HIGHLIGHTS OF ASTRONOMY

INTERNATIONAL ASTRONOMICAL UNION

UNION ASTRONOMIQUE INTERNATIONALE

HIGHLIGHTS OF ASTRONOMY

VOLUME 8

AS PRESENTED AT THE XXth GENERAL ASSEMBLY OF THE IAU, 1988

EDITED BY

D. McNALLY

University of London Observatory, London, U.K.

KLUWER ACADEMIC PUBLISHERS

DORDRECHT / BOSTON / LONDON

The Library of Congress Cataloged This Serial Publications as Follows:
71-159657

ISBN-13:978-0-7923-0281-0   e-ISBN-13:978-94-009-0977-9
DOI: 10.1007/978-94-009-0977-9

Published on behalf of
the International Astronomical Union
by
Kluwer Academic Publishers, P.O. Box 17, 3300 AA Dordrecht, The Netherlands.

Kluwer Academic Publishers incorporates
the publishing programmes of
D. Reidel, Martinus Nijhoff, Dr W. Junk and MTP Press.

Sold and distributed in the U.S.A. and Canada
by Kluwer Academic Publishers,
101 Philip Drive, Norwell, MA 02061, U.S.A.

In all other countries, sold and distributed
by Kluwer Academic Publishers Group,
P.O. Box 322, 3300 AH Dordrecht, The Netherlands.

printed on acid free paper

TABLE OF CONTENTS

PREFACE   D. McNally   xv

INVITED DISCOURSES

1. **HALLEY'S COMET**

   Part I: Ground-based Observations   R.M. West   3
   Part II: Space Studies   V.I. Moroz   17

2. **THE RISE AND FALL OF QUASARS**   M. Schmidt   33

3. **GALAXY FORMATION AND DARK MATTER**   M. Rees   45

JOINT DISCUSSIONS

1. **NEW DEVELOPMENTS IN DOCUMENTATION**   67
   AND DATA SERVICES FOR ASTRONOMERS

   Chairman and Reporter: G.A. Wilkins

   1.   Introduction   69
   2.   Developments in primary publishing   71
        (Contributors: S. Mitton; H. Abt; G.A. Wilkins;
        C.O. Jaschek)
   3.   Developments in information retrieval   77
        and distribution
        (Contributors: J. Rey-Watson; B.M. Lasker,
        P.M.B. Shames & L. Butler; B.G. Marsden;
        J.-L. Halbwachs; C.R. Benn; R.M. Shobbrook)
   4.   Developments in data archiving and retrieval   83
        (Contributors: C.O. Jaschek; O.B. Dluzhnevskaya;
        C.R. Benn; M. Rushton; F. Ochsenbein; F.M. Spite)
   5.   The changing role of astronomical libraries   87
        (Contributors: G.A. Wilkins; S. Stevens-Rayburn;
        E. Bouton; E. Lastovica; A.-M.M. de Narbonne;
        R.M. Shobbrook; J. Dudley)
   6.   Summary   93
        Acknowledgements   95
        References   96

Additional paper:

The selection of scientific and   J. Dudley   97
technical records for permanent
retention

2.   <u>FORMATION AND EVOLUTION OF STARS IN BINARY SYSTEMS</u>   101

Chairman and Editor: R.C. Smith

New observational clues on binary formation in the galaxy	D.W. Latham	103
A search for planetary-mass companions to nearby stars	B. Campbell	109
Spectroscopic binaries among low-mass pre-main sequence stars	R.D. Mathieu	111
Binary frequency among pre-main sequence stars in Taurus and Ophiuchus	M. Simon	117
Brown dwarfs in binary systems	B. Zuckerman	119
Cloud collapse and fragmentation	A.P. Boss	123
Criteria for collapse and fragmentation of rotating clouds	S.M. Miyama	127
Mathematical status of the fission theory	N.R. Lebovitz	129
Numerical simulations of fission	R.H. Durisen S. Yang & R. Grabhorn	133
The J vs M relation for binary stars	J.E. Tohline	137
The formation and evolution of binaries in globular clusters	F. Verbunt	139
Binaries from unstable triples. Dynamical processes of formation	J.P. Anosova	143
"Undisturbed" evolution in binaries	J. Andersen	145
Disturbed binaries: the early phases	J.P. de Greve	149
Binary Wolf-Rayet stars	B. Hidayat K.A. van der Hucht	153
Progress of common envelope evolution	R.E. Taam	155
Millisecond pulsars	J.H. Krolik	161
Evolution of cataclysmic binaries	B. Paczynski	167
The AM Her period spike	J.P. Lasota	173
A new progenitor model of type Ia supernovae	I. Hachisu M. Kato & H. Saio	175
Concluding remarks	V. Trimble	177

3. SUPERNOVA 1987A IN THE LARGE MAGELLANIC CLOUD   181

Chairman: V. Trimble, Co-chairmen: W. Liller & J.C Wheeler
Editor: W. Liller

Editorial	W. Liller	183
Supernova 1987A: light curves and their interpretation	R.M. Catchpole	185
Evidence for asymmetries in SN1987A	M. Karovska L. Koechlin P. Nisenson C. Papaliolios C. Standley	193
Gamma-ray lines from SN1987A and interpretation	E.L. Chupp	199
Interpretation of the CO bands of supernova 1987A	C.M. Sharp P. Höflich	207
Three dimensional hydrodynamical simulation of type II supernova	M. Nagasawa	213
NLTE calculations of hydrogen line profiles for SN1987A	W. Schmutz	215
Non-equilibrium thermal X-ray emission in the early phase of supernova remnant	H. Hanami T. Yoshida	217

Additional papers:

Effects of the soft X-ray burst from SN1987A on its circumstellar medium	P. Lundqvist C. Fransson	223
Neutrinos: detection and interpretation	L.N. Alexeyeva	229

4. THE COSMIC DUST CONNECTION IN INTERPLANETARY SPACE: 239
    COMETS, INTERSTELLAR DUST AND FAMILIES
    OF MINOR PLANETS

Chairman and Editor: J.M. Greenberg

From interstellar dust to comet dust and interplanetary particles	J.M. Greenberg	241
What are families of minor planets?	Y. Kozai	251
IRAS dust bands and the origin of the zodiacal cloud	S.F. Dermott P.D. Nicholson	259
Spatially varying optical properties of the zodiacal dust	S.S. Hong S.M. Kwon	267
What we know about families of asteroids	V. Zappala'	273
Comets, meteorites and interplanetary dust	D.E. Brownlee	281
Dynamics and spatial shape of short-period meteoroid streams	P.B. Babadzhanov Yu.V. Obrubov	287
Cometary dust and zodiacal light connection	A. Dollfus	295
Dust from the comets	T. Mukai	305

The origin and physical characteristics   D. Olsson-Steel   313
of meteoroids

5.   ATOMIC AND MOLECULAR DATA FOR ASTROCHEMISTRY   321

Chairman and Editor: P. Smith

Atomic and molecular data for diffuse   E.F. van Dishoeck   323
cloud chemistry

Ultraviolet, visible, and infrared   J.H. Black   331
spectroscopy of interstellar molecules

Microwave spectroscopy of astrophysical   W.M. Irvine   339
molecules

Some salient features of evolving models   S.P. Tarafdar   345
of interstellar clouds   S.K. Ghosh
    K.R. Heere
    S.S. Prasad

Molecules in circumstellar envelopes   A. Omont   357

Atomic and molecular data for stellar physics   R.A. Bell   365

Chemistry in dense interstellar clouds/   T.J. Millar   369
Data requirements

Chemistry in shocks   T.W. Hartquist   375

Chemical effects of interstellar grains   D.A. Williams   383

The volatile composition of comets   H.A. Weaver   387

Atmospheres of planets and their satellites   D.F. Strobel   395

6.   DISKS AND JETS ON VARIOUS SCALES IN THE UNIVERSE   397

Chairman and Editor: J. Dyson

The far-infrared (IRAS) excess in Roberts 22   M. Parthasarathy   399
and related objects

Recent observations of the beams in SS433   R.C. Vermeulen   403

Large scale jets in Class I and Class II   G.V. Bicknell   409
radio sources and quasars

Synchrotron thermal instabilities   G. Bodo   417
and radio filaments in the lobes of Cygnus A   A. Ferrari
    S. Massaglia
    E. Trussoni

Gravitation and jet induced velocities   M. Whittle   423
in the narrow line region of active galaxies

Two-flow model for extragalactic radio jets   H. Sol   429
    E. Asseo
    G. Pelletier

7.   THE HUBBLE SPACE TELESCOPE - STATUS AND PERSPECTIVES   433

     Chairman and Editor: G. Miley

     The science program   N.A. Bahcall   435
     of the Hubble Space Telescope
     Hubble Space Telescope   E.J. Weiler   441
     second generation instrument selection
     Wide field/planetary camera-II   J. Trauger   443
     for the Hubble Space Telescope
     The Space Telescope Imaging Spectrograph   B.E. Woodgate   445
     (STIS)
     The next generation: G.D. Illingworth 449
     an 8-16 m space telescope
     Space astronomy - The next thirty years   M.S. Longair   455

                    JOINT COMMISSION MEETINGS

1.   FOR MILLIARCSECOND OR BETTER ACCURACY   463

     Chairman and Editor: P.K. Seidelmann

     1.   Introduction   P.K. Seidelmann   465
     2.   Observational accuracies   469
          (Contributors: M. Shao; P. Bender; J.H. Taylor)
     3.   Theoretical Developments   472
          (Contributors: H. Kinoshita & J. Souchay; J. Wahr;
          N. Capitaine; S. Aoki)
     4.   Computational considerations   476
          (Contributors: E.M. Standish Jr.; M. Feissel; B.D. Yallop;
          C.A. Murray; H. Schwan)
     5.   Working Group Reports   482
          (Contributors: R.L. Duncombe; B. Morando;
          J.A. Hughes)

2.   SOLAR AND STELLAR CORONAE   501
     (In honour of Gordon Newkirk Jr.)

     Chairmen and Editors: E.R. Priest & R. Falciani

     1.   G. Newkirk's contribution to coronal studies   J.A. Eddy   503
     2.   Structure of the Solar Corona   T. Sakurai   513
                                                         E. Hiei
     3.   Coronal heating: theoretical ideas   J.V. Hollweg 517

4.	An update on X-ray emission from stars	R. Rosner	521
5.	Solar and stellar winds	G.L. Withbroe	525
6.	Coronal instabilities	G. Einaudi	529
7.	Accretion disk coronae	M. Kuperus	535
8.	Solar and stellar flares	A.O. Benz	539

3. **HIGH ANGULAR RESOLUTION IMAGING FROM THE GROUND**   543

Chairmen: J.E. Baldwin and J. Davis
Editor: J. Davis

Introduction	J. Davis	545
Principles of imaging using arrays	T.J. Cornwell	547
Is the imaging problem identical in all wave bands?	J.E. Baldwin	549
Review of linked array instruments	R.D. Ekers	551
Very long baseline interferometry	J.M. Moran	553
Millimeter wave interferometry	D. Downes	555
Meter wave interferometry	G. Swarup	557
Long baseline optical interferometry	S.T. Ridgway	559
Speckle interferometry	J.C. Christou	561
Infrared long baseline interferometry	W.C. Danchi M. Bester P.R. McCullough C.H. Townes	563
Active control and adaptive optics for optical interferometers	F. Merkle	565
Galactic and extragalactic applications	G.B. Field	567
The application of optical arrays to solar system and stellar problems	H.A. McAlister	569
Optical interferometry: summary and perspectives	P.J. Léna	571

4. **MOLECULES IN EXTERNAL GALAXIES**   573

Chairmen: F. Combes, N.Z. Scoville & J. Young
Editor: F. Combes

The molecular spiral structure in M51 derived from CO(J = 2 - 1) line observations	M. Guélin S. Garcia-Burillo R. Blundell J. Cernicharo D. Despois H. Steppe	575

Molecular cloud spiral arms and results from tidal interaction modeling   A. Hjalmarson   579

CO in NGC4438 and tidal stripping in the Virgo cluster   F. Combes
C. Dupraz
F. Casoli
L. Pagani   581

CO observations of the central region of NGC4258   Y. Sofue   583

The correlation of CO and IR emission from galaxies; what does it tell us?   F. Verter   585

Can galactic GMCs be identified from l-v diagrams?   D.S. Adler
W.W. Roberts, Jr.   587

Warm gas and spatial variations of molecular excitation in the nuclear region of IC342   A. Eckart
D. Downes
R. Genzel
A.I. Harris
D.T. Jaffe
W. Wild   589

Recent CO(2-1) observations of galaxies with the CSO   A.I. Sargent
T.G. Phillips
D.B. Sanders
N.Z. Scoville   591

Molecules in galaxies: results from Bell Laboratories   A.A. Stark   593

CO in M82 and other middly active galaxies   R. Wielebinski   595

Molecular clouds in dwarf irregular galaxies   C. Henkel   597

Molecular clouds in the Large and Small Magellanic Clouds   M. Rubio   599

CO in early type galaxies   T. Wiklind
C. Henkel   601

The ratio of H_2 to HI gas in infrared luminous galaxies   I.F. Mirabel
D.B. Sanders   603

Molecular gas in galactic nuclei   N.Z. Scoville   605

5. SPECTROSCOPY OF INDIVIDUAL STARS IN GLOBLULAR CLUSTERS AND THE EARLY CHEMICAL EVOLUTION OF OUR GALAXY

609

Chairman: G. Cayrel de Strobel
Editors: G. Cayrel de Strobel & M. Spite

(The transactions of this Joint Commission Meeting will be published by the Imprimerie de l'Observatoire de Paris, 92195 Meudon Principal Cedex, France)

Summary   G. Cayrel de Strobel   611

6. **STELLAR PHOTOMETRY WITH MODERN ARRAY DETECTORS**   615

Chairman and Editor: F. Rufener

Introduction and basic references   F. Rufener   617
for stellar photometry with CCD

CCD imagers for astronomy: J.C. Geary   623
past problems and future hopes

The CCD mosaic project by ESO and INSU/   S. d'Odorico   629
Toulouse Observatory   J.-L. Prieur

Ground-based photometric calibration   D.A. Hunter   631
of the Space Telescope CCD camera   H.C. Harris
                                                     W.A. Baum
                                                     J.H. Jones
                                                     T.J. Kreidl

Some factors affecting the accuracy   P.B. Stetson   635
of stellar photometry with CCDs

CCD data taking modes and   S. Djorgovski   645
flatfielding problems   M. Dickinson

High precision crowded field photometry   P. Linde   651

Analytical approximation   O. Bendinelli   657
of long-exposure point spread functions   G. Parmeggiani
and their use   F. Zavatti

Photometric data archives   C.O. Jaschek   663

7. **STAR CLUSTERS IN THE MAGELLANIC CLOUDS**   665

Chairman: P. Demarque

See Transactions XXB of the IAU under report of Commission 37.

ADDITIONAL JOINT COMMISSION MEETING

SYSTEMATIC OBSERVATIONS OF THE SUN   669
(In honour of Helen Dodson Prince)

Chairmen and Editors: J.C. Pecker and P. Wilson

Summary   671

1.   Observations   672
      (Contributors: P. McIntosh; H. Snodgrass; Z. Mouradian;
      K. Harvey; R. Altrock; P. Simon; J.-P. Legrand;
      G. Alissandrakis; H. Neckel; P. Petropoulos & X. Poulakis;
      M.H. Gokhale; K.R. Sivaraman; J. Pap)

2. Modeling implications   675
    (Contributors: P. Wilson; P. Gilman)
3. Future work   677
    (Contributors: W. Livingston; K. Zwaan; E. Hiei;
    L. Paterno)

ADDITIONAL CONTRIBUTIONS

The microwave background radiation: G. de Zotti   681
recent advances and new problems   L. Toffolatti

Submillimeter spectrum of the   T. Matsumoto   689
cosmic background radiation

The status of Big Bang nucleosynthesis   H. Reeves   693
in July 1988

Author Index   697

PREFACE

It is the customary practice to report the major events of a General Assembly -the Invited Discourses, Joint Discussions and Joint Commission Meetings in Highlights of Astronomy. Vol. 8 reports the highlights of the XXth General Assembly of the International Astronomical Union, 1988 August 2-11, Baltimore, USA. The present volume contains the 3 Invited Discourses and papers presented at 7 Joint Discussion Meetings and 6 Joint Commission Meetings. Two Joint Commission Meetings will be reported elsewhere -JCM5 Spectroscopy of Individual Stars in Globular Clusters and the Early Chemical Evolution of our Galaxy (in summary only here, published by the Imprimerie de l'Observatoire de Paris) and JCM7 Star Clusters in the Magellanic Clouds (see Transactions of the IAU, Vol.XXB, report of Commission 37).

I am most grateful to the authors of the invited discourses R.M. West and V.I. Moroz, M. Schmidt and M. Rees for sending me the manuscripts so promptly. I am also indebted to the Chairmen of the Joint Discussion and Joint Commission Meetings for their organisation of the meetings and for the assembly of their material for publication. Unfortunately the deadline for receipt of manuscripts coincided with an extended postal strike in France which seriously hindered the preparation of the volume for publication.

The organisation of the XXth General Assembly was in the capable hands of my predecessor Dr. J.-P. Swings (1985-1988) and it was his prerogative to choose additional papers for publication in this volume: 5 such additional papers are included -2 included with the report of JD3 "Supernova 1987A in the Large Magellanic Cloud" and the remaining 3 on the microwave background radiation and cosmic nucleosynthesis at the end of the volume.

I have followed the precedent set in Vol. 7 in that the volume has no subject index. There is an extensive table of contents and author index which should make tracking down any particular contribution straightforward.

It is my pleasure to record my appreciation of the work by Dr. Swings in organising the XXth General Assembly of the Union, to the Authors and Meeting Chairmen for their work in producing the manuscripts as near the publication deadline as is possible in an imperfect world and, in particular, to Mrs. M. Orine of the Paris Secretariat for her considerable help in the final editing of this volume.

IAU-UAI Secretariat
98bis, bd Arago
F-75014 PARIS
France

Derek McNALLY
General Secretary , IAU
1989 January 10

INVITED DISCOURSES

HALLEY'S COMET (Part I): Ground-based Observations

Richard M. West
European Southern Observatory
Karl-Schwarzschild-Str. 2
D-8046 Garching bei München, FRG

ABSTRACT. Since the recovery in October 1982, an extensive, international programme to observe Comet Halley with ground-based instruments has been coordinated by the International Halley Watch (IHW), and a comprehensive archive is now in the final phases of preparation. The observations were carried out at more than 150 observatories and with all available methods. A special effort was made to support the space missions during the comet encounters in early March 1986. Whereas the spacecraft provided detailed in-situ measurements over a short time interval, ground-based observers have so far followed the development of the comet over a period of nearly six years, and a number of spectacular events near the nucleus and in the tail have been documented in great detail. These observations still continue. This article gives an overview of the most important results obtained from the ground and also mentions the prospects for further observations with large telescopes during the next years.

1. Comet Halley and cometary science

Quite a few excellent books have recently been written about Comet Halley and there is no need to describe once more its role in the history of humanity. Still, an account of the impact of this object on astronomy and associated space sciences during the recent apparition can best be appreciated after a brief glance at some of the results obtained at the time of earlier passages. Cometary science has since long been intimately connected to Comet Halley and it received a tremendous impetus through the intensive observational and interpretational efforts during 1985-86.

Comet Halley belongs to a select group of ten known comets with periods near 70 years. However, five of these have only been seen during a single apparition and among the others, one (Westphal 1913 VI) was not seen at its last perihelion passage and may have ceased being active; one (Brorsen-Metcalf 1919 III) was only seen at two apparitions and the last two (Pons-Brooks 1954 IV; Olbers 1956 IV) have only been observed during three apparitions. In fact, Halley is by far the best known, periodic comet and it has been observed over more than two millennia; it also exhibits virtually all the characteristics of an active comet, from nuclear outbursts to rapid changes in the tail.

More or less detailed astronomical observations are now available at 30 apparitions since *240 B.C.* The earliest data come from China and it is not unlikely that improved orbital extrapolations further back in time will eventually uncover even earlier observations in the ancient annals of that country. However, "modern" studies may be said to start with a drawing by Hevelius on 8 September 1682, showing a prominent, curved jet in addition to a tail. Soon after followed the work by Halley (1705), who recognized that the bright comets in *1531, 1607* and *1682* had similar orbital elements and must therefore be the same object.

3

D. McNally (ed.), Highlights of Astronomy, Vol. 8, 3–16.
© *1989 by the IAU.*

The passage in *1759* - first predicted by Halley - was above all characterized by its crucial importance for the confirmation of Newton's theory of gravitation and also the improved accuracy of *astrometric* observations, due to better instruments and measuring methods. With the help of the frantic work by J.-J. Lalande and his assistant Madame N.-R. de la Brière, in November 1758 A.-C. Clairault was able to announce the predicted perihelion time as 13 April 1759, i.e. only one month too late. But the comet was accidentally recovered on 25 December 1758 by an amateur, J.G. Palitzsch near Dresden; this was also the first telescopic recovery. Ch. Messier found it independently on 24 January 1759, and a certain lack of international cooperation is illustrated by the fact that he was forbidden by his superior, J.-N. Delisle, to announce this until three months later. However, in the meantime Messier went on to produce an excellent series of astrometric positions.

Although Encke (1820) had earlier introduced a non-gravitational term into his orbital computations for the short-period comet that now carries his name, none of the Halley predictions for the *1835* apparition took this effect into account. It is during this apparition that the modern, *(astro)physical* study of comets was inaugurated by several series of detailed drawings of the inner coma, which were made in the course of painstaking, visual observations by F.W. Bessel, F.G.W. Struve, J.F.W Herschel and others; reproductions from the original papers may be found in the atlas by Donn et al. (1986). They documented a pronounced asymmetry and also rapid changes in the emitting light cones, almost exactly 150 years before the spacecraft observations of dust jets.

Another step from *qualitative* descriptions towards *quantitative* measurements was achieved at the *1910* apparition. A new generation of powerful telescopes in combination with the photographic emulsion as detector and information storage medium, resulted in a wealth of useful observations. The photographic recovery of the comet by Max Wolf, at 3.5 A.U. from the Sun, was followed up by at least 1600 photographs, more than 3000 astrometric positions and several dozen photographic spectra. The organization of coordinated observations was suggested by E.E. Barnard already in the 1890's and a "Comet Committee of the Astronomical and Astrophysical Society of America" was created. A "Circular Respecting Observation of Halley's Comet" in 1910 established well defined goals and invited cooperation. Several temporary observing stations were set up in order to get continuous coverage, for instance in Hawaii. However, the coordination failed, because manpower and funding were unable to handle the unexpectedly large number of observations and also because some of the most active observatories decided to proceed independently. In the end, the only comprehensive study was that of Bobrovnikoff (1931), lately supplemented by an impressive collection of 1909-11 Halley photographs (Donn et al., 1986).

Improved orbital computations began to appear in print more than 10 years before the next perihelion passage in *1986*, and the first recovery attempts were undertaken already in 1977. However, the first definite image was obtained on 16 October 1982 with the Palomar 5 m telescope, immediately followed by the 3.6 m CFH and 4 m KPNO telescopes and soon after by the Danish 1.5 m telescope at ESO. The advance of astronomical technology over the past centuries is interestingly illustrated by the interval between recovery and perihelion: 1531 (30^d, recovery magnitude 4^m), 1607 (36^d, $2.^m5$), 1682 ($31^d, 2.^m5$), 1759 (78^d, 8^m), 1835 (104^d, $?^m$), 1910 (231^d, 16^m) and 1986 (1212^d, $24.^m5$). Monitoring of the nucleus continued until the comet entered the active state by developing a dusty coma late in 1984 and soon thereafter the presence of CN was detected spectroscopically. Table 1 summarizes some of the major events.

TABLE 1: CHRONOLOGY OF MAJOR HALLEY EVENTS 1982 - 91

Date			r(A.U.)	Event
1982	Oct.	16	11.05	Recovery with 5-m Palomar telescope
1983	Jan.	10	10.53	1^m variation: rotation or activity?
1984	Feb.	4	7.96	First spectrum: reflected sunlight
1984	Sep.	25	6.14	6" coma visible
1984	Oct.	22	5.91	Photometry shows coma contribution
1984	Oct.	30	5.84	30" coma in slit spectrum
1985	Feb.	17	4.84	First spectra with CN and [OH] emission
1985	Jul.		3.3	First radio detection of OH
1985	Nov.	12	1.76	First photos with ion tail
1985	Nov.	27	1.54	Minimum distance to Earth at $\Delta = 0.62$ A.U.
1985	Dec.	9	1.36	Two $3° - 4°$ ion tails
1986	Jan.	10	0.87	Major Disconnection Event (D.E.) in tail
1986	Feb.	9	0.59	Perihelion
1986	Feb.	15	0.60	First photo after perihelion
1986	Feb.	late	0.65	Multiple dust tails; antitail
1986	Mar.	6	0.79	Vega-1 flyby at $d = 8890$ km
1986	Mar	8	0.82	Suisei flyby at $d = 151000$ km
1986	Mar.	8-10	0.83	Major D.E.
1986	Mar.	9	0.83	Vega-2 flyby at $d = 8030$ km
1986	Mar.	11	0.86	Sakigake encounter at $d = 6.99 \cdot 10^6$ km
1986	Mar.	14	0.90	Giotto flyby at $d = 596 \pm 2$ km
1986	Mar.	20-22	1.01	Major D.E.
1986	Mar.	25	1.07	ICE closest approach at $d = 28 \cdot 10^6$ km
1986	Apr.	11	1.33	Minimum distance to Earth at $\Delta = 0.42$ A.U.
1986	Apr.	11-12	1.34	Major D.E. in tail
1986	May	6		η-Aquarids meteor stream
1986	May			Major sunward spike
1986	Jun.	14	1.82	Last photo showing ion tail
1986	Jul.	23	2.75	Last visual detection of tail (0.3°)
1987	Feb.	2	4.84	Last spectrum with CN and C_3 emission
1987	Apr.	1	5.38	Last CCD-image showing faint tail (> 4')
1987	Apr.	22	5.57	Possible outburst; m(total) = 13.6
1987	Nov.	26	7.35	Strong condensation (m = 19.6 within 5")
1988	Feb.	23	8.01	m(total) = 17 within 40"
1988	May		8.6	Asym. coma; dust cloud > 50"; m(nucleus) = 23.1
1989	Feb.		10.4	Predicted m(nucleus) $\sim$ 24
1990	Feb.		12.5	Predicted m(nucleus) $\sim$ 25
1991	Feb.		14.3	Predicted m(nucleus) $\sim$ 26

2. Coordination of observations during the present apparition

The current apparition represents a milestone in cometary research, not only because of the spacecraft encounters, but also because of the successful, world-wide coordination of the entire ground-based observational effort. The value of this vast undertaking has been clearly demonstrated and it will undoubtedly serve as a model for any similar, future programmes in other fields.

The coordination only became possible after the establishment of the *International Halley Watch (IHW)* with the following main goals: *to encourage and support any scientifically valid means of studying the comet; to coordinate activities among ground-based disciplines and with flight projects; to further standardization and to produce a complete Halley archive with all properly documented data.* The IHW was conceived under the auspices of NASA in 1979; Lead Centers at JPL in Pasadena and at Dr. Remeis Sternwarte in Bamberg were created in 1980, and a Steering Group was set up. Regional Centres were set up in various countries, among others in Japan, P.R.China, USSR and U.K. In 1982, the IAU recognized IHW as the sole world-wide coordinator of ground-based Halley observations and the same year close contacts were established between IHW and the Interagency Consultative Group (IACG) that coordinated the Halley spacecraft projects.

In order to achieve the stated goals, and to avoid the problems that had derailed earlier attempts at coordination, the IHW organised itself into eight disciplines, specified by the investigation technology: *Astrometry, Infrared spectroscopy and radiometry, Large-scale phenomena (tail studies), Near-nucleus studies, Photometry and polarimetry, Radio studies, Spectroscopy and spectrophotometry* and *Meteor studies.* A unique feature of the IHW is the *Amateur observation network* which was organized directly by the lead centers. 26 astronomers were appointed "Discipline specialists" with responsibility for the coordination within their respective disciplines. Direct contacts were established to observers all over the world, and the organisational success is reflected by the enthusiastic response: in the end, more than 1000 astronomers and several hundred amateurs from more than 50 countries actively participated in the IHW. Large amounts of data have now been collected (Table 2), which together with those from the spacecraft will soon become available in a unified "Halley Archive". It will comprise a total of ~ 22 Gbytes and is expected to appear in 1990 in the form of computer-readable Compact Discs, in printed form (pending funding) and also as a computer-retrievable data base. A "trial" archive with observations of P/Crommelin during its 1983-84 apparition was prepared by Sekanina and Aronsson (1985).

Net	No. of data	Observing period
TABLE 2: IHW ANTICIPATED DATA		
Astrometry	7000	1982 Oct. - continues
Near-nucleus	6000	1982 Oct. - continues
Large-scale	7000	1982 Oct. - continues
Photometry and polarimetry	55000	1982 Oct. - continues
Spectroscopy and spectrophotometry	2500	1984 Feb. - 1987 Feb.
Infrared spectroscopy and radiometry	2000	1984 Dec. - 1988 Feb.
Radio studies	2200	1985 Jan. - 1986 Aug.
Meteor studies	13500	1985 Oct. - continues
Amateur observations	13000	1985 - 1988

In astrophysical terms, the goal of the IHW has been to provide the observational data for the fullest possible description of the cometary phenomena and their temporal and spatial variations. This in turn allows conclusions about the constitution of the coma and the processes near the nucleus, as well as the interaction between the dust and gas tails with the interplanetary medium. The primary aim of cometary physics, namely the study of the nucleus itself, its structure, composition and evolutionary history, is only indirectly possible by ground-based observations; the present break-through in this area is of course due to the in-situ observations from spacecraft. However, the encounters were of relatively short duration and all took place at about the same heliocentric distance. The full benefit of the high-resolution spacecraft data can only be achieved when they are compared with the long time-series gathered by ground-based techniques. For the first time, accurate calibrations of these series in terms of production rates of many individual atomic and molecular species have now become possible, greatly improving the prospects of an accurate quantitative understanding of the evolution during the pre- and post-perihelion phases. From the recovery at the record heliocentric distance of 11 A.U., through the perihelion at 0.6 A.U., and out again to 8.6 A.U. in May 1988, and with the prospect of additional data during the coming observing seasons, the present apparition of Comet Halley has brought nothing less than a revolution in cometary science.

3. Ground-based observations 1982 - 1988

The preparations for observing Halley resulted in a number of conferences, the Proceedings of which reflect the various aspects of pre-Halley cometary research (Ponnamperuma, 1981; Véron et al., 1982; Wilkening, 1982; Gombosi, 1983; Carusi and Valsechi, 1985).

Detailed presentations of the initial results from space and ground may be found in the Nature Supplement of 15 May 1986 (Vol. 321, pages 259 - 366) and also in the Proceedings of two major conferences, the 20th ESLAB Symposium on "Exploration of Halley's Comet" held in Heidelberg in October 1986 (Battrick et al., 1986; Grewing et al., 1988) and the Symposium on the "Diversity and Similarity of Comets" held in Brussels in April 1987 (Rolfe and Battrick, 1987). The reader is referred to these volumes for details and only an overview of the major ground-based results, including some observations made from orbiting satellites, rockets and aircraft, will be attempted here. In view of the importance of the space encounters, special emphasis will be given to the interval 5 - 15 March 1986, and for practical reasons, the observations will be presented by IHW discipline.

3.1 ASTROMETRY

Astrometric measurements of Comet Halley with reasonable accuracy are available since 1607 (Kepler, Hariot and Longomontanus). The non-gravitational forces are now known to arise from the jet effect of outgassing from the rotating nucleus, and they are therefore dependent on several parameters, including the activity level, heliocentric distance and rotational state. For the orbital linkage over several apparitions, they constitute an important source of uncertainty - in the case of Halley, a delay of ~ 4 days per orbital period - although semi-empirical modelling (Marsden, 1985) has been reasonably successful. For instance, Brady and Carpenter (1971) linked observations from 1682 (13 positions), 1759 (243), 1835 (1517) and 1910 (3085) and obtained the perihelion time $T_o = 1986$ Feb. 9.39474 ET. In a critical study, Yeomans (1977) selected the best 885 positions, including 9 from 1607, and followed the orbit back to the year 837; based on those from 1759 - 1911 alone, he predicted $T_o = 1986$ Feb. 9.6613 ET. Including observations after the recovery and up to April 1985, Landgraf (1986) obtained Feb. 9.45891. In April 1986, after the encounters, the actual value was found to be Feb. 9.45888 ET (Morley, 1986).

In addition to the interest in the orbital motion of Comet Halley *per se*, the data provided by the IHW Astrometry Net were used by the spacecraft centers for navigation. Also for this reason, an additional effort was made to increase the accuracy of the measured positions and hence the predicted, spatial position of the comet at the time of the encounters. For instance, the 1835 visual measurements were re-investigated, using more precise, modern positions of the reference stars and many of the 1909-11 plates were remeasured (Morley, 1984; Röser, 1987), significantly improving the accuracy. Special, astrometric catalogues with higher density and accuracy than the SAO Catalogue were prepared along Halley's path by means of new astrographic plates obtained at various observatories in the USA and USSR. A conference on "Cometary Astrometry" (Yeomans et al., 1984) provided detailed recommendations for observers and measurers.

The result is gratifying: about 7000 positions from at least 143 observatories have until now been communicated to the IHW Astrometry Net. The r.m.s. residual is of the order of 1 arcsec, and more than 85% of the data are within 3 arcseconds. During the critical period from mid-February until the encounters in early March 1986, several southern observers daily provided the spacecraft centers with accurate positions within a few hours of the observations. This ground-based support provided crucial data for the navigation before the Pathfinder data became available from the Vega spacecraft.

A major problem for cometary astrometry has always been the presumed off-set between the center of the diffuse image of the inner coma, as seen on short-exposure photographic plates, and the actual position of the comet's nucleus; this is known as the *light shift* and reflects the asymmetric light distribution near the nucleus, because the dust is preferentially released on the side which is illuminated by the Sun. A comparison of the Vega- and Giotto-sightlines to the nucleus with the computed orbit from ground-based observations, for the first time allowed a direct determination of this bias as ~ 1100 km ($\sim 1.''5$) at the time of the encounters.

3.2 NEAR-NUCLEUS STUDIES

Three circumstances particularly contributed to the success of the near-nucleus observations. First, the advent of CCD's, which are optimally suited for this purpose (sensitivity, dynamic range, field). Secondly, the IHW Near-Nucleus Net specified and procured a standard set of optical filters, centered on the spectral bands of the most prominent molecules and radicals in the coma (CN, C_3, CO^+, C_2 and H_2O^+) and also at three continuum wavelengths for comparison and dust studies. And thirdly, greatly improved image processing techniques now make it possible to isolate and quantitatively measure even very faint, transient features like gas and dust jets.

A major result: there are discrete sources of activity, on the sunlit side only. In a series of papers before the 1986 observations, Sekanina and Larson (1986) described how they digitized and processed many of the 1910 photographs. With a radial/rotational shift-difference algorithm, they demonstrated an amazing amount of morphological detail in the coma, including spiral *dust jets*, that "unwind" from the nucleus and evolve into expanding envelopes on a time scale of days. The jets are generated by ejection from discrete sources, enabling the authors to determine the nucleus spin vector, a provisional rotation period of ~ 2.2 days, and also to draw a rough topological map of 13 dust vents. The general picture was confirmed by the spacecraft images. Post-perihelion, short photographic exposures in red light showed major day-to-day changes in nuclear activity (West et al., 1986), in support of a non-uniform distribution of vents on the surface.

Similar processing of CCD frames obtained before and after the perihelion passage in 1986 also showed nuclear dust jets. In addition, pinwheel-shaped *gas jets* were discovered in the CN-images (A'Hearn et al., 1986; Hoban et al., 1988) by using a radial renormalization technique. Gas jets were also seen in CCD-exposures behind an [O I] filter, identical with the one in the Giotto Halley Multicolour Camera, but due to the strong dust background near 700 nm, they could not be seen in the [H_2O^+]-filter (Cosmovici et al. 1988). The gas jets do not coincide spacially with the dust jets, and it has been suggested that the organic CHON particles may be the source of the gas by photosputtering.

Another phenomenon, expanding *coma shells*, which was first seen on 1910 photographs by Larson and Sekanina (1985), was studied in detail by Schlosser et al. (1986) on photographs taken in the light of CN. At least 15 spherical shells were observed from Feb. 17 to Apr. 17 and the expansion rate appeared to decrease with heliocentric distance. From the extrapolated ejection-times, a periodicity of 53.4 ± 0.9 hours was found. However, a rediscussion of this material by Celnik and Schmidt-Kaler (1987) casts some doubt on the correct identification of the shells from night to night. A comparison with the onset times of nuclear activity, observed photometrically by Schleicher et al. (1986), indicates that brightening in CN is observed at the time of the release of a new CN shell. Moreover, backwards extrapolation of the outward motion of condensations in the tail, as seen in the light of CO^+, shows that within the uncertainty, plasma release may also occur at the same time.

Some reports about possible splitting of the nucleus may have referred to bright ejecta.

3.3 LARGE-SCALE PHENOMENA

The temporal changes in the dust and gas tails of a comet reflect the complex interaction between the material released from the nucleus and the interplanetary medium. For such studies, continuous monitoring is of the greatest importance, and it is therefore most gratifying that a large number of observers with access to wide-angle cameras decided to participate in the IHW. They were rather well distributed by geographical longitude and by latitude from Antarctica to northern Norway, and the Pacific Ocean "gap" was bridged by well-equipped expeditional observers on several islands.

The large-scale phenomena commenced with the appearance of a short ion tail by mid-November 1985. Several ion-tails became visible in early December and the first, major disconnection event (D.E.) took place around January 10, 1986. After perihelion, Comet Halley displayed a broad, fanshaped dust tail with seven or more subtails and also several ion tails. A predicted antitail was observed in late February. A spectacular D.E. was recorded on March 8 - 10. The tail reached a maximum length of about $75 \cdot 10^6$ km in mid-April, although the visibility was hampered by the bright Milky Way background. The ion tail disappeared by mid-June 1986 and the last CCD-image to show a pronounced tail was obtained in April 1987. The coma has persisted at least until May 1988 and a strong nuclear condensation became visible in late 1987; since March 1988, it has been experiencing 1^m night-to-night variations, proving that we are at last beginning to see the nucleus again.

About 20 D.E.'s have been observed; at least the major one on March 8 - 10 can unambiguously be connected to the passing of a magnetic sector boundary, detected by the magnetometers on Vega-1 at March 7.87 (Niedner and Swingenschuh, 1987). This is the first direct confirmation of the suspected cause of D.E.'s and underlines their importance for the study of the solar wind.

A long series of wide-field photographs was made by W.Liller on the remote Easter Island. At La Silla in Chile, H. Pedersen and his associates obtained wide-field CO^+ CCD images (resolution 30") during no less than 57 consecutive nights after Feb. 17; the combination of more than 600 individual frames into a video-sequence is nearing completion. Also from this site, a group from the Bochum University obtained narrow-band CN, CO^+, H_2O^+, N_2^+ and CO_2^+ photos over the same period (cf. Celnik and Schmidt-Kaler, 1987). On these, the authors also measured the motion of a large number of tail knots, humps, kinks, etc. and found that the outward velocity increases quadratically with the distance from the nucleus. More than 500 amateur photos were collected and studied in a similar way in Japan (Saito et al., 1987). Of particular value is the large number of photos obtained at the time of the spacecraft encounters, which allows a detailed study of the development of the features which were observed in-situ.

The impressive fan tail of dust in late February was immediately interpreted by Sekanina (1986) as a series of discrete synchrones from isolated dust outbursts near the time of perihelion; the average time interval was $\sim$ 52.5 hours. A sunward spike of dust was observed from April 28 to June 7 up to a distance of 700,000 km and could be explained as due to the ejection of very small particles (Sekanina et al., 1987).

Another large-scale phenomenon, the Ly-α halo, was observed several times from sounding rockets (Opal et al., 1987) and a 2.2 day "breathing" period was found from the data returned by the Suisei UV-imager.

3.4 PHOTOMETRY AND POLARIMETRY

Photoelectric photometry was made in many places through sets of 8 standard IHW interference filters, centered on the major emission lines and the adjacent continuum. These sets were procured and distributed by the IHW Photometry Net; a general calibration allows direct transformation of the measured fluxes to column densities and then to production rates by a suitable model. Observations through diaphragms of different sizes show the radial distribution of the emitting species. As a main result, the photometrically measured H_2O production (from H_2O^+) agreed well with that found in-situ and Halley's production levels could be compared with those of other bright comets from recent years, notably Bennett (1970 II), Kohoutek (1973 XII) and West (1976 VI). Moreover, the deduced rates for OH and CN were compared with radio measurements of OH and HCN.

Among others, long series of observations were made at Mauna Kea (Piscitelli et al., 1986) and by Sterken and collaborators (1987) in New Zealand and Chile. Of particular interest is the 7.4-day periodic variation in the C_2, OH and CN-fluxes which was first described by Millis and Schleicher (1986), and which was also found in the HCN radio data (Schloerb et al., 1987).

The overall brightness of the comet was estimated by many amateur visual observers and a uniform series of measurements was made with the IUE Fine Error Sensor (FES). In general, the photometric behaviour was similar to what was observed in 1910-11, including the typical asymmetry, reflecting the larger, more persistent post-perihelion activity (Green and Morris, 1987).

Linear polarisation data are available from a number of observers from late 1985 to mid-1986, showing a maximum value near 25% and the reversal at a scattering angle near 160°. Post-perihelion circular polarisation was measured by observers at Taricha (Bolivia) and La Silla (Chile). From a comparison with the in-situ dust particle fluxes measured by the impact sensors on the Vega and Giotto spacecraft, Lamy et al. (1987) conclude that the polarimetry is compatible with rough, moderately absorbing silicate grains having a density decreasing with radius, and also with rough graphite grains, or a mixture of the two.

3.5 SPECTROSCOPY AND SPECTROPHOTOMETRY

This IHW net went into action already in early 1984, when a rather noisy spectrum was obtained of the 23^m object (Belton et al., 1987). The spectrum showed only reflected sunlight until the first weak CN emission lines appeared in Feb. 1985 at r = 4.8 A.U. (Wyckoff et al., 1985). From then on, several dozen observers collected the richest spectral material ever obtained for a single comet. This includes a long series of UV-spectra from IUE (e.g. Feldman et al.,1987) and ASTRON (Boyarchuk, et al., 1987), supplemented by spectra from several rocket flights (Opal, 1987).

As Halley brightened, it became possible to gradually increase the spectral resolution, culminating in R=100000 spectra during March/April 1986 (Arpigny et al., 1987) that allowed a complete separation of the [O I] and NH_2 blend at 630 nm and the resolution of the isotopic lines of $^{12}C_2$ and $^{13}C^{12}C$ in the C_2 (0,0) band at 515 nm. Similar spectra of the CN(0,0) R-branch by Wyckoff et al. (1988) have lead to a determination of the $^{12}C/^{13}C$ isotope abaundance ratio of 65 ± 9 while the solar system value is 89. A very large number of spectra was obtained with a variety of telescopes and equipment by Wyckoff and associates (1986). They were able to isolate pure ionic spectra with improved image processing techniques and possibly detected CO_2^+.

A number of observers used the well-established objective prism technique. In addition to slot and long-slit spectra, 2-dimension spectra (aperture plate) were obtained, for instance by Jockers et al. (1987), allowing an investigation of the simultaneous spatial distribution of several ions in the coma, not possible with other techniques.

The spectral data provided a detailed view of the time variations of the various atomic and molecular species in the coma. The spectra which were obtained near the spacecraft encounters could be directly compared to the in-situ data on the radial distribution of the major gaseous components in the coma and good agreement was found in most cases. Similarly, comparisons with CCD-images and photometry done through the IHW filters were useful to ensure the overall data compatibility.

3.6 INFRARED SPECTROSCOPY AND RADIOMETRY

Several very successful programmes were carried out in the infrared region. Ground-based observations in the near-infrared bands from high-altitude sites were complemented with important data, also at longer wavelengths, which were gathered during several flights of the Kuiper Airborne Observatory (KAO).

Comet Halley was regularly monitored from Sept. 1985 until June 1986, mostly in the standard JHKLMNQ infrared photometric bands (1.2 - 20 μm), and through diaphragms of varying sizes. These measurements allow an analysis of the temperature, production rate, particle size and composition of the dust in the coma. Large variations in the amount of dust were observed, reflecting jet activity also seen in the visual region, but there was no evidence of changes in the nature of the dust over this period. The infrared polarisation was measured in the J, H and K bands and pointed to a two-component dust population (Brooke et al., 1987).

Within infrared spectrophotometry, a major event was the first, unambiguous detection of H_2O emission in a comet at 2.65 μm; the line was seen in high-resolution spectra from a Fourier Transform Spectrometer (FTS) onboard KAO. This difficult observation was only possible when the line was Doppler-shifted sufficiently into the wings of the strong terrestrial absorption line, in December 1985 and March 1986. Expansion velocities of about 1 km/sec were measured (Larson et al., 1987). The measured H_2O production rate was about 10^{30} mol/sec in March 1986, confirming water as the dominant species. Other results from the KAO include a measurement of the ortho-to-para H_2O-ratio; a value of 2.2 - 2.3 was found, indicating a nuclear spin temperature $\sim$ 25 K. OH rotational emission was detected at 119 μm. A search for methane emission was negative and yielded an upper limit for the production rate at about $4 \cdot 10^{28}$ mol/sec. The silicate feature at 10 μm was easily detected, and a new emission feature at 28 μm may be attributed to dust emission.

In the 3 - 4 μm band, emission was detected from formaldehyde (H_2CO) at 3.6 μm; other lines were from CO_2, CO and possibly OCS. Two unidentified emission features at 3.28 and 3.37 μm are thought to originate in the C-H bond in unsaturated and saturated hydrocarbon molecules, respectively, and are also present in the spectra of interstellar dust clouds. It is not known whether the emitter is gaseous or solid; in the first case, the relative, global abundance of carbon in the coma would be no less than 50% of the water abundance. However, the emission could also be thermal, from small dust grains (Encrenaz, 1988).

3.7 RADIO STUDIES

The radio studies of Comet Halley commenced as early as January 1985 when the Nancy radio telescope began to monitor the 1665 and 1667 MHz OH lines in a programme that lasted until the end of August 1986 (Gérard et al., 1987). The first detection was made in early July 1985, almost simultaneously with this telescope and also with the NRAO 43 m antenna. The OH production and kinematics were studied; a peak rate of $3 \cdot 10^{29}$ mol/sec was reached in January 1986 and ΔV increased from 2 km/sec in November 1985 to 3 km/sec in March 1986. A possible 7.1 day variation in the January 1986 data was not seen in those from March and April.

More than 30 radio observatories participated in the Halley observations. Of particular interest is the long series of measurements of the HCN rotational emission at 3.4 mm which was made with the 13.7 m antenna at the Five College Radio Astronomy Observatory (Schloerb et al., 1987). The variations followed the 7.4 day period of the C_2-flux, discovered by Millis and Schleicher (1986).

Several groups attempted to detect more complex molecules. However, the radio searches for HNC, HC_3N, CH_3CN, HCO^+, OCS and CO were negative. Radar contact was established with the Arecibo 300 m antenna in late November 1985, but it is doubtful that any scintillation enhancement from Halley's plasma tail was seen during occultation observations of compact radio sources in early 1986 (Ananthakrishnan et al., 1986).

Many of these results are extensively discussed in the Proceedings of a follow-up meeting on "Cometary Radio Astronomy" (Irvine et al., 1987).

3.8 METEOR STUDIES

Comet Halley is the parent body of two meteor streams, the Orionids with peak activity near October 21, and the η Aquarids around May 3-5. Visual and radar observations were therefore stepped up in October 1985 and May 1986, respectively, although fresh injections of dust particles could not yet be expected to be observed on these dates (Babadzhanov et al., 1987). Indeed, even less meteors were observed than in previous years (Hajdukova et al., 1987), but these observations serve as a useful comparison for the coming seasons. Spectra of some Orionid meteors show them to be essentially similar to meteors from other showers; the brightest are produced by meteoroids with masses of several grams.

3.9 AMATEUR CONTRIBUTIONS

Astronomy is one of the few sciences where the amateur still plays an important role and cometary astronomy is a classic amateur domain. In recognition of this potential, the IHW created an Amateur Net and a comprehensive Manual was issued (Edberg, 1983); it was also translated into several other languages, including Japanese.

The coordination bore fruit in several ways. First, it resulted in about 10000 magnitude estimates, 700 drawings of the visual appearance, about 2000 photographs, many of high quality and very useful for the large-scale investigations, and about 50 spectra. Secondly, and equally important, the Halley programme taught many amateurs how to make scientifically useful observations and left them well prepared for future work. The feeling of being involved in "real science" was of great satisfaction to many amateur participants and a good advertisement for astronomy in contemporary society.

4. Conclusions and outlook

Never before have so many astronomers been mobilized for the observations of a single object and never before has such a rich material been obtained with so many different methods. The organization of the observations of Comet Halley by the IHW has been an unequalled success and long after the initial studies are over, important research will continue on the basis of the enormous archive. The intensive collaboration between space- and ground-based observers has blazed the trail for future projects, also outside the present field.

But the ground-based Halley campaign is not yet over and some of the IHW Nets will continue to produce new observations for some time. Long CCD exposures still permit a morphological study of the coma and therefore of the diminishing activity. They also provide information about the varying brightness of the nucleus. Astrometric positions can be determined as long as the nucleus can be detected and may eventually throw more light on the troublesome non-gravitational forces.

After so many new discoveries, it is almost ironic that the major outstanding problem is the rotational state of the nucleus. It was impossible to determine unambiguously the rotation period from distant, pre-perihelion photometry alone, but the study of jets (1910) and outbursts (1986) seems to favour a period near 52 hours. This period would also appear compatible with the Vega and Giotto views of the nucleus. However, a 7.4 day periodicity was found in coma photometry and radio measurements soon after perihelion and since then several authors have tried to reconcile the data by introduction of the effects of nutation and precession; a comprehensive summary is given by Sekanina (1987) and the formal theory is described by Kamél (1988).

It is not obvious that the rotation of the nucleus will be clearly reflected by major changes in the coma and it may therefore be futile to attempt a linkage between the early pre-perihelion data and measurements of the well-developed coma in 1985-86. However, the knowledge of the rotational state and therefore of the attitude during the spacecraft encounters is of course of crucial importance for the determination of the three-dimensional shape of the nucleus and the surface topology. Some observers with access to large telescopes have therefore decided to continue photometric measurements of the "bare" nucleus at large heliocentric distances. The attempts in 1987 were unsuccessful because of a persistent, opaque coma and only from March 1988 can the nucleus again be seen and measured during excellent seeing, as a central light spike in the inner coma.

As part of this renewed effort, observations have been performed with the Danish 1.5 m telescope at the ESO La Silla Observatory. A composite image of Comet Halley was prepared from more than 50 individual CCD frames, obtained during 19 nights in April - May 1988 and totalling 11^h15^m exposure time (West and Jørgensen, 1988). At this time the heliocentric distance was 8.6 A.U.. The 23^m cometary nucleus is seen in an asymmetric "inner" coma, surrounded by a much larger, "outer" coma, more than 50", or 300,000 km across, cf. Fig.1. The brightness of the nucleus varies by more than 1^m; it was not yet possible to determine the period unambigously. Subtracting the composite image from its own rotationally symmetric mean, an fan-shaped area of enhanced surface brightness at the 27.5^m per square arcsecond level is seen in the SE quadrant, i.e. opposite the direction

14

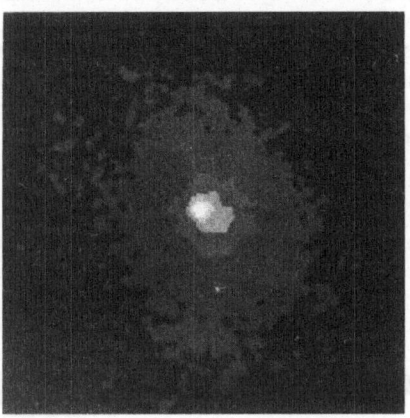

Figure 1: Comet Halley in April-May 1988; see text

to the Sun. These findings indicate that some dust is still being released into the coma and from there it is lost into interplanetary space via a dust tail, albeit exceedingly faint.

Hopefully more measurements of this type will follow in 1989 and, by use of the largest ground-based telescopes, also in 1990 and possibly in 1991 as well. After that the HST may take over. Not only may such observations provide a firmer basis for the determination of the rotation of the nucleus; they also carry cometary research further into the outer reaches of the solar system than ever before attempted.

Acknowledgements

It is pleasure to thank several colleagues, in particular P.B. Babadzhanov, S.Edberg, H.E. Jørgensen, R.L. Newburn, J. Rahe, Z. Sekanina, P. Wehinger, Ya.S. Yatskiv and D.K. Yeomans, for very useful comments on a draft of the present article.

References

A'Hearn, M.F., Hoban, S., Birch, P.V., Bowers, C., Martin, R., Klinglesmith, D.A., 1986, Nature, **324**, 649

Ananthakrishnan, S., Manoharan, P.K., Venugopal, V.R., 1987, Nature, **329**, 698

Arpigny, C., Magain, P., Manfroid, J., Dossin, F., Danks, A.C., Lambert, D.L., 1987, Astron. Astrophys., **187**, 485

Babadzhanov, P.B., Obrubov, J.A., Pushkarev, A.N., Hajduk, A., 1987 Bull. Astron. Inst. Czechosl., **38**, 367

Battrick, B., Rolfe, E.J., Reinhard, R. (eds.), 1986, ESA SP-250, Paris

Belton, M.J.S., Spinrad, H., Wehinger, P.A., Wyckoff, S., Yeomans, D.K., 1987, Astron. Astrophys., **187**, 569

Boyarchuk, A.A., Grinin, V.P., Zvereva, A.M., Sheikhet, A.I., 1987, Pisma v Astron. Zh., **13**, 228

Bobrovnikoff, N.T., 1931, Lick Obs. Bull., **17** Pt. II

Brady, J.L., Carpenter, E., 1971, Astron. J., **76**, 728

Brooke, T.Y., Knacke, R.F., Joyce, R.R., 1987, Astron. Astrophys., **187**, 621

Carusi, A., Valsecchi, G.B. (eds.), 1985, "Dynamics of comets: their origin and evolution", Proc. IAU Coll. 83, D.Reidel Publ. Company

Celnik, W.E., Schmidt-Kaler, Th., 1987, Astron. Astrophys., **187**, 233

Cosmovici, C.B., Schwarz, G., Ip, W.-H., Mack, P., 1988, Nature, **332**, 705

Donn, B., Rahe, J., Brandt, J.C., 1986, "Atlas of Comet Halley 1910 II", NASA SP-488, Washington D.C.

Edberg, S.J., 1983, IHW Amateur Observers Manual for Scientific Comet Studies, Parts I and II, JPL 83-16, Pasadena

Encke, J.F., 1820, in Berliner Astron. Jahrbuch f. 1823, 222

Encrenaz, Th., 1988, "La physique cometaire aprés l'exploration de la cométe de Halley", Annales de Physique, in press

Feldman, P.D. et al., 1987, Astron. Astrophys., **187**, 325

Gérard, E., Bockelée-Morvan, Bourgois, G., Colom, P., Crovisier, J., 1987, Astron. Astrophys., **187**, 455

Gombosi, T.I. (ed.), 1983, "Cometary Exploration", Proc. Int. Conf. 15 - 19 November 1982, Hung. Ac. Sci., Budapest

Green, D.W.E., Morris, C.S., 1987, Astron. Astrophys., **187**, 560

Grewing, M., Praderie, F., Reinhard, R. (eds.), 1988, "Exploration of Halleys' Comet", Springer-Verlag, Berlin - New York (pp.1-936

Hajdukova, M., Hajduk, A., Cevolani, G., Formiggini, C., 1987, Astron. Astrophys., **187**, 919

Halley, E., 1705, "Astronomiae cometicae synopsis", Philosoph. Transact. Royal Soc. London, **24**

Hoban, S., Samarasinha, N.H., A'Hearn, M.F., Klinglesmith, D.A., 1988, Astron. Astrophys., **195**, 331 also in Astron. Astrophys., **187**, 1987)

Irvine, W.M., Schloerb, F.P., Tacconi-Garman, L.E. (eds.), 1987, Proc. NRAO Workshop No. 17, 24-26 Sept. 1986, Green Bank

Jockers, K., Geyer, E.H., Rosenbaumer, H., Hänel, A., 1987, Astron. Astrophys., **187**, 256

Kamél, L., 1988, Astron. Astrophys., **193**, L4

Lamy, P.L., Grün, E., Perrin, J.M., 1987, Astron. Astrophys, **187**, 767

Landgraf, W., 1986, Astron. Astrophys., **163**, 246

Larson, H.P., Mumma, M.J., Weaver, H.A., 1987, Astron. Astrophys., **187**, 391

Larson, S.M., Sekanina, Z., 1985, Astron. J., bf 90, 823

Marsden, B.G., 1985, "Dynamics of Comets: Their Origin and Evolution", eds. Carusi, A. and Valsecchi, G.B., Astrophys. Spac. Sci. Library, **115**, 343

Millis, R.L., Schleicher, D.G., 1986, Nature, **324**, 646

Morley, T., 1984-6, Giotto Flight Dynamics Report, Vols. I-IV, ESOC, Darmstadt

Niedner, M.B.Jr., Schwingenschuh, K., 1987, Astron. Astrophys., **187**, 103

Opal, C.B., McCoy, R.P., Carruthers, G.R., 1987, Astron. Astrophys., **187**, 320

Piscitelli, J.R., Tholen, D.J., Hammel, H.B., 1986, ESA SP-250, Vol. III, 499

Ponnamperuma, C. (ed.), 1981, "Comets and the Origin of Life" Proc. 5th College Park Colloquium 29 - 31 October 1980, D.Reidel Publ. Company

Röser, S., 1987, Astron. Astrophys. Suppl. Ser., **71**, 363

Rolfe, E.J., Battrick, B. (eds.), 1987, ESA SP-278, Paris

Saito, T., Saito, K., Aoki, T., Yumoto, K., 1987, Astron. Astrophys., **187**, 201

Schleicher, D.G., Millis, R.L., Tholen, D., Lark, N., Birch, P.V., Martin, R., A'Hearn, M.F., 1986, ESA SP-250, Vol. I, 565

Schloerb, F.P., Kinzel, W.M., Swade, D.A., Irvine, W.M., 1987, Astron. Astrophys., **187**, 475

Schlosser, W., Schulz, R., Koczet, P., 1986, ESA SP-250, Vol. III, 495

Sekanina, Z., 1986, IAU Circ. No. 4187

Sekanina, Z., 1987, Comet Science Team Preprint Series No. 106, JPL, Pasadena

Sekanina, Z., Aronsson, M., 1985, Archive of Observations of Periodic Comet Crommelin (1983-84 apparition), JPL, Pasadena

Sekanina, Z., Larson, S.M., 1986, Astron. J., **92**, 462 (Paper IV)

Sekanina, Z., Larson, S.M., Emerson, G., Helin, E.F., Schmidt, R.E., 1987, Astron. Astrophys., **187**, 645

Sterken, C., Manfroid, J., Arpigny, C., Astron. Astrophys., **187**, 523

Véron, P., Festou, M., Kjär, K. (eds.), 1982, "The Need for Coordinated Ground-based Observations of Halley's Comet" Proc. ESO Workshop 29 - 30 April 1982, ESO, Garching

West, R.M., Jørgensen, H.E., 1988, submitted to Astronomy & Astrophysics

West, R.M., Pedersen, H., Monderen, P., Vio, R., Grosbøl, P., 1986, Nature, **321**, 363

Wilkening, L. (ed.), 1982, "Comets", Univ. Arizona Press, Tucson

Wyckoff, S., Lindholm, E., Wehinger, P.A., Peterson, B.A., Zucconi, J.-M., Festou, M.C., 1988, Astrophys. J. (in press)

Wyckoff, S., Wagner, R.M., Wehinger, P.A., Schleicher, D., Festou, M., 1985, Nature, **316**, 241

Wyckoff, S., Wehinger, P.A., Spinrad, H., Belton, M.J.S., 1986, ESA SP-250, Vol. I, 311

Yeomans, D.K., 1977, Astron. J., **82**, 435

Yeomans, D.K., West, R.M., Harrington, R.S., Marsden, B. (eds.), 1984, "Cometary Astrometry" Proc. of ESO/IHW Workshop 18 - 19 June 1984, JPL, Pasadena

HALLEY'S COMET (Part II): Space Studies

Vassili I. Moroz
Space Research Institute
Academy of Sciences of USSR
Moscow 117810, USSR

ABSTRACT. An international armada of spacecraft encountered Comet Halley in March 1986. The present article gives a brief overview of this unique event during which a cometary nucleus was seen as a spatially resolved object for the first time. It is a very dark body, the shape is irregular and the structure is inhomogeneous; only the sunward side is active. Many parent molecules were identified in the coma, including complicated hydrocarbons. There was also organic matter in the comet dust particles. Extensive studies were made of the complicated plasma environment of the comet.

1. Cometary missions

An impressive armada of spacecraft encountered Comet Halley in March 1986 (Fig. 1). The largest of them were Vega-1 and Vega-2. They had two goals: (1) studies of Venus by means of balloons and landers and (2) fly-by studies of Comet Halley. In June 1985 both spacecraft successfully delivered the landers and balloons at Venus, made a gravitational swing-by manoeuvre near this planet and were directed towards Comet Halley. The Vega mission was an international project. Although the spacecraft themselves were controlled by the Soviet Union, the scientific programme and the payloads were coordinated by the International Science and Technical Committee (CIST), representing scientific institutions in nine countries. Academician Sagdeev was the president of the CIST.

Also the Giotto project was international. It was set up by the European Space Agency (ESA). Giotto was the first attempt of ESA to visit deep space. The Japanese "Planet-A" project included two nearly identical spacecraft - "Suisei" ("Comet") and "Sakigake" ("Pioneer"). They both studied the solar wind/comet interaction, and Suisei also provided UV-imaging of the hydrogen coma.

Giotto encountered the comet after the two Vega spacecraft. Its trajectory was corrected by the use of accurate data for the comet's orbit which were obtained by Vega. This also became an area of efficient international collaboration. Fast processing, transfer of the information from the Vegas and almost real-time calculations of new ephemerides (taking also into account new ground-based astrometric data) was the subject of the very successful, international "Pathfinder" programme.

Important observations of Comet Halley were provided by two space telescopes in near-Earth orbits (IUE and Astron), sounding rockets and also by the UV-spectrometer from Pioneer Venus orbiter.

How can this remarkable accumulation of space efforts be explained? There are several reasons: 1) Every apparition of Comet Halley is an important astronomical event because

17

D. McNally (ed.), Highlights of Astronomy, Vol. 8, 17–31.
© *1989 by the IAU.*

this is the brightest periodical comet; its physical properties are more typical of a young comet than of a periodical one; 2) periodical comets are the only ones suitable for close-up studies by spacecraft; 3) spacecraft give a unique possibility of approach to the cometary nucleus and to investigate it as a spatially resolved object, 4) cometary nuclei are probably samples of pristine materials and their study opens new paths towards the understanding of the origin of the Solar System.

The main scientific results of the missions to Comet Halley are contained in the Proceedings of several conferences (COSPAR XXVI, Toulouse, 30 June-11 July 1986. Adv. Space Res., Vol. 5 (No. 12), 113-121; ESLAB Symposium on the Exploration of Halley's Comet, Heidelberg, 27-31 October 1986; Symposium on the Diversity and Similarity of Comets, Bruxelles, 6-9 April 1987) and in many journals and papers. There were some special issues on the subject: Nature (Vol. 321, No. 6067, 1986), Kosmicheskije Issledovanija (Vol. 25, No. 5, 1987 - in Russian), Astronomy and Astrophysics (Vol. 187, 1987). The most important topics in these collections of papers are: *nucleus, gas, dust and plasma.*

Below we will outline the most important achievements, emphasizing those results which give new information about the nucleus. These are the close-up images, the measurements of IR radiation and also the studies of the gas and dust composition which are indirect, but unique sources of data about the chemical composition of the nucleus. The quantity of published original papers is very large and we will include in our references only a small part of them. There are comprehensive reviews devoted to some selected topics with more detailed references, for instance by A'Hearn (1988) about cometary nuclei and by Galeev (1987) about plasma events.

2. The nucleus: Images and IR-radiometry

Early in the morning of March 6, 1986, hundreds of people were at their working stations at IKI (Moscow) and at the Soviet Deep Space Network station in Evpatoria (Crimea). The culmination of this huge project was approaching. Vega-1 was about to encounter Comet Halley.

The nucleus comes nearer and nearer. The comet now looks phantastic. There is a broad cone (fan) with a bright spot at its top and the nucleus inside. The fan is directed towards the Sun, opposite to the orientation of the usual cometary tail. The reason is clear. We see the innermost part of the coma, virtually unattainable for observations before. The cone is formed by dust grains driven out by evaporated volatiles from the sunlit side of the nucleus towards the Sun. Only at a distance of many thousands of kilometers (outside our field of view) the trajectories become curved, due to solar radiation pressure and they reverse their direction, thereby forming the dust tail.

Now there are a few minutes left before the pericenter passage. A huge stone with blurred edges is visible on the displays. This is the nucleus surrounded by dust. Frames follow quickly, one after another; "the stone" is turning before the enchanted audience. It is like a flight above the nucleus on-board an aircraft. The goal has been achieved - for the first time, scientists have seen a cometary nucleus.

The interpretation of these images was not a simple task because of the screening effect by the dust. However, the analysis showed that the true boundaries of the nucleus were visible through the surrounding dust. So the structure, shape, size and reflectivity of the nucleus could be determined.

The nucleus was imaged by the two spacecrafts at different rotation phases of the nucleus and this gives a possibility to reconstruct a 3D model and to define its size (Fig. 3) and also to evaluate its period of rotation and the direction of its axis (Sagdeev et al. 1986a, 1986b, 1986c).

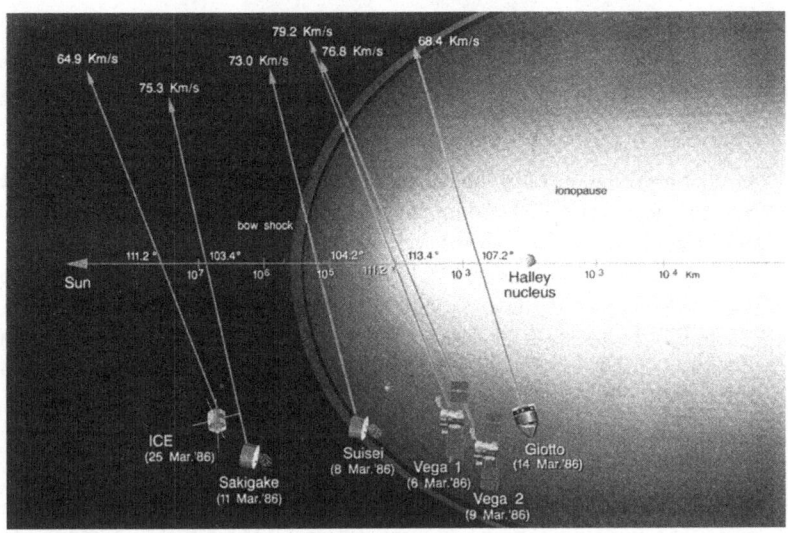

Figure 1: Six spacecrafts met Comet Halley near the descending node of its orbit. Three of them (Vega-1, Vega-2 and Giotto) penetrated deeply into the dust coma. The relative speed of the encounter was $\sim 70 - 80$ kms^{-1} and cometary dust grains were extremely dangerous for the spacecraft; for this reason, protective screens covered the most critical systems by Vega-1 and Vega-2. However, it was decided not to fly too close to the nucleus and minimum distances of 8900 and 8000 km were chosen. The Giotto spacecraft was more compact and better protected. Its closest approach to the nucleus was only 600 km.

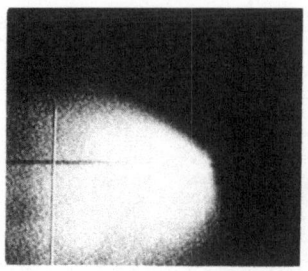 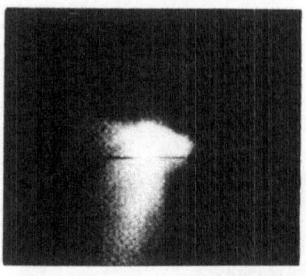 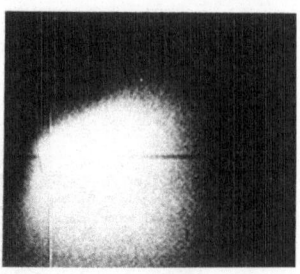

Figure 2: This is how Comet Halley looked from a spacecraft flying through its coma: a) The distance to the nucleus d = 29,000 km; the Sun is on the left, the nucleus is in the right part of the bright dust fan directed towards the Sun; the frame size is 228 × 308 km. b) d = 8,000 km; the Sun is above the plane of the image (phase angle ∼ 30°); frame size 62 × 84 km; the nucleus, surrounded by dust, is visible. c) The Sun is to the right, the nucleus is to the left; the distance and the frame size are the same as in a). These examples are taken from the collection of images obtained by Vega-2 (Sagdeev et al., 1986a). No processing was done on the images shown here.

Excellent images were obtained by the Halley Multicolour Camera on Giotto (Keller, et al., 1986a, 1986b, 1988). The data about shape and size are in general agreement with Vega's imagery. The imaging conditions were different: the almost constant (near 100°) phase angle compared to the fast change in the phase angle in the case of Vega. This made it possible to combine many Giotto images for contrast enhancement (Fig. 4), but it is impossible to make a 3D reconstruction using the Giotto data alone.

Let us summarize the main conclusions which result from the studies of the Vega and Giotto images:

1. The nucleus is a single, solid body. Its shape is irregular (like a "potato"); the size is approximately 8 × 8 × 16 km and the volume is ∼ 500 km^3.

2. The reflectivity is very low, the geometric albedo is ∼ 0.04. Thus the nucleus of Comet Halley is one of the darkest bodies in the Solar System.

3. Dust is emitted from the sunlit side of the nucleus only. Its flow is spatially inhomogeneous (and time-variable). There are some regions - a few percent of the entire surface - which eject narrow dust jets, but a weak dust outflow also exists outside these jets. The theory predicted a thin (∼ 1 km), near-surface layer of dust in which particles are accelerated by the gas from zero to almost final velocity. Now we have evaluated the average optical depth of this layer on the sunlit side of the nucleus to about 0.1.

4. There is a complicated pattern of surface features, including ring structures, bright spots, etc.

5. The period of rotation of the nucleus is about 2.2 days. Almost the same value was obtained from the analysis of the outburst activity periodicity observed from the Earth (Sekanina and Larson, 1986). The inclination of the equator to the orbital plane is ∼ 30°. The direction of rotation is retrograde, like the orbital motion. Some of the

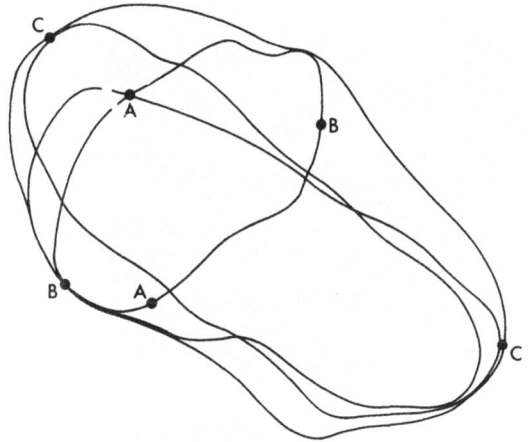

A - A: 7.5 ± 0.8 km
B - B: 8.2 ± 0.8 km
C - C: 16.0 ± 1.8 km

Figure 3: 3D model of the nucleus of Comet Halley, based on processed images obtained by Vega-1 and Vega-2 (Sagdeev et al., 1986 b).

ground-based studies suggest a rotation period near 7.4 days (Millis and Schleicher, 1988) or even 14.6 days (Festou et al., 1988); these authors proposed a model with rotation around the long axis with a period of 14.6 days and precession with a 2.2 day period around another axis inclined by 84° to the first one.

The TV imaging of the nucleus was supplemented by measurements of its thermal infrared radiation. This was done by the infrared radiometer which was part of the IKS instrument on Vega-1 (Nikolsky et al., 1987). The result was unexpected: a surface temperature of ~ 370 K was estimated in the vicinity of the subsolar point! This fact led to difficulties in understanding the surface structure of the cometary nucleus if we take into account that we need a large amount of water ice there. There is a strong outflow of gas that escapes the nucleus and H_2O is the main constituent. Consequently, water ice is one of the important components of the nucleus.

To provide the observed flux of H_2O molecules (see below), a substantial part of the illuminated surface (not less than 25%) must be icy. A important fraction of the received solar energy is spent on evaporation of this ice. Evaporation leads to cooling and quantitative estimates show that the ice surface temperature in this case should be ~ 200 K. But the surface is hot and black. How can this strong discrepancy be explained?

Two hypotheses may possibly solve this difficulty:

1. The ice is covered by a black, porous layer with low thermal conductivity. We can see the same for spring snow in towns. The outer surface of such a thin layer absorbs solar energy and transfers it down to the ice boundary, where the temperature is indeed equal to ~ 200 K. Owing to its porosity, the black layer may be transparent for a gas flow from below. This is not a new idea. The possibility of the presence of such a layer ("mantle") has been discussed for many years (cf. Markovich, 1988). There have been predictions, however, that such a layer would not remain at the relatively small heliocentric distance where the spacecraft encounters with the comet took place.

2. The surface of the nucleus has a horizontally inhomogeneous ("spotted") structure. There are some cold areas with high gas and dust production and some warm ones

Figure 4: Composite image of the nucleus and the inner coma, as obtained by an analysis of the data obtained by the Giotto Multicolour Camera (Keller et al., 1988). The Sun is to the left. Craters and other surface details are visible near the terminator.

of lower activity. This model has its own difficulties: "open ice", if it exists in cold areas, must be dark. No one has yet presented quantitative evidence that this is really possible. So some specific properties of the surface layer in cold spots are probably implied, even by this model. It is hardly open ice without a more or less pronounced black mantle.

A strong argument for the second model is the thermal lag of about 1.5-2 hours between the regions of the maximal temperature and the subsolar point (Nikolsky et al., 1987). It is not compatible with any reasonable, horizontally homogeneous model.

3. Neutral and Ionized Gas

Many cometary molecular and atomic emission lines and bands were recorded during the ground-based observations. But most of the corresponding atoms and molecules (like H, OH, C_2, CH, CN, NH_2, etc.) cannot be the actual constituents of cometary nuclei. They are definitely secondary species and they must originate from primary, "parent" molecules. These parent gases, present in a frozen state inside the nucleus, are supposed to be its major constituents. Molecules such as H_2O, CO_2, CH_4, NH_3 and also many other and more complicated ones have been discussed. However, all of these are multiatomic molecules, radiating mainly in the IR range, and therefore not available for observations from the ground.

Studies of Comet Halley by spacecraft provided the real break-through in the solution of this difficult problem. Now we have for Comet Halley a long list of actually observed parent molecules (see Table 1). Infrared spectrometers (IKS for 2.5 - 4.8 μm and TKS IR channel for 0.8 - 1.7 μm ranges) were used on the Vega missions and mass-spectrometers (NMS, IMS, PICCA) were on-board Giotto.

Some of the results are presented in Fig. 5. The most intense bands in the IKS range are located near 2.7 and 4.3 μm. They belong to H_2O and CO_2, respectively. The same molecules were successfully measured by the neutral mass spectrometer (NMS) on Giotto. The results from both experiments are in excellent agreement. They showed that in the gaseous coma, H_2O is the major constituent among the parent molecules, and that the CO_2/H_2O-ratio is $\sim 3\%$. The H_2O band (2.7 μm) in the spectrum of Comet Halley was detected already several months before the Vega and Giotto encounters from the high-altitude US Kuiper Airborne Observatory (Weaver et al., 1987); it was observed with extremely high resolving power ($\lambda/\Delta\lambda \simeq 10^5$).

TABLE 1: PARENT MOLECULES DETECTED IN COMET HALLEY

Gas	Q_i/Q_{H_2O}	Q_{max}	Line-spectr.	Mass-spectr.
H_2O	1	10^{30}	V1:IKS [1]	G:NMS [3]
		$1.6 \cdot 10^{30}$	V2:TKS [2]	
CO_2	0.03	$3 \cdot 10^{28}$	V1 [1]	G:NMS [3] (upper limit)
CO^*	0.04	$4 \cdot 10^{28}$	V1 [1]	G:NMS (upper limit)
			IUE [4]	
CH-X	0.5	$5 \cdot 10^{29}$	V1 [1]	
H_2CO	0.05	$5 \cdot 10^{28}$	V1 [1]	
$(H_2CO)_5$	qual. ident.			G:PICCA [5]
CH_4^{**}	0.02	$2 \cdot 10^{28}$		G:IMS [6]
NH_3^{**}	0.01-0.02	$(1-2) \cdot 10^{27}$		G:IMS [6]
HCN	~ 0.001	10^{27}		radioastronomy [7]

*   Total Q_{CO} is higher, this is only the part which is directly connected with the nucleus.
**   No direct identification, conclusion is based on IMS data taking into account models.
References: [1] Combes et al. (1986); Moroz et al. (1987); [2] Krasnopolsky et al. (1988); [3] Krankowsky et al. (1986); [4] Festou et al. (1986); [5] Hübner, W.F. (1987); [6] Allen et al. (1987); [7] Schloerb et al. (1987).

A broad feature, consisting of a few overlapping peaks, is seen between 3.2 and 3.6 μm (Fig. 5). The central wavelength 3.4 μm corresponds to the CH-band in complex organic molecules (hydrocarbons). This feature was confirmed soon after the Vega mission by ground-based observations; fortunately, the terrestrial atmosphere is transparent in this part of the spectrum. A definitive identification of the detected band is not yet possible, but a mixture of saturated and unsaturated hydrocarbons is presumed. Approximate estimates show that their mixing ratio can be larger (up to several tens of percent), if the band is generated in a resonance emission by molecules in the gas phase. However, it is also possible that 3.4 μm band is radiated in the process of UV excitation of large molecules or, partly or completely, as thermal radiation from dust grains.

24

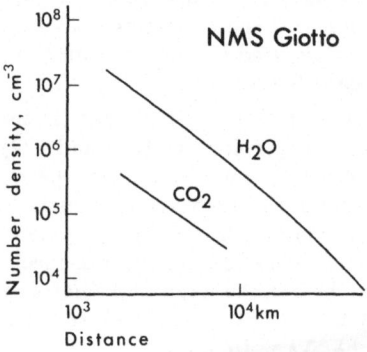

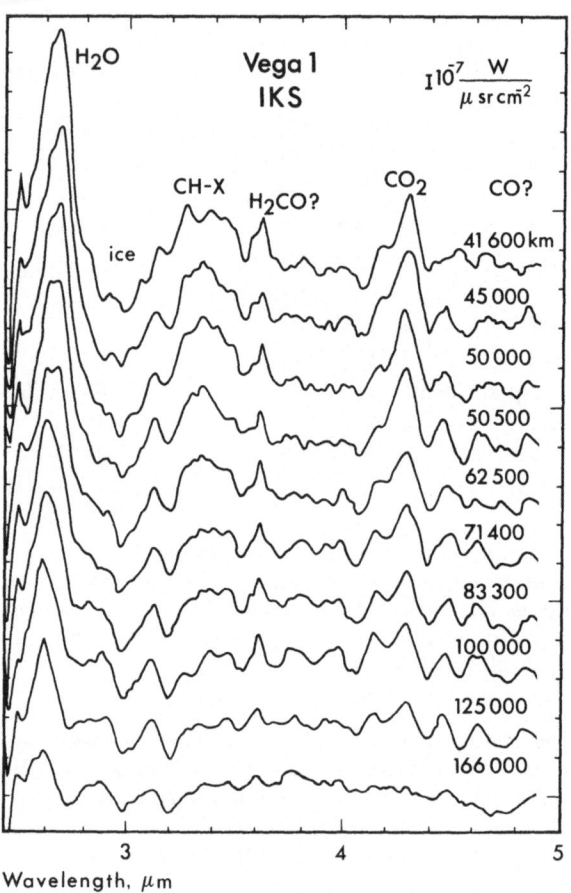

Figure 5

A narrow peak near 3.6 μm has been tentatively identified with formaldehyde H_2CO. According to Hübner (1987), the PICCA mass-spectrometer on Giotto shows the presence of more complicated molecules, including H_2CO as a part of polyformaldehyde $(H_2CO)_5$.

The ion mass-spectrometer on Giotto provided many identifications of ionospheric constituents. Taking into account chemical models of the coma, an accurate analysis of the ion data shows that it definitely contains neutral NH_3 and very probably CH_4 (Allen et al., 1987); both are classical candidates to the list of parent molecules.

Carbon monoxide CO may also be one of the parent molecules. However, part of its flow is generated not by the nucleus itself, but by the so-called "distributed source", i.e. dust or other molecules; in the latter case not all of the CO molecules are parent.

Some of the parent molecules, CO_2 and especially CO, NH_3, CH_4, have very low condensation temperatures, much lower than H_2O. Thus, the nucleus of Comet Halley could not have solidified inside the orbit of Neptune. This imposes certain limitations on theories of the origin of comets.

The low condensation temperature, and consequently the high pressure of saturated vapor at small distances from the Sun, could in principle explain the time-variable outburst activity of comets.

The time dependence of the overall gas production rate is now better known for Comet Halley than for any other comet. It shows a strong asymmetry relative to perihelion (Fig. 6). One of the important applications of this dependence is an estimate of the mass of the nucleus. Gas ejected from the nucleus creates a jet force and influences the motion of the comet; in a sense it behaves like a rocket. Rickman (1986) and Sagdeev et al. (1987a, 1988) have independently tried to determine the mass; values from 0.5 to $2 \cdot 10^{17}$ g are given by the first, and from 1 to $8 \cdot 10^{17}$ g by the second author. The value $2 \cdot 10^{17}$ g is within the limits given by both. It corresponds to a density of 0.5 gcm^{-3}. The uncertainty is large and several poorly known factors affect this value. However, the average density is most likely lower than 1 and higher than 0.1 gcm^{-3}. We can also evaluate a nominal "decay time" for Comet Halley, as M and dM/dt are known; it is $\sim 5 \cdot 10^4$ years. In reality, it could be longer, since comets are supposed to gradually reduce their activity in the course of their lifetime.

The neutral gas coma consists of gas that has escaped from the nucleus. It can be treated as a cometary gas atmosphere. Contrary to planets, this atmosphere is not in hydrostatic equilibrium and the number density only obeys the continuity condition. The number density near the surface is $\sim 3 \cdot 10^{13}$ and the pressure is ~ 1 μbar. This is equivalent to a good laboratory vacuum. However, the matter in this near-vacuum is still sufficient to generate the entire array of remarkable cometary events.

The outermost part of the cometary atmosphere is its hydrogen coma. It extends over $\sim 10^7$ km, consists of hydrogen atoms and emits the resonance $L\alpha$ line at $\lambda 1216$ A. This hydrogen coma was detected with sounding rockets during earlier observations of other comets. Among all members of the Halley armada, only Suisei was provided with a device for observations of the coma of Comet Halley. These observations commenced about 5 months before the encounter. It was discovered by Suisei that the H-coma is "breathing" and that the intensity varies by a factor of 2 with a 2.2 day period (Kaneda et al., 1986). The explanation is clear: it reflects the rotation of the nucleus and the horizontal inhomogeneity of the surface activity.

Neutral atoms in the coma create ions by means of photoionization. Their number density is low near the nucleus, it increases with distance, passes through a maximum at some distance and decreases further out. The first cometary ions were detected by the spacecraft, $\sim 10^7$ km from the nucleus.

The interplanetary space is not empty. Solar ions move from the Sun with a speed

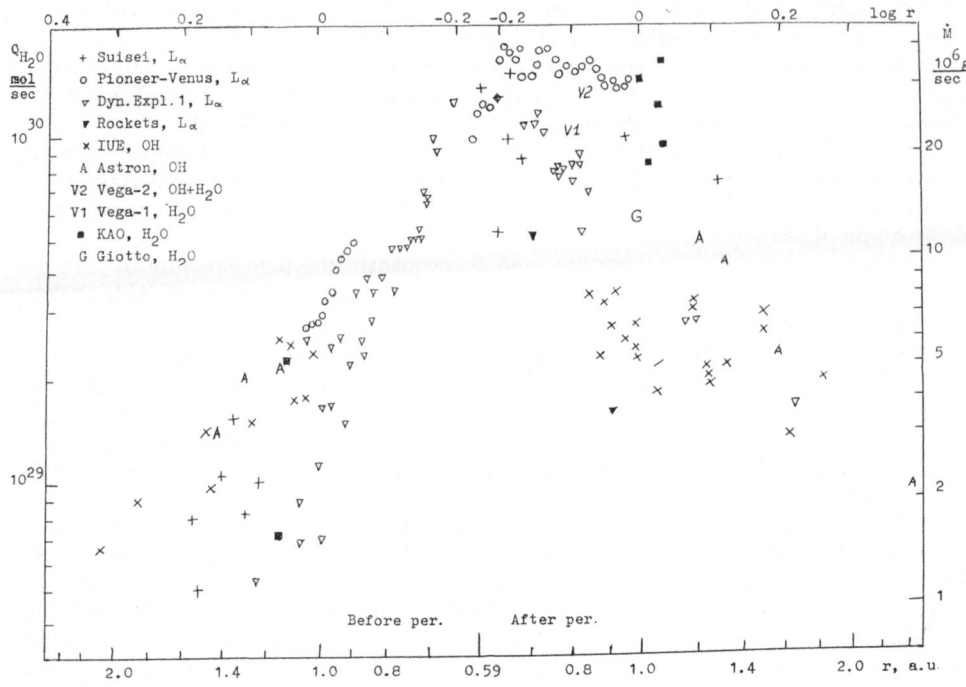

Figure 6: Gas production rate of Comet Halley as function of distance to the Sun before and after perihelion. Abscissa: heliocentric distance; ordinate: gas production rate in mol/s (left scale) and g/s (right scale). There is a pronounced asymmetry relative to the perihelion; seasonal effects might be a possible explanation. The data are from flux measurements of three different species: H_2O [Kuiper Airborne Observatory - Weaver et al. (1987); Vega-1 - Moroz et al. (1987); Vega-2 - Krasnopolsky et al. (1987,1988); Giotto - Krankowsky et al. (1986)]; OH [IUE - Feldman et al. (1987); Astron - Boyarchuk et al. (1986)]; H [Suisei - Kaneda et al. (1986); Pioneer-Venus - Stewart et al. (1987); Dynamics Explorer 1 - Graven and Frank (1987); sounding rockets - Opal et al. (1987)].

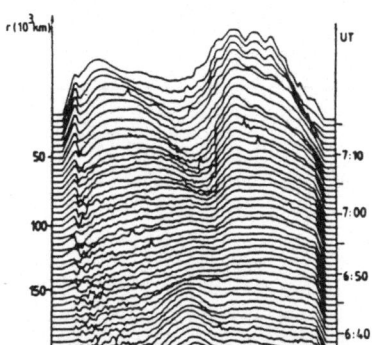

VEGA-2 PLASMAG-1 RAM-ANALYSER 1MIN AVERAGES

Figure 7: The ion energy spectrum changed during the approach of Vega-2 towards the central region of Comet Halley. The solar wind ions dominate at distances more than ~100,000 km, the cometary ions at smaller distances (Gringauz et al., 1986).

of 400 kms^{-1}. This is the "solar wind". Solar ions encounter the ones from the comet. It was expected that a shock wave would arise in this process. Indeed, the spacecraft intersected the shock wave at a distance of $\sim 10^6$ km from the nucleus. It was "smoother" than predicted by theoreticians. The same property of the shock boundary was observed 6 months earlier by ICE at Comet Giacobini-Zinner.

The speed of the solar wind is diminished, the magnetic field is amplified and a turbulence forms behind the shock wave boundary. The number density of cometary ions increases when we move towards the nucleus. At the distance $\sim 10^5$ km, "quantity converts to quality": within this boundary, the cometary ions dominate in the chemical composition of the plasma (Fig. 7). Such a relatively distinct "chemical boundary" was not predicted by any theory. It has appropriately been named "cometopause". It was observed that the ion flux inside the cometopause is subjected to short-time periodic variations; this phenomenon has not yet been explained.

The next important boundary is the so-called "contact surface" through which the magnetic field does not penetrate (Fig. 8). The contact surface was detected by Giotto at ~ 5000 km from the nucleus. The magnetic field dropped from 60 nT to zero when Giotto flew through this contact surface.

4. Dust

There is a lot of dust in the coma. A simple explanation of this fact is more or less obvious: the volatiles in the nucleus are mixed with refractory grains and push them outside during the evaporation. Consequently, the measurements of the composition of the dust grains have told us about the nature of the refractory material inside the nucleus.

New and important information about this subject was obtained by the dust impact mass-spectrometers PUMA on Vega (Sagdeev et al., 1987) and PIA on Giotto. A target

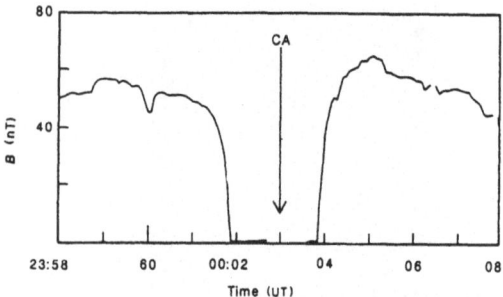

Figure 8: Magnetic "hole" inside the comet (Neugebauer et al., 1986)

of chemically pure silver was placed at the entrance of the mass-spectrometers. A dust particle converts to a small cloud of ions when impacting on this target. The ion cloud enters the mass-spectrometer, allowing the measurement of its elemental composition. Only small dust grains with the masses in the 10^{-16}-10^{-14} g interval were studied in this way, but there were many of them. For example, PUMA registered about 2000 cometary dust particles.

There were particles with predominance of light elements, carbon or metals. Sometimes peaks of H, C, O, N were visible simultaneously. The analysis of these so-called CHON particles led to the next interesting conclusion: 1) these particles have a C/Si relative abundance which is substantially higher than in any meteorites of known classes; 2) nitrogen is present in all particles with large C/Si-ratio, but this is possible only if both C and N are constituents of organic molecules. Thus we have here the second evidence that there is organic matter within the cometary nucleus. Greenberg qualitatively predicted the existence of CHON particles on the basis of his model, describing the nucleus as a "snow-ball" that consists of interstellar dust grains. He favoured a complicated structure of such grains (see below).

The existence of CN-jets correlated with dust (ground-based observations) is probably explained by the outgassing of CHON particles. This correlation between gas and dust is one of the hitherto unknown cometary properties which were detected during this apparition of Halley.

Many other, less complicated particle detectors were used during the cometary missions in order to measure in-situ number densities and mass distributions - on Vega (Masets et al., 1987; Vaisberg et al.; 1987, Simpson et al., 1987) and on Giotto (McDonnell et al., 1987). The number of small grains is higher than what was predicted from earlier ground-based remote observations. A summary of the physical properties of cometary dust is given in Table 2. There is evidence of destruction of dust particles on their paths. It means that at least the larger particles are fragile. Their density is probably much less than 1 gcm^{-3}. Many examples of fragile, fine-structured interplanetary particles (so-called Brownly particles) have been collected in the terrestrial stratosphere. Their cometary origin is now almost proven. A direct calibration of the dust particle detectors was not possible and for this reason the estimates of the dust production rate for Comet Halley can only be given with an accuracy of half an order of magnitude. The dust-to-gas ratio was probably between 0.1 and 0.5 at the time of the Vega and Giotto encounters.

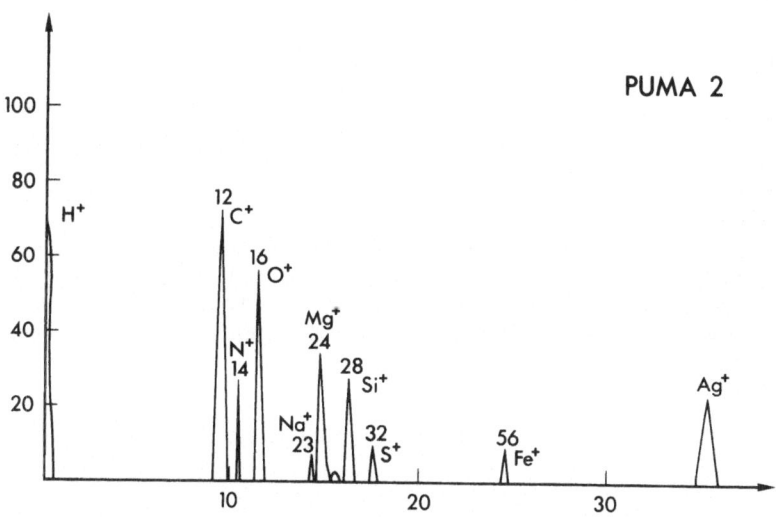

Figure 9: One of the mass-spectra obtained by the particle impact mass-spectrometer PUMA on Vega-2. This example shows that a complicated mixture of silicate, metallic and organic material is present in some cometary particles.

TABLE 2: SOME PROPERTIES OF COMETARY DUST

(a) **Cumulative mass distribution**: $r = 8100$ km (Vega-2)
$3 \cdot 10^{-16} < m < 10^{-10}$ g $\qquad$ $F(>m) \propto m^{-0.3}$
$10^{-10} < m < 3 \cdot 10^{-7}$ g $\qquad$ $F(>m) \propto m^{-0.85}$

(b) **Optical "activity"**
90% of scattered solar light is provided by
particles with $m > 10^{-12}$ g

(c) **Average density** of particles with $m > 10^{-12}$ g
$\sim 0.3 \mathrm{gcm}^{-3}$ $\qquad$ uncert. factor ~ 2

(d) **Velocity**

$m(g)$	$u_d(\mathrm{kms}^{-1})$
10^{-12}	0.75
10^{-8}	0.5

a - Masets et al. (1987); b,c,d - Krasnopolsky et al. (1987)

5. Concluding remarks

The nucleus

Let us now summarize what we know about the nucleus of Comet Hally, taking into account the information obtained by the studies of gas and dust. It is a single solid body; it cannot be a collection of gravitationally bound or unbound small components. Water ice is the main volatile constituent. There are many other, less abundant constituents, the most important of them are probably organic molecules as in volatiles and/or refractory compounds. The average density of the nucleus is low, probably less than 1 gcm^{-3}.

A model of the cometary nucleus as a single, solid body consisting of a mixture of volatiles and meteoritic matter was proposed by F. Whipple (1950). His model has been confirmed by the Halley missions. In addition, some new important properties were found:

1. the presence of the organic matter in different forms,

2. a pronounced inhomogeneity of the surface properties,

3. a low albedo,

4. a high temperature in some areas,

5. specific properties of the surface layer, and

6. a low density.

Different ideas about the internal structure of the nucleus are now being discussed and two of them will be mentioned here. The first was elaborated by Greenberg (1982) who described the nucleus as a package of interstellar grains. These are supposed to be small cylinders, a few microns in length. Each of them contains all the components of the nucleus: silicate core, organic internal mantle, an icy external mantle and even submicron particles within the latter.

But this is only one of the hypotheses under discussion. Another example is an icy glue model proposed by Gomboshi and Houpis (1986). It describes the nucleus as a collection of many massive silicate stones connected by ice layers of varying thickness. This concept explains very well the linear shape of the active regions on the surface of the nucleus. There are other models, but these two are probably the extreme cases. We could hardly decide between them without cometary sample return missions. Such plans are discussed very enthusiastically in the wake of the successful Halley missions. This is one of the most important new lines in the future studies of the Solar System.

The place of the comets in the Solar System.

Comets almost definitely come from Oort's cloud. It contains $10^{11} - 2 \cdot 10^{12}$ cometary nuclei and is located at $2 \cdot 10^4$ AU from the Sun. It is the most distant component of the Solar System. If the mass of Comet Halley's nucleus is typical and if the number of such nuclei in Oort's cloud is $2 \cdot 10^{12}$ then its mass is ~ 100 M$_\oplus$, i.e. 1/4 of the total mass of all planets. The angular momentum could then be of the order of one magnitude higher than that of the planets. These two simple ideas were mentioned recently by Marochnik and Mukhin (1987). They lead to a new understanding of the possible place of comets in the Solar System. It may be much more important than was supposed before.

References

References to papers presented at the 20th ESLAB Symposium on the Exploration of Halley's Comet, Heidelberg, eds. by Battric B., Rolf, E.J., Reinhard R., ESA SP-250, Noordwijk, The Netherlands, ESA Publ. Div., 1986, are marked (H). Most of the papers presented at this symposium were also published in Astron. and Astrophys., Volume 187. In these cases references are given to the appearance in this journal.

A'Hearn, M.F., 1988, Ann. Rev. Earth Planet. Sci., **16**, 273
Allen, M. et al., 1987, Astron. and Astrophys., **187**, 502
Boyarchuk, A.A. et al., 1986, (H), Vol.3, 193
Combes, M. et al., 1986, Nature, **321**, 266
Feldman, P. et al., 1987, Astron. and Astrophys., **187**, 325
Festou, M.C. et al., 1986, Nature, **321**, 361
Festou, M. et al., 1987, Astron. and Astrophys., **187**, 575
Galeev, A.A., 1987, Astron. and Astrophys., **187**, 12
Gomboshi, T.I. and Houpis, H.L.F., 1986, Preprint KFKI-1986-23/C, Budapest
Graven, J.D. and Frank, L.A., 1987, Astron. and Astrophys., **187**, 351
Gringauz, K.I. et al., 1986, Geophys. Res. Lett. **13**, 613
Greenberg, J.M., 1982 in Comets, ed. Wilkening, L.L., Tucson, University of Arizona Press, 131
Hübner, W.F., 1987, Science, **237**, 628
Kaneda, E. et al., 1986, Geophys. Res. Lett., **13**, 833
Keller, H.U. et al., 1986a, Nature, **321**, 320
Keller, H.U. et al., 1986b, (H) Vol.2, 347
Keller, H.U. et al., 1988, Nature, **331**, 333
Kissel, Y. et al., 1986, Nature, **321**, 336
Krankowsky, D. et al., 1986, Nature, **321**, 326
Krasnopolsky, V.A. et al., 1987, Astron. and Astrophys., **187**, 707
Krasnopolsky, V.A. et al., 1988, Astron. and Astrophys. (in press)
Markovich, M.Z., 1958, Bull. of Inst. of Astrophys. and Phys. (Dushana, USSR), **28**, 25 (in Russian)
Marochnik, L.S. and Mukhin, L.M., 1987, Preprint IKI Pr-1319, Moscow (in Russian)
Moroz, V.I. et al., 1987, Astron. and Astrophys., **187**, 513
Millis, R.L. and Schleicher, D.G., 1986, Nature, **324**, 646
Neugebauer, F.M. et al., 1986, Nature, **321**, 352
Nikolsky, Yu.V. et al., 1987, Kosmich. issled., **25**, 793 (in Russian)
Opal, C.B. et al., 1987, Astron. and Astrophys., **187**, 320
Rickman, H., 1986, Proc. of an ESA Workshop held at the University of Kent at Canterbury, UK, 15-17 July 1986, xxx
Sagdeev, R.Z. et al., 1986a, Nature, **321**, 262
Sagdeev, R.Z. et al., 1986b, (H) Vol.2, 307
Sagdeev, R.Z. et al., 1986c (H) Vol.2, 335
Sagdeev, R.Z. et al., 1987a, Pisma v Astron. Zh., **13**, 621 (in Russian)
Sagdeev, R.Z. et al., 1987b, Kosmich issled., **25**, 856 (in Russian)
Sagdeev, R.Z. et al., 1988, Nature, **331**, 340
Schloerb, F.P. et al., 1987, Astron. and Astrophys., **187**, 47
Sekanina, Z. and Larson, S.M., 1986, Nature, **321**, 357
Stewart, A.I.F. et al., 1987, Astron. and Astrophys., **187**, 351
Weaver, H.A. et al., 1987, Astron. and Astrophys., **187**, 411
Whipple, F., 1950, Astrophys. J., **111**, 375

THE RISE AND FALL OF QUASARS

Maarten Schmidt
Palomar Observatory
California Institute of Technology
Pasadena, CA   91125, USA

It is a great honor for me to speak to you tonight on the subject
of quasars. In the past twenty-five years, an enormous amount has been
learned about these objects and the field has become very technical. I
recognize that there are many people in the audience who are not professional astronomers and I hope not to disappoint them. At the same
time my colleagues may recognize a few things that I touch upon. They
will also hopefully forgive me for leaving out many attributions: many
hundreds of astronomers have contributed to the field.

1. DISCOVERY OF QUASARS 1960-63

The discovery of quasars happened during a concentrated effort in the
1950s and 1960s to obtain optical identifications of extragalactic radio
sources. Many of the strong radio sources were identified as giant
elliptical galaxies. Minkowski's success in confirming the identification of 3C 295 with a galaxy of redshift 0.46 was a major achievement
in 1960.

It was arround that time that Tom Matthews, at Caltech, obtained an
accurate position of the radio source 3C 48. Allan Sandage took a plate
of the field in September 1960, which showed a stellar object with faint
nebulosity at the radio position. He also obtained the first spectra of
the stellar object which contained strong broad emission lines that
could not be identified. Photometry showed that the object had an ultraviolet excess and that it varied in brightness over several months. In
the announcement in the March 1961 issue of Sky and Telescope, the possibility that it was a galaxy was dismissed in favor of an interpretation in terms of a peculiar local star.

In 1961 and 1962 Tom Matthews concentrated on the radio positions
of radio sources of small angular diameter in the hope of detecting
galaxies in very distant clusters. Without knowing it in advance, he
attempted in those two years optical identifications of as many as
seven radio sources that eventually turned out to be quasars. Some of
the optical objects were misidentifications, including 3C 273 which was
identified with a galaxy. In the other cases, the optical objects were

33

D. McNally (ed.), Highlights of Astronomy, Vol. 8, 33–44.
© 1989 by the IAU.

stellar, but the spectra which I obtained defied interpretation.

The source 3C 273 provided the clue to the mystery of the radio stars. Hazard et al. had been observing lunar occultations of the source in 1962, showing that the source was double. One component coincided with a bright star, of magnitude 13, and the other with a jet-like feature. I suspected that the bright star was a foreground object un-related to the radio source. Essentially with the idea of eliminating the star from consideration, I took some short-exposure spectra of it in December 1962. Surprisingly, the star showed a number of broad emission lines.

It took another six weeks for the mystery to be resolved. Hazard had written up the occultation results for publication and suggested that I write a companion article about the optical spectrum. While writing the manuscript, I took another look at the spectra. I noticed that four of the emission lines showed a pattern of increasing strength and increasing separation from blue to red. For some reason, I decided to construct an energy level diagram based on these lines. The results seemed to contradict the regular spacing of the lines. To check on that, I proceeded to take the ratio of the wavelength of each of the lines to that of the nearest Balmer lines, which do have a regular spacing pat-tern. The first ratio, of the line at 5630 Å to H-beta, was 1.16. The second ratio was also 1.16. When the third ratio was 1.16 again, it came as a flash that I was looking at a Balmer spectrum redshifted by 16 per cent.

A little while later, Jesse Greenstein and I were able to identify the emission lines observed in 3C 48 at a redshift of 37 per cent. The impact of the realization that fairly bright star-like objects could have large redshifts was stunning. We soon had intense discussions about what caused these large redshifts and within a week came to the seemingly reckless conclusion that these "stars" were at distances of thousands of millions of the light years. Considering the controversy about quasar distances that was to develop later, I believe in hindsight that we did all right in the time span of a week. I will come back to the interpretation of the redshifts presently.

2. OPTICAL PROPERTIES

In the first few years following the discovery of their redshifts, there was little understanding of the quasar phenomenon, as was well expressed in George Gamov's exhortation in the May 25, 1964 issue of Newsweek:

> Twinkle, twinkle, quasi-star
> Biggest puzzle from afar
> How unlike the other ones
> Brighter than a billion suns
> Twinkle, twinkle, quasi-star
> How I wonder what you are.

As a consequence, their definition had to be established empirically from observed properties listed below.

2.1 Star-Like Image

Quasars were originally required to have a dominant star-like component. This came from the observation that quasars look at first sight just like stars. In practice, that meant that much of the light came from a diameter of less than one arcsecond. However, nebulosity such as seen around the nucleus of 3C 48 is probably the rule.

2.2 Emission Lines

The emission lines shown by quasars are Lyman-alpha (1216 Å), C IV (1549 Å), C III (1909 Å), and Mg II (2798 Å). Quasar spectra also show the Balmer lines of hydrogen and sometimes forbidden lines as seen in planetary nebulae. The lines are typically 50-100 Å wide, corresponding to velocities around 5000 km s^{-1}.

2.3 Large Redshifts

Recognition of the line pattern is usually unambiguous, so there is rarely any major uncertainty about a quasar's redshift. The redshifts range from a few per cent to several redshift is 4.43 for a quasar found by Warren et al. in Cambridge. This redshift is as large as a bullet shot with a velocity that is 93.4 per cent of the speed of light. The total number of quasars with known redshift is around 4000. Their total number in the universe is of the order of millions.

2.4 Variability

Variability, such as first seen in 3C 48, turned out to be common in quasars. The observations of 3C 273 go back one hundred years--they show at times rapid variations, with a time scale of about a month. This means that the size of the light emitting region cannot be larger than a light month. In some quasars, and recently in 3C 273, even faster variations have been observed, producing upper limits to the size of the objects of the order of light weeks or light days.

3. INTERPRETATION OF REDSHIFTS

The interpretation of the redshifts of quasars has given rise to much controversy. I believe that this was partly caused because quasars were discovered too early. In the early days, I used to discuss four hypotheses for the redshift. We will recall these briefly. I will discuss the cosmological hypothesis, which is accepted by most astronomers today, last.

3.1 Local Explosion

This early proposal by Terrell posed that quasars were all ejected by our own Galaxy. Velocities of ejection would be large, with many at 80

or 90 per cent of the velocity of light. From our position about
30,000 light years off the center of the Galaxy, we would expect to see
angular motion on the sky unless the objects are very far away. No
angular motions were observed, requiring a minimum age of the explosion
of about 10 million years. My own arguments against this hypothesis
were mostly based on the total mass of the ejected quasar cloud. One
might also expect in this case to see quasars around other galaxies--
which should give rise to clustering of quasars and to blueshifts of
quasars, both of which are not observed.

3.2 Gravitational Redshift

This interpretation was one that Greenstein and I considered carefully
after the discovery of redshifts. On the basis of spectroscopic and
dynamical arguments, we were able to show that the mass of quasars in
this hypothesis had to be larger than that of the most massive galaxies
known.

3.3 No Physical Interpretation

This was an extraordinary option, mostly based on Arp's finding of
associations on the sky between quasars and galaxies at much smaller
redshifts. These associations according to Arp show that quasars are
at much smaller distances than would correspond to their cosmological
redshift distance. If the cosmological redshift interpretation is
correct, these associations on the sky must be just an accidental effect
of projection of foreground objects against background objects. Much
of the debate about this issue has centered on whether statistical
arguments for or against the physical reality of these associations
were valid.

3.4 Cosmological Redshift

If the redshifts of quasars are cosmological, then they are very
luminous, highly compact objects. In early discussions of the nature
of the redshifts, I tended to argue that the alternative interpretations
led to objects that were at least as exotic as cosmological quasars. We
can, however, show that quasars are at cosmological distances even ig-
noring their large redshifts!
      The argument is based on the counts of quasars as a function of
magnitude. Quasar surveys conducted by a number of teams of astronomers
have shown that the number of quasars rises by a factor of 8 per magnitude, from magnitude 15 to 19, as shown in Figure 1. It can easily be
shown that objects having a uniform distribution in local Euclidean
space show a rise in numbers of a factor 4 per magnitude. Therefore,
the space density of quasars must increase with distance, if they reside
in local Euclidean space. Since quasars here are isotropically distributed, this means that we are located in a density minimum of quasars.
This situation is in conflict with the Copernican principle, which states
that it is very unlikely that the human observer is at a special location
in the universe. We can only escape the Copernical condemnation if we

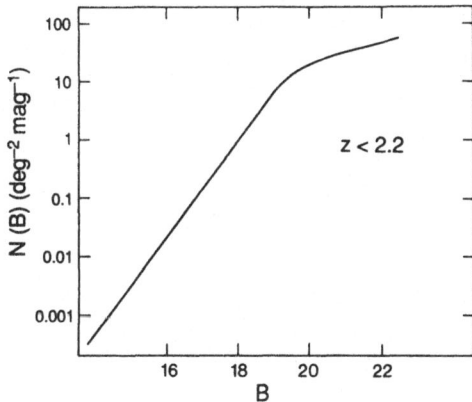

Fig. 1. Counts of quasars with redshifts less than 2.2, as a
function of blue magnitude B. The curve represents the work
of many astronomers.

put quasars at distances so large that their light takes most of the age
of the expanding universe to reach us. In that case, fainter more distant quasars reflect the situation in the universe at earlier epochs.
Since we accept that the expanding universe shows evolution, the increasing density toward earlier epochs is no objection. Ryle used this
argument in the 1950s to show that extragalactic radio sources had to be
very distant, even before most of their redshifts had been measured. I
believe that this line of reasoning provides the most direct support for
the cosmological redshift of quasars.

4. NATURE OF QUASARS

On the cosmological interpretation of the redshifts, quasars are at distances from 1 to 13 million light years, assuming a Big Bang model of
the universe with an age of 15 thousand million years. Their luminosities
range from 10^{11} to 10^{14} solar luminosities, while the size of the central
luminous part is probably less than a light week as we mentioned earlier.
Compared to the most luminous galaxies, quasars are up to a thousand
times more luminous, yet their size is 100,000 times smaller! It is not
surprising then that for many years the relation between quasars and
galaxies remained unclear, in particular since no quasars were found in
rich clusters of galaxies.
      A useful clue has been provided by the nebulosity, such as seen
around 3C 48. Optical imaging of quasars at modest redshift has shown
that they are all surrounded by nebulosity and it is now believed that
all quasars are resident in a host galaxy. The dominant light of the
quasar makes such studies difficult. It is expected that the Hubble
Space Telescope with its high spatial resolution will make a major contribution to the study of host galaxies of quasars.
      The spectral energy distribution of quasars is remarkably broad.

Each octave of frequency or wavelength contributes a roughly equal amount of energy all the way from gamma-ray or X-ray energies to infrared or radio wavelengths. Individual quasars differ strongly in their behavior at radio wavelengths. Only a small fraction emit strong radio radiation, such as shown by 3C 48 and 3C 273. Essentially all quasars are strong X-ray emitters.

The spectral energy distribution shows a number of peaks that suggest a number of components. Studies of variability of the different components are very useful in the diagnostics of a complex situation; these require coordinated ground-based and space observations.

The enormous energy output of quasars, coupled with their very small size, requires a compact, massive object. An attractive hypothesis is that the central body is a black hole of about 1000 million solar masses. Material falling into the black hole from the surrounding galaxy would be responsible for the enormous energy output of quasars. The total size is about equal to the orbit of Saturn in our solar system. The emission lines are formed in clouds at greater distances, of a few light years.

From the beginning, our understanding of quasars has been hampered by the fact that the central object could not be resolved optically. How well would we understand the Andromeda nebule with its spiral arms, cepheids, clusters, rotation, etc., if it presented itself to us as a point source?

At radio wavelengths, the situation is more favorable, for two reasons. First, radio quasars often do show structure on a scale of seconds of arc. Jets are usually prominent, sometimes stretching over enormous distances into the intergalactic medium. Second, radio astronomy has succeeded in creating images that show detail a thousand times finer than optical images. This is achieved by using radio telescopes located continents apart as an interferometer. This allows studies on a scale of light years, close to the central engine of quasars.

The resulting radio maps of quasars show variations from month to month. Generally, a component appears that moves away from the central source at high speed. In fact, the speed naively calculated is usually five to twenty times the speed of light, hence the term superluminal motion. Three radio blobs in 3C 273 moved apart 25 light years in the three year interval 1977-1980! Since speeds larger than the velocity of light are frowned upon, this finding at first seemed to argue strongly against the large distances that follow from the cosmological redshifts. It may be shown, however, that relativisitc ejection of material in a direction close to the line of sight to the observer leads to superluminal motion. This does not mean that the frequent occurrence of superluminal motion among quasars is fully understood. The orientation of these sources relative to the observer is the subject of much discussion.

5.   QUASARS AS PROBES

Besides the emission lines observed in quasar spectra, they also exhibit absorption lines, mostly at redshifts well below that of the emission lines. A single quasar can show absorption lines at many different redshifts: these are formed by gas clouds at different locations on our line of sight to the quasar. There are two main types of absorption-line systems. The first one shows both hydrogen lines and absorption lines of heavier elements such as carbon, silicon, iron, etc. These systems are caused by galaxies that are on the line of sight to the observer. The second type of absorption system consists only of the Lyman-alpha line of hydrogen. They are observed at wavelengths below that of the Lyman-alpha emission line. These systems are probably gas clouds that have not formed stars yet and therefore have not experienced metal enrichment through stellar evolution, supernova explosions, etc. The only evidence for the existence of these pure hydrogen gas clouds is in the absorption spectra of quasars.

Quasars have also led to the first observation of the gravitational lens effect that was expected on the basis of Einstein's General Theory of Relativity. The lens effect typically manifests itself in multiple images of the quasar, with each image having an identical spectrum and redshift. The different images reach us through different paths around the lens. Since the light travel time along the two paths may differ by many years, any time variation in the spectrum of quasars would show up as a difference in the spectra of the images. This makes the distinction between a pair of quasars at the same redshift and a true lens effect sometimes difficult in individual cases. Most of the lenses are combinations of a galaxy in a cluster and the cluster itself. Standard theory requires that the number of images must be odd. Ironically, most if not all cases show an even number of images--perhaps the last image is very faint or very close to the quasar. Here, again, the Hubble Space Telescope with its high spatial resolution will add much to our knowledge.

6. SPACE DISTRIBUTION OF QUASARS

You will recall that we discussed earlier the steep increase of the counts of quasars with magnitude and how that increase could be used to argue against the local hypothesis. I would now like to discuss the space distribution of quasars based on the cosmological interpretation of their redshifts. In order to do this, we have to find and count quasars, i.e., to take their census, before we can derive their space density.

Let me first explain how we search for quasars. In the 1960s, it was noticed that all quasars identified from radio sources had a blue color, or so-called ultraviolet excess. This turned out to be a property of all quasars of redshift less than 2.2, regardless of their radio emission. As an example, consider the Palomar-Green survey, started by Green by taking two-color exposures of 266 fields in the northern sky with the Palomar 18-inch Schmidt telescope. The millions of objects on these exposures were digitally recorded and those of the appropriate blue color selected. This produced a sample of several thousand : objects, of which individual spectra were taken with the 60 and 200-inch telescopes at Palomar. This long-term study produced the PG catalog, which is a complete sample over 10,714 square degrees of

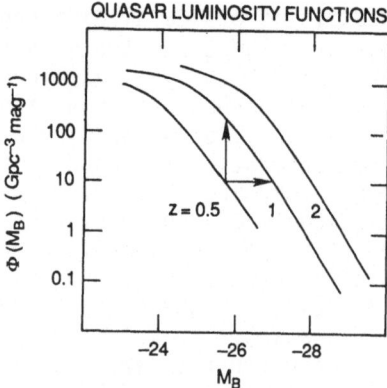

Fig. 2. Schematic representation of quasar luminosity functions based on a review by Boyle et al. Density evolution is represented by the vertical arrow, luminosity evolution by the horizontal arrow.

around 1700 blue objects to an average limiting blue magnitude of 16.2. Among them were 115 quasars which constitute the Palomar Bright Quasar Survey.

Another important survey is that by Boyle et al., which covers 4.2 square degrees and is complete to a magnitude of 20.9. The method of selection was essentially the same as for the Bright Quasar Survey. Since the fields were relatively small, spectra could be taken with a fiber-fed spectrograph allowing dozens of spectra to be observed simultaneously.

With the date from these suveys and many others in hand, the space density of quasars of given luminosity at given redshift can be derived. Before I show the results, let me emphasize that they reflect many years of effort and telescope time by many groups of astronomers.

Since quasar numbers depend on absolute luminosity, we express quasar space densities in terms of the luminosity function, which represents the space density per interval of luminosity. We see from Figure 2 that at given redshift, say 0.5, the luminosity function declines steeply with increasing luminosity: quasars of the highest luminosity are very rare. We also see that the luminosity function of quasars at different redshifts is different. At given luminosity, say absolute magnitude -26, the space density increases by a factor of about 130 from a redshift of 0.5 to 2 (cf. Fig. 3).

An alternative way of looking at these luminosity functions is in terms of a horizontal shift, as illustrated in Figure 2. The shift between the luminosity functions at redshift 1 and 2 is about a magnitude. In other words, the change can be understood if all quasars brightened by about a magnitude between redshifts 1 and 2. In fact, all the luminosity functions can be understood if quasar brightened by about 5 magnitudes, or a factor of 100, from redshift zero to redshift 2.

These two alternative ways of looking at quasar evolution are usually labeled density evolution and luminosity evolution. Much discussion has been going on lately about the relative merits of these points of view. Luminosity evolution could be understood if quasars are long-lived, that is they all formed at or before redshift 2 and all dimmed uniformly by 5 magnitudes to the present. It is easy to show that if quasars radiate 10 per cent of the rest-mass energy of the accreted mass, then under the hypothesis of luminosity evolution the typical quasar would have a mass of around 10^{11} solar masses. This is much larger than the mass deduced from the dynamics of the emission-line region of quasars, which is around 10^9 solar masses.

It is essential for the hypothesis of luminosity evolution that the luminosity function of Seyfert galaxy nuclei be added to the quasar luminosity function since at large redshifts many of these would brighten to become quasars. This creates another problem. Seyfert galaxies contribute at X-ray energies about 30 per cent of the total X-ray background at 2 keV. If they brighten as required in the luminosity evolution scenario, their X-ray contribution would exceed that of the observed background.

For these reasons, I prefer to think in terms of density evolution, with quasars typically living only ten or a hundred million years. In this scenario, quasars were more numerous at larger redshifts because their birth rates were higher at earlier cosmic times.

7. THE REDSHIFT LIMIT

Even though the space density of quasars of given luminosity rises sharply with redshift, there has been a suspicion for a long time that quasars of very high redshift may not exist. To illustrate, consider the quasar KPS 2000-030 with a redshift of 3.78. Its optical magnitude is around 18. Such a quasar at a redshift of 5 would be about magnitude 19, and therefore easily observable, yet not a single one has been found yet. In order to investigate this, we have to look carefully into the ways quasars of large redshift are discovered.

Quasars at redshifts beyond 2.2 do not have blue colors any more, therefore a different search technique has to be used. The Cambridge group headed by Hewitt uses many different colors and separates out those that show color combinations unlike those of the majority of the stars. Slit spectra are taken of all the candidates. Quasars with colors similar to those of stars will be missed in this technique.

An alternative method is the objective prism or grism technique. Each object in the field is recorded as a short, low-resolution spectrum. Candidate quasars are selected by searching for an emission line in the short spectra. This technique has been used for many years using photographic plates and visual inspection of the spectra. Even though around a thousand quasars have been found this way, the resulting space densities are of little value as a census of quasars, because the selection effects can not be quantified.

Several years ago, Schneider, Gunn and I decided to start a program using this method in an objective way. We employ CCDs as detectors, and search for emission lines in the digitally recorded spectra by computer. Soon it developed that we needed to cover larger areas of the sky to discover a substantial number of quasars. This was a problem, because a CCD at the 200-inch telescope covers less than

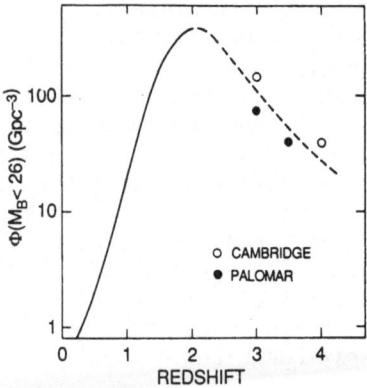

Fig. 3. Co-moving space density of quasars more luminous than $M_B = -26$, as a function of redshift. The points represent preliminary results of two searches for high redshift quasars.

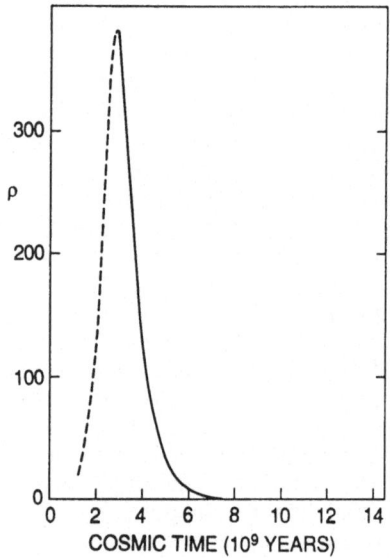

Fig. 4. Co-moving space density of quasars more luminous than $M_B = -26$, versus cosmic time.

one-hundredth of a square degree. We resolved the problem by stopping the tracking of the telescope: the array of four CCDs are read out exactly at the sidereal rate with which objects move across them. The exposure time is very short, of course, generally less than a minute. In a 7-hour exposure a long strip of sky can be observed with a total area of 14 square degrees.

There are many details of this approach that warrant mentioning; suffice it to say that one night's surveying records tens of thousands of short spectra, and that the computer search for emission lines produces typically hundreds of candidates. In the course of this work we have found two quasars with redshifts over 4.

We now have tentative results for one survey of 14 square degrees, in which we found 44 quasars. Among these quasars, 15 have a redshift larger than 3. Since the candidate search was done by computer algorithm, the selection effects pertaining to this survey can be evaluated quantitatively. It turns out that our survey essentially covered quasars of absolute mangitude brighter than -26 at redshifts above 2.7. This is the same luminosity for which we found an increase in co-moving space density by a factor of 130 from redshift 0.5 to 2. The points to the right of Figure 3 are from our group and from first results announced by the Cambridge group. I want to emphasize that our results are still tentative. Also, I am presenting the Cambridge results in a different fashion than they did.

It is clear from these results that the co-moving space density of quasars brighter than -26 is reaching a maximum at a redshift of 2 or 2.5 and that it then declines rapidly. Some of our colleagues have suggested that we are missing high redshift quasars in our census as a consequence of dust absorption in the distant universe. This appears unlikely, since extragalactic radio sources, not affected by dust absorption, also peak at a redshift of 2.

At this stage it is important to realize that at increasing redshifts we are looking back to progressively earlier epochs. It makes sense, then, to replace the redshift by the appropriate cosmic time: 0 for the Big Bang and 15 thousand million years for the present. Also, I will use a linear scale for the density rather than the logarithmic scale used here.

The curves drawn in Figure 4 correspond precisely to those in Figure 3. Here for the first time, we see the rise and fall of quasars. Quasars started appearing perhaps a thousand million years after the Big Bang, reached a sharp maximum at 3 thousand million years, and since that time have declined to almost nothing. Clearly, the fireworks has long been over. The fact that we see the height of the fireworks at all is due to the finite velocity of light, which allows us to look back to these early epochs.

The sharp maximum must signify an extraordinary event in the early universe. Now, as you know, astronomers have been looking for many years for the epoch of galaxy formation. Theoreticians place this event anywhere in the range of redshift 2-10. When galaxies start forming, the collapse of the inner parts must provide plenty of material to accrete onto any black hole present in the center of the galaxy, producing a bright quasar. Perhaps the black holes were formed in the same period. In any case, I would argue that the peak of quasar activity and the formation of galaxies are probably related, directly or indirectly.

I am sure you realize that his scenario is very speculative on my part. For that reason, I am awaiting with some trepidation Professor Martin Rees' judgment on these matters in his Invited Discourse on the formation of galaxies next Monday night.

GALAXY FORMATION AND DARK MATTER

Martin Rees
Institute of Astronomy
Madingley Road
Cambridge, CB3 0HA.

INTRODUCTION

My aim in this general talk will be to air some questions, rather than offer firm answers, because the most basic questions about galaxies are indeed still unresolved. In particular:

1. We do not know why such things as galaxies should exist at all — why these assemblages of stars and gas with fairly standardised properties are the most conspicuous large–scale features of the cosmos.

2. About 90% of the mass associated with galaxies is hidden. The luminous stars and gas contribute only about a tenth of the gravitating material inferred from dynamical arguments. What the rest consists of is still a mystery.

3. It is unclear why the nuclei of some galaxies flare up, and release the colossal amount of non–stellar radiation emitted from quasars and radio galaxies, as discussed by Maarten Schmidt in his discourse.

We are perplexed about these issues, just as 50 years ago our predecessors were perplexed about the nature of stars. But some of us are hopeful that the physical processes underlying galaxies are coming into focus, and can at least be seriously addressed. I must apologise in advance to specialists on this topic for the 'broad brush' and inevitably distorted exposition I shall be giving.

In their already-classic book on galactic dynamics, Binney and Tremaine (1987) make the point that galaxies are to astronomy what ecosystems are to biology. They are not only dynamical units, but chemical units as well. The atoms we are made of come from all over our Milky Way galaxy, but few come from other galaxies. The ecological analogy reflects other features of galaxies: their complexity, ongoing evolution, and relative isolation.

Single stars, the individual organisms in the galactic ecosystem, can be traced from their birth in gas clouds through their lifecycle. And we have come to understand why *stars* exist with the general properties we see. The question why *galaxies* exist is less straightforward than the equivalent question for stars. Galaxies formed at an earlier and remote cosmic epoch. We don't know how much *can* be explained in terms of ordinary processes accessible to study now, and how much has its causes in the earliest universe.

There is an elaborate taxonomy for galaxies, but the most obvious categories are disks and spheroids or ellipticals. There is a well–known cartoon model, dating back about 30 years, to account for this basic morphological distinction. Suppose

45

D. McNally (ed.), Highlights of Astronomy, Vol. 8, 45–64.
© *1989 by the IAU.*

that a galaxy started life as an irregularly–shaped gas cloud contracting under gravity, and that the collapse of such a gas cloud were highly dissipative, in the sense that any two globules of gas that collided would radiate their relative kinetic energy and merge (Figure 1). The end result of the collapse of such a cloud would be a rotating disk. This is the lowest energy state that the cloud can reach if it does not lose or redistribute its angular momentum. On the other hand, stars do not collide with each other, and are unable to dissipate energy in the same fashion as gas clouds. So the *rate of conversion of gas into stars* could be the crucial feature determining the type of galaxy that results. Elliptical galaxies would be those in which the conversion is fast, so that most stars have already formed before the gas has had time to settle down in a disk. Disk galaxies result when the star formation is delayed until the gas has already settled into a disk. According to this traditional picture, disk galaxies are those with slower metabolism, which have not yet got so close to the final state in which essentially all the gas is tied up in low mass stars or dead remnants.

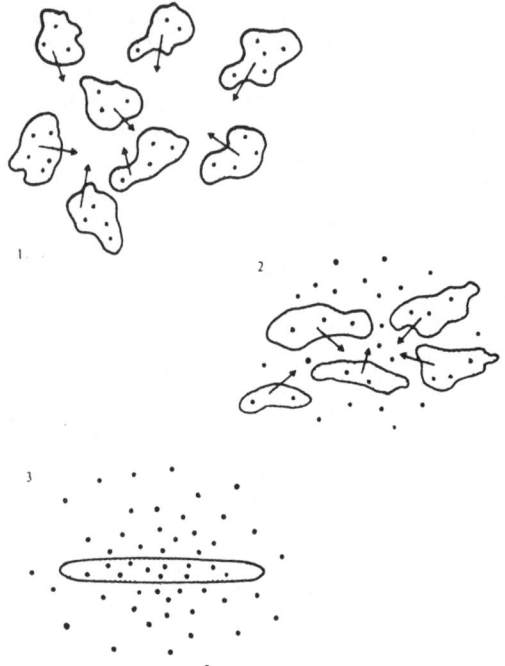

Figure 1. *'Cartoon' showing three stages in the traditional picture of protogalactic collapse*

WHAT IS SPECIAL ABOUT GALACTIC DIMENSIONS?

This story has many inadequacies, and I'll return to some of them later. In particular, there is no scale in the picture. Is there any physics that singles out clouds of galactic dimensions, just as, since Eddington and Chandrasekhar, we have known the natural scale of stars? All we have for galaxies is a simple but suggestive physical argument. Two timescales are important in determining how a self–gravitating

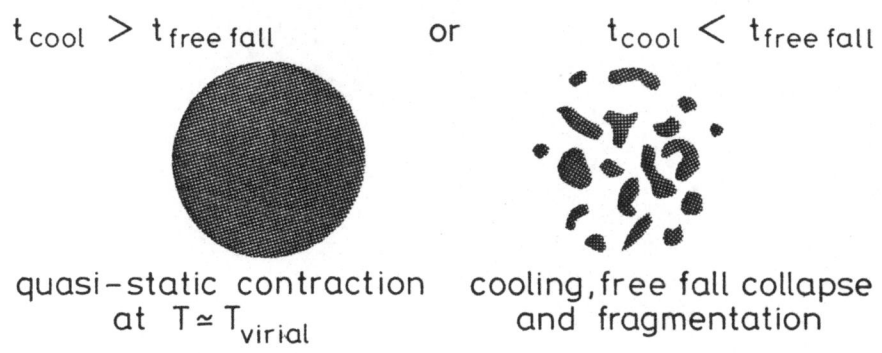

two modes of contraction

$t_{\text{cool}} > t_{\text{free fall}}$ or $t_{\text{cool}} < t_{\text{free fall}}$

quasi-static contraction
at $T \simeq T_{\text{virial}}$ cooling, free fall collapse
and fragmentation

Figure 2. *Cooling and contraction of self-gravitating gas clouds.*

gas cloud evolves. The first of these is the dynamical or freefall time, which is of order $(G\rho)^{-\frac{1}{2}}$, its precise value depending on the geometry of the collapse. The second is the radiative cooling timescale. This depends on the gas temperature T_g, and can be written $T_g/\rho\Lambda(T_g)$ where Λ can be calculated from atomic physics.

If t_{cool} exceeds $t_{\text{dynamical}}$, a cloud of mass M and radius r can be in quasi-static equilibrium, with the gas at the virial temperature. But if $t_{\text{cool}} < t_{\text{dynamical}}$ such equilibrium is impossible (Figure 2). The cloud cools below the virial temperature and undergoes freefall collapse or fragmentation. We would expect clouds to collapse and fragment in the fashion depicted in Figure 1 only if they enter the part of $M - r$ plane where cooling is faster than freefall. A simple calculation shows that this criterion involves a characteristic mass–independent radius of order 75 kpc and a characteristic mass M_{crit} of order $10^{12}\,M_{\odot}$. Clouds less massive than M_{crit} will readily fragment, but above M_{crit} fragmentation is impossible unless the cloud contracts until its radius is below r_{crit}. This characteristic mass and radius, consequences of straightforward physics (Figure 3 *overleaf*), feature in many cosmogonic schemes as at least setting an upper limit to the scale of galaxies.

Eddington claimed that a physicist on a cloud–bound planet could have predicted the properties of the gravitationally–bound fusion reactors that we call stars. But these simple considerations don't suffice to predict galaxies, even with hindsight. This is because any true explanation of galaxies must involve setting them in a cosmological context.

THE COSMOLOGICAL CONTEXT

In a memorable invited discourse at the Patras IAU General Assembly, Zel'dovich (1982) discussed the hot big bang model, which he opined was as sure as that the Earth goes round the Sun. We may not all quite share his exuberant certitude. But most of us regard the hot big bang as the 'best buy' cosmology, more than 50% likely to be essentially correct. According to this picture everything emerged from a universal thermal soup which was initially smooth, and almost featureless,

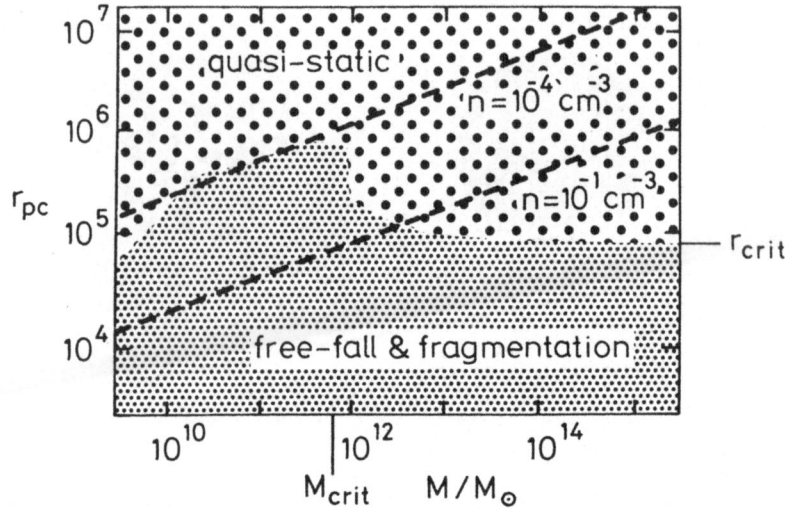

Figure 3. *The quasi–static and free–fall regions (cf. Fig. 2) are here presented in a mass–radius plot. M_{crit} and r_{crit} should set characteristic upper limits to the dimensions of galaxies. Clouds with $M \gg M_{crit}$ would be quasi–static unless at very high densities (cf. gas in clusters of galaxies).*

but not quite. There were, we don't really know why, small fluctuations from place to place in the expansion rate. Structures emerged via gravitational instability as over–dense regions lagged more and more behind the universal expansion, and eventually condensed out as embryo galaxies and clusters.

Theorists trace back the history of the hot big bang over 60 decades of logarithmic time. The events and stages in the cosmic expansion are summarised in Figure 4, which goes back to the earliest era, the intellectual habitat of the 'gee whiz' fringe of particle physicists. For our present purposes the uncertain details are irrelevant. It may, though, be conceptually useful to divide cosmic history into 3 parts. For the first 40 decades the microphysics is uncertain. When the universe cools below 10 MeV and the density falls below nuclear density, the microphysics become straightforward. Initial irregularities, owing their origin to the first era, amplify via gravitational instability, and things become less straightforward when the first of these condense out. Then we confront a set of new difficulties. The physics is just Newtonian gravity and gas dynamics, the easy bits of Landau and Lifshitz, but the complications are those of non–linearity. The 'recent' universe is hard to understand for the same reason that weather prediction is difficult.

The types of difficulty that one faces in studying the early and late phases were amusingly contrasted in a recent article by the distinguished Canadian relativist Werner Israel. The early universe was like the challenge of chess, he said, but the later non–linear stages were like mud wrestling. Maybe, but it's glorious mud and some of us have wallowed contentedly in it for years.

A key question is how much can be explained by processes occurring at the

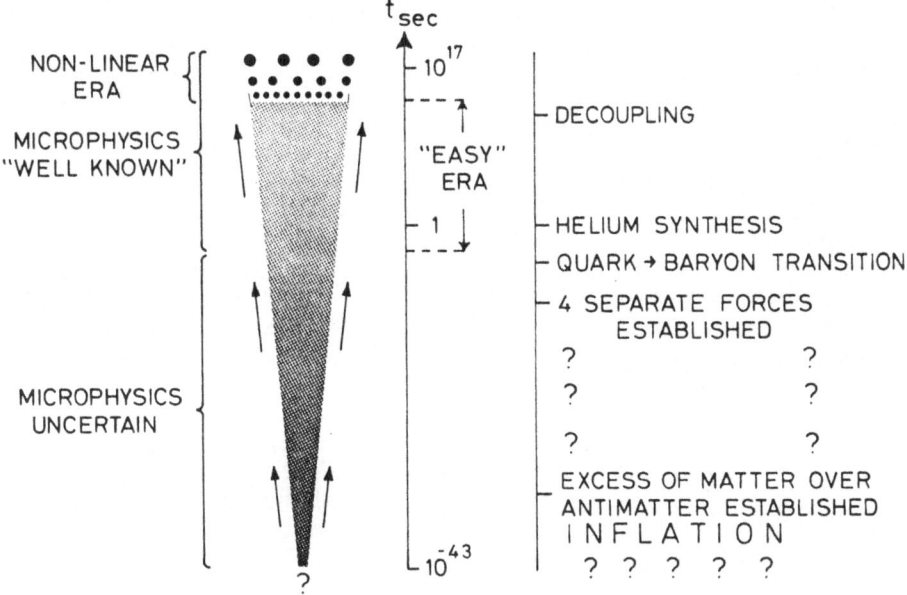

Figure 4. *Stages in the evolution of the 'standard' big bang model universe*

range of epochs accessible to astronomers, and how much has to be attributed to the 'chess playing' era.

The main types of relevant data are morphological classifications (dating back to Hubble); correlations between luminosity, velocity dispersion and size; and the statistics of galaxy clustering. Any quantitatively satisfactory theory must explain these things. We have no generally agreed theory yet. Indeed, as Saslaw has put it, 'if galaxies didn't exist, we'd have no problems explaining the fact'. Moreover, the seekers for any such theory must first face a most embarrassing circumstance: this is the *dark matter problem,* evidence that 90% of the mass of galaxies is unaccounted for, and takes some unknown form.

DARK MATTER

The evidence for dark matter dates back more than 50 years, but has firmed up since the classic papers of Einasto, Cowsik and Saar and Ostriker, Peebles and Yahil, both published in 1974. Vera Rubin (1985) reviewed this (now compelling) evidence in her discourse at the last IAU.

The masses inferred from relative motions of galaxies in apparently bound groups and clusters exceed by a factor 10 those inferred from the internal dynamics of the luminous parts of galaxies. This apparent discrepancy could be resolved if galaxies were embedded in extensive dark haloes. The halo hypothesis can be checked in some edge–on disk galaxies, where emission from gas can be observed out at radii far exceeding the extent of the conspicuous stellar disk. The mass of this gas is itself negligible, but rotation velocities derived from its spectral lines do not fall off as $R^{-\frac{1}{2}}$, as would be expected if the gas were orbiting a mass distribution concentrated at much smaller radii. Instead the velocity remains almost constant, implying that the mass within radius R is proportional to R out to 80 kpc in some cases. Direct lower limits on the mass–to–light ratio in the outlying parts of some galaxies exceed 300 solar units.

In some elliptical galaxies also, the mass seems to increase proportional to R out to large R. In M87, such evidence comes from globular cluster orbits, and, still further out, is inferred from the X-ray temperature and profile of the diffuse gas. On a larger scale, we have evidence from clusters of galaxies, along the lines first discussed by Zwicky and Sinclair Smith in the 1930s. Many independent lines of evidence point towards the existence of dark matter (these are summarised in Figure 5). This has as good a claim to be termed a paradigm shift as any development one can think of in modern astronomy.

The dynamically–inferred dark matter, though ten times the luminous matter, still amounts to only 10 or 20 per cent of what is required for a closed universe: the corresponding value of the density parameter Ω, the ratio of the actual density to the cosmological critical density, is 0.1 or 0.2. I shall return later to the question of whether there could be enough dark matter to make $\Omega = 1$.

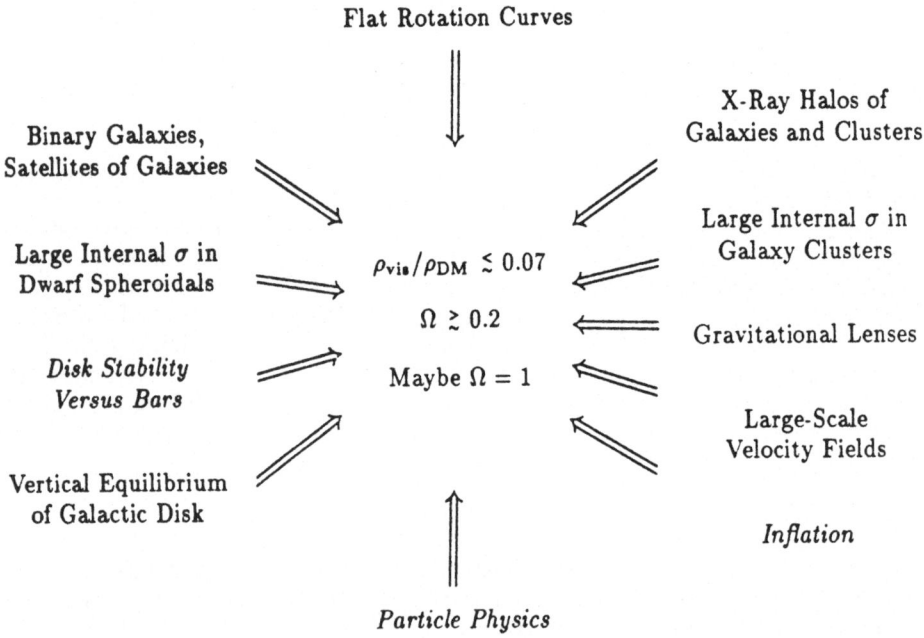

Figure 5. *The various lines of evidence for dark matter (diagram due to J. Kormendy)*

THE NATURE OF THE DARK MATTER: BARYONIC OR NOT?

What is the halo dark matter? The first possibility that comes to mind is faint stars or stellar remnants. Figure 6, due to Carr, Bond and Arnett (1984), quantifies the maximum hidden contribution to Ω that could be made by stars or their remnants in the mass range from 10^{-2} to $10^8\,M_\odot$. There are two tenable dark matter candidates: very low mass stars, below $0.1\,M_\odot$, or the remnants of very massive stars. Ordinary stars above $0.1\,M_\odot$ would contribute too much background light unless they had all evolved and died, leaving dark remnants. But the remnants of ordinary massive stars of $10-100\,M_\odot$ would produce too much material in the form of heavy elements. Stars with core masses above $200\,M_\odot$ end their lives, via the pair production instability, by collapsing rather than exploding. These very massive objects, (VMOs for short), do not eject heavy elements, and leave black hole remnants. Such objects, if they constitute our own galactic halo, can't however exceed $10^6\,M_\odot$ each, because otherwise dynamical friction, whereby a hole transfers energy to lighter stars close to its path, would have led to excessive thickening of the galactic disk.

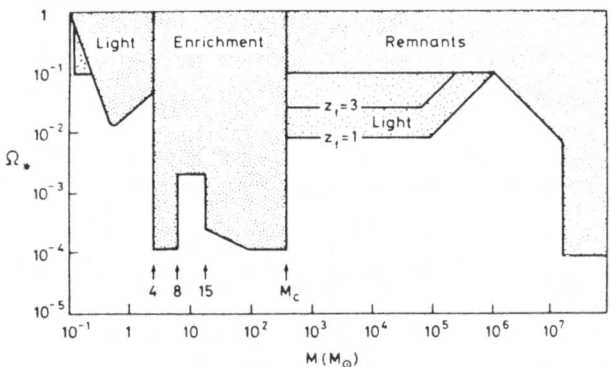

Figure 6. *Constraints on the fraction of the critical density that could be present in stars or stellar remnants of various masses. The stars are presumed to have formed at some redshift z_f. Possible candidates for the dark matter are low mass stars (brown dwarfs or 'Jupiters') or very massive objects (VMOs).*

Is it likely or unlikely that a forming galaxy should convert most of its mass into either ultra–low mass stars or objects heavier than a few hundred suns? We don't understand enough about star formation, even close at hand in for instance the Orion nebula, to be confident in saying how the initial mass function might be affected by intense background radiation, absence of heavy elements, lack of magnetic fields, and the rest. Theory therefore cannot arbitrate reliably between low mass and VMO options (Figure 7 *overleaf*).

Can we learn from observations about what the dark matter is? Low mass objects would be perhaps detectable in the infrared: the nearest would be less than a parsec away, with high proper motions. There are two handles on VMO remnants. They might reveal their presence by accretion on passage through interstellar clouds. Also, they imply that galaxies would be bright when young –

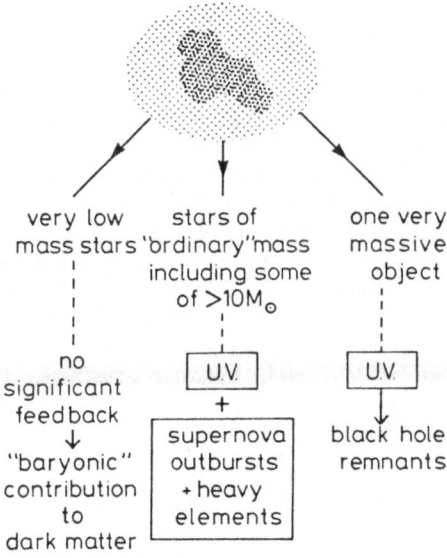

Figure 7. *In many cosmogonic schemes, star formation would be initiated in baryonic clouds of around 10^6 M$_\odot$, but we have no firm theoretical basis for deciding the characteristic mass, or the IMF, of these first stars.*

there are constraints from the sky brightness, and from the faint galaxy counts, but the quantitative interpretation of these limits depends on the uncertain redshift of galaxy formation.

One way of detecting compact dark objects, and discriminating between the Jupiter and VMO options, is by searching for evidence of gravitational lensing. The probability of seeing lensing due to an object in our own halo is only about 10^{-6}. But the cross–section for effective lensing is proportional to distance, so there is, perhaps surprisingly, much more chance of detecting objects in the haloes of galaxies half way out to the Hubble radius (Figure 8). The probability that a compact source at redshift $z > 1$ is significantly microlensed by objects along the line of sight is of order Ω, independent of the individual lens mass involved (Refsdal 1970, Press and Gunn 1973). The *angular separation* of the images, proportional to (lens mass)$^{\frac{1}{2}}$ is however a diagnostic of the masses. For masses above 10^5 M$_\odot$, very long baseline radio interferometers provide adequate resolution. We could probably already exclude $\Omega = 1$ in such objects.

For brown dwarfs of below 0.1 M$_\odot$, the angular scale is less than a micro arcsecond. This cannot be directly resolved by any technique, until optical interferometers are deployed in space. There is nevertheless a genuine prospect of detecting lensing of this kind because of the variability that would ensue if the lens were to move transversely (*e.g.* Gott 1981). An object at the Hubble distance moving at $100\,\text{km}\ s^{-1}$ takes only a few years to traverse a micro arcsecond. The image structure and time variation are more complicated if the line of sight passes through, for example, a galactic halo, thereby encountering an above–average column density of dark matter. Several objects may then contribute to the imaging, yielding a frosted glass effect, whose pattern, though too small to be seen directly, would vary on a timescale of months or years.

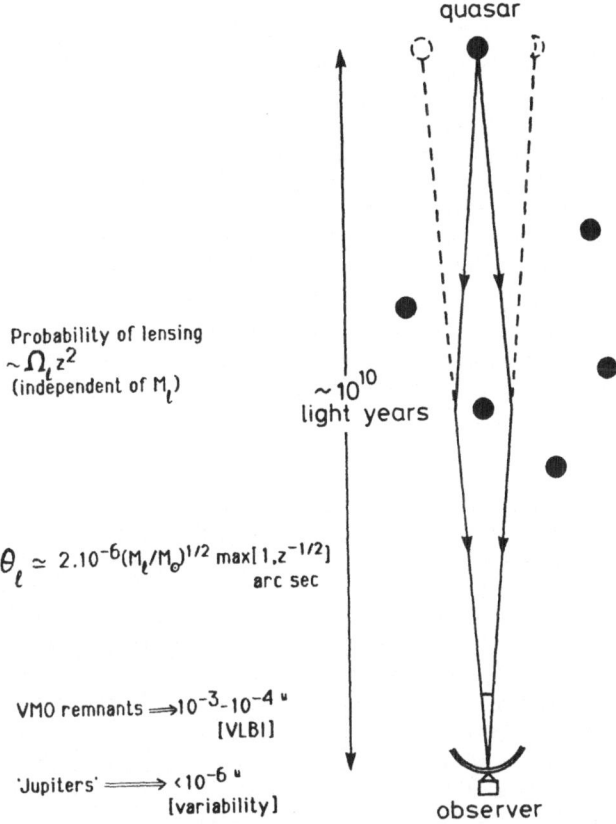

quasar

Probability of lensing
$\sim \Omega_\ell z^2$
(independent of M_ℓ)

$\sim 10^{10}$
light years

$\theta_\ell \simeq 2.10^{-6} (M_\ell/M_\odot)^{1/2} \max[1, z^{-1/2}]$
arc sec

VMO remnants $\Longrightarrow 10^{-3} \text{-} 10^{-4}$ "
[VLBI]

'Jupiters' $\Longrightarrow < 10^{-6}$ "
[variability]

observer

Figure 8. *Properties of gravitational microlensing by a population of compact objects along the line of sight to a cosmologically-distant quasar.*

Those I've just discussed are the 'dull man's' options for dark matter. The big bang may have left not just baryons and radiation, but other species as well, which may contribute to Ω. In the standard big bang model, neutrinos are almost as abundant as microwave background photons, outnumbering baryons by around 10^9. Their mass would only need to be a few eV to make them dynamically important. More than 15 years ago Cowsik and McClelland (1973) and Marx and Szalay (1972) conjectured that neutrinos could provide the dark mass in galactic haloes and clusters. At that time the suggestion was not followed up very extensively. But by the 1980s physicists had become more open-minded about non-zero neutrino masses. A change in theoretical attitude, coupled with experimental claims that the electron neutrino had a mass around 36 eV (Lyubimov *et al.* 1980), stimulated astrophysicists to explore scenarios for galaxy formation in which neutrino clustering and diffusion played a key rôle. More recently, other kinds of non-baryonic matter have also been considered.

Provided that we know the mass and annihilation cross-section for any species of elementary particle, we can in principle calculate how many survive from the big bang, and the resultant contribution each species makes to Ω. Progress in experimental particle physics may therefore reveal a particle which must contribute significantly to Ω, unless we abandon the hot big bang theory entirely.

Neutrinos have the virtue of being known to exist, but particle physicists are inventive, and have come up with a long shopping list of relics that *might* exist. The most theoretically–favoured option is some kind of electrically neutral weakly interacting massive particles, WIMPs for short. These have attractive cosmogonic consequences which I'll come back to in a moment. What is perhaps more remarkable is that such particles may be looked for in the lab.

If our Galactic Halo were composed of WIMPs with individual masses of a few GeV, they would be swarming through this room, with a density of $10^5 \mathrm{m}^{-3}$ and speeds of around 300 km s^{-1}. Collision cross–sections are small, but whenever a WIMP collided with an atomic nucleus, the nucleus would recoil with a similar velocity, and an energy around a keV. The collision rates depend on the physical details and the target nucleus, but are in the range 1 - 1000 events per day per kilogram of detector.

These collision events may be detectable by a variety of cryogenic techniques in a low background environment, at quite modest cost. Such experiments are being planned in various countries (see Primack *et al.*, 1988, for a review). Ingenious schemes for detecting a halo background of exotic particles are surely among the most worthwhile and exciting high risk experiments in physics or astronomy today – potentially as important as those that led to the discovery of the microwave background in the 1960s. A null result, with just upper limits, would surprise nobody. On the other hand, such experiments could reveal new particles, as well as determining what 90% of our universe consists of. Because the detection is sensitive to velocity, they would even reveal the halo's velocity dispersion and rotation. The mean velocity of halo particles relative to the detector would change during the year, owing to the Earth's motion round the Sun. The resultant annual modulation, with an amplitude of a few per cent and a peak in June, would be an unambiguous signature discriminating against spurious background.

DARK MATTER AND GALAXY FORMATION

A less direct line of attack on pinning down the dark matter entails exploring the consequences of each option for galaxy formation. If it is dynamically dominant, then non–baryonic matter plays a key rôle in the process whereby small primordial perturbations evolve into protogalaxies and clusters.

The key parameter is the spectrum of density fluctuations, the rms amplitude as a function of mass scale, at the recombination epoch, z = 1000. Density contrasts on all relevant scales amplify at the same rate thereafter, so the first bound systems to arise via gravitational instability will have mass scales for which this amplitude peaks. The spectrum depends on what is imprinted initially, possibly modified by preferential damping of smaller scales before recombination.

The left–hand panel in Figure 9 shows the spectrum expected if the universe is dominated by neutrinos with masses 10 or 20 eV. These are moving sufficiently fast that everything is homogenised on scales at least up to $10^{14} \mathrm{M}_\odot$. The first bound systems would then be superclusters, and galaxies would result from some kind of secondary fragmentation process.

On the right is shown a 'white noise' spectrum, with amplitude larger for smaller scales. Here we have a hierarchical 'bottom–up' cosmogony, with the emergence first of subgalactic scales, then galaxies, and then clusters. (There may then

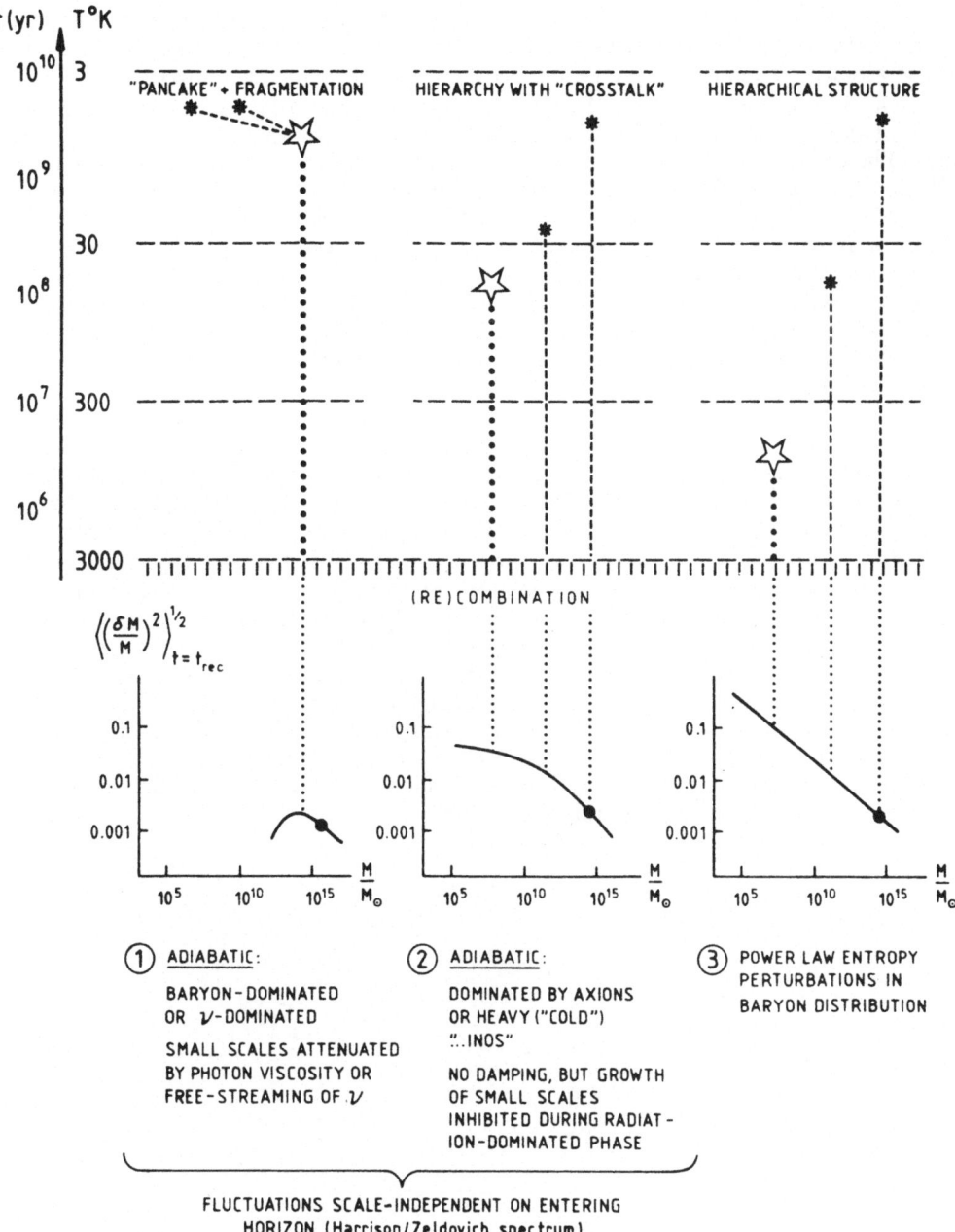

Figure 9. *Cosmogonic scenarios corresponding to three different spectra for the post-recombination density perturbations. See text for further explanation.*

be an interesting complication: radiative or explosive output from the first small bound objects could create secondary large–scale inhomogeneities that swamp those already present. We then have neogony as well as palaeogony.)

At recombination, when the universe was 10^6 years old, the microwave background shifted redward of the visible band, and the universe entered a literal dark age. The universe remained a simple place until the first bound systems condensed. We don't know when 'first light' occurred. The dark age may have been brief, as in the right–hand panel of Figure 9, or it could have lasted a billion years if the left–hand panel is closer to the truth. We remain more confused and ignorant about this phase of cosmic history than many seem to be about the first 10^{-35} seconds.

Let us focus now on the middle panel in Figure 9. The fluctuation spectrum here has the shape unambiguously calculable for WIMPs, or for any non–baryonic dark matter that is 'cold', in the sense that the individual particles move too slowly for damping due to free–streaming to occur, as it does for neutrinos. This 'cold dark matter' spectrum is nearly flat for small masses, so the typical fluctuation of $10^6 \, M_\odot$ would collapse no earlier than the epoch corresponding to z = 10. The build–up of structure is hierarchical, in the sense that smaller scales tend to form earlier. However, because of the flat spectrum, there would be complicated 'cross talk' between many different scales. The 3σ peaks in the density distribution on galactic scales, $10^{11} \, M_\odot$, would have the same amplitude as more typical peaks of mass $10^6 \, M_\odot$, and would therefore collapse at the same time. It is consequently hard to analyse, either analytically or numerically, even the purely dynamical and non–dissipative aspects of the clustering. However, those studies that have been done are encouraging, in that when the amplitude of the fluctuations is normalised so as to match the data on galaxy clustering, the finer scale disposition of the dark matter closely reproduces the sizes and profiles of individual galactic haloes. An example of how such clustering develops, based on simulations of Frenk *et al.* (1985), is shown in Figure 10. Figure 11 shows the final spatial disposition of the dark matter for a slightly different model.

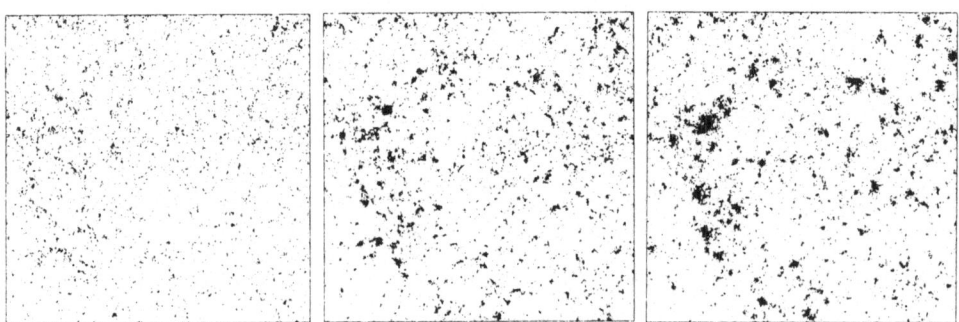

Figure 10. *Three stages in the evolution of non–dissipative gravitational clustering within a comoving cubical volume, for an initial spectrum with $\langle (\delta M/M)^2 \rangle^{\frac{1}{2}}$ proportional to $M^{-\frac{1}{3}}$. If the right–hand panel is taken to represent the present epoch, then the middle panel is z = 0.9 and the left–hand panel z = 3.5.*

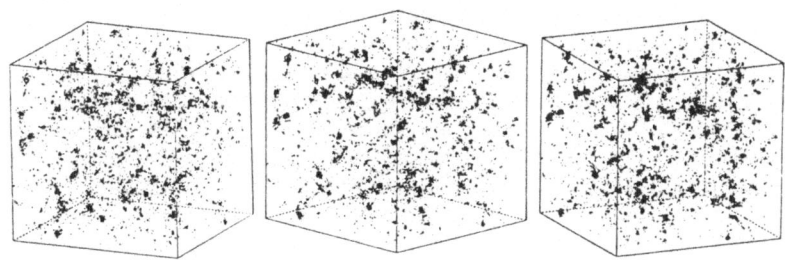

Figure 11. *Three views to illustrate the spatial structure within a simulated cubical volume of the expanding universe. In this model the initial fluctuations were Poissonian, with amplitude proportional to $M^{-\frac{1}{2}}$.*

Even though the dark matter may be dynamically dominant, it manifests itself only gravitationally. To predict what the universe would actually look like in this model — the luminosity function of galaxies and how they are clustered — we need to develop more understanding of several physical processes. Baryons are presumed to condense in virialised haloes of dark matter in the mass range 10^8 – $10^{12} M_\odot$. For larger masses, dissipative cooling may be inefficient for the reason mentioned earlier (*cf.* Figure 3). Below $10^8 M_\odot$ the potential wells may be too shallow to capture primordial gas. The mass distribution of isolated virialised systems can in principle be learned from N-body simulations. But even if the dissipationless clustering of the dark matter is accurately known, the fate of the baryonic component, how much gas falls into each potential well and how much is retained, involves complex gas dynamics. We need also to understand how the baryonic component behaves during mergers. If we trace back the history of the large haloes in Figure 10, half have experienced a merger since z = 2.

Theoretical fashions are often transient. But the cold dark matter model (Peebles 1982, Bond and Szalay 1983, Blumenthal *et al.* 1984 and references cited therein) has survived for more than 5 years (though there are two possible problems with it that I'll mention later). Insofar as it can account for galaxies and their haloes it offers circumstantial support for the idea that the dark matter is in WIMPs or axions. But this evidence is only circumstantial. The nature of the dark matter is still an open question. I am personally agnostic and would bet 25% on Jupiters, 25% on black holes, 25% on WIMPs or other cold dark matter, leaving the remaining 25% for things not yet thought of.

The late Professor Redman of Cambridge, warmly remembered by senior IAU members as a no–nonsense observer with little taste for speculation, once claimed that any competent astrophysicist can reconcile any theory with any set of facts. An even more cynical colleague extended this claim, asserting that the astrophysicist often need not even be competent. Dark matter theorists are perhaps exemplifying Redman's theorem and its extension. All things considered, the existence of dark matter is unsurprising. There are all too many forms it could take, and the aim of theorists and observers alike must be to narrow down the range of options. What is encouraging is that various lines of observations, experiments, and theoretical modelling should over the next few years do just this.

It would be specially interesting if we could, by astronomical methods, discover some fundamental particle which has been predicted by theorists. If such particles turned out to account for the dark matter, we would however have to view the galaxies, the stars, and ourselves, in a downgraded perspective. Copernicus dethroned the Earth from any central position. Early this century, Shapley and Hubble demoted us from any privileged location in space. But now even baryon chauvinism might have to be abandoned: the protons, neutrons, and electrons of which we and the entire astronomical world are made could be a kind of afterthought in a cosmos where photinos or neutrinos control the overall dynamics. Great galaxies could be just a puddle of sediment in a cloud of invisible matter ten times more massive and extensive.

IS THE UNIVERSE FLAT ($\Omega = 1$)?

Dark matter is relevant to the future of the entire universe, to eschatology. Will the universe eventually recollapse to a big crunch, or will it expand forever? This depends on whether or not the density parameter Ω exceeds unity.

In an influential review published back in 1974, Gott, Gunn, Schramm, and Tinsley summarised the arguments bearing on Ω. They concluded that the dynamical evidence favoured a value 0.1 or 0.2, and noted that if this matter were all baryonic, the lower end of the range was compatible with the value favoured by big bang nucleosynthesis, for a Hubble time of 2×10^{10} years, (a value consistent with the ages of the globular clusters, *etc.* .) Much new evidence has accumulated since 1974, especially on cluster dynamics and element abundances, and some relevant theoretical issues have been refined and elaborated. But if one were to update Gott *et al.*'s discussion, their net conclusion would not change much.

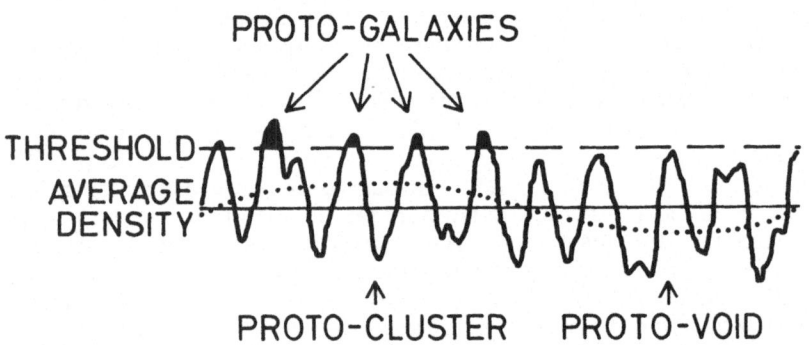

Figure 12. *If galaxies form from exceptionally high–amplitude peaks in a gaussian density field, they display enhanced clustering (or biasing) because the probability of a high peak is sensitive to whether or not there is a positive contribution to the amplitude from longer–wavelength modes.*

The attitudes of theorists, however, seem to have changed markedly. This is partly because non–baryonic matter is now taken much more seriously, seeming in some ways almost a natural expectation. But the other new element in the discussion is the concept of inflation. This resolves some stubborn paradoxes in a rather natural way. In particular, it suggests why the expansion rate is so fine tuned that our universe has neither collapsed long ago, nor is expanding too fast for galaxies to have condensed. Inflation indeed suggests that the fine tuning should be so precise that Ω has almost exactly the value unity. If Ω is indeed 1, the balance of argument tilts in favour of non–baryonic dark matter, because standard big bang nucleosynthesis favours a value for Ω_{baryon} of around 0.1.

The fact that direct dynamical evidence suggests a lower Ω implies that, if Ω is actually unity, the galaxies must be more clustered than the overall mass distribution. Voids must not be as empty as they look. The efficiency of bright galaxy formation could be a sensitive function of the density or the depth of the potential well. Biasing might at first sight seem just an ad hoc contrivance introduced by theorists to save the philosophically attractive $\Omega = 1$ model when confronted with apparently conflicting evidence. But there are physical reasons for expecting it in the cold dark matter cosmogony. Bright galaxies would be more clustered than the mass for the same reason that in an ocean swell the highest waves come in groups: peaks are more likely to be exceptionally high if they are superimposed on a very large scale positive fluctuation (an incipient cluster) rather than in an incipient void (Figure 12). Nonetheless, the case for $\Omega = 1$ comes primarily from theoretical prejudice and has no really direct observational support.

LARGE–SCALE STRUCTURE

A word now about large–scale structure. Initial fluctuations imprinted in the early universe cannot 'know' what is special about galactic mass. They will spread up to larger scales – though the amplitude must fall off with increasing scale, because of the Universe's overall observed homogeneity. We may get cleaner evidence on these fluctuations from the bigger scales, because these have not yet been confused by non–linear and dissipative effects. Hence the interest in clustering and superclusters. The well–known Lick counts whose clustering properties have been analysed extensively by Peebles and his co–workers, are now complemented by data from the UK southern sky survey. Maddox and Efstathiou, using the APM machine in Cambridge, have now studied the galaxy correlation function and clustering data for this survey.

Objective statistical tests for large–scale clustering are sorely needed. Astronomers react to the data rather as to ink–blot psychological tests. Some see filamentary features, bubbles, or sheets. Others see only gaussian fluctuations, the contrast enhancement of the bright galaxies being perhaps enhanced by biasing.

Redshifts are now available for larger samples of galaxies, and apparent sheet–like structures are seen when the data are plotted with redshift as a radial coordinate. However, one cannot tell whether such a feature is a physically thin sheet, or a thick over–dense region expanding slower than the Hubble flow with a small velocity difference between front and back. What is really wanted for probing the dynamics are the velocities *relative to* the Hubble flow: the motions induced by the gravitational fields of clusters, or perhaps by giant explosive events. For this one needs not just redshift, but an independent measure of distance. This is just becoming possible for some samples. We shall soon know more about the reality of large–scale motions ('cosmic plate tectonics') and whether clusters really have large velocities relative to the Hubble flow. Also, are the edges of voids expanding faster than the Hubble flow? If not, the total density in voids cannot be much lower than outside.

Another line of evidence comes from the microwave background isotropy. This is amazingly precise, upper limits now being a few times 10^{-5}. The universe was therefore quite smooth at z = 1000, the redshift where, in most models, photons were last scattered. But perturbations grow by only a factor a 1000 at most since that epoch. Can we then reconcile the existence of conspicuous large–amplitude inhomogeneities in the present universe, on the scales of superclusters and voids, with the smoothness at the recombination epoch which these microwave limits imply – with the lack of gravitational or Doppler perturbations due to incipient clusters? We still have only upper limits, but these are now stringent enough to exclude some options. Theorists still tell the microwave observers, as they have been telling them for the last fifteeen years or more, that another factor of two improvement will be crucial.

THE EPOCH OF GALAXY FORMATION AND THE FIRST QUASARS

The microwave background is our most direct link with very high redshifts, but Maarten Schmidt reminded us in his discourse that quasars are now seen with redshifts exceeding 4. Their light set out when the universe was less than a fifth of its present scale. The corresponding age is model–dependent, but for $H = 50\,\mathrm{km}\ s^{-1}\ \mathrm{Mpc}^{-1}$ and $\Omega = 1$, it is only around a billion years. Maarten Schmidt also displayed the dramatic peak in quasar activity at redshifts of 2 or 2.5. [It is an anti–anthropic irony that the most interesting time to be an astronomer was before the Earth formed.]

At least a few galaxies must therefore have formed, and evolved to the stage when a runaway catastrophe occurs in their nuclei, at a redshift exceeding 4. This is a severe problem for top–down models involving neutrinos. It is also a constraint on hierarchical models, especially the CDM model, in which galactic masses aggregate rather late. Only very exceptional CDM peaks would be on galactic scales at these redshifts. It is hard to quantify how much of a problem the data already pose for the CDM model. We need to know how big a galaxy has to be to 'host' a quasar, and how many generations of quasars there are. But the embarrassment threshold of CDM advocates will certainly need to rise if the redshift barrier gets pushed much beyond 5.

One *could* take the apparent quasar cut–off setting in beyond redshifts 2.5 as corroboration of the CDM model. Or one could attribute the peak to something else. And there is an argument, robust and insensitive to details, that an important stage of galaxy formation, the formation of disks, must have happened late, at a redshift less than 3, even if spheroids formed much earlier.

Angular momentum cannot be stored in the big bang. Protogalaxies would have had random shapes, and non–zero quadrupole moments. They would then have acquired angular momentum near the time of turnaround via tidal torques (Figure 13). This process imparts, however, only 10% of what is needed for rotational support. That is a problem unless there is a long 'lever arm', whereby the angular momentum could have been acquired when the protogalaxy had a ten times larger radius than now. The gas that makes a 10 kpc disk would then have to have fallen in from beyond 100 kpc. That infall would have taken more than a billion years, implying recent disk formation even if spheroids formed earlier (Figure 14).

Galaxies are generally isolated, but mergers occur today. We see many interacting galaxies. Moreover, mergers would have been much more frequent at larger redshifts. The next series of slides (not, unfortunately, reproducible in the

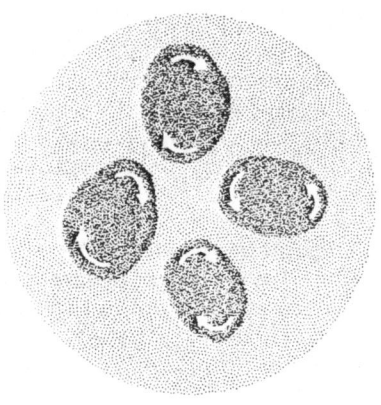

Figure 13. *Protogalaxies would have non–zero quadrupole moments, and their mutual gravitational interactions near the time of turnaround would have imparted angular momentum to each.*

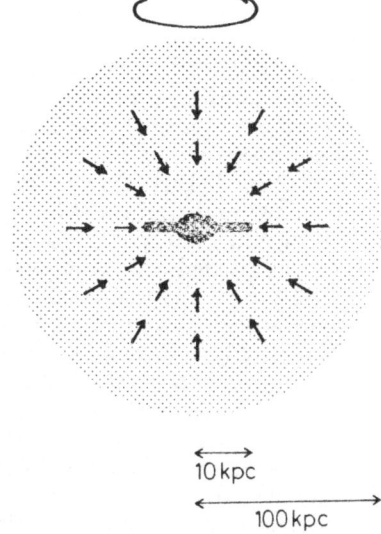

Figure 14 *Tidal torques typically impart less than 10 per cent of the rotational velocity needed for centrifugal support. The material that ends up in a disc of radius 10 kpc must therefore have fallen in from $\gtrsim$ 100 kpc. This infall timescale is $\sim 10^9$ yrs, so the formation of discs cannot have been completed before the epoch corresponding a redshift of 2 or 3.*

written version) shows a merger of a disk and small elliptical, from simulations by Hernquist at Princeton. Many galaxies must have experienced such events.

'NORMAL' GALAXIES

Quasars may involve an atypical subset of galaxies. So we must be cautious about inferring anything about typical galaxies from the quasar redshift distribution. The same is true of radio galaxies, because they are exceptional too. [We heard just today, incidentally, that the galactic redshift record had been broken, with a newly discovered object at z = 3.8.] Until recently, hardly anything was known about *ordinary* galaxies sufficiently far back in time for evolutionary changes to really show up. But large telescopes and more sensitive detectors are changing this. Images can now go so faint that there are 100,000 galaxies per square degree, and counts can be compiled down to fainter than 26 magnitude. These counts cannot be uniquely modelled in the absence of knowledge of the redshift. But models suggest that the dominant faint galaxy population may be being seen at the stage when they are acquiring disks. We must await the sharper images the ST will give to test this hypothesis. Next–generation telescopes should give us snapshots of galaxies at different redshifts, (different epochs), thereby allowing us to check how galactic evolution actually occurred.

These faint galaxies vastly outnumber quasars and radio sources, which could mean one of two things. *Either* a very small fraction of galaxies have long–lived active nuclei; *or* more do, but the activity represents a relatively brief phase in each galaxy's life history. Maarten Schmidt favoured something closer to this second option, because if individual quasars were too long–lived, they would build up to unacceptably large masses. If there were many generations of quasars, we would expect that dead quasars, massive black holes now starved of fuel, should lurk in the nuclei of many nearby galaxies. Recently, just such evidence has emerged, primarily from the work of Dressler, Kormendy, Richstone and Tonry. The stars near the centre of the Andromeda galaxy have a rotation and velocity dispersion revealing a central spike, and seem to be orbiting a dark compact object which at least fits the description of a black hole of $10^7\,M_\odot$. Similar effects are seen in M32 and the Sombrero galaxy as well.

These holes could be reactivated, perhaps as radio galaxies or Seyferts, if the galaxy were disturbed by a merger. Otherwise they would be quiescent, but not quite. Now and again a star would wander so close that tidal forces ripped it apart. We would then see a flare persisting for as long as it took the debris to be swallowed or expelled, maybe a year or so. Searches for such a phenomenon would be a crucial test of the reality of these quiescent black holes.

There is darkness at the centre of even the most familiar galaxies. Moreover, 90% of the gravitating stuff that binds them may be a dark relic of the hot early phases of the big bang, whose elucidation transcends the physics we understand. Even normal galaxies point to new links between the cosmos and the microworld, depicted here in my final slide (Figure 15).

I argued earlier that the mundane physics of gas cooling and Newtonian collapse singles out a galactic mass and lengthscale, so that a favoured mass need not be imprinted *ab initio*. But there must have been some initial fluctuations. Otherwise the universe would still be amorphously uniform, with no galaxies, no stars, and no astronomers. There is still no agreed understanding of why the universe combines the small–scale roughness needed to initiate galaxies with the large–scale uniformity that has allowed it to expand smoothly for 10 billion years. This must await the ultimate synthesis depicted 'gastronomically' in Figure 15.

The problems of large–scale cosmogony are so intermeshed that we will not really solve any until the whole picture comes into sharper focus. For instance, we cannot test theories of galaxy formation and evolution until we understand the gas dynamics of star formation, and the possible rôle of active nuclei, as well as the exotic physics of the initial fluctuations.

The empirical data — observations in all wavebands, and laboratory experiments as well — are burgeoning and all advancing the subject. And theorists are injecting a range of not necessarily compatible ideas whose vector sum at least pushes the subject forward. Hubble's great book, 'The Realm of the Nebulae', concludes with these words. 'With increasing distance our knowledge fades and fades rapidly. Eventually we reach the dim boundary, the utmost limits of our telescope.

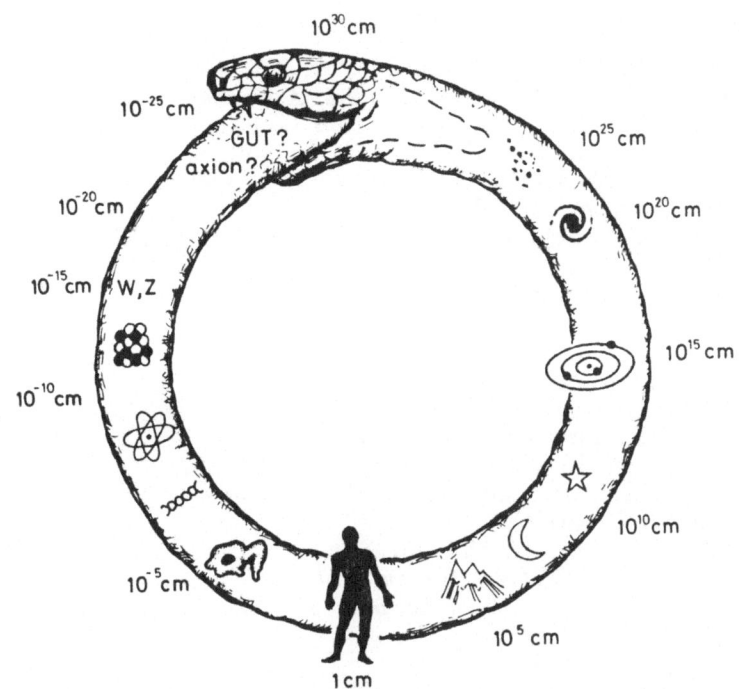

Figure 15. *The everyday world is determined by atomic structure; stellar evolution depends on nuclear physics. Recent ideas, accordingly to which galaxies and clusters may be held together by particles that are relics of the ultra–early universe, suggest further links between the 'micro'– and the 'macro'–world – the left and right segments of this picture.*

There we measure shadows, and we search among ghostly errors of measurement for landmarks that are scarcely more substantial. The search will continue. Not until the empirical resources are exhausted need we pass on to the dreamy realm of speculation.'

This search *has* continued as more powerful telescopes and detectors have been deployed. Observers have colonised the speculators' former territory, and theory itself now has a speculative range undreamt of by Hubble's contemporaries. The origin of the nebulae, and the emergence of cosmic structure, are still mysterious but the key questions are at least in clearer focus.

I am grateful to many colleagues for discussions and collaboration on topics mentioned in this talk, and to Judith Moss for her careful preparation of the typescript.

REFERENCES *

Binney, J. and Tremaine, S. *Galactic Dynamics* (Princeton U.P.).

Blumenthal, G.R., Faber, S.M., Primack, J.R. and Rees, M.J. 1984. *Nature* **311**, 517.

Bond, J.R. and Szalay, A.S. 1983. *Astrophys. J.* **274**, 443.

Carr, B.J., Bond, J.R. and Arnett, W.D. 1984. *Astrophys. J.* **277**, 445.

Cowsik, R. and McClelland, J. 1973. *Astrophys. J.* **180**, 7.

Einasto, J., Kaasik, A. and Saar, E. 1974. *Nature* **250**, 309.

Frenk, C., Davis, M., Efstathiou, G.P. and White, S.D.M. 1985. *Nature* **317**, 595.

Gott, J.R. 1981. *Astrophys. J.* **243**, 140.

Gott, J.R., Gunn, J.E., Schramm, D.N. and Tinsley, B.M. 1974. *Astrophys. J.* **194**, 543.

Lyubimov, V.A., Novikov, E.G., Nozik, V.Z., Tretyakov, E.F., and Kozek, V.S. 1980. *Phys. Lett.*, **394**, 266.

Marx, G. and Szalay, A.S., 1972. *Proc. Neutrino 72, Technoinform, Budapest*, p.191.

Ostriker, J.P., Pebles, P.J.E. and Yahil, A., 1974. *Astrophys. J. Letters* **193**, L.1.

Peebles, P.J.E., 1982. *Astrophys. J. Letters* **263**, L.1.

Press, W.H. and Gunn, J.E., 1973. *Astrophys. J.* **185**, 397.

Primack, J.R., Seckel, D. and Sandoulet, B., 1988. *Ann. Rev. Nucl. and Particle Sci.* (in press).

Refsdal, S., 1970. *Astrophys. J.* **159**, 357.

Rubin, V., 1985. *Highlights of Astronomy* **7**, 27.

Zel'dovich, Y.B., 1982. *Highlights of Astronomy* **6**, 29.

* It is not feasible to list original references to *all* topics touched on in this 'discourse'. This list primarily includes just the literature specifically mentioned in the text.

JOINT DISCUSSIONS

1. NEW DEVELOPMENTS IN DOCUMENTATION AND DATA SERVICES

FOR ASTRONOMERS

Scientific Organising Committee

G.A. Wilkins (Chairman)

O.B. Dluzhnevskaya, B. Hauck, C.O. Jaschek, S. Mitton, F.M. Spite,
P.A. Wayman, G. Westerhout

Supporting Commission: 5

NEW DEVELOPMENTS IN DOCUMENTATION AND DATA SERVICES FOR ASTRONOMERS

Report on the proceedings of Joint Discussion 1 at the 20th General
Assembly of the International Astronomical Union at Baltimore,
Maryland, U.S.A., on
1988 August 3

G. A. Wilkins (President, IAU Commission 5)
Royal Greenwich Observatory, Herstmonceux Castle
Hailsham, East Sussex, England BN27 1RP

ABSTRACT. The meeting provided a forum in which those responsible for
documentation and data services in astronomy could interact with users
of these services in an endeavour to establish how new techniques for
the distribution and retrieval of information may be used to best
advantage by astronomers. The report contains summaries of each of the
short papers that were presented during the four sessions and brief
accounts of the ensuing discussions. The principal topics were: the
preparation and publication of papers and reports; bibliographic data
services; electronic-mail facilities; archiving of current
observational data; the changing role of libraries; and the maintenance
of the historical record of astronomical activities.

1. INTRODUCTION

1.1   Background to the Joint Discussion

During the short period of less than three years since the 19th General
Assembly of the International Astronomical Union (IAU) in Delhi, India,
there have been two areas of rapid advance that have had major impacts
on documentation and data-service activities in astronomy. Many
authors now use "desk-top publishing" techniques to produce
camera-ready copy that is comparable in appearance with the output of
commercial photo-typesetting equipment, while 'electronic mail' is now
widely used for rapid, cheap communication within and between many, but
not all, countries of the world. Other areas of growth in recent years
include the distribution of preprints rather than of reprints of
papers, and information retrieval by remote interrogation of large
bibliographic data-bases. The direct recording in digital form of
observational data has also become much more common and is leading to
the need for new procedures for the archiving of data; at the same time
astronomers are becoming more aware of the need for greater attention
to the conservation of astronomical plates and documents. It was
therefore very appropriate that these and other related matters should
be discussed at the 20th General Assembly of the IAU in 1988 at a
meeting that brought together astronomers from many different
commissions with librarians and other providers of the information
services that are used by astronomers.

D. McNally (ed.), Highlights of Astronomy, Vol. 8, 69–96.
© 1989 by the IAU.

The busy schedule of the IAU General Assembly meant that only one day
(Wednesday, August 3) could be allocated to such a meeting, and so it
was preceded by IAU Colloquium No. 110 on "Library and Information
Services in Astronomy". This was held in Washington, D.C., on July
29-30 and at the Goddard Space Flight Center on August 1. (The
original intention had been to hold a workshop at the U.S. Naval
Observatory, but the numbers wishing to attend soon proved to be so
large that it was decided to hold it at the nearby Dupont Plaza Hotel
and it was necessary to arrange a more formal programme than had been
originally envisaged.) The programme of Colloquium 110 covered a wider
range of topics than that of Joint Discussion 1; in particular, more
attention was given to the problems of librarians and astronomers in
countries that do not have access to the advanced computer and
communication systems that are now available in highly-developed
countries. Financial support was provided by the IAU and by other
American organisations and individual librarians, and so several
librarians and others from less developed countries were able to attend
the Colloquium and, in some cases, the General Assembly as well.
Participants in the Colloquium who were not able to attend all of the
General Assembly were invited to attend the Joint Discussion in order
that they could interact with astronomers who had not attended the
Colloquium; about 20 such persons took part in the Joint Discussion,
some of them as speakers.

The Joint Discussion thus served as the valuable follow-up to
Colloquium 110 and it also provided an introduction to more detailed
discussions during the subsequent meetings of IAU Commission 5 on
Documentation and Astronomical Data. It also created a greater
awareness amongst astronomers of the functions of Commission 5, and it
stimulated some to attend its meetings which were necessarily held
simultaneously with other scientific meetings on specialists topics.

The following report on the Joint Discussion contains summaries of the
four sessions, each of $1\frac{1}{2}$ hours, which consisted of a series of short
prepared contributions and periods for general discussion. Comments
made during the period of general discussion have usually been reported
immediately after the contribution to which they referred; editorial
insertions of additional comments or information have been indicated by
[enclosure]. In several cases the speakers have submitted papers on
their topics for publication in the Proceedings of Colloquium 110
(Wilkins and Stevens-Rayburn 1989), which contains other papers and
reports of discussions on these and related topics. The subsequent
meetings of Commission 5 are reported in Volume 20B of the Transactions
of the IAU, where will also be found the text of a resolution on the
improvement of publications that was adopted by the General Assembly
during its final session on August 11.

1.2   Introduction to the Joint Discussion

The chairman of the first session, Professor Patrick Wayman, a former
General Secretary of the IAU, welcomed the participants to the Joint
Discussion, who numbered about 200.

The chairman of the Scientific Organising Committee, George Wilkins, then summarised the principal aim of the Joint Discussion as the transfer of information and ideas between the providers and the users of documentation and data services in astronomy. It would act as a follow-up to IAU Colloquium 110 and as an introduction to the meetings of Commission 5 on the following days. There would be four sessions:

Session 1 on developments in primary publishing would be concerned with economic issues and with new techniques that would affect the long-term character of the astronomical literature and of the author's role in the publication process. Consideration would also be given to the revision of the IAU Style Manual which, it is hoped, will be of assistance in the preparation of astronomical papers for journals, as well as of reports and papers of IAU publications.

Session 2 on developments in the retrieval and distribution oɪ information will provide information about the availability and use of bibliographic databases that cover fields of interest to astronomers and also about the use of computer networks to provide rapid communication between individuals and to 'publish' urgent announcements and other such information. The project to develop a thesaurus of astronomical terms to improve the quality of information retrieval would also be described and discussed.

Session 3 on developments in data archiving and retrieval would include a review of data-centre activities, but would be primarily concerned with new systems for the archiving of new observational data. The problem of the designation of the sources to which the data referred would also be discussed.

Session 4 on the changing role of astronomical libraries would provide a review of the discussions at IAU Colloquium 110 on the current problems of astronomical libraries and on the ways in which they are developing new techniques and activities in response to the availability of new facilities. Finally, there would be consideration of the task of maintaining the historical record through the archiving of correspondence, manuals and other unpublished records, plates, tapes and discs; this would include guidance on the selection, indexing and conservation of such records.

Participants who gave prepared contributions and others who spoke during the discussions were invited to submit short papers or notes for use in the compilation of the proceedings.

2   DEVELOPMENTS IN PRIMARY PUBLISHING

2.1   The publisher's viewpoint

2.11   S. Mitton: Economic issues in astronomical publishing. Simon Mitton, of the Cambridge University Press (CUP), indicated that he would be concerned mainly with the factors that affect the prices of astronomical books; many of them apply also to journals. He would base his statistics on the experience of CUP, but he felt that they would be typical of other major publishing houses that aimed to cater for astronomy.

There was a steady decline in the size of the core market from 1965 until about 1981, and this appears to have been largely due to physics libraries being unable to afford to take books in peripheral areas; another factor is that research grants are often now insufficient to allow for the purchase of extra books. There are now about 550 university libraries and astronomical organisations that buy astronomy books on a regular basis. The number of copies bought by individuals from their own pockets may run from only 100 up to several thousand in the case of very popular books. The availability of photocopiers may have reduced the number of copies bought by individuals, but on the other hand publishers should not complain of photocopying if they have set the prices on the basis that libraries will be the only purchasers of a particular book. The contributions towards the sale price of a book are typically as follows:

Physical manufacture (typesetting, printing and binding)	25%
Fees/royalties	10%
Bookseller discount	35%
Marketing, distribution, advertising and overheads	25%
Surplus/profit	5%

These proportions vary from book to book and sometimes the bookseller discount is higher than 35% especially for books that are imported. Some publishers may look for a higher surplus than 5%, possibly to 15%. Those who consider that book prices are too high because of profits made by the publisher should invest in their shares! (The Cambridge University Press is a not-for-profit organisation.) The typical price for a 400-page book will depend primarily on the number of copies that the publisher expects to sell and may be illustrated as follows:

Specialised monograph (typeset)	approximately 1000 copies	$75-120
Conference proceedings (camera-ready copy)	400-1000 copies	$50-80
Graduate textbook	2500 copies over 3 years	$35-45
Undergraduate textbook	5000 copies over 3 years	up to about $25
Popular books	10^4-10^5	$35 maximum

The popular or mass-market books are bought mainly by amateur or armchair astronomers, and it can be said that they, to some extent, subsidise the professional market. The price of a book may be affected by various factors; for example, although authors of papers in conference proceedings do not receive royalties, the publisher may make a contribution to the cost of the organisation of the conference and these expenses are recovered from the sales of the proceedings. Inclusion of colour plates may increase significantly the cost of production. Advertising may be used sometimes to reduce the prices of conference proceedings, for example. It is not expected, however, that the use of new technologies, such as desk-top publishing, will reduce the prices significantly. The costs of composition have already been significantly reduced over the past 10 years by the introduction of new technology and by the elimination of restrictive practices. The effect of desk-top publishing techniques will be mainly to increase the quality of the product; competition between publishers helps to keep down the prices and the editorial boards of journals can help to control costs, but it should be noticed that societies, such as the Royal Astronomical Society, do want to generate a surplus from the sales of their publications in order that they may build up cash reserves against future periods when they may wish to sell the journal at a loss. It may be noticed that this is shown in the published accounts of the Royal Astronomical Society.

In the future, one can see an increased use of desk-top publishing, but this will require that standards be introduced for disc size and formats and so on; the variety of typefaces used should also be limited. Electronic mail will also be increasingly used for the submission of papers and in order to speed up the refereeing processes. It is still expected that copyediting will necessary in order to remove at least the worst of the inconsistencies that are so common in camera-ready copy today.

2.12 Discussion. The first questioner asked why authors should get less than 10% in royalties while the booksellers margin was 35% or more, and he then went on to ask whether booksellers are really necessary when most of the sales are to libraries. In response Mitton said that it would be impossible to increase the royalties without putting up prices sharply; attempts had been made to cut out booksellers by the use of direct mail-order, but it was found that many librarians prefer to go through an agent or bookseller in order to save their own time; an agent will deal with many different publishers and yet present the librarian with only one invoice; this could result in a saving of staff required in the library. A. Ratnakar pointed out that this would be less important in developing countries; he had often tried and failed to place direct orders with the publisher. Mitton commented that there are difficulties in importing books into India because of the bureaucracy; if books are sent directly through the mail then this also leads to problems of customs and losses.

2.2 New techniques for the submission and preparation of papers.

2.21 H. Abt: The use of new techniques in editorial procedures. Helmut Abt, the editor of the Astrophysical Journal, considered the effects on the publication of journals of the introduction of new techniques in each of the chronological steps in reviewing and publication.

(a) Submission. For speed and reliability, express mail is the best current method. With regular mail we have lost manuscripts mailed between American cities, and about 0.1% of the manuscripts from abroad are lost in transit. Even airmail can take up to one month between the USSR and USA. Facsimile transmission has marginal quality at present and is not yet suitable for long manuscripts, although the quality is adequate for review copies.

Computer-readable diskettes are not yet competitive because (1) they still require much work by the publishers, (2) they require much more work by the authors, (3) there are still too many competing software systems, and (4) papers are written too infrequently (about one paper worldwide per author per year) for authors to remember the system details, although that is not true of manuscript secretaries.

An excellent current method is the use of an optical character-recognition machine made by Kurzweil, which scans and reads typescripts in any typeface. An operator follows the interpretation for about two pages to monitor for confusing letters (e.g., one and lowercase L); then the machine scans one page per minute and records it in a computer-readable file. This method requires the least work and consistency of the authors. The operator edits this file for composition by a phototypesetter. The American Institute of Physics uses it for 40% of its manuscripts, namely the non-mathematical ones.

(b) Reviewing. In the future, manuscripts will be sent to referees by electronic mail (e-mail), saving about a week due to mailing. At present about 10% of the referee reports are sent by electronic mail, and the fraction is increasing rapidly.

(c) Transmission to publishers. At present this is done by express mail for the reason of reliability.

(d) Copyediting is still necessary because many authors do not recognise the need for consistency. Currently copyedited manuscripts are discussed with authors by telephone for the Ap.J. Letters and are mailed to the main journal authors. In coming years the latter should be done by telefax or e-mail.

(e) Publication. To save shelf space, future publication may be on both paper and compact disks (CD), with the latter used for older issues. CD publication will not be much cheaper because only the costs for paper and mailing are reduced. The eventual method will probably be to have individual accepted papers entered, when ready, into a central memory bank for immediate access by readers. Printed issues of journals may then no longer be produced

2.32 G. A. Wilkins: Author's viewpoint of new techniques. Wilkins expressed the view that the use of desk-top publishing techniques for the preparation of camera-ready copy for papers for conference proceedings is producing results of high quality but, on the other hand, such techniques make much more demands on the author: for example, he needs to have much more awareness of typographical matters, of the detail associated with copyediting, and of various systems of computer software. The extra effort may well be justified if the proceedings are published more quickly, but the author will be frustrated if it is necessary for him to revise his paper extensively before publication in order to meet the standards normally expected in journal publications. Many authors will be reluctant to spend all the time and effort that is required to learn and use these new techniques, and moreover it seems unlikely that they are likely to result in significantly lower prices for journals. The question therefore arises as to whether greater use of these techniques should be demanded.

2.33 G. A. Wilkins: IAU Style Manual. Wilkins then went on to draw
attention to the new IAU Style Manual that he is preparing and that is
to be published in the Transactions of the IAU with the reports on the
meetings at Baltimore. The Manual is intended to provide a general
guide on the writing of astronomical papers and reports, and on copy
preparation for authors, editors, referees and even publishers. The
Manual is much more detailed than previous versions of the IAU Style
Book since it covers many points that would normally be dealt with by
the copyeditor of a journal. Particular attention is, however, paid to
such matters as units and designations of astronomical objects that are
primarily the responsibility of the author; it also contains other
information that it is hoped will be useful to readers as well as to
authors. The drafts have been circulated quite widely, and several of
the editors of the principal journals have commented on it. It is
hoped that most of the recommendations will apply to all the principal
astronomical journals and not just to IAU publications. One principal
point of disagreement concerned the form of references; in particular,
Wilkins considered that references should include the title of the
paper and should give the full internationally-agreed abbreviations for
the titles of journals: however, several of the editors considered
that this would take up too much space and that titles should be
omitted and very abbreviated forms should be used for the principal
journals. These matters were discussed further at a later session of
Commission 5 on editorial policy.

2.34 Discussion. David Lide, the president of CODATA, shared the
concern about whether authors should be expected to act as copyeditors
and compositors. The computerised typesetting of tables in large
central facilities was very valuable, but it was clear that the
standards of presentation have gone down very rapidly. Most authors
are unwilling to learn all the nuances of style and to take into
account all the standards necessary to achieve a fully satisfactory
result. He noted that the publisher's copyeditor now works at a
terminal and no longer uses blue pencil on a manuscript. On the other
hand, Bob Hanisch thought that authors were not reluctant to learn
detail, but that the main problem is the lack of standards; it would be
possible to prepare standard templates in computer software to match
the style of the journals. H.-U. Daniel, who is responsible for
Astronomy and Astrophysics in Springer-Verlag, stated that his company
had developed a package that prompts the author about layout and other
details; he did, however, recognise that some decline in quality is to
be expected. The instructions that secretaries need to follow cover
about 5 pages. Astronomy and Astrophysics is willing to accept
manuscripts that have been produced by TEX; the situation will be
improved when MATHOR is available since the display on the screen
corresponds to what will be printed. He accepted, however, that
savings in the cost of typesetting by the use of such techniques will
not be large; in any case the costs of printing and binding make up a
large proportion of production costs. Alan Batten considered that the
biggest barrier to a scientist acting as copyeditor is his or her own
attitude: many do not consider that the things that copyeditors attend
to are important; even those of us who do are not very good at
copyediting; the new ways of book production force the author to be his
own copyeditor and we need more than ever to regard editors and
publishers as our partners and not as our adversaries. Abt in reply to

Hanisch considered that, in fact, good progress had been made in the development of common formats for journals. Barry Lasker said that it is important to distinguish between formal publication and informal communication, such as is now provided by computerised bulletin boards, and he was concerned that this distinction might be blurred if papers were submitted to a central database and not published in printed form. Several participants spoke in favour of the inclusion of the full title of papers in the list of references, and it was pointed out that very long titles could easily be abbreviated by the use of leaders ... . In order to try to obtain a better indication of the general feeling of the meeting on this matter, the chairman asked for a show of hands of those who were in favour of the inclusion of titles, and those who did not feel that they were sufficiently useful to justify the extra space; about 50 of those present expressed a preference for titles, while 30 did not want them.

2.4 Publication of data in printed form

The chairman stated that Professor Jaschek of the Data Centre at Strasbourg had intended to be present but had been unable to attend at the last moment. He had, however, supplied a summary of the remarks that he intended to make; the chairman then read the following short paper:

2.41 C. O. Jaschek: Journal policies and alternative procedures for the publication of data. The principal reasons why we publish on paper are that (a) paper is cheap, (b) paper provides a long lasting support, and (c) paper provides an immobile version (i.e. unchangeable) of the author's script. In general we find that there are no problems with small collections of data, e.g. 500 observed colours or 300 proper motions, but problems do occur with larger data sets. Page charges make compression necessary, and this may produce almost illegible results in microprint or suppression of important information, such as dates of photometric observations.

There also exist routine observations, like double-star measurements or minor-planet positions, which are sometimes not accepted by journals because they only contribute to later work and cannot be used for immediate interpretation. Besides this, we have the compilation catalogues - such as Mermilliod's photoelectric catalogues of some ten thousand stars - which journals simply refuse to publish. Such catalogues are only available on magnetic tape, and what appears in print is, at most, an announcement. Finally, the problem becomes essentially unsolvable with huge amounts of data like those coming from instruments in space or from radiotelescopes. Often these are published as 'observatory publications', but this decreases their availability.

As a result, valuable data are lost from general circulation; various remedies have been proposed.

(a) Microfiche. This is a cheap and durable support, lasting at least 40 years. Microfiche were never popular with astronomers because they need a reading device. For long compilation catalogues they are nevertheless the best solution. For shorter data lists - say less than 6000 lines - they are cumbersome.

(b) <u>Unpublished data archives</u>. The IAU Commission 27 on Variable Stars keeps such an archive on machine-printed pages. The problem here is that most people do not know of the existence of these archives, nor about their contents. As a consequence the archives are very sparsely used.

(c) <u>Data journals</u>. Data journals exist in other sciences, as in physics and chemistry, where the data flow is larger than in astronomy; they include <u>only</u> data and some technical details. Since Jaschek was convinced that such a journal would be successful he attempted to sell the idea, and he even found a publisher. Unfortunately, astronomers (usually the happy variety of colleagues who do not try to publish long data series) were against it, firstly because it would mean one more journal to buy, and secondly because if the data were not properly discussed, one could perhaps get some data of lower quality published. So in the end he gave up.

An intermediate solution would be to set a limit to the number of publishable data and convince authors that they should agree to have their observations printed on microfiche <u>and</u> provide their observations for data centres on tape or diskette. If observations are kept only on microfiche or only on tape, we have difficulties of access in the first case and no permanent record left in the second. Please observe that magnetic tape is perishable - five years or so - and that the often recommended remedy, namely a digital optical disc, has no guaranteed lifetime. Only one manufacturer, to his knowledge, has indicated a lifetime of 20-30 years.

In summary, this means that we will have to continue with printed journals for short collections of data, but that we must put the larger data collections into data centres. Indeed data centres were set up just for that purpose.

2.42 <u>Discussion</u>. There was no time for discussion of Jaschek's contribution.

3. DEVELOPMENTS IN INFORMATION RETRIEVAL AND DISTRIBUTION

The chairman of Session 2 was Professor Hauck of the University of Lausanne, Switzerland.

3.1 Retrieval of Bibliographic Information.

3.11 <u>J. Rey-Watson: Bibliographic databases</u>. Joyce Rey-Watson, librarian at the Center for Astrophysics (Cambridge, Mass., USA), first of all pointed out that until recently on-line retrieval of bibliographic information about astronomical topics has been largely limited to the use of databases intended primarily for physics, and information about particular objects could only be obtained if the name of the object was included in the title or abstract of the paper. This situation has been dramatically changed by the development of SIMBAD (Sets of Identifications, Measurements and Bibliography for Astronomical Data) by the Data Centre of the Observatory of Strasbourg (CDS). The database is held on a computer in Paris but is accessible from the USA on a transatlantic link provided by NASA; a terminal for demonstration was available in the exhibition associated with the General Assembly. Searches are automatically carried out for all known names of each object, even though the enquirer need specify only one name.

The second development of particular importance to astronomers is the agreement between the Astronomisches Rechen Institut (ARI) at Heidelberg and the Fachinformationszentrum (FIZ) at Karlsruhe to the effect that Astronomy and Astrophysics Abstracts will be available on-line in the Physics Briefs database from the beginning of 1989.

Other major databases that are accessible to astronomers for on-line searches include:

INSPEC: This is based on Science Abstracts A (Physics) and goes back to 1969;

NASA/RECON: An aerospace database developed by NASA that is available through NASA and ESA-IRS: it dates back to 1962;

SCISEARCH: This is based on the Science Citation Index and is used in a different way to trace the development of a particular topic from earlier references to the present day.

Other aids in bibliographic searching include:

Current Contents: this appears weekly and contains the tables of contents of many astronomical journals.

Dissertation Abstracts Online: this includes all American theses back to 1961.

Astronomy and Astrophysics Monthly Index: published in California in both printed and machine readable form.

Lunar and Planetary Bibliography: this is available on-line at the Lunar and Planetary Institute (Houston, Texas); it covers the Moon, planets, asteroids, comets, meteorites, and space colonisation and utilisation.

Finally, Rey-Watson indicated that she would be glad to provide additional information on request.

3.12 <u>Discussion</u>. N. G. Roman enquired whether cross-referencing by position is possible in SIMBAD. Rey-Watson stated that the enquirer was able to specify limits between which the coordinates of objects of interest should lie.

3.2   Electronic bulletin boards and similar services

3.21 <u>B. M. Lasker, P. M. B. Shames and L. Butler: Public access computer services</u>. Barry Lasker, of the Space Telescope Science Institute, considered that public access computing services, based on the existing and rapidly developing networks that interconnect astronomical institutions, already offer the community significant resources for information exchange; and the potential for growth at minimal cost is limited only by the imagination and enthusiasm of the scientific users. The requirements for such services are (1)

widespread connectivity for both direct logins and for mail handling,
(2) friendliness to users regardless of their level of sophistication
with computers, (3) active, responsible management of the scientific
materials, and (4) effective management of the computer resources.
Some networks, such as BITNET, provide e-mail facilities, but not login
access facilities. It has been found that most users read but do not
contribute to the bulletin boards, and that astronomers are usually
very different from the software community and demand a high level of
service. The service should be available at all times and the bulletin
board should be controlled. Remote interactive working can be very
expensive unless the system is efficient and is used effectively.

One example of this kind of service is the Astronomy Information System
(ASTIS), which has been established at the Space Telescope Science
Institute in order to provide open access to bulletin boards,
databases, information, and a directory of astronomers. Access to the
ASTIS services is supported on SPAN, the ARPA Internet, and the
international packet-switching (PSS) networks so as to be within reach
of the widest possible community.

The initial implementation contains material on SN 1987A, the Space
Telescope Science Data Analysis System, and directory entries for over
700 individuals and astronomy sites (as collected from individual
contributions and from the activities of the RGO and the AAS). A
project to expand the service to include Space Telescope news, proposal
information, and bulletin boards on other astrophysical topics is in
progress.

3.22 Discussion. P. W. Hill stressed that users must be careful not
to waste time and money when working interactively over PSS networks.
P. Wehinger considered that bulletin boards need contributions from
users.

3.23 B. G. Marsden: The Computer Service of the IAU Central Telegram
Bureau and Minor Planet Center. Brian Marsden, of the
Harvard-Smithsonian Center for Astrophysics, described the development
and current status of the computer services now provided by the IAU
Central Telegram Bureau and by the Minor Planet Center, of which he is
the Director. The impetus for providing a mechanism for computer
distribution of the IAU Circulars was Comet IRAS-Araki-Alcock.
Announced just one week before its passage only 0.03 au from the Earth
in May 1983, this comet was extensively discussed in airmailed
Circulars that failed to reach the majority of recipients until after
it had passed! By January 1984 appropriate arrangements had been made
on one of the VAX computers at the Harvard-Smithsonian Center for
Astrophysics, and during that first year of operation of the Computer
Service there were some two dozen subscribers, all but two of them
located in North America.

Initially, use of the Computer Service required the subscriber to
contact the VAX by modem. In addition to providing the IAU Circulars,
the system had a facility for the exchange of messages. There was also
a useful feature that allowed the caller to extract orbital elements of
comets and minor planets from the Minor Planet Center's files and to
calculate eleven points in an ephemeris - at an interval between 0.1
day and 20 days.

XMODEM and KERMIT file-transfer protocols were available, and late in 1985 the system was transferred to a MicroVAX computer essentially dedicated to the work of the Central Telegram Bureau and Minor Planet Center. E-mail capability become possible by BITNET, but because of its frequent unreliability, it was decided to continue to provide the electronic IAU Circulars only to those who actually logged in for them. A new addition to the Computer Service allowed the caller to find all the known comets and minor planets within a 5-degree range of right ascension at a particular time. Submitted as a batch-job running while the caller made use of other features, this calculation enables him to establish whether a suspected comet discovery was new or whether a possible nova might be a particular minor planet.

By the beginning of 1987 the Computer Service boasted a steady 40 or so subscribers, but it was still the case that 90% of them were in North America. With the discovery of the LMC supernova there was a dramatic change. The possibility of connecting with SPAN suddenly became a reality. Modems were relegated to amateur use as professional callers could "set host" from other VAX computers. The number of subscribers tripled over the course of a single month, and a second MicroVAX was added in the expectation that one or the other computer would be available at all times. A separate login allowed the SPAN user to e-mail the IAU Circulars to a specific VAX node and username.

Automatic e-mail distribution of the IAU Circulars to subscribers who can conveniently be reached via SPAN or BITNET was finally organized in July 1988. About half of the Computer Service users seem to be in this category. It still happens that messages get lost (or, at least, substantially delayed), but subscribers are able to log in and read any that they have not received.

Although some may think this to be an unnecessary relic from the days of the absolute dominance of the printed word, the Computer Service still always circulates complete IAU Circulars, rather than individual items as they become available,. With only some 20% of the subscribers participating in the computer dissemination, and all of these restricted to the nations of the first world, there is certainly no plan to discontinue the printed Circulars, which should still be regarded as the "official" version.

It is certainly to be hoped that, by the time of the next IAU General Assembly, astronomers in many more countries will be able to participate in the Computer Service. Most of these astronomers currently rely on telex for rapid communication, and while some changes in the dissemination by telex of the most urgent data are planned, it would still be expensive to distribute complete IAU circulars in this way. New features are planned for incorporation in the Computer Service, and further suggestions are very welcome.

3.24 <u>Discussion</u>. Lasker wondered whether the use of the search facilities might impose a high load on the system; Marsden replied that a MicroVAX is largely dedicated to the Service and is able to meet the demands on it. A. Ratnakar commented that the Service is not available in countries like India. In reply to a question, Marsden said that

there are small fixed monthly charges for the use of the Computer
Service in addition to the subscription to the mail service for the
circulars.

3.3   Directories of organisations and astronomers

3.31   J.-L. Halbwachs: International directories prepared at
Strasbourg. Jean-Louis Halbwachs, of the Centre de Donneés Stellaire
(CDS) at Strasbourg, said he was speaking on behalf of André Heck, who
had developed his first directory after finding that his publisher
could not distribute publicity material properly. The International
Directory of Astronomical Associations and Societies contains over 2000
entries for 65 countries; the seventh edition will be published in
1988. He later produced the International Directory of Professional
Astronomical Institutions, and its first edition contains a wide range
of information about 2200 organisations in 75 countries; the second
edition will be published in 1989. (See references.) It is expected
that new editions will be published by CDS at a frequency of two years;
enquiries should be addressed to Dr. A. Heck at CDS.

3.32   Discussion. Hill wondered about the problems of obtaining
up-to-date reliable data; Halbwachs replied that Heck had made every
effort to check the accuracy and completeness of the data. Hauck
considered that it would be useful to have the names of the astronomers
of each institution, but Halbwachs felt that this would be too long.
Rey-Watson felt that a frequency of 2 years is adequate for a
directory. It was generally agreed that it would be very useful if the
directories were to be made available on SIMBAD, since entries could be
updated when new information was obtained.

3.33   C. R. Benn: World e-mail directory prepared at Herstmonceux.
Chris Benn, of the Royal Greenwich Observatory, first of all spoke
about the revolution in communications that had been brought about by
electronic mail which is quick, cheap, convenient and informal. As a
new system it still has problems, such as the complicated syntax of
addresses for routing messages. Several national and international
computer networks are linked together, but some parts of the world are
inaccessible. He and R. Martin had felt the need for a directory and
had prepared one with about 3000 entries for astronomers and 200
entries for organisations; it includes an introductory guide to the
technique. Copies of the first edition have been distributed widely
without charge, and the directories are on-line on the VAX computer at
the RGO. He suggested that user names should have a standard form,
that e-mail addresses should be given for authors of papers in
journals, and that computer networks should also be used for the
dissemination of news and for software exchange.

3.34   Discussion. Lasker pointed out that other countries were
accessible through the PSS network, and that BITNET had links to some
places not shown on the map presented by Benn; Wehinger said that a
link to Moscow will be available soon. There were several comments
about user names: some systems allow an alias to be used; some node
managers do not allow user names to be publicised because of fear of
unauthorised access by hackers; and on some systems a Postmaster is
obligatory and allows messages to be sent to individuals whose user
names are not known. Gösta Lyngå said that he was a keen user of

e-mail and very happy about the RGO directory for which he thanked
Benn. He drew attention to the problem that arises when a node breaks
down. The mail is often returned with insufficient explanation,
whereas one would expect that another routing would be automatically
chosen. Benn recognised the problem, but said that the response is
different in different networks. The chairman closed the discussion by
pointing out that such problems could be discussed further at a session
of Commission 5 on e-mail on August 5.

3.4 Aids to information retrieval

3.41 R. M. Shobbrook: The IAU Thesaurus: A way to more-effective
information retrieval. Robyn Shobbrook, the librarian of the
Anglo-Australian Observatory, introduced her talk by giving a
definition of a thesaurus: a compilation of words and phrases showing
synonymous, hierarchical and other relationships and dependencies, the
function of which is to provide a standardised vocabulary for
information storage and retrieval systems.

Astronomers and librarians have been experiencing difficulties in
keeping up with the amount of published literature. The astronomer
tries to keep abreast in his particular field and the librarian in the
management, control and retrieval of scientific information. The 1980s
have seen a revolution in the methods for information storage and
retrieval; in particular the use of the on-line database is now well
established. The speed of processing information for storage has been
embraced by all but with little thought given to how we shall achieve
the effective and high-precision recall of documents.

An international team of eight librarians, coordinated by Robyn
Shobbrook has compiled a preliminary draft version of preferred terms
in astronomy and astrophysics. The aim of the thesaurus is to
standardize terminology for use in automated and manual library systems
and in the allocation of keywords for scientific papers destined for
publication. The keyword list has been compiled from a number of
authoritative sources according to the American National Standard for
thesaurus construction. It is designed to show the interrelationships,
as well as synonymous terms, for a given subject, and each term has
been assigned an American Institute of Physics classification number.

The librarians firmly believe the best road to success in information
retrieval from automated systems is provided by vocabulary control.
Contrary to belief, free-text or natural-language searching does not
lead to high-precision recall. Consistency and integrity of the
on-line catalogue can only be achieved with a controlled vocabulary.
With today's technology it is possible to maintain the best of both
worlds for high-precision recall. The controlled vocabulary is used to
index the major concepts of a given document over and above the natural
language used within the document.

When the IAU thesaurus has been sufficiently refereed and tested by
scientists and librarians over at least the next year or two, it is
hoped that the editors of the major journals will adopt the list and
that astronomy librarians around the world will find they have a useful
reference tool to assist not only with the input of data into manual or

automated systems but to realise high-precision recall when retrieving information. The ultimate goal will be a cost saving in time and resources for both scientists and librarians.

3.42 Discussion. In her talk Shobbrook drew attention to the costs that would be incurred in completing the project and to the need for the assistance of astronomers in checking the drafts for completeness and accuracy and in resolving the differences of opinion on the choices of preferred terms. Discussion on these and other points had to be deferred to a session of Commission 5 held on August 5.

4. DEVELOPMENTS IN DATA ARCHIVING AND RETRIEVAL

The chairman of session 3, Dr. Gart Westerhout, Director of the U. S. Naval Observatory, introduced the session by pointing out that in the past there were relatively few occasions when photographic plates were used by persons other than those who had made the observations; nevertheless it was recognised that the plates should be kept and made available for further study. Now we have the problem of what to do with the enormous quantities of data that are now being obtained directly in computer-readable form; three approaches to this problem are discussed in this session, but first there are two papers about the activities of data centres that are endeavouring to make accessible the data that have already been published.

4.1 Developments at astronomical data centres

4.11 C. O. Jaschek: Developments at CDS and cooperating data centres. In the absence of Jaschek, Westerhout read out, with interjected comments, the summary submitted by Jaschek in advance of the meeting.

Jaschek commented on each of the following common objectives of the data centres:

(a) to increase the number of catalogues;

(b) to encourage the constitution of specialized data bases;

(c) to extend the present network of data centres; and

(d) to integrate the new archives of basic observational data in the network of data centres, which hold mainly compiled catalogues.

(a) At present we have an efficient organization of catalogue exchange between the three major data centres, those in the US, USSR and France, and certain countries like the UK and the German Democratic Republic. A catalogue arriving at anyone of the three data centres is circulated to the others; the numbering system is kept uniform, so that it is immaterial to which centre one applies for data. What is less efficient is the way catalogues arrive at the data centres. One of his activities is to write to authors of long lists or catalogues to ask them to send copies of their tapes. Since manuscripts are usually produced on tapes or diskettes, it should be easy to get them, but because of the delay before he can write the letter (after the catalogue has appeared in print) he often receives the discouraging answer that in the meantime the diskette has been re-used and that the catalogue has disappeared. This implies that somebody has to put it again on tape. Such difficulties could be overcome by better cooperation.

(b) One of the nicer aspects of data centre work is the knowledge one gains of underdeveloped areas of astronomy and of what could be done to improve the situation. When one has located such a problem area one starts looking for an expert capable and willing to do the job. This is usually a hard job, which does not always end happily. What many people have not realized is that the simple fact of collecting all information on a small part of astronomy is a powerful way to get acquainted with problems and to spot new questions. So one should encourage - and this is specially valid for locations where astronomy is developing - the formation of groups to produce specialized catalogues. Such a catalogue often also constitutes a good showcase for the activities of the group!

(c) Despite the fact that networks function across political borders, the transmission of large data sets is now done almost exclusively by mail. When a parcel crosses a national border its delivery may be delayed by customs. What we could do to minimize such problems is to create subcentres for certain regions or countries to act as distributors for their country and/or geographical region - South America, South Asia, and Islamic countries, for instance. It is clear that whenever astronomy has grown sufficiently it needs a data centre!

(d) Integration of data centres with archives is a clear tendency, which will probably be solved in a natural way through networks. It has already started with the forthcoming integration of NASA and ESA space data into the existing data networks, but it is equally clear that we should not stop at space data. We have, for instance, large archives at optical observatories and at radio and infrared facilities, and it is urgent that these be integrated too. It should become possible to know, for example, who has observed 3C 243, with what technique and when.

So in the end we may expect that towards the end of this century any astronomer may, from his own terminal, have access to all data accumulated up to now by his colleagues and his predecessors. Perhaps you might say that this is just a glorious vision, but this is what we are working for.

4.12 O. B. Dluzhnevskaya: Soviet Astronomical Data Centre. On the invitation of the chairman, Olga Dluzhnevskaya of the Astronomical Council of the Academy of Sciences of the USSR, presented a brief report on the activities on the Soviet Astronomical Data Centre in Moscow. It was established in 1981 to serve the USSR and other socialist countries, and it also provides information to other countries, such as Brazil, Mexico, Sweden, Egypt and India. Catalogues on magnetic tape are exchanged with the Strasbourg Data Centre, and a connection to SIMBAD is established about once a week. It holds about 500 catalogues, of which 32 are from the Soviet Union, including catalogues on variable stars and on star clusters and associations; extensions of these catalogues are in preparation. All the large observatories in the USSR have their own databanks, and it is hoped to establish a computer network soon. [A short paper on "the network of the centres of astrometrical data in the USSR", by L. I. Chunseva and others at the Pulkovo Observatory, was received before the meeting, but was not presented; two of the five centres are primarily concerned with satellite tracking data (including laser-ranging data), while others hold astrometric data on stars, solar-system bodies and earth-rotation.]

4.2   New programmes for the archiving of observational data

4.21   C. R. Benn: The RGO system for the archiving of data from La Palma.   Benn described briefly the principal features of the system developed for the archiving of current observational data from the three telescopes (1-metre, 2.5-metre, 4.2-metre) on La Palma for which RGO is responsible.   The data archive is growing at the rate of about 10 magnetic tapes per week and it is seen as a major national facility. The tapes have FITS headers, and there is a complementary catalogue that lists the objects observed and additional information about the time, telescope, detector, and weather; this observation catalogue is accessible remotely for searching.   A statistical study for 1984-86 (Benn and Martin 1987) showed that the telescopes were actually recording data for an average of about 4 hours per night and that very few observations were made at zenith distances greater than 45°.   Such information is useful for assessing the effectiveness of the operational and maintenance procedures for the telescopes and it could also be useful for planning and design.   Optical discs will be used for storage as soon as appropriate standards have been developed.   It is also intended to extend the system to include the archiving of the reduced data.   Eventually an international union catalogue of observations should be formed, and the data from past observations should be made widely accessible.

4.22   Discussion.   M.-C. Lortet enquired about the identification of the observed objects and was told that about 10% of the names used by the observers were insufficient to identify the object easily.   It was pointed out that most planetary observing would be excluded if telescopes were not able to reach zenith distances greater than 45°.

4.23   M. Rushton: The STScI Data Management Facility.   Minick Rushton, of the Space Telescope Science Institute, said that the purposes of the STScI Data Management Facility were to provide a local archive of data from the Hubble Space Telescope and to support their distribution to other archives.   The archive includes also data from ground-based observations; for example, 1500 Sky Survey plates have already been digitised.   The mission is expected to last 15 years and to produce about 4.8 gigabytes per day.   Optical discs provide the main storage medium.   The facility will be independent of any particular device or manufacturer, and so it uses FITS standards for data formats, standard software packages and modular architecture.   Much of the system is now operational.

4.24   Discussion.   H. Friedman enquired whether data compression techniques were being used for image processing; Rushton answered in the negative, but said he would welcome information about them.

4.25 F. Ochsenbein: The ESO archive project. Francois Ochsenbein, of
the European Southern Observatory based at Garching, FRG, said that the
main reasons for creating and maintaining an archive of the data
obtained at the ESO telescopes are (a) to keep an historical record of
observed objects, (b) to reuse the data for other purposes, and
therefore avoiding the duplication of observations, and (c) to allow
researches based on the accumulated material.

The archive is made of two parts:

1. Raw observed data, with a complete description of each image
   and the related calibration data, are stored on slow access
   media, using the FITS format (Wells et al. 1981). The media
   which will be used (tapes, cassettes or optical disks) are not
   yet defined, but could be a combination of several media. Only
   non-proprietary data will be available on request.

2. The catalogue of observations is organised as a data-base, and
   will be accessible for enquiries at the ESO computer facilities
   in Garching and over computer networks. This catalogue
   contains sufficient information to locate and estimate the
   suitability of archived data; it also includes the title of the
   observation as it was submitted and accepted.

The archiving policy recently defined at ESO (Van der Laan 1988) grants
to the observing team a proprietary period of one year, which may be
extended on request in special cases. The contents of the catalogue,
i.e. the list of the observed targets and the title of the observing
run, will however normally become public immediately after the end of
the observations.

A test over one year of EFOSC (ESO Faint Object Spectrograph and
Camera) observations on the 3.6-m telescope was performed to ensure the
feasibility of the project. The reliability of some key parameters
(telescope position, timing, target names) was checked, and some
improvements in the acquisition systems will be implemented to ensure a
maximal reliability of the archived material.

4.26 Discussion. In response to an enquiry about whether the weather
conditions are recorded, Ochsenbein said that this is under discussion.
The statement that some vendors of optical discs have claimed lifetimes
of up to 200 years was viewed with some scepticism, and it was pointed
out that appropriate readers might not be readily available after a
much shorter period. The cost of maintaining the archives is likely to
be large. Ochsenbein was closely questioned about the imposition on
observers of standardised procedures for recording details of the
observing runs, including calibrations; H. Dickel commented that it is
clear that standard designations ought to be used for identifying the
objects observed.

4.3 Designation of objects for data archiving and retrieval.

4.31 F. Spite: Problems of designation and some recommendations.
Francois Spite, of the Observatoire de Paris at Meudon, said that
ideally designations should be short, unambiguous and informative, but
in practice short designations often proved to be ambiguous and failed

to identify objects clearly. In SIMBAD (see section 3.11) a standard form of designation is used; it consists of a group of up to four alphanumeric characters, a space and a composite number (including leading zeros and signs where appropriate). Even this form is inadequate for multiple stars, aggregates of objects, extended sources observed in different wavelengths, and stars in clusters and in galaxies; in particular, the identification of objects in the Magellanic Clouds poses an imminent major problem.

Some negative and some positive recommendations can be made on the basis of past experience;

do not introduce a new name for an object that already has an unambiguous name;

do not abbreviate abbreviations when forming names for objects without names;

avoid strange forms of designation;

do not use roman numerals or special characters;

do give leading zeros and signs rather than spaces or no space;

do use a typeface (e.g., OCR-B) that differentiates between zero and the letter O and between 1 and the lower-case letter L; and

give magnitude scales where charts are used to identify objects.

These recommendations should be well known, but they are often ignored.

4.32 <u>Discussion</u>. Lyngå emphasised the desirability of keeping existing designations. The chairman then closed the session as time had run out. [A new statement of recommendations for the designation of astronomical objects (radiation sources) outside the Solar System was prepared during the General Assembly (see Trans. IAU 20B) for wide dissemination.]

5. THE CHANGING ROLE OF ASTRONOMICAL LIBRARIES

5.1   Review of IAU Colloquium No. 110

5.11 <u>Introduction</u>. The Chairman of the session, George Wilkins, explained that it had been decided to review the proceedings of IAU Colloquium No. 110 on Library and Information Services in Astronomy, instead of holding the panel discussion announced in the posted programme. He would review the programme for the Colloquium and at appropriate places he would interrupt the review to ask some of those present to speak about particular aspects that had not been discussed during the earlier sessions of this discussion. Each of these contributions would then be followed by an open discussion.

5.12 G. A. Wilkins: Review of the programme of IAU Colloquium No. 110.
The first session on the publication and acquisition of books and
journals was opened with a keynote address by Helmut Abt on the future
of astronomical literature, and then the views of publishers on the
economics and pricing of astronomy books and journals were expressed by
G. Kiers of the IAU publisher (Kluwer) and P. Boyce of the American
Astronomical Society. Then the problems of international acquisitions
were discussed by a panel from many different parts of the world. This
drew attention to the problems in obtaining publications that are
experienced by librarians in developing countries, as well as by those
in western countries.

The second session on searching for astronomical information started
with a review of different ways of choosing words for use in searching
for information. The working group of Commission 5 has produced a
vocabulary of terms for use as keywords. Various classification
schemes are in use but these need updating regularly. A project to
develop a structured thesaurus of astronomical terms is in progress
(see section 3.41). Representatives of several indexing and
abstracting services and of providers of bibliographic databases
described their products; it was stated that Astronomy and Astrophysics
Abstracts will be made available on-line by the Fachinformationszentrum
(FIZ) in Karlsruhe. Techniques for obtaining the documents once they
had been identified were also discussed and progress on the development
of union lists of the holdings of astronomy serials was described. The
techniques for the use of computer networks and electronic mail were
described and on-line demonstrations of the use of SIMBAD and other
databases were provided during the evenings as well as during the day.

The third session on the handling and the use of special format
materials revealed the great increase in the use of preprints and the
corresponding decrease in the provision of reprints and of observatory
publications, particularly in western countries. It was noted that
reprints and observatory publications are still considered to be of
particular value to developing countries as they provide a medium for
the exchange of publications with other institutions. The use of
non-printed materials such as microfiche, but particularly magnetic
tape, was also discussed. The fourth session was devoted to a visit to
the Library of Congress following lunch in the cafeteria in the Madison
Building; the introductory tours started in the parts of the main
building open to the public, but then the participants in the
colloquium were taken around various other parts of the Library of
Congress that are not normally accessible and were given talks about
different aspects of the work of the Library. Then all the
participants came together again for talks on the work of the Science
and Technology Division and on the optical-disc pilot project, which
has demonstrated the great value of this medium not only for the
storage of information but also as a means of very fast retrieval of
information from journals and many other sources.

The third day began with a short session on the conservation of
historical materials. A representative of the Library of Congress
showed graphically how much damage is caused by improper handling of
books and gave advice on how they ought to be treated. Archival
activities at the Royal Greenwich Observatory and the Konkoly
Observatory were described, and then attention turned to a large

variety of other library activities. There was a lively discussion on
the role of astronomers in libraries and, in particular, on the
relative merits of expertise in astronomy or in library science in
providing library and information services. Other speakers discussed
the ways in which they are using computers for a variety of jobs in
libraries, such as cataloguing and word-processing, as well as for
information retrieval. Problems of deciding what correspondence and
unpublished documents ought to be kept, provided another topic of
discussion. The provision of support for remote observatories was one
of several topics that were discussed in small groups outside the main
sessions, and on which brief reports on the general conclusions were
given to the main meeting. A panel discussion on resource sharing and
cooperative activities proved to be of great interest and it is hoped
that it will lead to action that will be of benefit to libraries in
developing countries. The Saturday-afternoon session concluded with a
series of short summaries and reviews by the chairmen of the previous
sessions, and then Pat Molholt, Associate Director of Libraries at the
Renssaeler Polytechnic University, took as the subject of her closing
address a look at the future of astronomical libraries; she gave the
participants a dramatic illustration of the way in which computers
might be used as sources of artificial intelligence as well as of
information.

The last day of the Colloquium was spent at the Goddard Space Flight
Center. In the morning there were several talks by members of the
staff of GSFC and by other participants about various aspects of the
organisation of data centers, the archiving of observational data, and
access and retrieval techniques. Participants were then given the
opportunity to see some of the facilities of GSFC for data activities,
for libraries and for the construction and testing of spacecraft.

5.13 S. Stevens-Rayburn: The handling of preprints. Sarah
Stevens-Rayburn, librarian at the Space Telescope Science Institute
(STScI), pointed out that preprints are replacing reprints as the
primary means by which astronomers endeavour to ensure that their
papers are brought promptly to the attention of their peers in other
institutions. As an aid to astronomers who wish to know what preprints
have been received, the librarians at the Space Telescope Science
Institute and the National Radio Astronomy Observatory (NRAO) cooperate
in preparing a bi-weekly list of preprints. This list is available for
on-line searching and is also distributed on a world-wide basis, either
electronically or on paper. Final references are added as soon as they
are known; but "preprints" that are received after the paper has been
published are not listed! Institutions are requested to send their
preprints to STScI or NRAO promptly.

5.131 Discussion. The ensuing questions and comments demonstrated both
the general interest in the list and its value for a wide range of
purposes. Each title is kept in the on-line file; many papers remain
unpublished. Details are changed if it is clear that an early version
of a paper has been revised and a new preprint issued. It is sometimes
obvious that one paper is being submitted to several different journals
in slightly different forms. A lot of work is required to maintain the
file and it is not feasible to extend the amount of detail given. It

was suggested that authors should ensure that copies of their preprints are sent to the host institution when the papers use observations made on an instrument at that institution.

5.14 E. Bouton: Library support for remote site. Ellen Bouton, librarian at the National Radio Astronomy Observatory (NRAO), said that the discussions had revealed many problems in meeting the needs of astronomers at remote sites (this term also includes locations where groups are separated from the library facilities of the headquarters institution even if the place is not a remote one). The budget usually drastically limited the number of books and journals that may be purchased, and the stock is often in the care of an astronomer or an administrator who has other more pressing duties and no training in librarianship. Sometimes the HQ librarian never visits the remote site, and usually only infrequent visits are possible. There are often difficulties in getting material to the site quickly; diplomatic pouches and visiting astronomers are sometimes used instead of regular mail. The problems and the partial solutions are quite varied and the general feeling is that better solutions must be sought.

5.141 Discussion. A. E. Wright spoke of the situation at the Parkes Observatory, where there are only two astronomers on site and 2 or 3 visiting astronomers at any one time,. There are problems of space as well as of budget in holding an adequate stock of journals. He wondered, however, whether consideration has been given to replacing the printed copies by access to electronic copies. Abt said that his concept of a database for papers would provide this facility, but he considered that it was at least 10 years away. Lyngå wondered whether microfiche would be a viable alternative, even though astronomers are normally reluctant to use this medium; a good reliable reader/printer would be required, so that, for example, finding charts could be reproduced at full scale.

5.15 E. Lastovica: Resource sharing. Ethleen Lastovica, librarian at the South African Astronomical Observatory, pointed out that the essential theme of the IAU Colloquium 110 was resource sharing, but a panel of four members and a moderator had discussed resource sharing as it relates to sharing of information between those who have it, and those who do not. The reasons for not having information available when a user requires it are varied. Library and information-centre budgets world-wide are restricting the buying power of libraries, but generally, western countries are better endowed than others. Moreover, libraries in Eastern-Bloc countries, Asia and developing countries often have their foreign acquisitions restricted by currency regulations and political considerations. Several suggestions were put forward.

(a) Lists of duplicate publications not required in a library must continue to be offered to others. These lists were considered of prime importance to many libraries that have a limited stock. It was suggested that a central depository for surplus publications could be established.

(b) A forum for international exchange of information should be built up. Here it was felt that the Physics-Astronomy-Mathematics Division of the Special Libraries Association could play a role through its quarterly P.A.M. Bulletin. It was decided to circulate the next issue of the Bulletin to all who attended the Colloquium.

(c) An International Directory of Astronomical Resources should be compiled as an aid to knowing what is available in astronomical libraries and information centres. Access to these resources could then be made via interlibrary loans, or electronic mail, or in machine-readable form.

5.151 _Discussion_. Olga Dluzhnevskaya drew attention to the situation in the USSR, where there are about 50 astronomical institutes. Only a few of them are able to buy books and equipment from abroad; for example, the main Lenin Library has a budget of only about $2000 per year for all foreign literature, and the Sternberg Astronomical Institute is allowed only $200 for foreign books. The international exchange of books, as well as of serials, is very important for such libraries. Dockers commented that such exchanges are not attractive in the west since so few astronomers are able to read Russian. There is a move towards giving longer abstracts in English, but this is not always possible.

5.16 A.-M. M. de Narbonne: Meeting the future needs of astronomers for information. Anne-Marie Motais de Narbonne, the librarian of the Paris Observatory, considered that, although new tools, such as computers, e-mail and databases, are changing the work of librarians, even greater changes may be expected from the evolution of the needs of astronomers as a result of the introduction of new methods in astronomical research. Astronomical research and, hence, astronomical information have been expanding for a long time; astronomical libraries should have expanded as well but they also need to evolve. Progress comes more and more from new connections with other fields of science. The closer the border the more fruitful the research. Librarians can help astronomers to make these connections by providing new sources of information. These new sources can be found in specialised databases and libraries which are not organised with an astronomical point of view. The direct use of such sources appears to be difficult for astronomers, even for those quite familiar with astronomical sources of information. Librarians are generalists and information scientists, they know how to bridge the gap with other sciences.

Other changes in astronomical research include the higher and higher concentrations of instruments providing astronomical data, the increasing quantities of data, and the greater sharing of instruments and data. These changes constitute a challenge for libraries since information is more and more scattered and local needs become greater and greater. Astronomical libraries will face these new demands through closer and closer cooperation. Libraries will share information in the way that astronomers share instruments and data, so that local needs will be satisfied through the astronomical library network. The success of IAU Colloquium 110 is a recent demonstration of the common will of astronomical librarians to work in this way. Such a network needs to be set up, maintained, and improved continuously. For that purpose librarians need the cooperation of

astronomers. This implies, of course, enough money and staff because
cooperation requires time and money, but above all astronomers should
also be involved in this information exchange programme to make it
successful. Finally, she made a plea to the astronomers present to
talk to their librarians.

5.17 R. M. Shobbrook: The Astrobungler's Guide. Robyn Shobbrook gave
an example of the way in which librarians are responding to the new
methods of working. There have been three separate organisations on
Siding Spring Mountain: the Australian National University (ANU), the
Anglo-Australian Observatory (AA), and the UK Schmidt Telescope Unit
(UKSTU). [There is also a small unit for the tracking of
earth-satellites; the AAO has now taken over responsibility for the
Schmidt Telescope.] She has produced the Astrobungler's Guide (taking
the name from the nearby Warrumbungle Mountains) to give information
about library-related facilities on the site (including photocopiers);
the guide describes the loan policies and procedures, as well as what
is available and where it is located. This guide has proved to be very
popular with the astronomers on the site and with visiting astronomers.

5.2   The maintenance of the historical record

5.21 J. Dudley: the archiving of astronomical records. Janet Dudley,
formerly the Librarian and Archivist at the Royal Greenwich
Observatory, gave a very detailed review of problems that are
associated with the task of ensuring that appropriate records of recent
astronomical activities are available to the historians of the future.
She also discussed briefly the conservation of those records of past
activities that have been kept. (An extended summary of her
presentation is given separately after this report.)

The basic problems are organisational in character because very few
astronomical institutions have adequate systems for controlling the
large amount of paper that is produced. In some cases papers are
regarded as the private property of the scientist concerned or the
papers concern more than one institution; in such cases it is very easy
for documents that would be of great interest to be lost. Then it must
be noted that there is a very large variety in the types of records
that are generated: correspondence, records of meetings, descriptions
of equipment, legal records, results of observations, and many other
types. In some cases the records may not necessarily be on paper; they
may be on magnetic tape or even as three-dimensional models. It is
important to have a policy about what is to be kept and not merely what
may be thrown away. It is necessary to consider whether the aim is to
preserve the information or the document itself. It must be recognised
that documents may have to be kept for varying periods before a final
decision about whether to keep or reject them may be taken; hence there
should be a regular schedule for the review of papers. There should be
a consistent policy about what should be kept and the policy should be
a neutral one: records of failures as well as successes should be
kept. It is important that the persons carrying out the appraisal of
the documents should keep a record of what has been done and of why
particular decisions have been taken. Such a policy should ensure that
the records that are kept will have long-term value. Particular
problems arise when the information is generated or stored on magnetic
media such as tapes and discs; it is probable that the archivist will

need to transfer such information into new formats in periods of five to ten years to ensure that the information remains accessible. Many documents are copied on to microform in order to reduce the amount of storage space required but, as yet, it is not known how long such materials will remain readily legible. Many documents produced in the 19th century are decomposing rapidly because of the acidic nature of the paper. Even today many important documents are not produced on archival quality paper.

It is also necessary to consider how best to ensure that the documents remain accessible to prospective users. It is not sufficient merely to store the documents, they should be listed and indexed so that the information in them can be retrieved. Each institution does have the responsibility for looking after its records; it has to bear in mind that once a document has been destroyed it is lost forever, while a decision to keep a document is also a decision to incur further expense. Such decisions should not be left to chance.

5.22 Discussion. Wayman stated that the IAU does have a policy of keeping its records. Abt said that he wanted to throw away the records of the Astronomical Journal when he took over, but this was compared to burning the library at Alexandria and the University of Chicago decided that the records should be archived. Even now, many consider that the reports of referees are confidential and should not be archived for others to study later. Referees are therefore given the opportunity to indicate whether or not they are willing for their reports to be kept. At present 200 000 sheets accummulate each year and he had found it necessary to rent a storage locker in Tucson. At that point the chairman decided that the time to close the Joint Discussion had come, and he thanked all those who had contributed by presenting papers or by participating in the Discussion, and he expressed the hope that all participants had found the Discussion to be of interest and value.

6   SUMMARY

6.1   Overall view

The Joint Discussion brought together many astronomers, librarians and other specialists who would not normally attend the meetings of Commission 5 and it led to a valuable exchange of information and views about new developments in astronomical documentation and information services. In this respect it achieved its principal objective. It is hoped that this report will be read by an even wider community and that others will be encouraged to use the new techniques and facilities and to contribute to their development. It is too early to judge whether the Joint Discussion achieved its ultimate objective of bringing about a closer relationship and better understanding between the providers and the users of these new astronomical services.

The contributions and discussions showed clearly that the new techniques are potentially of great value. In some cases, however, further development and evaluation is required before it will be possible to recommend their widespread adoption. Some of the benefits,

problems and tasks that were identified during discussions are
highlighted in the following paragraphs. The list is by no means
complete, but it should provide a guide for future action.

6.2 Publications

The transmission of copy by national postal services is subject to
unjustifiable delays and losses in some cases, but it is unlikely that
reliable computer networks will be available to replace the post in all
countries in the near future.

The transmission of copy in magnetic or electronic forms may save time,
but there is a need for standardisation of procedures and format.

The availability of the IAU Style Manual should make it easier for
conscientious authors to produce copy in a consistent style, but
copyeditors will still be necessary if publishers are to maintain high
standards in their journals and books. More effort should be made to
encourage astronomers to use the existing recommendations that are
intended to give greater clarity and precision in the specification of
physical quantities, in the statement of bibliographic references and
in the designation of astronomical objects.

6.3 Retrieval of bibliographic references

The SIMBAD database is a powerful new tool for the retrieval of
information about astronomical objects, and there is a heartening
prospect that Astronomy and Astrophysics Abstracts will soon be
available on-line for the retrieval of references. Many astronomers
will, however, need more guidance about the availability and techniques
of use of these and the many other databases from which they may wish
to retrieve references of interest to their work.

The development of an IAU Thesaurus should provide a useful aid to
improve the completeness and precision of retrieval of information;
assistance from astronomers is required to ensure that the thesaurus is
up-to-date and reliable.

The astronomical community should also take action to revise the
Universal Decimal Classification scheme 52 for astronomy in order to
ensure that librarians and information specialists are no longer forced
to use an incomplete and misleading scheme when indexing astronomical
books and papers.

6.4 Communications

General-access computing services and electronic-mail, using the
networks that now connect many astronomical institutions, will
revolutionize both the general distribution of information and the
communications between individual astronomers. The procedures need to
be simplified and greater uniformity in addressing systems is
desirable. Directories of organisations and individuals need to be
extended, and they should be made available on-line as well as in
printed form. The needs of institutions that are not connected to
these networks must not be overlooked.

6.5   Numerical databases

The development of a system of large astronomical databases at a few centres, such as Strasbourg, Washington and Potsdam, continues, but the new networks will make possible general access to specialist databases that are each held at only one institution. The prompt archiving of new observed and reduced data in digital form will make such data available to much wider communities than in the past, when the main storage media were the photographic plate and eventually the printed page. Such systems could be used at small observatories as well as at the major observatories and institutions, but the full benefits will only be obtained if common procedures for access can be developed.

6.6   Designations

The failure of many astronomers to follow the IAU recommendations on designations when compiling new lists of objects, or when referring to objects in published lists, is a matter of considerable regret. There is a need for greater vigilance by editors and referees, but this is not sufficient. Astronomers must be convinced of the value of following the recommendations. More effort is needed to develop a general system that can be applied in an appropriate manner whenever new types of object are discovered, when new techniques of observation are introduced, or when new major surveys are made.

6.7   Libraries

The availability of desk-top computers, communications networks, and major databases has already had a major impact on the work of many librarians. These new facilities have both aggravated and compensated for the diminishing budgets of many libraries. They also both require and make possible greater cooperation between librarians in different institutes. Commission 5 should play a greater role in stimulating and assisting such cooperation, and especially in ways that will assist librarians in developing countries.

The maintenance of the historical record of the activities of an institution is still an important task but very rarely is direct provision made for it. This task often falls to the librarian who must decide or recommend what is to be kept and take steps to see that the documents as well as the library books are conserved for posterity. Commission 5 in cooperation with Commission 46 should do more to provide guidance to librarians on this specialist task. They will need the support and assistance of astronomers if a full record of the activities as well as the results of 20th century astronomers is to be kept for posterity.

7   ACKNOWLEDGEMENTS

I would like to thank Elizabeth Roemer for providing me with a copy of her comprehensive set of notes on the discussions, Chris Benn for taking notes on Session 1, and all the other participants who have provided summaries of their contributions to the Discussion. I would also like to thank my secretary Annette Hedges for her assistance during the preparations for the meeting and for preparing the camera-ready copy for this report.

REFERENCES

Benn, C. R., Martin, R., 1987.   Isaac Newton Telescope observations
    1984-6. Q. J. Roy. Astron. Soc. 28, 481-496.

Benn, C. R., Martin, R., 1988.   World electronic-mail guide [for
    astronomy]. Herstmonceux, UK: Royal Greenwich Observatory.

Grosbol, P., Harten, R. H., Greisen, E. W., Wells, D. C., 1988.
    Generalized extensions and blocking factors for FITS. Astron.
    Astrophys. Supple. Ser. 73, 359-364.

Harten, R. H., Grosbol, P., Greisen, E. W., Wells, D.C., 1988.   The
    FITS tables extension. Astron. Astrophys. Suppl. Ser. 73, 365-372.

Heck, A., Manfroid, J., 1988.   (7th ed.) International directory of
    astronomical associations and societies (IDAAS 1988). Obs.
    Strasbourg: Publ. Spec. CDS 10.

Heck, A., Manfroid, J., 1989.   (2nd ed.) International directory of
    professional astronomical institutions (IDPAI 1989). Obs.
    Strasbourg: Publ. Spec. CDS 12.

Van der Laan, H., 1988.   ESO archiving policy. The Messenger 52, 3-4.

Wells, D. C., Greisen, E.W., Harten, R. H., 1981.   FITS: a flexible
    image transport system. Astron. Astrophys. Supple. Ser. 44,
    363-370.

Wilkins, G. A., Stevens-Rayburn, S., 1989.   Library and Information
    services in astronomy. Proceedings of IAU Colloquium No. 110.
    Washington, D.C., 1988 July 27-August 1. Washington, D.C: U.S.
    Naval Observatory

THE SELECTION OF SCIENTIFIC AND TECHNICAL RECORDS FOR PERMANENT
RETENTION

Janet Dudley
Royal Signals & Radar Establishment
Malvern, Worcestershire
UK   WR14 3PS

ABSTRACT. A small fraction of the enormous volume of records that are
generated by research and development activities should be selected for
permanent retention because of their potential value to future
scientists, engineers, historians and others. The records are
generated for many different purposes and in many different forms (not
all are documents). There are, however, some basic principles that
should be used in appraising the records so that the most appropriate
of them are selected according to identifiable criteria.

1. THE PROBLEMS

The problems of the reviewing and archiving of records do not have a
particularly high profile with either the administrators or the
scientists and engineers in institutions that produce scientific and
technical records. The first question anyone interested in, or
responsible for, such matters is likely to be asked is 'why bother?'.
Perhaps a quotation from the plinths of two statues outside the
National Archives in Washington will suggest the answer: "The Past is
Prologue. Learn from the Past".

The suggestions and ideas which follow are based on some years of
experience working in England in various scientifically-oriented
government departments whose records are subject to the Public Records
Acts 1959 and 1962. It is thus an English experience and should be
viewed as such, although the principles and criteria suggested are
applicable across a far wider range of records.

Some of the more important problems associated with building a
collection of records for permanent retention are:
    (a) Status: the records of a particular institution may be
governmental records, state records, company records, private records
or a mix of any or all of these.
    (b) Organisation: many institutions producing scientific and
technical records have inadequate (or non-existent) registries.
Individuals keep their own unorganised collections of papers and
various groups within the institution keep files of, mainly,
administrative material, which may be duplicated many times over by
other groups. Identifying the main or source file on a particular
subject under these circumstances is virtually impossible. In such an
uncontrolled environment indiscriminate and uninformed destruction is

D. McNally (ed.), Highlights of Astronomy, Vol. 8, 97–100.
© 1989 by the IAU.

also much more likely to take place.

(c) Ownership: individual scientists tend to consider records relating to activities in which they are involved as 'theirs'. There is a tendency for records to migrate from one institution to another as people change jobs. The result is that some institutions do not have a complete, permanent record of the work performed by and/or funded by that institution. Intellectual property rights are of considerable importance in such circumstances and need to be addressed on a national - and an international - basis.

(d) Size: the quantity and range of scientific and technical records is enormous, as is the variety of their quality.

(e) Selection: who, within a particular institution, is responsible for the selection of records? How is it done and what criteria are used?

It is really the last problem which is discussed in the remainder of this paper. Reviewers must always bear in mind that the destruction of a record worthy of permanent retention is irreversible but, equally, the cost of preserving a record unworthy of permanent retention is high and will continue indefinitely. How can a reasonable balance be achieved between these two opposite truths?

2. CRITERIA FOR SELECTION

Before considering what criteria may be applicable to the selection for retention from a collection of scientific material, two points should be stressed since the originators of such records may not always recognise them.

(a) Records are seldom used in isolation, but are frequently consulted in conjunction with others, often from other departments within the same institution or from other, completely separate, institutions.

(b) Records are frequently used for research apparently unconnected with the purpose for which they were created.

Some criteria for selection are obvious and applicable to virtually any institution:

(a) Records needed for the conduct of the business of an institution - the operational or administrative needs - must be kept for at least as long as they are necessary to the efficient running of that business.

(b) Records with a legal value: deeds, contracts, etc.

(c) Records which must, by statute, be retained.

Other criteria are more evaluative and applicable only to certain institutions. They can best be generically described as identifying records with a potential research value:

(d) Records cited in institutional histories, particularly published histories; such records are often concerned mainly with the results of research and development.

(e) Raw data, particularly observational or experimental data, that are the basis for reduced (or smoothed) data and published material. Such data may exist as images on photographic plates or as other types of analogue records; now such data are usually recorded directly on

magnetic tapes or discs.
(f)   Records of legislative activities.
(g)   Records of notable events and people, causes célèbres,
controversies etc, usually known as associative records.
(h)   Demographic, statistical and quantitative records.
(i)   Photographic and pictorial records.
(j)   Audio and audio-visual records.
(k)   Internally-generated training material.
(l)   Committee agendas, circulated papers and minutes.
(m)   Three-dimensional objects, such as models.

All such types of records should be considered as candidates for
permanent retention, but such material should be appraised so that the
significant can be identified and preserved and the ephemeral marked
for eventual (but not necessarily immediate) destruction.

3. APPRAISAL

The appraisal of records should be a matter of deciding what the
institution needs to keep, not what it can afford to destroy. Whether
or not existing files of papers can be weeded should also be decided at
an early stage. Weeding is time consuming and staff expensive: a
generally acceptable method in the UK is that a file is reviewed paper
by paper until one is identified as worthy of permanent preservation.
The whole file is then marked for retention without the remainder being
reviewed. The removal of significant papers from a file so that the
remainder of that file can be destroyed is not considered good archival
practice.

In the UK, the Public Record Office (PRO) reckons to retain between 2%
and 10% of government records and anyway retains everything dated prior
to 1660. In the business field the figures are between 1% and 9%.
No-one, however, believes that the correct 10% is saved, but the
adoption of the following principles should help to ensure that the
most appropriate material is retained.
(a)   Decide what should be kept, not what can be thrown away.
(b)   Make a first appraisal as early as possible in the life of a
record (and not more than 5 years after its creation) and regularly
thereafter (possibly at five-yearly intervals).
(c)   Be neutral: no organisation is totally successful and records
must be of failures as well as successes.
(d)   Be consistent: always use the same criteria to appraise the
records.
(e)   Document what is done, so that future decisions can be based on
the same criteria and so that future historians know why a particular
decision was made.

The report by Haas et al (1985) contains descriptions of the types of
records that are to be found in academic and commercial organisations
and gives suggestions for what should be retained and what may be
discarded.

4. THE VALUE OF THE RECORDS

If these principles are borne in mind during the review processes they should ideally lead to the records selected for permanent preservation having an identifiable value:

(a)   evidential (usually setting a precedent or recording provenance);

(b)   informational (conveying an action, a fact or an opinion);

(c)   instructional (describing policies and procedures);

(d) educational (illustrating the actual development of concepts and methods); and

(e)   cultural (showing the social context of the activities).

The value of records to a particular institution will vary according to the activities and needs of that institution. It is, however, vital that all institutions recognise that their records are important, both to them (in the short term), and to a wider audience of historians, social researchers and so on (in the medium and long term). Only by recognising this fact will they see the necessity of investing time and money in the organisation and control of their records. But those responsible for such organisation and control must also remember that no archivist can, or should, keep everything. Appraisal and review is essential if a useful and usable collection is to be maintained for future generations. There is as much merit in a positive decision to destroy records as there is in a positive decision to preserve. It is vital to maintain a proper and logical perspective, and to this end a paper by Rapport (1981) is essential reading, if only for its thought-provoking qualities.

REFERENCES

Haas, J. K., Samuels, H. W., Simmons, B. T., 1985. Appraising the records of modern science and technology: a guide. USA: Massachusetts Institute of Technology; distributed by the Society of American Archivists (Chicago).

Rapport, Leonard, 1981. No grandfather clause: reappraising accessioned records. American Archivist 44 (2), 142-150.

2. FORMATION AND EVOLUTION OF STARS IN BINARY SYSTEMS

Scientific Organizing Committee

R.C. Smith (Chairman)

J. Andersen, P.A. Charles, I. Hachisu, B. Hidayat,
I. Iben, K.-C. Leung, M. Plavec, S.M. Rucinski, V. Trimble

Supporting Commisions:

26, 29, 30, 34, 35, 37, 42, 44 & 48

NEW OBSERVATIONAL CLUES ON BINARY FORMATION IN THE GALAXY

David W. Latham
Harvard-Smithsonian Center for Astrophysics
60 Garden Street
Cambridge, Massachusetts 02138, U.S.A.

ABSTRACT. New observations of binaries are beginning to provide new clues on the formation and evolution of binary and multiple systems in a variety of stellar populations in the Galaxy. New orbital determinations are shedding light on the frequency and orbital characteristics of binaries in the disk and the halo of our Galaxy, both in clusters and the field. These results support the view that the formation of binaries involving solar-mass stars is relatively independent of the stellar environment. Evolutionary effects can have a major influence for close binaries with periods up to at least ten days, with a strong dependence on the age of the population. Progress towards determining the frequency of low-mass companions and planetary systems is promising but still very limited.

1. Introduction

Binaries are everywhere. They occur in every stellar environment in the Galaxy, whether in the field or in clusters, whether in the disk or in the halo. In the field, binaries are common in metal-rich and metal-poor populations alike, and are found among the oldest stars in the Galaxy and among the most recently formed.

To use the observed frequency and orbital characteristics of binaries to learn about their formation, one must take into account changes in these characteristics which may have resulted from evolutionary effects, such as the disruption of wide binaries by stellar encounters, circularization of the orbits and synchronization of the rotation for close binaries by tidal effects, and orbital changes due to mass transfer as one or both members of the binary undergoes stellar evolution. One might hope to avoid many of these evolutionary effects by studying binaries among stars that are just forming onto the main sequence. For the time being this approach is not feasible, because only nine spectroscopic orbits are available for pre-main-sequence stars (Mathieu 1988). Of course, to investigate the effects of metallicity and environment on binary formation, one may not be able to avoid working with samples where evolutionary effects are important.

In this brief introductory review I attempt to mention the areas of

103

D. McNally (ed.), Highlights of Astronomy, Vol. 8, 103–107.
© *1989 by the IAU.*

observational research where new results are having an impact on our views of binary formation in the Galaxy. Many of the new results for spectroscopic binaries have been made possible by the development of new instruments capable of mass-producing accurate radial velocities of faint stars (Philip and Latham 1985). On the astrometric side, the development of speckle interferometry techniques promises to close the classical period gap between spectroscopic and astrometric orbits (McAlister et al. 1987).

2. Binaries in the Galactic Disk

2.1. SOLAR-TYPE STARS IN THE FIELD

The frequency and characteristics of solar-type binaries in the disk have been reviewed nicely by Abt (1983). Recently Morbey and Griffin (1987) showed that many of the low-amplitude spectroscopic orbits derived by Abt and Levy (1976) were spurious. However, when 15 binaries were removed from the sample in the period range 0.1 to 10 yr, the incompleteness correction increased in rough compensation, so the overall frequency of binaries remains unchanged, albeit with larger uncertainties in the affected period range (Abt 1987).

2.2. STARS IN OPEN CLUSTERS

The frequency of binaries as a function of period for stars on the main sequence of the Hyades, the best studied open cluster, is indistinguishable from the Abt and Levy (1976) results for field dwarfs (Mathieu et al. 1985), supporting the view that the formation of binaries is the same in open clusters as in the field. In the Hyades, all the orbits with periods longer than 5.7 days are eccentric, while those with shorter periods are circular (Mayor and Mermilliod 1984), due to the effects of tidal circularization over the 0.8 Gyr lifetime of the cluster. Discounting this evolutionary effect, there is no evidence that the Hyades binaries were formed with a different distribution of eccentricities than the field binaries.

In the old open cluster M67, the binaries have a spatial distribution that is more concentrated to the center of the cluster than the single stars (Mathieu and Latham 1986). This is thought to be a result of dynamical relaxation of the more massive systems to the center of the cluster, and is not a reflection of the spatial distribution at the time of formation.

3. Binaries in the Galactic Halo

3.1. HALO DWARFS IN THE FIELD

Recent surveys of proper-motion stars have turned up hundreds of new high-velocity low-metallicity dwarfs which belong to a true Galactic halo population but happen to be passing through the local solar neighborhood (c.f. Fouts and Sandage 1986, Carney and Latham 1987). The radial velocities of these stars have been monitored for several years, and already dozens of new spectroscopic orbits have been solved (Latham et al.

1988). Although this survey for binaries is not yet complete, the preliminary indication is that binaries are just as frequent in the halo as in the disk. All the halo binaries with periods shorter than about 13 days have circular orbits, presumably due to tidal circularization over the lifetime of the Galaxy. For the halo binaries with periods longer than about 13 days, the distribution of orbital eccentricities is indistinguishable from that for the disk binaries. As far as we can tell from these limited results, the formation of binaries was the same in the halo as in the disk. In particular, we have not yet detected any effects of low metal abundances on the formation of binaries.

3.2 HALO GIANTS IN THE FIELD

There are several long-term efforts underway to monitor the radial velocities of halo giants, such as the stars on Bond's (1970, 1980) lists of metal-poor red giants, and the first few orbital solutions should be published soon. One should expect two main evolutionary effects: all binaries with periods less than roughly a year should have circular orbits because the expanded convection zone in the giants make the tidal mechanisms much more efficient; and the shortest period binaries should be removed from the sample as the result of stellar evolution and mass transfer. Otherwise one expects the binary population among the halo giants in the field to be consistent with the binaries among the halo dwarfs, from which the giants must have evolved.

3.3 GIANTS IN GLOBULAR CLUSTERS

Gunn and Griffin (1979) suggested that binaries are underabundant among globular cluster giants, and this impression has been supported by subsequent work. Although searches for spectroscopic binaries in globular clusters are finally uncovering a few candidates (Pryor et al. 1988; Pryor et al. 1989), so far only one cluster star, a giant in Omega Centauri, has an orbital solution (Mayor 1988). Did main-sequence binaries form much less frequently in globular clusters than in the field, perhaps because of the high-density environment, or have various evolutionary effects and stellar encounters removed the short- and long-period binaries from the observed sample of giants? The resolution of this question may require binary surveys of dwarfs in globular clusters, a rather daunting observational undertaking.

4. Low-Mass Companions

When binaries and multiple systems are formed, how is the available mass divided up among the components? Does the distribution of secondary masses continue right down into the range of planetary companions, or does the formation of a planetary system require a distinctly different process? Tantalizing progress has been made on three observational fronts - photometric, astrometric, and spectroscopic - but so far only the surface has been scratched in this field.

4.1. PHOTOMETRIC SEARCHES

Searches for the infrared signature of cool low-mass companions are beginning to produce candidates for very low mass stars and even brown dwarfs below the hydrogen-burning limit of 0.08 solar masses (Forrest *et al*. 1988, Zuckerman 1988). These observations are difficult, and some of the announced candidates for brown dwarf companions have not been confirmed, for example VB8B (McCarthy, Probst, and Low 1985; Perrier and Mariotti 1987).

An important result from IRAS has been the detection of far-infrared excesses for dozens of stars (Backman 1988). In a few cases the infrared signal has been spatially resolved, indicating the sources are extended dust clouds, and in at least one case the cloud is flattened as would be expected for a disk geometry seen edge on. In a couple of cases it can be seen that there is an inner gap in the cloud inside a few tens of AU. Is this evidence for the formation of planets in the inner regions of giant dust clouds around these stars?

4.2. ASTROMETRIC ORBITS

Despite many years of hard work, most of the claims for astrometric orbits showing brown dwarf or even planetary companions are marginal at best. However, there are a few cases where the evidence seems reasonably compelling, such as the brown dwarf companion of LHS1047 (Ianna *et al*. 1988).

4.3. SPECTROSCOPIC ORBITS

Spectroscopic searches have the advantage that they are sensitive to close companions with short orbital periods, independent of the distance to the system. The disadvantage is that the viewing angle of a spectroscopic orbit cannot be determined from the radial-velocity data alone, thus introducing a fundamental ambiguity in the interpretation of any single orbital solution. A very low velocity amplitude may simply be the result of the orbit being viewed nearly face on.

Several instruments capable of measuring stellar radial velocities with a precision of about 10 m/s are now in operation. The results from a modest survey of 18 stars at the CFHT suggest that perhaps half of the stars have excess velocity variation that may be due to the presence of low-mass companions of roughly a few Jupiter masses (Campbell 1988). However, in only one case has even a tentative orbit been solved.

Two stars which have been observed intensively for many years because of their use as radial velocity standards have now been shown to have low-amplitude orbits. One of the Griffin standards, HR152, has a companion which would be 30 or 40 Jupiter masses if the orbit (McClure *et al*. 1985) is being viewed edge on, while an IAU standard, HD114762, has a companion that might be as small as 10 Jupiter masses (Latham *et al*. 1989). The orbital period for HD114762 is 84 days, and the primary star is similar to the sun, which would place the companion at the same distance as the orbit of Mercury.

Recently Marcy and Moore (1988) have reported a spectroscopic orbit for the M3 dwarf Gliese 623. They use various approaches to estimate that the companion has about 70 Jupiter masses.

There is some hope that large spectroscopic surveys for monitoring radial velocities will eventually provide some information on the frequency of low-mass companions. However, this approach can only reach the upper range of planetary-sized objects, and can not hope to detect earth-sized planets in reasonable orbits.

REFERENCES

Abt, H. A. 1983. *Ann. Rev. Astron. Astrophys.* 21, 343.
Abt, H. A. 1987. *Astrophys. J.* **317**, 353.
Abt, H. A. and Levy, S. G. 1976. *Astrophys. J. Suppl.* **30**, 273.
Backman, D. 1988. Paper presented at the *Toronto International SETI Conference.*
Bond, H. E. 1970. *Astrophys. J. Suppl.* **22**, 117.
Bond, H. E. 1980. *Astrophys. J. Suppl.* **44**, 517.
Campbell, B. 1988. Paper presented at this Joint Discussion.
Carney, B. W. and Latham, D. W. 1987. *Astron. J.* **93**, 116.
Forrest, W.J., Skrutskie, M.F., and Shure, M. 1988. *Astrophys. J.* 33, L119
Fouts, G. and Sandage, A. 1986. *Astron. J.* **91**, 1189.
Gunn, J. E. and Griffin, R. F. 1979. *Astron. J.* **84**, 752.
Ianna, P. A., Rohde, J. R., and McCarthy, D. W., Jr. *Astron. J.* **95**, 1226.
Latham, D. W., Mazeh, T., Stefanik, R. P., Mayor, M., and Burki, G. 1989. *Nature*, submitted.
Latham, D. W., Mazeh, T., Carney, B. W., McCrosky, R. E., Stefanik, R. P., and Davis, R. J. 1988. *Astron. J.* **96**, 567.
Marcy, G. W. and Moore, D. 1988. Preprint.
Mathieu, R. D. 1988. Paper presented at this Joint Discussion.
Mathieu, R. D. and Latham, D. W. 1986. *Astron. J.* **92**, 1364.
Mathieu, R. D., Stefanik, R. P., and Latham, D. W. In *IAU Coll. No. 88, Stellar Radial Velocities*, A. G. D. Philip and D. W. Latham, eds., L. Davis Press, Schenectady, p. 385.
Mayor, M. and Mermilliod, J.-C. 1984. In *IAU Symp. No. 105, Observational Tests of Stellar Evolution Theory*, A. Maeder and A. Renzini, eds., Reidel Publishing, Dordrecht, p. 411.
Mayor, M. 1988. Private communication.
McAlister, H. A., Hartkopf, W. I., Hutter, D. J., Shara, M. M., and Franz, O. G. 1987. *Astron. J.* **92**, 183.
McCarthy, D. W., Probst, R. G, and Low, F. J. 1985. *Astrophys. J. Lett.* **290**, L9.
McClure, R. D., Griffin, R. F., Fletcher, J. M., Harris, H. C., and Mayor, M. 1985. *Publ. Astron. Soc. Pacific* **97**, 740.
Morbey, C. L. and Griffin, R. F. 1987. *Astrophys. J.* **317**, 343.
Perrier, C. and Mariotti, J.-M. 1987. *Astrophys. J. Lett.* **312**, L27.
Philip, A. G. D. and Latham, D. W. 1985. *IAU Coll. No. 88, Stellar Radial Velocities*, L. Davis Press, Schenectady. See especially papers 2-4.
Pryor, C., Latham, D. W., and Hazen, M. L. 1988. *Astron. J.* **96**, 123.
Pryor, C., McClure, R. D., Hesser, J. E., and Fletcher, J. M. 1989 In *Dynamics of Dense Stellar Systems*, D. Merritt, ed., Cambridge Univ. Press, Cambridge, in press.
Zuckerman, B. 1988. Paper presented at this Joint Discussion.

A SEARCH FOR PLANETARY-MASS COMPANIONS TO NEARBY STARS

Bruce Campbell
Department of Physics and Astronomy
University of Victoria
P.O. Box 1700
Victoria, B.C.
Canada   V8W 2Y2

ABSTRACT. Nine of 18 stars observed with a high precision radial velocity technique show long term, low level variations which imply the presence of companions in the range of 1 to 10 Jupiter masses. These companions could represent the tip of the planetary-mass spectrum.

1. INTRODUCTION

High precision radial velocities have been measured for a sample of solar-type stars at the Canada-France-Hawaii telescope since 1981. Precise velocities are obtained by using an absorption cell containing HF gas to impose reference absorption lines on the stellar spectrum. This eliminates the systematic errors which limit the precision of conventional velocity measurements. Random errors are minimized by obtaining very high signal-to-noise ratio spectra, typically 1000:1. Details of this technique may be found in Campbell et al. (1986) and Campbell, Walker, and Yang (1988; hereafter Paper I).

2. ANALYSIS OF THE RADIAL VELOCITIES

To assess whether there are significant long-term trends present in the velocities, we fit the data for each star with alternately a straight line, and a parabola. This yields the first and second derivatives of the velocities, as well as the formal 1 sigma uncertainties in each of these quantities. We conclude that a significant velocity variation is present if either the slope or curvature differ from zero at more than the 3 sigma level.

Of 17 stars analyzed by this procedure, 8 are significant velocity variables. The slope or curvature significance values are shown for these stars in Table 1. We attribute these low level velocity variations to the presence of low mass companions.

The eighteenth star in our sample, which cannot be analyzed in the same way, is γ Cep. This star shows an obvious change in velocity over the past seven years of about 2000 m s^{-1}, probably due to a stellar

D. McNally (ed.), Highlights of Astronomy, Vol. 8, 109–110.

companion. However, on top of this gross trend there is a small, oscillatory variation. A simultaneous fit of a parabola and a sine-wave to the data yields a best-fitting sine-wave of period 3.1 years, and amplitude 22.5 m s^{-1}. We ascribe this variation to a third body in the system.

TABLE 1. Velocity variables

Star	Significance (sigmas)	M_C sin i (Jupiters)	
		Lower Lim.	Upper Lim.
ε Eri	4.5	1.0	4.5
36 UMa	4.8	1.6	12.7
β Vir	3.3	0.9	9.6
β Com	3.6	1.0	9.0
61 Vir	3.8	0.8	7.2
β Aql A	4.1	1.0	12.7
η Cep	3.0	1.1	19.4
61 Cyg A	3.0	0.7	4.0
γ Cep	M_C sin i = 1.6 Jupiters		

3. LIMITS TO COMPANION MASSES

Even though we do not know orbital periods in most cases, it is possible to place upper limits on the masses of companions by combining our radial velocity results with astrometric information. Most of our program stars have been observed astrometrically, and none shows a convincing perturbation. Combining this with limits to the velocity variations (see Paper I for details) yields the upper limits to the companion masses shown in Table 1. These show a large range because they depend on the parallax, but average around 10 Jupiter masses.

Lower limits to the companion masses can be derived from the observed range in velocities. The lower limits are all about 1 Jupiter mass. For γ Cep, the third body has M_C sin i of 1.6 Jupiter masses.

In summary, nine of 18 stars observed show significant low level velocity variations which imply companions of 1 to 10 Jupiter masses. Companions in the traditional brown dwarf range of 10 to 80 Jupiter masses are conspicuously absent, which implies that those we have detected could represent the tip of the planetary mass spectrum. If so, then ≳50% of the stars in the Galaxy could have planetary companions.

REFERENCES

Campbell, B., Walker, G. A. H., Pritchet, C., and Long, B. 1986, in Astrophysics of Brown Dwarfs, ed. M. C. Kafatos, R. S. Harrington, and S. P. Maran (Cambridge: Cambridge University Press), p. 37.
Campbell, B., Walker, G. A. H., and Yang, S. 1988, Ap. J., 331, 902.

SPECTROSCOPIC BINARIES AMONG LOW-MASS PRE-MAIN SEQUENCE STARS

Robert D. Mathieu
Department of Astronomy
University of Wisconsin
Madison, Wisconsin 53706 USA

1. Introduction

Although the unusual nature of T Tau was noted over four decades ago, the first orbit for a low-mass pre-main sequence (PMS) spectroscopic binary was not determined until Mundt *et al.* (1983) serendipitously discovered the double-lined nature of V826 Tau. To some degree the paucity of spectroscopic binary detections may be attributed to the faintness of such stars and the consequent difficulty in obtaining high resolution spectra; the first high-precision radial-velocity survey was that of Herbig (1977). Advances in radial-velocity measurement technology now permit relatively easy velocity measurements of PMS stars. However, V826 Tau was also one of the first discovered members of the naked T Tauri (NTTS) class of PMS stars (Walter 1987) and its discovery as a spectroscopic binary foreshadowed a prevalence for binary detection among this population. In this short paper we review the present observational status of PMS spectroscopic binaries and present several initial results and thoughts for consideration.

2. The Roster of PMS Spectroscopic Binaries (August 1988)

At present, the roster of low-mass PMS spectroscopic binaries with orbit determinations stands at nine.These are listed in order of increasing period in Table 1, where the orbital periods, eccentricities, mass ratios for the double-lined cases, status of the binaries as classical T Tauri stars (CTTS) or NTTS and the relevant references are presented. We have also included in the list the star 045251+3016 as at present the data show a systematic increase in velocity of 15 km/sec over 800 days; this is clearly a long-period system for which an orbit determination will require further monitoring. Finally we include the double-lined eclipsing binary EK Cep. Arguably, EK Cep might not be considered a PMS binary as the primary is a 2.0 M_O star on the ZAMS. However, Popper (1987) argues that the 1.1 M_O secondary is still contracting to the main sequence and hence we include it here. We note that numerous other authors have suggested velocity variability for additional PMS stars (e.g., Herbig 1977, Bouvier *et al.* 1986, Hartmann *et al.* 1986, Edwards *et al.* 1988). However, as later observations have not always corroborated earlier indications of variability (e.g., Hartmann *et al.* 1986) we have taken the conservative position of assigning binary status only to those stars with clear evidence for orbital motion.

Several trends are evident in Table 1. The distribution of spectroscopic binaries among PMS populations and the orbital eccentricity distribution are discussed further below. The high frequency of both double-lined binaries and short-period systems are

D. McNally (ed.), Highlights of Astronomy, Vol. 8, 111–115.
© *1989 by the IAU.*

TABLE 1. Roster of PMS Spectroscopic Binaries

	$\underline{P}$ (d)	$\underline{e}$	$\underline{q}$	Population	Reference
155913-2233	2.4237	0.00		NTTS	MWM
V4046 Sgr	2.43	0.00	1.	CTTS	BDLR
V826 Tau	3.9063	0.00	1.	NTTS	MUN
EK Cep	4.42782	0.109	1.8	?	TPO
160905-1859	10.401	0.18		NTTS	MWM
AK Sco	13.6093	0.469	1.	CTTS	AND
P1540	33.73	0.12	1.3	NTTS	MM
162814-2427	35.95	0.49	1.1	NTTS (?)	MWM
162819-2423	89.0	0.5		NTTS (?)	MWM
160814-1857	144.5	0.24		NTTS	MWM
045251+3016	>800			NTTS	MWM

AND = Andersen *et al.* 1989; BDLR = Byrne 1986, de la Reza *et al.* 1986; MM = Marschall and Mathieu 1988; MUN = Mundt *et al.* 1983; MWM = Mathieu, Walter and Myers 1989; TPO = Tomkin 1983, Popper 1987

certainly the result of selection biases. Only the NTTS binaries, excluding P1540, are the product of a comprehensive radial-velocity survey (Mathieu, Walter and Myers 1989; MWM). Of these seven binaries, only two are double-lined, a frequency not significantly different from that found among solar-mass field stars. Similarly, most of the stars in that study have only been observed over 2-3 yr, biasing the sample to shorter periods.

3. Issues and Results

3.1 Spectroscopic Binary Frequency Among PMS Stars

If the number of binaries are conserved after reaching the PMS stage of evolution, than the frequency of binaries among all PMS stars of a given mass at any given age should be comparable to that found among the field population. This need not be true within subgroups of the PMS population. For example, the presence of a close binary companion will likely alter the distribution and evolution of circumstellar material; indeed it has been conjectured (Walter *et al.* 1988) that the action of such companions may be one process by which CTTS evolve into NTTS. Comparison of CTTS, NTTS and field short-period binary frequencies may provide insight into such processes.

The evidence to date indicates that the spectroscopic binary frequency among the NTTS is indistinguishable from that in the field. In total, eight spectroscopic binaries have been found among the NTTS. Considering only the shortest period (P < 100 days) binaries in the Taurus-Auriga and Ophiuchus star-forming regions, MWM find a frequency of $10 \pm 4\%$ (5 out of 49), to be compared to 13% among the field solar-mass stars (Abt and Levy (1976), modified as Morbey and Griffin (1987) and Abt (1987)).

Interestingly, among the CTTS there are only two confirmed spectroscopic binaries, V4046 Sgr and AK Sco. (The classification of AK Sco as CTTS is based only on its infrared excess and light variability; CTTS emission lines are not present (Andersen *et al.* 1989, Brown and Walter, priv. comm.)). Several high-precision radial-velocity surveys of CTTS, with total numbers of stars comparable to those for the NTTS, have been completed and other spectroscopic binary candidates have been suggested (references in Sec. II). Unfortunately sufficient numbers of velocity measurements to confirm orbital motion have not yet been obtained for any of these candidates and conclusions regarding binary frequencies must be drawn from less secure analyses of velocity measurement distributions. A high-precision radial-velocity survey of CTTS in the Taurus-Auriga, NGC 2264 and λ Orionis star-forming regions is ongoing at the Harvard-Smithsonian Center for Astrophysics (in collaboration with Hartmann, Latham and Stahler). Approximately two thirds of the nearly 100 program stars have spectra at 5200A permitting high quality velocity measurements; the remainder provide lower quality or no measurements due to emission, veiling or rotation. All stars have been observed in at least two observing seasons and typically in three or more. Among those stars providing high-quality velocity measurements, the largest peak-to-peak velocity variations have been at only the 10 km/sec level. The lack of higher amplitude velocity variability, as well as double-lined systems, is in notable contrast to an essentially identical study of NTTS (MWM) and may indicate a lower frequency of the shortest period binaries among the CTTS. However, the observations per star, as with previous studies, are still too few to permit a definitive conclusion. Nonetheless, these preliminary results and the paucity of known short-period binaries among the CTTS are intriguing; the implications regarding the evolution of circumstellar material and accretion disks in binary systems are sufficiently important that a substantial continued observational effort is warranted.

While the CTTS spectroscopic binary frequency remains uncertain, two important conclusions can still be drawn. First, V4046 Sgr and AK Sco are clear evidence that the presence of close stellar companions does not necessarily preclude T Tauri characteristics. Second, given a short-period binary frequency among the NTTS of only 10%, the action of stellar companions at separations of less than 1 A.U. on circumstellar material is not the dominant mechanism converting CTTS into NTTS. Whether companions at wider separations might also be effective needs to be investigated, but at first glance it seems unlikely given the substantial frequency of wide binaries among the CTTS (Simon (this volume), Cohen and Kuhi 1979).

3.2 Masses of PMS Stars

To date the only dynamical mass determination of a low-mass PMS star is that of $1.12 M_O$ for the secondary of EK Cep. A preliminary analysis by Popper (1987) finds the position of the secondary in the theoretical H-R diagram to be between the $1.0 M_O$ and $1.25 M_O$ evolutionary tracks of Iben at an age of 2×10^7 yr. Increasing the number of fundamental mass determinations, particularly for binary components with ages of 10^6 yr or less, must be a primary goal of future spectroscopic binary work.

3.3 The Epoch and Coevality of Formation of Stars in Binary Systems

Young stars are typically dated through reference to PMS stellar evolutionary models for single stars. As yet, the relevance of these models for the components of short-period binaries has not been investigated. With this caveat, the ages of the NTTS binaries range from somewhat less than 10^6 yr to 10^7 yr (references in Table 1). Thus by ages of 10^6 yr close binary systems exist and indeed are indistinguishable from those in the field except for the PMS nature of the components and their eccentricity distribution (discussed below). Although the numbers are small, the age distribution of the NTTS binaries does not differ from that of the entire NTTS population (MWM; Walter et al. 1988). V4046 Sgr and AK Sco both have ages of $5-10 \times 10^6$ yr, older than typical CTTS.

While the coevality of formation of binary components is often taken as a given, it is subject to observational test. For short-period binaries, double-lined PMS binaries represent a valuable tool in this regard. At the simplest level, two coevally formed stars of the same mass should have the same photospheric luminosity and spectra. Interestingly, the three binaries listed in Table 1 with q~1, V4046 Sgr, V826 Tau and AK Sco, have apparent brightness ratios (in limited spectral ranges) of 2, 1-1.15 and 2.0-0.5, respectively (references in Table 1). Only V826 Tau can be straightforwardly interpreted as having coevally formed components. However, the remarkable variation in the brightness ratio of AK Sco is interpreted by Andersen *et al.* (1989) as evidence for substantial extinction gradients on size scales comparable to the binary orbit and the system acts as a warning, once again, against interpreting apparent brightness differences as intrinsic luminosity differences, at least among CTTS binaries.

A somewhat more sophisticated analysis has been applied to the NTTS double-lined binary Parenago 1540 by Marschall and Mathieu (1988), who deconvolved the composite light and obtained age estimates for each component. Interestingly, using the PMS evolutionary tracks compiled in Cohen and Kuhi (1979) they find the primary and secondary ages to differ by a factor two.The most straightforward interpretation of this result is that the two stars in fact formed noncoevally in a bound system. An alternative interpretation derives from the fact that the center-of-mass space velocity of P1540 deviates from the Trapezium cluster velocity by four times the cluster velocity dispersion; apparently the binary is presently escaping from the cluster. This suggests that P1540 may have been recently involved in a close stellar encounter, during which exchange of an original binary component with an encountered star may have occurred. Given that star-formation regions have formation timespans of up to 10^7 yr, this would easily explain the noncoevality. A more conservative, but perhaps wiser, conclusion is that such analyses are beginning to seriously test the accuracy of PMS isochrones. Similar analyses of a sample of double-lined PMS binaries will provide a benchmark for testing the PMS evolutionary tracks and the coevality of formation.

3.4 Evolution of Binary Orbits

This important issue can best be introduced by noting that two solar-mass stars in an orbit of 2.5 day period are separated by 10 R_o. This separation is comparable to the radii of solar-mass stars at the stellar birthline. It is difficult to avoid the conclusion that either the single-star PMS evolutionary tracks are not valid for close binary components or the orbital separations of at least short period systems decrease with time. Furthermore this argument shows that the issue of orbital evolution and interaction between binary components is not restricted to the earliest stages of the star-formation process; it remains relevant into the PMS phase. The relative timescales of the decrease in stellar radii and orbital semi-major axis will dictate the nature of any interaction between two binary components during the PMS phase. Observation of the youngest and shortest-period systems should provide the best insight into the nature of early binary evolution. At present we have yet to find a binary substantially younger than 10^6 yr, by which age the stellar radii are such that solar-mass stars lie well within their Roche radii even for orbital periods of only a few days. Hence it is not surprising that at present the binary 155913-2233 with a period of 2.4 days shows no evidence of interaction between the stars. On the other hand, V4046 Sgr - also having a period of 2.4 days - has both strong emission activity and large infrared excesses. V4046 Sgr is one of the rare cases of a CTTS isolated from any dark cloud or molecular material; Herbig (1978) has suggested interaction between the two stars as one explanation for the continued emission activity. Better observational insight into the evolution of binary separation during the PMS phase will improve with the discovery of binaries of younger age than those presently known.

That orbital evolution does occur however is indicated by the orbital eccentricity distribution shown in Table 1. The binaries with periods of less than 4 days have circular orbits while those with longer periods have eccentric orbits. Indeed this eccentricity distribution represents one of the most solid observational facts that any theory of binary formation must explain: among all young stellar systems (Hyades, Pleiades, α Per and the PMS population), with only rare exception, every low-mass binary with a period greater than 4-6 days has an eccentric orbit while every binary with shorter period has a circular orbit. This is a remarkably robust result in the face of a presumed variety of formation conditions for each binary. Perhaps the most straightforward explanation would be to attribute the circular orbits to circularization processes occurring relatively late in the star-formation process when the differences in the initial conditions have been erased. In this picture, the cutoff period of 4 days between the circular and eccentric orbits becomes a critical benchmark for the evolution of internal stellar structure, stellar radius and orbital separation prior to 10^6 yr. However complex that evolution may be, the process must ultimately be capable of producing a 4 day period binary with a circular orbit. It is worth noting that in 10^7 yr two solar-mass stars with ZAMS internal structure can only circularize orbits with periods of less than two days (Mathieu and Mazeh 1988). That tidal circularization is more effective during the PMS stage is likely due to larger stellar radii and deeper convective zones.

Drs. J. Andersen, A. Brown, M. Slovak, F. Walter and H. Zinnecker are gratefully acknowledged for both insightful discussions and contributions of information in advance of publication.

Abt, H.A. 1987, Ap.J., **317**, 353.
Abt, H.A. and Levy, S.G. 1987, Ap.J. Supp., **30**, 273.
Andersen, J., Lindgren, H., Hazen, M.L. and Mayor, M. 1989, in preparation.
Bouvier, J., Bertout, C., Benz, W. and Mayor, M. 1986, Astron. Astrophys., **165**, 220.
Byrne, P.B. 1986, Irish. A.J., **17**, 290.
Cohen, M. and Kuhi, L.V. 1979, Ap.J.Supp., **41**, 743.
de la Reza, R., Quast, G., Torres, C.A.D., Mayor, M., Meylan, G. and Llorente de
    Andres, F. 1986, in New Insights in Astrophysics, ESA SP263, 107.
Edwards, S., Cabrit, S., Strom, S., Heyer, I., Strom, K. and Anderson, E. 1987, Ap.J.,
    321, 473.
Hartmann, L.W., Hewett, R., Stahler, S. and Mathieu, R.D. 1986, Ap.J., **309**, 275.
Herbig, G.H. 1977, Ap.J., **214**, 747.
Herbig, G.H. 1978, in Problems of Physics and Evolution of the Universe (Academy of
    Sciences of the Armenian SSR, Yervan), p. 171.
Marschall, L.A. and Mathieu, R.D. 1988, A.J., in press.
Mathieu, R.D., Walter, F.M. and Myers, P.C. 1989, in preparation.
Mathieu, R.D. and Mazeh, T. 1988, Ap.J., **326**, 256.
Morbey, C.L. and Griffin, R.F. 1987, Ap.J., **317**, 343.
Mundt, R., Walter, F.M., Feigelson, E.D., Finkenzeller, U., Herbig, G. and Odell, A.P.
    1983, A.J., **269**, 229.
Popper, D.M. 1987, Ap.J., **313**, L81.
Tomkin, J. 1983, Ap.J., **271**, 717.
Walter, F.M. 1987, P.A.S.P., **99**, 31.
Walter, F.M., Brown, A., Mathieu, R.D., Myers, P.C. and Vrba, F.J. 1988, A.J.,
    96, 297.

BINARY FREQUENCY AMONG PRE-MAIN SEQUENCE STARS IN TAURUS AND OPHIUCHUS

M. Simon
Astronomy Program
State University of New York at Stony Brook
Stony Brook, N.Y. 11794, U.S.A.

ABSTRACT. The lunar occultation technique applied in the IR offers a powerful means of identifying binaries among obscured young stars. Our program has revealed binaries with separations from 1 to 100 AU in the Taurus and Ophiuchus star forming regions to about K=9 mag. To date, 29 objects have been observed; 6 were discovered to be binaries. The observed binary frequency is about half that expected from the binary statistics of a comparable sample of field stars. The discrepancy is probably attributable to our insensitivity to binary systems with secondary mass much less than that of the primary.

Binaries among young stars are being identified by optical spectroscopy (Mundt *et al.* 1983, Mathieu 1988), by IR speckle interferometry (*e.g.* Dyck *et al.* 1981, Chelli *et al.* 1988) and by lunar occultations in the infrared (Simon *et al.* 1987). In the infrared K band the IR occultation technique offers a powerful means of identifying binaries among obscured young stars to K~ 9 mag and angular separations from 0.005 to 0.5 arc sec. The lower limit is set by the signal to noise of the data and the instrumental effects that smooth the Fresnel pattern (Simon *et al.* 1987). The upper limit is set by the photometer aperture, typically 4 to 6 arc sec. One can observe only at the dark lunar limb.

The moon passes over both Taurus and Ophiuchus, two of the nearest star forming regions. Because the Taurus star forming region is so extensive, occultations of the young stars within it occur each year. The opportunities to observe occultations of the young stars in Ophiuchus are more limited. The most recent occultations occurred in 1985-6 and the next series will be in 1991-3.

We have to date observed 29 objects, 19 in Taurus and 10 in Ophiuchus. The systems discovered to be binaries are DF, FF, HQ, and Haro 6-8 in Taurus and SR-12 and ROX 31 in Ophiuchus. That DF Tau and Haro 6-8 are binaries with separations, in the projection of the occultations, of ~ .011 and ~ .560 arc sec respectively are recent findings; the other systems are described by Simon *et al.* 1987.

A central question is how the binary statistics of the young stars and field stars compare. For now, we consider the binaries in Taurus and Ophiuchus together; when the sample is larger,

D. McNally (ed.), Highlights of Astronomy, Vol. 8, 117–118.
© *1989 by the IAU.*

it will be possible to analyze the regions separately. The stars in our sample are of late spectral type. The F3 - G2 dwarfs studied by Abt (1983) are the most comparable field star sample. Abt's results are in terms of binary period. Assuming 1 $M_\odot$ total mass for each binary system and 150 pc as representative distance to both star forming regions, our angular resolution range 0.005 to 0.5 arc sec corresponds to systems with periods in the range 10^2 to 10^6 days. Spectroscopic binaries tend to have periods shorter than 100 days. Speckle observations at K have a resolution limit of $\sim .120$ arc sec, or in the spirit of these estimates, systems with periods exceeding 10^5 days. The occultation technique nicely fills this gap. Using Abt's result for the 10^2 to 10^6 day range of periods, and correcting for both orbital projections and the distribution of occultation directions, we estimate that about 40% of the systems observed, or 12 out of 29, should be binaries.

Our sample is still small but it seems that we are detecting a smaller number of binaries than expected. The reason may lie in the mass distribution of the secondaries. If it proves similar to that of the "bifurcation" sample identified by Abt (1983), then, as we approach the faint end of our flux limited sample, we will start to be insensitive to systems with secondary masses much less than that of the primary. Data for more systems, providing information on the observed flux ratios will help resolve the situation.

A thoroughly studied sample of a significant number of young star binaries will yield values of the fundamental stellar parameters, provide data essential to the determination of the initial mass function, and contribute to our understanding of the formation and evolution of multiple star systems. We look forward to progress in these areas in the next few years as the moon sweeps southward through Taurus and again across Ophiuchus.

I am grateful to J. Benson, W. Chen, R. Howell, A. Longmore, M. Smith, B. Wilking, and G. Wright for observing occultations for this program and thank W. Chen and D. Peterson for discussions and help. This research was supported in part by NSF Grant AST 85-14337.

References

Abt, H.A. 1983, Ann. Rev. Astr. Ap., 21, 343.
Chelli, A. *et al.* 1988, Astr.Ap., *in press.*
Dyck, H.M. *et al.* 1982, Ap.J.Lett., 255, L103.
Mathieu, R.D. 1988, *these proceedings.*
Mundt, R. *et al.* 1983, Ap.J., 269, 229.
Simon, M. *et al.* 1987, Ap.J., 320, 344.

BROWN DWARFS IN BINARY SYSTEMS

B. Zuckerman
Astronomy Department
University of California, Los Angeles
Los Angeles, CA   90024-1562
U.S.A.

ABSTRACT.   Searches for brown dwarf companions to known stars have revealed a number of tantalizing candidates.

1.   INTRODUCTION

Until 1983, stars were the only objects that we knew with certainty were in orbit about stars other than the Sun.   Then IRAS discovered clouds of particulate matter in orbit around Vega and many other nearby main-sequence stars.   Even more recently, since 1987, various reports of potential brown dwarf and large planetary mass companions to nearby stars have appeared.   (None of the pre-1987 claims have ever been confirmed.)

The purpose of this brief review is to summarize some of the recent searches for brown dwarf companions to nearby stars.   The upper mass limit to brown dwarfs is about 0.08 $M_\odot$, which is also the lower mass limit to main-sequence stars.   Below this mass, objects do not fuse protons into helium nuclei in their interiors.   The lower mass limit to brown dwarfs, which is also the upper mass limit to large planets, is less well defined.   One reasonably popular choice is 0.01 $M_\odot$ (or 10 $M_{Jupiter}$), because above this mass objects, i.e., brown dwarfs, can fuse deuterium nuclei into helium nuclei in their interiors.   Models of brown dwarfs indicate that, following a short period of deuterium burning, they will have radii within a factor of 2 or so of that of Jupiter (e.g., Nelson et al. 1986).

2.   INFRARED SEARCH TECHNIQUES

Most of the serious IR searches for brown dwarfs have involved either imaging with the new two-dimensional array cameras, or photometry of white dwarf stars.   A list of the most ambitious of the array searches known to us is presented in Part IA of Table 1.   The Rochester search has been described by Skrutskie et al. (1986) and the Arizona search by Rieke and Lebofsky was described briefly by Liebert and Probst (1987, page 490).   The Rochester search revealed one low-luminosity object,

D. McNally (ed.), Highlights of Astronomy, Vol. 8, 119–122.
© 1989 by the IAU.

Gliese 569B, which Forrest et al. (1988) interpret as a possible brown
dwarf. Since brown dwarfs are expected to be cool and of low
luminosity during most of their lifetimes (e.g., Nelson et al. 1986),
and since Gℓ569B is neither especially cool nor of low luminosity, the
brown dwarf interpretation rests on some very shaky theory of the
evolution of low-mass stars and brown dwarfs. Gℓ569B is compared with
some other very low luminosity stars in Table 2 and Figure 2 of Becklin
and Zuckerman (1988) and considered in some detail by Zuckerman and
Becklin (1989).

It is also possible to search for brown dwarfs near white dwarf
stars with a photometric technique first exploited by Probst (1983a,b).
Subsequent investigations by Kumar (1987), Zuckerman and Becklin
(1987a,b, 1989), and Becklin and Zuckerman (1988), have probed down
into the brown dwarf mass range. Two brown dwarf candidates have been
found near a total of about 120 white dwarfs that have been searched to
date.

The first candidate appeared as excess infrared emission from
Giclas 29-38 (Zuckerman and Becklin 1987b). Although a reasonable case
can be made for a brown dwarf source for this excess radiation
(Greenstein 1988), it is not possible to rule out a swarm of dust
particles or asteroid-like objects in orbit around G29-38 as the source
of the IR excess. Recent observations by Tokunaga and Becklin (1988,
private communication) with CGAS on the IRTF failed to reveal any sharp
spectral features between 1.9 and 3.6 μm. As dust grains might carry a
feature at 3.3 to 3.4 μm (see, e.g., the observations of Comet Halley
reported by Knacke et al. 1986), and a cool brown dwarf might show deep
atmospheric absorption features due to H_2O (Berriman and Reid 1987),
the CGAS data are a bit perplexing. Recent photometric data at 4.8 μm
by Becklin, Tokunaga, and Zuckerman (1988, private communication), when
combined with 10 μm data reported by Tokunaga et al. (1988), indicate
that the IR excess at G29-38 can be well fitted from 2 to 10 μm by a
black body of about 1200°K.

The second candidate brown dwarf, GD165B, was discovered to lie
about 4" from the white dwarf GD165 (Becklin and Zuckerman 1988). If
GD165B is a true companion to GD165, and not merely a very red
background object (which seems improbable), then their separation is
about 120 A.U. GD165B is certainly not a dust cloud, since it appears
to be stellar and is well-resolved from GD165. Although, at about
2100°K, GD165B is cooler than any known dwarf star, it is not so cool
that it can be classified unimpeachably as a brown dwarf. According to
the cooling models of Nelson et al. (1986), GD165B is a brown dwarf
with a mass that lies between 0.06 and 0.08 $M_\odot$. But the models of
D'Antona and Mazzitelli (1985) would classify GD165B as a star at the
very bottom of the main-sequence (mass about 0.08 $M_\odot$) rather than as a
brown dwarf.

3. OPTICAL AND RADIO TECHNIQUES

The search techniques listed in Part II of Table 1 are sensitive to
much lower mass objects than are the infrared methods. Indeed, at
present, techniques IIC and IID are about the only methods capable of

detecting planets with Earth-like masses. Eclipses of white dwarfs by orbiting brown dwarfs are, potentially, yet another detection technique. However, a recent exciting report of just such an eclipsing system (I.A.U. Telegram No. 4648) apparently will be retracted (R. Rubin and B. Zuckerman 1988, private communication).

The most secure detection of an object with a mass in the brown dwarf range (Fruchter et al. 1988) was achieved with technique IID. However, very likely, this brown dwarf (mass $\gtrsim$ 20 $M_{Jupiter}$) was originally a bona fide low-mass star that is now being destroyed by its nearby pulsar primary. Technique IIB has revealed only one potential brown dwarf candidate to date. Reported at this Joint Discussion by Dr. Latham, the mass of the unseen companion to a G-type primary star is $\gtrsim$ 10 $M_{Jupiter}$. The actual mass of the companion depends on the unknown inclination angle of the orbit. A back-of-the-envelope calculation indicates a 1% chance that the orbit is sufficiently unfavorably inclined that the mass of the companion is greater than 0.08 $M_\odot$, i.e., stellar. Since Dr. Latham and collaborators have examined thousands of stars, we should not be surprised if the companion is, indeed, stellar.

4. SUMMARY

In spite of many searches utilizing various techniques, at this moment there is only one object known whose mass unquestionably lies in the domain of brown dwarfs. This object, the companion to an eclipsing millisecond pulsar, is most unusual and, very probably, began its life as an ordinary star. Other plausible candidates have been identified during the past year and the field is ripe for new discoveries.

References

Becklin, E.E. and Zuckerman, B. 1988, Nature, in press, December 1988.
Berriman, G., and Reid, N. 1987, M.N.R.A.S., 227, 315.
D'Antona, F., and Mazzitelli, I. 1985, Ap. J., 296, 502.
Forrest, W.J., Skrutskie, M.F., and Shure, M. 1988, Ap. J., 330, L119.
Fruchter, A.S., Stinebring, D.R., and Taylor, J.H. 1988, Nature, 333, 237.
Greenstein, J.L. 1988, A.J., 95, 1494.
Knacke, R.F., Brooke, T.Y., and Joyce, R.R. 1986, Ap. J., 310, L49.
Kumar, C.K. 1987, A. J., 94, 158.
Liebert, J., and Probst, R.G. 1987, Ann. Rev. Astr. Ap., 25, 473.
Nelson, L.A., Rappaport, S.A., and Joss, P.C. 1986, Ap. J., 311, 226.
Probst, R.G. 1983a, Ap. J., 224, 237.
_____ 1983b, Ap. J. Suppl., 53, 335.
Skrutskie, M.F., Forrest, W.J., and Shure, M. 1986, in Astrophysics of Brown Dwarfs, ed. M.C. Kafatos, R.S. Harrington, S.P. Maran (Cambridge, Cambridge University Press), p. 82.
Tokunaga, A.T. et al. 1988, Ap. J., 332, L71.
Zuckerman, B., and Becklin, E.E. 1987a, Ap. J., 319, L99.
_____ 1987b, Nature, 330, 138.
_____ 1989, in preparation for Ap. J.

Table 1

Searches for Brown Dwarfs as Companions to Known Stars

I. INFRARED TECHNIQUES

A. TWO-DIMENSIONAL IMAGING CAMERAS (Mostly Main-Sequence Stars)

Camera/ Telescope	Approximate Number of Stars Searched	Brightness Limit at 2.2 μm
Rochester/IRTF	60	15 mag
Arizona/90"	50	15.5
NOAO/1.3 and 2.1 m	40	16
IRCAM/UKIRT	10	17

B. PHOTOMETRICALLY (All White Dwarfs)

Telescope	Approximate Number of Stars Searched	Brightness Limit at 2.2 μm
IRTF (and a few UKIRT)	120	16 mag

II. OPTICAL AND RADIO TECHNIQUES

Technique	Number of Stars Searched
A. Classical Astrometry	Many
B. Radial Velocity Variations	Dozens
C. Variations in Optical Periods of ZZ Ceti White Dwarfs	A Handful
D. Variations in Radio Periods of Pulsars	Many

Notes to Table 1:

For the infrared techniques IA and IB, the outer search radius at each
  star is typically 100 to a few hundred or even a few thousand A.U.
For A the inner radius is a few tens to a few hundred A.U.
For B the inner radius is that of the white dwarf primary.
Techniques IIC and IID are sensitive to companions with Earth-like
  and even smaller masses.

CLOUD COLLAPSE AND FRAGMENTATION

A. P. BOSS
DTM, Carnegie Institution of Washington
5241 Broad Branch Road, N.W.
Washington, D.C. 20015
U.S.A.

ABSTRACT. Interstellar clouds are thought to undergo a rapid phase of collapse in the process of contracting to form stars. Break-up during this collapse phase is termed *fragmentation*. Computer codes capable of calculating the hydrodynamics of cloud collapse in three spatial dimensions have been used to study the fragmentation process. Fragmentation into binary or multiple protostellar systems is the preferred outcome of collapse; only very slowly rotating, high thermal energy clouds, or clouds starting from power-law initial density profiles, avoid fragmentation and form single stars.

1. Introduction

This review focuses on the aspects of interstellar cloud collapse of most interest for the formation of binary systems. Break-up of a cloud during its dynamic, self-gravitational collapse toward stellar densities is termed *fragmentation*, in contrast to *fission*, which refers to break-up of a quasi-equilibrium, rapidly-rotating configuration (see reviews by Durisen and Lebovitz, this volume). Fragmentation intrinsically involves the nonlinear, time evolution of a self-gravitating, three dimensional (3D) fluid undergoing rapid, dynamical collapse, and as a consequence is not amenable to detailed analytical study. However, substantial progress in understanding fragmentation has come from the use of numerical codes developed for this problem. A brief summary of protostellar dynamics and thermodynamics is given, followed by the results of the standard test case for 3D codes and a description of the outcomes of isothermal and nonisothermal 3D collapse.

2. Protostellar Dynamics and Thermodynamics

Once interstellar clouds become sufficiently dense, exterior ionizing radiation is unable to penetrate to the cloud center, leading to loss of the magnetic field support that appears to dominate diffuse clouds. Shielding from external radiation also allows the cloud temperature to decrease to $\sim 10 \text{K}$. The combination of decreased magnetic and thermal support is thought to produce a phase of dynamic collapse, where self-gravitational forces overwhelm thermal and magnetic forces. This first collapse phase occurs nearly isothermally, until densities of $\sim 10^{-13}$ g cm^{-3} are reached, whereupon the center is dense enough to trap IR radiation that previously escaped. Thereafter the evolution at the center is more nearly adiabatic, or more properly, *nonisothermal*, to stress the fact that at least initially, continued radiative losses preclude the assumption of strict adiabaticity. Because of rising thermal pressure, an *outer* core forms, and this core begins to collapse itself once temperatures high enough to dissociate molecular hydrogen ($\sim 2000 \text{K}$) are reached. The second

123

D. McNally (ed.), Highlights of Astronomy, Vol. 8, 123–126.
© *1989 by the IAU.*

collapse leads to formation of the *inner* core with densities comparable to stellar densities (~ 1 g cm^{-3}). Tscharnuter (1987) has shown that the inner core can disappear and reappear in a series of explosive rebounds (driven by reassociation and dissociation of molecular hydrogen). However, we are interested here in binary formation, which means we must consider the collapse of rotating clouds, and Tscharnuter's 'hiccups' are stabilized by rotation. Larson (1972) showed that when rotation is included, axisymmetric (2D) protostellar collapse can lead to flattened structures (rings) that are likely to fragment in a fully 3D calculation.

3. Standard Test Case

Because of the absence of analytical solutions for 3D collapse, one of the most important means of testing the accuracy of 3D codes is through mutual comparisons on a standard test case. Boss and Bodenheimer (1979) presented results obtained with two separate finite-difference (FD) codes applied to the same initial conditions. They found good agreement, both qualitatively and quantitatively: with both codes, the initial cloud collapsed (isothermally) and formed an intermediate bar that fragmented into a binary protostar. The properties of the binary thus formed are typical of all succeeding calculations: e.g., each binary member contains about 15% of the total mass and spins with a specific angular momentum reduced by about a factor of 20 from that of the initial cloud. However, when Gingold and Monaghan (1981) used their smoothed-particle hydrodynamics (SPH) code to study the standard test case, a different evolution ensued: the binary decayed back into a bar. Efforts to resolve this disagreement failed (Bodenheimer and Boss 1981; Gingold and Monaghan 1982), until a refined SPH method (Monaghan and Lattanzio 1985) coupled with a greatly increased number of particles led to an evolution that is very similar to that of the FD codes (Monaghan and Lattanzio 1986). The independent SPH code of Miyama, Hayashi, and Narita (1984) also produces results similar to that of the FD codes with initial conditions close to that of the standard test case. This mutual agreement between four independent numerical codes using greatly differing techniques gives strong support for the credibility of these 3D collapse calculations.

4. Isothermal Fragmentation

All published results for 3D isothermal collapse are summarized in Figure 1. Both FD calculations (Narita and Nakazawa 1977; Boss and Bodenheimer 1979; Tohline 1980; Boss 1980; Bodenheimer, Tohline, and Black 1980; Różyczka, Tscharnuter, and Yorke 1980; Boss 1981a,b) and SPH calculations (Larson 1978; Wood 1982; Gingold and Monaghan 1983; Miyama, Hayashi, and Narita 1984) are included. Figure 1 shows whether fragmentation (formation of two or more fragments) resulted from isothermal collapse of an initially uniform density cloud with initial ratios of thermal to gravitational energy (α_i) and rotational to gravitational energy (β_i) designated by location on Figure 1.

Clouds that collapse from high α_i, β_i initial conditions do not fragment, but instead undergo relatively little collapse before settling into a diffuse, rotating, isothermal equilibrium state, termed a *Bonnor-Ebert ellipsoid*, the triaxial analogue of the Bonnor-Ebert isothermal sphere. Clouds with lower α_i, β_i undergo a sustained collapse that nearly always leads to fragmentation into a binary or multiple system (the one calculation at $\alpha_i = 0.1$, $\beta_i = 0.05$ was only taken to 1.0 free fall times, insufficiently far to allow fragmentation to occur). The oblique line shows the fragmentation criterion ($\alpha_i \times \beta_i < 0.12$) advanced by Hayashi, Narita, and Miyama (1982) on the basis of the stability of 2D rotating, isothermal equilibrium models. Figure 1 shows that Hayashi's criterion represents well the results for rapidly rotating clouds, but the criterion appears to fail for slowly rotating clouds (see below and the one model with $\alpha_i = 0.55$ and $\beta_i = 0.02$).

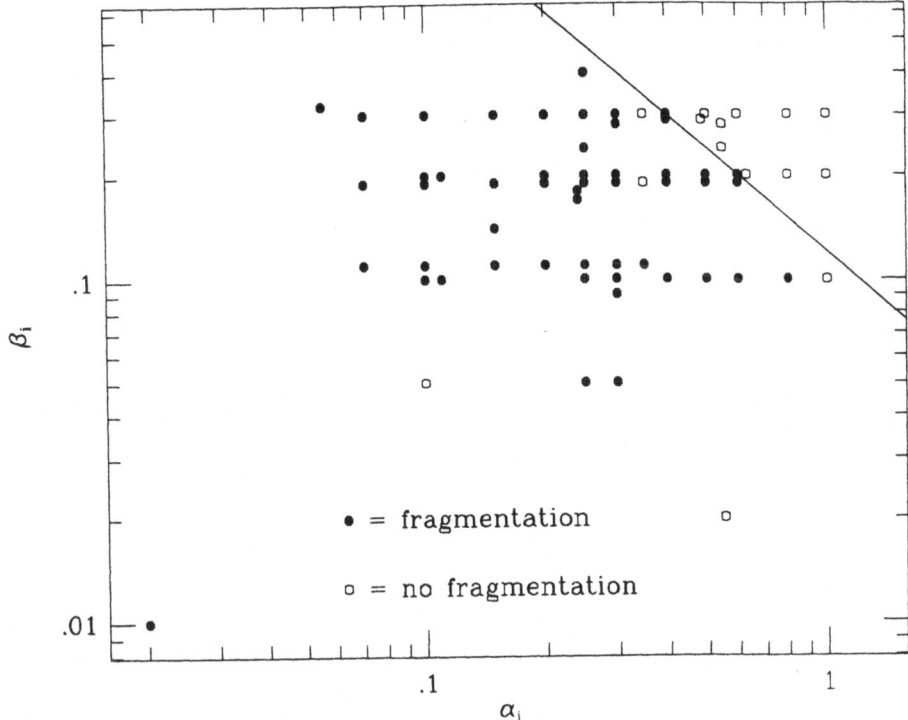

Figure 1. Results of all 3D isothermal collapse calculations.

It should be noted that while the FD and SPH codes largely agree on whether fragmentation should occur, agreement on the number of fragments produced may not always be so good. In general, FD codes produce smaller numbers of fragments than SPH codes for a given set of initial conditions. This is undoubtedly because FD codes model a continuum that resists fragmentation (through numerical viscosity), while SPH codes model a swarm of N particles that can become N fragments if they are separated sufficiently far apart. Reality is hopefully somewhere in between, and this point alone emphasizes the need to continue to compare results obtained with both types of codes.

5. Nonisothermal Fragmentation

In the previous section we asserted that very slowly rotating clouds are unlikely to undergo fragmentation. Isothermal calculations are unable to properly address this point, because slowly rotating clouds collapse to nonisothermal densities prior to undergoing rotationally-induced fragmentation. Calculation of evolution through this phase requires detailed thermodynamics and 3D radiative transfer. Boss (1985) used a 3D Eddington approximation code to show that very slowly rotating clouds collapsing from high α_i initial conditions do *not* undergo fragmentation, though they may become bar-like. Hence these initial conditions appear to lead to single star formation.

There are two other physical situations in which binary formation is prevented. One is the collapse of progressively smaller mass clouds, which is equivalent to initiating collapse at progressively higher densities. Increased initial densities means less collapse can occur before nonisothermality sets in and thermal pressures prohibit fragmentation; the minimum protostellar mass that can be produced through the fragmentation of a collapsing cloud is $\sim 0.01 M_{\odot}$ (Boss 1986). The other is the collapse of clouds that are initially strongly centrally condensed; clouds with power-law initial density profiles cannot fragment into binaries (Boss 1987). In contrast, changes in the dust grain opacity by factors of 3 to 100 have relatively little effect on protostellar fragmentation (Boss 1988a).

6. Conclusions

This review has shown that fragmentation during protostellar collapse into a binary or multiple system is a natural, even preferred outcome, strongly suggesting that fragmentation is a likely means for explaining the formation of binary stars. Fragmentation also appears to be uniquely capable of explaining the ubiquity and wide range in dynamical properities of binary systems (see Bodenheimer 1978 and Boss 1988b for details).

References

Bodenheimer, P. 1978, *Ap. J.*, **224**, 488.
Bodenheimer, P., and Boss, A. P. 1981, *M.N.R.A.S.*, **197**, 477.
Bodenheimer, P., Tohline, J. E., and Black, D. C. 1980, *Ap. J.*, **242**, 209.
Boss, A. P. 1980, *Ap. J.*, **237**, 866.
—— 1981a, *Ap. J.*, **246**, 866.
—— 1981b, *Ap. J.*, **250**, 636.
—— 1985, *Icarus*, **61**, 3.
—— 1986, *Ap. J. Suppl.*, **62**, 519.
—— 1987, *Ap. J.*, **319**, 149.
—— 1988a, *Ap. J.*, **331**, 370.
—— 1988b, *Comments Ap.*, **12**, 169.
Boss, A. P., and Bodenheimer, P. 1979, *Ap. J.*, **234**, 289.
Gingold, R. A., and Mongahan J. J. 1981, *M.N.R.A.S.*, **197**, 461.
—— 1982, *M.N.R.A.S.*, **199**, 115.
—— 1983, *M.N.R.A.S.*, **204**, 715.
Hayashi, C., Narita, S., and Miyama, S. M. 1982, *Prog. Theor. Phys.*, **68**, 1949.
Larson, R. B. 1972, *M.N.R.A.S.*, **156**, 437.
—— 1978, *M.N.R.A.S.*, **184**, 69.
Miyama, S. M., Hayashi, C., and Narita, S. 1984, *Ap. J.*, **279**, 621.
Monaghan, J. J., and Lattanzio, J. C. 1985, *Astr. Ap.*, **149**, 135.
—— 1986, *Astr. Ap.*, **158**, 207.
Narita, S., and Nakazawa, K. 1977, *Progr. Theor. Phys.*, **59**, 1018.
Różyczka, M., Tscharnuter, W. M., and Yorke, H. W. 1980, *Astron. Ap.*, **81**, 347.
Tohline, J. E. 1980, *Ap. J.*, **235**, 866.
Tscharnuter, W. M. 1987, in *Physical Processes in Comets, Stars, and Active Galaxies*, eds. E. Meyer-Hofmeister, H. C. Thomas, and W. Hillebrandt (Berlin: Springer-Verlag), p. 96.
Wood, D. 1982, *M.N.R.A.S.*, **199**, 331.

Criteria for Collapse and Fragmentation of Rotating Clouds

Shoken M. Miyama
Department of Physics
Kyoto University
Kyoto, 606
Japan

ABSTRACT. Three dimensional computations for processes of collapse and fragmentation of rotating gases are performed and criteria for fragmentation of rotating clouds in the cases of the various equation of state are obtained.

We are considering formation processes of stars in binaries and multiple systems. As one of these processes, we are studying fragmentation processes of rotating clouds or adiabatic cloud-cores. Using three-dimensional numerical computations, we obtain the criteria for fragmentation of rotating clouds.

In order to concentrate our attention on physics of fragmentation processes,
we make the following assumptions and simplifications. The equation of state
(EOS) is assumed to be polytropic as $P = K\rho^\gamma$, where K and γ are a constant number and the adiabatic constant, respectively. The initial state of a rotating cloud is assumed to be a spherical gas with uniform density ρ_0 and constant angular velocity Ω_0, but there are small and random density perturbations,
$\Delta\rho/\rho \leq 0.03$.

From above assumptions, the initial state of a cloud is parameterized by two
nondimensional parameters as $\alpha_0 = T/|W_g|$ and $\beta_0 = E_k/|W_g|$ where T, E_k and
W_g are thermal energy, rotational energy and gravitational energy, respectively.

We summarize the computational results in the case of $\gamma = 7/5$ in Figure 1,
where at the position of the initial state in the $\alpha_0 - \beta_0$ plane, we plot the final
stage of the computations. In this figure, we understand easily the dependence of
the final state on the initial α_0 and β_0. In this case, in the low α_0-valued and high
β_0-valued region in the $\alpha_0 - \beta_0$ plane the initial model fragments into pieces. In the
higher α_0-valued region, the cloud does not fragment but a spiral mode appears. In
the more upper region, the cloud only shrinks and the nonaxisymmetric mode
does not appear. The final stages of the computations depend also on the value of
β_0. The computations are performed in the cases of other EOS. We find that in the
gases with $\gamma > 4/3$, as the initial value of β_0 becomes larger, the fragmentation
occurs more easily. This dependence on β_0 is very different from the cases of
isothermal collapse (Miyama, Hayashi and Narita, 1984)

In order to interpret the numerical results, we estimate the flatness of a rotating disk formed in a collapsed gas. Then using the stability analysis of the rotating
infinite sheet, we obtain the criterion of fragmentation. Hence if the flatness of
the disk is greater than some value, say 4π, the disk is assumed to be unstable

127

D. McNally (ed.), Highlights of Astronomy, Vol. 8, 127–128.

for fragmentation. It is to be noticed that the value 4π is determined because in the isothermal sheet the growth rate is largest at that value for any angular velocity as long as the system is unstable and this value depends weakly on the EOS. Then, we obtain the criteria for fragmentation as $\alpha_0\beta_0^{4-3\gamma} = constant$. For typical value of γ, they are $\alpha_0 = 0.064\beta_0(\gamma = 5/3)$, $\alpha_0 = 0.090\beta_0^{1/5}(\gamma = 7/5)$, $\alpha_0 = 0.098(\gamma = 4/3)$ and $\alpha_0 = 0.15\beta_0^{-1}(\gamma = 1)$. These criteria agree the numerical results well. Especially the difference of the critical curves between the cases of $\gamma > 4/3$ and $\gamma < 4/3$ is well interpreted by this criteria.

Then the quantity $\alpha_0\beta_0^{4-3\gamma}$ is a very important value to determine the evolution of a rotating cloud. In the case of isothermal clouds, this value is proportional to square of $c_s J/GM^2$ (c_s, M and J are sound velocity, the total mass and the total angular momentum) and in the case of adiabatic gases, this value is also described only by the constants of motion (in this case, entropy, J and M). Hence, we consider the criterion obtained here are applicable to the general initial data, because that quantity is very fundamental for rotating gases.

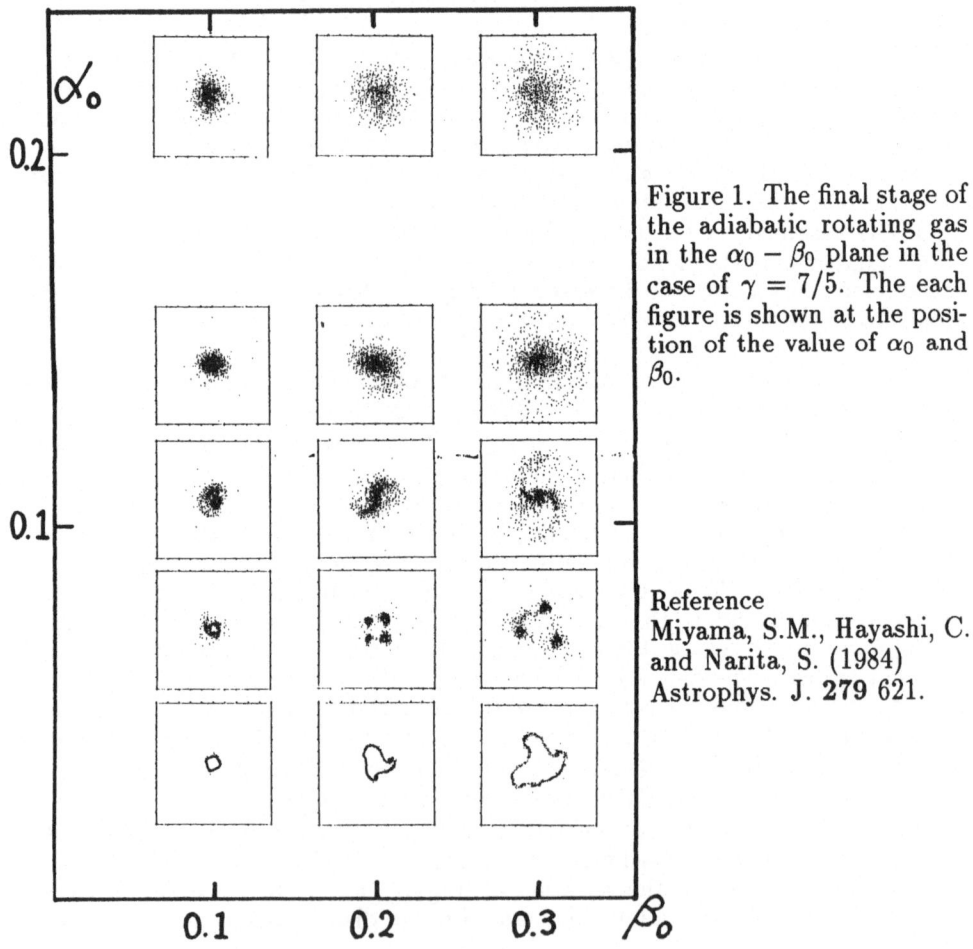

Figure 1. The final stage of the adiabatic rotating gas in the $\alpha_0 - \beta_0$ plane in the case of $\gamma = 7/5$. The each figure is shown at the position of the value of α_0 and β_0.

Reference
Miyama, S.M., Hayashi, C. and Narita, S. (1984) Astrophys. J. **279** 621.

MATHEMATICAL STATUS OF THE FISSION THEORY

N. R. LEBOVITZ
The University of Chicago
5734 So. University Ave.
Chicago, IL 60637, USA

ABSTRACT. A review is given of the current mathematical and theoretical status of
the possibility of the formation of binary stars by fission. It is emphasized that none of the
competing theories is completely deductive: each begins with a preconception of how
fission might occur, and considers mathematical problems that bear on the viability of that
preconception. The classical fission theory turned out to lack internal consistency. The
present author's version of the theory, revised from the classical theory in that viscosity
plays a negligible rather than a dominant role, appears to be internally consistent and
compatible with the idea that close binaries with mass ratios near one may form in this way.
The version of the theory described elsewhere in this volume by Durisen, which fails to
produce binaries, is based on a preconception of the route to fission different from the
classical one. Its failure to produce binaries therefore reflects only on that route, and not
on other possible routes to fission.

1. Introduction

It has become common to use the term 'fragmentation' to refer to the breakup of a rapidly
collapsing gas cloud and the consequent formation of binary or multiple systems; articles
in this volume by Boss and Miyama are related to fragmentation. It occurs in the earliest
phase of star formation, that of dynamical collapse, and is characterized by a single
timescale (the dynamical timescale). Its mathematical and numerical treatment is via a
standard initial-value problem.

By contrast, the word 'fission' refers to the formation of a binary star from a single, rotating
star during a phase of slow, or secular contraction. The present article and that by Durisen
are devoted to this subject. Two timescales (the dynamical and the contraction
timescales) characterize the pre-main-sequence phase when fission is contemplated. A
standard initial-value problem is impractical because of the widely separated timescales
(but Lucy [1977] has attempted such a treatment anyway). A quasistatic approximation,
like that used in the theory of the evolution of spherical stars, is the logical approach, but
there are serious mathematical obstacles to applying it in the circumstances relevant to
fission. It has so far been implemented only in the context of idealized models of uniform
density, as described below in section 2.

The traditional, quasistatic approach toward fission envisages the following sequence of
events (or preconception). Initially, the figure is axisymmetric, or nearly so. In
consequence of a slow loss of energy through radiation, the axisymmetric figure

129

D. McNally (ed.), Highlights of Astronomy, Vol. 8, 129–131.
© *1989 by the IAU.*

becomes progressively more flattened, resulting in a progressively larger fraction of its energy in the form of kinetic energy of rotation. When the ratio of kinetic to gravitational energy reaches a critical value, the axisymmetric figure becomes unstable to a symmetry -breaking perturbation carrying the circular cross-section into an elliptical one. Further evolution takes place quasistatically along a nonaxisymmetric (or bar-shaped) family on which the equatorial cross-section lengthens progressively in one direction. This continues until this nonaxisymmetric family itself becomes unstable. The nature of this secondary instability and of the family bifurcating at this point is crucial for the theory. One imagines that instability is initiated by a perturbation tending to concentrate matter toward the ends of the bar, resulting in a dumbbell shape. This process continues as evolution proceeds until the matter connecting the two ends of the dumbbell has thinned to the vanishing point, and one is left with a pair of binary stars orbiting each other.

There is nothing in this preconception of binary-star formation that would preclude the use of realistic models in checking the various assertions. However, the mathematical difficulties remain formidable, and, as a result, the only models in which attempts at verification have been made are ellipsoidal figures of uniform density and perturbations therefrom. The work described elsewhere in this volume by Durisen uses realistically stratified models, but in a preconception different from that described above and (in my opinion) not compelling; we discuss this further in section 3 below.

2. Ellipsoidal Figures

We turn now to a brief description of the classical fission theory and of its revised form, wherein one attempts to verify the preconception of fission described above in the ellipsoidal context. A fuller description of these topics is given by Lebovitz (1987).

2.1 THE CLASSICAL VERSION

Here the assumption is made that viscosity is important. This has the effect of enforcing a uniform rotation and influences the stability properties of the figures as well. The axisymmetric family consists of Maclaurin spheroids, which become unstable when the ratio of kinetic to gravitational energy reaches a certain critical value. The family of Jacobi ellipsoids bifurcates from this critical point, and becomes the stable, bar-shaped family along which further evolution takes place. The Jacobi family in turn becomes unstable when the bar reaches a critical elongation. Up to this point of instability along the Jacobi family, all the assertions of the preconception above are verified (in fact, they were originally abstracted from this model). Beyond this point, however, the facts fail to agree with the preconception. The perturbation initiating the instability of the critical Jacobi figure does not shape it into a dumbbell. More seriously, the family of figures bifurcating from this critical point is itself unstable: there are no stable, quasistatic figures just beyond the critical Jacobi figure. Hence quasistatic evolution ceases at this point and motion must take place (at least for a while) on the dynamical timescale. The precise behavior of the figures at this point has never been fully resolved.

2.2 THE REVISED VERSION

Here viscosity is neglected, on the ground that this assumption is nearer to the truth than the opposite assumption of the classical version. A wider variety of ellipsoidal configurations is now possible (the Riemann ellipsoids, in which there are fluid motions beyond those of rigid-body rotation). The initial evolution again takes place along the Maclaurin family (approximately). Again an instability along this family signals a shift in the evolution onto a family of ellipsoids with three unequal axes. This family itself subsequently becomes unstable. While details differ, there is a rather complete analogy with the classical version up to this point. However, there is an important difference beyond this point. On the revised version, the perturbation initiating the instability carries the ellipsoid into a dumbell shape. Moreover, the family of figures bifurcating at the critical point is stable beyond the critical point, allowing quasistatic evolution to continue.

3. Conclusions

The principal conclusion is that, by removing the artificial assumption of a dominating viscosity from the theory, one finds a pattern of behavior supporting the preconception of fission as described above. There remain a number of mathematical details to confirm before this picture is mathematically solid (Lebovitz [1987]), but it appears at this writing that the impasse that affected the classical version of the fission theory is overcome on the revised version.

The theory suffers from an obvious defect: the models are uniformly dense whereas stars are centrally condensed. One may wonder whether the qualitative conclusions drawn from these idealized models would persist in the context of realistically stratified models.

The work described elsewere in this volume by Durisen employs realistically stratified models and extensive numerical computation, but also suffers from defects, in the nature of its preconception of fission. One of the elements of this preconception is that fission should be the immediate and direct outcome of the instability of the axisymmetric figure. Since the initiating perturbation carries the figure into a bar (not a dumbbell), it is difficult to understand the motivation for this idea. In fact the computations do not support it. One can construe this as confirming that the preconceived route to fission is not possible; one need not construe it to mean, as some have, that fission is not possible.

References

Lebovitz, N. (1987) Geophys. Astrophys. Fluid Dynamics **38**, 15.
Lucy, L. (1977) Astron. J. **82**, 1013.

NUMERICAL SIMULATIONS OF FISSION

R. H. DURISEN, S. YANG, and R. GRABHORN
Department of Astronomy
Swain Hall West 319
Indiana University
Bloomington, IN   47405
U.S.A.

ABSTRACT. Recent hydrodynamic simulations indicate that gravitational torques due to spirals arms suppress binary fission in rapidly rotating polytropic fluids.

1. Introduction

Self-gravitating equilibrium states of rotating fluids are subject to a variety of secular and dynamic instabilities and bifurcations. Many scenarios have been suggested whereby evolution through these instabilities or bifurcations might produce a binary or multiple body system from a single body by growth of nonaxisymmetric distortions. I will refer to this hypothetical process as "binary fission".

2. Numerical Simulations

2.1. METHODS

In recent years, numerical hydrodynamic simulations in three dimensions have been used to explore the regime where axisymmetric equilibrium states are known to be dynamically unstable to barlike distortions. Contributions can be distinguished by the polytropic index n of the fluids considered and by the nature of the numerical hydrodynamics scheme, either smoothed particle (SPH) or finite difference (FD): n = 3/2, FD and SPH (Tohline et al. 1985, Durisen et al. 1986); n = 1/2 and 3/2, SPH (Durisen and Gingold 1986); n = 0.8 to 1.8, FD (Williams and Tohline 1987, 1988); n = infinity (isothermal), SPH (Miyama et al. 1988). In all these studies, axisymmetric equilibrium models with small nonaxisymmetric perturbations are loaded into the hydro codes as initial conditions.

133

D. McNally (ed.), Highlights of Astronomy, Vol. 8, 133–135.
© *1989 by the IAU.*

2.2 EXAMPLE

Figure 1 illustrates two stages from an evolution with n = 3/2. The rotation parameter $T/|W|$ is 0.33 initially, where T is the rotational kinetic energy and W is the gravitational potential energy. By 2.5 rotations, the initial seed bar perturbation has grown into a two-armed spiral, and the central region has almost undergone binary fission. The binary never pinches off, however, because the outer spirals exert gravitational torques which extract orbital angular momentum. A short while later the lumps of the protobinary crash together to form a stable, barlike equilibrium which resembles a Riemann S-type ellipsoid. The final state is a detached ring and disk of material surrounding the central bar. The gap between the bar and ring is the position of orbital corotation with the tumbling bar.

2.3 GENERAL RESULTS

From n = 1/2 to infinity, independent of code and initial perturbation, simulations exhibit the following common features: 1.) Two-armed spirals grow dynamically near and above $T/|W| \approx 0.27$, in agreement with classic bar mode results when n = 0. 2.) The trailing spirals transport angular momentum outward and suppress binary formation. 3.) As n increases for fixed $T/|W|$, more mass and angular momentum are ejected, and the central regions experience more significant contraction. 4.) As $T/|W|$ increases for fixed n, higher order structure and more long-lived binary fragments can appear.

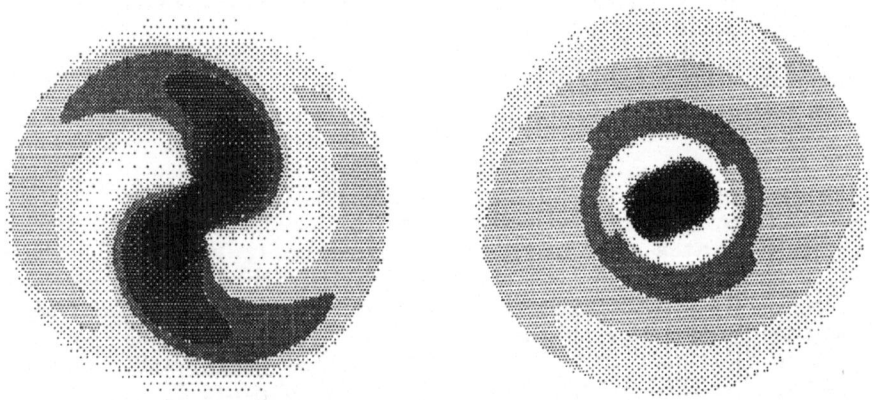

Figure 1. Grey scale representations of the density in the equatorial plane for an n = 3/2, $T/|W|$ = 0.33 polytrope. Black signifies the highest densities, and white denotes essentially zero density. The fluid rotates in a clockwise sense. The left and right panels corresponds to about 2.5 and 5.0 central initial rotation periods, respectively.

3. Implications

The simulations to date all produce a form of fission, in that a ring or disk of material is ejected, but "binary" fission is generally suppressed by gravitational torques. Unfortunately, the finitely unstable initial equilibrium conditions of these simulations are not physically realistic. All binary fission scenarios involve slow evolution in equilibrium from stable conditions toward and through an instability or bifurcation point. Simulations of this type are currently being pursued. However, a simple physical argument suggests that "binary" fission will still be suppressed. Sufficiently rapid rotation to induce instability and bifurcation requires differential rotation for n > 0.8. It is differential rotation that induces the trailing spiral arms and consequent torques. This seems difficult to escape for ordinary stellar compressibilities (n ≳ 3/2).

The investigations reported here use centrally condensed configurations with similar angular momentum distributions. Studies are now underway by various groups to explore a wider variety of equilibrium models, including tori and protostars. Preliminary results indicate significant differences in stability properties.

Although it is becoming clear on observational grounds that low mass stars rotate too slowly to fission during pre-main sequence contraction, fission instabilities may well play an important role during the protostellar accretion phase, when rapidly rotating equilibrium states of various sizes and forms can be expected to occur.

Durisen, R.H., and Gingold, R.A. (1986) "Numerical Simulations of Fission", in W.K. Hartmann et al. (eds.), Origin of the Moon, Lunar & Planetary Inst., Houston, pp. 487–498.

Durisen, R.H., Gingold, R.A., Tohline, J.E., and Boss, A.P. (1986) "Dynamic Fission Instabilities in Rapidly Rotating n = 3/2 Polytropes", Astrophysical J., 305, 281–308.

Miyama, S.M., Narita, S., Kiguchi, M., and Hayashi, C. (1988) "Nonaxisymmetric Instability of Rotating Isothermal Equilibria", Astrophysical J., in press.

Tohline, J.E., Durisen, R.H., and McCollough, M.L. (1985) "Linear and Nonlinear Stability of Rotating n = 3/2 Polytropes", Astrophysical J., 298, 220–234.

Williams, H.A., and Tohline, J.E. (1987) "Linear and Nonlinear Dynamic Instability of Rotating Polytropes", Astrophysical J., 315, 594–601.

Williams, H.A., and Tohline, J.E. (1988) "Circumstellar Ring Formation in Rapidly Rotating Protostars", Astrophysical J., in press.

THE J vs. M RELATION FOR BINARY STARS

J. E. TOHLINE
Department of Physics and Astronomy
Louisiana State University
Baton Rouge, LA   70803-4001
U.S.A.

ABSTRACT. For a given mass M and environmental temperature T, there is a well-defined angular momentum J_{max} above which physical systems cannot exist as self-gravitating entities. The quantity $J_{max} \propto M^2 T^{-1/2}$. Observations of J and M in young binary systems should put useful constraints on the temperature of the medium from which they formed.

Brosche (1963) was the first to publicize the idea that a plot of log J vs. log M for a wide range of astronomical systems shows a strong correlation fit nicely by a line of slope +2, i.e., dlog J/dlog M = 2. In his initial analysis, Brosche identified this relationship in a plot that contained systems ranging in mass from planet-satellite systems orbiting the Sun to the local supercluster of galaxies--a range covering more than 20 decades in mass and more than 40 decades in J. (If this plot is extended in angular momentum over an additional 60 decades--i.e., covering a total of more than 100 decades in J --the point defined by the Planck mass and ℏ falls very nearly on the same line than runs through the points in Brosche's diagram!) It behooves us to understand the physical origin of this universal dlog J/dlog M relation.

A number of discussions of the dlog J/dlog M relation can be found scattered through the literature over the past twenty-five years. Many (Carrasco et al. 1982, and references cited therein; Trimble 1984) have focused on systems having a limited mass or period range--such as normal binary star systems--for which data correlations often show dlog J/dlog M closer to +5/3 than to +2. Our discussion will be confined to Brosche's more universal slope.

Because very slowly rotating objects or, ideally, zero angular momentum systems fall well below the data shown in Brosche's Fig. 2, it is unreasonable to adopt the line of slope +2 as an absolute correlation obeyed by all physical systems. Instead, it is preferable to identify from Brosche's paper an upper envelope that Nature sets on J(M) for all systems. With this in mind, we should not only be concerned with explaining the slope but also the absolute location of the line (i.e., its y-intercept) in the log J-log M plane.

It appears as though both the position and slope of this upper

137

D. McNally (ed.), Highlights of Astronomy, Vol. 8, 137–138.
© 1989 by the IAU.

envelope can be explained in terms of the characteristic sound speed--or temperature--of the medium from which astronomical systems form under the influence of gravity. For a given total mass and sound speed c, there is an angular momentum J_{max} above which a system cannot become, or cannot exist as, a self-gravitating entity. The limit is set by (Tohline and Christodoulou 1988)

$$J_{max} = f \frac{G}{c} M^2 , \qquad (1)$$

where G is the gravitational constant and $f \approx 0.1$ is a dimensionless coefficient. (Alternatively, for a given M and c, the maximum allowed orbital period $P_{max} = 2\pi f^3 GM/c^3$.) This limit on J can also be ascertained from Chandrasekhar's (1961) dispersion relation for gravitational instability in a uniformly rotating, infinite homogeneous medium. Clearly, for fixed c, this relation demands an upper envelope that is a line of slope +2 in a log J - log M diagram. Furthermore, the <u>value</u> of J_{max} is reasonable. All of Brosche's data lies below the line defined by a sound speed of 0.3 km s^{-1}, indicating that the angular momentum for all these systems has been limited by environments warmer than 10 K.

Using relation (1), it is interesting to examine data sets having a restricted range of masses or orbital periods, such as the spectroscopic binaries discussed by Trimble (1984). Virtually all binaries shown in her Fig. 8 having P > 1000 days lie below the envelope set by a temperature T = 1000 K (c = 3 km s^{-1}). This strongly suggests that these "long period" binary systems formed in an environment having T $\geq$ 1000 K. The short period (P < 2 days) systems shown in Trimble's Fig. 8 are confined below a line defined by relation (1) and a temperature $\sim 10^6$ K, reflecting the much warmer environment in which the stellar components of these systems reside.

Models that have been developed to explain the dynamical process of star formation have generally included standard physical processes which are believed to be important in dictating what the gas temperature $T(\rho)$ is as a function of the gas density ρ during cloud collapse. Direct checks of this $T(\rho)$ relation are precluded by the large optical depth in protostellar clouds and/or the spatial resolving power of our present day instrumentation. An indirect test of the environmental temperature at which binary systems form may be available, however, through relation (1). Careful observational documentation of the J vs. M relationship among the youngest binary star systems can be used to put useful constraints on physical models of star formation.

This work has been supported by the U. S. National Science Foundation through grant AST 87-01503.

REFERENCES

Brosche, P. 1963, Zeitschrift für Astrophysik, 57, 143.
Carrasco, L., Roth, M., and Serrano, A. 1982, Astron. Ap., 106, 89.
Chandrasekhar, S. 1961, Hydrodynamic and Hydromagnetic Stability, Dover, New York.
Tohline, J. E. and Christodoulou, D. M. 1988, Astrophys. J., 325, 699.
Trimble, V. 1984, Astrophys. Space Sci., 104, 133.

THE FORMATION AND EVOLUTION OF BINARIES IN GLOBULAR CLUSTERS

Frank Verbunt
Max Planck Institut für Extraterrestrische Physik
D-8046 Garching bei München
Federal Republic of Germany

Abstract. The number density of stars in the cores of globular clusters is high enough for close encounters between stars to be frequent. These encounters may lead to the formation of binaries. Those binaries which do not easily form via the evolution of primordial main-sequence star binaries, and are therefore rare in the galactic disk, can be common in globular clusters. Examples of such binaries are the low-mass X-ray binaries. Such binaries may evolve into radiopulsars.

1. Introduction

Binaries have long escaped detection in globular clusters. In 1987 the binary period was finally found for two X-ray sources (Table 1). Subsequently, five radiopulsars were found in globular clusters, three of which are in a binary (Table 1). These binaries are thought to have evolved from low-mass X-ray binaries, and the single radiopulsars may originate in binaries too. Several other binaries were found via accurate radial velocity measurements of cluster giant stars /8/. I limit myself here to the discussion of the formation and evolution of binaries containing a neutron star. The formation of other binaries may follow a similar pattern. More detail, and references, are given in /11-13/.

Table 1. *Parameters for binaries containing neutron stars, and for single radiopulsars in globular clusters. X = X-ray source, PSR = radiopulsar.*

source	P	P_{orb}	e	cluster	ref
X1820 $-$ 30		864s		NGC 6624	/10/
X2127 $+$ 12		0.35d		M 15	/5/
PSR1821 $-$ 24	3.1ms			M 28	/6/
PSR0021 $-$ 72A	4.5ms	1924s	0.33	47 Tuc	/1/
PSR0021 $-$ 72B	6.1ms	7-95d		47 Tuc	/1/
PSR1620 $-$ 26	11.1ms	191.4d	0.025	M 4	/7/
PSR2127 $+$ 12	110.7ms			M 15	/15/

139

D. McNally (ed.), Highlights of Astronomy, Vol. 8, 139–142.
© 1989 by the IAU.

2. Formation mechanisms

A star is deformed by the gravitational pull when another star passes nearby. The deformation energy is taken from the relative kinetic energy, and exceeds it at a distance d less than about three times the radius of the deformed star, for masses of $\sim M_\odot$, and velocities ~ 10 km/s. The two stars are then bound /2/. Because the energy in the deformation is proportional to d^6, a rough estimate of the deformation energy gives a pretty accurate estimate of the closest distance required. In this way, a neutron star passing close to a main-sequence or giant star can be tidally captured by it.

Another possibility is an exchange encounter between a binary and a single neutron star, leading to the formation of a temporary triple system, from which one of the original binary members escapes, leaving the other two stars in a binary /4/. Because the number of binaries is small, most low-mass X-ray binaries in globular clusters are probably formed via tidal capture /13/.

3. Stellar content of the core of a globular cluster

To determine what stars will participate in close encounters, one assumes that all stars in a globular cluster were formed simultaneously, with a distribution of masses m according to an initial mass function (IMF) $dN(m) = C_o m^{-1-x} dm$. The normalization constant C_o and the slope index x can be determined by counting main-sequence stars, which are still unchanged in the cluster. The more massive stars have evolved, the most massive ones have left neutron stars, the less massive ones have left white dwarfs. The number of neutron stars can be estimated by extrapolating the IMF, but two problems arise: i) the large extrapolation required (from the main-sequence stars at $\lesssim 0.8 M_\odot$ to neutron-star progenitors at $\gtrsim 8 M_\odot$) introduces a large uncertainty, ii) from observations of radiopulsars in the galactic disk it appears that many of them are born with velocities exceeding the escape velocity from a globular cluster. Thus, the number of neutron stars in a globular cluster is very uncertain, even if the IMF is well studied /13/.

Around the turnoff mass, stars are evolving into (sub)giants, and their number can be found with help of the IMF. Although the number of giants is small, their size is large, and they account for a sizable fraction of the tidal captures. Virtually all close encounters occur in the cluster core, since the density is highest there. Mass segregation must therefore be taken into account: the most massive stars will be concentrated towards the cluster center, which enhances close encounters involving these stars, including the (sub)giants. Depending on the concentration of the cluster, the fraction of captures by giants varies between about 10 % and 30 % /14/.

4. Tidal capture of neutron stars, and their consequences

Figure 1 gives a schematic overview of the most common capture processes involving

DIFFERENT CAPTURE PROCESSES AND THEIR CONSEQUENCES

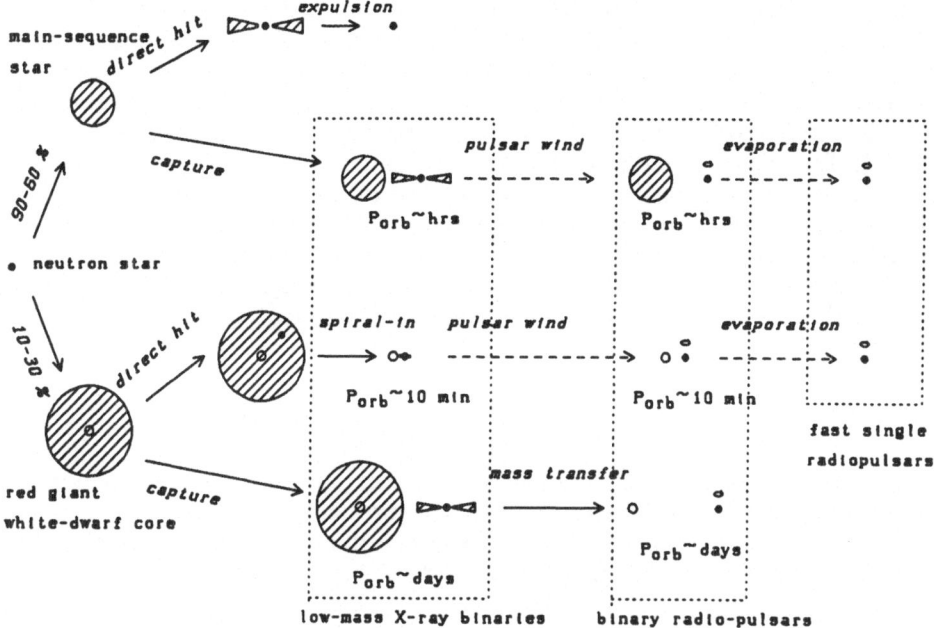

Figure 1. Schematic overview of the formation via tidal capture of X-ray binaries, and of their subsequent evolution. After /11/.

neutron stars, and of subsequent occurrences.

A neutron star can be captured by either a main-sequence star, or by a (sub)-giant. A direct hit on a main-sequence star destroys it, and after a short intermediate phase where some gas still floats around the neutron star, the neutron star emerges pretty much as it was. In a slightly wider encounter, a binary may be formed. Loss of angular momentum, due to gravitational radiation or magnetic braking by stellar wind, drives the two stars together, and causes mass transfer. The ensuing X-ray binary has P_{orb} ~1-8 hr. It appears that interruption of the mass transfer can cause the neutron star to switch on as a radiopulsar, which subsequently evaporates its companion /3/. Thus a binary radiopulsar is formed, and, after complete evaporation, a single radiopulsar.

A direct hit of a neutron star on a giant, may cause it to spiral in, until the released energy expells the envelope, leaving the white-dwarf giant core in close orbit around the neutron star. Gravitational radiation drives the two stars together until mass transfer stars at P_{orb} ~ 10 min. Again, it is possible that the neutron star switches on as a radiopulsar, and evaporates its companion. A wider encounter with a giant leads to the formation of a binary with P_{orb} ~1-50 days. Mass transfer in such a binary is caused by expansion of the giant radius. After exhaustion of the giant's envelope, a wide binary with a radiopulsar is left.

Comparison with the observed sources (Table 1) shows that one X-ray binary (in NGC 6624) originated in a direct hit between a neutron star and a giant, and

142

another one (in M 15) in a wider encounter, followed by some loss of angular momentum. Evolution of a wide X-ray binary has led to a radiopulsar in M 4 and one in 47 Tuc. The origin of the 1924.3s binary in 47 Tuc probably involved a spiral-in in a giant, but the origin of its high eccentricity is a mystery. The single radiopulsar in M 28 may have evaporated an earlier companion, but the pulsar in M 15 rotates too slowly for such a process. Wide radiopulsar binaries like those in M 4 and 47 Tuc live long, and are subject to encounters with single cluster stars /9/. Such encounters may induce eccentricity, which explains the e of the M 4 binary, and leads to the prediction that the wide binary in 47 Tuc has a high e. More rare encounters can liberate the neutron star from a wide binary by exchanging it with a single cluster star. This may well be the origin of the single radiopulsar in M 15, a cluster with a dense core, in which encounters are frequent.

5. Summary.

Tidal capture of neutron stars by main-sequence or (sub)giant stars explains the variety of X-ray sources and radiopulsars in the cores of globular clusters. It is not possible to predict the absolute number of sources, due to the large uncertainty in the number of neutron stars in globular clusters.

References

1. Ables, J.G., Jacka, C.E., McConnell, D., Hamilton, P.A., McCulloch, P.M. & Hall, P.J. 1988, *IAU Circ.#* 4602.
2. Fabian, A.C., Pringle, J.E. & Rees, M.J. 1975, *Mon. Not. R. astr. Soc.* **172**, 15p.
3. Fruchter, A.S., Stinebring, D.R. & Taylor, J.H. 1988, *Nature* **333**, 237.
4. Hills, J.G. 1976, *Mon. Not. R. astr. Soc.* **175**, 1p.
5. Ilovaisky,S.A., Aurière, M., Chevalier, C., Koch-Miramond, L., Cordoni, J.P. & Angebault, L.P. 1987, *Astron. Astrophys.* **179**, L1.
6. Lyne, A.G., Brinklow, A., Middleditch, J., Kulkarni, S.R., Backer, D.C. & Clifton, T.R. 1987, *Nature* **328**, 399.
7. Lyne, A.G., Biggs, J.D., Brinklow, A., Ashworth, M. & McKenna, J. 1988, *Nature* **332**, 45.
8. Pryor,C.P., McClure, R.D., Hesser, J.E. & Fletcher, J.M. 1987, *BAAS* **19**, 676.
9. Rappaport, S., Putney, A. & Verbunt, F. 1988, *Astrophys. J.* submitted.
10. Stella, L., Priedhorsky, W. & White, N.E. 1987, *Astrophys. J. (Letters)* **312**, L17.
11. Verbunt, F. 1988a, in: *The physics of compact objects, theory vs. observation*, eds. N.E. White and L. Fillipov, Pergamon Press, Oxford, p.529.
12. Verbunt, F. 1988b, in: *The physics of neutron stars and black holes*, eds. Y. Tanaka, ISAS, Tokyo, in press.
13. Verbunt, F. & Hut, P. 1987, in: *The origin and evolution of neutron stars*, IAU Symp. # 125 Nanjing P.R. China, eds. D.J. Helfand and J.H. Huang, Reidel, Dordrecht, p.187.
14. Verbunt, F. & Meylan, G. 1988, *Astron. Astrophys.* in press.
15. Wolszczan, A., Middleditch, J., Kulkarni, S.R. & Fruchter, A.S. 1988, *IAU Circ.#* 4552.

BINARIES FROM UNSTABLE TRIPLES. DYNAMICAL PROCESSES OF FORMATION

Joanna P. Anosova
Astronomical Observatory
Leningrad State University
Leningrad, U.S.S.R.

The dynamical processes of formation, evolution and disruption of binaries may be effectively studied by computer simulations in the N > 3-body gravitational problem. As a result of analysis of these investigations of diverse authors, the classification of the dynamical processes of formation of wide and close binaries may be proposed (see Table 1). This Table shows the following general processes: I-triple approaches of the single bodies; II-approaches of binaries with single bodies; III-escape from physical triples. The actions of these processes, and kinetics of a frequency of binaries in general field were studied at the Astronomical Observatory of the Leningrad State University (1965-1988) by computer simulations in the three-body problem. More than 3.10^4 orbits with negative total energy $E < 0$ and 5.10^4 with $E > 0$ have been run on the computers. The film "Dynamical evolution of triple systems" was produced. Part I of this movie shows the evolution of the unstable non-hierarchical triplet as well as the processes of formation, evolution, and disruption of temporary wide and final close binaries inside the physical triples. Part II of film presents in detail the trajectories of the bodies on the triple approaches of "fly-by" - and of "exchange"-types. The triple approach of "fly-by"-type results often in an escape from triple as well as the formation of final close binary. The triple approach of "exchange"-type consists as a rule of a few close double approaches of bodies and rarely results in an escape from triplet, it results in formation of temporary wide binary inside triplet. Part III of movie presents the trajectories of the different-mass bodies: an escape of the minimum-mass body, the intermediate-mass body, and the maximum-mass body as well as a formation of binaries with different-mass components.

[In Dr Anosova's absence, the film was introduced and presented by Dr R S Harrington, US Naval Observatory.]

D. McNally (ed.), Highlights of Astronomy, Vol. 8, 143–144.

REFERENCES:

1.Anosova J.P.1986,Astrophys. and Space Science, 124,217.
2.Anosova J.P.1988,In "The Few Body Problem", Proc. IAU Coll.96, ed Valtonen, M.J., Reidel, Dordrecht, Astrophys.and Space Sci. Library,140.27.

TABLE 1. Processes of formation of wide and hard binaries. E is the total energy of the triplet, E* the energy of the closest bodies.

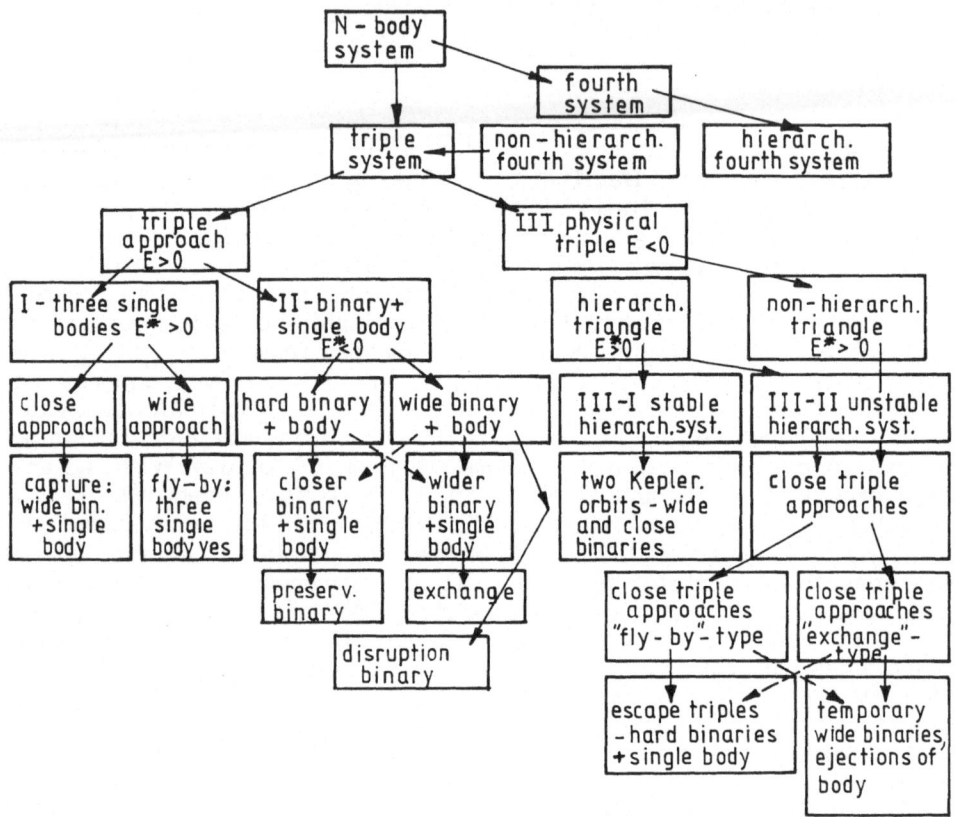

"UNDISTURBED" EVOLUTION IN BINARIES

J. Andersen
Copenhagen University Observatory, Denmark, *and*
Harvard-Smithsonian Center for Astrophysics
60 Garden Street
Cambridge, MA 02138, U.S.A.

ABSTRACT. The use of binary systems as tools for testing models for single-star ("undisturbed") evolution is briefly reviewed. Recent successes and directions for future work are discussed.

1. INTRODUCTION

In many binary systems, interactions between the components are weak enough that the stars can be assumed to evolve initially just as they would have done if they were single. This is not a trivial observation, because some of these binary systems offer possibilities for testing current theory of stellar evolution in ways which are impossible in single stars. Moreover, this first stage provides the initial conditions for the later evolution of some binaries into much more exotic objects, and for constructing models of these advanced phases of evolution.

A comprehensive review of the subject cannot be given in the time and space available here. I shall therefore focus on a few highlights of recent results in the field, drawing heavily on work carried out by a group of colleagues centered at Copenhagen Observatory, who have concentrated on determinations of accurate data for this purpose from observations of suitable detached, double-lined eclipsing binaries.

2. TESTING STANDARD MODELS OF SINGLE-STAR EVOLUTION

Given the initial mass, chemical composition, and age of a star, theory in principle predicts its current mass, radius, temperature or luminosity, and emitted spectrum. Only for the Sun can all these parameters be independently and accurately determined and compared with the best current evolutionary models. However, ambiguities arise from uncertainties in, especially, convection theory and computed opacities. Consequently, the solar constraints are primarily used to fix certain adjustable parameters in the models, notably the mixing-length parameter, α, and the helium abundance, Y. One may then compare, e.g., the observed and predicted mass-radius or mass-luminosity relations, using masses, radii,

145

and luminosities determined from binary studies (Straižys and Kuriliene, 1981; Balona, 1984).

Unless very precise mass and radius data are used in such studies, however, observational scatter may obscure the evolutionary effects and the latter be overlooked. In fact, log g as determined directly from the basic data M and R is a very sensitive indicator of evolution, even *within* the main-sequence (MS) band (Nordström, 1988); evolutionary effects of only 5% of the width of the MS can be distinguished in the most accurate data (masses and radii precise to ~1% or better). Still, even with first-rate masses and radii, fundamental ambiguities remain when the composition and age of the stars are not well known.

In stars other than the Sun, the actual age and, usually, the helium abundance cannot be determined independently of the stellar models one wants to test. Close binaries, however, must satisfy the additional constraint that the ages and initial compositions of the components be identical. In stars of near-solar heavy-element abundance Z, the obvious further assumption is for the helium abundance to be also essentially solar. Thus, if the metal abundance of a binary can be determined, e.g. spectroscopically, only its age remains unknown among the fundamental parameters. If evolutionary models with mixing-length parameter and helium abundance normalized to the Sun and with the observed Z can reproduce all observed properties of two substantially different binary components for the *same* age, a non-trivial test has been made. Evidently, the test becomes even stronger if agreement can be demonstrated for binaries with a variety of different metal abundances.

3. RECENT RESULTS

The first detailed test of this kind was made by Popper and Ulrich (1984), using the binary V818 Tau (HD 27130). Its membership of the Hyades provides both its metal abundance, distance, and the information that both stars are on the ZAMS. Popper and Ulrich showed that solar-calibrated stellar models matched the luminosity of the primary satisfactorily; a low helium abundance (Strömgren et al., 1982) was inconsistent with the binary data. This analysis was refined by Schiller and Milone (1987) on the basis of accurate masses and radii of both stars.

A stronger test of the models is possible in the system AI Phe, in which the more massive component (1.24 $M_\odot$) has already evolved onto the lower giant branch while the other, only 3% less massive star has just left the main sequence. A complete determination of the masses, radii, luminosities, and metal abundance ([Fe/H] = -0.14) of the system by Andersen et al. (1988a) led to remarkably close agreement with solar-normalized VandenBerg (1983, 1985) standard models for these observed parameters *and* the solar helium abundance: Derived ages for the two stars agreed to ±1%. It could be further concluded that convective core overshooting had no observable effect in these stars. It was noted that the same models, for the Hyades metal abundance, also matched the properties of V818 Tau. It was finally *predicted* that the very similar-mass system UX Men should be more metal-rich than the Sun, as those stars were much cooler than the models matching AI Phe.

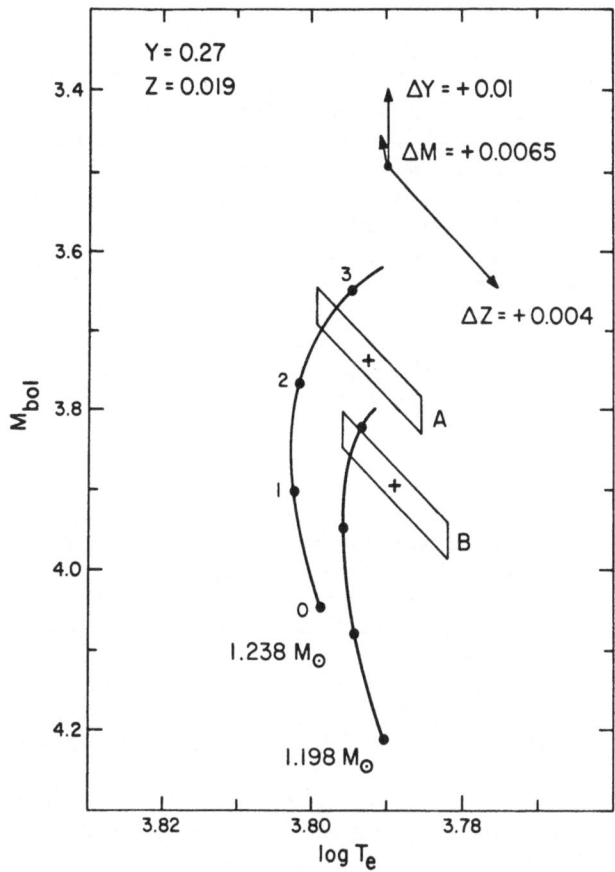

Fig.1. Evolutionary tracks and observed locations for the components of UX Men (Andersen et al., 1988b). Ages (Gyr) are labeled along the tracks; arrows show the effects of 1 σ changes in the parameters.

   This prediction was confirmed by the subsequent similar study of UX Men by Andersen et al. (1988b) who found [Fe/H] = +0.04. The theoretical HR diagram in Fig. 1 compares the theoretical tracks for the observed masses and metal abundance, and the solar helium abundance, with the observed location of the components as derived from temperatures and radii. As will be seen, the agreement is again quite satisfactory.
   From these comparisons, we draw the following conclusions: First, standard VandenBerg models with a solar helium content can now successfully predict, *at the 1% level*, all observed properties of binaries with Z from 0.012 (AI Phe) over 0.019 (UX Men) to 0.024 (V818 Tau) as well as the Sun. Second, this is only possible with the newest Los Alamos opacities; models using older (Cox-Stewart) opacities require a significantly lower helium abundance to fit the data. Third, even at this level of refinement, errors in the masses and radii have now ceased to be the limiting factor; the main uncertainty in the comparison with theory is due to that of even the best modern metal abundance determinations.

4. PROBLEM AREAS AND FURTHER WORK

While these recent successes in using binary systems as "laboratories" in which to study "undisturbed" stellar evolution are quite encouraging, not all problems have been solved and certainly not all work completed. A few main problem areas and directions for the future are listed below.

For very massive (O) stars, similarly informative comparisons are not currently possible, because temperature calibrations are quite uncertain and metal abundances difficult to determine. In addition, the observable consequences of interesting features of stellar models in this domain (convective overshooting, mass loss) are quite subtle in the main-sequence phase. A small discrepancy between observed and predicted *apsidal-motion* constants for standard models may offer additional clues.

In some well-established cases, two stars in different binaries have essentially equal masses and radii, but quite different temperatures, both in B stars (Popper, 1987) and in A stars (Andersen et al., 1984). Present experience suggests that the first hypothesis to explore is that of significantly different metal abundances. Only after such observations have been made should more exotic explanations in terms of helium abundance variations or non-standard evolution be invoked.

Finally, Habets and Zwaan (1988) have shown again how the observed stellar (synchronous or non-synchronous) *rotation* of binary components can be used to probe the evolution of internal structure of the stars.

Our overall conclusion is, however, that the greatest observational need is now for more *abundance* data to match the known masses and radii.

The long-standing pleasant cooperation of my colleagues and support of this project from the Danish Natural Science Research Council, the Danish Board for Astronomical Research, European Southern Observatory, and the Carlsberg Foundation are gratefully acknowledged.

REFERENCES

Andersen, J., Clausen, J.V., Gustafsson, B., Nordström, B., and Vanden-Berg, D.A. (1988a) Astron. Astrophys. 196, 128.y

Andersen, J., Clausen, J.V., and Magain, P. (1988b) Astron. Astrophys., in press.

Andersen, J., Clausen, J.V., Nordström, B.: (1984) Astron. Astrophys. 134, 147.

Balona, L.A. (1984) Monthly Notices Roy. Astron. Soc. 211, 973.

Habets, G.M.H.J. and Zwaan, C. (1988) Astron. Astrophys., in press.

Nordström, B. (1988) Astrophys. J., in press.

Popper, D.M. (1987) Astron. J. 93, 672.

Popper, D.M. and Ulrich, R.K. (1986) Astrophys. J. (Lett.) 307, L61.

Schiller, S.J. and Milone, E.F. (1987) Astron. J. 93, 1471.

Straižys, V., and Kuriliene, G. (1981) Astrophys. Space Sci. 80, 353.

Strömgren, B., Olsen, E.H., and Gustafsson, B. (1982) Publ. Astron. Soc. Pacific, 94, 5.

VandenBerg, D.A. (1983) Astrophys. J. Suppl. Ser. 51, 29.

VandenBerg, D.A. (1985) Astrophys. J. Suppl. Ser. 58, 711.

DISTURBED BINARIES : THE EARLY PHASES

J.P. DE GREVE
Astrophysical Institute, V.U.B.
Pleinlaan 2
B 1050 Brussels
Belgium

ABSTRACT. The disturbance of components and system parameters by mass transfer during and at the end of core hydrogen burning, as well as at the onset of core helium burning is reviewed. Emphasis is given to the influence of the evolution of the gainer and to the variation of the chemical composition of the surface.

1. Introduction

This paper is restricted to "disturbances" in the sense of mass transfer interaction. It does not deal with disturbances due to magnetic braking or influence of tidal interaction on the rotational behaviour of the components. More general aspects of the latter have been reviewed by Savonije and Papaloizou (1985). The importance of such interactions to clarify apparent anomalies is illustrated by the explanation of the discrepancy between the asynchronous rotation and the circularised orbits in systems containing F, G, or K-type components (Habets and Zwaan, 1988), through the increase of the moments of inertia (Rutten and Pylyser, 1988) and magnetic braking.
 There are at least 2 reasons to focus attention to the early disturbance of a binary system:
 1. Calculations of close binary evolution, involving the structure of the two components now give a better insight to the occurrence of contact phases and more detailed information on the remnant system (Nakamura and Nakamura, 1984, De Greve, 1986, Packet, 1988). This, in turn, serves to a better evaluation of the further evolution, more especially the advanced common envelope evolution.
 2. Implementation of a detailed nuclear network in the computer code results in information of the variation of surface abundances of elements such as carbon and nitrogen during mass transfer (De Greve and Cugier, 1988). Analysis of high dispersion spectra supply us with observational counterparts. From the comparison we may hope to get a better understanding of the process of mass transfer and, subsequently, of mixing processes involved in accretion of matter.

149

D. McNally (ed.), Highlights of Astronomy, Vol. 8, 149–152.
© *1989 by the IAU.*

2. Interaction near the Terminal Age Main Sequence (TAMS)

For a discussion of the interaction of a binary sysem at the end of core hydrogen burning of the primary component (late case A) or at the onset of hydrogen shell burning (case B), we follow the evolution of two detached systems, V539 Ara (6.1 M_O+5.25 M_O, P=3.17d) and QX Car (9.2 M_O+8.5 M_O, P=4.48d). Both systems were extensively studied within the framework of the Copenhagen program on accurate determination of absolute dimensions (Andersen, 1983, Andersen et al., 1983). For both systems sequences were computed with and without convective overshoot of the convective core (respectively given by the characters R and S). Their evolution is extensively described by De Greve (1988).The larger radius of the "overshooting" models forces them into a late case A evolution, whereas the models with classical convective cores evolve into case B interaction. Obviously, overshooting results in less extreme mass ratios and, consequently, in smaller final periods (De Greve, 1986).

The evolution of the surface abundance of hydrogen, carbon and of the carbon to nitrogen fraction is summarized in Figure 1. It was derived using the results obtained by De Greve and Cugier (1988), assuming thermohaline mixing in the envelope of the mass gaining star when hydrogen depleted layers are accreted on top of the latter. It shows clearly that abundances lower than cosmic values, but larger than those expected from simple unperturbed accretion of layers of the primary (Cugier and Hardorp, 1988), are expected during the slow mass transfer phase. Such intermediate values were recently derived for a number of Algol systems (see De Greve and Cugier, 1988, for references).

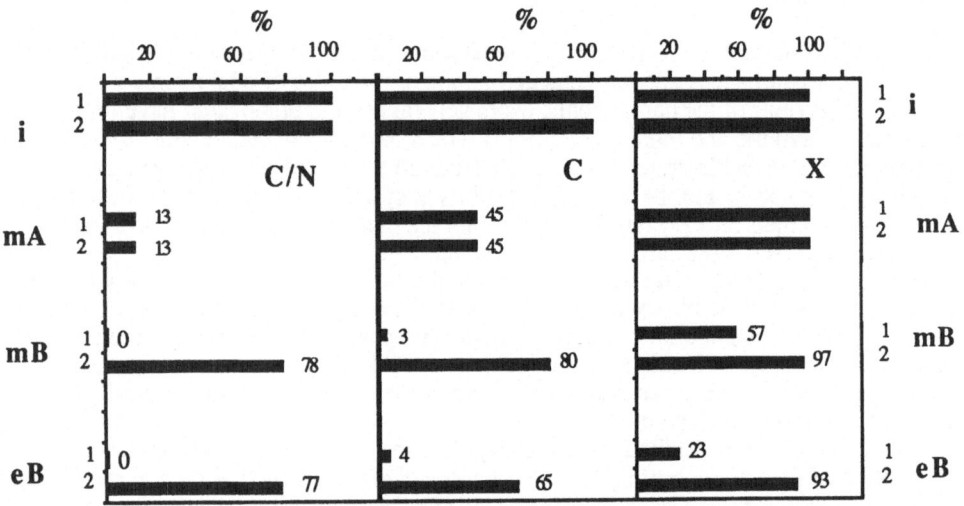

Figure 1. Proportional variation of the surface abundances of carbon-to-nitrogen (C/N), carbon (C) and hydrogen (X), during case AB of mass transfer of the system 9.2 M_O+ 8.5 M_O, P=4.48 d (QX Car), assuming thermohaline mixing when hydrogen depleted layers are accreted by the secondary star. The numbers 1 and 2 refer to the components. The following phases are given:
i= initial state, mA= mid case A, mB= mid case B, eB= end case B.

With regard to the semidetached state, systems with comparable mass such as V356 Sgr (12.1 M_0+4.7 M_0, P=8.9d; Popper, 1980), BF Cen (8.7 M_0+3.8 M_0, P=3.7d; Andersen et al., 1988) and perhaps even u Her (7.8 M_0+2.8 M_0, P=2.05d; Van der Veen, 1985) may have evolved from systems like V549 Ara and QX Car.

3. Mid case A interaction

For the low mass range, the evolution has been investigated by Ron Webbink (1976). For masses of the primary around 12 M_0 Nakamura and Nakamura (1984, 1987a,b) found a complex set of scenarios, depending on the initial conditions. One of their important results is the occurrence of reversed mass transfer during the main sequence interaction. The origin of it is shown in Figure 2, for a medium mass system calculated by us (De Greve and Packet, 1988). The Figure gives the evolution of the central hydrogen content by mass as a function of time. The fast accretion first leads to a rejuvenation of the gainer, but during the slow semidetached phase, nuclear burning proceeds at a faster rate than in the primary star. The net result is a faster evolution compared to the primary. Before the end of core hydrogen burning, the secondary fills its own critical surface and reversed mass transfer starts. In this particular system the same situation is repeated once more, again reversing the roles of loser and gainer. This time the outer critical surface is filled and a common envelope phase starts while both components are still on the main sequence.

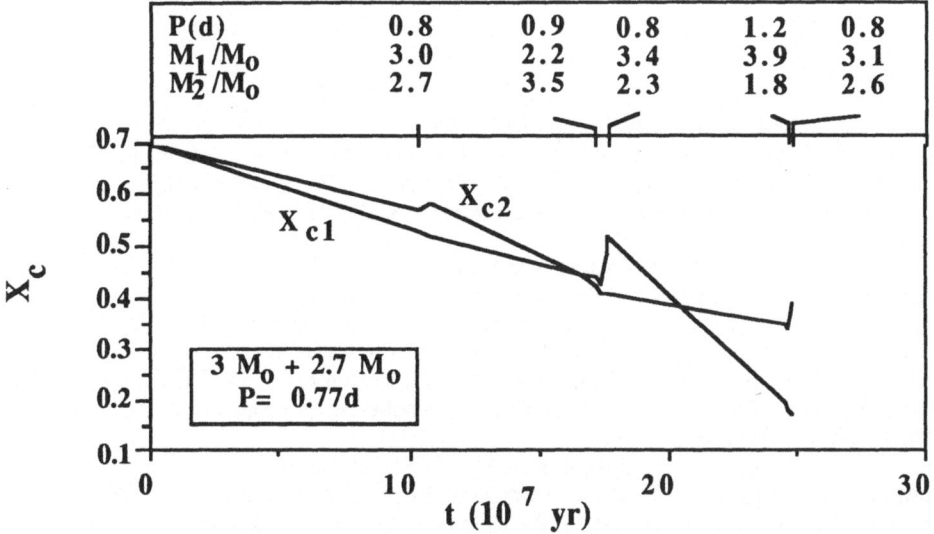

Figure 2. Evolution of the central hydrogen content (by mass) for the two components of an interacting binary system (case A of mass transfer). Characteristics of the system parameters are shown on top.

The net result for systems evolving through mid case A interaction is that the mass ratio and the period are confined to small ranges (resp. 0.3 to 2 and 1 to 2.5 days). Nakamura and Nakamura (1984) indicated what kind of systems follow the interrupted interactive evolution, as a function of initial separation d (expressed as the fraction of the possible main sequence range of values, ZAMS: d=0, TAMS: d=1), and mass ratio. Reversed mass transfer occurs for systems with d=0.7 (q=1) to d=0.5 (q=0.5). Practically all scenarios lead to overflow of the outer critical surface during the interactive phase. Only systems with d>0.5 and q~1 survive the interactive phase for some time. Giants with helium or carbon-oxygen companions only result from systems with d larger than 0.8 (case AB and B).

4. Late case B mass transfer.

Masses and characteristics of systems evolving through mass transfer around the moment of helium ignition in the core, were extensively studied by Iben and Tutukov (1985, and references therein). One of their important results was the derivation of the mass and composition of the degenerate dwarfs, formed by such interaction (probably valid for all case B interactions):

1. In the primary mass range 8.8 to 10.6 M_O, oxygen-neon dwarfs of mass 1.1 to 1.4 M_O are formed.
2. Below 2.3 M_O, down to 1 M_O, helium white dwarfs are formed down to a minimum mass of 0.13 M_O.
3. Carbon-Oxygen dwarfs of masses 0.3 M_O to 1.1 M_O originate from primaries of 2.3 to 8.8 M_O.

References

Andersen, J.: 1983, Astron. Astrophys.118, 255.
Andersen, J., Clausen, J.V., Nordstrom, B., Reipurth, B.: Astron. Astrophys. 121, 271.
Andersen, J., Clausen, J.V., Gustafsson, B., Nordstrom, B., Vandenberg, D.A.: 1988, preprint "Absolute dimensions of eclipsing binaries: XIII. AI Phoenicis: A case study in stellar evolution".
Cugier, H., Hardorp, J.:1988, Astron. Astrophys. (in press).
De Greve, J.P.: 1986, Space Sc. Rev. 43, 139.
De Greve, J.P.: 1988, Astron. Astrophys. (in press).
De Greve, J.P., Cugier, H.: 1988, Astron. Astrophys. (in press).
De Greve, J.P., Packet, W.: 1988, in 'Algols', A.H. Batten (ed.), IAU Coll. 107.
Habets, G.M.H.J., Zwaan, C.: 1988, preprint "Asynchromous rotation in close binary systems with circular orbits".
Iben, I., Tutukov, V.: 1985, Astrophys. J. Suppl. Ser. 58, 661.
Nakamura, M., Nakamura, Y.:1984, Astrophys. Space Sci. 104, 367.
Nakamura, M., Nakamura, Y.:1987a, Astrophys. Space Sci. 134, 161.
Nakamura, M., Nakamura, Y.:1987b, Astrophys. Space Sci. 134, 219.
Packet, W.: 1988, Ph. D. thesis, V.U.B., Brussels.
Popper, D.M.: 1980, Ann. Rev. Astron. Astrophys. 18, 115.
Rutten, R.G.M., Pylyser, E.:1988, Astron. Astrophys. 191, 227.
Savonije, G.J., Papaloizou, J.C.B.: 1985, in 'Interacting Binaries', P.P. Eggleton and J.E. Pringle (eds.), D.Reidel Publ. Co., Dordrecht, Holland, p.83.
Van der Veen, W.E.C.: 1985, Astron. Astrophys. 145, 380.
Webbink, R.F.: 1976, Astrophys. J. Suppl. Ser. 32, 583.

BINARY WOLF-RAYET STARS

B. Hidayat, Observatorium Bosscha, ITB, Lembang, Indonesia
K. A. van der Hucht, SRON Space Research Utrecht, The Netherlands

ABSTRACT. The distribution of the WR cluster members is interpreted in
terms of their association with the galactic spiral-arm pattern. It is
suggested that in a limited volume the number fraction of WR binaries is
influenced more by magnetic field properties of the spiral arms, than by
function of the galactocentric distance.

INTRODUCTION
     There is strong evidence that WR stars evolve from WNL to WC,
through mass loss of single massive stars (Maeder, 1987; van der Hucht
et al., 1988). A controlling factor of the WR population in galaxies is
metallicity (Smith, 1988). This factor also causes variation in the
subtype distribution of WR stars (van der Hucht et al., 1988).
     Paczynski (1967) for the first time recognized the importance of
mass exchange in a close-binary as a channel for producing WR star. This
channel, in contrast with single star mode of WR formation, is found to
be independent of metallicity (Maeder, 1982).
     The binary percentage in the solar neighborhood is on the average
37% (van der Hucht et al., 1988), 29% in LMC (Breysacher, 1980) and
perhaps 100% in SMC. A percentage of 50% would be quite normal. It is
therefore interesting to look for other evidence for the cause of binary
WR production, in the case of high percentages.

THE DATA
     Lundstrom and Stenholm (1984) show that $\sim$30% of the known galactic
WR stars are clusters and associations members. Van der Hucht et al.
(1988) indicate that eleven of the 43 WR cluster and association members
are double-line spectroscopic binaries and thus form a significant
fraction.
     In order to find a local galactic environment that may affect the
formation of binary WR stars, the galactocentric distribution of the
clusters and their relationship with the WR subtypes and their binaries
are reexamined. By isolating the factor of metallicity, one may be able
to deduce other galactic environment parameters which are operating to
produce binary WR star. For this purpose 17 clusters within 2.5 kpc from
the sun (up to 3 kpc the number is 24) are studied. In this limited
volume we expect a uniform metallicity. Within 2.5 kpc from the sun, we
are essentially dealing with only 2 branches of galactic spiral arms,
namely the Local and the Sagittarius and Carina arms (Bok, 1983). All
the clusters under discussion are found in these arms. The age spread
of the clusters in d < 2.5 kpc is limited within $6.0 \le \log T \le 6.7$ yr.
This may indicate that the clusters have similar physical characteristics.

153

D. McNally (ed.), Highlights of Astronomy, Vol. 8, 153–154.
© 1989 by the IAU.

154

RESULTS AND CONCLUSIONS

It is found that WN 7 stars in clusters are more abundant than earlier WN and WC subtypes. This can be attributed to the age of the sample clusters, which favour WN types - as these stars have not had enough time to evolve into later evolutionary phases yet. The overall pattern of the distribution of WR subclasses as a function of distance from the clusters's centers similar to that obtained by Lundstrom and Stenholm (1984), but there exist an unexplained feature if one breaks down the data into WN and WC populations. Here the WN 6 and 7 stars, which are more massive than other subtypes, are found at larger distances from the clusters centres as compared to the earlier, presumably less massive, subtypes.

Table I shows the breakdown of the number of WR stars which belong to the Local arm and the Sagittarius arm. It can be seen that the number ratio of WR binaries to WR stars in the Local arm is larger (4/7) than in the Sagittarius arm (2/8). Since the clusters are at the same galactocentric distance, the different number ratios should be attributed to another parameter than metallicity. A

Table 1. Clusters in the Local and Sagittarius Arms having WR stars

Type	Local		Sagittarius	
	N	$N_{bin.}$	N	$N_{bin.}$
WN	3	2	6	1
WC/WO	4	2	2	1

different IMF may be the cause (Conti et al., 1983; Garmany, 1984). While this could perhaps explain the excess in WN stars in the Sagittarius arm (6) over that in the Local Arm (3), it is unclear how it could account for the fraction of binaries.

Zinneker (1982) proposed a hypothesis that the fraction of binary systems may depend on the local strength of the mean interstellar magnetic field. Therefore the difference in the binary frequency in clusters, which are associated with two different galactic arms, but are at rather similar galactocentric distance may be attributed to the difference of magnetic properties in the two arms.

We thank G. Admiranto for his help. B.H. thanks Unesco ROSTSEA and the Leids Kerkhoven-Bosscha Foundation for their financial assistance.

REFERENCE

Bok, B., 1983, H.N.Russel Lecture, Publ.Steward Obs., No. 435.
Breysacher, J., 1980, Astron. Astrophys. Suppl. 43, 203.
Conti, P.S., Garmany, C.D., de Loore, C.W.H., and Vanbeveren, D., 1983, Ap. J. 234, 303.
Lundstrom, I. and Stenholm, B., 1984, Astron. Astrophys. Suppl. Ser. 58, 163.
Maeder, A., 1982, Astron. Astrophys. 105, 149.
Maeder, A., 1987, Astron. Astrophys. 173, 247.
Smith, L., 1988, Ap. J. 327, 128.
Van der Hucht, K.A., Hidayat, B., Admiranto, A.G., Supelli, K.R., and Doom, C., 1988, Astron. Astrophys. 199, 217.
Zinneker, H., 1982, in "Binary and Multiple Stars as Tracers of Stellar Evolution", Z. Kopal and J. Rahe (eds.) Dordrecht, Reidel, P. 115.

PROGRESS OF COMMON ENVELOPE EVOLUTION

R. E. TAAM
Department of Physics and Astronomy
Northwestern University
Evanston, Illinois 60208
U.S.A.

ABSTRACT.   The current understanding of the common envelope binary
phase of evolution is presented.   The results obtained from the
detailed computations of the hydrodynamical evolution of this phase
demonstrate that the deposition of energy by the double core via
frictional processes is sufficiently rapid to drive a mass outflow,
primarily in the equatorial plane of the binary system.   Specifically,
recent calculations suggest that large amounts of mass and angular
momentum can be lost from the binary system in a such a phase.   Since
the time scale for mass loss at the final phase of evolution is much
shorter than the orbital decay time scale of the companion, the
tranformation of binary systems from long orbital periods (> month) to
short orbital periods (< day) is likely.   The energy efficiency factor
for the process is estimated to lie in the range between 0.3 and 0.6.

1.   INTRODUCTION

From the existence of short period binary systems containing a compact
object one can infer that there were prior evolutionary phases where
mass and angular momentum were lost.   Examples of such systems are
cataclysmic variable binaries (with orbital periods, P, ranging from 2
to 14 hours), binary nuclei of planetary nebulae (with periods ranging
from 2.3 to 16 hours), binary X-ray sources (the most extreme example of
which is 4U 1820-30 with an orbital period of 11.08 min.; Stella,
Priedhorsky, and White 1987), and binary radio pulsars (PSR 1913+16, P =
7.7 hours; PSR 0655+64, P = 1.03 days).   The fundamental difficulty in
forming such systems is clear when it is recognized that the progenitors
of the compact components must have had a radius much larger than the
present day orbital separation of these systems.   A possible resolution
to this difficulty involves the relaxation of the assumption of
corotation between the spin and orbital motion (see, for example,
Paczynski 1976).   The lack of corotation results in processes which can
lead to the ejection of substantial amounts of mass and angular momentum
via a common envelope or double core stage.   For an overview of this
evolutionary phase and for references to previous work see the papers by
Bodenheimer and Taam (1986) and Taam (1988).

155

D. McNally (ed.), Highlights of Astronomy, Vol. 8, 155–159.
© 1989 by the IAU.

A binary system can enter into the common envelope stage by a variety of evolutionary paths. For example, consider a binary system with a red giant companion for which the mass transfer process is unstable. Since matter can be transferred from the red giant to its companion on a time scale much shorter than the thermal time scale of the red giant (see Paczynski and Sienkiewicz 1972), the mass accreting companion will likely expand to fill its own Roche lobe after only a small amount of mass ($\sim 0.01 M_\odot$) is accreted. It is likely that the rotation of the giant loses synchronism with the orbital motion because the time scale to achieve solid body rotation is longer than this rapid evolutionary phase. Evolution to the common envelope stage can also result if, at the onset of mass transfer, the red giant is not rotating synchronously with the orbital motion (Counselman 1973). For this case, corotation between spin and orbital motion is lost when the ratio of the spin moment of inertia of the giant is greater than 1/3 the moment of inertia of the binary. This circumstance occurs when the mass ratio of the binary system exceeds about 5 since there is insufficient angular momentum available in the orbit to achieve synchronism. Another evolutionary path to the common envelope stage involves the physical collision of two stars, a possible aftermath of a tidal encounter in a globular cluster stellar system (Verbunt 1987).

2. RESULTS

The detailed hydrodynamical calculations of the common envelope phase have been largely exploratory in nature since the hydrodynamic and thermodynamic processes occur over a wide range of time scales and length scales. Although the complexities of the intrinsic problem hamper rapid progress, a few results can be given based upon general considerations. Specifically, if the entire common envelope is to be ejected before the companion reaches the boundary of the red giant core, then the following conditions must be satisfied (see Taam, Bodenheimer, and Ostriker 1978; Taam 1984): (1) there must be sufficient energy released from the orbit to unbind the envelope, (2) the energy lost from the orbit is directly transferred to the hydrodynamic mode; that is, the time scale for ejection must be shorter than the time scale for the removal of energy by transport mechanisms, and (3) the ejection must be rapid with respect to the orbital decay time scale.

Some recent work describing the mass ejection phase has been reported by Livio and Soker (1988) in the three dimensional approximation and by Taam and Bodenheimer (1989) in the two dimensional approximation. Although the mass ejection process must be described in three dimensions during the early evolutionary stages of the orbital decay (see Livio and Soker 1988), it can be adequately modeled in two dimensions during the later stages (see Bodenheimer and Taam 1984; Taam and Bodenheimer 1989).

The interaction between the star and the envelope of the giant can be described in terms of a Hoyle and Lyttleton (1939) and Bondi and Hoyle (1944) type accretion picture. Within this framework, the kinetic energy of the relative motion between the secondary and common envelope is converted into thermal energy in a shock and a gravitational drag is exerted on the secondary forcing the orbit to decay. Because this energy is deposited at such a rapid rate (greater than 10^7 $L_{\odot}$) the energy is not transported efficiently toward the stellar surface by radiative diffusion or by convection, but, instead, is converted into kinetic energy of motion. However, the mass is not ejected uniformly over a spherical volume, but is concentrated to the equatorial plane over a half angle of about 13 degrees (Taam and Bodenheimer 1989). Such an outflow geometry is favored due to the presence of steeper density gradients in that direction. Although the angular momentum lost from the binary orbit is redistributed in the radial direction by the matter outflow leading to only a slight spin up of the common envelope near the vicinity of the secondary, most of this angular momentum is carried out of the system by the ejected matter. Toward the final phase of the hydrodynamical evolution (P < day) the mass outflow is so severe that the density in the vicinity of the companion has declined sufficiently that the drag luminosity drops and, consequently, the orbital decay time scale increases rapidly. At this point the spiral in time scale is more than an order of magnitude longer than the mass loss time scale and more than 75% of the mass of the common envelope has been ejected. If we can extrapolate the results of Taam and Bodenheimer (1989), it appears likely that the entire envelope will be ejected and that the spiral in process will stop.

The general results that can be gleaned from these calculations can be summarized as follows. Multi-dimensional effects are important in the evolution leading to the preferential ejection of matter along the equatorial plane within a greater half angle for larger energy input rates. In addition, the mass ejection process is nearly adiabatic since the time scale on which energy is deposited into the common envelope is much shorter than the energy transport time scale by either radiation or convection. Thus, most of the orbital energy is converted into the kinetic and potential energy of the outflowing matter. However, the energy is not distributed uniformly over the common envelope, but rather in the equatorial plane with the matter accelerated to greater than the escape speed. Thus, the conversion of orbital energy to mass loss is only moderately efficient (~ 30% to 60%).

3. CONCLUSIONS

Within the framework of the double core hypothesis, it has been found that the ejection of the entire common envelope is very likely. The transformation of long period binary systems to short period systems via this evolutionary stage is, thus, confirmed. However, our studies do not predict the relationship between the final orbital period for which the system emerges from the common envelope phase and the initial period just prior to the spiral in process. It is quite possible that this aspect of the problem will depend upon the detailed response of the nuclear burning shells of the red giant to the degree and efficiency of material circulations induced in the inner regions. For example, if hydrogen rich matter is mixed into the helium burning region and carbon nuclei from the helium rich region is mixed into the hydrogen burning shell, the energy generation rates in the nuclear burning shells may be enhanced to the extent that the nuclear energy release may significantly aid in the ejection of matter during the terminal stage. On the other hand, if the circulations do not mix combustible fuels into the burning regions, then the extensive mass loss will eventually cause the nuclear burning to be extinguished. In either case, once the mass above the white dwarf core is reduced below some critical value ($< 0.001 - 0.01$ $M_\odot$ the radius of the primary remnant will shrink (in order to maintain hydrostatic equilibrium) and the common envelope phase will terminate.

A second major area of study requiring attention involves the response of the secondary to the conditions within the red giant envelope. Webbink (1988) has inferred from the observations of the cool components of planetary nebulae binary nuclei that the secondary may emerge from the common envelope evolution relatively undisturbed. This suggests that the secondary component accretes or loses very little mass. Along the same lines, the investigation by Kato (1982) indicates that it is likely that there are phases in the evolution where either accretion or evaporation are possible. Clearly, future work in these areas will help clarify our understanding of this important phase of binary evolution.

This research has been supported in part by NSF under grant AST-8608291.

REFERENCES

Bodenheimer, P., and Taam, R. E. (1984), 'Double Core Evolution II. Two Dimensional Hydrodynamic Effects', Ap. J., 280, 771-779.
Bodenheimer, P., and Taam, R. E. (1986), 'Common Envelope Evolution', in J. Truemper, W. H. G. Lewin, and W. Brinkmann (eds.), The Evolution of Galactic X-Ray Binaries, Reidel, Dordrecht, p. 13-24.
Bondi, H., and Hoyle, F. (1944), 'On the Mechanism of Accretion By Stars', M.N.R.A.S., 104, 273-282.
Counselman, C. C. (1973), 'Outcomes of Tidal Evolution', Ap. J., 180, 307-314.
Hoyle, F., and Lyttleton, R. A. (1930, 'The Effect of Interstellar Matter on Climatic Variation', Proc. Cambridge Phil. Soc., 35, 405-415.
Kato, M. (1982), 'Mass Loss and Mass Gain of Stars Immersed in a Supermassive Star', P.A.S.J., 34, 173-182.
Livio, M., and Soker, N. (1988), 'The Common Envelope Phase in the Evolution of Binary Stars', Ap. J., 329, 764-779.
Paczynski, B. (1976), 'Common Envelope Binaries', in P. Eggleton, S. Mitton, and J. Whelan (eds.), IAU Symp. No. 73, The Structure and Evolution of Close Binary Systems, Reidel, Dordrecht, pp. 75-80.
Paczynski, B., and Sienkiewicz, R. (1972), 'Evolution of Close Binaries VIII. Mass Exchange on the Dynamical Time Scale', Acta Astr., 22, 73-91.
Stella, L., Priedhorsky, W., and White, N. E. (1987), 'The Discovery of a 665 Second Orbital Period From the X-Ray Source 4U 1820-30 in the Globular Cluster NGC 6624', Ap. J. Letters, 312, L17-L21.
Taam, R. E. (1984), 'Evolution of Binary Systems Into Transient Sources', in S. E. Woosley (ed.), High Energy Transients in Astrophysics, AIP Press, New York, p. 1-10.

Taam, R. E. (1988), 'Aspects of Common Envelope Evolution', in K. C. Leung (ed.), Critical Observations vs. Physical Models for Close Binary Systems, Gordon and Breach, New York, pp. 365-370.

Taam, R. E., and Bodenheimer, P. (1989), 'Double Core Evolution III. The Evolution of a 5 Solar Mass Red Giant With a One Solar Mass Companion', Ap. J., in press.

Taam, R. E., Bodenheimer, P., and Ostriker, J. P. (1978), 'Double Core Evolution I. A 16 Solar Mass Star With a 1 Solar Mass Neutron Star Companion', Ap. J., 222, 269-280.

Verbunt, F. (1987), 'Formation of Ultra-Short Period Binaries in Globular Clusters', Ap. J. Letters, 312, L23-L25.

Webbink, R. F. (1988), 'Late Stages of Close Binary Systems--Clues to Common Envelope Evolution', in K. C. Leung (ed.), Critical Observations vs. Physical Models for Close Binary Systems, Gordon and Breach, New York, pp. 403-446.

MILLISECOND PULSARS

J.H. KROLIK
Department of Physics and Astronomy
Johns Hopkins University
Baltimore MD 21218
USA

ABSTRACT. Millisecond pulsars are intrinsically interesting because they illustrate some of the most extreme physical conditions to be found anywhere in the Universe, and because their evolution exhibits several stages of great drama. It had been widely believed for several years that spin-up of an old neutron star by accretion from a close stellar companion explained their fast rotation, but the absence of companions in several cases cast doubt on that picture. This spring a millisecond pulsar in a close binary was discovered in which the companion appears to be evaporating, thus reconciling the existence of lone millisecond pulsars with the standard picture. Ongoing observations of this new system, and complementary calculations, promise to answer many of the questions remaining about this dramatic phase in stellar evolution.

1. Why Study Millisecond Pulsars?

At first blush, it might seem odd to give such close attention to a sub-class of a sub-class. Radio pulsars are a small fraction of all neutron stars, and the known population of millisecond pulsars is a tiny minority ($\sim 10^{-2}$) of radio pulsars. Yet, once described, the attractive properties of millisecond pulsars become immediately clear, for they exhibit some of the most extreme physical conditions, and one of the most dramatic evolutionary histories, of any class of astronomical objects.

To begin, they are among the most stable clocks anywhere. Whereas normal pulsars change period at a long-term average rate $\dot{P} \sim 10^{-15}$, for millisecond pulsars the typical $\dot{P} \sim 10^{-19}$ (Backer 1987). Such extreme long-term stability pushes even the best laboratory clocks to their limits. On shorter timescales, i.e., days, they are also extremely regular: typical timing excursions over a single day are generally smaller than 500 ns.

At the same time, their extremely fast rotation rate pushes them into very exotic physical conditions. Like other neutron stars, their gravitational potential is mildly relativistic: $GM/(rc^2) \sim 10^{-1}$. However, in contrast to ordinary neutron stars, millisecond pulsars rotate near break-up: $\Omega/\Omega_{break-up} \simeq 0.1 - 0.6$, rather than $\sim 10^{-3}$. In fact, they rotate so rapidly that they exceed the stability limit for the axisymmetric Maclaurin spheroids, and become triaxial. The resulting gravitational radiation enforces a lower bound on the period of $1 - 1.3$ ms.

Finally, in order to spin down at all, they must couple electromagnetically to the outside world, and so possess a fairly strong magnetic field: $\dot{P} \sim 10^{-19}$ implies $B \sim 10^9$ G for a moment of inertia $\sim 10^{45}$ gm cm^2. Because, as I will discuss later, the age of these neutron stars is

161

D. McNally (ed.), Highlights of Astronomy, Vol. 8, 161–165.
© *1989 by the IAU.*

probably $\gtrsim 10^9$ yr, in contrast to the typical age of ordinary radio pulsars $\sim 10^6$ yr, mechanisms as yet unknown must exist to preserve the field against Ohmic decay.

2. Best Bet Evolutionary History

There is a general consensus that the most likely progenitor systems for millisecond pulsars are low-mass X-ray binaries (Alpar, *et al.* 1982). A number of arguments support this conclusion, but perhaps the strongest is that one expects the neutron star in these X-ray binaries to ultimately be spun up to millisecond periods. When material is accreted from a disk onto a magnetized collapsed star, it brings along the angular momentum it had in its Keplerian orbit just outside the magnetospheric boundary. In principle, the resulting change in neutron star spin could have either sign. Balance is achieved only when:

$$P \simeq 2 L_{38}^{-3/7} B_9^{6/7} \mathrm{ms},$$

where L_{38}, the luminosity in units of 10^{38} erg s^{-1}, is typically order unity in low-mass X-ray binaries. If the spin rate initially is slower, continued accretion decreases P until enough mass has been transferred to achieve equilibrium. Because the specific angular momentum at the magnetospheric boundary is roughly twice the specific angular momentum of a neutron star rotating at break-up, the star must accrete only $\sim 0.05\times$ its initial mass in order to achieve a millisecond period. At the accretion rate that would produce 10^{38} erg s^{-1}, that takes only $\sim 10^7$ yr.

Other arguments are consistent with this story. The relatively weak magnetic field found in the millisecond pulsars is consistent with the observed absence of pulses from low-mass X-ray binaries. Likewise, the comparatively large number of millisecond pulsars in globular clusters (about a third of all those known: Kulkarni 1988) comports well with the relatively large number of low-mass X-ray binaries in globulars.

However, prior to this spring several large gaps remained in this picture. Most importantly, there was no plausible explanation for what happened to the companion in the solitary millisecond pulsars. In addition, if the accretion rate diminished gradually, then the same mechanics which spun up the neutron star during the period of high accretion rate would spin it down as the accretion rate fell.

A possible answer to these questions was provided in a pair of papers written last winter (Ruderman, Shaham, and Tavani 1988; Ruderman, Shaham, Tavani, and Eichler 1988). They suggested that the increasing spin rate of the accreting neutron star in a low-mass X-ray binary could, through a variety of possible mechanisms, produce a substantial luminosity in high energy photons. Those energetic photons striking the companion would then heat its surface above the escape temperature, ultimately evaporating the entire companion in a relatively short time. Thus, accretion would stop quickly, avoiding the spin-down problem, and the companion star would be eliminated, explaining the observed solitary millisecond pulsars within the context of the low-mass X-ray binary model.

3. PSR 1957+20

3. 1. Observations and Immediate Inferences

Remarkably enough, within a few months of the submission of these papers, an example of just this sort of process was discovered (Fruchter, Stinebring, and Taylor 1988: FST). PSR 1957+20 has the second-fastest known period—1.6 ms—but more importantly, it is in a close binary with a period of 9.2 hr and a projected semi-major axis (of the neuton star's orbit) of 2.7×10^9 cm. The orbit is very nearly circular, for the eccentricity is no more than 10^{-4}

(Fruchter 1988). From these numbers (and an assumed neutron star mass of $1.4 M_\odot$), we may infer that:

$$M_c \simeq 0.024/\sin i\, M_\odot \qquad a_c \simeq 1.6 \times 10^{11} \text{ cm} \qquad R_L \simeq 2 \times 10^{10} \text{ cm}$$

where M_c is the companion mass, a_c is its semimajor axis, and R_L is the radius of its Roche lobe, assuming synchronous rotation (as seems quite likely, considering the very small eccentricity). It is particularly instructive to compare R_L to the radius of a cold degenerate H-rich star of this mass: $R_d \simeq 8 \times 10^9 \sin^{1/3} i$ cm. Other than the very small companion mass, these numbers taken by themselves are not so remarkable.

What makes this pulsar truly special is that it *eclipses*. At an observing frequency of 430 MHz, the eclipse lasts for 0.83 hr, almost a tenth of the orbital period. Thus, the radius of the optically thick (to radio photons) region of the companion is $\simeq 5 \times 10^{10}$ cm, or $\simeq 2.5 \times R_L$!

We are immediately forced to the conclusion that the companion is losing mass at a substantial rate. Matter outside the companion's Roche lobe cannot stay near it, and yet the densest part of this material orbits the neutron star in association with the companion because the eclipse is quite well-defined. Therefore, the eclipsing material must be continually replenished by mass-loss from the companion. From the magnitude of fluctuations in pulse time delays just outside eclipse, FST estimated that the electron density $n_e > 10^6$ cm^{-3} (interestingly, these fluctuations are much stronger during egress than ingress); if the plasma moves with a velocity comparable to the orbital velocity (this is certainly a lower limit), then the mass loss rate $\dot{M}_c > 10^{-14} M_\odot$ yr^{-1}. Although this is an interesting rate, it is far too small to affect even such a low mass companion in a Hubble time. It is likely, however (see §3.2) that $\dot{M}_c$ is considerably greater than this lower limit.

If this mass loss is being driven by heating of the companion's surface, the visible light from the system should be modulated at the orbital period (again assuming synchronism). In fact, this is just what is seen (Fruchter, Gunn, Lauer, and Dressler 1988: FGLD; Kulkarni, Djorgovski, and Fruchter 1988; Djorgovski and Evans 1988; van Paradijs, *et al.* 1988). At maximum light, $m_V = 20.3$ mag, while minimum light is at least 0.85 mag, and possibly as much as 3 mag, fainter. Moreover, maximum and minimum occur at the correct phases relative to the pulsar eclipse, giving an absolute confirmation to the optical identification.

From the optical data, even more may be learned about the system (Djorgovski and Evans 1988; FGLD; van Paradijs, *et al.* 1988). Although the extinction is a bit uncertain, it may be bounded between 1 and 2 mag, so that the colors at maximum light correspond to a color temperature of $\simeq 5800$ K. Combining the extinction correction and the distance estimated from the dispersion measure ($\simeq 0.9$ kpc), the radius of the optical photosphere is found to be between 4.9×10^9 cm and 1.4×10^{10} cm, with the most likely value of 1.1×10^{10} cm. This number is in very good agreement with the *a priori* supposition that the companion should be just slightly larger than a cold degenerate star of that mass.

Furthermore, if the color temperature is a good approximation to the effective temperature, and we have an estimate of the stellar radius, we know its absolute luminosity. It is $\simeq 5 \times 10^{31}$ erg s^{-1}. As of yet there is only an upper limit on the pulsar spin-down rate: $\dot{P} < 3 \times 10^{-20}$ (Fruchter 1988), but we may use it, along with theoretical predictions of the neutron star's moment of inertia to estimate what fraction of the strong dipole wave radiated by the pulsar directed at the companion (assuming isotropic radiation) is converted to heat at the surface of the companion. Uncomfortably for theorists, $L_c/L_{pulsar} > 0.1$, using the best estimate of the stellar radius and the upper limit on $\dot{P}$ (FGLD; van Paradijs, *et al.* 1988). If the measured $\dot{P}$ proves to be much smaller than 3×10^{-20}, we may be driven to consider nuclear matter equations of state which allow larger moments of inertia than had hitherto been considered.

3. 2. Theory and Future Work

It will come as no surprise when I remark that there are already almost twice as many theoretical as observational papers on this object (Cheng 1988; Eichler and Levinson 1988;

Kluźniak, *et al.* 1988; Michel 1988; Phinney, *et al.* 1988; Rasio, *et al.* 1988; van den Heuvel and van Paradijs 1988), with more certain to appear in the near future. The main problems under consideration so far have been: What is the density of the eclipsing matter outside the companion's Roche lobe? And, how is the power in the low frequency wave from the pulsar transformed so that it can penetrate to the surface of the companion and drive the mass loss?

The answer to both of these questions remains controversial. A plausible answer to the former is given by Phinney, *et al.* (1988). They argue that the strong wave from the pulsar can be treated as a relativistic fluid so that the boundary between the radio-opaque matter escaping the companion and the transparent relativistic fluid can be determined simply by momentum balance. Because we know the answer (from the eclipse geometry), it is possible to work backwards to find the momentum flux in mass loss from the companion. Given a typical velocity of order the orbital speed, one then finds that the density is six orders of magnitude greater than in the region producing the timing fluctuations just outside eclipse. The corresponding $\dot{M}_c$ is then $\sim 10^{-8} M_\odot$ yr^{-1}, which will destroy the companion in a mere 2×10^6 yr.

Not all models suppose that the low frequency wave and the stellar mass loss interact hydrodynamically (*e.g.*, Rasio, *et al.* 1988 and Michel 1988). Fortunately, there is a clear observational test: the hydrodynamic picture predicts a contact discontinuity between the relativistic and ordinary fluids which defines the boundary between transparent and radio-opaque regions, while the others predict a more gradual transition. If the eclipse duration at other radio frequencies is the same as at 430 MHz, then the density gradient is sharp; if the duration decreases at higher frequencies, then the gradient is more gradual. VLA measurements will soon provide an answer.

The situation is even vaguer with respect to how the low frequency power penetrates to the companion's surface. If the mass loss takes place as a plasma, then the low frequency wave cannot possibly reach the companion. Somehow its power must be "up-converted" into higher frequency photons with greater penetrating power. Various authors have suggested mechanisms for producing photons ranging from the ionizing ultraviolet to MeV γ-rays, but all the mechanisms are little more than guesses. It is possible, if the photons are more energetic than a few keV, and if a large enough fraction is directed away from the companion, that they might be directly observable. Whether or not that is the case, more detailed calculations of the thermal and hydrodynamic conditions in the base of the stellar wind can also constrain the energy of the penetrating photons because heat deposited deep inside the stellar atmosphere will be wasted in thermal radiation, while if the photons are absorbed at too high an altitude the wind's sonic point will occur in a region of such low density that the mass loss rate will be too small (Cheng 1988; Eichler and Levinson 1988).

Beyond these questions, work is also likely to soon begin on simulating the dynamics of the wind from the companion in order to understand the duration and asymmetry of the eclipse, as well as on modeling the detailed shape of the optical light curve.

In a final *coup de théâtre*, it is also possible to see the effects of the low frequency wave in the interstellar medium far outside the binary. Kulkarni and Hester (1988) have obtained an Hα image of the pulsar region which shows a nebula with a distinct cometary shape several arcseconds across. They interpret the line emission as due to a bow shock supported by the pressure of the relativistic wind driven by the low frequency wave. To order of magnitude, the separation between the head of the cometary nebula and the pulsar agrees with the prediction made assuming a spin-down luminosity corresponding to $\dot{P} \sim 3 \times 10^{-20}$, an external density of ~ 1 cm^{-3}, and a pulsar space velocity ~ 100 km s^{-1}.

References

Alpar, M.A., Cheng, A.F., Ruderman, M.A., and Shaham, J., 1982, *Nature* 300, 728.

Cheng, A., 1988, submitted to *Ap. J.*

Djorgovski, S. and Evans, C., 1988, submitted to *Ap. J. Lett.*

Eichler, D. and Levinson, A., 1988, submitted to *Ap. J. Lett.*

Fruchter, A.S., 1988, private communication

Fruchter, A.S., Gunn, J.E., Lauer, T.R., and Dressler, A., 1988, *Nature* 334, 686.

Fruchter, A.S., Stinebring, D.R., and Taylor, J.H., 1988, *Nature* 333, 237.

Kluźniak, W., Ruderman, M., Shaham, J., and Tavani, M., 1988, *Nature* 334, 225.

Kulkarni, S.R., 1988, in *Physics of Neutron Stars and Black Holes*, ed. Y. Tanaka (Institute of Space and Astronautical Science: Tokyo)

Kulkarni, S.R., Djorgovski, S., and Fruchter, A.S., 1988, *Nature* 334, 504.

Kulkarni, S.R. and Hester, J.J., 1988, submitted to *Nature*

Michel, F.C., 1988, submitted to *Nature*

Phinney, E.S., Evans, C.R., Blandford, R.D., and Kulkarni, S.R., 1988, *Nature* 333, 832.

Rasio, F.A., Shapiro, S.L., and Teukolsky, S.A., 1988, submitted to *Nature*

Ruderman, M., Shaham, J., and Tavani, M., 1988, *Ap. J.* in press.

Ruderman, M., Shaham, J., Tavani, M., and Eichler, D., 1988, *Ap. J.* in press.

van den Heuvel, E.P.J. and van Paradijs, J., 1988, *Nature* 334, 227.

van Paradijs, J., Allington-Smith, J., Callanan, P., Charles, P.A., Hassall, B.J.M., Machin, G., Mason, K.O., Naylor, T., and Smale, A.P., 1988, *Nature* 334, 684.

EVOLUTION OF CATACLYSMIC BINARIES

BOHDAN PACZYŃSKI
Department of Astrophysics
124 Peyton Hall
Princeton University
Princeton, NJ 08544, USA

ABSTRACT. The minimum period of hydrogen rich cataclysmic binaries at about 80 minutes is due to evolution driven by gravitational radiation. The nature of the gap between 2 and 3 hour periods is less clear. Magnetic braking may be responsible, but this would be in contradiction with the hibernating nova scenario. It is interesting that contact binaries of W UMa type are subject to little or no magnetic braking.

The origin and evolution of cataclysmic binaries has been revied recently by King (1988). Here I shall discuss a few problems that are either new or controversial.

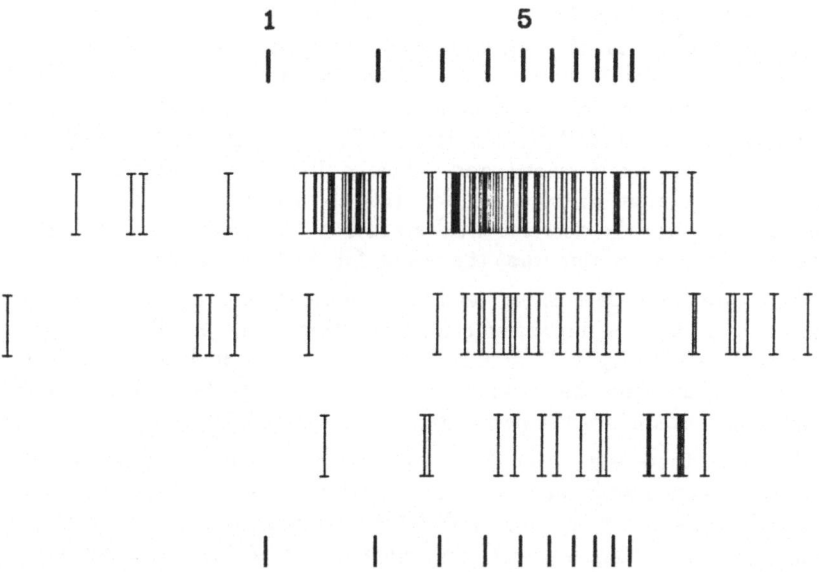

Figure 1. The distribution of binary periods is shown in a form of three spectra: for cataclysmic variables (upper), for low-mass X-ray binaries (middle), and for related systems (lower) following Ritter (1987). The comparison spectra (top and bottom) are labeled with the binary period (in hours). Four cataclysmic variables with periods below 80 minutes are hydrogen deficient.

167

D. McNally (ed.), Highlights of Astronomy, Vol. 8, 167–171.

One of the important observations that theory has to explain is the distribution of binary periods. It is presented in Figure 1 as a "spectrum" of periods, each system shown as a vertical line. The cut-off in the distribution of periods of cataclysmic binaries near 80 minutes is understood in terms of evolution driven by gravitational radiation (Paczyński 1981, Paczyński and Sienkiewicz 1981, Rappaport, Joss and Webbink 1982). The origin of the gap between 2 and 3 hours is less clear. The most popular explanation is binary evolution driven by magnetic braking (Eggleton 1976, Verbunt and Zwaan 1981, Spruit and Ritter 1983, Verbunt 1984, and many others). In order to have a gap between 2 and 3 hours, systems above the gap must be transferring mass at a rate between 2×10^{-9} and $10^{-8} M_\odot$ yr^{-1} (Paczyński and Sienkiewicz 1983, Verbunt 1984, Hameury et al. 1987), more or less as indicated by the observations (Patterson 1984).

The most serious problem with this high mass transfer rate is that accretion of hydrogen rich matter onto a white dwarf primary leads to mild shell flashes, not to nova explosions (Kutter and Sparks 1980, Prialnik et al. 1982, Fujimoto 1982). A very attractive solution to this problem was proposed by Shara and his associates (Shara et al. 1986, Prialnik and Shara 1986, Shara 1988). They find that very old novae, like WY Sge (Nova 1783) and CK Vul (Nova 1670) are 10 to 100 times fainter than the more recent "old novae". They propose that between the eruptions novae are "hibernating", with mass transfer rate and luminosity reduced by orders of magnitude. The proposed space density of such objects is about $10^{-4} pc^{-3}$, i.e. one out 1000 nearby G, K, or M dwarfs should be a hibernating nova. The time averaged mass transfer rate is reduced to about $10^{-10} M_\odot$ yr^{-1}, and hydrogen ignition on the accreting white dwarf is explosive.

It is well established that mass transfer in all kinds of semidetached binaries changes on various time scales. Among cataclysmic variables this is well known for polars, i.e. AM Her type objects (Liebert and Stockman 1985 and references therein). The observed spin-up times for DQ Her type systems are 2×10^6 years (DQ Her), and 3×10^5 years (H2252-035) (Lamb and Patterson 1983). This is much shorter than the expected life time of any cataclysmic binary, and indicates that the currently observed mass transfer rates are higher than average. Notice, that if there are large fluctuations in the mass transfer rate then the accreted nuclear fuel is likely to ignite during the high mass transfer episode, no matter what the reason for the fluctuations is.

There are very interesting and far reaching consequences of the hibernation hypothesis, at least in its extreme form mentioned above. The average mass transfer rate of only $10^{-10} M_\odot$ yr^{-1} requires angular momentum loss at the level provided by gravitational radiation alone. There is no need for magnetic braking. In this case, a new explanation is needed for the period gap between 2 and 3 hours.

Of course, there was never direct evidence that magnetic braking actually operates efficiently in cataclysmic binaries. One should notice that magnetic braking is much less efficient in the contact systems of W UMa type than it is among their predecessors, as the contact systems are so much more numerous than their detached progenitors in the field (Ruciński 1985, Ruciński and Seaquist 1988), as well as in NGC 188 (Kałużny and Shara 1987, Kałużny 1988). The activity related to surface magnetic fields is smaller in W UMa systems than expected for stars with such short rotation periods (Rucinski 1985). Therefore, it is not obvious why the magnetic braking scenario, based on single or detached systems, should work for semidetached cataclysmic binaries any better than for contact W UMa stars. The origin of the gap remains unclear.

The issue of what fraction of their lifetimes novae are hibernating is likely to be resolved observationally, by systematic search for short period binaries among nearby G, K, and M dwarfs. Such searches can be conducted in at least two ways: with photoelectric radial velocity measurements using technique developed by Griffin (1988 and references therein) and also adopted by Mayor (1980, CORAVEL), and with accurate CCD photometry. The expected radial velocity amplitudes are in the range $100 - 300\, km\, s^{-1}$, while ellipsoidal light variations of the secondary components filling up their Roche lobes are in the range of $0.1 - 0.2\, mag$. Binary periods from 1 to 10 hours are expected.

Patterson (1984) noticed that the death rate of novae exceeds by an order of magnitude the birth rate of short period cataclysmic binaries. This discrepancy is even larger if most novae hibernate. If the discrepancy is real then there must be a process that destroys cataclysmic binaries. One may envision a scenario where any given cataclysmic binary transfers only a fraction of the secondary's mass during its life time. In this picture cataclysmic variables would be born at all binary periods, and would spend most of their rather short life close to that period. The gap between 2 and 3 hours may indicate that systems in that range are most prone to self-destruction. Run-away angular momentum loss during a very long and slow nova eruption may be responsible (MacDonald 1986).

It is not likely that theory alone can answer the question of whether a typical cataclysmic variable evolves over a large or small range of binary periods. In particular, do binaries evolve from above the gap to below the gap? It is most likely that large surveys of nearby stars will determine the distribution of hibernating as well as detached binaries, and will help to resolve this issue.

Apart from the gap in their period distribution there are two unusual groups of cataclysmic binaries. There are 7 systems with periods between 0.0788 and 0.0798 days, and 6 systems in the range 0.1396 - 0.1406 days. 40 binaries are known to have periods between 80 minutes and 2 hours, i.e. between 0.0556 and 0.0833 days. Assuming uniform distribution the a priori probability of finding 7 systems with periods within 0.001 days anywhere between 80 minutes and 2 hours is 0.02. There are 34 binaries known with periods between 3 and 4 hours, i.e. between 0.125 and 0.1667 days. For a uniform distribution the probability of finding 6 systems with periods within 0.001 days somewhere between 3 and 4 hours is 0.02. These two groups may be just due to chance.

Hameury et al. (1988a, 1988b) proposed that the group of 6 AM Her type binaries with periods near 0.079 days may be explained within a magnetic braking evolutionary scenario. In order to explain the narrowness of the period range for this group they had to assume that the rate of angular momentum loss was identical for all systems, to within 1% or so. Considering the complicated nature of stellar magnetic fields and stellar winds, such standardization would be most extraordinary and very important. It will be interesting to see if the grouping persists when the number of known short period cataclysmic binaries has doubled.

The nature of the 2 to 3 hour gap is not clear for cataclysmic binaries, but its existence is beyond any doubt. The low mass X-ray binaries seem to have a cut-off at about 3 hours, but its reality is not so obvious (cf. Figure 1). In any case, there are hydrogen rich systems with periods below 4 hours (UY Vol, Schmidtke and Cowley 1987, GR Mus, Motch et al. 1987), in spite of the theoretical prediction that none should exist below 5 hours (Phinney et al. 1988). One system, Cyg X-3, increases its binary period of

4.8 hours on a time scale of only 5×10^5 years, making it a likely candidate for a progenitor of the recently discovered eclipsing millisecond pulsar 1957+20 (Fruchter et al. 1988, Molnar 1988). It is not clear what process drives this rapid evolution or whether it applies to other low mass X-ray binaries. The existence of "on" and "off" states and of various types of transient and "X-ray nova" phenomena makes the picture even more confusing. It is impossible at this time to say anything definite about time-averaged mass transfer rates or the variety of evolutionary scenarios for low mass X-ray binaries.

There is another very confusing issue related to cataclysmic binaries and to progenitors of Type I supernovae. In one recent search for short period double white dwarfs, none was found in a sample of 44 (Robinson and Shafter 1987), while other groups found some short period pairs of white dwarfs (Saffer, Liebert and Olszewski 1988, Bragaglia et al. 1988). It is not clear what fraction of all white dwarfs is double or whether they are good candidates for the progenitors of Type I supernovae (Tutukov and Yungelson 1979, Webbink 1979, Iben and Tutukov 1984, Paczyński 1985). If future searches establish that the fraction of double white dwarfs is small, the most likely explanation will be a shortage of progenitors for such systems, i.e. main sequence binaries with periods of few years and mass ratio close to one. A survey of about 80 late B-type binaries by Wolff (1978) has revealed only one such system. However, the components of long period binaries are likely to be rapid rotators, so it may be very difficult to discover low amplitude radial velocity variations. Considering all the uncertainties, the deficiency of short period double white dwarfs is not alarming for the theory of evolution of close binaries, but it may be uncomfortable for those who seek the progenitors of Type I supernovae.

It is a great pleasure to acknowledge many discussions with Dr. Virginia Trimble. This project was supported in part by the NSF Grant AST87-18432.

References

Bragaglia, A., Greggio, L., Renzini, A., and D'Odorico, S. 1988, *The Messenger, No. 52*, p. 35.

Eggleton, P. P. 1976, *IAU Symp.* **73**, p. 209, Eds: P. P. Eggleton, S. Mitton, and J. A. J. Whelan, (D. Reidel Publ. Co.).

Fruchter, A. S., Stinebring, D. R., and Taylor, J. H. 1988, *Nature,* **333**, 237.

Fujimoto, M. Y. 1982, *Ap. J.,* **257**, 767.

Griffin, R. F. 1988, *The Observatory,* **108**, 90.

Hameury, J. M., King, A. R., Lasota, J. P., and Ritter, H. 1987, *Ap. J.,* **316**, 275.

Hameury, J. M., King, A. R., Lasota, J. P., and Ritter, H. 1988a, *Mon. Not. R. astr. Soc.* **231**, 535.

Hameury, J. M., King, A. R., and Lasota, J. P. 1988b, *Astron. Ap.,* **195** , L12.

Iben, I. Jr., and Tutukov, A. V. 1984, *Ap. J. Suppl.,* **54**, 535.

Kałużny, J. 1988, private communication.

Kałużny, J., and Shara, M. M. 1987, *Ap. J.,* **314**, 585.

King, A. R. 1988, *Quarterly J. R. A. S.,* **29**, 1.

Kutter, G. S., and Sparks, W. M. 1980, *Ap. J.,* **239**, 988.

Lamb, D. Q., and Patterson, J. 1983, *IAU Coll. No 72*, p. 229, Eds: M. Livio and G. Shaviv, (D. Reidel Publ. Co.).

Liebert, J., and Stockman, H. S. 1985, in *Cataclysmic Variables and Low-Mass X-Ray Binaries*, p. 151, Eds: D. Q. Lamb and J. Patterson, (D. Reidel Publ. Co.).

MacDonald, J. 1986, *Ap. J.*, **305**, 251.

Mayor, M. 1980, *Astron. Ap.*, **87**, L1.

Molnar, L. A. 1988, *Ap. J. (Letters)*, **331**, L25.

Motch, C., Pedersen, H., Beuermann, K., Pakull, M. W., and Courvoisier, T. J.-L. 1988, *Ap. J.*, **313**, 792.

Paczyński, B. 1981, *Acta Astron.*, **31** 1.

Paczyński, B., and Sienkiewicz, R. 1981, *Ap. J. (Letters)*, **248**, L27.

Paczyński, B., and Sienkiewicz, R. 1983, *Ap. J.*, **268**, 825.

Paczyński, B. 1985, in *Cataclysmic Variables and Low-Mass X-Ray Binaries* , p. 1, Eds: D. Q. Lamb and J. Patterson, (D. Reidel Publ. Co.).

Patterson, J. 1984, *Ap. J. Suppl.*, **54**, 443.

Phinney, E. S., Evans, C. R., Blandford, R. D., and Kulkarni, S. R. 1988, *Nature,* **333**, 832.

Prialnik, D., Livio, M., Shaviv, G., and Kovetz, A. 1982, *Ap. J.*, **257** , 312.

Prialnik, D., and Shara, M. M. 1986, *Ap. J.*, **311**, 172.

Rappaport, S., Joss, P. C., and Webbink, R. F. 1982, *Ap. J.*, **254**, 616.

Ritter, H. 1987, *Astron. Ap. Suppl.*, **70**. 335.

Robinson, E. L., and Shafter, A. W. 1987, *Ap. J.*, **322**, 296.

Ruciński, S. M. 1985, in *Interacting binary stars*, p. 113, Eds: J. E. Pringle and R. A. Wade, (Cambridge University Press).

Ruciński, S. M., and Seaquist, E. R. 1988, *A. J.*, **95**, 1837.

Saffer, R. A., Liebert, J., and Olszewski, W. 1988, *Ap. J.*, **334**, ...

Schmidtke, P. C., and Cowley, A. P. 1987, *Astron. J.*, **93**, 374.

Shara, M. M. 1988, 20th General Assembly, IAU, Joint Discussion II, this volume.

Shara, M. M., Livio, M., Moffat, A. F. J., and Orio, M. 1986, *Ap. J.*, **311**, 163.

Spruit, H. C., and Ritter, H. 1983, *Astron. Ap.*, **124**, 267.

Tutukov, A. V., and Yungelson, L. R. 1979, *Acta Astron.*, **29**, 665.

Verbunt, F. 1984, *Mon. Not. R. astr. Soc.*, **209**, 227.

Verbunt, F., and Zwaan, C. 1981, *Astron. Ap.*, **100**, L7.

Webbink, R. E. 1979, in *White Dwarfs and Variable Degenerate Stars*, Ed.: H. M. Van Horn and V. Weidemann (New York: University of Rochester Press), p. 426.

Wolff, S. C. 1978, *Ap. J.*, **222**, 559.

THE AM Her PERIOD SPIKE

J.P. LASOTA
Département d'Astrophysique Relativiste et Cosmologie
Observatoire de Paris, Section de Meudon
92195 Meudon Principal Cedex, France

ABSTRACT. The AM Her period spike is due to the resumption of mass transfer after evolving through the period gap. This implies that (i) at the spike period all secondary stars are the same (ii) most AM Her white dwarfs have a unique mass $\simeq 0.6 - 0.7 M_\odot$ (iii) nova explosions on magnetic white dwarfs remove exactly the mass accreted between outbursts (iv) the newly discovered AM Her system at 127min has a mass $\simeq 1.3 - 1.4 M_\odot$ (v) the maximum magnetic field for most AM Her's is $\leq 4 \times 10^7$ G.

There is a large, highly significant (Morris *et al.* 1987) accumulation (6 out of 15) of AM Her systems in the period interval 113.5 – 114.8 minutes. Recently Hameury *et al.* (1988a) proposed an explanation for this in terms of the resumption of mass transfer after these systems have crossed the period gap. When accretion resumes at a period of about 2 hr the fully convective secondary star undergoes a short (Kelvin-Helmholtz time) episode of adiabatic expansion, until it is sufficiently far out of thermal equilibrium that it can contract under the influence of mass loss. The orbital period first increases by 2 to 3 min, reaches a maximum, and decreases. This enhances the discovery probability at this period: (i) the mass transfer rate $-\dot{M}_2$ is about twice the value for a star close to the main sequence contracting in response to mass loss, and (ii) the expansion and subsequent contraction of the secondary lengthens the time spent near the initial period. To give the observed period "spike" requires most systems to have similar parameters when mass transfer resumes at the lower edge of the period gap: otherwise the initial value of P_0 would not be the same and the spike would be smeared out. The secondary stars are naturally similar if the gap is *caused* (Rappaport *et al.* 1983, Spruit & Ritter 1983) by them becoming fully convective. Thus at the spike, $M_2 \simeq 0.2 \ M_\odot$. Further, the period spike implies constraints on the primary masses which have been examined by Hameury *et al.* (1988b). Higher masses cause a narrower period gap and hence a higher value P_0 of the initial period below the gap. The value of the primary mass for systems at the period spike depends on the fate of the matter accreted during the secular evolution. Hameury *et al.* (1988c) show that for magnetic systems nova outbursts remove (almost) exactly the mass accreted between explosions. The white dwarf masses therefore stay constant. Using the fact that the newly discovered AM Her system EXO033319-2554.2 has a period of 127min they conclude that most AM Her systems have a *unique* mass $\simeq 0.6 - 0.7 M_\odot$, while EXO033319-2554.2 has a significantly larger white dwarf mass

173

D. McNally (ed.), Highlights of Astronomy, Vol. 8, 173–174.
© *1989 by the IAU.*

($\simeq 1.2 - 1.3 M_\odot$). This prediction was subsequently supported by radial–velocity measurements (Beuermann *et al.* 1988).

Most AM Her binaries have white dwarfs magnetic fields (Schmidt *et al* 1986) $B_1 \leq 3 \times 10^7$ G, whereas isolated white dwarfs can have fields up to at least 5×10^8 G (Schmidt 1987). Since most properties of magnetic binaries depend on the magnetic *moment*, the upper limit on magnetic *field* has been difficult to understand. However, if one neglects the small class of massive systems (like EXO033319-2554.2) we may regard AM Her systems as having effectively a unique white dwarf mass. With R_1 essentially fixed at $\simeq 8 \times 10^8$ cm, the limit on B becomes a limit on moments μ_1 through $\mu_1 = 5 \times 10^{33} B_7$ G cm^3 where B_7 is B_1 in units of 10^7 G. Hameury *et al.* (1988d) have shown that, moments $\mu_1 \geq 2 \times 10^{34}$ G cm^3, (i.e. fields stronger than $B_7 \simeq 4$) imply catastrophic angular momentum losses $\dot{J}$ via magnetic braking involving the *white dwarf* magnetic field (Schmidt *et al.*1986). These drive the secular evolution so rapidly that the systems enter a prolonged period gap at $P \sim 5$ hr, remaining as detached systems for a time of order the Hubble time. Some of these systems may in fact be observed as apparently isolated, rapidly rotating and highly magnetic white dwarfs (Schmidt 1987), or, for very old systems, as *non*–magnetic white dwarfs with brown dwarf companions.

References

Beuermann, K., Thomas, H.C. & Schwope, A., 1988. *Astr. Astrophys.*, **195**, L15.

Hameury, J.M., King, A.R., Lasota, J.P. & Ritter, H., 1988a. *Mon. Not. R. astr. Soc.*, **231**, 535.

Hameury, J.M., King, A.R. & Lasota, J.P., 1988b. *Astr. Astrophys.*, **195**, L12.

Hameury, J .M., King, A.R., Lasota, J.P. & Livio, M., 1988c. *Mon. Not. R. astr. Soc.*, , .submitted

Hameury, J .M., King, A.R. & Lasota, J.P., 1988d. *Mon. Not. R. astr. Soc.*, , .submitted

Morris, S.L., Schmidt, G.D., Liebert, J., Stocke, J., Gioia, I. & Maccacaro, T., 1987. *Astrophys. J.*, **314**, 641.

Rappaport, S., Verbunt, F. & Joss, P.C., 1983. *Astrophys. J.*, **275**, 713.

Schmidt, G., Stockman, H.S. & Grandi, S.A., 1986. *Astrophys. J.*, **300**, 804.

Schmidt, G., 1987. *Mem. Soc. Astron. It.* **209**, 227.

Spruit, H.C. & Ritter, H., 1983. *Astr. Astrophys.*, **124**, 267.

A NEW PROGENITOR MODEL OF TYPE Ia SUPERNOVAE

Izumi Hachisu[1], Mariko Kato[2], and Hideyuki Saio[3]

[1]Department of Aeronautical Engineering, Kyoto University
Kyoto 606, Japan
[2]Department of Astronomy, Keio University, Yokohama 223, Japan
[3]Department of Astronomy, University of Tokyo, Tokyo 113, Japan

ABSTRACT. A new progenitor model of Type Ia supernovae (SNe Ia) is proposed. The model consists of a carbon-oxygen white dwarf (0.8-1.2 M⊙) and a low-mass red giant star (0.8-1.5 M⊙) with a helium core (0.2-0.4 M⊙). When a red giant fills its inner critical Roche lobe and its mass transfer rate exceeds a critical value, a common envelope state is realized. Then the mass accretion rate onto the white dwarf, i.e., the mass transfer rate is tuned up to be $\dot{M} = 8.5 \times 10^{-7}(M_{WD}/M⊙ - 0.52)$ M⊙ yr^{-1}, where M_{WD} is the mass of the white dwarf. This rate is high enough to suppress the hydrogen shell flashes, but too low for carbon to be ignited off-center. When the carbon-oxygen core mass grows to the Chandrasekhar limit during the common envelope phase, a Type Ia supernova explosion is expected to occur.

PROGENITOR MODEL

The carbon deflagration model of carbon-oxygen white dwarfs has succeeded in explaining the various observational aspects of Type Ia supernovae (SNe Ia). It has been extensively argued that double carbon-oxygen white dwarf systems are the progenitor of SNe Ia (e.g., Iben and Tutukov 1984; Webbink 1984). Since the scenario was proposed, some suspicions on the scenario have been presented (e.g., Saio and Nomoto 1985; Robinson and Shafter 1987). Although the above suspicions cannot reject the possibility that double carbon-oxygen white dwarfs are the site of SNe Ia, the rate of SNe Ia coming from the double white dwarf progenitor is probably much smaller than that expected by Iben and Tutukov and by Webbink. In this sense, it is worthwhile to search for another way to Type Ia supernovae.

We assume that the progenitor of SNe Ia is a close binary system consisting of a carbon-oxygen (C+O) white dwarf and a low-mass red giant star. This assumption is consistent with the observational fact that SNe Ia appear everywhere in spiral galaxies and even in elliptical galaxies. When the giant fills its inner critical Roche lobe and the mass transfer rate exceeds a critical value, i.e.,

D. McNally (ed.), Highlights of Astronomy, Vol. 8, 175–176.

$$\dot{M}_{cr} = 8.5 \times 10^{-7}(M_{WD}/M\odot - 0.52) \ M\odot \ yr^{-1}, \tag{1}$$

where M_{WD} is the mass of the white dwarf, a common envelope is formed. The mass transfer rate cannot be determined only by the condition of the mass losing star because of the pressure gradient effect in the common envelope. Then the mass transfer rate is equal to the rate with which the mass accreting component fills/overfills its Roche lobe. In such a situation, the mass transfer rate must be tuned up to be $\dot{M}_2 = -\dot{M}_{cr}$. With this accretion rate the envelope remains to be a red giant size. It should be noted that the mass of the extended envelope around the C+O white dwarf is as small as 10^{-5} M⊙ or less for $M_{WD} > 1$ M⊙ because the radius of the extended envelope around the white dwarf ranges from a few to several tens solar radii (it is bluer than the Hayashi track).

When the mass ratio is too large, the separation decreases and finally the common envelope overfills the outer critical Roche lobe. Then the helium core and the C+O core spiral-in in the envelope. We do not regard this case as the progenitor of SNe Ia. When the mass ratio is close to or somewhat larger than unity, the separation decreases first and then increases because the mass ratio is reversed. Then the outer critical Roche lobe overflow does not occur and the mass of the C+O white dwarf can grow to the Chandrasekhar limit. This rate ($\dot{M}_{cr}$) is high enough to suppress the hydrogen shell flashes, but too low for carbon to be ignited off-center. When the mass reached the Chandrasekhar limit, a carbon burning ignites at the center to trigger a SN Ia explosion. When the mass transfer rate becomes smaller than the critical value before the C+O white dwarf mass reaches the Chandrasekhar limit, the common envelope disappears. Further, when the mass accretion rate decreases to less than 10^{-7} M⊙ yr^{-1}, hydrogen shell flashes occur (nova systems) and it blows off almost all of the accreted material and no SNe Ia are expected to occur.

If the low-mass star is a red giant (helium core mass is about 0.40 M⊙), SN Ia explosion is possible for red giant mass of $M_{2,0} > 0.8$ M⊙. For $M_{2,0} > 1.5$ M⊙, the possible parameter range for $M_{WD,0}$ is small because of the outer critical Roche lobe overflow.

Type Ia supernova is characterized by no hydrogen lines. Just before the supernova explosion, mass of the hydrogen-rich envelope around the C+O white dwarf becomes as small as 10^{-7} M⊙. We believe that such mass of the hydrogen is too small to be observable. Assuming that the initial masses ranging from 3 to 8 M⊙ can leave C+O white dwarf larger than 0.8 M⊙, we obtain the rate of 0.012 yr^{-1} in our Galaxy. The full paper of our scenario will appear elsewhere (Hachisu, Kato, and Saio 1988).

REFERENCES

Hachisu, I., Kato, M., and Saio, H. 1988, Ap. J., submitted.
Iben, I. Jr., and Tutukov, A. V. 1984, Ap. J. Suppl., 54, 335.
Robinson, E. L., and Shafter, A. W. 1987, Ap. J., 322, 296.
Saio, H., and Nomoto, K. 1985, Astr. Ap., 150, L21.
Webbink, R. F. 1984, Ap. J., 277, 355.

CONCLUDING REMARKS

VIRGINIA TRIMBLE
Astronomy Program   and   Physics Department
Univ. of Maryland   Univ. of California
College Park MD 20742   Irvine CA 92717 USA

1. THE PAST

In one sense our subject is an old one, dating back more than two centur_
ies to the work of John Michell on statistics of visual star pairs. In‾
another sense, it is both young and rapidly growing. The first meeting
devoted exclusively to binary systems took place in 1966 in Uccle, Belgium, followed by the first IAU Joint Discussion on the subject in 1967
and IAU Colloquia 6 and 18 in 1969 and 1972. From this average of less
that one meeting with published proceedings per year, our gatherings
have proliferated to about five per year in the mid 80's. I am inclined
to suspect that the topic of formation and evolution of binary stars is
now too broad to fit into any one meeting, room, day, or mind.

Over these 21 years, both the subject and its practitioners have e-
volved a good deal. The 1969 Elsinore colloquium, for instance, had 21
official participants, 12 of whom are still publishing in binary star
astronomy. Topics of extensive discussion there included activity in
old novae, models for Beta Lyrae, evidence for mass loss (including the
first rocket UV data), definition of RS CVn stars by Popper, data on
masses of Algols, and the first persuasive models for Algol formation
via conservative mass transfer from Kippenhahn & Weigert, Paczyński, and
Plaveç et al.

Some years down stream, the 1983 NATO workshop on interacting binaries had 115 participants, at least 98 of whom published something on
the subject in 1986-87. Much of the discussion focussed on departures
from conservative mass transfer, including common envelope binaries, the
loss or transfer of angular momentum during mass loss and transfer, and
the nature and persistence of contact. Other hot topics were the physics
of outbursts in RNe and DNe, statistical issues (both the distribution
of P, a, e, q, etc. and the question of which objects and phenomena real_
ly are systematically associated with binaries), and formation mechan-‾
isms. Many of these are clearly still with us, A. Underhill's remarks
having reminded us that there is not yet full concensus on the importance
of binaricity even among Wolf-Rayet stars, though explaining WRs was one
of the early triumphs of conservative mass transfer.

Somehow in the same time frame, the present author has evolved from
gate-crasher, through contributed papers and invited reviews, to conclud_
ing remarks, and expects shortly to be asked to give after dinner talks.‾

D. McNally (ed.), Highlights of Astronomy, Vol. 8, 177–180.
© *1989 by the IAU.*

2. THE PRESENT

2.1 Observations

The full importance of results presented at this JD will become obvious
only some years in the future. From the perspective of a few hours'
hindsight, however, two major and several minor observational points and
about four theoretical areas seem noteworthy. First, 1988 is the year
in which pre-main-sequence binaries finally became common enough to look
at statistically. M. Simon's lunar occultation sample (6 of 29 stars
double with $\Delta\theta$ = 0.01-0.'5 and Δm = 1-2), R. Mathieu's pre-MS spectroscopic systems (10 orbits, some separations less than T Tau accretion
disc sizes), and R. de la Reza's isolated binary T Tau collectively lead
to the impression that pre-MS binaries are about as common at various
separations as MS binaries, apart from a shortage of short-period true
T Tau's. This deficit can be blamed both on the difficulty in getting
good photospheric velocities for the stars and on the transience of the
phase: inevitably two stars separated by about their own disc sizes will
spiral together and expel the discs. Mathieu noted that better pre-MS
evolutionary tracks are needed to confirm coevality (or lack of it) in
his pairs.

A second observational highlight is the proliferation of pairs with
mass ratios less than 0.1. D. Latham introduced us to several such systems, including his own HD 114762, where the companions are arguably in
the brown dwarf mass range. The 0.05 M_0 secondary of AA Dor was probably still less massive on the main sequence, according to B. Paczyński.
Zuckerman's IR companions to WDs (10 of 100 searched) are presumably also
BDs or late M's. B. Campbell's 9 systems (of 18 late MS stars studied)
are, on the other hand, in the planetary range, with companion masses of
1-10 Jupiters, and no brown dwarfs. If this is the tip of the planetary
iceberg, then half or more of pop I late MS stars have solar systems.
Given the difference in parent populations, I do not see any real contra
diction between Campbell's and Latham's results. Self-evidently, then,
we do not yet have the data to decide whether two discrete physical processes are needed to produce low-mass-ratio binaries and planetary systems respectively, but I would bet the answer is yes.

Other neat new things on the observational front, and the people
who remarked upon them, include (a) the filling-in of the gap in period
and separation by radial velocity spectrometer and speckle techniques (D.
Latham), (b) the circularization of 4^d orbits within 10^6 yr (R. Mathieu)
(c) the non-coplanarity of multiple systems (K.D. Borne), (d) the prolif
eration of main sequence pairs with good enough data to confirm that the
components share both composition and birthdays (J. Andersen), (e) the
remarkable present faintness of the novae of 1437 and 1670 near M_v = +10
(M. Shara), and (f) the high mass of the white dwarf member of the CV
(EXO 033319-2554.2) that trespasses on the period gap (J.P. Lasota).
Finally, preprints and rumors (noted by M. Shara and R. Webbink) abound
with, at last, a few short-period double degenerates, one at Steward Ob
servatory and two at Bologna, the latter in company with two WDs that
have close M dwarf companions, within a sample of 20 WDs.

2.2. Theory

Recent theoretical advances divide naturally into four areas: formation from dense gas, formation from pre-existing stars, early system evolution, and later system evolution. Formation ab initio might occur by either fission or fragmentation of a rotating gas cloud, the former being a quasi-equilibrium process, the latter a dynamical collapse one. In summary, fission (R.W. Durisen, N.R. Lebovitz) is out, and fragmentation (A.P. Boss, S.M. Miyama, J. Tohline) is in. Even with differential rotation, quasi-equilibrium processes spin off only low-mass arms and discs, not comparable components (though the residual central tri-axial bars still need to be followed through further contraction and may yet prove interesting). Fragmentation, on the other hand, systematically leaves two (or more) comparable lumps over a well-defined range of initial rotation energy, thermal energy, and degree of central concentration (not too much!). Calculations cannot yet predict how many systems of each separation, mass ratio, etc. should form, but it does seem possible to produce the full known range of properties, and heirarchical processes can occur.

Pre-existing stars can give rise to binaries through disruption of triples (J. Anosova, on film) and via capture and exchange in dense environments like cluster cores (F. Verbunt). Verbunt noted that tidal capture and exchange of a neutron star for an MS component in an existing binary are competing contributors of low mass X-ray systems in globular clusters and that the spun-up NS can be liberated by a second exchange as well as by evaporating its companion.

A number of interesting points about early system evolution defy logical ordering. They include (a) the necessary shrinkage of pre-MS or bits to get the closest MS systems (R. Mathieu), (b) the use of circularization time as an age criterion (D. Latham), including the implication that pre-MS circularization must reach 4^d period systems in 10^6 yr, even though MS calculations get only to $2^d.4$ in that time (R. Mathieu), (c) the explanation of asynchronous MS rotators as overshooting when the stars contract after pre-MS synchronization (C. Zwaan), (d) the fact that $q \sim 1$ small-a systems can fill their Roche lobes and retract more than once while still MS stars (because mass transfer increases the size of the recipient's convective core and so raises central hydrogen content, decreasing radius, J.-P. de Grève), (e) the undetectability of convective overshoot effects on structure of main sequence O binaries, despite its later importance (J. Anderson), and (f) the possible effects of magnetic fields in the formation of binary Wolf-Rayets (B. Hidayat).

Leading to later evolutionary phases come R.E. Taam's important simulations of the common envelope phase. He finds that ejection is largely equatorial, occurs in 1-10 yr, and has efficiency of only 0.3-0.6. The systems need to be followed further to decide whether M_2 accretes or ablates and whether some pairs will merge before ejection is complete.

Before the JD began, I was inclined to think that the second mass transfer phase (CVs, X-ray binaries, etc.) was now about as well understood as the first phase was at the time of IAU Symp. 73 (1975). This illusion persisted through M. Shara's discussion of hibernating novae, J.P. Lasota's explanation of the AM Her period spike (and the associated spike in B values (because higher ones evolve catastrophically fast), P.C. Joss's confession that the shortest-period CVs and LMXRBs push very hard on scenarios both for formation and for driving adequate mass trans fer, and J. Krolik's revelation that the group whose scheme for forming isolated msec pulsars predicted systems like 1957+20 are no longer sure their evaporation mechanism works very well. Illusion shattered over B. Paczyński's conclusions that magnetic winds turn off in close systems (from the absence of W UMa progenitors in NGC 188), that hibernation may conflict with observed numbers of related systems, that the CV period gap is not understood, and that CVs may simply evolve through a short P range and then die. Finally, attempts to model type I supernova progenitors as RG+WD common envelope systems (I. Hachisu), which turn out to be unacceptably bright, or as double degenerates, which will explode if mass transfer is strongly super-Eddington (R. Webbink) but which don't seem to exist, lead one to the conclusion that what we need is a class of SNI progenitors with no detectable properties at all.

3. THE FUTURE

There is a sense in which binary star evolution is a solved problem: if we are told the values of M_1, M_2 and a and that e = 0 at t = 0 and are given rules for rates of mass loss from the system, angular momentum loss from the system, and angular momentum transfer between orbit and components, all as a function of time, then there exist calculations that can predict the future state of the system and what it should look like. Approaching from the other side, we seem to find that most kinds of systems we see, from RS CVn stars to millisecond pulsars, can be modeled somewhere in one or more of the scenarios.

But there is a catch. There is clearly underlying physics that determines all these things: initial masses and separations from the processes of star formation; circularization from tidal and perhaps magnetic interactions; loss of mass and angular momentum from single-star winds, common envelope processes, and probably other things we have not thought of yet; exchange of angular momentum via tides, accretion, magnetic fields, etc. This underlying physics largely eludes us, its products being represented by adjustable parameters in the calculations. Admittedly, our present understanding of single star evolution shares many of these problems.

At the end of the 1983 workshop, I bid the participants au revoir or the equivalent in 11 languages, predicting that we would all gather again in 1987. We are a year late (but Serbo-Croatian, Danish, Russian, and Turkish -- Allahaismarladik -- have been added to the list), which suggests 1992+1 for the next stock-taking. By then we can reasonably expect that fragmentation calculations will be predicting statistics of main sequence systems, that common envelope simulations will indicate which systems eject and which merge, that the samples of pre-main-sequence and low-mass-ratio binaries will have expanded further still, and, if we are very lucky, that someone will finally have found a type I supernova progenitor.

3. <u>SUPERNOVA 1987A IN THE LARGE MAGELLANIC CLOUD</u>

<u>Scientific Organising Committee</u>

V. Trimble (Chairman), W. Liller & J.C. Wheeler (Co-chairmen)

R.D. Cannon, M.W. Feast, D.K. Nadyozhin, K. Nomoto, R.Z. Sagdeev,
J. Trümper, A.J. Turtle, W. Wamsteker, R.E. Williams, L. Woltjer

<u>Supporting Commissions:</u>

27, 28, 34, 35, 40, 44, 47 & 48

EDITORIAL

William Liller
Instituto Isaac Newton
Ministerio de Educacion
Casilla 8-9, Correo 9
Santiago, Chile

Joint Discussion III, <u>Supernova 1987A in the LMC</u>, and the subsequent Supernova Working Group meeting produced a total of thirty-two papers by investigators from twelve countries. Organized by the out-going Chairperson of the Supernova Working Group, Virginia Tribmle, these two meetings gave IAU members attending the XXTth General Assembly a timely update on the status of understanding of this immensely important event.

Out of the discussions that took place during and after the Working Group meeting came one very strong recommendation first put forward to me by J. Craig Wheeler and seconded by a number of others:

The observations of SN1987A should be fully archived

and in much the same way as those of Periodic Comet 1982i (Halley) have been. It should be remembered that Comet Halley returns to our neighborhood every 76 years; we cannt expect another bright supernova to appear so soon.

Writing on behalf of the Supernova Working Group, I urge that interested parties organize the effort needed to collect and publish all the available observational data on SN1987A. This work should be started immediately; the longer we wait, the greater the chance will be that observations will be misplaced, misinterpreted, forgotten, or lost.

To be more specific, I have suggested, noting the extremely fine x-ray and neutrino observing programs carried out by Japanese investigators, that it would be highly appropriate for scientists of that nation to head this archival project. The Supernova Working Group very much hopes that our eastern colleagues will accept this difficult but important challenge.

D. McNally (ed.), Highlights of Astronomy, Vol. 8, 183.
© 1989 by the IAU.

SUPERNOVA 1987A: LIGHT CURVES AND THEIR INTEPRETATION

R M Catchpole
South African Astronomical Observatory (SAAO)
P O Box 9, Observatory 7935
Cape, South Africa

The Type II supernova SN1987A which occurred in the LMC is the brightest and most completely observed supernova ever recorded. Objective prism and UBV observations were made of the blue supergiant progenitor Sanduleak $-69°202$ and indicate that the visual absorption lies in the range $0.4 < A_V < 0.6$. Furthermore, the distance to the LMC is known in absolute units with a precision of about $\pm$ 15% (m-M = 18.45, Feast 1988) which combined with the above data and subsequent photometric observations permits detailed comparison with theory.

Within 107 minutes of the Kamiokande IMB neutrino event the region of the supernova was being observed by Albert Jones, although it was not until 0.8 days after the event that the supernova was officially discovered by Shelton. The first photoelectric observation was made at 1.1 days, by William Allen (1988) a New Zealand amateur. Other observations made during the first two days, have been conveniently tabulated by Arnett (1988). During this time the supernova steadily brightened in V, although theory predicts that it was rapidly fading bolometrically and cooling, after the intense heating that occurred when the shock wave reached the stellar surface about 3 hours after core collapse.

As of August 1988, there is still a large body of unpublished photo-electric data so that it is premature to undertake a comprehensive review. This paper will be confined to a discussion of bolometric fluxes, based on the broad band photometry obtained by the CTIO/ESO and SAAO observers. The other major body of published UBVRI photometry is by the MSSSO observers (Dopita et al. 1988) and is in good agreement with the SAAO data. The relevant papers are listed in Table 1.

There is a large body of narrow band photometry in various systems which could prove valuable for comparison with theoretical model atmosphere calculations for the supernova. The narrow band widths of the filters make absolute flux calibration very much easier than for the broad band photometry.

The fine error sensor (FES) on the IUE satellite, freed as it is from clouds and seeing, gives excellent temporal coverage of SN1987A. However the FES sensitivity peaks near B but stretches all the way to 9000 Å, which means that its effective wavelength will change in a complicated way as the SN evolves from a hot continuum to an emission line spectrum, making quantitative interpretation very difficult.

185

Table 1
Photometry Papers discussed below

	Days since core collapse	Photometric Coverage
CTIO		
Hamuy et al. 1988	1 to 177	U to I
Suntzeff et al. 1988	188 to 476	U to I
ESO		
Bouchet et al. (1987a)	6 to 21	J to L
Cristiani et al. (1987)	2 to 27	U to I
Bouchet et al. (1988)	21 to 231	J to Q (20μm)
SAAO		
Menzies et al. (1987)	1 to 50	U to L
Catchpole et al. (1987a)	51 to 134	U to L
Catchpole et al. (1987b)	135 to 260	U to M (4.8μm)
Whitelock et al. (1988)	261 to 385	U to M

Fig. 1 shows the variation with time of the various broad band colours measured from SAAO. Note the rapid decline in the brightness at U which is interpreted as partly due to the decreasing photospheric temperature and partly due to the dramatic increase in the UV line opacity. The effect in the IUE short wavelength bands was even more marked and their contribution to the bolometric flux can be safely ignored after the first few days. Also note the difference in linear decline rates at different wavelengths. Fig. 1 illustrates the

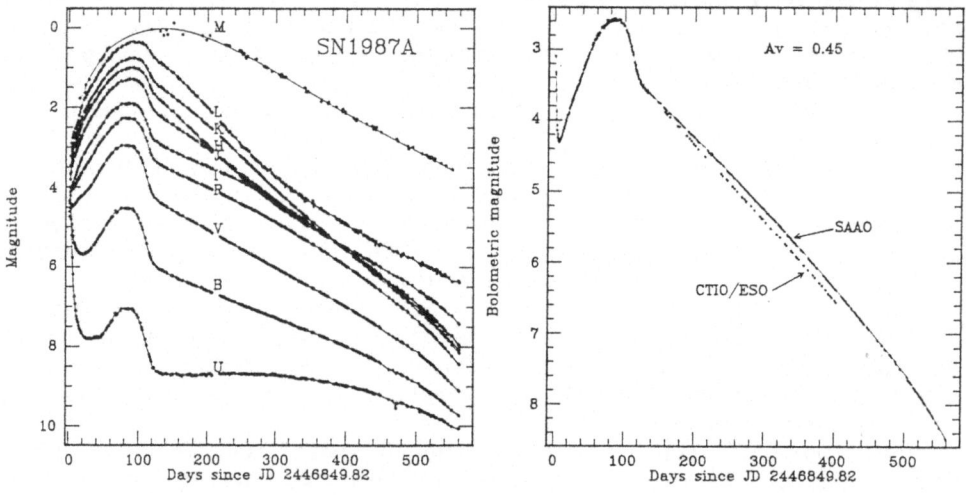

Fig. 1. The SAAO magnitudes as a function of time

Fig. 2. The bolometric magnitudes as a function of time

importance of combining the photomet., to form a bolometric curve rather than trying to compare individual magnitudes with models.

There are various ways in which the individual magnitudes, after conversion to fluxes, can be integrated to give the bolometric luminosity. In Fig. 2 the SAAO curve is based on a spline fit to U to M data where M values have been taken from a smooth curve of K-M against time. Also shown are magnitudes based on fluxes kindly supplied by Suntzeff and Bouchet (private communication) based on data obtained at ESO and CTIO. To allow comparison the SAAO bolometric magnitudes are calculated using $A_V = 0.45$, the value adopted by ESO/CTIO workers. The agreement between the two data sets for the first 140 days is very close although during this time the ESO/CTIO results include SAAO J to L photometry so that the two data sets are not entirely independent. Beyond day 140 differences arise between the two data sets largely because of differences in sensitivity in the I bands (Hamuy et al. 1988) at the two observatories. Menzies (1988) has shown that the wavelength sensitivity of the CTIO I band is much narrower than the SAAO band which results in the CTIO observers not including flux from the strong CaII emission lines at 8600 Å.

Two things are striking about the bolometric light curve, firstly it is quite unlike that of any other supernova in showing a second maximum and secondly it is very smooth on a time scale of a few days. Recently Young and Branch (1987) and Schmitz and Gaskell (1988) have pointed out examples of other supernovae that may have had similar light curves to SN1987A.

The difference between SN1987A and a more typical type, II if there is such a thing, is illustrated in Fig. 3 in a comparison with SN1969L, which after an initial rapid decline showed a distinct plateau phase before continuing to decline more slowly. The fact that the two SN light curves are very similar during the linear decline phase is probably fortuitous as we would expect Type II supernova progenitors to cover a wide range of masses with a possible corresponding range for the mass of ^{56}Ni created in the explosion.

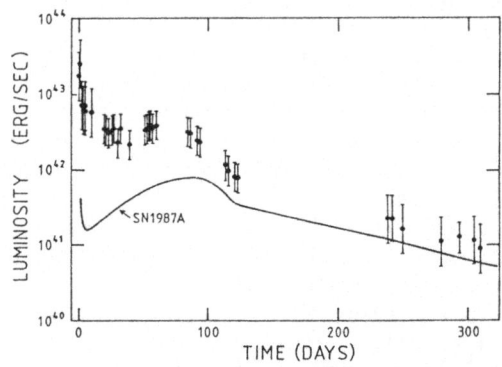

Fig. 3. SN19871 compared with SN1969L. The data is from Ciatti et al. (1971) & Kirshner et al. (1973) as presented in Weaver & Woosley (1980).

The unique and unexpected shape of the light curve, rapid colour evolution and the short time interval between the arrival of the neutrinos and visual brightening for SN1987A are all apparently consequences of it arising from a compact blue, rather than an extended red giant, progenitor.

188

The behaviour of the light curve can be divided into 5 phases which are illustrated symbolically in Fig. 4. Phase 4, which lasts for 143 days between days 122 and 265 after core collapse, shows a very closely linear decline in bolometric magnitude that in turn corresponds to an exponential decay in flux. The e-folding time of the decline depends on the method of integration and slightly on the adopted reddening but is very close to the 111.26 day mean life for radioactive decay of ^{56}Co. Fig. 5 summarizes the e-folding times derived by different methods of integration and for different values of the reddening. The behaviour of individual colours is also shown. The slope of the bolometric curve can be thought of as the flux weighted mean of these individual colour slopes. The close agreement between the observed decline rates and the ^{56}Co mean life provides convincing evidence that ^{56}Co and its progenitor ^{56}Ni are the major sources of energy in the late time light curves of type II supernovae. This was predicted by Pankey (1962) and modelled for SN1987A by Woosley et al. (1988), Nomoto et al. (1988) and others. The light curve also indicates that any contribution from other

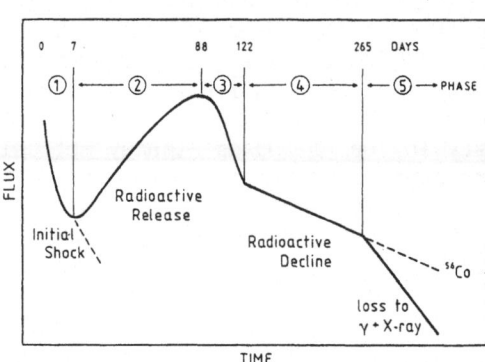

Fig. 4.   Symbolic light-curve for SN1987A illustrating the 5 phases and their starting time in days.

energy sources such as a pulsar was not significant during phase 4. The phase 4 light curve can be used to calculate the mass of ^{56}Ni produced in the initial explosion if we assume that only a negligible amount of the ^{56}Co decay energy came out as γ and x-rays. This assumption is justified by Kumagai et al. (1988) who show that 3% of the flux is emitted as γ and x-rays at this time. For $A_v = 0.6$ and a distance modulus = 18.5 a mass of ^{56}Ni = 0.08 Mθ is derived. If we integrate the bolometric flux curve until day 265 and compare the total flux (9.1 x 10^{48} erg) with that generated by radio-active decay in the same time interval we find the decay energy exceeds the radiated energy by 51% which is increased to 66% if we follow Woosley's theoretical prediction that up to day 40 the energy is derived from the initial shock wave. This excess energy is comparable to the kinetic energy of the ^{56}Co material and is probably responsible for mixing this material higher into the ejecta. This in turn can explain both the smooth shape of the light curve and the early appearance of γ and x-rays (Nomoto et al. 1988). Once we have accepted that ^{56}Ni and ^{56}Co are major sources of energy we can then readily explain the remaining phases of the light curve. Phase 1 is defined by a rapid decline in brightness and redward change in all colour indices while during phase 2 the supernova slowly climbs to maximum brightness, something that at the time was totally unexplained and unexpected.

During phase 1 and well into phase 2 the flux distribution is well approximated by a blackbody which allows us to define a photospheric

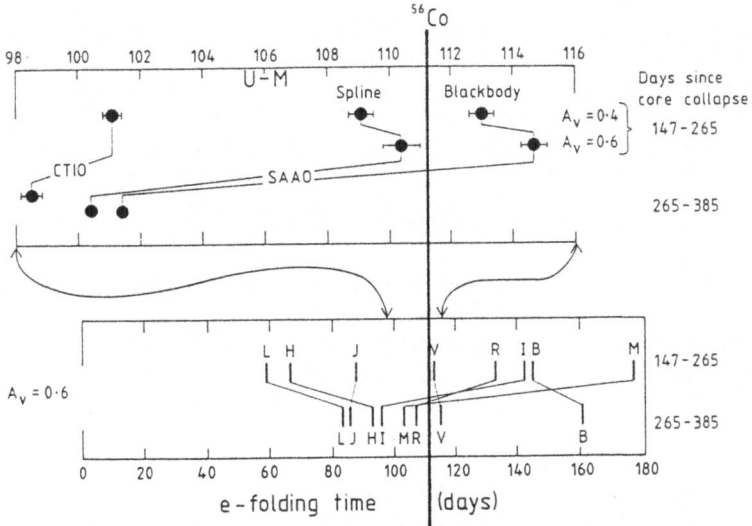

Fig.5. The e-folding time in days for various methods of integration and sets compared with ^{56}Co decay. The upper panel is for bolometric magnitudes, the lower panel for individual SAAO magnitudes.

radius and temperature. The variation of these with time are shown in Fig.6. Branch (1987) has shown that the density gradient within the photosphere can be deduced from the slope of the Log Radius against Log Time curve. The slope changes on day 5 and corresponds to a change of density from $\rho \approx R^{-11.7}$ to $\rho \approx R^{-5.0}$. The definition of a photosphere combined with radial velocity measurements of faint lines formed at the photosphere allow the distance of SN1987A to be calculated. Using this method Branch (1987) derives a modulus of 18.7±0.2 in good agreement with the generally adopted value of 18.5±0.15, demonstrating the validity of the method for distance determination. The temperature initially declines rapidly and then remains constant at the value appropriate to that of H recombination showing that the position of the photosphere is defined by the sharp increase in opacity as one crosses from the neutral to ionized hydrogen.

The decline in brightness during phase 1, that lasts until day 7, is caused by the rapid decline in photospheric temperature being more important than the increase in curve enters phase 2 and the SN

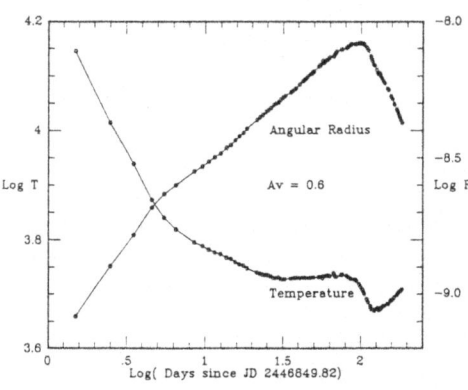

Fig.6. Variation of blackbody temperature and radius with time from the SAAO data.

starts to brighten the increasing radius of the photosphere more than compensates for the much slower decrease in temperature. Woosley et al. (1988) predict that for about the first 40 days the SN is entirely powered by energy deposited in the outer envelope by the passage of the initial shock wave. Thereafter the steady rise in brightness to maximum light on day 88 is a consequence of the photosphere which, although expanding in length coordinates, is falling back in mass coordinates and progressively liberating radiation generated by radioactive decay. The slope of the light curve during phase 2 and the time of maximum brightness both impose constraints on models of the explosion in terms of the degree of mixing, the mass of the expanding envelope and the energy of the explosion.

During Phase 3, from day 88 to day 122, the brightness declined steeply. It was during this time that an excess of flux first appeared at 4.8μm and rapidly increased in strength. Longer wavelength photometry indicated that this excess was confined to the M band and it was identified with emission from the fundamental vibration rotation band of CO. The width and velocity of the CO lines showed that they arose in the envelope and indicated that the excess flux should be added into the current energy budget of the SN. In some supernovae an infrared excess has been observed and attributed to a light echo of the original UV flash. In which case the energy should not be included in the current energy budget. In SN1987A a weak excess of flux beyond 4μm, of uncertain origin, was noted by Bouchet et al. (1987b) as early as day 59 but amounted to less than 1% of the total flux.

Phase 5 which commenced on day 265, corresponds to an increasingly rapid decline in brightness in the U to M bolometric flux as more of the ^{56}Co energy escapes as γ rays and x-rays.

The CTIO/ESO data is not in complete agreement with this interpretation as their curve always falls below the SAAO curve during the linear decline and does not show a change of slope between phases 4 and 5. The differences are more clearly illustrated in Fig. 7 where the SAAO and CTIO/ESO data are compared directly with the ^{56}Co e-folding

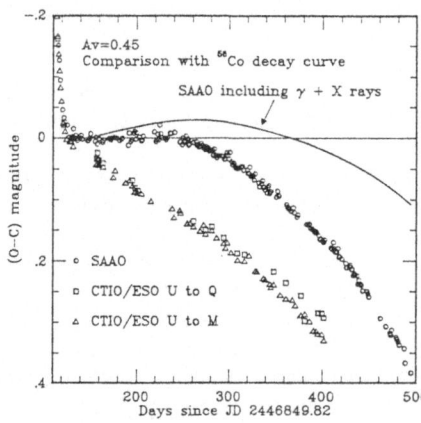

Fig. 7. Differences between the CTIO/ESO and SAAO bolometric magnitudes and the ^{56}Co decay curve are given as a function of time. The continuous curve shows the effect of adding the γ and X-ray flux to the SAAO curve.

curve (111.26 day). The CTIO/ESO observers show two curves one for U to M (0.3 to 4.8μm) and one for U to Q (0.3 to 20μm). The similarity of the two slopes shows the discrepancy does not arise from the failure of SAAO to include data out to 20μm. About 0.05 of the difference between the SAAO and ESO/CTIO residuals may be caused by the different methods of integration used at the two observatories while the main difference arises from differences in the I band discussed above.

In order to determine the true bolometric flux from the supernova all the flux from outside the U to M region must be included. Up to the present (August 1988) the only other important source of energy is in the γ and x-rays. X-rays in the 16-30keV range have been detected from SN1987A since day 130 and ^{56}Co line emission at 847keV and 1238keV has been measured since about day 180. Unfortunately before this data can be combined with the U to M data we must include the Compton scattered but usually unobserved flux between 16 and 847keV which must be modelled. This has been done by Kumagai et al. (1988) and the results of adding their predicted fluxes to the SAAO data are also shown in Fig. 7. This has the effect of reducing the deviation of the CTIO data from the ^{56}Co decay curve while the SAAO data now lies within ± 3% of the curve for 300 days after day 140. In view of quite large uncertainties in the γ and x-ray data the overall agreement is rather satisfactory.

Uncertainties in determining the absolute bolometric flux can be divided into astrophysical and observational uncertainties. The main astrophysical uncertainties are in the distance modulus and inter-stellar reddening. The problem of whether or not to include any infrared excess in the energy balance has not yet arisen. The two major observational problems are undoubtedly the incompleteness of coverage of the spectrum by the broad bands and the choice of the most appropriate integration technique. In the early days, when the super-nova still showed a strong continuum, interpolation between adjacent bands introduced relatively small errors. As the spectrum shows more emission lines and the continuum fades the problem becomes much worse and it is possible for individual lines to lie between bands and not be measured at all. The omission of Paschen α at 1.876μm between the H and K bands may have caused the bolometric flux to be underestimated by a few percent from day 200. Also the band heads of the CO fundamental band lie outside to blue of the M band. As more spectrophotometry becomes available it should be possible to improve our knowledge of the bolometric curve. However it must be emphasized that this will be an iterative procedure as spectrophotometry also has calibration problems. Other observational problems are an accurate knowledge of both the filter transmissions and their absolute flux calibrations.

Future photometric possibilities include possible dust formation, evidence for a pulsar and the search for further evidence of light echoes.

I am indebted to the entire SAAO staff and visitors who have contri-buted to observations of SN1987A and to Stan Woosley, Ken Nomoto and David Arnett for their enlightening preprints on which much of the theoretical discussion here is based.

REFERENCES

Allen, W H, 1988.   Publ. Var. Star Sect., RASNZ.
Arnett, W D, 1988.   Astrophys. J., 331, 377.
Bouchet, P, Moneti, A, Slezak, E, Le Bertre, T, & Manfroid, J, 1988.
    ESO preprint No. 592.
Bouchet, P, Stanga, R & Le Bertre, T, 1987a. Astr. Astrophys., 177,L9.
Bouchet, P, Stanga, R, Moneti, A, Le Bertre, T, Manfroid, J, Silvestro,
    G & Slezak, E, 1987b.   ESO Workshop on the SN1987A. ESO Confer-
    ence and Workshop Proceedings No. 26, p.79, ed. Danziger, I J,
    Garching.
Branch, D, 1987.   Astrophys. J., 320, L23.
Catchpole, R M & Menzies, J W et al., 1987a. Mon. Not. R. astr. Soc.,
    229, 15p.
Catchpole, R M, Whitelock, P A & Feast, M W et al., 1987b.   Mon. Not.
    R. astr. Soc., 231, 75p.
Ciatti, F, Rosino, L, & Bertola, F, 1971. Mem. Soc. Astr. Ital. 42,
    163.
Cristiani, S, Babel, J & Barwig, H, 1987.   Astr. Astrophys., 177, L5.
Dopita, M A, & Dawe, J A, et al. 1988.   Astr. J., 95, 1717.
Feast, M W, 1988.   SAAO preprint No. 599.
Hamuy, M, Suntzeff, N B, Gonzalez, R, & Martin, G, 1988.   Astr. J.,
    95, 63.
Kirshner, R P, Oke, J B, Penston, M V, & Searle, L, 1973.   Astrophys.
    J., 185, 303.
Kumagai S, Itoh, M, Shigeyama, T, Nomoto, K, & Nishimura, J, 1988.
    Astr. Astrophys., 197, L7.
Menzies, J W, 1988.   SAAO preprint No. 598.
Menzies, J W & Catchpole, R M et al. Mon. Not. R. astr. Soc., 227, 39p.
Nomoto, K, Shigeyama, T & Hashimoto, M, 1988.   Lecture Notes in
    Physics, 305, 319.
Pankey Jr, T, 1962.   Ph.D. Thesis, Howard University, Washington DC.
Schmitz, M F & Gaskell, C M, 1988.   Proc. of the fourth George Mason
    Astrophysics Workshop, p.112, ed. Kafatos, M & Michalitsianos,
    A G.   Cambridge University Press.
Suntzeff, N B, Hamuy, M, Martin, G, Garnez, A, & Gonzalez, R, 1988
    preprint.
Weaver, T A, & Woosley, S E, 1980.   Ann. NY Acad. Sci., 336, 335.
Whitelock, P A, Catchpole, R M, Menzies, J W & Feast, M W, et al.,
    1988.   Mon. Not. R. astr. Soc., in press.
Woosley, S E, Pinto, P A & Ensman, L, 1988.   Astrophys. J., 324, 466.
Young, T R & Branch, D, 1987.   Nature, 333, 305.

EVIDENCE FOR ASYMMETRIES IN SN1987A

M. KAROVSKA, L. KOECHLIN, P. NISENSON,
C. PAPALIOLIOS, and C. STANDLEY
Harvard-Smithsonian Center for Astrophysics
60 Garden Street, Cambridge, MA 02138

ABSTRACT. Results of observations carried out during the past year and a half after the explosion of SN1987A (February 23 1987), show strong evidence for asymmetry in the expanding shell. The departure from spherical symmetry in SN1987A has important implications for further detailed modeling of supernovae explosions, light curves and spectra.

INTRODUCTION

The supernova SN1987A in the LMC provides an exceptional opportunity to observationally test the assumption of spherical symmetry which has been widely used in the theoretical models of the supernova explosions. The assumption of spherical symmetry has also been made in attempts to use supernovae as cosmological distance indicators (Baade method). If the supernovae are not spherically symmetric, the distances estimated will be too high. We present here the evidence for asymmetries in SN1987A obtained using spectroscopic, polarimetric and speckle interferometric observational techniques.

OBSERVATIONAL RESULTS

I. *SPECTROSCOPY*

Early spectroscopic observations of the SN1987A (March, 1987) obtained by Hanuschik and Dachs (1987) revealed the presence of a double-peaked feature in the Hα region. A bump appeared first in the transition region between the Hα absorption trough and the maximum emission about 20 days after the explosion. An equally displaced component from the rest wavelength of the Hα in the LMC was observed a few days later. A similar double-peaked structure was also observed in several H I lines in the optical and the infrared (Phillips, 1987; Danziger et al, 1987) between the 20th and

193

D. McNally (ed.), Highlights of Astronomy, Vol. 8, 193–198.
© 1989 by the IAU.

the 80th day after the explosion. The origin of this
spectral feature is probably related to some departure from
spherical symmetry of the supernova shell, but the presence
of two equally displaced components indicates that the
source must have some global symmetry (Lucy, 1987).

Infrared spectra obtained 120 to 250 days after core
collapse by Danziger et al (1987) show a redshift bias for
the emission components of strong P Cygni lines in the
velocity range of 500-1400 Km/s. A similar effect has been
observed by Terndrup et al (1988) in the optical and the
infrared for the peaks of the H I, Ca II and [Fe II]
emission lines: the peaks are redshifted by 500-700 Km/s in
the rest system of SN1987A. A possible interpretation of
the observed effect is that it arises from a geometrical
asymmetry of the line-forming region.

The peaks of the [O II] $\lambda\lambda$ 6300Å, 6363Å forbidden
emission lines which became visible in the supernova
spectrum around 130 days after the explosion did not show
any redshift bias (Terndrup, 1988). On the other hand, the
high-dispersion spectra of the top of the [O II] doublet
obtained at the Anglo-Australian Observatory (April 1988,
AAO Newsletter #45) show the presence of small fluctuations
which occur at the same line-of-sight velocities for the two
lines. These observations may indicate that several clumps
are present in the expanding envelope, or alternatively,
that there are radial 'fingers' of material with a higher
velocity of expansion.

II. *POLARIMETRY*

Shapiro and Sutherland (1982) proposed an observational
test of the spherical symmetry of the supernovae based on
the fact that the light of an unresolved, aspherical
supernova atmosphere (scattering dominated) is linearly
polarized. This test was performed on SN1987A, and the
results of several polarimetric and spectropolarimetric
observations obtained at different epochs after the
explosion indicate that this supernova is not spherically
symmetric.

Some of the earliest polarimetric observations of the
SN1987A (10-11 days after the explosion) were carried out by
Schwarz and Mundt (1987). The results of the narrow band
polarimetry obtained in the peaks and the troughs of several
spectral lines (Hα, Hβ, Hγ, Na D) show variation in
polarization across line profiles. Jeffery (1987)
interpreted the observed polarization structure as arising
from an asymmetric, expanding, scattering atmosphere. He has
carried out Sobolev-Method radiative transfer calculations
with an ellipsoidal model for the SN1987A photosphere and
shell in order to fit the polarimetric data. The best fits
to the data were obtained for an axis ratio of the shell
(minor axis over major axis) between 0.6 and 0.8.

Multicolor linear polarimetric observations of SN1987A
(U,B,V,R, and I filters) have been obtained between February
28 and April 29, 1987 by Méndez et al (1988). The
measurements made during the first month after the explosion
show polarization of 0.1-0.5 % at position angle of about
25⁰ (or at 205⁰ since there is a 180⁰ ambiguity). After the
first month, the position angle remained almost unchanged in
U and B filters, but an additional component became visible
in V, R, and I filters. The wavelength dependence of the
polarization measured during the first 30 days indicates
that the scattering is dominated by Thompson scattering. The
appearance of the second component after 30 days has been
interpreted as due to the change of absorption in the total
opacity, although the possibility that it arises from a
completely different polarizing mechanism has not been
excluded. In order to interpret the measurements obtained
during the first month, Méndez et al (1988) propose a model
which is based on the assumption that the outer layers of
the supernova shell are ellipsoidal. Results obtained imply
that the axis ratio of the shell is between 0.8 and 0.9.

Results from the spectropolarimetric measurements
obtained by Cropper et al (1988) between February and July
1987, also suggest that the SN1987A is asymmetric. The
polarization has been measured in the spectral range between
350 nm and 900 nm including the regions around Hα and Ca II
lines. The measurements indicate that the position angle of
the principal scattering axis is wavelength dependant. For
example, the principal scattering axes for the Hα and the Ca
II triplet appear to be perpendicular. These results have
been interpreted using two different supernova polarization
models, one proposed by Shapiro and Sutherland (1982) and
the other by McCall (1984). Assuming that the scattering
atmosphere has a shape of an oblate or prolate spheroid, the
best fit to the data has been obtained for the axial ratio
of 0.7 to 0.9.

The observations obtained by Cropper et al (1987) yield
another important result. When all polarimetric measurements
are plotted in the Stokes-parameter QU plane, the locus
defines a band, the axis of symmetry of which is at a
position angle of about 200⁰. This axis is nearly aligned
with the direction (P.A.=194⁰ ± 2⁰) of the bright companion
to the SN1987A observed in March and April 1987 by Nisenson
et al (1987) and Matcher et al (1987) using speckle
interferometry techniques. Also, the position angle of this
axis is almost identical with the position angle of the
major axis of the elongation in the supernova images (P.A.=
200⁰ ± 5⁰) obtained between June (1987) and April (1988)
using speckle imaging techniques (Karovska et al, 1988a).

The spectra of the polarized radiation recorded
between 140 and 360 days after the explosion, using
spectropolarimetry at AAO show broad redshifted features
corresponding to the emission lines (Bailey, 1988). The

observed polarization structure has been interpreted in
terms of electron scattering in an asymmetric, expanding
atmosphere. The polarization position angle measured since
July 1987 was about 110° (in the red wing of Hα,after
correcting for the effects of interstellar polarization).
This is shifted by 90° from the measurements of the first
few months after the explosion. The position angle of the
electron scattering axis seems to be aligned with the minor
axis of the elongation in the supernova speckle images
(P.A.=110° ± 5°).

III. *SPECKLE INTERFEROMETRY*

   High angular resolution speckle interferometry
observations of SN1987A obtained on March 25 and April 2,
1987 revealed a bright source separated from the supernova
by approximately 60 mas (Nisenson et al, 1987). Data were
recorded with the two-dimensional photon counting detector
(PAPA) on the Cerro Tololo Interamerican Observatory (CTIO)
4 m telescope using several narrow bandpass (10 nm) filters.
The supernova 'companion' was observed at the position angle
of 194° in Hα, at 533 nm, and at 450 nm with the magnitude
differences of 2.7, 3, and 3.5-4 respectively. no companion
was seen in the data recorded on the comparison stars
(unresolved by the telescope).
   This bright source was observed again 12 days later (
Meikle et al 1987) at AAO 4 m telescope at the same position
angle and at nearly the same distance from the supernova. A
magnitude difference of 3 has been measured at 658.5 nm.
Successive speckle observations, starting with June 1987,
did not yield new detection of this source for a magnitude
difference smaller than 4 (a magnitude difference of 4 was
the upper limit for certain detection). This indicates that
the source had faded faster then the supernova itself.
Attempts to interpret the speckle data have not resulted in
a definitive explanation of the nature of this bright
source. However, there is additional evidence that the
supernova companion source may well have been physically
associated with the supernova itself. The source lies on an
axis almost aligned with the principal symmetry axis of the
supernova envelope as defined from the early polarization
observations, and also by the elongation in speckle images
of the SN1987A observed between June 1987 and April 1988.
   Speckle interferometric observations from March-April,
1987 did not show any convincing evidence of an asymmetry in
the supernova shell. However, spectroscopic and polarimetric
observations show evidence that the supernova shell was
asymmetric at this early epoch after the explosion. The
asymmetry was not detected by speckle interferometry because
of the small size of the supernova at that epoch (≤ 12 mas,
Karovska et al 1988b), and because of the relatively low
signal-to-noise in the data.

Successive speckle interferometry observations made at
CTIO 4 m telescope between 95 and 411 days after the
explosion showed unambiguously that the supernova is
asymmetric (Karovska et al, 1988a). The asymmetry was
observed at four epochs (day 95-98, 265-267, 370-373, and
409-411) at several different wavelengths between 450 nm and
850 nm including Hα line and Ca II triplet. The position
angle of the major axis of the elongation in the images was
200° ± 5°. The images appeared to be somewhat brignter in
the south-west direction. The variation of the position
angle as a function of wavelength was less than 10°.

Measurements of the minor and the major axis of the
elongated images obtained at several different wavelengths
show that the supernova size increased substantially since
June 1987. The largest change was detected in the images
obtained from the data recorded near the Hα maximum emission. The minor axis increased from 6 mas in June 1987 to 20
mas in April 1988. During the same epoch the major axis
increased from 10 mas to 30 mas. The ratio between the minor
and the major axes of the elongated images is between 0.6
and 0.9, depending on the wavelength of observation. The
ratio between the axes has not diminished since June 1987.

CONCLUSION

A number of spectroscopic, polarimetric and speckle
interferometric observations of the SN1987A show that this
supernova is not spherically symmetric. The asymmetry is
not negligible and should be included in theoretical models.
This can be crucial for the interpretation of other
observational results, for example the early emergence of
the X-rays and gamma rays and the evolution of their spectra
as a function of time (Grebenev and Sunyaev, 1987).

Current observational results do not give definitive
answers to the questions: When did the asymmetry appear?
What caused it? Although some of the observations showed
evidence for an asymmetry as early as few days after the
explosion, it is unclear whether the SN1987A was initially
asymmetric or the departure from the spherical symmetry
appeared after the explosion. Rotational effects, presence
of strong magnetic fields, and appearance of nonspherical
instabilities during the core collapse are some of the
potential sources for an initial departure from spherical
symmetry. The departure from the spherical symmetry in the
SN1987A envelope can be also caused by an initial asymmetry
in the progenitor envelope. The progenitor envelope could
have been rotationally flattened due to angular momentum
transfer from a binary companion during the stellar
evolution (Chevallier and Soker, 1988). Another possibility
is that the asymmetry appeared after the collapse due to the
growth of nonspherical instabilities in the expanding

envelope.

Further observations of SN1987A may provide more information and help us to determine the cause of the asymmetry. If the source of the asymmetry is not peculiar to SN1987A, a major revision of the supernova models and an assessment of the reliability of the distances derived using Baade method would be necessary.

This work has been partially supported under grant AFOSR-86-0103 and the Smithsonian Institution Scholarly Studies and Research Opportunities grant programs.

REFERENCES

Bailey, J. 1988, *preprint.*
Chevalier, R.A., and Soker, N. 1988, *preprint.*
Cropper, M., Bailey, J., McCowage, J., Cannon, R.D., Couch, W.J., Walsh, J. R., Strade, J.O., and Freeman, F. 1988, *M.N.R.S.,* 231, 695.
Danziger, I.J. et al 1987,*Fourth George Mason University Workshop in Astrophysics*, "SN 1987A in the LMC", eds. M. Kafatos and A.G. Michalitsianos.
Grebenev, S.A., and Sunyaev, R.A. 1987, *Soviet Astr. Letters*, 13, 1042.
Hanuschik, R.W. and Dachs, J. 1987 , *Astron. & Ap. (Letters)*, 177, L4.
Jeffery, D. J. 1987, *Nature*, 329, 419.
Karovska, M., Koechlin, L., Nisenson, P., Papaliolios, C., and Standley, C. 1988a, *IAU Circ. No 4604.*
Karovska, M., Koechlin, L., Nisenson, P., Papaliolios, C., and Standley, C. 1988b, *Ap. J., in print*
Lucy, L.B. *1987,* Fourth George Mason University *Workshop in Astrophysics*, "SN 1987A in the LMC", eds. M. Kafatos and A.G. Michalitsianos.
McCall, M.L. 1984, *M.N.R.S.,* 231, 695.
Meikle, W.P.S., Matcher, S.J., and Morgan, B.L. 1987, *Nature*, 329, 608.
Méndez, M., Clocchiatti, A., Benvenuto, O.G., Feinstein,C., and Marraco, H.G., *preprint.*
Nisenson, P., Papapliolios, C., Karovska, M., and Noyes, R. 1987, *Ap. J. (Letters),*320, L15.
Phillips, M.M.1987, *Fourth George Mason University Workshop in Astrophysics*, "SN 1987A in the LMC", eds. M. Kafatos and A.G. Michalitsianos.
Schwarz, H.E., and Mundt, R. 1987, *Astron.&Ap.,* 177, L4.
Shapiro, P.R., and Sutherland, P.G. 1982,*Ap. J.,*263,902.
Terndrup, D.M., Elias, J.H., Gregory, B., Heathcote, S.R., Phillips, M.M.,Suntzeff,N.B., and Williams, R.E. 1988, *Proceedings of the Astronomical Society of Australia.*

GAMMA-RAY LINES FROM SN1987A AND INTERPRETATION

E. L. CHUPP
Physics Department and
Institute for Earth, Oceans and Space
University of New Hampshire
Durham, New Hampshire 03824, USA

ABSTRACT. Gamma-ray lines from the decay of ^{56}Co in the SN1987A remnant have been detected by satellite and balloon experiments. The observations directly confirm the basic theoretical tenet that ^{56}Ni was explosively synthesized in the aftermath of core collapse of the blue super giant Sanduleak - 69 202. The flux level of the ^{56}Co lines at 847 keV and 1238 keV, from the Gamma-Ray Spectometer (GRS) on the *SMM* satellite, is consistent with a constant level from 1987 August through 1988 May. The early appearence of the γ-ray lines and the continuum reported by the *Ginga* and *Mir-Kvant* satellite experiments require mixing, or clumping of a few percent of the newly synthesized ^{56}Ni in the expanding envelope. Results from balloon experiments, which are in the preliminary stage of analysis, do not give clear evidence for γ-ray line shifts from rest energies with a limit $\Delta E/E < 0.002$, nor is there definitive evidence, at this time, of line splitting, but γ-ray line widths are clearly wider than instrument resolution at $\Delta E/E \leq 0.02$. The continuum (50-500) keV/line ratio has stayed approximately constant at 8 from 1987 August to 1988 May indicating the origin of the continuum from Compton down-scattering of deeper lying ^{56}Co and lines from ^{56}Co at smaller γ-ray optical depths. Future balloon flights of the most sensitive high-resolution spectrometers, planned for 1988 November and 1989 April, are expected to be able to detect weaker γ-ray lines at a flux level of $\sim 10^{-4}$ γ cm^{-2} s^{-1} for the 847 keV line. SN1987A was just close enough to the Earth to confirm theory with existing instruments. In this report we review only γ-ray line results, since the continuum observations have been reviewed by Professor Trümper in this session.

1. Gamma-Ray Line Observations

The first detection of γ-ray lines from the remnant of SN1987A was made using the Gamma-Ray Spectrometer (GRS) on board the Solar Maximum Mission (*SMM*) satellite (Matz et al. 1988). In Figure 1, we show the satellite-Earth geometry for solar and LMC observations. Any extraterrestrial source of γ rays detectable by the GRS may be occulted by the Earth (in each orbit), depending on the source direction and the orientation of the orbital plane. For SN1987A observations occultation periods last for ~ 30 days with ~ 20 days during which no occultation of the source occurs. During the periods when the LMC is not occulted by the Earth any SN γ rays can reach the GRS, penetrate the side shield, and be detected. Figure 2 shows a schematic of the GRS with

199

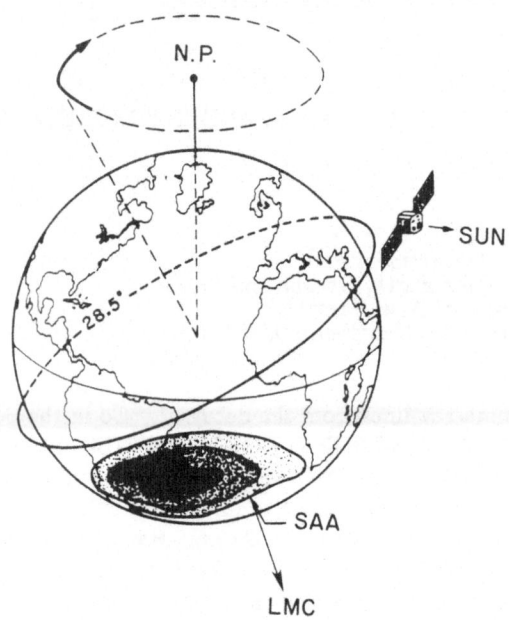

Figure 1.   The geometry is shown for SN1987A viewing by the *SMM* Gamma Ray
Spectrometer (GRS).

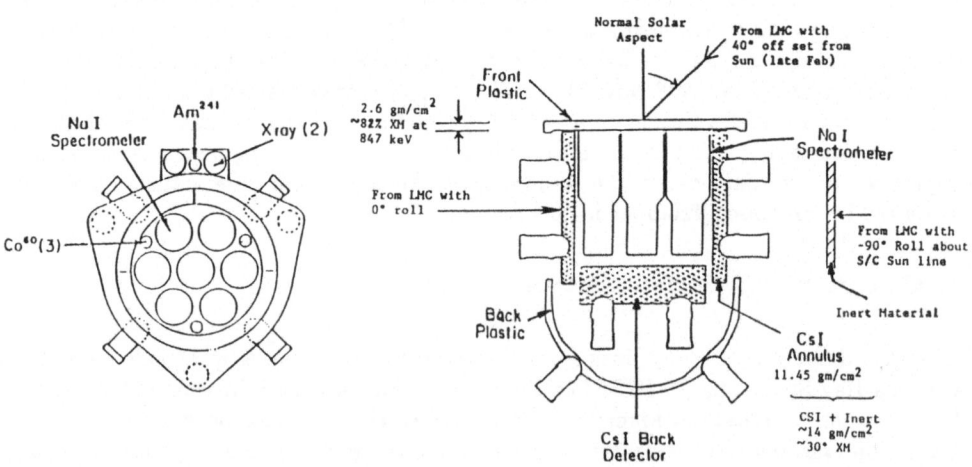

Figure 2.   The schematic drawing for the *SMM* GRS is shown along with absorbers
for different SN1987A viewing.

top and side views. The side view shows the amount of absorber that γ rays from the LMC must penetrate. For example, the 847 keV line is attenuated by about a factor of 3, giving an effective area of ~ 40 cm^2. The basic technique used with GRS for determining the γ-ray line flux is to form a net γ-ray spectrum by subtracting, in each orbit, the occulted (or background) spectrum from the spectrum obtained when the LMC is exposed and accumulating these for several days or weeks. This procedure eliminates spectral features present in unsubtracted spectra which are due to induced radioactivity in the spacecraft and instrument. However, there is unequal exposure to atmospheric radiation in the two spectra so a net atmospheric residual remains after subtraction.

These background subtracted spectra are then examined for evidence of excess γ-ray line flux at the expected ^{56}Co energies of 847 keV, 1238 keV and 2599 keV. To determine the specific line flux values the net spectra are fitted in the regions near a given line with a power law continuum and a Gaussian peak with a width fixed at the instrument's resolution including background lines. The best fit yields the line fluxes and the parameters of the power law continuum. Figure 3 shows the intensities of the 847 keV and 1238 keV lines after the event resulting from approximately 30 day data accumulations. Beginning with the accumulation starting in early 1987 August flux increases in the 847 keV and 1238 keV lines appear, followed by a continuing succession of positive flux enhancements through 1988 May. Matz et al. (1988) have shown that prior to 1988 August and for the seven years preceding SN1987A, there is no significant evidence for flux enhancements at the ^{56}Co line energies. The first and last SMM GRS results are listed in Table 1.

In 1987 May, NASA inaugurated its Souther Hemisphere balloon campaign (Riegler et al. 1988) to study the ^{56}Co γ-ray line emissions from the SN1987A remnant. The balloon experiments were carried out at atmospheric depths, ranging from ~ 5 g cm^{-2} to ~ 3 g cm^{-2}. To view the LMC the γ-ray telescopes must point off the zenith direction by some angle, θ_z, depending on the balloon's location. This requires a secant θ_z correction (typically $\sim 40\%$) to the atmospheric thickness any SN γ-ray must penetrate to reach the detector. The geometry (greatly exaggerated) for balloon observations of the LMC, from two Southern Hemisphere locations, is shown in Figure 4; i.e. a launch from Alice Springs, Australia (S 23°) and from Antarctica (S 79°). The zenith angles shown for the two cases pertain only to the case of maximum elevation of the LMC source or at culmination.

The first positive observation of a ^{56}Co line from the balloon campaigns was made by the Lockheed/Marshall Space Flight Center in late 1987 October, using a large high-energy resolution γ-ray spectrometer (Chase et al. 1988). The flux for the 847 keV line is quoted as $(10 \pm 2.8) \times 10^{-4}$ γ cm^{-2} s^{-1}, where the experimenters quote the errors as statistical. It is also of significance that these observers do not report a positive detection of the 1238 keV ^{56}Co line, but the flux cannot be greater than 0.4 times the flux of the 847 keV line (Chase et al. 1988). This result is not inconsistent with the positive report by SMM, because of the large errors involved (cf Table 1 and Figure 3).

The first image of SN1987A remnant in γ rays was made by the Caltech group, using a coded aperture telescope on 1987 November 19 (Cook et al. 1988). The first reports gave results on the continuum spectrum consistent with the Mir-Kvant results, (Sunyaev et al. 1987). In the energy intervals (801–833) keV and (1185–1281) keV, the integral flux

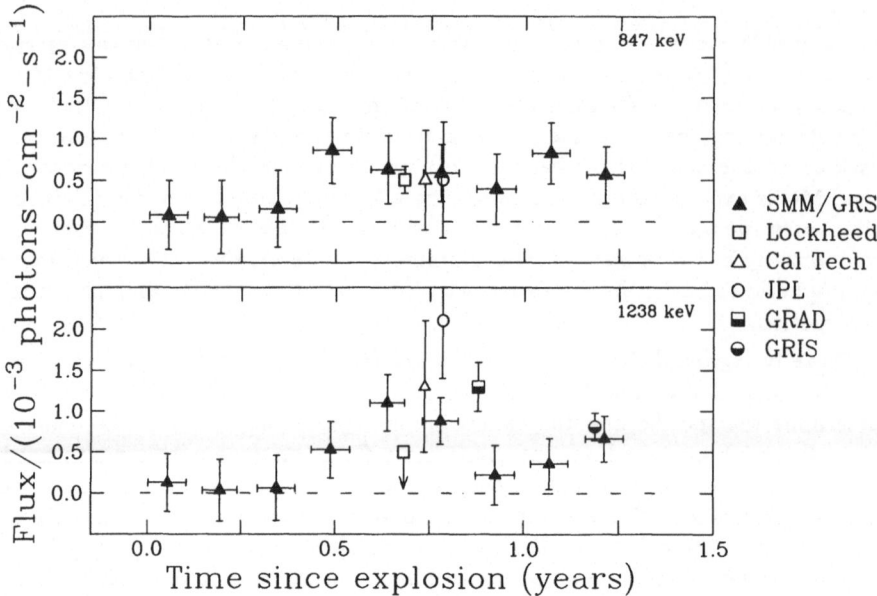

Figure 3.   The flux time history of the 847 keV and 1238 keV ^{56}Co lines is shown as measured by the *SMM* GRS. Flux values obtained by balloon experiments are also shown as discussed in the text.

Table 1.
SN1987A Gamma-Ray Lines -
Capabilities and Selected Fluxes

Instrument	Date(s)	Flux Limit (1 σ) (10^{-4})$_7$ cm^{-2} s^{-1}	Flux Reported (10^{-4})$_7$ cm^{-2} s^{-1}	Peak Position keV
SMM GRS	87 Aug 1 - Sept 7	847 (2) 1238 (1.6)	8.6±4 5.3±3.5	840±6 1239±10
LPARL/MSFC	87 Oct 29 - Oct 31	847 (3.5) 1238 (4.5)	10+2.8 -	838 - 850
CIT GRIP	87 Nov 19	847 (3.2) 1238 (4.9)	5±6 13±8	
JPL	87 Dec 7	847 (2.5) 1238 (3.9)	- 21±7	1240.8±1.7
UF/GSFC GRAD	88 Jan 8 - Jan 11	847 (8.6) 1238 (10.2) 2599 (12.3)	13±3 3.6±2.1	1239.6±1.5 2602.5±1.7 3.6±2.1
SMM GRS	88 Apr 22 - May 29	847 (1.7) 1238 (1.4)	5.6±3.4 6.6±2.9	840±6 1239±10
GSFC/BL/SD GRIS	88 May 2	847 (1.4) 1238 (1.7)	8.1±1.7	1234.9$^{+1.1}_{-1.3}$

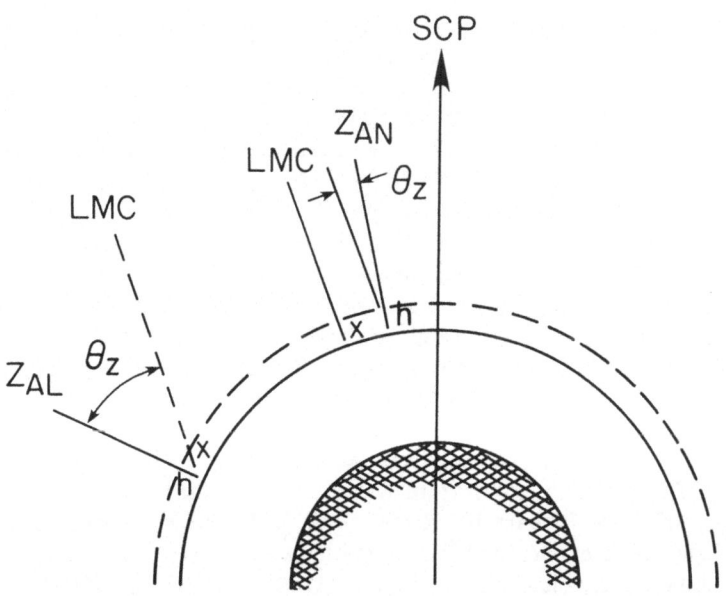

Figure 4. The geometry is shown for SN1987A viewing at culmination by balloon instruments at balloon altitudes in the Southern Hemisphere. The minimum zenith angle θ_z is shown.

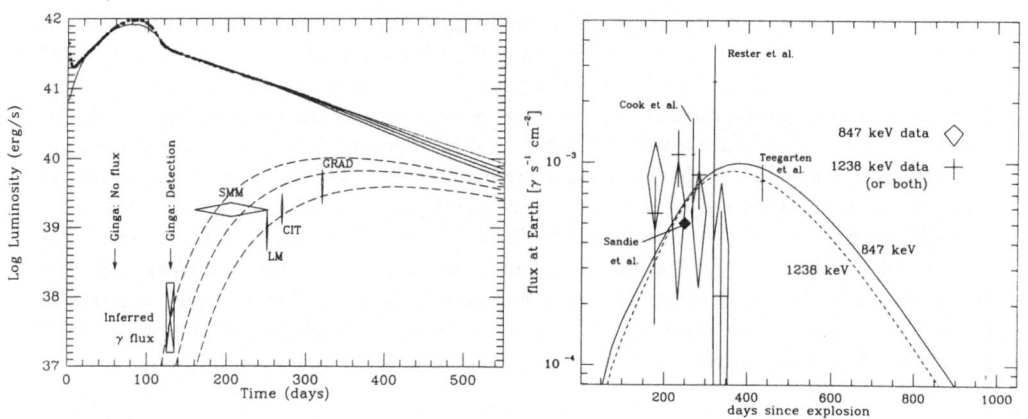

Figure 5. (left) The theoretical γ ray light curves are shown for ^{56}Co mixing models from Arnett and Fu (1988).

Figure 6. (right) The theoretical γ ray light curve is shown for Pinto and Woosley (1988) mixed model 10 HMM.

values are (1.1 ± 0.6) and (1.2 ± 0.7), respectively, in units of 10^{-3} photons cm^{-2} s^{-1}. It is important to remember that these flux values contain both continuum and any line contribution. When lines at the instrument width are fit to the spectrum, the resulting flux values (cm^{-2} s^{-1}) are $(5 \pm 6) \times 10^{-4}$ and $(13 \pm 8) \times 10^{-4}$ for the 847 keV and 1238 keV lines, respectively. These values (of low significance) are shown in column 4 of Table 1 and are consistent with the average *SMM* fluxes for the time period November 16 – December 22 of 1987, which are shown in Figure 3.

The next balloon borne spectrometer which observed the LMC region was launched by the JPL group on 1987 December 6 (Mahoney *et al.* 1988). A positive flux value was found only for a line at (1240.8 ± 1.7) keV at a level $(2.1 \pm 0.7) \times 10^{-3}$ photons cm^{-2} s^{-1}. This value is higher than the *SMM* value given above for the corresponding time interval but the reason for this discrepancy is not known. The flux for a line at 847 keV was $(5 \pm 7) \times 10^{-4}$ photons cm^{-2} s^{-1}, clearly an upper limit which is not inconsistent with the corresponding *SMM* average (see Figure 3).

A DOD supported payload (*GRAD*) was launched by the University of Florida/ Goddard Space Flight Center group from Antarctica (McMurdo) on 1988 January 8 and stayed aloft until late January 10. Rester *et al.* (1988) have reported the presence of strongly red-shifted components of ^{56}Co γ-ray lines at 847 keV, 1238 keV, 2599 keV. They also reported slightly blue-shifted peaks at (1239 ± 1.5) keV and (2602 ± 1.7) keV which are very near the rest energies for the 847 keV and 1238 keV lines. For direct comparison with the other results discussed in this report we discuss only the "blue-shifted" lines and leave the reader to the experimenter's report (Rester *et al.* 1988) for other details. Column 4 of Table 1 gives the reported flux values for the 847 keV line, which again is somewhat higher than the *SMM* average value for the corresponding time period. This experiment gave the only evidence for detectable flux for the 2599 keV line, but the experimenters' significance is $< 2\ \sigma$.

The largest germanium spectrometer (*GRIS*) used thus far for SN1987A observations was launched by the GSFC/Bell Laboratory/Sandia Laboratory consortium on 1988 May 2 (Barthemly *et al.* 1988). The analysis of the experimental results (at the time of this report), is still preliminary, however, a very significant ($\sim 5\ \sigma$) flux value for the 1238 keV line is reported at $((8.1 \pm 1.7) \times 10^{-4}\ \gamma$ cm^{-2} s^{-1}, with the error only statistical. This value is also consistent with the last *SMM* average value, shown in Table 1. The statistics were good enough in this flight to determine that the 1238 keV line width was $\sim (25 \pm 5)$ keV, centered at $1234^{+2.1}_{-2.3}$ keV (Teegarden 1988). These latter values are preliminary and include an estimate of systematic errors. Future flights of the largest high-resolution spectrometers will clearly be needed to measure the rate of decay of the ^{56}Co line fluxes and it is heartening to learn that NASA will include *GRIS* in its 1988 November Australia Campaign.

2. Spectrometer Capabilities

We now review the capabilities of the γ-ray spectrometers (to measure a line flux) under the actual experimental conditions encountered during SN1987A observations. The method used for this assessment is to calculate for each spectrometer, the 1 σ statistical limit (expressed in flux units) for a null flux observation.

This is

$$F_{min}(\text{Line}) = \frac{3}{S(cm^2)} \left(\frac{2\dot{B}\Delta E}{T_{S,B}} \right)^{1/2} e^{\mu_\ell x} \tag{1}$$

where $\dot{B}$ (counts s^{-1} keV^{-1}) is the measured background count rate, ΔE is the FWHM band width for a given line observation, $T_{S,B}$ is the source or background observation time (whichever is smaller) and $\exp(\mu_\ell x)$ is the correction factor for an observation made under an atmospheric thickness x(g cm^{-2}) where μ_ℓ(cm^2 g^{-1}) is the absorption coeffecient for a given line and finally $S(cm^2)$ is the effective area for a given line.

In Table 1, column 3 is shown the 1 σ statistical flux limits (in parenthisis) for the conditions of the first *SMM* observation in August 1987 and for several later balloon observations and for the last completed *SMM* observation. In column 4 of Table 1. are shown the line flux values reported for each observation along with the quoted error. These errors should be compared with the values in parenthesis in column three.

In using Equation (1) to calculate the flux limits shown in column three, the experimenters stated line width for each specific line with a quoted flux value was used. Also, some experimental flux errors include estimates for systematic errors with the exception of the LPARL/MSFC and the GSFC/BL/SL values. In most cases, the statistical 1 σ limits are well below the quoted flux errors so the significance of the flux values can be judged with some confidence. In two cases, the second and fifth entries, the quoted errors are somewhat below the statistical limits indicating that the corresponding flux values are probably not as significant as indicated. The general conclusion that one can make from Table 1 is that several experiments have confirmed positive flux values for the 847 keV and 1238 keV lines.

3. Comparison with Theory

It is now widely appreciated that the earlier than expected appearance of γ-ray lines in 1988 August required that some small fraction ($\sim$ 1%) of the ^{56}Co expected at that time had to be fully exposed with the greater portion of ^{56}Co at sufficient depths to explain the hard X-ray continuum spectrum (Sunyaev et al. 1987) by Compton down scattering of the lines. The time of appearance of detectable γ-ray line fluxes at 847 keV and 1238 keV was expected several months later than observed if the requisite amount of original ^{56}Ni was confined to a thin layer under the expanding and thinning envelope. Under these conditions one would expect to first see the down-scattered hard X-ray continuum and later the lines. The (near) simultaneous appearance of the lines and continuum suggests that some form of redistribution of the radioactive material is taking place, such as mixing, fragmentation, etc. Several theoretical calculations are now available that model the expanding envelope with various assumptions about mixing etc.. In Figure 5 we show an example of the γ-ray light curve of Arnett and Fu (1988), which corresponds to the case of initial production of 0.073 M$_\odot$ of ^{56}Ni and 7.5 M$_\odot$ of ejecta and for different choices for the density structure in the ejecta, giving the three different γ-ray light curves shown dotted in the figure. In all three cases it was assumed that mixing of ^{56}Co occurred out to a fractional radius of 0.4 into the ejecta. The corresponding bolometric light curves are also shown in the figure. Another example for a theoretical γ-ray light curve, with mixing of ^{56}Co in the envelope, is given by the 10 HMM model of Pinto and Woosley (1988) (see Figure 6). In this case the envelope model is 10 $M_\odot$ with 0.075$M_\odot$ of synthesized ^{56}Ni. The mixing is accomplished by radioactive heating of a small amount of the ^{56}Co which is accelerated to a high velocity, and then plows into the envelope and core. Only a few percent of the ^{56}Co needs to be redistributed to low γ-ray optical depths to drastically affect the γ-ray line light curve.

4. Conclusions

The basic result of this review is that *SMM* GRS and several balloon instruments have confirmed the detection of ^{56}Co γ-ray lines at 847 keV and 1238 keV. Further, the observations are consistent with a constant flux level for the two lines through 1988 May with $\bar{F}(1238)/\bar{F}(847) \simeq 1 \pm 0.5$. There is, in this authors' opinion, no evidence for a significant shift of the centroids of the lines from their rest energies. Some experimenters claim measurable, small blue shifts from the rest energies, but intercalibration of instruments is required. At this point we conclude that $(\Delta E/E_0) < 0.002$, corresponding to $v_- < 700$ km s^{-1}. There is, however, now good evidence that the 1238 keV line is broadened to ~ 25 keV (FWHM) about the rest energy. Finally, Gehrels *et al.* (1988) in a recent summary have pointed out that the ratio of the γ-ray continuum flux in the energy range (50–500) keV to the 847 keV flux has remained constant at about 8 through 1988 April. This ratio should be about 30 if all the ^{56}Co were at a constant depth. Further analysis of the balloon and *SMM* results is in progress.

5. References

Arnett, W.D. and Fu, A. (1988), Submitted to Astrophysical Journal.

Barthemly, S. *et al.* (1988), IAU Circular 4593.

Chase, L.F., Jr. *et al.* (1988), Submitted to Astrophysical Journal (Letters).

Cook, W.R. *et al.* (1988), Submitted to Astrophysical Journal (letters).

Gehrels, N., Leventhal, M. and MacCallum, C. J. (1988), N. Gehrels and
    G.H. Share (eds.) AIP Proc. Workshop on Nuclear Spectroscopy, of
    Astrophysical Sources, 1988, p. 87.

Mahoney, W.A. *et al.* (1988), Submitted to Astrophysical Journal (Letters).

Matz, S. M. *et al.* (1988), Nature **331**, 416; IAU Circular 4618,
    and N. Gehrels and G.H. Share (eds.) AIP Proc. Workshop on Nuclear
    Spectroscopy of Astrophysical Sources, 1988, p. 51.

Pinto, P.A. and Woosley, S.E. (1988), Nature **333**, 534.

Riegler, G.R. (1987), To be published in 'Proceedings ESO Workshop on the
    SN1987A,' MPE, Garching.

Rester, A.C. *et al.* (1988), Submitted to Astrophysical Journal (Letters).

Sunyaev, R.A. *et al.* (1987), Nature **330**,227.

Teegarden, B. J. (1988), Personal Communication.

INTERPRETATION OF THE CO BANDS OF SUPERNOVA 1987A

C.M. Sharp and P. Höflich
Max-Planck-Institute für Physik und Astrophysik
Institut für Astrophysik
Karl Schwarzchild Str.1, 8046 Garching, FRG

ABSTRACT. Model calculations for the interpretations of the CO bands in the IR are discussed. Spherical symmetry is assumed. Profiles of the physical quantities (i.e. temperature, mass, electron density, abundances, etc.) are given by NLTE models which allow for the representation of observed spectra of supernova 1987A in the optical and near infrared. C and O have to be chosen overabundant by a factor of 25 in the helium-rich layers, or a clumpy density distribution structure of a corresponding factor relative to the mean local value is needed. However, the comparison between the observed forbidden OI line at about 6300 A favours the first interpretation. A factor of about 3 has to be taken into account as uncertainty for the overabundance because of the assumptions of the background model. This implies evidence for strong mixing of the helium rich region with more central layers during the stellar evolution and/or the explosion.

1. Introduction

Many of the properties of the supernova 1987A in the Large Magellanic Cloud are known fairly well, e.g., the progenitor star had about a 20-25 M_0 at the main sequence (Arnett, 1987; Hildebrandt et al., 1987; Höflich, 1987; Woosley et al., 1987).

There are, however, many questions and uncertainties related to the stellar explosion and possible mixing of different layers. Strong evidence for mixing of the H and He layers and the innermost layers of the progenitor star has been found from the light curve as well as the observed spectra (Nomoto, 1988; Woosley, 1988; Höflich, 1988). There is also some evidence of mixing of the C and O region with the He shell during or after the explosion, but there are a number of difficulties in directly observing C and O because of the lack of unblended atomic lines and strong NLTE effects. However, the molecule CO is measured in the infrared (IAU Circulars 4457, 4468, 4484, 4500; Danziger et al., 1987) which allows good determinations of the enrichment of carbon and oxygen.

D. McNally (ed.), Highlights of Astronomy, Vol. 8, 207–211.

In order to investigate a possible mixing of the C- and O-rich layers, we have calculated the CO emission features. In the first section, the background model is described, followed by a discussion on how the CO opacities are computed; then its influence on the models for different abundances are considered and comparisons are made with the observed spectra.

2. The model construction

NLTE models of scattering-dominated photospheres of type II supernovae are used as the background models and have been applied to SN 1987A. They can be characterised by the assumption of radiative equilibrium, spherical geometry and homologeous expansion of an initial density and chemical distribution. These models allow for the representation of the observed spectra including the time dependence in the optical and nearer IR-wavelength range (see Höflich 1987, 1988 for more details of the models). The electron temperature, the particle density and chemical profile, the electron density and especially the NLTE ionization stages of CI and OI have been used in order to calculate the CO abundances and cross sections (see below) for a given model. We calculated the formation of CO by the assumption of LTE relative to CI and OI. This is a reasonable assumption because the particle densities in the CO-forming regions are quite high ($\approx 10^{11}$ cm^{-3}; see Höflich, 1988), and consequently the CO formation is very fast. In the optically thin regions, it is assumed that once CO is formed, it will not be destroyed because of its high dissociation energy of 11 eV. This is a reasonable assumption because of the decrease of the collisional rates due to the decreasing density, and because of the geometrical dilution of the radiation field. The retroaction of the CO opacities on the NLTE model has been neglected, because most of the C and O ($\leq 90\%$) is in CI or OI in the relevant regions of the envelope. However, a possible cooling effect of CO on the local electron temperature should be noted.

3. Models for the CO molecule

As CO is the most tightly bound diatomic molecule and carbon and oxygen are relatively abundant, CO is usually the second most abundant molecule in stellar atmospheres and the interstellar medium after H$_2$. The IR vibration-rotation spectrum of CO is one of the simplest molecular spectra, so the monochromatic absorption coefficient can be rapidly computed using approximate techniques that are sufficiently accurate for our purposes here.

Due to its high dissociation energy, the Morse approximation is good for many of the vibrational levels. With this method the first order anharmonic correction to the harmonic oscillator is allowed for, with higher order corrections being neglected, and an efficient analytic approximation is used to calculate the vibrational band strengths (Sharp, 1988). However, higher order anharmonicity is allowed for in computing the band frequencies. Fortunately, as the ground electronic state of CO is $^1\Sigma$, the rotational fine structure of each band consists of only a single P-branch ($\Delta J = -1$) and R branch ($\Delta J = +1$). The total line strength is obtained from the Boltzmann factor for the temperature corresponding to the collisional temperature, partition function, vibrational band strength and the rotational Hönl-London factor. Non-rigidity and the coupling of vibration and rotation is allowed for in calculating the rotational levels and line frequencies, but is neglected in determining the line strengths. Each rotational line is broadened with a Gaussian profile dependent on the temperature and some assumed turbulent velocity.

The region of the spectrum that is of interest is divided up into several thousand equally spaced intervals of widths approximately comparable to the line widths. Each vibration-rotation band is computed in a progression starting with the fundamental and its "hot" bands, i.e. $\Delta v=1$, then the first overtone and its "hot" bands, i.e. $\Delta v=2$ etc. In general each profile will cover several intervals, and using the error function, the mean absorption in each interval is obtained.

4. Discussion of the synthetic spectra and applications to the SN

An atmospheric model is used as the background which allows for the representation of the optical and near-infrared spectra observed on 2 October 1987 ($T_{eff} = 4900$ K; $R_{photosphere} = 7.8\ 10^{14}$ cm; particle density $N(R_{photosphere}) = 1.5\ 10^{11} cm^{-3}$; model IX, see Höflich, 1988). The profiles of the density, electron temperature and ionization and of the excitation stages of all ions are assumed to be given by this NLTE model. The typical tempweratures in the CO-forming region are in the order of 4000 to 4500 K.

The influence of the carbon and oxygen abundances on the emitted IR spectra are investigated (Figure 1). The sensitivity of the band strength on the abundances should be noted. Below abundance of about 10 times solar, the 1st overtone system of CO can be assumed to be optically thin. However, note the finite optical thickness for higher

Figure 1: The synthetic IR spectra labelled by the assumed over-abundance of C and O relative to the solar photosphere are given as calculated for the background model (see text).

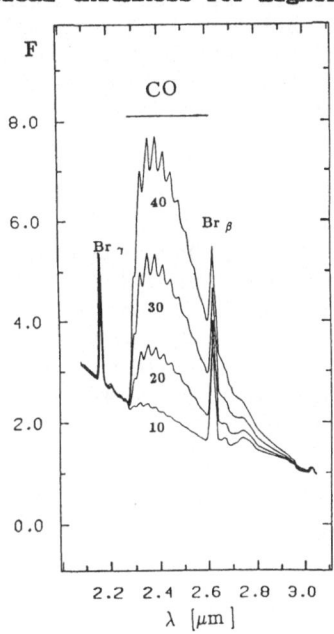

Figure 2: The optical depths of the first overtone system of CO at the Thompson optical depth of 1 (curve 1) and the maximum ratio of CO/(C+CO) (curve 2) are given as a function of the abundance relative to hydrogen in the solar photosphere as calculated for the background model (see text).

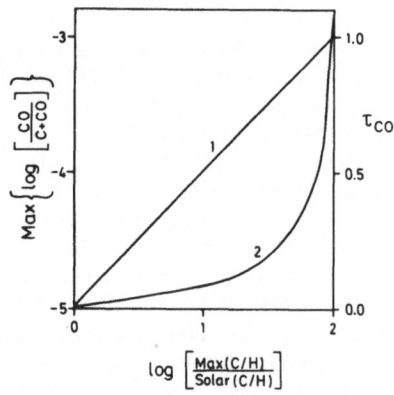

abundances. The first overtone bands are formed near the stellar photosphere in the parameter range we are interested in. Therefore, the temperatures in these layers are not strongly affected by the cooling due to this molecule.

However, the fundamental system is formed at much larger distances at which the temperature is influenced by molecular cooling. Therefore, we restrict our discussion here to the overtone bands. Note, in addition, that most of the carbon and oxygen are in the atomic states (curve 2, Figure 2). Therefore, the CO-particle density depends on the particle density of C and O only. The given abundances may also be interpreted as a locally higher density, e.g., the overabundance of 25 of CO can also be produced by local density fluctuations of the corresponding order. This would have only small influence on the calculated spectra in the optical wavelength range.

A comparison of the observed and calculated spectra (Figure 3) show good agreement for an overabundance in the order of 25 times solar in the layers just above the photosphere. This overabundance may either be interpreted as mixing of these layers with the carbon- and oxygen-rich layers that were originally located somewhat deeper in the star, or this may be due to density fluctuations. However, the observed forbidden OI line in the optical wavelength range favours the first explanation as the main effect (Höflich, 1988).

Figure 3: IR spectrum as observed by ESO between 1 and 6 Oct. 1987 (upper curve shifted by two units) in comparison with the reddened, synthetic spectrum (lower curve, 25 * solar C and O, $E_{B-V}=0.15$). In addition, some line identifications are given.

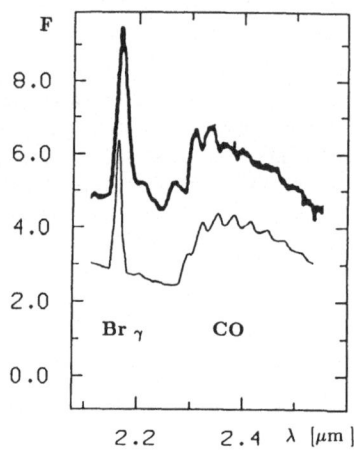

5. Summary and conclusions

We have demonstrated that the CO emission of SN1987A as observed after about 7 months is well explained by a model which allows for the representation of the optical spectra if a higher particle density of carbon and oxygen by a factor of about 25 is assumed. This may be either due to an overabundance of C and O in the helium rich layers or due to local temporary density fluctuations. However, the optical spectrum of the forbidden OI feature favours the first interpretation, and therefore indicates some mixing of the helium rich stellar layers with those below. Additionally, we already know that the H and He layers as well as the inner layers are mixed (Nomoto, 1988; Woosley, 1988; Höflich, 1988a). We can conclude that all layers have to undergo some mixing and a detailed determination of the chemical profiles by the observed spectra may give important information concerning the typical mixing scales. However, the observed OI line profiles (Dopita, personal communication, 1988), the FeII and CoII lines indicate some inhomogeneities of the density structure (Ericson, personal communication; Hilldebrandt *et al.*, 1988). Therefore, some contribution caused by the clumpiness of the matter of our CO determination has to be expected.

We have also to stress the limitations of our anaylses. The assumption of the negligible retroaction of the cooling of the molecules for the relevant regions certainly breaks down if the corresponding features are formed well above the photosphere. This is true in these conditions for the fundamental bands. Therefore, a discussion of the line ratios of the different systems needs consistent models (including the molecular cooling). Such models will enable us to determine the chemical profiles of C and O more accurately. This will be discussed in a forthcoming paper.

6. References

Arnett, W.D. 1987, *Astrophys. J.*, **319**,136.

Danziger, I.J., Bouchet, P., Fosbury, R.A.E., Gouiffes, C.; Lucy, L.B., Moorwood, A.F.M., Oliva, E., and Rufener, F. 1987. George Mason Conference, October 1987.

Hilldebrandt, W., Höflich, P., Truran,J.W., and Weiss, A. 1987, *Nature*, **327**, 597.

Hilldebrandt, W., Höflich, P., Janka, H.- T., and Monchmeyer, R. 1988. To appear in *Proceedings of the ESO-CERN Workshop*, June 1988.

Höflich, P. 1987, in "ESO-workshop on SN1987A", ed. I.J.Danziger, Garching, p.447.

Höflich, P. 1988, in *Proceedings of the Astronomical Society of Australia*, Canberra, June 1988, in press.

Nomoto, K. 1988, preprint.

Sharp, C.M. 1988, *Astron. Astrophys. Suppl. Ser.* **72**, 355.

Woosley, S.E. 1988, preprint.

Woosley, S.E., Pinto, P.A., and Ensman,L. 1987, *Astrophys.J.*, **324**, 466.

THREE DIMENSIONAL HYDRODYNAMICAL SIMULATION OF TYPE II SUPERNOVA

—— *Mixing and Fragmentation of Ejecta* ——

MIKIO NAGASAWA
National Astronomical Observatory
Mitaka, Tokyo 181
Japan

ABSTRACT. Adiabatic supernova explosions of polytropic stars are investigated by a three dimensional Smoothed Particle Hydrodynamics. The evolution of thermal point explosions is almost spherically symmetric in a global sense, but they are found to be unstable against Rayleigh-Taylor instabilities. The typical unstable wavelength, which grows in the nonlinear stage, is comparable to the thickness of the spherical shell. As a result, we find a porous density structure on the expanding shell. These results suggest the clumpiness of the ejecta of supernova explosions. The accompanying mixing motion in the expanding shell can explain the rapidly rising light curve of SN1987A. Because it may mix up the energy source ^{56}Ni towards the outer layers of supernovae.

1. Introduction

In hydrodynamical studies of supernova explosions, there have been several suggestions that expanding spherical shells are unstable against non-spherical perturbations. The deflagration fronts in Type I supernovae are not always spherical but showed finger-like shapes (Müller and Arnett 1986) . Tomisaka and Ikeuchi (1983) studied the stability of spherical shells against the non-radial perturbations and found that there are Rayleigh-Taylor instability modes for short wavelength perturbations.

Recently there are several observations which suggest non-spherical motion in the ejecta of SN1987A. The first is the observation of large linear polarizations, which suggests that the supernova ejecta itself is not spherical but prolate or oblate with a certain aspect ratio (Barret 1987) . The other is the observation of X-ray from SN1987A. It is suggested that the ^{56}Ni is mixed toward the surface due to convection or Rayleigh-Taylor instability (Itoh et al. 1987) .

D. McNally (ed.), Highlights of Astronomy, Vol. 8, 213–214.
© *1989 by the IAU.*

2. Numerical Models

There have been several simulations of axially symmetric explosions (Bodenheimer and Woosley 1983) . In such restricted systems, some important non-axisymmetric instability modes may have been omitted from the beginning. We try to study supernova explosions including non-axisymmetric modes (Nagasawa et al. 1988) .

We have developed our SPH code to be able to solve the energy equation to treat the adiabatic problem with arbitrary equation of state, not restricted for polytropic gas. Our code has resolutions fine enough to study the criterion for gravitational instabilities or to simulate the analytic shock solutions. (Nagasawa and Miyama 1987) . We solved the adiabatic explosion of polytropic stars.

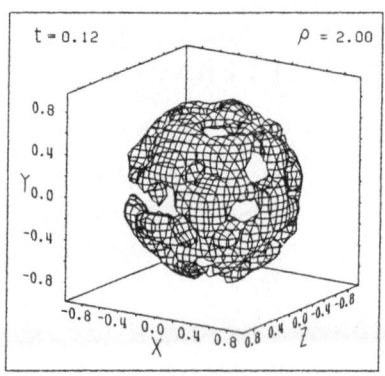

Fig.1. The 3D equidensity surface shows the void structure although the general motion is a spherical expansion.

Because of the large explosion energy, self-gravity during the explosion has been neglected.

3. Results

Thermal point explosions induce spherical shell expansions similar to that in one dimensional calculations and the Rayleigh-Taylor instability raises the density fluctuations in that shell. First in the linear stage, the small size of fluctuations appear on the contact discontinuity. The typical unstable wavelength, in the nonlinear stage, is comparable to the thickness of the spherical shell. The initial fluctuation is introduced as a numerical noise when we construct the 3-D equilibrium spheres by relaxation. The ejected gas, observed with a certain density level, seems very clumpy. The efficiency of the disruption does not increase with the explosion energy because the fast expansion has a shorter time before the break-up beyond the initial radius. The convection between shock front and contact discontinuity causes effective mixing in the shell. While, due to the phase cancellation of complicated fragments, there is no dominant contribution to the degree of polarization.

References

Barret, P., 1987. *ESO Workshop on SN1987A*, 174.

Bodenheimer, P., and Woosley, S., 1983. *Asrtrophys. J.* , **269**, 281.

Itoh, M., Kumagai, S., Shigeyama, T., Nomoto, K., and Nishimura, J., 1987. *Nature*, **330**, 233.

Müller, E., and Arnett, W.D., 1986. *Asrtrophys. J.* , **307**, 619.

Nagasawa, M., and Miyama, S.M., 1987. *Prog. Theor. Phys.*, **70**, 747.

Nagasawa, M., Nakamura, T., and Miyama, S.M., 1988. *Publ. Astron. Soc. Japan*, **40**, No.6.

Tomisaka, K., and Ikeuchi, S., 1983. *Publ. Astron. Soc. Japan*, **35**, 187.

NLTE CALCULATIONS OF HYDROGEN LINE PROFILES FOR SN1987A

Werner Schmutz
Joint Institute for Laboratory Astrophysics
University of Colorado and National Bureau of Standards
Boulder, Colorado 80309-0440

The synthetic line profiles presented here are the result of a series of models:
- Woosley's (1988) calculation of the stellar explosion yield the **luminosity**, the **density** and the **velocity** structure of the expanding SN as a function of time (Woosley's model 10HM was used)
- subsequent Monte Carlo simulations of the radiation transport in the expanding photosphere are based on the atmosphere structure given by the model above for a given time. The calculations take into account tens of thousands of lines and yield the line blanketed continuum flux and the **temperature** structure.
- The previous two models specify the physical condition of the expanding shell. With a third model the NLTE populations of 10 H and 12 He levels are determined and the emergent hydrogen lines profiles are obtained. The method of solving the NLTE problem is described by Wessolowski et al. (1988) and references therein.

The resulting hydrogen line profiles are given in Fig. 1. They are the result of essentially **ad initio** calculations: the free parameters are already determined by Woosley's calculations. The agreement of the theoretical profiles with the observed ones has to be called not too bad, keeping in mind how they are obtained: the ad initio modeling. But certainly, the synthetic profiles do not reproduce the observed ones. The question is, why not? In order to get an answer, several model parameters have been varied:
- the number of levels of the atomic hydrogen model
- the helium abundance
- the temperature outside the continuum formation region
- the luminosity (i.e. temperature structure in deeper layers)
- the treatment of line blanketing in the NLTE calculations
- the density structure
With the exception of the last point, all these variations resulted only in negligible to moderate changes of the line profiles, which were clearly not able to explain the observed discrepancy. However, the variation of the density structure caused a change of the desired order of magnitude. This might be the explanation why Höflich (1988)

215

D. McNally (ed.), Highlights of Astronomy, Vol. 8, 215–216.
© *1989 by the IAU.*

216

is able to reproduce the observed hydrogen line profiles. However, Höflich's calculations are based on a completely different density structure than the one resulting from Woosley's model-calculation.

This work was done in collaboration with Dave Abbott and Bob Russell. The financial support is acknowledged to the Swiss NF and the NSF Grant AST-8802937.

References
Höflich,P.: 1988, ... Coll. 108, 288
Wessolowski,U., Schmutz,W., Hamann,W.-R.: 1988, A. & A. 194, 160
Woosley,S.E.: 1988, Ap. J. 330, 218

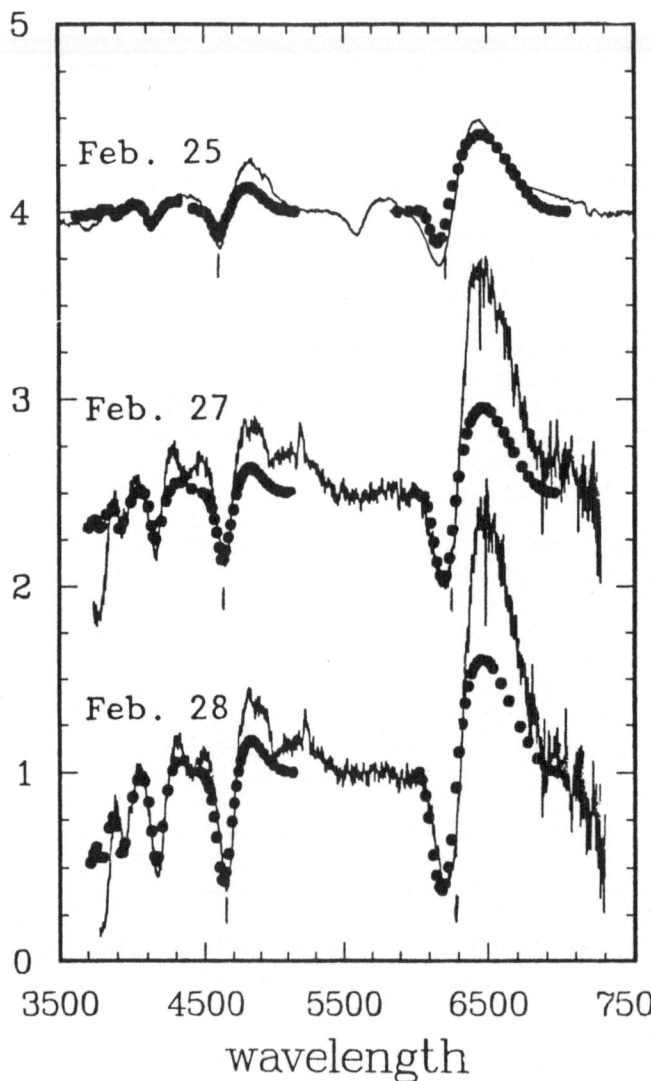

Figure 1. Normalized spectra of SN1987A for three dates: 4:00 UT Feb. 25, 0:40 UT Feb. 27. and 0:30 UT Feb. 28. Superimposed are the synthetic hydrogen line profiles (dots), calculated for the corresponding dates. The tick-marks below the H alpha and H beta profiles indicate the blue-shifted wavelength of the line corresponding to the expansion velocity at the location where the continuum is formed (Rosseland optical depth 2/3).

NON-EQUILIBRIUM THERMAL X-RAY EMISSION IN THE EARLY PHASE OF SUPERNOVA REMNANT

Hitoshi HANAMI and Tatsuo YOSHIDA
Department of Physics, I.F.S., University of Texas
Austin, TX 78712-1060
Department of Physics, Hokkaido University, Sapporo 060, Japan

Abstract. The X-ray emission from the interaction of the ejecta with circumstellar matter (CSM) for SN1987A is investigated. The electron and the ion temperatures seem to be in non-equilibrium in the early phase of a SN remnant. We have studied the two-temperature model in the early phase of SNR and discuss the X-ray emission from SN1987A and its CSM.

1. Introduction

It is implied that the density of the CSM of SN1987A is lower than that of SN1979C and SN1980K from radio observations (Turtle *et al*.1987). The progenitors of SN1979C and SN1980K seem to be red supergiants which had formed the dense CSM by winds. Indeed, from the positional coincidence and the observation by IUE, the blue supergiant called Sanduleak 69 202 is identified as the progenitor of 1987A (Gilmozzi *et al*. 1987). Then it was predicted that Ginga Observatory could not detect the X-ray emission from SN1987A because the density of CSM estimated by radio observation is low. However, since day 130 after the explosion, the thermal emission has been observed by Ginga (Dotani *et al*. 1987). Itoh *et al*. (1987) and Masai *et al*. (1987) have expected that the X-ray emission is enhanced when the shock hits the dense CSM formed in the red giant stage.

In the present paper we give a simple analytic method of getting the information about the CSM density from the X-ray observations. We estimate the electron temperature and the X-ray luminosity, which depends on the electron temperature, in a wide range of the parameter of the CSM density $\dot{M}/V_W$. Our simple analytic methods consists of two parts which are self-similar analysis for the dynamical properties and two-temperature model for the thermal process.

2. Analytic Method for the Two Temperature Model in the Early Phase of SNR

2.1 Self-similar solutions for the interaction region.

In early phase ejecta collide with the CSM, and the interaction region

217

with shocked ejecta and shocked CSM, which are separated by the contact discontinuity, is formed between outer and inner shocks. Chevalier (1982) investigated the interaction of the expanding ejecta with CSM by similarity analysis. Following him, we assume that the expanding ejecta and the CSM have power law density profiles which are represented by the density index n and s respectively. From dimensional analysis, the radius of the contact discontinuity which separates shocked ejecta and shocked CSM shells is given by

$$R_C = (Ag^R/q)^{1/(n-s)}t^{1/L}, \quad L = (n-s)/(n-3), \quad (1)$$

where g and q are constants related to the explosion energy E and mass M of ejecta and the density of the CSM. If the CSM is formed by steady mass loss with a constant terminal velocity, then s=2 and q = M/(4 V_W). A is a constant which can be determined from the result by this analysis. By using suitable similarity variables, we can get the structure of the density, the pressure and velocity in the shocked interaction region.

2.2 Two-temperature model

Itoh (1987) pointed out that the one-temperature model of SNRs cannot explain the observed X-ray temperature for young SNRs. He studied the two-temperature model that the electron temperature is not equilibrated with the ion temperature behind the shock front in the sedov phase. Here, in order to investigate the X-ray emission in the early phase, we extend the above one-temperature model to a two-temperature model.

We assume that the post-shock electrons are heated only through Coulomb collisions with the ions. The evolution of the normalized electron temperature $g_e = T_e/T$ for a fluid element represented in Lagrangian coordinate is given by

$$3/2 \ln((1+g_e^{1/2})/(1-g_e^{1/2})) - g_e^{1/2}(g_e+3) =$$

$$\ln A/80 \, n_i/T^{3/2}(t-t_o) = f(t) \quad (2)$$

where we use cgs units. T_e is the electron temperature; T is the mean temperature. Eq. (2) means that the electron temperature is determined by heating time $t-t_o$ and the postshock conditions. We can derive a useful form of the function f(t) rewritten by a temporal function with R_c and a function of similarity variables. Furthermore, in order to obtain g_e from f(t), Eq. (2) must be inverted. We can use the convenient explicit form with f(t) for g_e found by Cox and Anderson (1982),

$$g_e = 1 - \exp[-(5/3f)^{0.4}\{1+0.3(5/3f)^{0.6}\}] \quad (3)$$

Then we obtain the spatial distribution of the electron temperature in the interaction region at any time in the early phase. As an example, we show in Fig. 1 the spatial distribution of the electron and the mean temperature by the solid line and the dashed curve for the case s=2 and n=7. The labelling numbers indicate $\dot{M}/V_W$ normalized with 10^{-6} M_e/yr and 1000 km/s ($E=10^{51}$ erg and $M=1M_{\odot}$). The average electron temperature in inner shell and outer shell of the interaction region is also obtained. In the almost phase of non-equilibrium state, $f(t) \ll 1$. Then g_e is nearly equal to $(5/3f)^{0.4}$ which can be shown as the asymptotic form of Eq.(3) in the limit $f(t) \ll 1$. Then the time evolution of T_e is given by

$$\langle T_e \rangle_{s=2} \propto t^{-0.4} \quad . \tag{4}$$

By the use of the average electron temperature obtained above, we can estimate the X-ray luminosities of the outer and inner shells. Here we use the following approximate expressions: When $0.01\text{keV} < T_e < 1\text{kev}$ bound-bound transitions dominant the X-ray radiation process; then the cooling function is given by

$$A_L(T_e) = 9.26 * 10^{-24} T_e^{-1} \quad \text{erg cm}^{-3} \text{ s}^{-1} , \tag{5.1}$$

where T_e is in units of 1 keV and coronal ionization equilibrium is assumed. When T_e is higher than 1 keV, thermal bremsstrahlung dominates and the cooling function is represented by

$$A_L(T_e) = 5.9 * 10^{-24} T_e^{-1} \quad \text{erg cm}^{-3} \text{ s}^{-1} . \tag{5.2}$$

We obtain the luminosities by putting $\langle T_e \rangle$ into Eqns.(5.1) and (5.2). Then we compare them with the observed one.

3. Discussion of the CSM of SN1987A and Conclusions

As an example of the application of our analysis we mainly discuss the X-ray emission from 4 keV to 12 keV observed in the region of SN1987A by the X-ray astronomy satellite Ginga. [The hard X-ray (12-200 keV) emission observed by Ginga and Kvant is not discussed here. It can be explained by Compton scattering process of γ-rays from ^{56}Co decay (Sunyaev et al. 1987; Itoh et al. 1987)]. The observed spectrum on 3 September 1987 can be fitted by a composite of a thermal bremsstrahlung spectrum with temperature of 4 keV and a flat spectrum. If the X-ray emission above 12 keV is from Compton degradation of γ-rays, the observed spectrum between 4-12 keV is fitted by only a thermal bremsstrahlung spectrum with electron temperature of 10-12 keV. The luminosity in the range 4-12 keV was about $7 * 10^{36}$ erg/s.

If X-rays in the above range are emitted by the collision of the ejecta with the CSM, we can estimate $\dot{M}/V_w$ by comparing the observed luminosity with the estimated luminosity and compare the temperature expected from the observed spectrum with the average temperature derived here.

For the density profile of the outer region of the ejecta, Colgate and McKee (1987) have found that the explosion of a star with polytropic index results in the outer density distribution of the ejecta whose density profile index n is approximately 7. For the CSM the index of the density profile is 2, assuming a steady mass loss rate and a constant terminal velocity. Therefore, we mainly discuss the case s=2 and n=7. In this case the X-ray luminosities due to thermal bremsstrahlung are written for the outer and inner shells as

$$L_{ffout} = 1.1 * 10^{30}(E^2/M)^{-1/5}(\dot{M}/V_W)^{12/5} \ ty^{-1} \ erg/s$$

and

$$L_{ff \ in} = 6.0 * 10^{30}(E^2/M)^{-1/5}(\dot{M}/V_W)^{13/5} \ ty^{-1} \ erg/s.$$

In the detector range of Ginga, the bremsstrahlung emission is dominant. The luminosity in the inner shell is larger than that in the outer shell because the density on the shocked ejecta is higher. The outer and inner average electron temperatures are given by

$$\langle T_e \rangle_{out} = 1.3(\dot{M}/V_W)^{-2/5} \ keV, \ and \ \langle T_e \rangle_{in} = 0.49(\dot{M}/V_W)^{-2/5} \ keV.$$

In Fig.2 the total luminosity L_x and the average electron temperature are expressed as a function of $\dot{M}/V_W$ where we take $ty=0.5$, $E_{51}=1$, and $M_1=1$. $ty=0.5$ is the time from the explosion to 3 Sept. 1987; L_x hardly depends on E and M. As shown in Fig.2, the value of $\dot{M}/V_W$ which corresponds to the luminosity observed by Ginga is 240. Then the outer shock radius is 0.02 pc. The average electron temperatures in the outer and inner shells are 16 keV and 5.7 keV respectively. We may say that these temperatures are in good agreement with the spectrum observed by Ginga.

However, Chevalier and Fransson (1987) have deduced $\dot{M}/V_W=16$ from the radio observation of SN1987A on days 2.1 and 3.1 after the explosion. If the ejecta continue to interact with the CSM of this low value, X-ray emission could not be detected by Ginga. Therefore, the above difference between values $\dot{M}/V_W$ estimated by X-ray and radio observations seems to mean that the CSM of SN1987A consists of two CSM formed by the winds of different phase. As one possibility the dense outer CSM was forced by the wind from the red supergiant. After the red stage is over, the star contracts and becomes a blue supergiant. Then the low density CSM is formed by the high velocity wind. As another possibility, the outer was formed by the couldlet-evaporating wind from the blue supergiant. In this case, a hot wind-blown bubble, trapped by the cloudlets around the progenitor, might be formed like nebulae around the WR stars (Hanami and Sakashita 1987). Then we conclude that when the outer shock passed the boundary of the two CSM of different densities and reached the outer CSM, the X-ray emission was detected by Ginga.

4. References

Chevalier, R.A. 1982, *Astrophys.J.* **258**, 790.
Chevalier, R.A. and Fransson, C. 1987, *Nature* **328, 44.**
Colgate, S.A. and McKee, C. 1969, *Astrophys.J.* **157**, 623.
Cox, D.P. and Anderson, P.R. 1982, *Astrophys.J.* **253**, 263.
Dotani, T. *et al.* 1987, *Nature* **330**, 230.
Gilmozzi, R. *et al.* 1987, *Nature* **328**, 318.
Hanami, H. and Sakashita, S. 1987, *Astron. Astrophys.* 181, 343.
Itoh, H. 1978, *Publ.Astron.Soc.Japan* 30, 489.
Itoh, H. 1979, *Publ.Astron.Soc.Japan* 31, 429E.

Itoh, H., Hayakawa, S., Masai, K., and Nomoto, K. 1987, *Publ.Astr.Soc. Japan* **39**, 529.

Itoh, H. *et al.* 1987, *Nature*, **330**, 233.

Masai, S., Hayakawa, S., Itoh, H. and Nomoto, K. 1987, *Nature* **330**, 233.

Turtle, A.J. *et al.* 1987, *Nature*, **329**, 38.

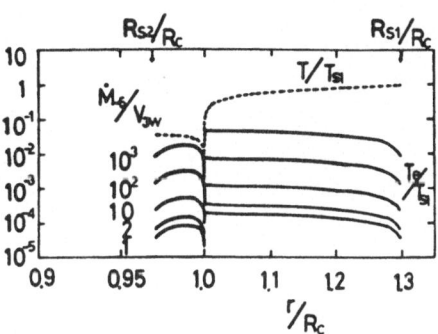

Fig.1. The spatial distributions of the electron normalized to the outer shock temperature, T_e/T_s are plotted (solid curves) against the radial distance. The normalized mean temperature T/T_s is also shown for the case s=2, n=7 (dotted curve).

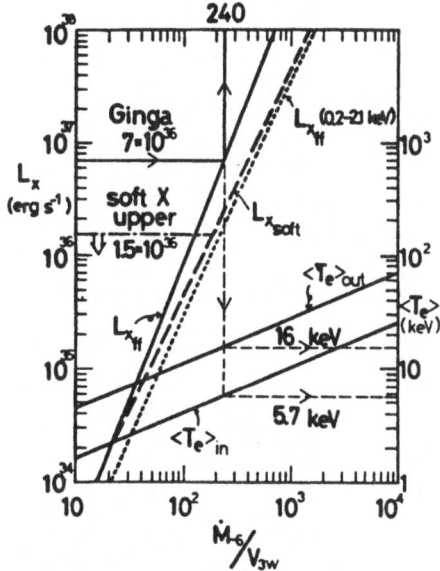

Fig.2. For the case s=2, n=7, ty=0.5, $E_{51}=1$, and $M_1=1$, the dependence of the X-ray luminosities on $\dot{M}/V_W$ is shown. The solid line represents the luminosity due to thermal bremsstrahlung between 0.2-2.1 keV. The dashed line shows the total soft X-ray luminosity with that due to bound-bound transitions. The outer and the inner average electron temperatures are also plotted as a function of $\dot{M}/V_W$.

Effects of the soft X-ray burst from SN 1987A on its circumstellar medium

Peter Lundqvist[1] and Claes Fransson[2]

[1]Lund Observatory, Box 43, S-221 00 Lund, Sweden
[2]Stockholm Observatory, S-133 00 Saltsjöbaden, Sweden

ABSTRACT: The observations of the narrow UV and optical emission lines from SN 1987A are modelled as reprocessed radiation in a shell around the supernova, heated and ionized by the soft X-ray and EUV radiation at the shock breakout. Constraints on the early soft X-ray burst are discussed, as well as the physical conditions and abundances in the shell.

1. Introduction

Since May 1987 the spectra of SN 1987A have shown a number of narrow emission lines in the UV (Fransson et al., 1989, hereafter F89). During the first ~400 days after the explosion, the luminosity of most of the lines (C III] λ1909, N III] λ1750, O III] λ1664 and N V λ1240) increased nearly linearly with time, while He II λ1640 and N IV] λ1486 were nearly constant. After ~400 days a decline of the line fluxes started (Sonneborn et al., 1988). Narrow optical [O III], He II λ4686 and Balmer emission lines have also been observed (Wampler and Richichi 1988). For a more detailed discussion of the observations, we refer to the paper by Panagia in this volume.

The observations provide important constraints on the parameters of the emitting gas. In F89, the electron density of the emitting gas was found to be ~2.6×10^4 cm^{-3}, and an analysis of the [O III] λλ 4959-5007/ 4363 ratio gave a gas temperature of ~5×10^4 K, ~300 days after the explosion (Wampler and Richichi 1988). A simple abundance analysis gave N/C and N/O ratios corresponding to ~37 and ~12 times solar, respectively (F89), indicating that the gas has been CNO-processed. The elapsed time between the explosion and the turn-over of the UV line fluxes indicates that the shell radius is ~5×10^{17} cm (Sonneborn et al. 1988), consistent with the optical determination by Wampler and Richichi.

In this paper we summarize some recent attempts to model this emission. A more detailed discussion of the constraints on the ionizing radiation, may be found in Fransson and Lundqvist (1988), and of the physical conditions of the emitting shell in Lundqvist and Fransson (1989).

D. McNally (ed.), Highlights of Astronomy, Vol. 8, 223–228.
© *1989 by the IAU.*

2. Modelling the narrow emission lines

Due to the high ionization potential of the UV line emitting ions (77 eV for N V), it was in F89 and Lundqvist and Fransson (1987) argued that the most likely source of excitation is the strong burst of EUV radiation expected to occur when the shock wave breaks through the surface of the supernova progenitor. This ionizes and heats the circumstellar gas around the progenitor. The emitting gas is likely to be in the form of a dense, thin shell at a distance, $\sim 5 \times 10^{17}$ cm from the supernova. The time evolution of the lines is then determined by a combination of light travel time effects and recombination of the gas. The *observed* emission at time t is

$$L_{obs} = \frac{c}{2 R_s} \int_{t_{min}}^{t} L_e(t_e) \, dt_e. \tag{1}$$

Here $L_e(t_e)$ is the total *emitted* luminosity at time t_e and $t_{min} = \max(\iota, t - 2 R_s/c)$. If the emissivity of the gas drops quickly after the initial ionization this becomes $L_{obs} =$ constant. If instead the emissivity is constant, $L_{obs} = c \, t \, L_e / 2 \, R_s$. Since the emissivity in the different lines is controlled by the recombination and cooling of the gas, a constant flux indicates either that the gas is recombining fast, or that the emissivity drops quickly because of cooling. A linear increase of the flux consequently implies fairly constant conditions of emission.

To test this scenario we have done detailed calculations of the temperature and ionization of the shell. Since the problem is time dependent, we have calculated the evolution of the physical quantities from the initial ionization until the final recombination of the shell. The evolution of the ionizing radiation is from hydrodynamic models of the explosion (Shigeyama et al. 1988; Woosley 1988; Arnett 1988). These show that immediately after the outbreak the effective temperature was $(2-5) \times 10^5$ K, after which the temperature decreased to less than 3×10^4 K in ~ 5 hours. The time integrated sum of the ionizing radiation from the photosphere has a form quite different from a blackbody, with an excess of soft photons.

As an illustration, we show in Fig. 1 the temperature and ionization in the middle of the shell as a function of time, together with the instantaneous and integrated (i.e. observed) line luminosities from the shell. For this model we have used the spectrum from the Shigeyama et al. 11E1Y6 model, a density of 2.6×10^4 cm^{-3}, and a shell distance of 5×10^{17} cm. The elements included are H, He, C, N, O, Ne and S with abundances He/H=0.1, N/C=5, N/O=2 and an overall metallicity of 0.3 times solar. The upper panel of Fig. 1 shows that the temperature immediately after the burst is 1.35×10^5 K. Hydrogen as well as helium is fully ionized, while most of the metals are in their helium like stages, like N

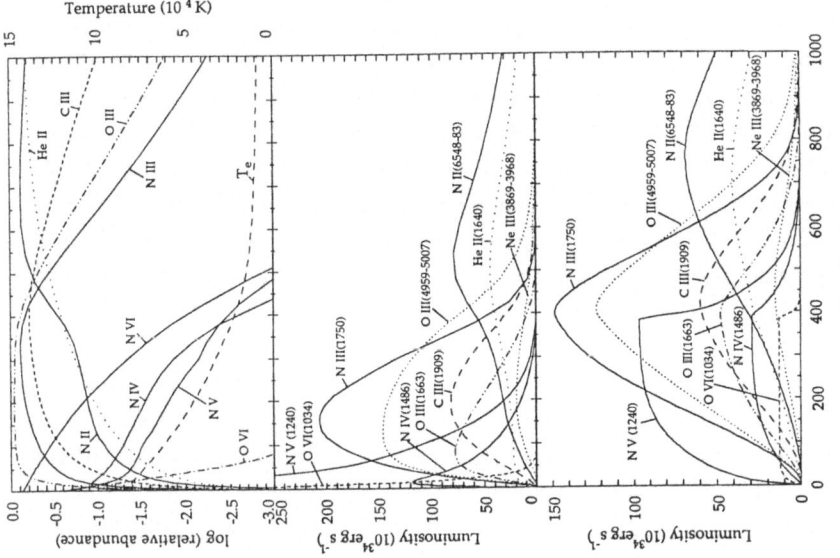

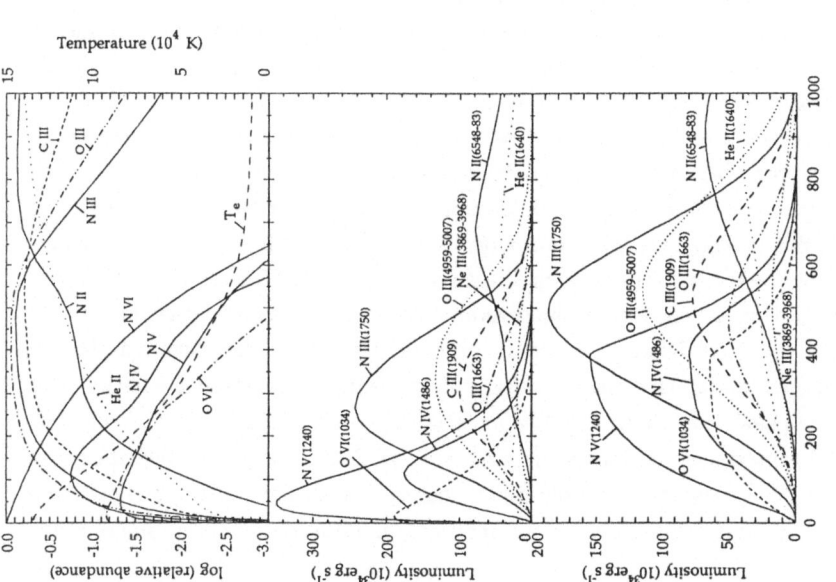

Fig. 1 and 2. Evolution of a shell ionized by the EUV burst from SN 1987A (see text for details). The upper panel shows the temperature and relative abundances in the shell as a function of time. Note the successive recombination of the nitrogen ions. The middle shows the *instantaneous* luminosities of the different lines and the lower the *observed* luminosities, integrated over the observable parts of the shell. In the left figure (Fig. 1) the ionizing flux has been taken from Shigeyama et al.'s 11E1Y6 model and in the right from Woosley's 10L model.

VI. Recombination, however, sets in fairly rapidly and a population of N V builds up, which then recombines to N IV and N III. The rate at which recombination proceeds is directly determined by the density of the gas, since the recombination time $t_{rec} = (\alpha_r n_e)^{-1}$, where α_r is the recombination coefficient. At a temperature of 10^5 K and a density of 2.6×10^4 cm^{-3}, the recombination time scales are 206 days for N VI to N V, 13 days for N V to N IV, 7 days for N IV to N III, and 14 days for N III to N II. These are thus short, or comparable, to the evolutionary time scale, and recombination is indeed important for the evolution. These time scales show that the gas would quickly become neutral if only recombination was involved. The gas temperature is, however, high enough for collisional ionization to prevent the gas from recombining more than to C III, N III etc., explaining why lines from these ions display the linear increase typical for a non-recombining ion. N IV, on the other hand, is less affected by collisional ionization and recombines, giving a roughly constant flux. This explains the qualitative behaviour of the observed fluxes in F89. The time scale determining the evolution is the longest of these time scales, i.e. the N VI to N V recombination time scale. The cooling also occurs on this time scale, and proceeds rather smoothly. At ~300 days the temperature is ~5.5×10^4 K, close to that determined from the [O III] lines. The ionic abundances of C III, N III and O III are all similar, until ~600 days. This is due to the importance of collisional ionization, which for these ions give a situation close to the coronal case. When the temperature decreases below ~3×10^4 K recombination of N III etc. sets in, and the different ions evolve differently on their respective recombination time scale. This calculation thus justifies the relations between the ionic and total abundances in F89. The middle panel of Fig. 1 shows the instantaneous values of the emission in the different lines. The N III] line e.g. has a very broad emissivity peak, due to the slow recombination, giving a nearly linear increase in the observed flux (lower panel). The emissivity of N IV] λ1486 is more concentrated to the early stages, and thus has a more constant observed flux.

Since the initial state of the gas is determined by the ionizing burst, the lines give constraints on this. In Fig. 2 we show an analogous calculation for the Woosley 10L model. Because of the lower radiation temperature (2×10^5 K), the initial ionization and temperature is lower, giving a more rapid recombination and thus a "flatter" light curve for N V. Also the final decay is faster. In order to reproduce the observed state of ionization, the spectrum must contain a sufficient number of photons above ~100 eV. This implies that the peak temperature must be more than ~$(3-6) \times 10^5$ K. The Shigeyama et al. 11E1Y6 model reproduces the increasing part well, while Woosley's models appear to have too low a peak temperature to reproduce the initial rise. This illustrates how these models can give unique information on the unobserved, but important early stages of the explosion.

We have also calculated models with different densities. Since both the cooling and recombination are governed by this, the time evolution and line ratios are sensitive to the density. It is encouraging that a density of $(2-3) \times 10^4$ cm^{-3}, very close to that derived from the C III] diagnosis, gives the best agreement with the observations. The total mass in the shell is somewhat sensitive to the ionizing spectrum, with a typical range of 0.02 - 0.05 M_o .

It is of great interest to determine the absolute metallicity, as well as the He/H ratio. Unfortunately, the errors in the observed Hβ and He II $\lambda4686$ fluxes are quite large, so the results should be taken with some caution. Preliminary results, however, suggest that the total CNO abundance is 0.2 - 0.3 times solar, while the He/H ratio is uncertain, but in the range 0.1-0.2, by number. The N/C and N/O ratios are for these models ~5 and ~2, respectively, confirming that the assumptions in F89 are reasonable approximations. These are within the errors given in F89.

Except for the observed UV, and optical [O III], He II and Balmer lines, there are also other lines that may be observable. In the far UV the C III $\lambda990$ and N III $\lambda970$ resonance lines are both strong. In the same way as the C IV $\lambda1550$ line they may, however, be unobservable due to interstellar absorption. The O VI $\lambda1034$ line is especially sensitive to the peak temperature of the radiation (Figs. 1 and 2). The possibility of actually observing this (Shull, priv. comm.) is thus extremely interesting. In the optical the [N II] $\lambda\lambda6548$-83, 5755 lines are expected to become strong after ~300 days. This coincides with the recombination from N III to N II. Also the [Ne III] $\lambda\lambda3869$–3968 lines can be fairly strong, as well as at late times [O II] $\lambda\lambda3726$-29.

3. Discussion

The observations of the narrow emission lines of SN 1987A have given us direct information about the structure of the circumstellar medium, which is of great importance for understanding the history of the progenitor. The radio observations by Turtle et al. (1987) were interpreted by Chevalier and Fransson (1987) as evidence for a wind with mass loss of 6×10^{-6} M_o yr^{-1} for a wind velocity of 550 km s^{-1}. The density of this gas at ~5×10^{17} cm is, however, far too small to explain these observations. The dense gas we observe in these emission lines probably only constitute a minor fraction of all the gas around the supernova. The total number of ionizing photons is ~10^{57} from the supernova, enough to ionize ~0.8 M_o , i.e. roughly 15-40 times that of the shell. The gas inside the shell is mainly shocked gas from the blue stellar wind and does not absorb many of the photons from the supernova. The gas outside the shell is due to the red supergiant (RSG) wind, with a density of ~20 $(r/10^{18}$ cm$)^{-2}$ $(\dot{M}/10^{-5}$ M_o yr$^{-1})$ cm^{-3}. Here $\dot{M}$ is the mass loss rate of the RSG. Of this gas, ~0.8 M_o will be ionized by the burst in a way similar to the shell, but then recombine on a much longer time scale (~10^3 years for N VI to N V). In addition, the radiation from the blue progenitor will ionize the hydrogen in the RSG wind, as long as $\dot{M}$ is less than ~2×10^{-5} $(R_s/10^{18}$ cm$)^{1/2}$ M_o yr^{-1}. Since the emission measure of this gas is a factor of ~25 $(\dot{M}/10^{-5}$ M_o yr$^{-1})^{-2}$ smaller, emission from this gas will be difficult to observe. The column density in ions like C IV, N V and Si IV etc. may, however, be large enough (10^{14}-10^{15} cm^{-2}) to contribute to the absorptions observed in the early spectra (de Boer et al. 1987; Dupree et al. 1987). Also from radio observations it is difficult to obtain any constraints on the total amount of gas lost by the progenitor. The free-free optical depth at a wavelength λ is given by $\tau_{ff}=6.1\times10^{-30} \lambda^2 T_5^{-1.5}$ EM, where EM is the emission measure and T_5 the gas temperature in 10^5 K. For a RSG wind with inner boundary R_{in} and a velocity of 10 km s^{-1}, the emission measure of the gas is only $\sim1.8\times10^{20}$ ($\dot{M}/10^{-5}$ M$_\odot$ yr^{-1})^2 ($R_{in}/10^{18}$ cm)$^{-3}$ cm^{-5}. It is thus easy to "hide" several solar masses of circumstellar gas, lost in the main sequence and red supergiant phases.

Acknowledgements: We have benefitted from discussions with Roger Chevalier, Ken Nomoto, Mike Shull, and Joe Wampler. This research has been supported by grants from the Swedish Natural Science Research Council and the Royal Swedish Academy of Sciences.

REFERENCES

Arnett, W.D. 1988, in Proc. of the George Mason Conference on SN 1987A, in press.

Chevalier, R.A., and Fransson, C. 1987, Nature, 328, 44.

de Boer, K.S., Grewing, M., Richtler, T., Wamsteker, W., Gry, C., and Panagia, N. 1987, Astron. Astrophys. 177, L37.

Dupree, A.K., Kirshner, R.P., Nassiopoulos, G.E., Raymond, J.C., and Sonneborn, G. 1987, Ap. J. 320, 597.

Fransson, C., and Lundqvist, P. 1988, in preparation.

Fransson, C., Cassatella, A., Gilmozzi, R., Kirshner, R.P., Panagia, N., Sonneborn, G., and Wamsteker, W. 1989, Ap. J., Jan. 1 (F89).

Lundqvist, P., and Fransson, C. 1987, in Proc. of the ESO Workshop on SN 1987A, ed. I.J. Danziger, p. 495.

Lundqvist, P., and Fransson, C. 1989, in preparation.

Sonneborn, G., et al., IAU Circ.

Shigeyama, T., Nomoto, K., and Hashimoto, M. 1988, Astron. Astrophys. 196, 141.

Turtle, A.J., Campbell-Wilson, D., Bunton, J.D., Jauncey, D.L., Kesteven, M.J., Manchester, R.N., Norris, R.P., Storey, M.C., and Reynolds, J.E. 1987, Nature, 327, 38.

Wampler, E.J., and Richichi, A. 1988, submitted to Astron. Astrophys.

Woosley, S.E. 1988, Ap. J. 330, 218.

NEUTRINOS: DETECTION AND INTERPRETATION

L.N. ALEXEYEVA
Institute for Nuclear Research
Academy of Sciences of the USSR
60th October Anniversary Prospect 7a
117 312 Moscow
U.S.S.R.

ABSTRACT. Observations of the neutrino burst from Supernova 1987A by water Cherenkov detectors (KAMIOKANDE II, IMB) and liquid scintillator detectors (Baksan, Mont Blanc) are reviewed. It is shown that neutrino signal from SN 1987A was observed. There are 24 events in three detectors (KAMIOKANDE II, IMB, Baksan) recorded at 7:35 UT. The average properties of the signal (effective neutrino temperature, total energy of neutrino emission, burst duration) are consistent with the general theoretical description of supernova explosions. Special attention is concentrated on individual characteristics of the signals detected and the available discrepancies of the model estimates. Time profile of the neutrino burst, estimates of effective neutrino temperatures and total neutrino energies, angular distributions of the events are discussed. These properties point out, probably, a more compound picture of the phenomenon. The more detail analysis of the experimental data is needed and all possibilities must be at least considered. Based upon the Baksan observations, an upper limit of 0.35 core collapse in the Galaxy per year (90% C.L.) is shown.

1. THE DETECTORS

There is no doubt that the detection of the neutrino signal from SN 1987A is the remarkable corroboration of the theory of supernova explosions.

The idea of searching for neutrinos from collapsing stars suggested in 1965 by G.Zatsepin [1] led to the development of specific underground detectors with a high content of hydrogen in their targets. There were four groups looking for neutrino burst associated with this supernova: the IMB collaboration [2], the KAMIOKANDE II collaboration [3], the Baksan telescope [4], and the LSD detector of USSR-Italy collaboration [5].

229

D. McNally (ed.), Highlights of Astronomy, Vol. 8, 229–237.
© 1989 by the IAU.

The first two detectors are water Cherenkov devices where Cherenkov light produced by relativistic charged particles measures by photomultiplier tubes. The other two detectors use liquid organic scintillator (C_9H_{20}) as a target of neutrino interactions. All of them are located underground at different depths. The main properties of the detectors and reported data are summarized in table 1.

Table 1. The detectors and reported candidates for the neutrino burst in association with SN 1987A.

Detector	Fiducial mass (t), target	Energy thresh (MeV)	Backgr rate (sec^{-1})	Number of events	Duration (sec)	Time (UT)
IMB	6800 (5000) H_2O	35	0.077	8	6.0	}7:35
KAM II	2140 H_2O	8.5	0.022	11	12.5	
Baksan	200 C_9H_{20}	10	0.034	5	9.1	
Mont Blanc	90 C_9H_{20}	5.5	0.012	5	7.0	2:52

Arrival times of events with relative accuracy $\lesssim 1$ msec and energies with energy errors $\sim 20\%$ are defined by each detector.

The basic interaction which can be observed by both types of detectors is reaction of $\bar{\nu}_e$ absorption by freetarget protons, $\bar{\nu}_e + p \rightarrow n + e^+$. Angular distribution of positrons produced must be isotropic. It is possible to detect some additional reactions of $\nu(\bar{\nu})$ interactions in water and scintillator. The most importent of them is neutrino-electron elastic scattering, $\nu_e + e \rightarrow \nu_e + e$. A recoil electron approximately conserves a neutrino direction and angular distribution of recoil electrons will be sharply anisotropic one, showing the neutrino direction. Total contribution of other interactions ($\nu_e(\bar{\nu}_e) + {}^{16}O \rightarrow {}^{16}F({}^{16}N) + e^-(e^+)$ for Cherenkov detectors, $\nu_e(\bar{\nu}_e) + {}^{12}C \rightarrow {}^{12}N({}^{12}B) + e^-(e^+)$ for scintillation detectors) to the total number of observed events is estimated to be small [6,7].

The values of energy threshold at the level of 50% detection efficiency and background counting rates are also shown in Table 1. Evidently, the best detector is the KAMIOKANDE II (K II) due to its large mass, low energy threshold and low background rate.

2. THE DETECTION OF THE NEUTRINO SIGNAL

The details of the discovery of Supernova 1987A have been

described elsewhere [8]. Three groups (table 1) reported the observations of $\bar{\nu}_e$ signals at 7:35 UT on February 23 [2-4]. The Mont Blanc group observed signal of 5 events at 2:52 UT [5]. Firstly, we shall discuss the second burst detected at 7:35 UT and then we shall return to the first burst.

The overall uncertainty in time is ± 1 min for the K II signal, is ± 50 msec for the IMB one and is (-54 sec,+2 sec) for the Baksan one. Within errors these three signals can be supposed to be simultaneous. Figure 1 shows comparative trigger efficiencies of all detectors. The trigger efficiencies of the Mont Blanc, the K II and the Baksan are rather close each other but the IMB can detect only a high energy tail of neutrino spectrum.

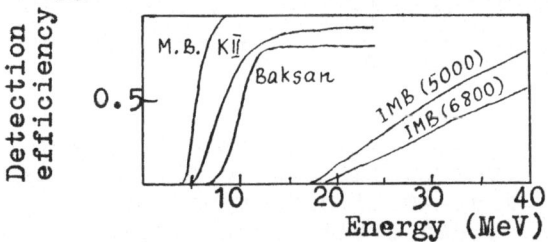

Figure 1. Comparative trigger efficiencies of all detectors.

Figure 2 shows the Poisson distributions for events within 10 sec interval detected in the period surrounding the time of Supernova. It is seen that background pulses of all detectors are well described by the Poisson law. There is usually no doubt that the KII signal and the IMB one were not originated by background. As regards the Baksan one, the chance probability of such signal to get into the one-minute interval respects to one occurrence per $\sim$20 years. So, we can conclude that these detectors have sampled the same source of the observed events.

Figure 3 shows the observed neutrino event rate per second normalized to the individual number of the observed events versus time. Figure 4 depicts the intergrated number of neutrino events, normalized in the same way, versus time. It is remarkable that just such general evolution of the neutrino emission was predicted by different model calculations of stellar core collapse and subsequent cooling of a nascent neutron star[9,10] :

1) the evolution of the neutrino emission is approximately described as an exponentially decaying signal with characteristic time $\sim$ 5 seconds;

2) the total duration of the neutrino signal is $\sim$ 20 sec;

3) the detected energies of the events are consistent with thermal neutrino spectrum and the effective neutrino temperature of 3-5 MeV (if a single temperature spectrum is supposed);

4) the total energy of the neutrino emission is $\sim 3 \cdot 10^{53}$;

5) the residue of Supernova is most probably a neutron star with a mass of $\sim$ 1.4 $M_\odot$.

The observation of the neutrino signal with the expected general characteristics is the great success of the theory and the experiment.

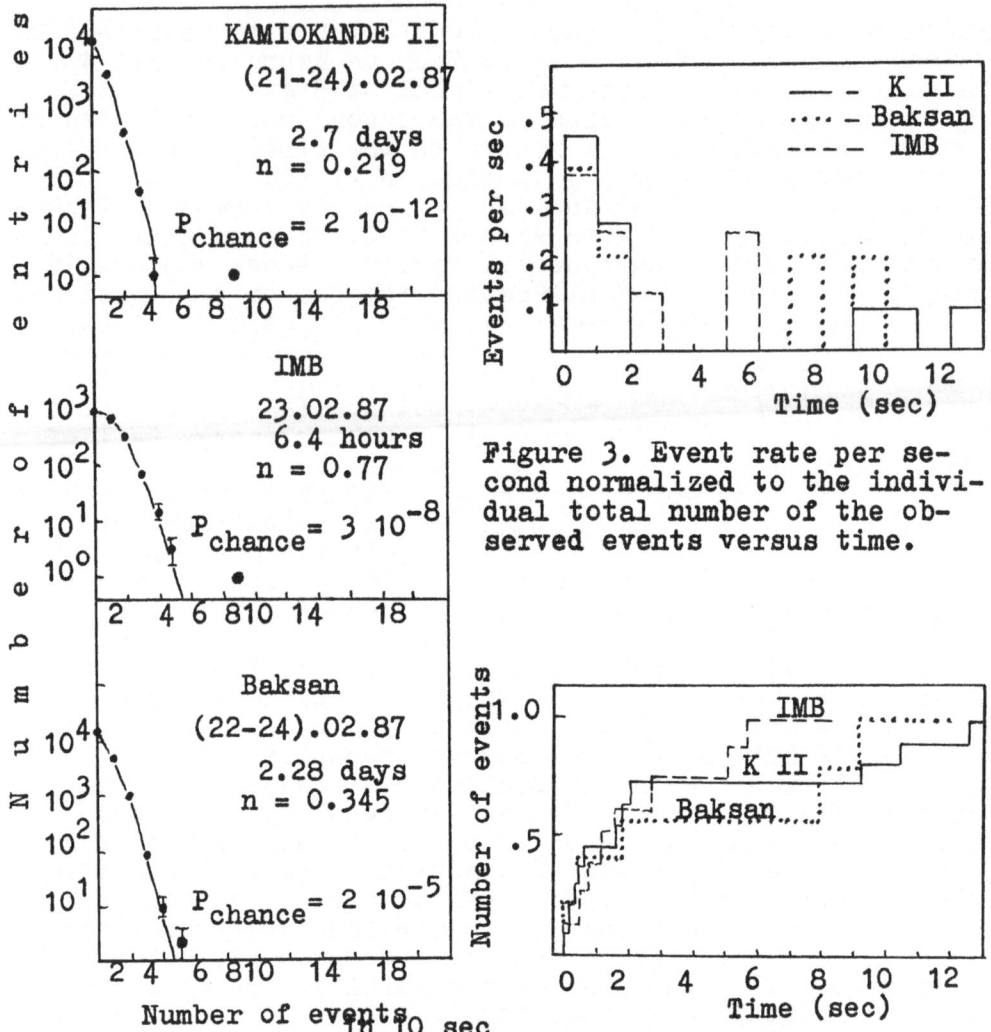

Figure 2. Poisson distributions for events within 10 sec intervals detected in the periods surrounding 7:35 UT on February 23.

Figure 3. Event rate per second normalized to the individual total number of the observed events versus time.

Figure 4. Integrated number of events normalized to the individual total number of the observed events versus time.

Now we shall discuss some facts most of which are ussually neglected with the reference to a "small-number statistic play". But we have the only supernova with the neutrino signal and the analysis of all, even small, facts is certainly desirable.

3. PROBLEMS

3.1. Time profile of the neutrino burst

What is the true time profile of the observed neutrino signal? Due to the absolute time inaccuracies ,we do not know it. Figure 5a shows the ensemble of the KII and the Baksan data as a function of time, setting t=0.0 to be the time of the first events. The events of both detectors show a bunch structure in time. There are the gaps of more than 7 sec in the KII signal and of 6 sec in the Baksan one between the second and the third bunches. Based upon a constant rate of

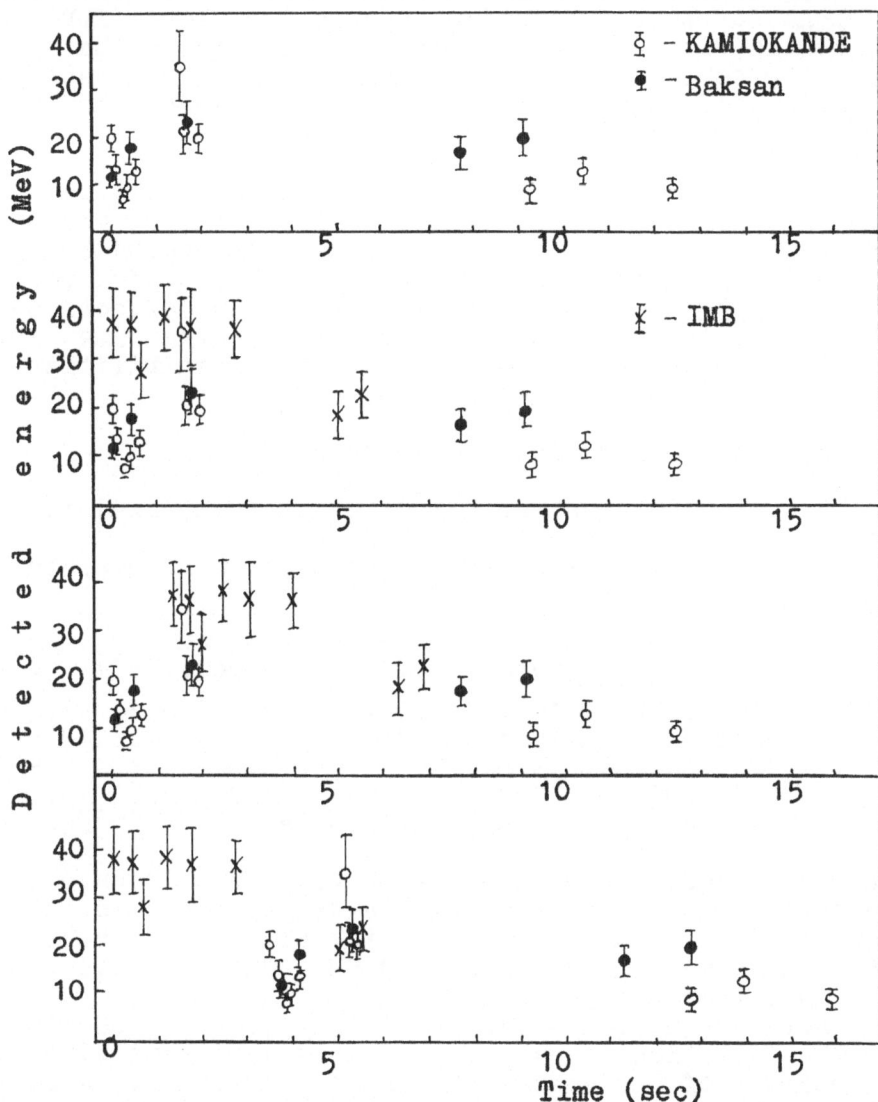

Figure 5. Time profiles of the events recorded by the KII and the Baksan (a) and by all three detectors (b), setting zero time to be the time of the first events. c) and d) depict possible time profiles of the observed signal.

11 events in 12.4 sec, the probability that a 7 sec interval would have zero events is $2 \cdot 10^{-3}$ [11].The same probability of the Baksan 6 sec gap is $3.5 \cdot 10^{-2}$ .So, the joint
probability is rather small. The first gap is statistically
less significant. Thus, the appearance, at least, of 6-7
gap in both data is the question which needs to be answered.
   The majority of authors superpose the IMB signal to the
KII one in the way shown on fig.5b [10]. Are there serious
reasons for such assumption? Due to the higher IMB energy
threshold, it seems more naturally to suppose that the IMB
sees the second bunch with higher energies of events
(fig.5c). If this picture is true, that means that the second gap is indeed the result of small-number statistic
play with low probability. But it is possible that the IMB
signal passes ahead of the KII and the Baksan (fig. 5d).
Outstripping can be about 3 sec. In this case the second
bunch of the IMB would coincide with the second bunch of
the KII-Baksan data, and the gap of 6-7 seconds remains
indeed empty. Perhaps, this superposition reflects better
the available data.Thus, the question on the time profile
of the neutrino signal is open and all possibilities need
to be considered.

3.2. Effective neutrino temperature

Derived temperatures for all data are summarized in table 2
[12].

TABLE 2. Derived effective neutrino temperatures
and total energies of $\bar{\nu}_e$ emission

Detector	Average detected energy (MeV)	Neutrino temper.(MeV)	$E_{\bar{\nu}_e}^{*)} \cdot 10^{52}$ (ergs)
KAMIOKANDE	16.7 ± 1.1	2.8 ± 0.3	5.8 ± 1.8
Baksan	19.4 ± 1.7	3.3 ± 0.4	18.6 ± 8.5
IMB (5000 t)	33.8 ± 2.9	4.5 ± 0.7	2.9 ± 1.1
IMB (6800 t)	33.2 ± 2.5	4.3 ± 0.6	3.2 ± 1.1

*)Distance to the star is adopted to be 50 Kpc

The estimates were obtained with the supposition of neutrino spectrum to be thermal single-temperature one. It is
seen that the temperatures for the KII and the Baksan are
much the same within errors,$\sim$ 3 MeV. The IMB data lead to
a higher temperature, $kT_{IMB} \sim 1.5 \, kT_{KII-Bak.}$.
   There are at least two possibilities to reconcile data.
i) The detection of events in IMB is very sensitive to an
inaccuracy of the energy threshold position.For example,
inaccuracy of $\sim$ 20% in the energy range 20-25 MeV results in
the substantial drop in the derived temperature,$\rightarrow$ 3.5 MeV.
ii) Another possibility is an assumption of neutrino spectrum in which high energy tail is considerably enhanced.

3.3. Total energy of neutrino emission

Table 2 shows also the derived total energies of $\bar{\nu}_e$ emission. The ratio of the values is (IMB:KII:Baksan) $\sim 0.5 : 1 : 3$. Why is the Baksan estimate so large and is the IMB one so small? The Baksan signal consists of 5 events instead of 1.6 predicted by standard model [12]. The probability of such discrepancy to be the result of small-number statistic play is $\sim 8\%$. To reconcile the data we may have to assume a neutrino spectrum with enhanced high energy tail once more.

The IMB small value of $E_{\bar{\nu}_e}$ can be partly caused by the same inaccuracy in the energy threshold position which was discussed in session 3.2. Figure 6 shows the derived total energy of $\bar{\nu}_e$ emission versus $kT_{\bar{\nu}_e}$. Dashed line illustrates the IMB effect of imaginary 20% inaccuracy of the energy threshold position.

Thus, the question on differences of the estimates of $kT_{\bar{\nu}_e}$ and $E_{\bar{\nu}_e}$, obtained for the detectors, needs to be considered in more detail.

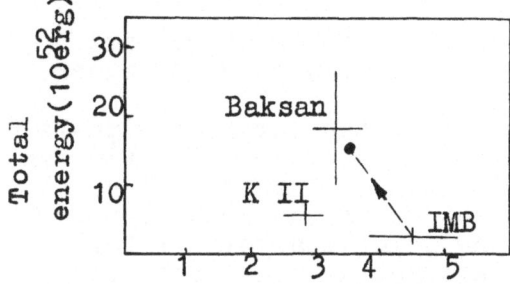

Figure 6. Derived total energy of $\bar{\nu}_e$ emission versus effective neutrino temperature.

3.4. Angular distributions of the events

Many authors have pointed out that the observed angular distributions, especially of the IMB events differ from the expected one and are quite puzzling [13]. The IMB group studied very careful all effects which could bias the expected distribution. They found the probability of the IMB distribution coming from a parent isotropic one to be only 5% [14].

3.5. The Mont Blanc signal

The Mont Blanc scintillation detector recorded a burst of 5 events within 7 seconds at 2:52 UT [5]. The chance background rate of such burst is ~ 0.7 per year. During the SN period two room temperature gravitational wave antennas installed at the Universities of Maryland and Rome were in operation [15]. Analysis of the data recorded by the Mont Blanc detector and by the antennas in the period of 2 hours roughly centred on the 5 burst shows correlation between data [15]. 14 Mont Blanc events instead of 2 expected by chance coïside with antenna peaks within time interval 1.2 ± 0.5 seconds. At present time the same analysis is performed using the data of the KAMIOKANDE II and the Baksan.

236

4. SUPERNOVA RATE LIMIT

The Baksan telescope observes Galaxy since June 1980 [4]. The
"live" observational time is 6.6 years. In accordance with
the standard collapse model we can expect about 35-50 events
if a distance to a star is 10 Kpc. We never see any pulse
burst which could be definitely interpreted as a collapse
neutrino signal. So, the upper limit on the collapse rate
in our Galaxy is $\zeta < 0.35$ per year (90% c.l.).

5. CONCLUSIONS

1)The neutrino signal from SN 1987A was observed. There are
24 events in three detectors, recorded at 7:35 UT on February 23.
2)The average derived characteristics of the burst are
consistent with the general theoretical picture of supernova explosions.
3)Some individual characteristics of the observed signals
 (time profile of the burst, differences in model estimates
of kT_{ν_e} and E_{ν_e} , angular distribution of the events) point
out, probably, more compound picture of the phenomenon.
4)The preliminary results of the joint analysis of the
Mont Blanc data and two gravitational antennas show the
correlation within two hours centred at 2:52 UT on February 23, which has a low level of chance probability.
5)Based upon the Baksan data obtaining during 6.6 years of
"live" observational time, the upper limit on Galaxy
collapse rate is $\zeta < 0.35$ per year (90% c.l.).
6)The approaching observation of the SN residue will help
us to understand the phenomenon in more detail.

References

[1] Domogatsky,G. and Zatsepin,G. Proc.9th ICRC,(1965)
     England, London, 2, 1030.
[2] Bionata,R. et al.(1987) Phys.Rev.Lett. 58, 1494.
[3] Hirata,K. et al.(1987) Phys.Rev.Lett. 58, 1490; (1988)
     Preprint UPR - 0150E, UT-ICEPP-88-03.
[4] Alexeyev,E. et al.(1987) JETP Lett. 45, 589; (1988) Phys.
     Lett. B205, 209.
[5] Aglietta,M. et al. (1987) Europhys.Lett. 3, 1315.
[6] Haxton,W. (1987) Phys.Rev. D36, 2283.
[7] Fukugita,M. et al. (1988) Preprint IASSNS - AST 88/25.
[8] Morrison,D. (1988) Preprint CERN/EP 88-9.
[9] Mayle,R. Wilson,J. and Schramm,D.(1987) Astroph.J.318,288;
     Nadyozhin,D. (1978) Astroph.Space Sci. 53, 131;
     Arnett,W. (1987) Astroph.J. 319, 136;
     Bruenn,S. (1987) Phys.Rev. Lett. 59, 938;
     Burrows,A.(1988) Preprint of the Steward Observatory
                    No. 815.

[10] Bahcall,J. et al.(1987) Preprint IASSNS-AST 87/8;
     Schramm,D. (1987) FERMILAB-CONF-87/161-A;
     Burrows,A. (1988) Preprint of Steward Observatory No 799.
[11] Kolb,E. et al. (1987) Phys.Rev. D35, 3598.
[12] Alexeyev,E. et al.(1988) Sov.Astron.Lett. 14, 41.
[13] Dar,A. (1988) Preprint TECHNION-PH-88-3;
     LoSecco,J. (1988) Preprint UND-PDK-88-4.
[14] Mattews,J. et al. (1987) Proc.U.of Minnesota Workshop;
     LoSecco,J. (1987) Preprint UND-PDK-88-3.
[15] Aglietta,M. et al. (1988) Report at Les Recontres de
     Physique de la Vallee d'Aoste, La Thuile, February 29.

4.THE COSMIC DUST CONNECTION IN INTERPLANETARY SPACE:

COMETS, INTERSTELLAR DUST AND FAMILIES OF MINOR PLANETS

Scientific Organising Committee

J.M. Greenberg (Chairman)

P.B. Babadzhanov, H. Fechtig, A. Hajduk, D. Hughes, Y. Kozai,
A.C. Levassur-Regourd, K. Mattila, T. Mukai, J.L. Weinberg

Supporting Commissions:

7, 15, 20, 21, 22, 34, 36, 49

FROM INTERSTELLAR DUST TO COMET DUST AND INTERPLANETARY PARTICLES

J. Mayo Greenberg
Laboratory Astrophysics, University of Leiden
Niels Bohrweg 2, 2333 CA   Leiden,
The Netherlands

ABSTRACT. Interstellar dust is described as consisting predominantly (by mass) of tenth micron (mean size) silicate core-organic refractory mantle particles which have evolved over galactic time scales of the order of 5 billion years. These particles were incorporated into comets and asteroids in the presolar nebula 4.5 billion years ago. The fragmentation of those primitive bodies gives rise to solar system debris which shows up as comet dust, zodiacal light, IDP's and meteorites. The chemical <u>and</u> morphological structure of comet dust is derived here as fluffy aggregates of interstellar dust. The chemical and morphological structure of the chondritic porous IDP's are then derived from comet dust which has evolved in the solar system. Zodiacal light particles are interpreted as various stages between comet dust and IDP's. Meteorites appear to be a side branch in the evolution from interstellar to solar system particles.

1. Introduction

How far removed in chemistry and morphology are the small particles in the solar system from their progenitors - the interstellar dust? Can we expect to find really close similarities? What we see at the present time in the form of comet dust, zodiacal light, meteors, IDP's, and meteorites must have originated in larger bodies which formed 4.5 billion years ago. Not only could these parent bodies have undergone significant metamorphosis in the aggregation stage but we might expect to find further changes to have occurred both within the parent bodies as well as from chemical and physical changes in the solar system following fragmentation. It is therefore at first glance almost inconceivable that close resemblances with solar system particles are traceable to the original interstellar dust. In some cases - as with comet dust and IDP's - it appears remarkably close.

2. Size, shape and composition of interstellar dust

Recent studies of the observations of so-called diffuse cloud dust (dust

D. McNally (ed.), Highlights of Astronomy, Vol. 8, 241–250.
© *1989 by the IAU.*

not in molecular clouds) in the ultraviolet have revealed the fact that there are three populations of dust (Greenberg and Chlewicki, 1983). There are elongated "large" grains of ~ 0.12 μm in mean radius which provide the major blocking of starlight in the visual. There are also very small carbonaceous particles of ≤ 0.01 μm in radius which produce a strong absorption feature at about 220 nm (Greenberg et al., 1987). In addition there is an independent population of ~ 0.01 μm silicate type particles. Large carbon bearing molecules (or very small particles) like PAH's (Greenberg, 1987) consume a small (≤ 5%) of the carbon.

The evolutionary picture of dust which is emerging is a cyclic one in which the particles, before being destroyed or going into solar system bodies, find themselves alternately over many cycles in diffuse clouds and in molecular clouds (Greenberg, 1982a, Greenberg, 1986, Schutte, 1988). A small silicate core captured within a molecular cloud accretes various ices and gradually builds up an inner mantle of organic refractory material which has been produced by photoprocessing of the volatile ices. Since the silicates which are formed in cool evolved stars are not crystalline, the elongation required for polarization in the 10 μm (Si-O stretch) band must be due to connected more-or-less spherical silicate beads. Representation of the core-mantle particles by concentric cylinders has been a mathematical convenience. The organic refractory mantles are subjected to the highest photoprocessing rates in the diffuse cloud phase - higher by factors of 10,000 or so than in the molecular cloud. Because of the cyclic evolution the organic refractory mantle on a grain is not a homogeneous substance but rather layered like the rings of a tree trunk in which the innermost layers have been the most irradiated and the outermost layer in the most recent molecular cloud phase is first generation organic refractory which is surrounded finally by lightly photoprocessed ices of which H_2O is the dominant component. Since further photoprocessing of organics leads to a greater and greater depletion of O, N, and H, the innermost layers are the most "carbonized" and the most non-volatile.

A schematic representation of grains in the various regions of space is shown in Fig. 1. Theoretical calculations of core-mantle

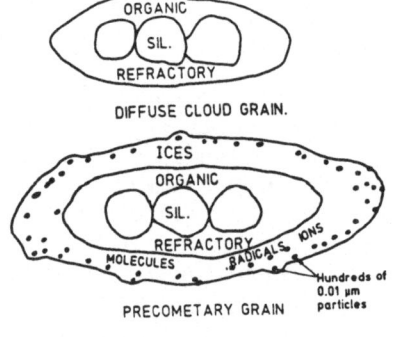

DIFFUSE CLOUD GRAIN.

PRECOMETARY GRAIN

Fig. 1. Interstellar grains are core-mantle structures.

Table 1

Suggested mass distribution of the principal chemical constituents of a cometesimal based on the dust model. Parentheses refer to very small particle components (a ≤ 0.01 μm). (Greenberg, 1988).

Component	Mass Fraction
Silicates	0.14+(0.06)
Carbon (carbonaceous) Nonvolatile Complex	(0.06)
Organic Refractory	0.19
H_2O	0.37
CO	0.05
Other Molecules + Radicals (H_2CO, CO_2, NH_3, OCN^-, HCO, S_2, CH_3OH...)	0.13

particles have been shown to match the oserved extinction and polarization as well as the albedo of interstellar dust (Chlewicki and Greenberg, 1988). In the final stage of cloud condensation we may expect that all remaining (condensable) molecules will have accreted onto the dust. In addition, the very small (≤ 0.01 µm) particles will be collected and trapped within the outer volatile icy mantle.

Our focus here will be on establishing a relationship between the chemical and morphological structure of presolar interstellar dust and comet dust, interplanetary dust, and meteorites.

3. Dust aggregation and morphology

In our solar system all of the planets and satellites have incorporated into their bodies at least the most refractory components of the interstellar dust which existed in the pre-solar nebula. Comets, appear likely to have preserved their original composition best including their volatiles not only because the volatile molecule S_2 may be traced back to the photochemical evolution of the interstellar dust (Grim and Greenberg, 1987) but also because of the observed CH_4/H_2O ratio (Larson et al., 1988).

As a first approximation, therefore, we consider a comet nucleus as if its chemical composition and morphological structure are directly related to interstellar dust. Table 1 shows the relative fractions of the various chemical constituents which have been obtained by an extrapolation from the molecular cloud dust phase (Greenberg, 1982b, 1983).

In forming the nucleus we assume that first clumps of grains form, and then clumps of clumps, and so on, until finally we reach the size of the comet nucleus. If we should start with the interstellar dust tightly packed and then remove all the volatiles (along with the trapped super small particles) the resulting mean density of the remaining core-organic refractory grains skeleton is about 0.5 g cm^{-3} (Greenberg, 1986). It is however observed that meteors (which are what is left after the original cometary volatiles have evaporated) have a characteristic density much lower than this, often being even less than 0.1 g cm^{-3}. This leads to a packing factor of 0.2; i.e., a comet is about 80% empty space! A model of such an open aggregate of 100 typical precometary grains is shown in Fig. 2a.

4. Comet dust

a) Size and fluffiness
The "unexpected" submicron sized particles detected by the Vega and Giotto impact systems extended down to the limit of 10^{-17} g. The interstellar core-mantle grains (without volatiles) have an average mass of $\approx 0.65 \times 10^{-13}$g and the bare (0.01 µm) particles have a mass of $\sim 10^{-17}$ g.

The spatial and temporal distribution of the masses and flux of dust particles measured by the dust counter and mass analyzer (DUCMA) on Vega 1/2 showed, among other things, that the lowest masses were the

244

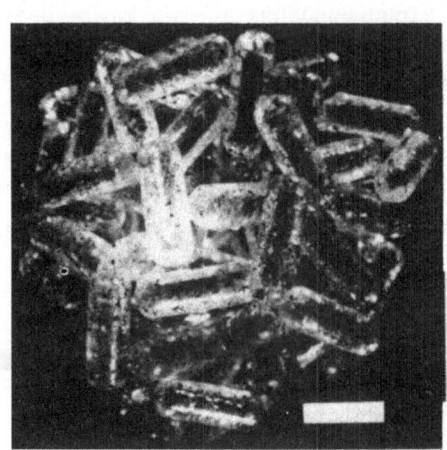

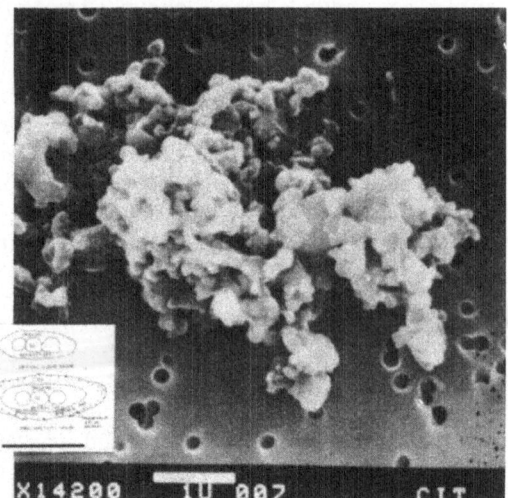

Fig. 2a: A piece of a fluffy comet: Model of an aggregate of 100 average interstellar dust particles each of which consists of a silicate core, an organic refractory inner mantle and an outer mantle of predominantly water ice in which are embedded the numerous very small (< 0.01 μm) particles responsible for the interstellar 216 nm absorption and the far ultraviolet extinction See Fig. 1). Each particle as represented corresponds to an interstellar grain ½ μm thick and about 1½ μm long. The mean mantle thickness corresponds in reality to a size distribution of thickness starting from zero. The packing factor of the particles is about 0.2 (80% empty space) and leads to a mean mass density of 0.28 gm cm^{-3} and an aggregate diameter of 5 μm.

Fig.2b: A highly porous chondritic IDP.
Note that the bird's nest particle (Fig.2a), the IDP (Fig.2b) and average interstellar core-mantle particle (Fig.2b insert) are equally scaled to 1 μm.

first particles encountered at the fringes of the coma. One of the explanations for this phenomenon by Simpson et al. (1986) is that some of the dust particles are "comprised of much smaller particles" from which pieces are shed which appear at great distances as the material which binds them sublimes; i.e., the initial dust is clumpy and resembles the bird's nest model (Fig. 2a without volatiles)

b) Chemical composition

The dust impact mass analyzers on Vega 1/2 (PUMA) and on Giotto (PIA) showed a predominance of the light elements H, C, O, N (organics) relative to the heavier elements Si, Mg, Fe (rockies) in the dust (Kissel et al. 1986a, 1986b).

Kissel and Krueger (K+K) (1987) have derived a molecular analysis of the comet dust and in particular its organic component. Masses between 2×10^{-15} and 10^{-11} were measured with the masses of most of the particles estimated to be in the range 10^{-12}-10^{-13} g with "systematic error within an order of magnitude". Their typical total relative atomic abundances in their molecules (of the organic refractory) show a significant lack of oxygen just as is predicted by the interstellar dust model. A four-fold enhancement of carbon was predicted relative to oxygen (Greenberg, 1982b, 1982c). The ratio of organics to silicate mass deduced by K+K is $m_{OR}/m_{Sil} \simeq 1:2$ which, not surprisingly, is less than that in the interstellar dust because of the expected evaporation of the less refractory organics at solar system temperatures.

c) Size plus chemical composition: The 3.4 and 10 µm excess emission

The 3.4 µm and 10 µm excess emission in comet dust provide evidence not only for the basic chemical ingredients - as given in the mass spectra - but also for the morphological structure (Greenberg, Zhao and Hage, 1989). It turns out that pure silicates no matter how small do not achieve high enough temperatures to produce the observed 10 µm emission. At, for example 1.11 AU the required temperatures needed to keep the total mass of the emitting particles at all reasonable is T > 430 K. Absorbing organic refractory mantles - such as those on interstellar silicate cores - are absolutely required to raise the compound grain temperatures high enough to make the 10 µm peak observable. Furthermore, the T > 430 K temperature constraint leads to a most probable silicate core radius ~ 0.05 µm and a mantle thickness $\geq$ 0.02 µm. i.e. an organic to silicate mass ratio $m_{OR}/m_{Sil} \simeq 0.9$ which, within the uncertainties, is like that deduced from comet dust mass spectra. If only such small particles (m $\leq 10^{-13}$g) could produce the 10 µm (and 3.4 µm) emissions their fluxes would have been more than 10,000 times higher than observed. It is, only by considering them to be in fluffy aggregates that the integrated fluxes come into reasonable resemblance to the particle impact detector data (McDonnell et al., 1987) - although still by a factor of about 25 too high for masses $\leq 10^{-9}$g.

6. Interplanetary dust

a) Zodiacal light

Interplanetary dust has classically been observed via its scattering of sunlight - the zodiacal light. The addition of infrared observations has revealed some significant physical distinctions between particles as a function of distance from the sun. Those which are within 1 AU scatter visible light much more effectively than those which are beyond 1 AU. At the same time, those which are farther out are more effective emitters of infrared radiation. This implies a difference in kind as well as number with increasing solar distance (Hong and Kwon, 1988). The most obvious explanation of this phenomena is that the radial decrease of the albedo of the zodiacal light particles is produced by a decrease in material density, just as the albedo of cometary dust is decreased because of its fluffiness. The interplanetary particle probe results of Pioneer 10/11 were also interpreted in terms of a radial decrease of particle density (Fechtig, 1984)

It has been suggested that the zodiacal light is predominantly produced by particles which started out as comet dust (Whipple, 1967).

The alternative point of view is that interplanetary particles result from asteroidal collisions (Olsson-Steel, 1988). Probably something in between may be true although, if some asteroids are just inert comets, the distinction may be academic. That asteroids play only a minor role as a dust source (Dohnanyi, 1976) appeared to be confirmed by the Pioneer 10/11 data which did not show any dust increase in the asteroidal belt (Humes et al. 1974). With the assumption that most interplanetary particles start out as fluffy low albedo comet dust particles (like that in Fig. 2a), Mukai and Fechtig (1983) proposed a mechanism by which solar heating would lead to a gradual compaction of the initially fluffy dust by evaporation of the volatiles in what they called "Greenberg particles" leading to more compact and higher visual scattering particles like the "Brownlee particles" (Fig. 2b).

b) IDP's

Although the mean density of the chondritic porous IDP's is low it is much higher than the initial cometary dust. But, as has been pointed out by Brownlee himself (Brownlee, 1988), there is no evidence of a bird's nest structure in the IDP's (Fig. 2b). What we see in Fig. 2b is an aggregate of more or less spherical particles of about 0.1 μ diameter whose infrared signature is that of silicates. When the interstellar dust is scaled like the IDP we see how its silicate core segments – which are hidden in the bird's nest model (Fig. 2a) – are like the silicates in the IDP. But where are the organic refractory mantles in the IDP's? In the original (interstellar dust) comet nucleus material the ratio of O.R. mass to silicate mass is given as about 1.5:1 (Table 1). However, already in the comet dust, the loss of the more volatile O.R. molecules has led to a reduction of this ratio by about a factor of 3 to about 1:2 (K+K). While the organic mantles are not "seen" in the IDP electron micrographs they become immediately apparent with Raman spectroscopy (Wopenka, 1988). It appears that every silicate particle is covered by some organic mantle. The fact that the mean silicate particle size is like that of the interstellar core pieces and each silicate or clump of silicates has an O.R. coating is certainly suggestive of the interstellar origin while the bird's nest morphological structure is lost because of the removal, during the passage from the comet to the earth, of a further part of the original comet dust O.R.

Additional indications for the cometary to interplanetary dust evolution may be seen in the lower density of meteors whose aphelion distances are beyond 5.4 AU as compared with those which spend more time closer to the sun (Verniani, 1973).

c) Meteorites

How do meteorites and their parent asteroidal bodies fit into the cosmic dust connection? Since the formation region for the asteroids was certainly at a higher temperature than that for comets we do not expect the interstellar dust to be nearly as well preserved. Within the framework of the theory of Ruzmaikina and Maeva (1988) the temperature of the pre-solar nebula relevant to the asteroidal belt was 250-300 K which was sufficient to evaporate all the dust volatiles while preserving a fraction of the organics. One factor which appears to provide a basis for believing the connection lies in the preservation (Brownlee, 1988; Huss and Alexander, 1987) of the pre-solar isotopic abundances of the heavy noble gases Ar, Kr and Xe in the carbonaceous component. These elements are presumed to have been trapped in the interstellar organic refractory mantles and retained during asteroid formation. Thus, although meteorites may be identified with the same interstellar dust ancestors as comets, they are like cousins rather then siblings.

Based on the observations of the largely amorphous, carbonaceous coatings in the Allende (C3V) meteorite (Green et al. 1971; Bunch and Chang, 1980; Bauman et al. 1973) Huss (1987) has suggested that the matrices in the parent bodies of the C3V, C3O, and type 3 ordinary chondrites probably accreted from presolar dust that had lost the icy mantles. On the other hand he proposed that CI (C1) chondrites and the matrices of C2 chondrites probably accreted as bulk samples of presolar dust with some icy mantles intact - almost cometary. Parent body heating (not characteristic of comets) then caused the icy mantles to react with the fine grained dust to produce the hydrothermal mineral assemblages now observed. The icy mantles in comet dust evaporate rather than melt so that, although we should not be surprised by seeing some resemblance between CP IDP's and CI chondrites, the differences should also not be a surprise - there are no hydrated silicates in low density IDP's. If IDP's are remants of comet dust they should more resemble the chemical and physical composition of the latter in which the H_2O evaporated rather than melted. There is a monotonic sequence of carbonaceous content from interstellar dust to comet dust to IDP's to meteorites.

In Fig. 3 we summarize the relationship between interstellar dust, interplanetary dust and meteors and meteorites as conceived of here.

7. Concluding remarks

We have to look to future space missions to recover comet material much more pristine than we can infer from flyby or even rendezvous missions. If the comet nucleus material can be retrieved from its depths and maintained intact cryogenically for laboratory studies, we may hope to study not only its atomic and molecular compositions but also its morphology. Microprobes are being developed (Bradley and Brownlee, 1986) which will make investigations possible of submicron structures. If it should turn out that the interstellar dust model is correct, individual grains whose mean lifetime before becoming part of a comet is about 5×10^9 yr will reveal cosmochemical evolution not only of the solar system but dating back a further 5 billion years before the earth's beginning - back to the earliest stages of the chemical evolution of the Milky Way. Dramatic differences in isotopic abundances could be expected on scales of microns. The next twenty to thirty years should be exciting ones indeed for studies of our origins.

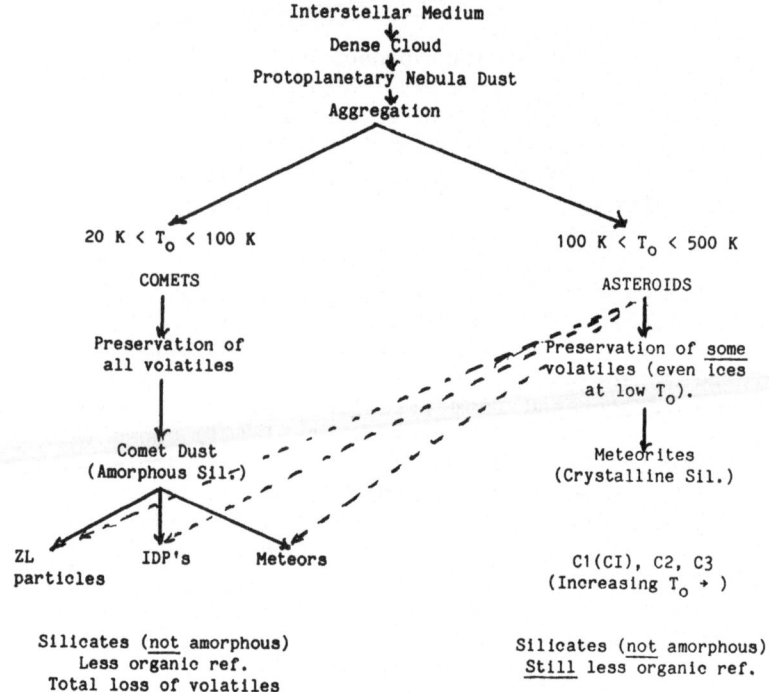

Fig. 3. Decrease of organics and increase of silicate crystallinity according to aggregation temperature, T_o, and thermal history.

8. References

Bauman, A.J., Devany, J.R. and Bollin, E.M. 1973, "Allende meteorite carbonaceous phase: intractable nature and scanning electron morphology, Nature, 241, 264-267.

Bradley, J.R. and Brownlee, D.E. 1986, "Cometary particles: Thin sectioning and electron beam analysis", Science, 231, 1542-1544.

Brownlee, D.E. 1988, "The composition of dust particles in the environment of Comet Halley", This volume.

Bunch, T.E. and Chang, S. 1980 "Carbonaceous chondrite phylosilicates and light element geochemistry as indicators of parent body processes and surface conditions, Geochim. Cosmochim. Acta 44, 1543-1577.

Chlewicki, G.C. and Greenberg, J.M., 1988, "Interstellar circular polarization and the dielectric nature of dust grains", Astrophys. J., submitted.

Dohnanyi, J.S. 1976, "Sources of interplanetary dust: asteroids" in Lecture Notes in Physics (eds. H. Elsässer and H. Fechtig, Berlin: Springer-Verlag) 48, 29.

Fechtig, H., 1984, "The Interplanetary dust environment beyond IAU and in the vicinity of the ringed planets", Adv. Sp. Res. Vol. 4 no. 9, 5-11.

Green, H.W. III, Radcliffe, S.V., and Hever, A.H. 1971, "Allende meteorite: a high voltage electron petrographic study", Science 172, 936-939.

Greenberg, J.M. 1982a, "Dust in dense clouds. One stage in a cycle", Submillimetre Wave Astronomy, eds. Phillips, D. and Beckman, J.E., Cambridge University Press, 261-306.

Greenberg, J.M. 1982b, "What are comets made of? A model based on interstellar dust", in Comets, ed. Wilkening, L.L., University of Arizona Press, 131-163.

Greenberg, J.M. and Chlewicki, G.C. 1983, "A far-ultraviolet extinction law: what does it mean?", Astrophys. J., 272, 563-578.

Greenberg, J.M. 1983, "Laboratory dust experiments - Tracing the composition of cometary dust", in: Cometary Exploration, ed. T.I.Gombosi (Hungarian Academy of Sciences) 23-54.

Greenberg, J.M. 1986, "Fluffy Comets", in: Asteroids, Comets and Meteors II, eds. C.-I. Lagerkvist, B.A. Lindblad, H. Lundstedt and H. Rickman, (Uppsala University Press) 221-223.

Greenberg, J.M. 1986, "Dust in Diffuse Clouds: One stage in a cycle", in: "Light on Dark Matter" ed. F P. Israel, (Reidel), 177-188.

Greenberg, J.M., de Groot, M.S. and Van der Zwet, G.P. 1987, "Carbon components of interstellar dust", in: Polycyclic Aromatic Hydrocarbons and Astrophysics, ed. A. Leger, L.B. d'Hendecourt and N. Boccara (Reidel) 177-181.

Greenberg, J.M., Zhao, Nansheng, Hage, J.I., 1988, Space Science Reviews, in press.

Greenberg, J.M. 1988, "The evidence that comets are made of interstellar dust" in "Comet Halley 1986 - Worldwide Investigations, Results and Interpretations", Ellis Horwood Ltd., Chichester, England (eds. John Mason and Patrich Moore), in press.

See "Polycyclic Aromatic Hydrocarbons and Astrophysics", 1987, ed. A. Leger, L.B. d'Hendecourt and N. Boccara (D. Reidel Pub.).

Grim, R.J.A. and Greenberg, J.M. 1987, "Photoprocessing of H_2S in interstellar grain mantles as an explanation for S_2 in comets", Astr. Astrophys. 181, 155-168.

Hong, S.S. and Kwon, S.M., 1988, "Spatially Varying Optical Properties of the Zodiacal Dust" In: Joint Discussion IV: The Cosmic Dust Connection.

Humes, D.H., Alvarez, J.M., O'Neal, R.L. and Kinard, W.H. 1974 "The interplanetary and near Jupiter meteoroid environments", J. Geophys. Res. 79, 3677-????.

Huss, G.R. and Alexander, C.Jr. 1987, "On the presolar origin of the "Normal Planetary" noble gas component in meteorites", J. Geophys. Res. 92, no. 134, E710-E716.

Huss, G.R. 1987 "The role of presolar dust in the formation of the solar system", Icarus.

Kissel, J., Brownlee, D.E., Büchler, K., Clark, B.C., Fechtig, H., Grün, E., Hornung, K., Igenbergs, E.B., Jessberger, E.K., Krueger, F.R., Kuczera, H., McDonnell, J.A.M., Morfill, G.M., Rahe, J., Schwehm, G.H., Sekanina, Z., Utterback, N.G., Völk, H.J. and Zook, H.A., 1986a, "Composition of comet Halley dust particles from Giotto observations", 1986, Nature, 321, 336-337.

Kissel, J., Sagdeev, R.Z., Bertaux, J.L., Angarov, V.N., Audouze, J., Blamont, J.E., Buchler, K., Evlanov, E.N., Fechtig, H., Fomenkova, M.N., von Hoerner, H., Inogamov, N.A., Khromov, V.N., Knabe, W., Krueger, F.R., Langevin, Y., Leonas, V.B., Levasseur-Regourd, A.C., Managadze, G.G., Podkolzin, S.N., Shapiro, V.D., Tabladyev, S.R. and Zubkov, B.V., 1986b, "Composition of comet Halley dust particles from Vega observations", Nature, 321, 280-282.

Kissel J. and Krueger, F.R. 1987, "The organic components in dust from Halley as measured by the PUMA mass spectrometer on board Vega 1", Nature 326, 755-760.

Larson, H.P., Weaver, H.A., Mumma, M.J. and Drapatz, S. 1988, "Airborne infrared spectroscopy of comet Wilson (19861) and comparisons with comet Halley", Astrophys. J. in press.

McDonnell, J.A.M. et al, 1987 "The dust distribution within the inner coma of comet P/Halley 1982i: encounter by Giotto's impact detection", Astron. Astrophys. 187, 719-741.

Mukai, T. and Fechtig, H. 1983, "Packing effect of fluffy particles. Planet Space Sci. 31, 655.

Olsson-Steel, D. 1988 "The origin and physical characteristics of meteoroids", this volume.

Ruzmaikina, T.V. and Maeva, S.V. 1988, "Process of formation of the protoplanetary disk", in COSPAR XXVII Helsinki proceedings.

Schutte, W. 1988, Ph.D. Thesis University of Leiden "The evolution of interstellar organic grain mantles" 1-295.

Simpson, J.E., Sagdeev, R.Z., Tuzzolino, A.J., Perkins, M.A., Ksanfomality, L.V., Rabinowitz, D., Lentz, G.A., Afonin, V.V., Ero, J., Keppler, E., Kosorokov, J., Petrova, E., Szabó, L. and Umlauft G., 1986, "Dust counters and mass analyser (DUCMA) measurements of comet Halley's coma from Vega spacecraft", Nature, 321, 278-280.

Verniani, F., 1973, "Physical parameters of faint meteors", J. Geophys. Res. 78, 8429-8462.

Whipple, F. 1967 "On maintaining the meteorite complex" in the Zodiacal Light and the Interplanetary Medium, ed. J.L. Weinberg NASA-SP 150, 409-426.

Wopenka, B. 1988, Earth and Planet Sci. Lett. 88, 221.

What are Families of Minor Planets?

Yoshihide KOZAI
National Astronomical Observatory
Mitaka, Tokyo 181
Japan

ABSTRACT. This paper describes how the families of minor planets were discovered and how minor planets are classified into families and shows several evidences that at least some families were originated by collisions of minor planets. In fact for major families there are much more fainter minor planets than in other regions and when the families were discovered by Hirayama in 1918 only one sixth of the total numbered minor planets belonged to families whereas now more than one third of them belong to the families. It is also emphasized that the classification into the families is theory dependent and that any theory available now is not accurate enough to derive stable quantities for the families to study their origins as the families are usually bounded partly by mean motion and/or secular motion commensurability regions in the phase space.

1. Discoveries of Families

Families of minor planets were discovered by Kiyotsugu Hirayama(1918), who was a professor of astronomy, the University of Tokyo, 70 years ago, when only 790 minor planets were known, as he noticed several clusterings of minor planets with similar values of the semi-major axes, a, the eccentricities, e, and the inclinations, i. Soon he realized that when the poles of the orbital planes of the minor planets in one clustering were plotted on a plane they were distributed nearly along a circumference with the center almost coinciding with the pole of Jupiter's orbital plane. He also found that the points drawn on a plane for the eccentricities and the longitudes of the perihelions, ω, instead of the inclinations and the longitudes of the ascending nodes, Ω, were also distributed nearly along a circumference with larger scatterings.

According to the linear classical theory of the secular perturbations of minor planets by assuming that both the eccentricity and the inclination are very small, the equations for $\xi = e \cos \omega$ and $\eta = e \sin \omega$, as well as $p = \tan i \cos \Omega$ and $q = \tan i \sin \Omega$ are linear differential equations with solutions expressed by the sum of a free oscillation and forced ones due to the secular perturbations in the corresponding orbital elements of the disturbing planets. Then ξ and η as well as p and q vary with time along circumferences with centers moving more slowly due to the forced oscillations. The radii of the circumferences are called the proper eccentricity and the proper inclination which are the amplitudes of the free oscillations. The diagrams which Hirayama drew show that ξ, η as well as p, q for the minor planets in the clustering are distributed along trajectories expressing the secular perturbations, although the phases are different from each other due to small deviations in the secular motions, that is, the angular velocities

251

of the circular motions.

Hirayama, therefore, concluded that since for the minor planets in the same clustering the semi–major axes, the proper inclinations and the proper eccentricities which are all stable quantities in the secular perturbation theory expressing averaged long–term behaviours of the motion take almost same values they had a common origin. And he called the clusterings as families of minor planets. By this way he identified five families, Themis, Eos, Coronis, Maria and Flora, and later he added the two families, Pallas and Phocaea(1923). Still they are the major families.

2. Identifications of Families

Groupings of minor planets into families were tried by several authors later on including Brouwer(1951). However, in fact the eccentricities and the inclinations for most of minor planets are not so small, and, therefore, the classical theory is not satisfactory to derive accurately stable quantities for them. Williams(1969) developed a new thory for secular perturbations and identified several more families according to his theory(1979). Kozai(1962 and 1979) modified the classical theory by introducing terms with integral multiples of $2g$ as arguments, where g is the argument of perihelion. Such terms are very important for minor planets with large eccentricities and inclinations.

For example, the inclination and the eccentricity of Pallas, the second minor planet discovered next to Ceres and a member of Pallas family, are $34.^{\circ}4$ and 0.234, respectively, and, therefore, are not at all small. By the theory of the author they change, respectively, between $28.^{\circ}7$ and $36.^{\circ}1$, and between 0.124 and 0.405 as periodic functions of twice the argument of perihelion. For any minor planet the variations of the eccentricity and the inclination with respect to the argument of perihelion can be estimated by assuming that all the disturbing planets move on the same plane along circular orbits, as under this assumption $\Theta = (1-e^2)^{1/2} \cos i$ is constant and the system of the equations of motion for the secular perturbations can be reduced to that of one degree of freedom.

In this paper the numbered minor planets are classified into the eight families listed in Table 1 with the ranges of the three stable quantities, namely, the semi–major axis, the value of the inclination to Jupiter's orbital plane at $2g = 0^{\circ}$, i_{min}, which is the minimum value, and Θ. Hungaria family, which has only few bright members, is the only one which is added to those in Hirayama's papers in Table 1, where the numbers of member minor planets shown in Hirayama(1923) and Brouwer(1951) as well as that by the author(1988) using 3840 minor planets are listed also.

TABLE 1. Eight major families

Family	a(AU)	Θ	i_{min}	Number of Members (1988)	(1923)	(1951)	v (km.s^{-1})	Δv
Hungaria	1.84–1.98	0.880–0.950	$19.^{\circ}0$–$28.^{\circ}0$	39	(1)	(8)	22	0.40
Flora	2.15–2.30	0.950–1.000	$00.\ 0$–$09.\ 6$	602	63	125	20	0.34
Phocaea	2.28–2.44	0.850–0.920	$20.\ 0$–$24.\ 0$	64	11	21	20	0.34
Maria	2.52–2.57	0.950–0.980	$13.\ 5$–$16.\ 0$	41	14	17	19	0.09
Pallas	2.60–2.80	0.770–0.830	$24.\ 0$–$31.\ 0$	15	3	6	18	0.33
Coronis	2.83–2.92	0.994–1.000	$01.\ 5$–$02.\ 6$	126	20	33	18	0.14
Eos	2.99–3.04	0.974–0.986	$09.\ 0$–$10.\ 7$	193	27	58	17	0.07
Themis	3.07–3.22	0.976–1.000	$00.\ 6$–$02.\ 5$	228	32	53	17	0.20

In Table 1 besides the ranges of the orbital parameters and the numbers of the members the velocity, v, and the scattering velocity, Δv, which are that for the circular motion with the mean value of the semi-major axes of the members and that producing the discrepancy of the semi-major axes, respectively, for each family are shown. The numbers in parenthesis for Hungaria family under (1923) and (1951) are estimated by the author by using the improved theory of the author(1962) whereas the numbers for Pallas family under (1923) and (1951) are the original ones in their papers by using the classical linear theory of the secular perturbations.

In Figure 1 a diagram showing the positions of i_{min} and a of the numbered minor planets is given for the range of a between 1.5 and 3.5AU(Astronomical Unit), the sizes of the points depending on Θ. In the Figure the eight families in Table 1 are clearly seen. Particularly, Maria, Coronis, Eos and Themis families are remarkable, as they are compact and dense. On the other hand, Hungaria, Flora, Phocaea and Pallas families are not so dense, however, they can be identified because of recognized boundaries where very few asteroids are found.

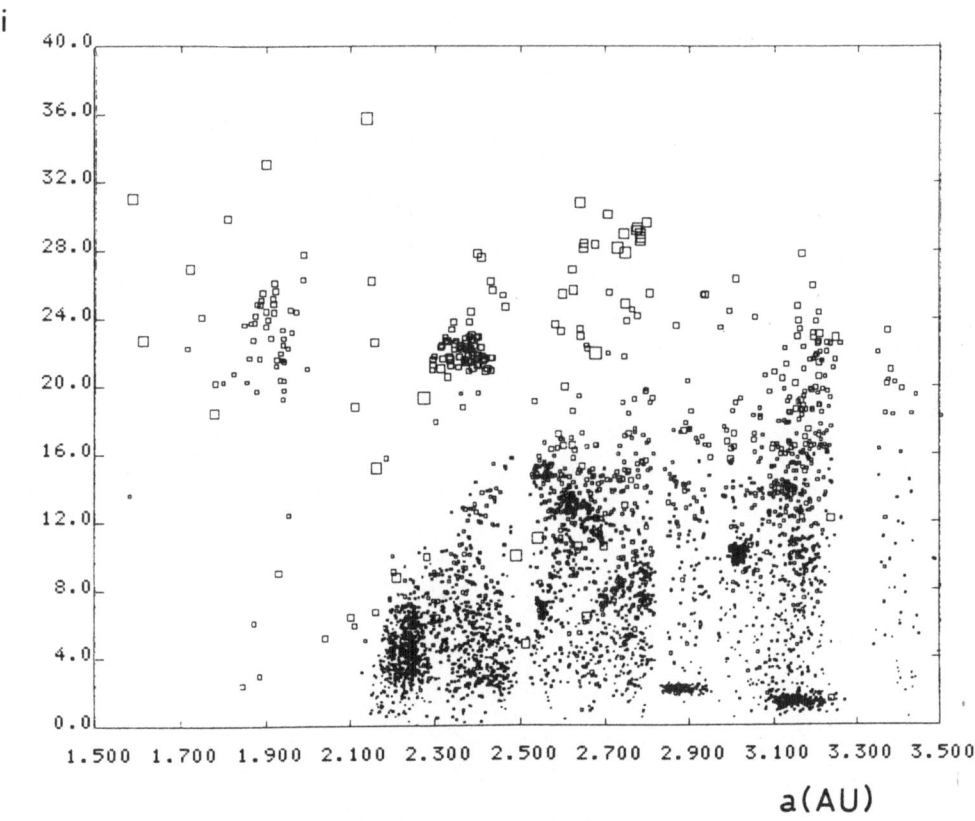

Figure 1. Diagram for the minimum values of the inclinations, i_{min}, and the semi-major axes, a, for numbered minor planets.

254

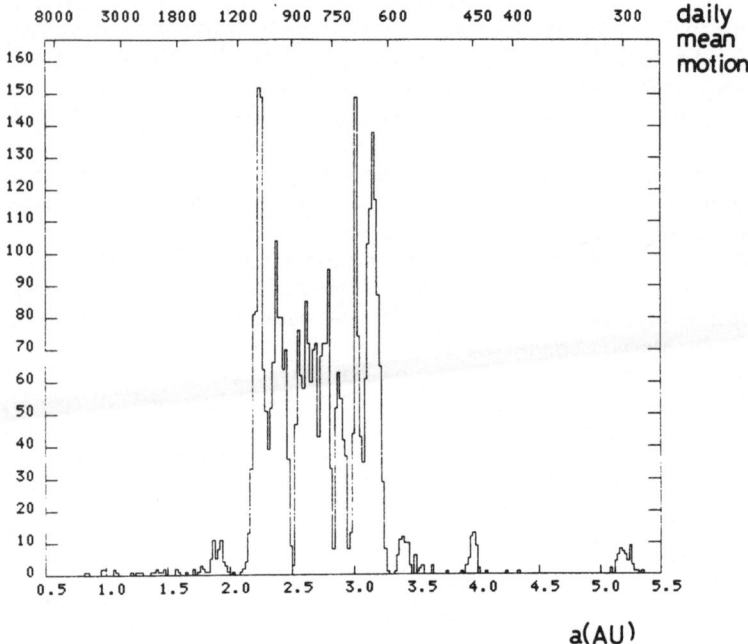

Figure 2. Number distribution with a for 3750 minor planets.

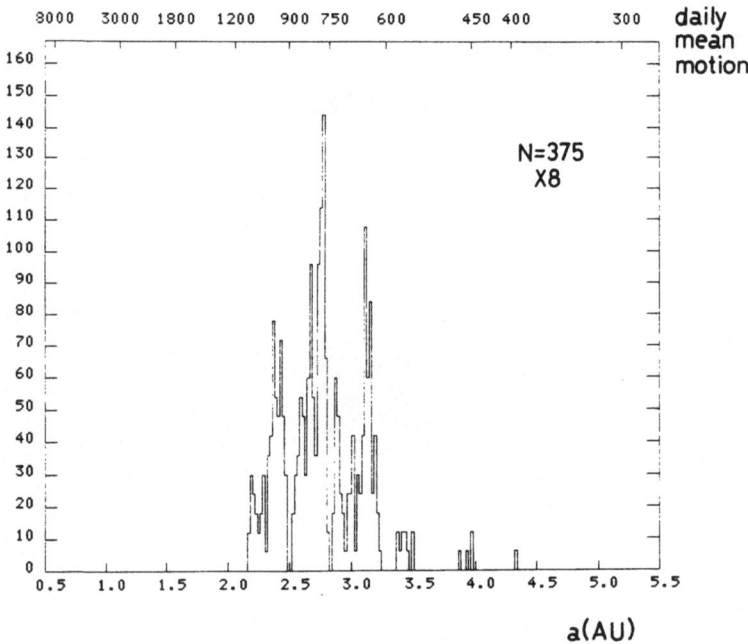

Figure 3. Number distribution with a for 375 minor planets.

3. Characteristics of Families

Other remarkable characteristics of the positions of the families in the figures are that almost all the families are bounded at least partly by so-called commensurable regions, either of mean motions with Jupiter or of secular motions of the longitudes of the perihelion or the node, where there are very few minor planets. For example, the left and upper boundaries of Flora and the lower boundaries of Phocaea and Pallas families correspond to secular commensurable regions as Williams(1969) pointed out.

As any secular perturbation theory does not hold for such mean motion and secular motion commensurable cases more general secular perturbation theories are needed to identify families and to derive more stable quantities to find out their origins, although to formulate such general theories is very difficult.

In Figure 2 and 3 number distributions of the numbered minor planets for the number up to 3750 and 375, respectively, are shown, although the numbers are multipled by 8 in Figure 3 for the 375 minor planets. It is noticed that generally in Figure 3 only bright minor plnets are included and here the highest peak is at 2.77AU, which is predicted by Bode's law. However, in Figure 2 there are other higher peaks, which correspond to positions of the major families. This means that there are much more fainter minor planets in the major families.

The same conclusion can be reached by comparing the numbers of member minor planets by Hirayama(1923) and that by the author as although the total number of the minor planets used in this paper is five times as large as that by Hirayama(1923) the ratio for the numbers for Flora family is roughly ten. Now more than one third of the numbered minor planets are members of the eight families, although in 1918 only one sixth were members. This means that nearly a half of newly discovered minor planets in the past 30 years belong to the families.

In Figure 4 brightness distributions for $B(1,0)$ (a kind of absolute magnitude in blue color) of minor planets belonging to the eight families as well as Trojan and Hilda groups and also those in some regions with semi-major axes similar to those of major families for comparison. The figure in the parenthesis in each diagram shows the number to be added to $B(1,0)$ to derive averaged apparent magnitudes at opposition.

These diagrams also show that there are much more fainter minor planets in the major families, particularly, for Flora family, which contains more than 600 members and the largest family and for which as the geocentric distance at opposition is small many fainter and smaller minor planets have been observed. And it is expected that when many fainter minor planets will be observed in future much more members will be added to Coronis, Eos and Themis families, for which as their distances from the earth are large even at oppositions fainter minor planets have not yet been discovered. Diagrams for non-member minor planets with the ranges of the semi-major axes similar to those for Coronis, Eos and Themis families and that outside of Flora family clearly show that there are more brighter minor planets and not so much fainter ones than those for corresponding family member minor planets.

In fact for Hungaria, Flora and Phocaea families there are only a few minor planets which are brighter than 12th magnitude in $B(1,0)$ and for Coronis, Eos and Themis families brighter than 10th magnitude. It must be noticed here that IRAS observed an enormous amount of micron-sized particles with orbits similar to those of the members of Eos and Themis families(Dermott et al, 1984).

Pallas family is the only one which has a very bright and large member, Pallas. However, it does not contain many fainter members. In fact the increase of the numbers compared with those by Hirayama(1923) and Brouwer(1951) is partly due to the theories adopted to derive the proper elements. The improved theory can identify more members

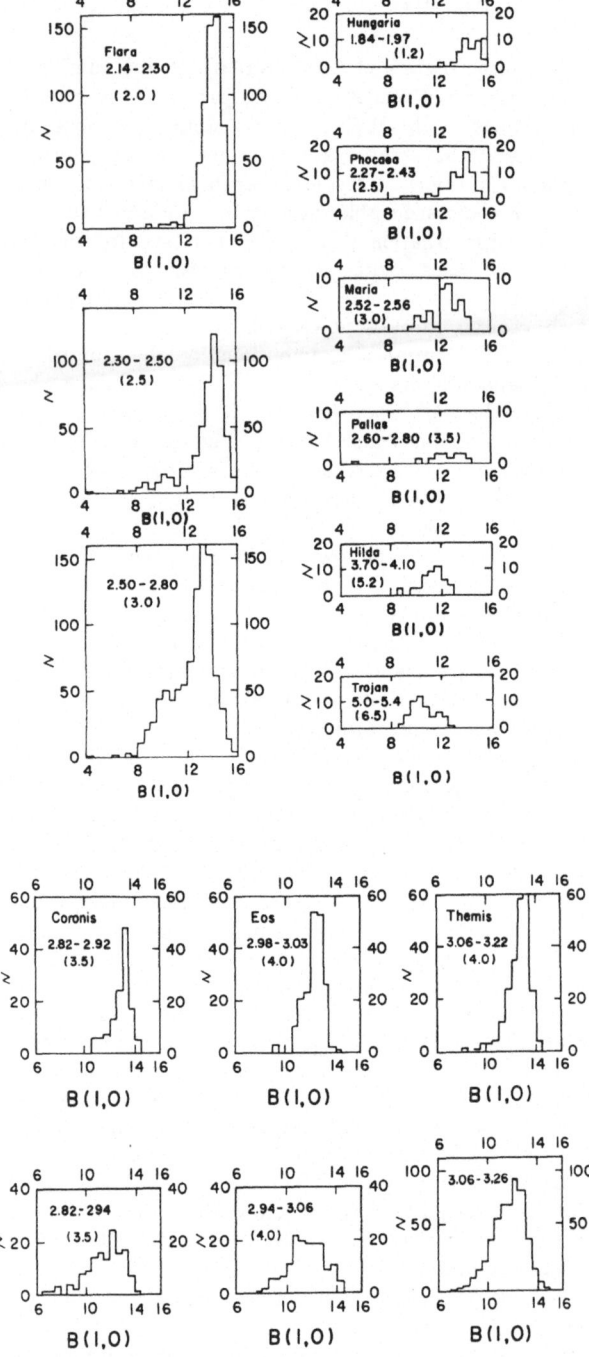

Figure 4.   Brightness distributions for families and others.

by taking into account the variations of the orbital elements with twice the argument of perihelion. Maria faimly has a peculiar brightness distribution which consists of a brighter and a fainter parts. It is expected for the two families more members will be discovered if high latitude sky will be surveyed systematically.

The brightness distributions for Hilda and Trojan groups are shown to find out any similarity to those of the families and it seems to the author that for Hilda group the diagram is not so different from those of some families showing some evidences of possible collisions.

4. Conclusions

In conclusion every data suggest that at least some of the families listed here were created by collisions among minor planets and, therefore, much more smaller particles produced during the formation of the families are expected to exist in interplanetary space. Still it seems to the author that since almost all the major families exist near mean motion and/or secular motion commensurable regions some resonance forces triggered the collisions and have kept the families in their present regions.

The author would like to stress, again, that any secular perturbation theories are not suffcient enough to produce very stable quantities for member minor planets of the families, and, therefore, it is very difficult to compute, for example, the original scattering velocity for each member after the possible collision accurately from the orbital data we have now. The scattering velocities in Table 1 which are computed from the present range of the semi—major axes of the member minor planets may not express the original scattering velocity, as the range of the semi—major axes are for most cases determined by mean motion commensurability zones.

Therefore, nothing can be said about the origin of the families very definitely now. Also it must be pointed out that even though collisions created the families listed here they cannot produce any meteorites found on the Earth. However, the author would like to mention that some of the craters on Mars were due to meteorites produced by the collisions which created Hungaria, Phocaea and Pallas families, as the perihelions of minor planets of the three families can approach to the orbit of Mars.

References

Brouwer, D. (1951) 'Secular variations of the orbital elements of minor planets', *Astron. J.*, **56**, 9–32.

Dermott, S.F., Nicholson, P.D., Burns, J.A. and Houck, J.R. (1984) 'Origin of the Solar System dust bands discovered by IRAS', *Nature*, **312**, 505–509.

Hirayama, K. (1918) 'Groups of asteroids probably of common origin', *Astron. J.*, **31**, 185–188.

Hirayama, K. (1923) 'Families of asteroids', *Japan, J. Astron. Geophys.*, 1, 55–93.

Kozai, Y. (1962) ' Secular perturbations of asteroids with high inclinations and eccentricities', *Astron. J.*, **67**, 591–598.

Kozai, Y. (1979) 'Secular perturbations of asteroids and comets', in R.L.Duncombe(ed.), *Dynamics of the Solar System*, 231–237. D. Reidel Publ. Company.

Williams, J.G. (1969) 'Secular perturbations in the Solar System', *Ph. D. Dissertation, Univ. California, Los Angeles*.

Williams, J.G. (1979) 'Proper elements and family memberships of the asteroids', in T. Gehrels(ed.), *Asteroids*, 1040–1063, Univ. Arizona Press.

IRAS DUST BANDS AND THE ORIGIN OF THE ZODIACAL CLOUD

S. F. DERMOTT and P. D. NICHOLSON
Center for Radiophysics and Space Research
Cornell University
Ithaca, NY 14853
USA

ABSTRACT. Previous discussions of the origin of the zodiacal cloud have attempted to distinguish between an asteroidal and a cometary source on the basis of collisional dynamics, that is, by calculating the rates of production and destruction of particles from the two possible sources. The uncertainties in these calculations are too large to permit a useful conclusion. The recognition that the solar system dust bands discovered by IRAS are probably produced by the gradual comminution of the asteroids in the major Hirayama asteroid families may allow us to estimate, with *comparative* confidence, the contribution to the zodiacal cloud of the asteroid belt as whole.

1. Introduction

Prior to the analysis of the data returned by IRAS, the origin of the particles that make up the zodiacal cloud was not known. While it was accepted that the source is either asteroidal or cometary, it was not possible to determine which of these two possible sources, if any, is dominant. While many attempts have been made to calculate the rate of production and destruction of asteroidal and cometary particles, the sources of uncertainty in these calculations are both large and numerous (see section 3). Two recent developments have provided important inputs to this problem.

Recent microscopic analysis of 200 interplanetary dust particles reported by Brownlee at this meeting (Schramm et al., 1988) have shown that while 45% of the particles (those classified as chondritic porous) are probably cometary, 37% of the particles (those classified as chondritic smooth) have characteristics (chemical alteration by liquid water) that suggest an asteroidal origin.

During its all-sky survey, IRAS discovered three prominent bands of warm (165 – 200 K) emission circling the sky at geocentric ecliptic latitudes of −10, 0 and +10 degrees (Low et al, 1984; Neugebauer et al., 1984). Low et al. (1984) suggested that the bands may be associated with dust derived from asteroidal collisions, while Sykes and Greenberg (1986) argued that random distintegrations of single asteroids in the 15 km diameter range are frequent enough to account for the IRAS observations. While these suggestions are important, neither does much to advance the debate. The fact that IRAS observed dust in the asteroid belt does not necessarily imply that the dust is asteroidal in origin. Numerous short-period comets have inclinations similar to those of the IRAS dust bands; the peak in the distribution of the observed inclinations is actually close to 10 degrees. Many of the observed short-period comets also have pericenters between 2 and 4 AU and, because of selection effects, we cannot rule out the possibility that a large number of short-period

259

comets with pericenters in the asteroid belt remain undetected. Furthermore, those comets that remain undetected because of low vaporization rates are the objects that are most likely to generate long-lived dust belts that extend around the sky.

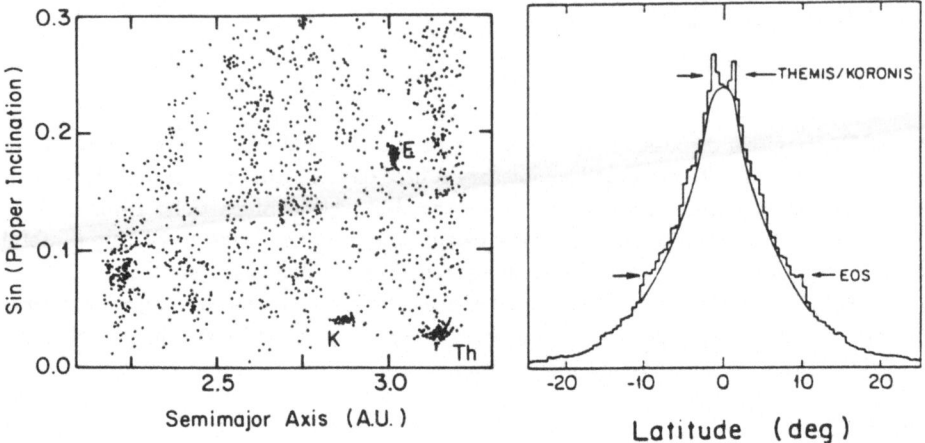

Figure 1. The proper inclination and semimajor axis of all asteroids listed in the TRIAD file are shown on the left. The clusters of asteroids at the points marked K, E, and Th are the Koronis, Eos and Themis asteroid families. The figure on the right shows a simulation of the latitudinal distribution of dust particles associated with the $\sim$ 2000 asteroids in the TRIAD file. The peaks marked Themis/Koronis and Eos arise from the clustering of inclinations associated with the three major families.

Dermott et al. (1984) pointed out (a) that the latitudes of the bands appear to coincide with the known latitudes of the three most prominent Hirayama asteroid families, that is, the Eos, Themis, and Koronis families (Fig. 1), and (b) that the expected *equilibrium* number density of particles in the 10 to $100\mu m$ size range associated with the families could be large enough to account for the IRAS observations. Dohnanyi (1969) has shown that despite the uncertainties in the physics of individual collisions, the *equilibrium* particle size-frequency distribution can be described with comparative confidence. In a series of papers, Dermott et al., (1984, 1985, 1986, 1988a) gave a detailed description of the expected geometry of the bands and described stringent tests of the asteroidal family hypothesis. The beauty of the asteroidal family model is that it is highly specific and clearly testable, although conclusive tests are far from simple. If the geometry of the bands could be related to the orbital elements of the families, then, since the particle production rate of the family members is proportionately no different from that of the rest of the asteroid belt, we would have conclusive evidence that asteroids are a significant source of the particles in the zodiacal cloud. Further outstanding problems that would then have to be addressed are the transport of particles from the asteroid belt to the Earth and the orbital eccentricities of the particles on arrival at the Earth. (Dermott and Gomes, 1988c).

2. Modelling the IRAS solar system dust bands

The data returned by IRAS showed, for the first time, that the structure of the zodiacal cloud is not entirely smooth. Some of the raw data used in our analysis of the newly discovered structure are shown in Figs. 2 and 3. The dust bands evident in these figures extend round the sky in unbroken bands that are approximately parallel to the ecliptic. A simplified view of the cross-section of a dust band is shown in Fig. 4. As a result of their vertical and radial harmonic oscillations, the particles spend more time near the outer edges of the boxes marked AABB than in their interiors, with the result that there are concentrations of particles in the corners of the boxes. Measurement of the apparent variations of the geometry of the bands, as seen from the Earth, with the time of the year and the elongation angle, ϵ, can be used to determine the orbital elements of the dust particle orbits.

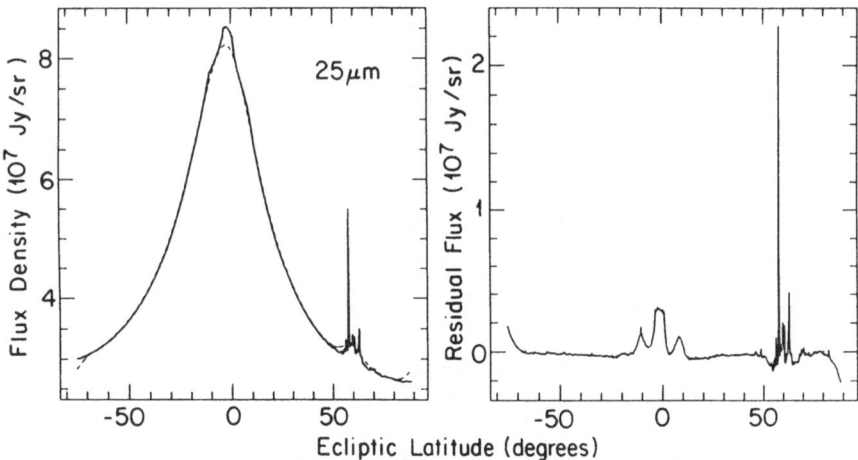

Figure 2. Fourier methods were used to separate the smooth, large-scale structure of the zodiacal background, shown on the left, from the narrower dust bands shown on the right. The short-scale structure evident at high latitudes is due to the galaxy.

There are two major considerations. (1) The latitudinal separation of the bands as seen from the Earth, β, varies with elongation angle, ϵ, and measurement of these variations leads to a determination of the inclination of the dust particle orbits and the heliocentric distances of the inner edges of the bands (the points marked P in Fig. 4). (2) The variation of the mean latitude of the bands with ecliptic longitude, or time of the year, due to the gravitational perturbation of the dust particle orbits by the planets leads to a determination of the forced inclinations and nodes and this also leads to a determination of the heliocentric distances of the bands (Fig 5).

The chief difficulties in interpreting the geometry of the bands are (1) the geocentric viewpoint of the IRAS telescope and (2) the large range of elongation angles sampled by IRAS. These difficulties cannot be completely removed by a simple coordinate transformation. Our geocentric viewpoint gives rise to small systematic displacements in the apparent locations of the bands which would not arise if the bands were viewed from the Sun. To overcome these and other difficulties we have constructed a three-dimensional numerical model that permits the calculation of the distribution of night sky brightness that would be produced by any particular distribution of dust particle orbits. This model includes the effects of planetary perturbations on the dust particle orbits, reproduces the exact viewing geometry of the IRAS telescope, and allows for the eccentricity of the Earth's orbit. The result is a model for the variation with ecliptic latitude of the brightness observed in a given waveband as the line of sight of the telescope sweeps through the model dust band at a constant elongation angle.

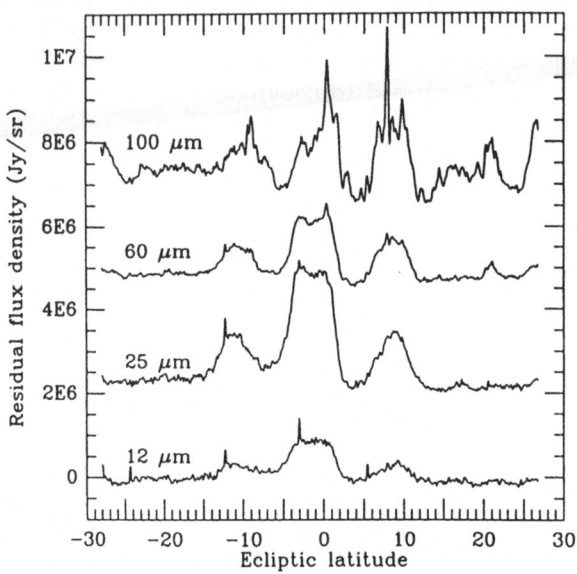

Figure 3. The dustbands seen in the four IRAS wavebands. The structure seen in the 100 μm waveband, although very similar to the structures seen in the other wavebands, is actually due to a region of star formation in the galaxy (Rowan-Robinson, private comm.) Some of the latter structure is also seen in the 60 μm waveband.

Some of our results are shown in Figs. 4 and 5. The parallax observations (Fig. 4) show that the inner edge of the "ten-degree" bands is close to 2.44 AU, while the inclination of the points P as seen from the Sun is 8.4 degrees. This is in *apparent* conflict with the asteroidal family model since the pericenter distance of the Eos band is to 2.8 AU while the mean inclination of the Eos family members is 10.2 deg.

In Fig. 5 an observed dust band profile is compared with the predicted profile for the Eos and Themis asteroid families. Several important comments can be made about this comparison (which is typical of the numerous other comparisons that we have made). (1) The observed latitudes of the peaks that we associate with the Themis family are in excellent agreement with the predictions based on secular perturbation theory. (2) The observed latitudes of the peaks that we associate with the Eos family are less than those predicted by one or two degrees (in good agreement with the discrepancy found from the parallax measurements). (3) The observed profiles are much broader and more rounded than the model profiles.

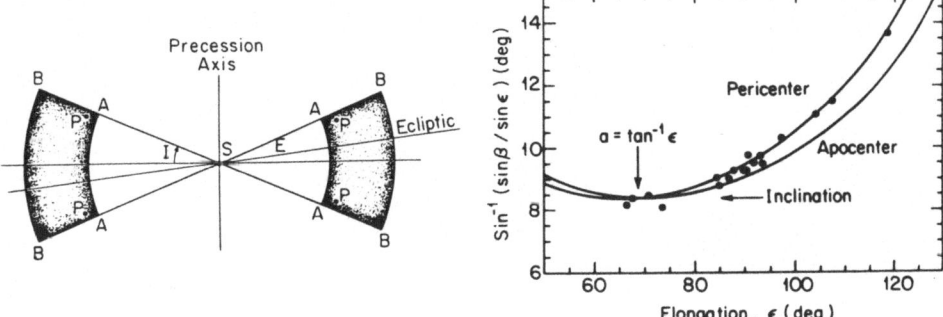

Figure 4. (Left): Cross-sectional view of a dustband. Particles are concentrated along the surfaces A (pericenter bands) and B (apocenter bands). The points P represent the positions of the apparent peaks in the latitudinal brightness variation as seen from the Earth. (Right): Variation of the geocentric ecliptic latitude β of the "ten-degree" dust bands with elongation angle, ϵ, as measured from IRAS HCON3 data. The curves P and A mark the bounds on β for a band with an inclination of 8.4 degrees, distance 2.44 AU, forced eccentricity 0.107 and forced pericenter of 32 degrees.

We have recently shown (Dermott et al., 1988b) that these discrepancies probably do not detract from the asteroidal family model. On the contrary, they may be evidence that particles formed in the asteroid belt spiral in towards the Earth. The model profile shown in Fig. 5 allows only for the dispersion in orbital elements displayed by the known family members. However, the thermal flux from the dust bands is due principally to those particles in the 10 to 100 μm size range and it is possible that the orbits of these particles decay due to Poynting-Robertson light drag before the particles are lost from the system (Dermott et al., 1986). If this is the case, then the particles in the dust bands will have significant dispersions in both their proper and their forced orbital elements. Our numerical model shows that dispersion of the orbital elements acts both to broaden a band and reduce its apparent inclination and we have suggested that the dispersion of the forced orbital elements due to the secular resonances at the inner edge of the asteroid belt may act to define the observed inner edges of the dust bands. Thus, by postulating that particles are transported out of the asteroid belt by P-R light drag we may be able to account for both the observed broadness of the dust band profiles and the apparent reduction in the inclination of the "ten-degree" bands that we associate with the Eos family.

3. The size of the source

Our numerical model also permits an accurate modelling of the total cross-sectional area of the particles associated with a given band. The observed areas are $\approx 2 \times 10^{19}$ cm^2. Since the orbits of the asteroids intersect, it is certain that they occasionally suffer high-velocity collisions and it is equally certain that the major asteroidal families, the Eos, Themis, and Koronis families, are the results of such collisions. However, the amount of dust that the disintegration of a single collision generates is highly uncertain. Consider the following, over-simplified, description of this type of calculation. If the size-frequency distribution of the particles in a given population is described by a power law of the form

$$\frac{dN(r)}{dln(r)} = \left(\frac{r}{r_o}\right)^{-3(q-1)}$$

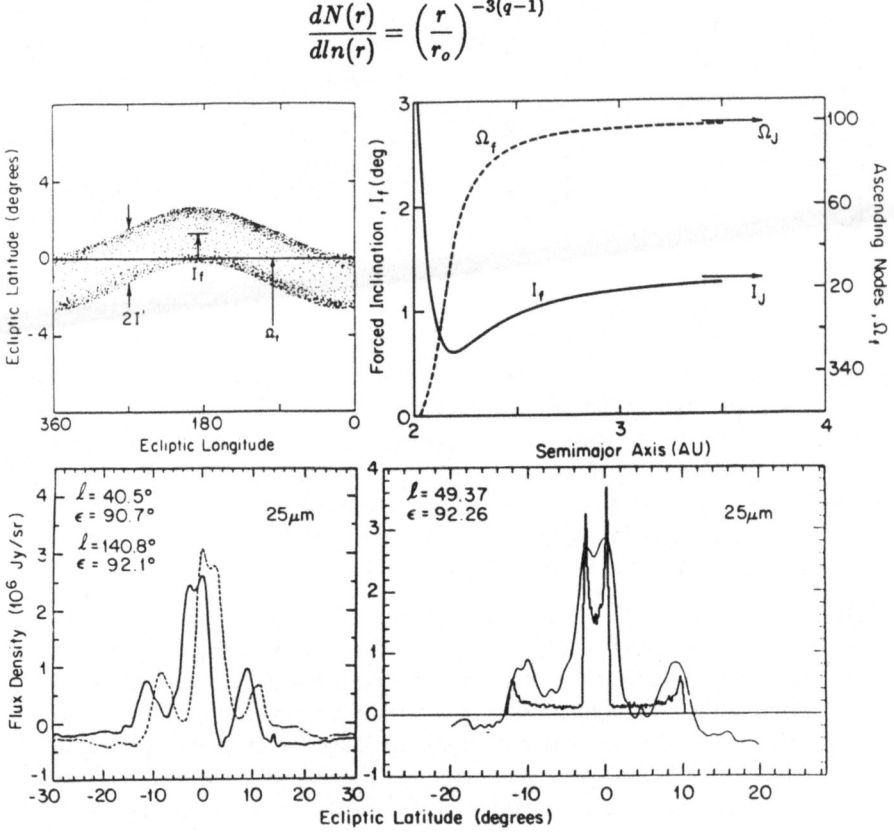

Figure 5. (Top left): Variation of the ecliptic latitude of a dust band with ecliptic longitude, as measured in a sun-centered coordinate system. The latitudinal width of the band at all longitudes is $2I'$, twice the proper inclination of the dust particle orbits. The amplitude of the the apparent sinusoidal variation in latitude is determined by the forced inclination, I_f. (Top right): Variation of the forced inclination I_f and the corresponding longitude of the ascending node Ω_f with distance from the Sun. (Bottom left): Two observations of the dust bands at different geocentric ecliptic longitudes showing the large variation in ecliptic latitude predicted by Dermott et al. (1984). (Bottom right): An observed dust band profile is compared with the prediction based on the distribution of orbits in the Eos and Themis families.

then four parameters are needed to specify the distribution. These are the scale factor r_o, the coefficient q, and the maximum and minimum particle radii, r_{max} and r_{min}. We can relate the surface area, A, of the dust to its total volume by calculating the equivalent radius, R_e, of that volume and by allowing that r_{max} is some fraction γ of R_e. The observed sizes of the asteroids in the major families suggest that $\gamma \approx 0.64$. The variation of A with q for various values of r_{min} is shown in Fig. 6.

If $q = 2$, then the disintegration of a single asteroid of diameter 15 km would provide the requisite area of dust (Sykes and Greeberg, 1986). However, since we have no way of knowing the appropriate value of q, and since varying q over a range as narrow as 1.8 to 2.2 produces a factor of 10^4 variation in A, we have little confidence that this type of prediction is useful. However, some semblance of confidence is restored if we consider the *equilibrium* size-frequency distribution. Dohnanyi (1969) has shown that despite the vagaries associated with the individual distintegrations the *equilibrium* size-frequency distribution is well-described by a power law with $q = 1.837$. If the Hirayama families represent equilibrium distributions down to some size as small as $\sim 100\mu m$, which we admit is debatable, then the equivalent radii needed to satisfy the observations are closely similar to the observed equivalent radii of the major families (Dermott et al., 1984).

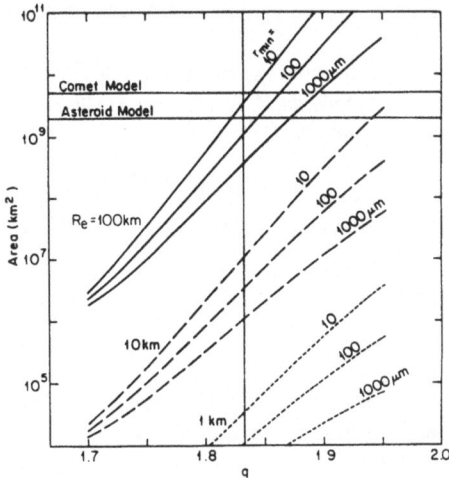

Figure 6. The total volume of material, as represented by the equivalent radius, R_e, needed to produce a given cross-sectional area in terms of the index of the power law distribution, q, and the lower cut-off in the particle size distribution, r_{min}. The vertical line refers to the equilibrium distribution, $q = 1.837$.

4. Conclusions

The arguments that we have presented for the asteroidal origin of a significant fraction of the particles in the zodiacal cloud are powerful since they are largely geometrical and depend only on physical processes, for example, secular perturbation theory, that are well understood. We have shown that the most prominent dust bands are probably derived from the major asteroidal families, that is, the Eos and Themis families, and Sykes (1988) reports that he has been able to detect a band that he associates with the Koronis family. However, to reconcile the observed shapes of the profiles, and the latitude of the "ten-degree" bands, with the predictions based on our accurate numerical models, we have to allow that the dispersions in the orbital elements of the very small dust particles ($r <$

$100 \mu m$) are larger than those of the parent asteroids. We have argued that this could be evidence for orbital decay due to Poynting- Robertson light drag and that the asteroids could be the source of "chondritic smooth" particles detected by Brownlee. The analysis of the available IRAS data is far from complete and we must expect that these arguments will undergo considerable refinement. Of particular interest, although not addressed in this short review, are the numerous faint bands discovered by Sykes (1988) that do not appear to be related to any known asteroidal families.

5. References

Dermott, S. F., Nicholson, P. D., Burns, J. A., and Houck, J. R. (1984) "Origin of the solar system dust bands discovered by IRAS", *Nature*, **312**, 505-509.

Dermott, S. F., Nicholson, P. D., Burns, J. A., and Houck, J. R. (1985) 'An analysis of IRAS' solar system dust band data', In *"Properties and interaction of interplanetary dust"*, eds. R. H. Giese and P. Lamy, Riedel, Dordrecht, 395-409.

Dermott, S. F., Nicholson, P. D., and Wolven, B. (1986) 'Preliminary analysis of the IRAS solar system dust data', In *"Asteroids, Comets and Meteors, II"*, eds. C-I. Lagerkvist and H. Rickman, Uppsala, pp. 583-594.

Dermott, S. F., Nicholson, P. D., Kim, Y., and Wolven, B. (1988a) 'The impact of IRAS on asteroidal science', In *"Comets to Cosmology"*, ed. A. Lawrence, Springer-Verlag, Berlin, pp. 3-18.

Dermott, S. F., Nicholson, P. D., Gomes, R., and Malhotra, R. (1988b). "Modelling the IRAS solar system dust bands", *Adv. Space Res.* (in press).

Dermott, S. F., and Gomes, R. (1988c). "Orbital decay of asteroidal dust particles due to Poynting-Robertson light drag and the origin of interplanetary dust particles" *Icarus* (submitted).

Dohnanyi, J. S. (1969) "Collisional model of asteroids and their debris", *J. Geophys. Res.*, **74**, 2531-2554.

Low, F. J. et al. (1984) "Infrared cirrus: new components of the extended infrared emission", *Astroph. J. (Letters)* **278**, L19.

Neugebauer, G. et al. (1984) "Early results from the infrared astronomical satellite", *Science*, **224**, 14-21.

Schramm. L. S., Brownlee, D., and Wheelock, M. M. (1988) "Major element composition of stratospheric micrometeorites", *Meteoritics* (in press).

Sykes, M. V. and Greenberg, R. J. (1986) "The formation and origin of the IRAS zodiacal dust bands as a consequence of single collisions between asteroids", *Icarus*, **65**, 51-69.

Sykes, M. V. "IRAS observations of extended zodiacal structures", *Astrophys. J. (Letters)*, **334**, (in press).

SPATIALLY VARYING OPTICAL PROPERTIES OF THE ZODIACAL DUST

S. S. HONG
Department of Astronomy, Seoul National University, KOREA
and
S. M. KWON
Space Astronomy Laboratory, University of Florida, U.S.A.

ABSTRACT. Analyses of both the zodiacal light in the visible and the zodiacal emission in the infrared have provided us with ample evidence to claim that the interplanetary dust particles are mixtures or coagulations of more than one constituents and their mixing ratios vary with the distance from the sun.

1. INTRODUCTION

Until quite recent years, most of what we know about the interplanetary dust particles came from the phenomenon of the zodiacal light, i.e., the scattered sunlight by the dust particles. The same dust particles ought to show up in the infrared by re-emitting the absorbed energy of the sunlight. Thus, such space experiments as the IRAS project (Hauser *et al* 1984) and the ZIP (Murdock and Price 1985) have added a new dimension to the study of the zodiacal dust cloud. In this contribution, we will critically review the recent analyses of the infrared zodiacal emission, and present evidence for supporting the multi-species nature of the zodiacal dust.

2. INVERSION OF THE IR BRIGHTNESS INTEGRAL

Brightness $Z(\varepsilon;\lambda)$ of the zodiacal emission measured at IR wavelength λ along the direction of elongation ε is given by the following integral of local volume emissivity $n(r)\sigma_{abs}(r;\lambda)B_\lambda[T(r)]$:

$$Z(\varepsilon;\lambda) = \int_0^\infty n(r)\sigma_{abs}(r;\lambda)B_\lambda[T(r)]dl, \qquad (1)$$

where l is the line-of-sight distance, B_λ represents the Planck function for the blackbody intensity, and $T(r)$ is the mean temperature of the dust particles at distance r from the sun. The multiple, $n(r)\sigma_{abs}(r,\lambda)$, of the dust number density $n(r)$ and the dust absorption cross section $\sigma_{abs}(r,\lambda)$, is called the volume absorption cross section (VAC). Please

267

D. McNally (ed.), Highlights of Astronomy, Vol. 8, 267–272.

note that the absorption cross section is considered here a function of both wavelength and heliocentric distance.

A methodology of inverting the IR brightness integral was developed to obtain the information on the r dependence of the VAC (Hong and Um 1987). A power-law

$$n(r)\sigma_{abs}(r;\lambda) = \zeta_0(\lambda) \left(\frac{r_0}{r}\right)^{\gamma}$$ (2)

of r with exponent γ was introduced for the spatial variation of the VAC, and another power-law with exponent δ, $T(r)=T_0(r_0/r)\delta$, for that of the mean dust temperature. Then, change of the integral variable l to scattering angle Θ transforms the brightness integral into the form

$$Z(\varepsilon;\lambda) = \frac{2hc^2}{\lambda^5} \frac{r_0\zeta_0(\lambda)}{\sin^{\gamma-1}\varepsilon} \int_{\varepsilon}^{\pi/2} \frac{\sin^{\gamma-2}\Theta \, d\Theta}{\exp|\alpha(\sin\varepsilon/\sin\Theta)^{\delta}|-1} ,$$ (3)

where $\alpha=hc/\lambda kT_0$ and other symbols have their usual meanings. Differentiating both sides of equation (3) with respect to elongation, they have

$$(\gamma-1)\cos\varepsilon = \frac{Z(\pi/2;\lambda)}{Z(\varepsilon;\lambda)} \left[\frac{d}{d\varepsilon}\ln Z(\varepsilon;\lambda)\right]_{\varepsilon=\pi/2} - \sin\varepsilon\frac{d}{d\varepsilon}\ln Z(\varepsilon;\lambda) + q(\varepsilon;\gamma).$$ (4)

Since $q(\varepsilon;\gamma)$ can be calculated numerically and $Z(\varepsilon;\lambda)$ is known from the observations, one may obtain the information on the unknown exponent γ from the solution to equation (4).

Our objective in implementing the IR inversion methodology [equation(4)] to the observations of the zodiacal emission is at knowing whether the derived value for γ is same for all elongations and wavelengths. If the dust optical property is spatially homogeneous, i,e., $\sigma_{abs}(r;\lambda)=\sigma_{abs}(\lambda)$ and a power-law with a fixed exponent describes $n(r)$, then the solution should give a constant γ for all wavelengths and elongations. If this, the simplest situation, were the case, the solutions for γ would draw a single straight line in the γ versus ε plane, running parallel to the ε-axis. As an alternative to the simplest case, if the dust property is spatially homogeneous as before but $n(r)$ doesn't follow a single power-law, then the solutions for γ should depend on elongation not on wavelength. In this alternative case, the inversion results would draw a same curve for different wavelengths. Keeping this line of reasoning in mind, let us look into the actual inversion results of the ZIP and IRAS data.

The ZIP experiments covered the ecliptic plane in two separate segments, one in the elongation range from 22° to 88° and the other from 138° to 180°. For the gap extending from 88° to 138°, Hong and Um (1987) smoothly connected the two segments, and they inverted the resulting ZIP data at 11 and 21 μm. For δ they used ½. Results of the inversion turned out to depend on both wavelength and elongation. In the γ *versus* ε plane, the solution for γ shows two curves of a similar shape for the two wavelengths.

The inversion result of the ZIP data over the gap may now be replaced by that of the IRAS, because the IRAS observed the zodiacal emission over the elongation range from 60° to 120°. Hauser and Vrtilek (1988) implemented the inversion methodology by Hong and Um (1987) to the IRAS data at 12, 25 and 60 μm. For δ they used two values, 0.36 and 0.5.

The γ values derived from the inversion with δ=0.36 are generally larger than those with δ=0.5, for all the three wavelengths. This may be understood as follows: For a given distribution of the zodiacal emission over the elongation, one needs to have a steeper decrease (larger γ) of the VAC with r in order to compensate the flattening (smaller δ) of the temperature variation. Sizes of the derived exponents for the VAC are thus closely related to the chosen value for δ. Because δ is not known precisely (discussed in section 3), numerical sizes of the exponents may not give us much information; our interest is rather in knowing whether they are independent of elongation and wavelength.

When δ=0.36 is used in the inversion, over the entire elongation range 60° < ε < 120°, the derived γ's for the three wavelengths are all different from each other. And there is a clear tendency of decreasing γ with increasing wavelength. The derived exponents also show an elongation dependence. In the first range of elongation 60° < ε < 80°, the exponents for all the three wavelengths decrease rapidly with increasing elongation; in the middle range 80° < ε < 100°, the three exponents become all constants of different sizes; in the last range 100°< ε < 120°, the exponent only for 12 μm decreases slowly with elongation and the other two exponents continue to be constants. When the larger value (0.5) is used for δ, general characteristics in the elongation dependence of the derived exponents are similar to the case of δ=0.36 with one exception, namely, the exponents for 25 and 60 μm become a same constant in the middle range of elongation. The decreasing tendency of γ with increasing wavelength is also evident in the first range of elongation, among the results with δ=0.5. We may summarize the inversion results of the IRAS data (Hauser and Vrtilek 1988) as follows: In the first and last ranges of elongation, the derived exponent for the VAC shows dependences on both elongation and wavelength. In the middle range, the exponent becomes independent of the elongation; but it may or may not be a function of wavelength, depending on the δ value.

Expectations based on the assumptions of the spatial homogeneity for the dust property and of the power-law relation for the VAC are not realized in the inversion results of the ZIP and IRAS. Thus, reasonings

based on *reductio ad absurdum* have led us to the following conclusions:
1. The absorption cross section of the zodiacal dust varies with
wavelength and heliocentric distance as well. 2. A power-law of r with
a fixed exponent is not enough to describe the spatial variation of the
VAC of the zodiacal dust.

3. SPATIAL VARIATION OF THE DUST TEMPERATURE

Dumont and Levasseur-Regourd (1988) devised an ingeneous way of determining the local grey temperature of the zodiacal dust. They showed
that if the local volume emissivity, $n(r)\sigma_{abs}(r;\lambda)B_\lambda[T(r)]$, decreases
with r as $a/r^3 + b/r^4$, then the brightness of the zodiacal emission at
an elongation ε is related to the value of the local emissivity at a
particular distance r_1 as follows:

$$Z(\varepsilon;\lambda) = n(r_1)\sigma_{abs}(r_1;\lambda)B_\lambda[T(r_1)] \left\{r_1^2/f(\varepsilon)\right\}, \qquad (5)$$

where r_1 is uniquely determined from ε, i.e., one-to-one correspondence
exists between them. Since $r_1^2/f(\varepsilon)$ has a length dimension, equation(5)
may be simply interpreted as a consequence of applying the *mean value
theorem* to the IR brightness integral given by equation (1). It should
be pointed out that $f(\varepsilon)$ is independent of a and b. If the absorption
cross sections are known at any two IR wavelengths, the local dust
temperature is easily determined by the ratio of the observed brightness
at the two wavelengths:

$$\frac{Z(\varepsilon;\lambda_i)}{Z(\varepsilon;\lambda_j)} = \frac{\sigma_{abs}(r_1;\lambda_i)}{\sigma_{abs}(r_1;\lambda_j)} \frac{B_{\lambda i}[T(r_1)]}{B_{\lambda j}[T(r_1)]}. \qquad (6)$$

Under the grey assumption for the absorption cross section, Dumont
and Levasseur-Regourd determined a run of local temperatures from the
12 and 25 μm data of the IRAS measurements at five different elongations.
The temperatures at the five locations over the range 1.1 AU < r < 1.3 AU
are found to vary as $r^{-0.33}$. Their analysis of the ZIP data at 11 and
21 μm gave a remarkably similar result $T(r) \propto r^{-0.32}$ over a somewhat
wider range 0.5 AU < r < 1.5 AU.

They noted the conflict between $\delta \approx 1/3$ and the grey assumption. To
reconcile the conflict, they relaxed the strict sense of the grey
assumption by assigning one constant value $\sigma_{abs}(r;VIS)$ to the absorption
cross section in the visible and another $\sigma_{abs}(r; IR)$ in the infrared.
With this relaxed version of the grey assumption, the departure of δ
from ½ was interpreted as an evidence that ratio of the cross sections
$\sigma_{abs}(r;VIS)/\sigma_{abs}(r;IR)$ varies as $r^{(2-4\delta)}$.

If the grey assumption holds true even in its relaxed version, the
spectral energy distribution of the zodiacal emission at any elongation
should follow the Planck curve [equation (5)]. However, observations

show spectral distributions that are significantly different from that of the blackbody. For example, from the same IRAS measurements of the zodiacal emission at a fixed elongation, one obtains widely different values of the color temperature, depending on the two wavelengths chosen for the color base. Local dust temperatures determined from the IRAS data at 25 and 60 μm are found to be significantly higher than the ones from the same data at 12 and 25 μm. Furthermore, they don't decrease smoothly with heliocentric distance, either. To a lesser extent, the same can be seen from the IRAS data at 12 and 60 μm. Therefore, at least some part in the departure of δ from ½ have been caused by the non-grey nature of the dust absorption cross section. The relation $r^{(2-4\delta)}$ may not holds as a whole, but it certainly proves that the dust property is not spatially homogeneous.

4. SPATIAL VARIATIONS OF DUST SCATTERING PROPERTIES

It can be shown that the brightness of the visible zodiacal light at a fixed elongation should decrease with the heliocentric distance of the observer as $r^{-(\nu+1)}$, if the scattering phase function is independent of location and the volume scattering cross section (VSC) varies with r as the power-law $r^{-\nu}$ with a fixed exponent ν (Giese and Dziembowski 1969). Space probes thus provide an excellent apportunity to determine the exponent for the VSC. Analysis of the Helios observations gave ν=1.3±0.05 for the region within the earth orbit (Leinert *et al* 1981); and that of the Pioneer suggested ν≃1.5 for the region outside the earth (Weinberg and Sparrow 1978). The difference between the indices for the two regions suggests that the power-law with a fixed exponent is too simple to describe the spatial variation of the VSC over the whole range of heliocentric distance.

Now, look at the scattering phase function. On the basis of a non-linear least squares analysis, Hong (1985) developed a methodology to derive the scattering phase function from the elongation dependence of the observed zodiacal light. Implementation of the methodology to the existing observations of the zodiacal light gave the scattering function. And the forward scattering part of the resulting function turned out to depend sensitively on the chosen value for ν. Only when ν≲1.1 is used, the observations yield scattering functions that have physically acceptable features near the forward direction. On the other hand, the Helios gave the tight bound, 1.3±0.05, to the exponent in the region that corresponds to small scattering/elongation angles. Thus the space probing and the analysis of the zodiacal light for the scattering function give conflicting results, but we have to remember that both approaches are vulnerable to the assumption of the spatially invarying scattering function. The discrepancy between 1.1 and 1.3 can be reconciled, if we suppose that the scattering phase function varies with the distance from the sun.

5. CONCLUSION

Scattering properties of the zodiacal dust particles in the visible and their absorption properties in the infrared are shown to vary with the heliocentric distance. Such spatial variations of the dust optical properties provide ample evidence for the multi-species nature of the zodiacal dust: the interplanetary dust particles are mixtures or coagulations of more than one constituents and mixing ratios of the constituents vary with the distance from the sun.

Acknowledgement: We thank Dr. Hauser and Dr. Vrtilek for communicating their inversion results with us. Partial supports were provided by the Basic Science Research Institute Program and by the US Air Force Office of Scientific Research.

REFERENCES

Dumont, R. and Levasseur-Regourd, A.-C. 1988, Astr. Ap., 191, 154.
Giese, R. H. and Dziembowski, C.V. 1969, Planet. Space Sci., 17, 949.
Hauser, M. G. *et al* 1984, Ap.J., 278, L15.
Hauser, M. G. and Vrtilek, J. M. 1988, a paper presented to the commission 21 meeting of the 20th IAU.
Hong, S. S. 1985, Astr. Ap., 146, 67.
Hong, S. S. and Um, I. K. 1987, Ap.J., 320, 928.
Leinert, C., Richter, I., Pitz, E., and Planck, B. 1981, Astr. Ap., 103, 177.
Murdock, T. L. and Price, S. D. 1985, A. J., 90, 375.
Weinberg, J. L. and Sparrow, J. G. 1978, in *Cosmic Dust* ed. by J. A. M. McDonell (New York: Wiley), p.75.

WHAT WE KNOW ABOUT FAMILIES OF ASTEROIDS

V. ZAPPALA'
Astronomical Observatory of Torino
I-10025 Pino Torinese
Italy

ABSTRACT. Asteroid families are considered for the most to represent fragments of collisional breakup of precursor bodies. If true, this offers the unique possibility to examine the interiors of large bodies and to study the processes of collision on a scale much larger than can be done in laboratory. Indeed, the general features of the mass distributions and of the ejection velocities of the family members can be interpreted in terms of collisional disruption of a parent body followed by self-gravitational reaccumulation on the largest remnant. However, several problems remain open: a) the degree of fragmentation in real families is generally lower than that observed for experimental targets; b) the relative velocities computed including also proper eccentricity and inclination differences are higher by about a factor 4 than those derived from semiaxes differences only; c) only very few of the presently proposed families have distributions of inferred mineralogies consistent with cosmochemistry. Further studies are needed, including better proper elements computation, classification methods, and new investigations on the physics of hypervelocity impacts.

1. INTRODUCTION

The very identification of a family via the analysis of "clusters" of orbital elements in the phase space, presents a number of difficulties and ambiguities due to the arbitrary nature of some key assumptions of the analysis, i.e., separating family members from the "field" objects, and to inaccurate or unreliable proper elements. The often produced divergent results were analyzed by Carusi and Valsecchi(1982). Moreover, although the idea of the origin of families by collisional breakup of a parent body is now widely accepted, the details of this process are not fully understood. While the potential of physical studies to test

273

and refine the collisional breakup theory is apparent, a wealth of information seems still to be hidden in the available data. In particular, a systematic coupling of orbital and physical data has to be more deeply investigated.

Obviously, one cannot exclude that some (or many) families have non collisional origin. In addition to the well known groups of Hungaria and Phocaea (Williams, 1971), other smaller groups can be separated from the field by secular resonances and can appear as true families. At the moment, however, this field of research is just at the beginning; therefore, the present paper will be devoted only to demonstrate that the collisional hypothesis is quite consistent with the data and that from these data one can extract some interesting information on the mechanism of collisional fragmentation. In addition, the major discrepancies which remain to be solved will be outlined.

2. MASS DISTRIBUTIONS

The knoweldge of the mass (or size) distribution of asteroids has been generally considered a powerful tool in understanding the evolutionary mechanisms which have been effective for the asteroid population as a whole. In particular, it is known that catastrophic collisions should result in a characteristic mass distribution of fragments. In this framework, the determination of the mass distribution of family members is crucial, since it allows direct comparison with laboratory experiments as well as numerical simulations of both the individual breakup process and the overall collisional evolution.

Gradie et al.(1979) made the first comprehensive attempt to "reconstruct" the parent bodies for some selected families. Fujiwara(1982) performed a detailed study of the mass distribution of the three "classical" Hirayama families (Koronis, Eos, and Themis), concluding that the three families were completely fragmented, but most of the fragments should have been reaccumulated by mutual gravitation, while the larger members could have rubble-pile structures, roughly fitting hydrostatic equilibrium figures. Zappala' et al.(1984) extended the analysis to the whole set of Williams'(1979) families. The first step of their work was to reconstruct the total mass of a family and, as a consequence, the mass of its parent body. They computed the missing mass of the unobserved smaller components using a differential mass distribution, with an assumed exponent (1.8) as suggested by the theoretical study of Dohnanyi(1971) for the whole sample of asteroids. Obviously, this procedure yelds only a crude estimate of the lower limit for the total mass of each individual parent body, but can be very useful in statistical analyses.

Zappala' et al. represented the mass distributions of specific families in terms of the "discrete mass distribution" introduced by Kresak(1977). Comparing the distribution tails, the best fit exponents, the mass ratios among the largest fragments, and the total masses of the precursor bodies, they found that - a part the very few largest fragments - the trend is quite similar among most of the families and it can be roughly fit by the usual exponent of about 1.8. A good agreement was also found with the results coming from laboratory experiments on hipervelocity breakups (Fujiwara, 1986).

The behaviour of the mass distribution among the largest bodies, in particular the mass ratios among the parent body, the largest fragment, and the second largest fragment, deserve further scrutiny. A few families show an unusual sudden mass drop from the largest and the second largest remnant, which is completely absent among catastrophic fragmentation experiments. This can be explained as a result of sub-catastrophic cratering impacts, which leave most of the parent body's mass intact, but also as a product of gravitational effects leading the reaccumulation of the slowest escaping fragments onto the largest remnant. The latter hypothesis is confirmed by the correlation existing between the M1/M0 ratio (M0=mass of the parent body, M1=mass of the largest remnant) and the size of the precursor body: larger mass ratios, implying more efficient reaccumulation, are associated to larger parent bodies. On the other hand, no correlation was found for the M2/M0 ratio (M2=mass of the second largest fragment), implying that no reaccumulation is effective for smaller remnants. The latter conclusion leads to a major discrepancy between the mass distributions of most families and the laboratory results. In fact, a scaling of the specific energy (E/M) from laboratory experiments to asteroid sizes predicts much more fragmentation for the asteroids than is seen: the specific energy necessary to disperse the fragments to infinity, overcoming the gravitational binding of the parent body, is considerably higher than the critical value for breakup observed in the laboratory. The problem is that any reasonable partition of energy would break a target body into innumerable tiny pieces, if the impact were sufficiently energetic to provide the kinetic energy necessary to disperse the fragments into a family. This dilemma could be resolved only if the effective strengths for asteroids were exceptionally high (Davis et al.,1985).

3. VELOCITY DISTRIBUTIONS

Another fundamental aspect of the families which can be compared with experimental data is the apparent ejection velocities of the fragments. In line of principle one could

derive the relative velocities from the differences between
semimajor axis, proper eccentricity, and proper inclination
of the family asteroids, and those of a reference body (for
which a natural choice is the largest asteroid of the family
under scrutiny). However, even at this preliminary stage,
there are some inescapable difficulties involving the
retardation of an ejected fragment due to self-gravitation
of the disrupted body; a possible further dispersion due to
subsequent breakups of the members; the dependences of the
velocities on some unknown angles at the moment of the
breakup event. Nevertheless, assuming to have reconstructed
quite accurately the mass of the parent body and that the
subsequent impacts should have affected only very small
"original" fragments, the problem of the unknown angular
elements at the time of breakup can be partially overcome by
using some mean value of the trigonometric functions or by
exploring the resulting velocities with various assumptions.
Obviously, this procedure cannot be taken into account in
order to understand the dynamical history of individual
families, but can be useful for statistical considerations.
This was the approach of Zappala' et al.(1984), who studied
the proper elements of Williams'(1979) families. For
obtaining the relative velocity components vS, vW, and
vT(S=along the direction toward the Sun, W= along the normal
to the orbital plane, and T=WxS) from the differences in
semimajor axis, eccentricity and inclination, they used the
classical Gauss' perturbation equations (see, e.g., Brouwer
and Clemence, 1961, p.299). Even with the most favourable
assumptions about the unknown angles quoted before, the
velocity distributions were found to be far from isotropy.
In fact, the r.m.s. values of vS and vW exceed by a factor 4
or 5 that of vT. This trend exists even for the three
largest classical families (Themis, Eos, and Koronis). There
is no obvious physical explanation for this result within
the collisional theory. Excluding at the moment any
cosmogonic rather than strictly collision origin for
families, one should point out that while vT depends mainly
on the difference in semimajor axis, vW and vS depend more
strongly on the differences between inclination and
eccentricity. Therefore, it is possible to ascribe the
asymmetry to poor reliability of the proper elements e' and
i'(a is generally a more reliable parameter). At least
within the linear theory, Carpino et al.(1986) confirmed
this hypothesis, by simulating some "synthetic" families and
performing numerical integrations for 10000 years. They
found that e' and i', as computed with the aid of the linear
theory, fluctuate widely in time, causing a systematic
"noise" in e' and i', artificially increasing the resulting
differences and thus the velocities vS and vW.The asimmetry
found by Zappala' et al.(1984) indicates that probably such
effect cannot be completely removed, even within a more

refined perturbation theory.

Based on these considerations, Zappala' et al.(1984),following Ip(1979) restricted their interpretation of family velocities to the velocity vT, which depends on the most "reliable" orbital parameter, the semimajor axis. The resulting value was multiplied by a factor 3**0.5 to account for the other two neglected components, assuming overall isotropy. The ejection velocity was computed by correcting the above velocity for the gravitational slowing down of the fragments escaping from the parent body.

From a plot of the mean ejection velocity of each family versus the size of the largest remnant, Zappala' et al. did not find any correlation. This result is consistent with the fact revealed by laboratory impact experiments that the ejection velocity depends mainly on the specific energy delivered to the target by the collision (Fujiwara and Tsukamoto, 1980); this quantity, in turn, depends on the impact velocity and on the projectile-to-target mass ratio, and both these parameters are not correlated with the target asteroid's size. The mean of the ejection velocity for the used sample of families resulted in 145 m/sec, which agrees well with the values found in the experiments for projectile-to-target mass ratio in the range 0.001 to 0.01 (assuming an impact velocity of about 5 km/sec); it is also remarkable that, according to Farinella et al.(1982), this range is precisely the same as that expected for the largest collision endured by all asteroids of size larger than about 10 km.

It is less easy to understand another result of this analysis: there are no velocities lower than 60 m/sec even for small target bodies, for which gravitational reaccumulation should be negligible. This result seems discrepant from experimental breakups, for which fragment velocities are generally lower for the same degree of fragmentation. Similar evidence about larger ejection velocities consistent with a moderate degree of fragmentation has been discussed in terms of the supposed catastrophic breakup of the saturnian satellite Hyperion (Farinella et al.,1983). This problem of velocity scaling may be related to the E/M scaling problem mentioned in the previous Section: in both cases, the apparent degree of fragmentation seems inadequate for the evident energy.

Even limiting the analysis to the vT component, it is possible to evidence some symmetry propriety of the ejection velocity field. The distribution of the differences in semimajor axis was investigated by Ip(1979) and extended by Zappala' et al.(1984). It was possible to distinguish between "symmetric" (or" dispersed") and "asymmetric" families, the latter ones showing most fragment on the same "side". Asymmetric families generally correspond to larger ejection velocities and to larger objects. Possibly this is again related to self-gravitation effects, which could amplify any initial anisotropy of the velocity field.

4. COMPOSITIONS OF FAMILY MEMBERS

Another fruitful way to study the origin of families and to investigate the collisional hypothesis is related to the mineralogy of the members of a given family. In the collisional assumption the inferred mineralogies must be consistent with a reliable cosmochemical model of the parent body.

Based on the TRIAD taxonomic classifications available (Bowell et al.,1979), Gradie et el.(1979) discussed the compositions for 47 Williams'(1979) families for which two or more members were classified. They concluded that while many of the more populous families are homogeneous, and consistent with the breakup of a homogeneous precursor body, a significant fraction of less populous smaller families are not. In addition, the families composed of dissimilar members are often difficult to explain in terms of the prevailing interpretations of mineralogy and cosmochemical models of parent bodies.

More recent studies have taken advantage of the refiniments in asteroid taxonomy and of the much larger database that has been compiled over the past decade. Bell (1988) concludes that there are only five families that seem to be well-established and composed of genetically related asteroids. He doubts the "reality" of a large fraction of the remaining families. On the other hand, Chapman(1987) performed a study which arrives at somewhat different -less pessimistic- conclusions. He confirms the distinctiveness and probable reality for the classical families (Nysa, Maria, Koronis, Eos, Themis, and subsets of the Flora family), and finds that several additional Williams families are compositionally distinctive, and a dozen more are probably distinctive although statistics are poor.

A more detailed review of the arguments of the present paper can be found in Chapman et al.(1988).

5. REFERENCES

Bell,J.F.(1988), submitted to Icarus

Bowell,E.,Gehrels,T. and Zellner,B.(1979)'Magnitudes,colors, types and adopted diameters of the asteroids',in T.Gehrels (ed.), Asteroids, University of Arizona Press, Tucson, pp. 1108-1129.

Brouwer,D. and Clemence,G.M.(1961) Methods of Celestial Mechanics, Academic Press, New York/London.

Carpino,M.,Farinella,P.,Gonczi,R.,Froeschle',Ch.,Froeschle', C.,Paolicchi,P. and Zappala',V.(1986)'The accuracy of proper orbital elements and the properties of asteroid families: Comparison with the linear theory', Icarus 68,44-76.

Carusi,A. and Valsecchi,G.(1982) 'On asteroid classification in families',Astron.Astrophys.115, 327-335.

Chapman,C.R.(1987)'Distributions of asteroids compositional types with solar distance, body diameter, and family membership'(abstract),Meteoritics 22, 353-354.

Chapman, C.R., Paolicchi,P., Zappala',V., Binzel,R. and Bell,J.F. (1988) 'Asteroid Families: Physical Properties and Evolution', submitted to R.P. Binzel, T. Gehrels and M. Matthews (eds.), Asteroids II, University of Arizona Press,Tucson.

Davis, D. R., Chapman, C. R., Weidenschilling, S. J. and Greenberg,R.(1985) 'Collisional history of asteroids : Evidences from Vesta and the Hirayama families',Icarus 62, 30-53.

Dohnanyi, J .W. (1971) 'Fragmentation and distribution of asteroids', in T.Gehrels (ed.), Physical Studies of Minor Planets, NASA-SP 267, Washington, pp.263-295.

Farinella,P.,Paolicchi,P.and Zappala',V.(1982)'The asteroids as outcomes of catastrophic collisions',Icarus 52,409-433.

Farinella,P., Milani, A., Nobili, A., Paolicchi, P. and Zappala', V. (1983) 'Hyperion: Collisional disruption of a resonant satellite', Icarus 54, 353-360.

Fujiwara, A. (1982) ' Complete fragmentation of the parent bodies of Themis, Eos and Koronis families', Icarus 52, 434-443.

Fujiwara, A. (1986) 'Results obtained by laboratory simulations of catastrophic impact', in D.R.Davis, P.Farinella, P.Paolicchi and V.Zappala' (eds.), Catastrophic Disruption of Asteroids and Satellites, Mem. S. A. It., Firenze, pp. 47-64.

Fujiwara, A. and Tsukamoto, A. (1980) 'Experimental study of the velocity of fragments in collisional breakup', Icarus 44, 142-153.

Gradie, J.C., Chapman, C.R. and Williams, J.G. (1979) 'Families of minor planets', in T. Gehrels (ed.), Asteroids, University of Arizona Press,Tucson, pp.359-390.

Ip, W.H. (1979) 'On three types of fragmentation processes observed in the asteroid belt', Icarus 40, 418-422.

Kresak, L. (1977) 'Mass content and mass distribution of the asteroid system', Bull.Astron.Inst.Czech. 28, 65-82.

Williams, J.G.(1971) 'Proper elements, families, and belt boundaries', in T.Gehrels(ed.), Physical Studies of Minor Planets, NASA-SP, Washington, pp. 177-181.

Williams, J.G. (1979) 'Proper elements and families memberships of the asteroids', in T. Gehrels (ed.), Asteroids,University of Arizona Press,Tucson,pp.1040-1063.

Zappala', V., Farinella, P., Knezevic, Z. and Paolicchi, P. (1984) 'Collisional Origin of the Asteroid Families: Mass and Velocity Distributions', Icarus 59, 261-285.

COMETS, METEORITES AND INTERPLANETARY DUST

D. E. Brownlee
University of Washington
Dept. of Astronomy
Seattle, Washington 98195
U.S.A.

ABSTRACT. Cometary debris of all sizes impacts the Earth but it is likely that only particles the size of dust survive atmospheric entry and are collected as meteoritic samples. Conventional meteorites and a substantial fraction of collected interplanetary dust particles appear to be asteroidal debris. Nearly half of the collected interplanetary particles have properties consistent with cometary material and resemble Halley dust that has lost the majority of its carbon and nitrogen. These particles might be aggregates of presolar grains and they provide some insight into the properties of interstellar grains.

1. INTRODUCTION

New information from the 1986 Halley flybys, meteorites, micrometeorites, and IRAS observations of interplanetary dust, have provided new insights into the relationship between meteoritic samples, comets and asteroids. The measured and inferred properties of cometary particles also provide new clues on the possibility that comets contain preserved presolar interstellar grains.

2. METEORITES

Conventional meteorites have been the traditional source of extraterrestrial material investigated in the laboratory. Meteorites, for the most part are primitive solar system materials and their chemical, mineralogical and isotopic properties provide a wealth of information on early solar system processes, materials and environments (Kerridge 1988). Conventional meteorites are large, critically important samples but they have a major limitation. They appear to be samples of only a small number of the solar system's comet and asteroid population. It likely that important classes of minor planets are not represented at all in meteorite collections.

Conventional meteorites are not a representative sample of the solar system's minor planet inventory because of selection effects associated with orbital transfer to Earth-crossing orbits and survival of hypervelocity entry into the atmosphere. Due to their size, meteorites can only be perturbed to Earth-crossing orbits by purely gravitational interactions. This is not a serious selection effect for cometary objects that are commonly released on Earth-crossing orbits, but it is important for asteroidal fragments. Meteorites from the asteroid belt reach the Earth after a series of gravitational perturbations. In the majority of cases the perturbations may initiate in chaotic zones associated with orbital resonances with Jupiter (Wisdom 1983). Wetherill (1985) has made a strong arguments that the the most common chondrite groups originate from the 3:1 Kirkwood gap, a very small region of the asteroid belt. The most serious selection effect on conventional meteorites is due to the stresses of atmospheric entry. To survive entry as an intact centimeter object, a meteoroid must strong. The peak atmospheric ram pressure experienced by centimeter objects results in disintegration of fragile materials and partial fragmentation of even the strongest stony meteorites. The most friable recovered meteorites are still rather strong rocks and are much stronger than cometary materials observed as typical meteors. Typical

D. McNally (ed.), Highlights of Astronomy, Vol. 8, 281–286.

cometary meteoroids in streams fragment when the dynamic ram pressure exceeds 10^5 dynes cm^{-2} (Verniani 1960) and yet the weakest chondrites have crushing strengths orders of magnitude higher. The most fragile stream meteors are the Draconids with strengths near 10^3 dynes cm^{-2} (Sekanina 1985). From atmospheric entry considerations it is clear that typical cometary meteoroids cannot become conventional meteorites. If comets do not contain strong components then no cometary samples would survive entry as particles larger than centimeter size.

An asteroidal origin for conventional meteorites is supported by mineralogical evidence that even the most chemically primitive meteorite classes have experienced at least mild metamorphism. The most volatile rich chondrites, those that most closely match solar abundances, show evidence for aqueous alteration. The type 1 (CI) chondrites even have conspicuous sulfate veins deposited from aqueous solution (Kerridge and Bunch 1979). Evidence for warm episodes in the history of all meteorites is strong evidence against origin from comets, bodies that sublime when heated above cryogenic temperatures. An additional argument that meteorites are derived from inner solar solar system bodies is that the abundance of implanted solar wind in brecciated chondrites is consistent with exposure conditions within 5AU (Anders 1975).

The evidence suggests that most and probably all meteorites are samples of asteroids but comparison with asteroid reflectance data shows that meteorites are not representative sample of the main belt asteroids. Over 70% of meteorites falls are ordinary chondrites (Wasson 1974) and yet spectral reflectance studies show that such materials are rare in the main belt asteroids (Gaffey and McCord 1979). The majority of asteroids have rather flat spectral reflectance curves similar to the carbonaceous chondrites. The carbon rich carbonaceous chondrites constitute only 2.6% of meteorite falls. In this regard it is highly significant that roughly 85% of the interplanetary dust particles that are collected and analyzed are more similar to the carbonaceous chondrites than any other meteorite group.

It is likely that there are no cometary materials that have become conventional meteorites. It also appears that the typical materials in the asteroid belt also do not commonly produce meteorites and are highly under represented in meteorite collections.

3. INTERPLANETARY DUST

3.1 Samples

Interplanetary dust (IDP) is particulate material that must include samples of the parentbodies that produce meteorites but they also appear to include samples of materials not contained in meteorite collections. Collected samples range in size from a few microns to a millimeter in diameter. The best preserved samples are the 10μm size particles collected from the stratosphere with U2 aircraft. Larger particles are collected from polar ice and deep sea sediments (Brownlee 1985). Most of the larger particles ($> 100\mu m$) melt during atmospheric entry but some properties such as elemental and isotopic composition are preserved.

3.2 Selection effects

Collected dust particles are expected to be a more representative sampling of the solar systems minor planet population because of less severe selection effects during orbital evolution and atmospheric entry. Orbits of dust particles decay due to Poynting-Roberston drag and all dust particles generated beyond 1AU can become Earth-crossers if they are not destroyed by mutual collisions. For this reason it is expected that all classes of solar system minor bodies should generate dust that eventually reaches the Earth. Once at the Earth there are of course atmospheric selection effects but the survival criteria are different than those applicable to centimeter and larger objects. A critical difference for survival of cometary matter is that small particles are not subjected to dynamic ram pressures as high as the crushing strength of cometary meteors. Particles of 10μm size decelerate from cosmic velocity above 90 km where the ambient air density is sufficiently low that the maximum ram pressure does not reach 10^5 dynes/cm^2, the typical strength of cometary meteors. Even Draconid particles should survive without fragmentation for sizes of 10μm and smaller. Larger and denser particles retain hypervelocity speeds into denser layers of the atmosphere and they are accordingly exposed to higher ram pressures.

3.3 Origin

Interplanetary dust particles are purged from the solar system by Poynting-Robertson drag and collisions. If the meteoroid complex is maintained in equilibrium against losses, new particles must be injected into the interplanetary medium on time scales of less than 100,000 years. The major sources of dust are asteroids and comets. Comets have traditionally been believed to be the dominant source of new particles because comets are the major source of meteors and until the IRAS observations, there was no observational evidence for the presence of dust in the asteroid belt. Due to light pressure effects, most of the dust released from long period comets travels on hyperbolic orbits and is lost from the solar system (Harwit 1963). Short period comets do release dust into bound orbits but the amount of observed dust from these objects appears to be insufficient to counter the 10^7 gs^{-1} loss rate due to collisions and the Poynting-Robertson effect (Kresak 1987). It is possible that much of the comet dust in the solar system was produced by extinct comets or that the major comet dust source is debris from progressive communition of larger cometary meteoroids (Kresak 1987) not directly observable in cometary comae.

For full interpretation of the cometary particles that are collected as micrometeorites, it is important to know whether the majority of particles are samples of only one or two comets or a large number of comets. Kresak's study (1987) indicates that over the past 200 years, particles from P/Encke have dominated the input to the zodiacal cloud and that over 90% of the total input has come from this comet as well as P/Halley, P/Brorsen, and P/Biela. Since the present dust population is the integral result of emissions over 10^4 to 10^5 years, prediction of the dominant sources of interplanetary comet dust is quite uncertain.

Some fraction of IDPs must be debris from the asteroids but the ratio of cometary to asteroidal interplanetary particles at 1AU is unknown. The discovery of IR emission from asteroid dust and the IRAS dust bands (Low et al. 1983) has lead to a renewed interest in asteroids as a source of interplanetary particles. Asteroids appear to be an important source of interplanetary particles that reach the Earth and it is possible that they are the major source (see Dermott and Nicholson in this symposium). The asteroid particles that spiral into 1AU should be relatively representative samples of the surface area of the asteroid belt but there may be a bias towards debris from asteroid families such as those that apparently produce the IRAS dust bands (Sykes and Greenberg 1986).

3.5 Types

The properties of micrometeorites collected in the stratosphere have been described elsewhere (Bradley et al. 1988, Mackinnon and Reitmeijer 1987, Sandford 1987 and Brownlee 1985). The particles can be classified into two broad groups that are primarily distinguished by mineralogical and morphological differences. One of the groups is dominated by hydrated minerals and the other is dominated by anhydrous minerals. Both particle types are black fine grained materials that generally match the elemental composition of carbonaceous chondrites within a factor of two for major and minor elements.

3.5.1. *The Hydrous Group.* This group, also called CS (chondritic smooth), is probably of asteroidal origin or at least it appears to be dominated by asteroidal particles. The CS particles are usually nonporous and have smooth, cracked, plately or fiberous surface textures. The dominant minerals are hydrated silicates and there are two subgroups one of which is dominated by serpentine and the other by smectite-like phases. Many of these particles contain evidence for aqueous alteration of the type that has altered the CI and CM (C1 and C2) carbonaceous chondrites. This evidence includes depletion of Ca (Schramm et al. 1988), presumably to water soluble phases, and the presence of carbonates and clusters of magnetite grains. The magnetite clusters are identical to those in CI chondrites formed by aqueous alteration. The evidence for aqueous alteration provides a strong argument against a cometary origin for particles in the CS class. It is possible that the progenitors of the CS particles were porous and fragile and that aqueous alteration produced the hydrated minerals, closed voids and strengthened the particles. Many of the CS particles are nearly identical to CI and CM chondrites and may be samples of the parentbodies of these meteorites.

3.5.2. *The Anhydrous Group.* This group (also called CP or chon dritic porous) is dominated by anhydrous minerals and is probably of cometary origin (Bradley and Brownlee 1986). Unlike all conventional meteorites many of the CP particles are porous with porosities that sometimes exceed 50%. The typical micro-porosity of chondrites is only a few percent. The open microstructures of CP particles is inconsistent with a history of compaction unless sublimable phases once filled void spaces. The porous aggregate structure of the CP particles is unique among meteoritic samples but it is consistent with the friable properties attributed to cometary meteors. The dominant constituents of CP particles are olivine, pyroxene, iron sulfide, glass and carbon. These materials occur as submicron equidimensional grains. Some of the submicron grains are solid microaggregates ("tarballs") of mineral grains less than $0.1\mu m$ in diameter. Carbon occurs as discrete submicron grains and as coatings on grains. Nearly of the carbonaceous matter is amorphous and generally resembles the aromatic rich kerogen found in carbonaceous chondrites. Graphite is exceedingly rare. Raman studies indicate the presence of carbonaceous layers on silicates that obscure the the characteristic Raman silicate bands (Wopenka 1988). Direct observation by electron microscopy of sectioned particles indicates however that such coatings are thin in the range of 100A.

4. COMET DUST

The 1986 Halley flybys provided the first direct compositional measurements of comet dust. The PIA and PUMA mass spectrometers flown on Giotto and the VEGA spacecraft provided thousands of time-of-flight mass spectra of individual micron and submicron particles (Kissel 1986a,b). The integral of these spectra implies a bulk elemental composition of Halley solids that is consistent with chondrites except that the carbon and nitrogen abundances are greatly enhanced. The abundances of these two elements relative to silicon appears to be approximately solar (Jessberger 1988). Roughly a third of the particles are dominated by H, C, N and O and are called "CHON". The most common particles are a mix of the light elements and the major chondritic "silicate" elements Mg, Si and Fe. The remainder of the particles contain only low carbon abundances and typically have compositions consistent with mixtures of silicates and sulfur. An important property of the silicate composition particles is that many of them are Fe free magnesium silicates. Presumably these are composed of forsterite or enstatite, minerals that condense at high temperatures from solar composition gas. The range in composition of micron and submicron Halley particles provides a powerful means of comparing Halley with other primitive solar system materials. The micron scale dispersion of Mg and Fe in Halley is clearly distinct from that observed in carbonaceous chondrites meteorites but it is a fairly good match with the anhydrous class of interplanetary dust (Jessberger *et al.* 1988; Lawler *et al.* 1988). Like the interplanetary particles, Halley dust shows a wide range of Mg/Fe ratios compared to those in the fine grained fraction of CI and CM meteorites.

   With the exception of C and N, the Halley particles have bulk compositions similar to primitive chondrites and interplanetary dust but the fine scale heterogeneity indicates that Halley is mineralogically distinct from the carbon rich chondrites and many of the hydrated types of IDPs. The anhydrous dust particles are the best match to the Halley data and their mineralogical composition is also consistent with fine structure seen in $10\mu m$ silicate emission observed in Halley (Bregman 1987). If indeed Halley is composed of anhydrous minerals this would imply that Halley is a mix of low temperature ice and volatiles mixed at fine scale with rather high temperature minerals such as pure Mg olivine. Condensation of this phase from the solar nebula occurs at temperatures above 1200K. The glaring discrepancy between the Halley measurements and collected samples is the abundance of carbon. The C/Si atomic ratio in Halley dust is estimated by Jessberger *et al.* to be 4.4, a value that is six times higher than CI chondrites and at least two and a half the mean value estimated for interplanetary dust. These differences could fundamental or is possible that much of the carbonaceous matter is Halley dust is relatively volatile and is lost in collected dust samples either during residence in the interplanetary medium or during atmospheric entry.

5. INTERSTELLAR DUST

Anomalous isotopic compositions of minor components isolated by chemical, mechanical and thermal means, indicate that meteorites contain at least trace quantities of presolar matter that has retained isotopic memories of presolar processes. The recent discovery of small isotopically anomalous SiC grains in the Murray meteorite is evidence that minute qualities of genuine presolar interstellar grains are preserved (Zinner *et al.* 1987). The presence of deuterium rich nuggets in IDPs is also evidence for presolar grains (McKeegan *et al.* 1985). If asteroidal meteorites contain interstellar dust is is likely that cometary matter should contain higher abundances of presolar solids and might even be dominated by such materials. Greenberg (1986) has proposed such a model and details what cometary matter composed of interstellar grains would be like. With the Halley results and the IDP data it is now possible to make rather detailed comparisons with the Greenberg model. A major prediction of the model was high carbon abundance in cometary grains. This is consistent with the in-situ Halley measurements. The laboratory dust studies do not reveal such high average carbon abundance but this is easily reconciled if part of the carbonaceous is moderately volatile. The model also predicts that carbon should occur as thick radiation processed mantles covering prolate silicate cores. Some of the Halley particles could have such a structure but there is no direct evidence that this is the case. Nearly a third of the particles are nearly pure silicate and could not have thick mantles. A significant fraction of the submicron particles are nearly pure CHON material and do not appear to have substantial silicate cores. The hundreds of collected interplanetary particles do not have the precise morphology predicted by Greenberg but it is possible that the differences could be minimized by either processing the comet or particles or by mild metamorphism of the model. The collected cp particles are aggregates of rather equidimensional submicron grains and prolate grains are relatively rare. The need for prolate interstellar grains is to provide alignment and polarization. Instead of solid rod-like grains proposed in the model, it is entirely possible that prolate interstellar grains could be end-to-end clusters of two or more equidimensional grains such as those found in interplanetary dust. The Raman measurements indicate that silicate grains in IDPs are covered with carbonaceous material but imaging of the grains shows that such films are typically less than 200A, thinner than predicted by the model. This difference could be due to evolutionary processes. An intriguing possibility is that some of IDP grains called "tar balls" could have properties similar to core-mantle grains. These grains do not consist of a single silicate core surrounded by a thick mantle but rather they contain hundreds of 100A size cores embedded in a carbonaceous mass.

6. FUTURE WORK

Results from the Halley flybys both corroborate the belief that comets are primitive materials and strengthen the notion that comets may be composed of presolar grains. Significant progress on the comet-interstellar grain connection can be made if proven samples of cometary material are subjected to laboratory analysis. The best samples would be collected directly from a comet nucleus such as is proposed by the Rosetta mission planned by the European Space Agency. The second best samples are interplanetary dust particles but it is imperative that means be developed for positive distinction of cometary and asteroidal samples. Information for such distinction should come from proposed missions such as the CRAF comet rendezvous proposed by NASA and the Cosmic Dust Collection Facility proposed for the NASA spacestation. At present it is important to persue the laboratory studies of probable cometary particles and to carefully evaluate data and models of cometary and interstellar grains. One of the obvious aspects to investigate is the implication that silicates in comets are anhydrous. Information on this problem can obtained for cometary and interstellar dust by careful study of the 10μm "silicate" feature. To provide new insight into early solar system processes it is also pertinent carfully investigate the differences between asteroid and comet samples. The most primitive asteroid samples contain hydrated minerals while it is possible that cometary silicates are anhydrous. This could be an indication that low temperature equilibration with solar nebula gas only occured in the inner solar system. It is also possible however, that hydrated phases in asteroids formed entirely by parent body processes such as chemical weathering of silicates in contact with ice or water. The ability to compare detailed isotopic, chemical and mineraolgical differences between asteroid and comet samples should provide fundamental insights into properties of the solar nebula over a wide range of solar distance.

7. REFERENCES

Anders, E. 1975, *Icarus*, **24**, 363.
Bradley, J. P. and Brownlee, D. E. 1986, *Science*, **231**, 1452.
Bradley, J. P., *et al.* 1988, in *Meteorites and the Early Solar System*, eds. J. Kerridge and M. Mathews, Univ. of Arizona press.
Bregman, J. D. 1987, *Astron. Astrophys.*, **187**, 616.
Brownlee, D. E. 1985, *Ann. Rev. Earth Planet. Sci.*, **13**, 147.
Sykes, M. V., and Greenberg R. 1986, *Icarus*, **65**, 51.
Gaffey, M. J. and McCord, T. B. 1979, in *Asteroids*, ed. T. Gehrels, U. Arizona press, pp 688-723.
Greenberg, J. M. 1982, in *Comets*, ed. L. L. Wilkening, U. Arizona Press.
Grun, E., *et al.* 1986, *Icarus*, **62**, 244.
Harwit, M. 1963, *J. Geophys. Res.*, **68**, 2171.
Jessberger, E. K. *et al.* 1988, *Nature*, **321**, 691.
Kissel, J., *et al.* 1986a, *Nature*, **321**, 280.
Kissel, J., *et al.* 1986b, *Nature*, **321**, 336.
Kerridge, J. F. 1988, *Meteorites and the The Early Solar System*, Univ. of Arizona Press.
Kerridge, J. F. and Bunch, T. E. 1979, in *Asteroids*, Ed. T. Gehrels, U. Arizona Press, pp. 745-764.
Kresak, L. 1987, *Pub. Astron. Inst. Czechoslovak Acad. Sci.*, **67**, 265.
Lawler, M. 1988, *Icarus*, in press.
Low, F. J. 1984, *Astrophys. J. Lett.*, **278**, 19.
Mackinnon, I. D. R. and Reitmeijer, F. J. M. 1987, *Rev. Geophys.*, **25**, 1527.
McKeegan, K. D. *et al.* 1985, *Geochim. Cosmochim. Acta*, **49**, 1971.
Sandford, S. A. 1987, *Fund. Cosmic Phys.*, **12**, 1.
Schramm, L. S., *et al.* 1988, *Meteoritics*, in press.
Sekanina, Z. 1985, *Astron. J.*, **90**, 827.
Sykes, M. V. and Greenberg, R. 1986, *Icarus*, **65**, 51.
Verniani, F. 1960, *Space Sci. Rev.*, **10**, 230.
Wasson, J. T. 1974, *Meteorites* (316 pages) New York: Springer Verlag.
Wetherill, G. W. 1985, *Meteoritics*, **20**, 1.
Wopenka, B. 1988, *Earth and Planet. Sci. Lett.*, **88**, 221.
Wisdom, J. 1983, *Icarus*, **56**, 51.
Zinner, E. 1987, *Nature*, **330**, 730.

DYNAMICS AND SPATIAL SHAPE OF SHORT-PERIOD METEOROID STREAMS

P.B.Babadzhanov and Yu.V.Obrubov
Astrophysical Institute of the Academy
of Sciences of Tajik SSR,
Dushanbe, 734670, USSR

ABSTRACT. At the early stage of evolution the meteoroid streams may be considered as elliptical rings of relatively small thickness. The influence of planetary perturbations can essentially increase the stream width and its thickness. As a result one stream may produce several couples of meteor showers active in different seasons of the year. 22 short-period meteoroid streams under review may theoretically produce 104 meteor showers. The existence of 67 is confirmed by observations.

1. INTRODUCTION.

The overwhelming majority of meteor streams are generally assumed to be formed by the process of cometary decay.The most effective process for the release of solid particles from a cometary nucleus is their ejection by sublimating gases when comets approach the Sun. Some asteroids may also be progenitors of meteoroid streams if we assume these asteroids to form numerous small fragments during mutual collisions. Irrespective of the source of meteoroid streams, the ejection velocities of particles from a parent body seem to be relatively small – from several m/s to 1 km/s. The ejection velocity of particles depends on the size of the cometary nucleus, its physical-chemical properties and on the distance from the Sun at ejection. The ejection velocity also depends on various meteoroid properties. Immediately after release the meteoroids are subjected to light pressure.

The effect of ejection velocity and light pressure produces the initial dispersion of orbital elements of meteoroids. In this case the semimajor axes and, perhaps, the eccentricities of meteoroids' orbits differ strongly. The difference in orbital orientation of meteoroids and parent body is negligible. So, at the initial stage of evolution a meteoroid stream is very flat, narrow at perihelion and broad at aphelion.

Planetary perturbations, in the general case, change all the orbital elements of meteoroids. Till the recent time when studing meteoroid stream evolution the orbital elements of all meteoroids were assumed to change in the same manner as those of the mean stream orbit. Such an approach caused the stream shape (or the initial dispersion of meteoroid orbits) to be independent of time. However, planetary perturbations may

287

essentially increase the dispersion of the orbital elements of meteoroids and, eventually, change the stream shape (Babadzhanov, Obrubov 1985, 1986, 1987 c).

In this paper we will show that the increase in dispersion of orbital elements of short-period meteoroid streams leads to the activity of several meteor showers produced by each stream.

2. DYNAMICS OF SHORT PERIOD METEOROID STREAMS.

In previous papers (Babadzhanov, Obrubov 1986,1987c) we showed that due to gravitation of large planets the long-period variations in orbital elements of asteroid Phaethon (a possible remnant of the Geminids' progenitor) and of stream meteoroids with semimajor axes from 1 to 1,7 AU are satisfactorily described by the following integrals of motion (Moiseev 1945; Lidov 1961):

$$(1-e^2) \cdot \cos^2 i = C1 = \text{const.,} \tag{1}$$

$$e^2(0.4 - \sin^2 i \cdot \sin^2 W) = C2 = \text{const.,} \tag{2}$$

where e is eccentricity, i is inclination and W is the argument of perihelion. C1 and C2 are constants which may be determined by the observed orbital elements at a given time. For example, for Phaethon C1=0.18 and C2=0.27.

If we add the condition of intersection of the meteoroid's orbit with the Earth's orbit:

$$a(1-e^2) \simeq 1 \pm e \cdot \cos W \tag{3}$$

to the formulae (1,2), then, from numerous orbits determined by a, C1, and C2 one may now identify the Earth-crossing orbits (Babadzhanov, Obrubov 1987 b,c). When the stream orbit lies in the ecliptic plane the stream may produce two showers if perihelion distance q<1 AU and only one shower if q≃1 AU. If i=0 and a combined solution of equations (1-3) gives only one value of e which the orbit may have during its evolution, then orbit may cross the Earth's one at four values of W.

2.1. Geminid stream.

Fig.1 presents secular variations of e, i and radii-vectors at the ascending (Ra) and descending (Rd) orbital nodes of Phaethon versus W. As follows from equations (1-3) and Fig.1 Phaethon's orbits, at which intersection with Earth's orbit occurs,have approximately similar inclinations and eccentricities. However their positions in space differ essentially. Each intersection may produce a stream radiant which differs essentially from the other ones.

Owing to the differences in the semimajor axes the variation rate in orbital elements of the Geminid meteoroids is different. The dispersion of orbits increases so that, eventually, after 20 000 yr it will embrace all four possible intersections. Thus, if the dispersion in meteoroid orbits is sufficiently large, and the arguments of perihelia

of individual orbits cover a range from 0° to 360° then the stream produces four annual meteor showers. The Geminid meteoroid stream produces
the nighttime pre-perihelion Geminid and Canis Minorid showers as well
as the post-perihelion Daytime Sextantids and δ-Leonids. Among these
four showers three are surely observed (Babadzhanov, Obrubov 1985, 1986,
1987 a,c).

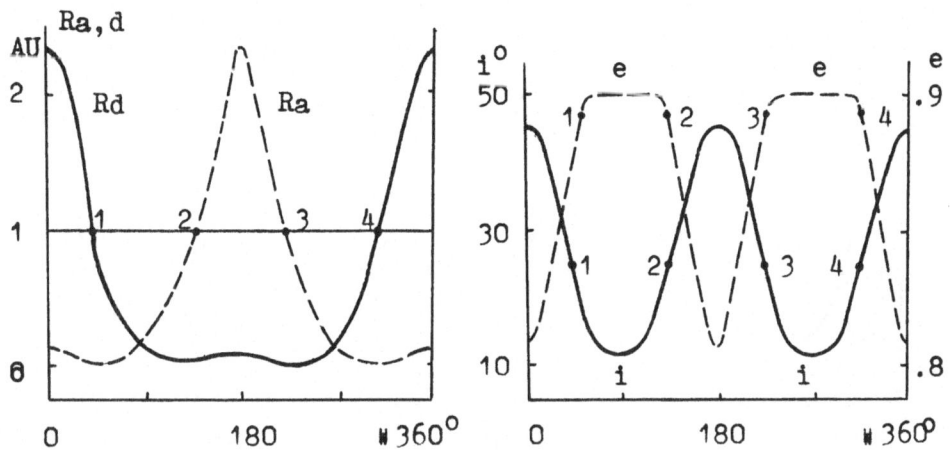

Fig.1 The dependence of the eccentricity, inclination and radii-vectors
at the orbital nodes of Phaethon on the argument of perihelion. The
intersection 1 corresponds to the δ-Leonids, 2 - Canis Minorids, 3 -
Daytime Sextantids and 4 - Geminids.

It should be noted that Geminid stream thickness of about 1 AU in
the regions of stream intersections by the Earth is a characteristic
feature of this stream.

2.2. Quadrantid stream

Consider now a more interesting case when the combined solution of
equations (1-3) gives two values of eccentricity at which a given orbit
crosses the Earth's orbit. We have come across this case for the first
time when studing the evolution of the Quadrantid stream. Two values of
the eccentricities at which the intersection may occur correspond to
four pairs of W. Hence, eight meteor showers are theoretically possible. For the Quadrantids this conclusion is confirmed by the calculations
of secular perturbations of the mean orbit (Babadzhanov, Obrubov 1987 c)
and by calculations of orbital evolution according to Everhart's method.
Fig. 2 shows the dependences of e, i, Ra and Rd on the argument of
perihelion. These dependences are based on calculations of secular
perturbations of the Quadrantid mean orbit by the Halphen-Goryachev
method. It is evident that this stream may produce four meteor showers
with small inclinations (i<30°) and perihelion distances (q < 0.1 AU) as
well as four showers with large inclinations (i ≃ 70°) and perihelion

distances q ≠ 1 AU. Qualitatively the same conclusion may be obtained from equations (1-3). These showers are x- Velids, Daytime Arietids, Southern and Northern δ- Aquarids, Quadrantids and Ursids, α- Cetids and Carinids (Table 2). It should be noted that the activities of all these showers (except of the x- Velids and Carinids) are confirmed by observations.

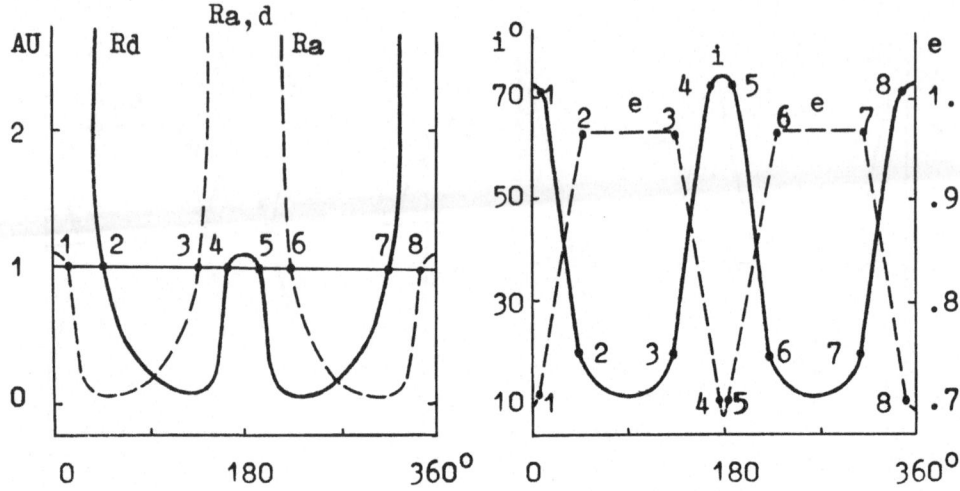

Fig. 2. The dependense of **e, i, Ra** and **Rd** of the Quadrantid orbit on **W**. The intersection 1 corresponds to x-Velids, 2 - Daytime Arietids, 3 - S. δ -Aquarids, 4 - Quadrantids, 5 - Ursids, 6 - Daytime α-Cetids, 7 - N. δ-Aquarids, 8 - Carinids.

In the Geminid stream the dispersion rate of orbits is mainly defined by the initial dispersion in semimajor axes. However, in the Quadrantid stream (because of the encounters of particles with Jupiter) the dispersion rate is defined also by the difference in meteoroid position in the same orbit. This conclusion is well illustrated by the results of Williams et al (1979) and also by our calculation using Everhart's method.

Investigation of the orbital evolution of 36 Quadrantid particles with semimajor axes a ranging from 2.7 to 3.2 AU using Everhart's method shows that the dispersion of meteoroid orbits causing the annual activity of all related meteor showers may be formed over 10-12 millenia.

2.3. Other Meteoroid Streams: The shape and number of meteor showers produced by these streams.

Simulation of the Geminid and Quadrantid stream evolution by the above methods shows that the qualitative features of these streams (the number of meteor showers, approximate orbits of meteoroids responsible for them etc.) may be obtained from equations (1-3). Furthermore, equations (1,2) permit us to estimate the shape which the stream may take under planetary perturbations. We have evaluated, firstly, a possible thickness of some short-period meteoroid streams at perihelion **Hq**,

aphelion **HQ** and middle part **Hm** of their orbits and, secondly, the possible number of meteor showers produced by each stream. These data are
given in Table 1. This Table also gives semimajor axes (in AU), maximum
and minimum of inclinations and the stream width L (in AU) along the
Earth's orbit computed from shower duration according to Cook (1973).
Nsh - the number of meteor showers which may be produced by the meteoroid stream.

Table 1. Semimajor axes, maximum and minimum of inclinations, thickness
and number of possible meteor showers for short-period meteoroid streams

No	Stream name	a AU	i max	i min	HQ	Hm	Hq	L	Nsh
1	Quadrantids	2.6-3.2	76	10	4.0	4.8	0.8	0.04	8
	δ -Aquarids							0.6	
2	δ-Cancrids	2.3	0	0	0	0	0	0.1	2
3	Virginids	2.6	10	3	0.5	0.4	0.03	1.2	4
4	δ-Leonids	2.6	12	6	1.0	1.7	0.14	0.7	4
5	Camelopardalids	1.5	8	7	0.5	0.4	0.23	0.4	4
6	σ-Leonids	2.4	2	1	0.1	0.2	0.00	0.9	4
7	δ-Draconids	2.8	38	9	1.6	2.6	0.31	0.4	6
8	μ-Virginids	3.1	24	10	2.0	1.5	0.36	0.7	4
9	α-Scorpiids	2.2	10	3	0.4	0.3	0.02	0.5	4
10	α-Bootids	2.6	30	17	2.6	2.0	0.44	0.5	4
11	Φ-Bootids	1.2	20	18	1.0	0.8	0.58	0.4	4
12	τ-Herculids	2.7	24	8	1.9	1.8	0.31	0.4	4
	Jun.Bootids	3.3							
13	() Ophiuchids	2.9	10	4	0.7	0.6	0.06	0.1	4
14	α-Capricornids	2.5	15	7	1.1	0.9	0.14	0.3	4
15	ι-Aquarids	1.8-2.4	19	5	1.1	0.9	0.14	1.1	4
16	ϰ-Cygnids	3.1	41	6	1.6	3.2	0.16	1.0	8
	χ -Scorpiids	3.1							
	Piscids	1.9-2.1							4
17	Taurids	1.9-2.6	16	2	1.0	0.7	0.12	1.8	4
	χ-Orionids	2.2							4
18	ϰ-Aquarids	3.2	3	1	0.1	0.2	0.01	0.3	4
19	Ann.Andromedids	3.2	22	4	2.1	1.7	0.30	0.8	4
20	Geminids	1.0-1.7	55	12	2.0	1.5	0.05	0.1	4
	D.Sextantids								
21	Dec.Phoenicids	3.0	16	9	1.5	1.2	0.27	1.0	4
22	δ-Arietids	2.1	3	2	0.2	0.2	0.05	0.1	4

Table 1 shows that the overwhelming majority of meteoroid streams can produce four meteor showers each. The investigation of the Encke meteoroid stream shows that eleven observable meteor showers (Table 2) are produced by this stream (Babadzhanov et al. 1987).

In addition to the Quadrantids the x-Cygnids and, perhaps, the δ-Draconids belong to the type of meteoroid streams which can produce eight showers each. This conclusion is confirmed by our calculation of the secular evolution of the mean orbits of these streams by the Halphen-Goryachev method.

Table 2. Theoretical and observed radiants of meteor showers.

	RA	Dec	Vg	Ls	RA	Dec	Vg	Ls	RA	Dec	Vg	Ls	RA	Dec	Vg	Ls
	Quadrantids(4)-.09				Carinids				Ursids(0)-.22				x-Velids			
Rt	227	51	41	282	151	-59	43	278	219	56	41	276	144	-55	43	276
Ro	230	48	41	283	?	?	?	?	223	62	38	281	?	?	?	?
	N.δ-Aquar.(4)-.11				S.δ-Aquar.(4)-.12				D.Arietids(3)-.07				α-Cetids(3)-.12			
Rt	336	-4	41	124	346	-13	40	133	49	24	40	80	49	12	42	76
Ro	337	-5	40	128	333	-16	41	125	43	23	37	77	44	12	39	78
	δ-Leonids(4)				ρ-Leonids(2)-.17				γ-Leonids(2)-.10				Aug.Cancr.(2)-.23			
Rt	159	19	21	338	152	2	21	338	148	23	21	136	142	6	21	136
Ro	159	19	23	338	161	7	21	342	150	20	22	152*	139	12	20	141
	α-Capricornids(4)				χ-Sagit.(2)-.20				ε-Aquar.(2)-.09				X-Capric.(1)-.26			
Rt	309	-10	23	127	313	-27	23	127	307	-10	23	305	312	-28	23	305
Ro	307	-10	23	127	290	-26	26	100	310	-7	23	315	314	-24	27	324
	S. ι-Aquarids (4)				N.ι-Aquarids(4)-.18				Ass.30 (3)-.13				Ass.16 (3)-.18			
Rt	335	-14	33	131	332	-7	33	131	16	3	34	35	13	10	34	35
Ro	333	-15	34	131	321	-8	33	120	13	-3	29	29	7	3	29	30
	N.Piscids(1)-.18				S.Piscids(4)-.25				May Ariet.(2)-.14				Ass 41 (3)-.18			
Rt	9	8	32	176	12	1	31	176	40	19	32	58	42	12	32	58
Ro	0	3	27	168	6	0	26	177	36	18	25	54*	33	9	29	52
	N.Taurids(4)-.10				S.Taurids(4)-.15				ζ-Perseids(2)-.13				β-Taurids(4)-.06			
Rt	51	23	32	219	53	15	31	219	86	27	32	101	86	20	32	101
Ro	58	22	29	230	50	14	27	220	84	24	29	102	86	19	30	96
	N.χ-Orion.(4)-.21				S.χ-Orion.(1)-.20				x-Aurig.(2)-.23							
Rt	73	26	32	239	74	19	31	239	109	26	31	121	108	18	32	121
Ro	84	26	25	258	85	16	26	259	94	28	24	102*	?	?	?	?
	Geminids (4)-.05				Canis-Minor.-.04				δ-Leonids				D.Sextan.(4)-.15			
Rt	112	32	34	260	109	12	34	260	169	16	34	194	161	-1	34	194
Ro	112	32	34	261	109	12	39	258	?	?	?	?	154	4	31	189

Summarizing the investigation of the dynamics and shape of short-period meteoroid streams one can say that a phenomenon of the formation of shower branches and twin showers is widely spread. 22 meteoroid streams given in Table 1 can theoretically produce 104 meteor showers. The existence of 67 is confirmed by observations. The results of a search for related meteor showers produced by six well-known short-period meteoroid streams are presented in Table 2.

Theoretical (Rt) and observed (Ro) coordinates of the geocentric radiant (right ascension **RA** and declination **Dec**), geocentric velocity **Vg** (in km/s) and solar longitude **Ls**, at which the shower activity occurs, are given in Table 2 for each shower. After shower name Table 2 gives the references to the catalogue where the observed orbit was found: (0)- Sekanina (1970), (1) - Sekanina (1973), (2) - Sekanina (1976), (3) - Kascheev et al. (1967), (4) - Cook (1973) and the values of D-criterion between the observed and theoretical orbits. Twin showers, found by Drummond (1982), and confirmed by the results of our investigation, are marked by an asterisk *.

Thus, owing to the differences in the ejection velocities and light pressure the initial dispersion of stream meteoroid orbits is produced. The influence of planetary perturbations can greatly increase the dispersion of orbital elements that causes a meteoroid stream to thicken essentially and leads to the activity of twin showers and northern and southern branches. In conclusion it should be noted that when estimating the volume, density and mass of a meteoroid stream it is necessary to take into account the stream thickness and the activity of all related showers.

REFERENCES

Babadzhanov P.B., Yu.V.Obrubov 1985, Report of Meetings of Comission 22, Transactions of IAU, XIX, p. 195

Babadzhanov P.B., Yu.V.Obrubov 1986, Dokl. AN SSSR, 290, 1, pp 53-57

Babadzhanov P.B., Yu.V.Obrubov 1987a, Handbook for MAP, 25, ed. R.G.Roper, US GPO, pp 2-10

Babadzhanov P.B., Yu.V.Obrubov 1987b, Dokl. AN Taj.SSR, 30, pp 486-491

Babadzhanov P.B., Yu.V.Obrubov 1987c, in Interplanetary Matter, eds. Z.Ceplecha, P.Pecina, Praha, pp141-150

Babadzhanov P.B., Yu.V.Obrubov, N.Makhmudov 1987, Sov. Comet Circular 373, pp 3-4

Cook A.F. 1973, in Evolutionary and Physical Properties of Meteoroids, eds. C.L.Hemenway, P.M.Millman, A.F.Cook, Washington D.C.,pp 183-191

Drummond J.D. 1982, Icarus, 49, pp 135-142

Kascheev B.L., V.N.Lebedinets, M.F.Lagutin 1967, Meteor Phenomena in the Earth's Atmosphere, Nauka, Moscow, 260 p.

Lidov M.L. 1961, Iskusstvennie Sputniki Zemli, 8, pp 5-45

Moiseev N.D. 1945, Trudy Gos. Astron. Inst. Mosk.Univ., 15, pp 75-99

Southworth R.B., G.S.Hawkins 1963, Smith. Contr. Astroph., 7, pp 261-286

Sekanina Z. 1970, Icarus, 13, pp 475-493

Sekanina Z. 1973, Icarus, 18, pp 253-284

Sekanina Z. 1976, Icarus, 27, pp 265-321

Williams I.P., C.D.Murray, D.W.Hughes 1979, Month. Not. R. Astron. Soc., 189, 2, pp 483-492

COMETARY DUST AND ZODIACAL LIGHT CONNECTION

A. DOLLFUS
Observatoire de Paris
92195 Meudon
France

ABSTRACT. Photopolarimetry of P/Halley coma characterizes, in the cloud of dust grains released by the nucleus, the presence of a significant population of large flakes, made of aggregated small and very dark grains. These globules are subsequently transported toward the Sun, with some modifications in texture, and concentrate in the inner part of the Solar System. Comets appear to be the source for the fluffy grains which are observed in the zodiacal light.

DUST GRAINS IN THE COMA OF P/HALLEY

When approaching the Sun, the nucleus of a comet releases gases and grains of solid material, because the increase of radiation heating from the Sun produces an evaporation of the ice and internal fracturations or explosions. The cloud of solid grains which are liberated into space scatters the solar light and is made visible around the nucleus, being telescopically observed as the cometary coma. The smallest grains are subjected to the solar radiation pressure which produces an additional force to the gravitation, radial to the Sun, they are moved outward and produce the cometary tail. Finally, these small grains spread out in the outer part of the solar system. We consider here the grains of larger size.

It was an assignment of the spacecrafts VEGA and GIOTTO which entered into the P/Halley coma cloud to record flux, mass and size distribution of these grains, directly in the vicinity of the nucleus. The dust impact-detectors CIS, PIA, DIDSY, SP and DUCMA placed on board these spacecrafts recorded essentially very small and sub-micron size grains. Mass distribution functions were derived (Mc Donnell et al., 1987 – Mazets et al., 1987 – Krasnapolsky et al., 1987 – Simpson et al., 1987).

However, the mass distribution deduced from these measurements does not exclude the presence of larger particules either if they have not been detected by the impactors (Green et al., 1987). The analysis of the spacecrafts deceleration when passing through the coma requires a flux of impacts by large particles (Edenhofer et al., 1987). The camera of spacecraft GIOTTO experienced attitude excursions explained by discrete impacts on the craft by grains with mass of at least a μg, corresponding to several hundreds μm in size (Curdt and Keller, 1988). The spacecraft SUISI was hit twice by large particles with masses of several mg and several μg, respectively (Uesugi, 1987), and spacecraft GIOTTO was disbalanced by a powerfull impact.

295

D. McNally (ed.), Highlights of Astronomy, Vol. 8, 295–304.

All these results suggest that the population of the solid particles which are ejected by the cometary nucleus is not limited to sub-micron or micron size grains, as was suggested by the impactors results alone, but must comprise a significant number of large pieces with sizes of at least 100 μm, or far more. In practise, the overall dust mass distribution is probably dominated by these large grains (Curdt and Keller, 1988).

The presence of large grains in the coma dust cloud is also supported by the Radar analysis. The ground-based Radar observation identified predominantly very large grains reaching the cm size range (Campbell and Harmon, 1988), and such was also the case for the comet IRAS-ARAKI-ALCOCK (Harmon et al., 1988).

CHARACTERIZATION OF THE COMA GRAINS BY TELESCOPIC PHOTOPOLARIMETRY

A way to characterize remotely some physical properties of a cloud of grains is by the technique of optical telescopic polarimetry. Polarization produces information about albedo, absorption, texture, size, aggregation and shape of grains (Dollfus, 1985 – Geake and Dollfus, 1986 – Wolff, 1981).

We implemented the method for P/Halley using the photoelectric photopolarimeter of Observatoire de Paris, between October and December 1985 at the 1 m. telescope at Meudon (France), and then in April 1986, with the 1.52 m ESO telescope at La Silla, Chile, (Dollfus and Suchail, 1987). A total of around 400 individual measurements were collected over 13 distinct nights covering a range of phase angle from 21.6° to 54.2°. The three Stokes polarization parameters Q/I, U/I and V/I were extracted with an accuracy of around 10^{-3} and nine maps of these parameters were constructed over the coma field.

Other polarimetric observations on P/Halley suitable for dust characterization were recorded by Bastien et al. (1986, 1987), Kikuchi et al. (1987), Kiselev et al. (1986), Le Borgne et al (1987), Brooke et al. (1987) and Vrba (1987). The instrument HOPE on-board spacecraft GIOTTO recorded in situ data (Levasseur-Regourd et al., 1986). All those measurements pertaining to the dust characterization have been analysed jointly in a workshop study in Paris on April 2-3, 1987, and synthetic linear polarization phase variation curves have been derived (Dollfus et al., 1988). The curves relevant to the present work are reproduced here in the Figures 1, 2 and 3.

Differences in the polarization signature are observed with the distance to the nucleus, which is also a function of time after the initial dust release into space. Figure 1 refers to the very bright envelope surrounding the nucleus in April 1986 with a radius not exceeding 100 km (Dollfus and Suchail, 1987), corresponding to a time after release of one or few minutes. Figure 2 is for the inner coma up to a distance of 1000 km (significantly more in the streamers) and corresponds to the recently ejected dust after they cleared up the bright envelope. Figure 3 corresponds to the older dust, after a travel over larger distances from the nucleus. This curve of Figure 3 pertains to a more advanced stage of cometary grains morphologic evolution after completion of the processes of modification due to evaporation of volatiles and re-shaping following immediately of the release into space.

INTERPRETATION OF THE POLARIZATION DATA

It was tempting to try an interpretation of the polarization on the basis of the optical effects by the grain population analysed by the impact detectors of the spacecrafts. Such an attempt was implemented by Mukai et al. (1987) who used the Mie theory with the grains size distribution given by the SP-2 instrument. They varied the two optical parameters n and k of the refraction index and found a reasonable fit with the curve of polarization at 0.63 µm with the precise values n=1.49±0.01 and k=0.03±0.004. But no realistic substances have such indices (Lamy et al., 1987). However, inspection of the phase dependent polarization curve of Figure 3 is immediately reminiscent of similar curves observed on other atmosphereless solar system objects, when they have their surfaces made of cohesive small grains. Such are Mercury (Dollfus and Aurière, 1974 – Gehrels et al., 1987), the lunar surface (Gehrels et al., 1964 – Dollfus and Bowell, 1971 – Dollfus and Geake, 1977), the asteroids (Zellner and Gradie, 1979 – Dollfus et al., 1989).

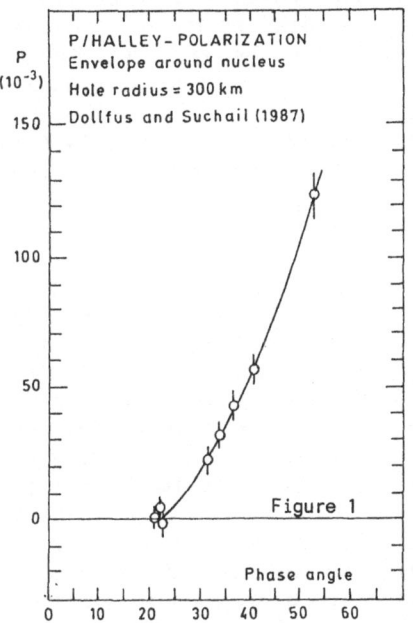

Figure 1

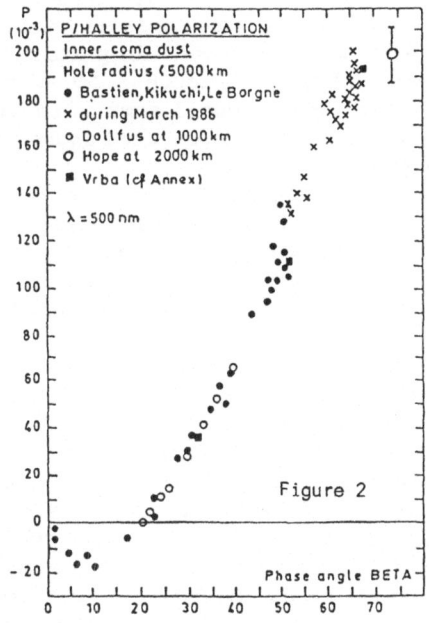

Figure 2

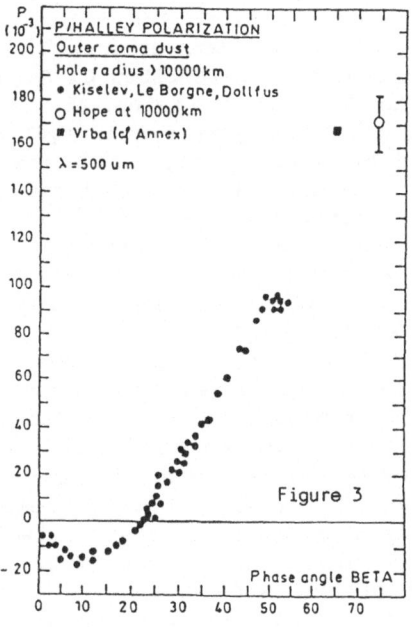

Figure 3

This similarity turns out to be an identity when we compared the coma curve with those for the darkest of these objects, which are among the C-type asteroids. The P/Halley dust grains polarization curve is exactly matched by the average of the four darkest asteroids presently analysed polarimetrically, which are 19 Fortuna, 54 Alexandra, 56 Meleta, and 84 Klio (Dollfus, 1988). The wavelength dependence fits as well. All these bodies have in common a same type of surface, which is made of a layer of grains, very dark and aggregated in a rough structure.

It is suggested that the solid particles which are responsible for the polarized light in P/Halley coma are large flakes made of aggregatred small and very dark grains.

This interpretation is entirely supported by the extensive laboratory studies on polarization by solid surfaces and samples which have been conducted at Observatoire de Meudon (Dollfus, 1971), and extended in a coordinated program involving the University of Manchester and the University of Arizona (Zellner et al., 1977a and b - Dollfus and Geake, 1977 - Geake et al., 1984 - Dollfus, 1985 - Geake and Dollfus, 1986). A large variety of terrestrial, meteoritic, lunar and artificial samples were analysed on different morphologies. Among this data, a very peculiar type of polarization behaviour emerges constantly. This typical family of curves characterizes surfaces which are made of a layer of small grains, cohesive and aggregated in complex rough structures. The polarization behaviour of P/Halley coma enters exactly into this specific category.

A limited number of polarization parameters are traditionally used to describe the phase dependence of the polarization by such solid surfaces (see ref. cited, specifically Dollfus, 1985, 1988). They are the minimum P_{min} of the degree of polarization curve, the inversion angle V_0 at which the degree of polarization changes sign and the slope h of the curve after the inversion angle. For the P/Halley polarization curve, all these three parameters fit with the values for very low albedo surfaces made of aggregated small dark grains. They are exactly reproduced, in particular, on a sample of meteorites Cl Orgueil when shifted through a 50 μm mesh leaving the smaller grains. At the microscope, the Orgueil sample exhibits opaque grennish grains of irregular shape, totally wrapped with aggregated small black particles, leafly textured (Dollfus, 1988).

The information is that the optical polarization which is observed in the P/Halley coma is essentially produced by dark flakes or globules, which are made of a rough aggregation of small and very black grains. These flakes have to be large enough to integrate the effect of this texture other many asperities, which imply diameters of perhaps several hundreds of μm (Dollfus, 1988).

This result is also supported by laboratory measurements on isolated single grains. Experiments were conducted either by the technique of microwave homothetic analogs (Zerull and Giese, 1974 - Schuerman, 1980) or with laser sources (Weiss-Wrana, 1983). Compact grains, in the size range of few tens micrometers in diameter, measured by Zerull and Geise (1984), do not fit al all; this is because compact textures do not permit the multiple exchanges of light between asperities or grains at the surface needed to produce the negative part of the polarization curve. Brownlee type grains, when they are made of very small aggregated particles producing a globule of around 4 μm in diameter (Zerull et al., 1980) does not fit well; the very deep negative polarization which is observed suggests particles of too small size. But a flying ash particle, irregular in shape, with a rough dark and particulate texture, measured by Weiss-Wrana (1983), fits properly with the cometary coma results (Dollfus, 1988).

It emerges again from these laboratory simulations that the polarization of the coma in the visible light is matched by grains far larger than the wavelength, very dark, with an extremely rough surface reminiscent of aggregates of small black particles.

GRAIN FLAKES AND SMALL PARTICLES

These globules, however, have to be intermixted with the mist of small particles sensed by the impact detectors of the spacecrafts. Occasionally, at the occurrence of specific ejection events by the nucleus, the population of small particles exhibits temporarily a dominant contribution to the polarization. Such is particularly the case of large phase angles and in the IR. Events of this type occurred during the course of the observation of P/Halley in 1985-1986 and they have been discussed (Dollfus et al., 1988 – Dollfus, 1988). For other comets, an outstanding case was comet West when its nucleus broke into four pieces on March 1976, with the release of a dense cloud of grains producing a very atypic polarization behaviour (Isobe et al., 1978).

More sensitive to the effect of the small grains is the polarization in IR. Brooke et al. (1987) attempted to explain their curves of polarization at 2.2 μm by a mixture of two types of grains. There is five free parameters however and, despite this flexibility, the author considers that an additive contribution by large rough particles is needed.

It was found also that, of six other comets analysed polarimetrically with sufficient accuracy, four of them, which were Bennett, Mrkos, Austin and West before breaking, were observed at a distances to the Sun smaller than 2 AU and produced the same polarization signatures as for P/Halley, which means dominated by the effect of large dark fluffy grains. The two other comets, which are Chernykh and Ashbrook-Jackson, were observed in excess of 2 AU from the Sun, and produced a polarization behaviour which is essentially dominated by the effect of small grains (Dollfus, 1988).

GRAIN EVOLUTION IMMEDIATELY AFTER RELEASE BY THE NUCLEUS

In the close vicinity of the nucleus, a bright envelope is observed at the telescope with almost the aspect of a star image. Its radius was estimated to be around 100 Km (Dollfus and Suchail, 1987). The phase dependent polarization curve of this bright feature is reproduced Figure 1. This curve departs from the specific polarization behaviour which characterizes dark rough solid surfaces. Although the cross-over angle of 22° remains compatible, the slope h after the inversion has a value of 2.8×10^{-3}/degree and the polarization increases with phase angle more and more steeply, a trend which is not observed on the specific rough particulate solid surfaces. The suggestion is made that ice is included in the immediate process of release by the nucleus, either as isolated grains intermixted with the population of dust grains and flakes, or as a cement or a coating, and that this volatile contribution is quickly removed by evaporation (Dollfus, 1988).

After they cleared up from this bright envelope, the large dark flakes disclose temporarily a slightly different polarization signature than at a large distances to the nucleus, as shown in the figure 2 compared to figure 3. It is tempting to suggest the effect of other volatile compounds, which vaporized more slowly with time (Wallis et al., 1987). Then the grains apparently stabilize in a more final stage.

FURTHER COMETARY GRAIN EVOLUTION

After the cometary grains escape the influence of the nucleus environment, they are subjected to solar gravitation and planetary perturbations with occasional collisions. For the largest grains and the flakes, the radiation pressure and magnetic field effects are negligible, but the Poynting-Robertson drag and the corpuscular drag are cumulating their influences. Detailed analysis of all these forces were derived by Gustafson and al. (1987 a, b). There is a decrease of the excentricity and of the semi-major axis, producing a spiral motion toward the Sun, and a concentration of the largest interplanetary grains in the ecliptic plane toward the Sun, which is made visible as the zodiacal light.

However, modification processes in the physical properties of the grains are at work during their travel into the solar system. Catastrophic collisions occur (Leinert, 1985 - Grun et al., 1985). Energetic ions and electrons accumulate damages (Strazulla et al., 1985 - Sandford, 1986). Aggregated grains in flake-like particles are slowly loosing their outer mantle, assumed to be organic, and the structures become denser and more compact (Fechtig and Mukai, 1985). In addition, there is a contribution to the interplanetary dust population by the solid particles ejected during collisions in the asteroidal belt, which supply compact grains.

The end-state of this double origin and evolution processes is an heterogenous population. The shape of the micro-craters at the surface of the lunar rocks indicates at least three types of particles, of respective densities 8 (irons), 3 (silicates) and 1 (aggregated flakes) gr/cm^3 (Fechtig, 1982 - Fechtig and Mukai, 1985).

ZODIACAL CLOUD CONNECTION

The next step of this evolution is a concentration in the inner zodiacal cloud. Photometry and polarimetry of the zodiacal light enable again, like for the cometary coma, to retrieve physical properties about the cloud itself (Weinberg, 1985 - Frey et al., 1974 - Leinert, 1975 - Leinert et al., 1981 - Levasseur-Regourd and Dumont, 1980 - Dumont, 1983 - Dumont and Levasseur-Regourd, 1985 - Giese et al., 1986), and about the individual grain properties.

Photopolarimetry indicates, in the zodiacal cloud, rather large, fluffy, low albedo grains (Giese et al., 1978 - Greenberg and Gustafson, 1981 - Weiss-Wrana, 1983 - Lumme and Bowell, 1985). The detailed analysis suggests an increase of albedo toward the Sun and a correlated decrease of the degree of polarization maximum (Dumont and Levasseur-Regourd, 1985 - Lumme and Bowell, 1985), in agreement with the in-situ measurements by the spacecraft Helios (Leinert et al., 1981). This fluffy structure of the zodiacal grains is reminiscent of the interplanetary "Brownlee particles" collected in the upper atmosphere. A link is suggested with our identification of flake-like clumps of dark particles in the cometary coma.

There is indication that the zodiacal grains are not identical to the cometary coma grains, however, being probably more compact and of less dark albedo. But the aggregated structure is essentially preserved. Meteoroid impacts at the surface of asteroids are not producing ejectas of this kind, and the processes at work in the interplanetary space may apparently not suffice to produce a large quantity of fluffy structures.

Our polarimetric analysis of P/Halley coma indicates that cometary nuclei are able to release into space a large amount of aggregated flake-like features. Comets appear to be the source, for the fluffy grains which are observed in the zodiacal cloud.

REFERENCES
Bastien P., Menard F., Nadeau R. (1986), "Linear polarization observations of P/Halley". Mon. Not. R. astron. Soc. 223, 827–834.
Bastien P., Drissen L., Menard F., St-Louis N., Nadeau R. (1987), "The dust around the nucleus of comet Halley". Proc. Symp. Diversity and Similarity of Comets", Brussels ESA SP-278.
Brooke T.Y., Knacke R.F., Joyce R.R. (1987), "The near infrared polarization and color of comet P/Halley". Astron. Astrophys. 187, 621–624.
Campbell D.B., Harmon J.K. (1988), "Radar observations of comet Halley" submitted to Ap. J.
Curdt W., Keller H.U. (1988), "Collisions with cometary dust recorded by the Giotto HMC camera", ESA Journal, 12, 189–208.
Dollfus A. (1971), "Physical studies of asteroids by polarization of the light", in "Physical studies of minor planets". T. Gehrels, ed. 95–116. NASA SP 267.
Dollfus A. and Bowell E. (1971), "Polarimetric properties of the lunar surface and its interpretation. Part. I – Telescopic observations". Astron. Astrophys. 10, 29.
Dollfus A., Aurière M. (1974), "Optical polarimetry of planet Mercury". Icarus 23, 465–482.
Dollfus A., Geake J.E. (1977), "Polarimetric and photometric studies of lunar samples". Phil. Trans. R. Soc. London 285, 397–402.
Dollfus A., (1985), "Photometric sensing of planetary surfaces". In "Advances in Space Research" Cospar conference, Gratz, Pergamon Press.
Dollfus A., Suchail J-L. (1987), "Polarimetry of grains in the coma of P/Halley – I. Observations". Astron. Astrophys. 187, 669–688.
Dollfus A., Bastien P., Le Borgne J.F., Levasseur-Regourd A.C., Mukai T. (1988), "Optical polarimetry of P/Halley : Synthesis of the measurements in the continuum". In Astron. Astrophys. (accepted for publication).
Dollfus A., (1988), "Polarimetry of grains in the coma of P/Halley : II – Interpretation" (Accepted Astron. Astrophys.).
Dollfus A., Wolff M., Geake J.E., Lupishko D.F., Dougherty L., (1989), "Photopolarimetry of Asteroids", chapter in "Asteroids II", T. Gehrels, Editor, Univ. Arizona Press.
Dumont R., (1983), "Zodiacal light gathered along the line of sight : The vicinity of the terrestrial orbit studied with photopolarimetry and with Doppler spectrometry". Planet Space Sci. 31, 1381–1387.

Dumont R., Levasseur-Regourd A.C., (1985), "Remote sensing of the zodiacal cloud along secants to Earth orbite", in "Properties and interactions of interplanetary dust" (Giese and Lamy, editors), Reidel Pub. Co., 207-213.

Edenhofer P., Bird M.K. et al. (1987), "Dust distribution of comet P/Halley's inner coma determined from the Giotto Radio Science Experiment". Astron. Astrophys. 187, 712-718.

Fechtig H., (1982), "Cometary dust in the Solar System", in "Comets" (Wilkening ed.), Univ. Arizona Press, 370-382.

Fechtig H., Mukai T., (1985), "Dust of variable porosity (densities) in the Solar System", in "Ices in the Solar System", (Klinger et al. editors), Reidel Pub. Co., 251-259.

Frey A., Hofmann W., Lemke D., Thun C., (1974), "Photometry of the zodiacal light with the balloon-borne telescope THISBE", Astron. Astrophys. 36, 447-454.

Geake J.E., Geake M., Zellner B. (1984), "Experiments to test theoretical models of the polarization of light by rough surfaces". Mon. Not. R. astr. Soc. 210, 89-112.

Geake J.E., Dollfus A. (1986), "Planetary surface texture and albedo from parameter plots of optical polarization data". Mon. Not. R. astr. Soc., 218, 75-91.

Gehrels T., Coffeen D., Owig D., (1964). "Wavelength dependence of polarization. III - The lunar surface". Astron. J. 639, 826.

Gehrels T., Landau R., Coyne G.V. (1987), "Mercury : Wavelength and longitude dependence of polarization", Icarus 71, 386-396.

Giese R.H., Weiss K., Zerull R.H., Ono T. (1978), "Large fluffy particles : A possible explanation of the optical properties of interplanetary dust". Astron. Astrophys. 65, 265-272.

Giese R.H., Kneissel B., Rittich U., (1986), "Threen-dimensional models of the zodiacal dust cloud : A comparative study", Icarus 68, 395-411.

Green S.F., McDonnell J.A.M., Perry C.H., Nappo S., Zarnecki J.C. (1987), "P/Halley dust coma grains or rocks ?". Symp. 6-9 April 1987, Brussels, Belgium ESA SP-278, 379-384.

Greenberg J.M., Gustafson B.A.S, (1981), "A comet fragment model for zodiacal light particles". Astron. Astrophys. 93, 35-42.

Grun E., Zook H.A., Fechtig H., Giese R.H., (1985), "Mass imput into and output from the meteoritic complex", in "Properties and interactions of interplanetary dust" (Giese and Lamy editors), Reidel Pub. Co., 411-415.

Gustafson B.A.S., Misconi N.Y. and Rusk E.T. (1987a), "Interplanetary dust dynamics. II - Poynting-Robertson drag and planetary perturbations on cometary dust". Icarus 72, 568-581.

Gustafson B.A.S., Misconi N.Y. and Rusk E.T., (1987b), "Interplanetary dust dynamics. III - Dust release from P/Encke : Distribution with respect to the zodiacal cloud". Icarus 72, 582-592.

Harmon J.K., Campbell D.B., Hine A.A. (1988), "Radar observations of comet IRAS-ARAKI-ALCOCK", submitted to Ap. J.

Isobe B., Saito K., Tomita K., Machara H. (1978), "Polarization of the head of comet 1976 VI West", Pub. Astron. Soc. Japan 30, 687-690.

Kikuchi S., Mikami Y., Mukai T., Mukai S., Hough J.H. (1987b). "Polarimetry of comet P/Halley". Astron. Astrophys. 187, 689-692.

Kiselev N.N., Pushnin P.A., Siklitsky V.I., Chernova G.P. (1986), "Polarimetry of comet Halley (1982 i)", Proc. 20th ESLAB Symposium, Heidelberg 27-31 nov. 1986, III, 29-30.

Kranapolsky V.A., Moroz V.I. et al, (1987), "Properties of dust in comet P/Halley measured by the Vega-2 three channel spectrometer". Astron. Astrophys. 187, 707-711.

Lamy P.L., Grun E., Perrin J.M. (1987), "Comet P/Halley : Implications of the mass distribution function for the photopolarimetric properties of the dust coma", Astron. Astrophys. 187, 767-773.

Le Borgne J.F., Leroy J.L., Arnaud J. (1987), "Polarimetry of comet P/Halley : continuum versus molecular bands". Astron. Astrophys. 187, 526-530.

Leinert C., (1975), "Zodiacal light - A measurement of the interplanetary environment". Space Sci. Rev. 18, 281-339.

Leinert C., Richter I., Pitz E., Planck B., (1981), "The zodiacal light from 1.0 to 0.3 A.U. as observed by the helios space probe". Astron. Astrophys. 103, 177-188.

Leinert C., (1985), "Dynamics and spatial distribution of interplanetary dust", in "Properties and interactions of interplanetary dust" (Giese and Lamy editors), Reidel Pub. co., 369-375.

Levasseur-Regourd A.C., Dumont R., (1980), "Absolute photometry of zodiacal light", Astron. Astrophys. 84, 277-279.

Levasseur-Regourd A.C., Bertaux J.L., Dumont R., Festou M., Giese R.H., Giovane F., Lamy P., Leblanc J.M., Llebara A., Weinberg J.L. (1986), "Optical probing of Comet Halley from the Giotto spacecraft". Nature 321, 341-344.

Lumme K., Bowell E., (1985), "Photometric properties of zodiacal light particles". Icarus 62, 54-71.

Mazets E.P., Sagdeev R.Z. et al., (1987), "Dust in comet P/Halley from Vega observations". Astron. Astrophys. 187, 699-706.

McDonnell, W.M. Alexander et al., (1987), "The dust distribution within the inner coma of comet P/Halley 1982 i : Encounter by Giotto's impact detectors". Astron. Astrophys. 187, 719-741.

Mukai T., Mukai S., Kikuchi S. (1987), "Complex refractive index of grain material deduced from the visible polarimetry of comet P/Halley". Astro. Astrophy. 187, 650-652.

Sandford S.A., 1986, "Solar flare track densities in interplanetary dust particles : the determination of an asteroid versus cometary source of the zodiacal dust cloud". Icarus 68, 377-394.

Schuerman D.W. (1980), "The microwave analog facility at SUNYA : Capabilities and current programs", pp. 227-232 in "Light scattering by irregularly shaped particles". D.W. Schuerman editor, Plenum Press.

Simpson J.A., Rabinowitz D. et al. (1987), "The dust coma of comet P/Halley : Measurements on the Vega-1 and Vega-2 spacecraft". Astron. Astrophys. 187, 742-752.

Strazulla G., Calcagno L., Foti G., Sheng K.L., (1985), "Interaction between solar energetic particles and interplanetary grains", in "Ices in the Solar System" (Klinger et al. editors), Reidel Pub. Co, 273-285.

Uesugi K. (1986), "Collision of large dust particles with Suisei spacecraft". Proc. 20th ESLAB Symp. Heidelberg, ESA SP-250, 219-222.

Vrba F. (1987), Document filed at the International Halley Watch Center.

Wallis M.K., Rabilizirov R., Wickramasinghe (1987), "Evaporating grains in P/Halley's coma". Astron. Astrophys. 187, 801-806.

Weinberg J.L., (1985), "Zodiacal light and interplanetary dust", in "Properties and interactions of interplanetary dust" (Giese and Lamy, editors), Reidel Pub. Co., 1-6.

Weiss-Wrana K. (1983), "Optical properties of interplanetary dust : comparison with light scattering by large meteoritic and terrestrial grains". Astron. Astrophys. 126, 240-250.

Wolff M. (1981), "Computing diffuse reflection from particulate planetary surface with a new function". Applied Optics 20, 2493-2498.

Zellner B., Gradie J. (1976), "Minor planets and related objects : XX - Polarimetric evidence for the albedos and compositions of 94 asteroids. Astron. Jl. 81, 262-280.

Zellner B., Leake M., Le Bertre T., Duseaux M., Dollfus A. (1977a), "The asteroid albedo scale. I - Laboratory polarimetry of asteroids". Proc. Lunar Sci. Conf. 8th, 1091-1100.

Zollner B., Le Bertre T., Day K. (1977b), "The asteroid albedo scale. II - Laboratory polarimetry of carbon-bearing silicates". Proc. Lunar Sci Conf. h, 1111-1117.

Zerull R.H., Giese R.H. (1974), "Microwave analogue studies", pp. 901-915 in "Planets, Stars and Nebulae studied with photopolarimetry", T. Gehrels Editor, Univ. Arizona Press.

Zerull R.H., Giese R.H., Schwill S., Weiss K. (1980), "Scattering by particles of non-spherical shape", pp. 273-282, in "Light Scattering by Irregularly Shaped Particles", D.W. Schuerman editor, Plenum Press.

DUST FROM THE COMETS

Tadashi Mukai
Kanazawa Institute of Technology
Nonoichi, Ishikawa 921, Japan

ABSTRACT.   We have found a similarity between the size spectra of the
observed interplanetary dust and the survived cometary dust,   referring
to   the dynamical behaviour of the dust leaving the comet. As a result,
we   can suggest that short-period comets with relatively   higher
eccentricities are major source of the interplanetary dust, especially
those with radii less than 10 μm. It is predicted, furthermore, that a
supply rate of the dust, which move on bound orbits after leaving the
comet,   becomes about 8x10 g/s from 85 short-period comets, and nearly
$3x10^4$ g/s from 101 long-period comets.

1. INTRODUCTION

One of the important questions for the interplanetary dust is where they
come from.   It is well known that the comets supply the dust in   the
solar   system.   Recently,   Comet Halley flybys have brought a   lot of
valuable information of the dust in the comet, such as the mass spectrum
of the cometary grains(e.g. McDonnell et al. 1987, and Mazets et al.
1987) and their chemical composition(e.g. Jessberger et al. 1986).
       The data of the dust near the earth's orbit are also surely
increasing.   The analysis of the particles collected in the upper
atmosphere of the earth provides us a powerfull tool to examine their
origin and evolution in the interplanetary space.   The size spectrum of
the interplanetary dust,   compiled based on the research of the lunar
microcraters plus in situ measurements of dust flux, is also available
to investigate their origin.
       In this paper,   I will study the interrelation between the dust in
the comet and those in the interplanetary space based on a comparison of
their size spectra.   Dynamical evolution of the dust ejected from   the
comet is mainly considered to estimate the size spectrum of the survived
cometary grains against an emission velocity of the grain from   the
cometary nucleus as well as   the solar radiation pressures.   As a
result,   it becomes possible to re-attack the old problem, i.e. how much
dust can be supplied from the comets into the solar system.

D. McNally (ed.), Highlights of Astronomy, Vol. 8, 305–312.
© *1989 by the IAU.*

2. SIZE SPECTRUM OF THE DUST

By using a variable mass density of the dust, decreasing with increasing radius, predicted in Lamy et al.(1987), the mass spectrum of the dust in comet Halley compiled by Divine and Newburn(1987) is converted to the size spectrum of the dust in the comet(see figure 1a). The data of the size spectrum derived from three different in situ data have shown slightly different features, but the existence of the smaller grains with radii s<0.1 μm has been proved.

On the other hand, figure 1b shows the size spectrum of the dust near the earth's orbit compiled by Grün et al.(1985)(GZFG in figure) and Le Sergeant and Lamy(1980)(LL in figure). Due to the difference of calibration of the microcrater analysis, the slopes of the grains with s<1 μm in both estimations are in disagreement.

From a comparison between figures 1a and 1b, we have a feeling that the dust with radii between 0.01 μm and 10 μm decreases during their dispersion from the comet to the interplanetary space. This is the initial motivation of the following study.

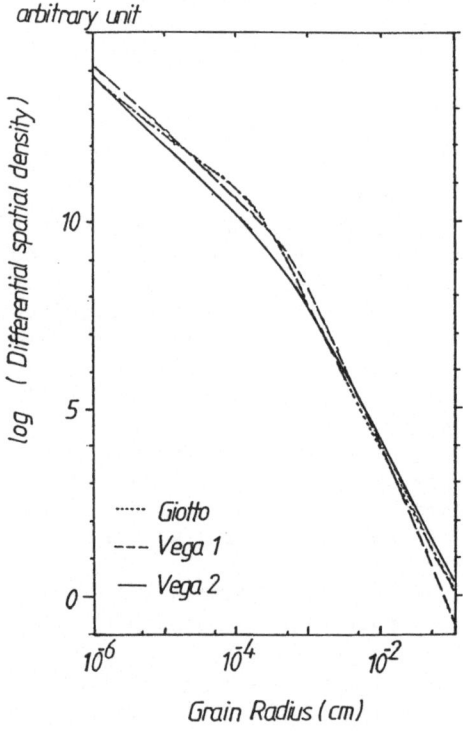

Fig.1a. Size spectrum of the dust detected in comet Halley.

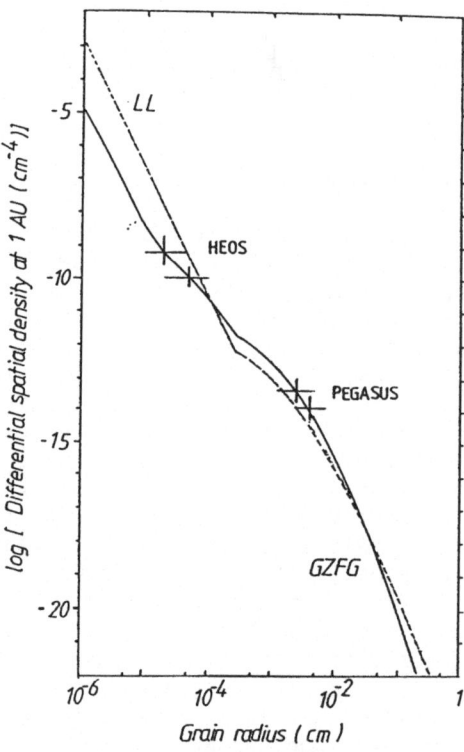

Fig.1b. Size spectrum of the dust near the earth's orbit. GZFG is Grün et al.(1985) and LL denotes LeSergeant and Lamy(1980).

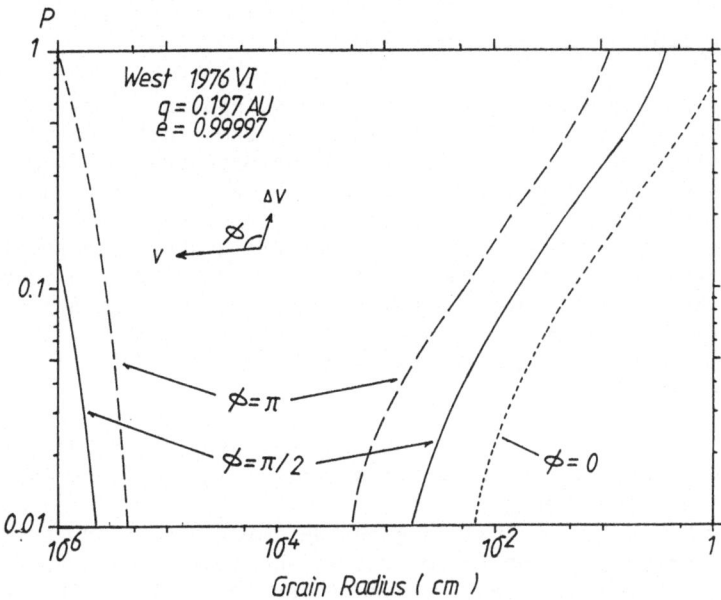

Fig.2a. Survival probability P of the dust ejected from long-period comet during its one orbit period. v denotes an orbital velocity of the parent comet and Δv means an emission velocity of the particle from the comet.

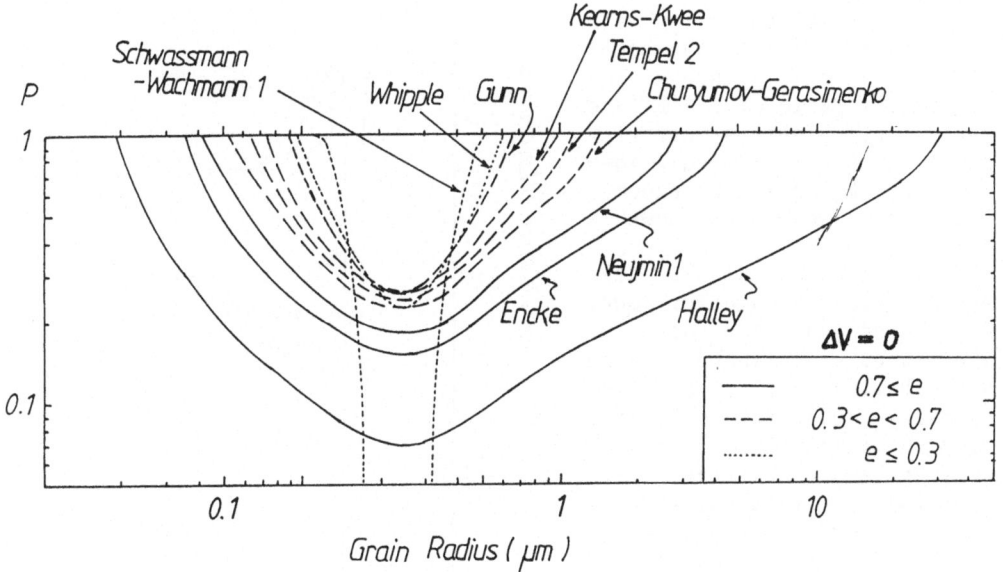

Fig.2b. The same as fig.2a, but those from short-period comets.

3. SIZE SPECTRUM OF SURVIVED COMETARY DUST

It is known that a part of the ejected grains cannot stay in the solar system after leaving the comet, because their orbital energy becomes positive due to the solar radiation pressures on them and their emission velocities Δv from the comet. Mukai(1985) has defined the survival probability P of the dust leaving the comet during one perihelion passage of the parent comet. Figure 2 shows the results of P, as a function of a grain radius s, computed for one long-period comet(comet West 1976VI) and 6 short-period comets. For a ratio of the solar radiation pressure to the solar gravity on the dust, the results for silicate given in Mukai(1985) were applied. In addition, the values of Δv came from an empirical formula in Mukai et al. (1985), estimated for comet Halley as functions of the grain radius and the sun-comet distance. These values of Δv, decreasing with increasing the grain radius, are roughly two order of magnitude smaller than the orbital velocity of the parent comet.

Figure 2 tells us that long-period comet scarcely provide the grains with radii less than 10 μm even in the case of backward ejection. Furthermore, it is found that for short-period comets, the depression in the curve of P near s=1 μm becomes wider and deeper as the eccentricity of the parent comet increases, and finally it would approach the feature found in long-period comet. If one considers only a perihelion ejection of the grain from the comet, no survived grains with radii smaller than some critical radius sc appear. The value of sc increases with increasing the sun-comet distance r. Since the allowable range of r for the comet with nearly circular orbit is very limited, then the depression in P becomes sharp and deep, as shown in figure 2.

We assume that the size spectrum of the cometary dust is independent of r. Therefore, the size distribution of survived cometary dust is estimated from a multiplication of P derived above by the size spectrum of cometary dust(Vega 2 data shown in figure 1a) .

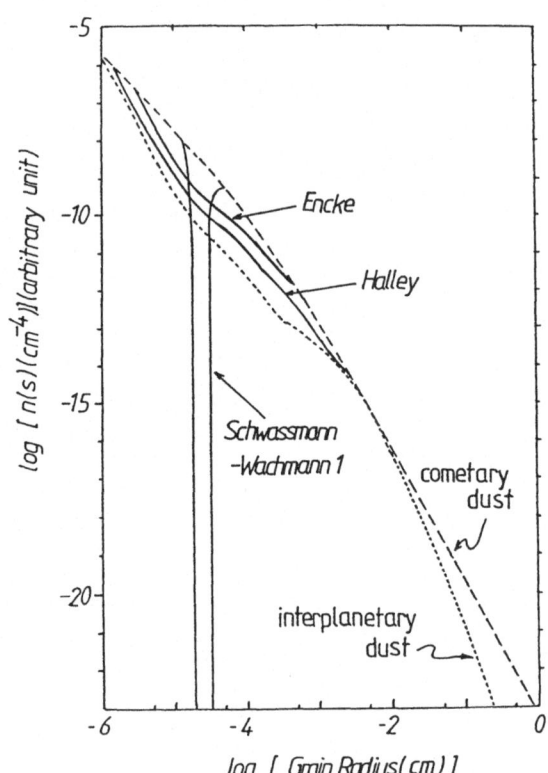

Fig.3a. Resulting size spectrum of survived dust ejected from short-period comets.

It is found that the resulting size spectrum of survived cometary dust ejected from, especially, short-period comets with larger eccentricities has a similar depression of the grains with radii between 0.01 μm and 10 μm to that found in the interplanetary dust(GZFG in figure 1b)(see figure 3).

On the contrary, the shape of the size spectrum expected for survived grains supplied from long-period comets and from short-period comets with nearly circular orbits is quite different, compared with that observed in the interplanetary dust. Since most of the asteroids have the nearly circular orbits, the survived dust released from the asteroids would have also sharp and deep depression feature in s<10 μm, although the process of dust ejection in the asteroids is unlike that in the comets.

Of course, we cannot conclude from this result alone that all of the

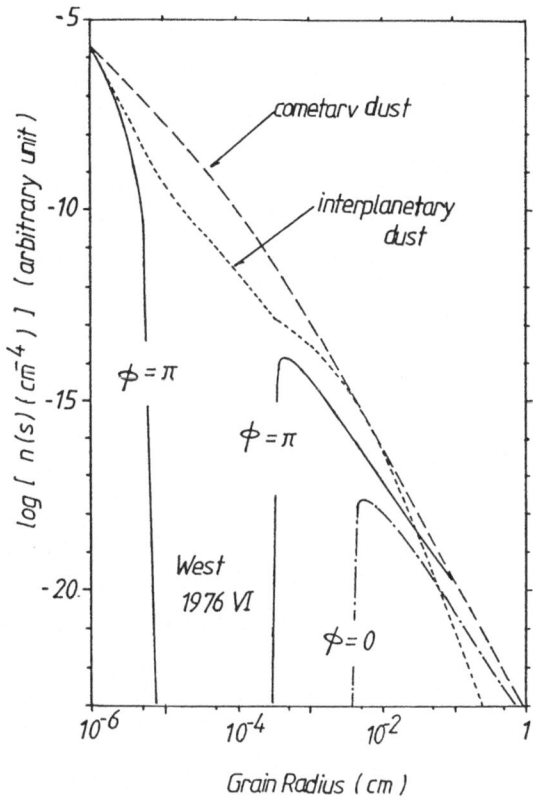

Fig. 3b. The same as fig.3a, but from long-period comet.

interplanetary dust come from short-period comets with relatively higher eccentricities. It might be problem to assume that all comets have the same size spectrum of the dust as that observed in comet Halley. Furthermore, a modification of the size spectrum of the dust in the interplanetary space, such as a production of smaller debris by mutual collisions of the interplanetary dust, should be taken into account in the future study. However, the similarity of the shape of the depression of the grains with radii between 0.01 μm and 10 μm discussed above strongly suggests that short-period comets play an important role for the supplier of the interplanetary dust, at least, those with radii less that 10 μm.

4. HOW MUCH DUST COME FROM THE COMETS?

A large fraction of the mass supplied from the comets is in the grains with radii larger than 100 μm. Therefore, the depression of the grains with radii less than 10 μm predicted above becomes out of the consideration when the total mass supplied from short-period comets is

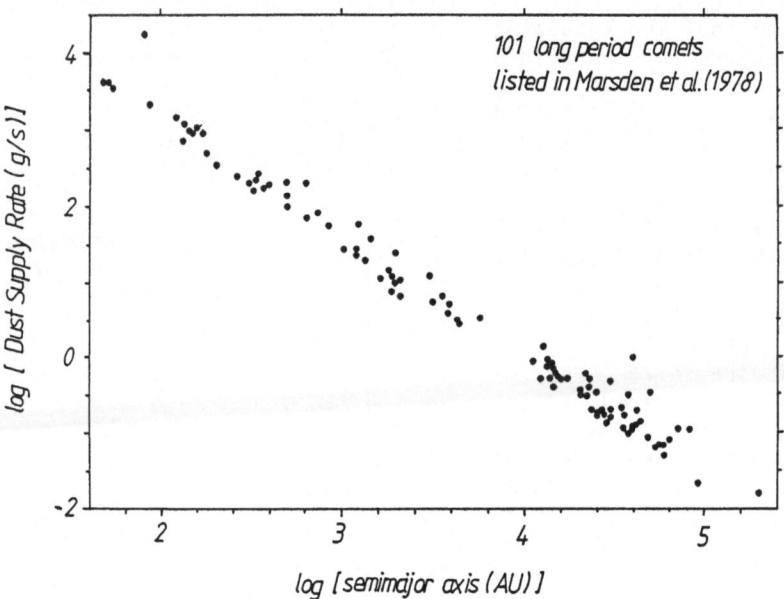

Fig. 4a. Mass supply rate of the dust from long-period comets.

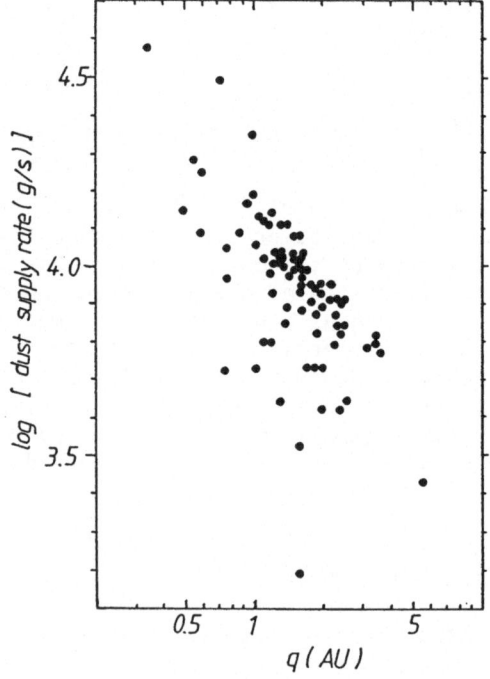

Fig. 4b. The same as fig.4a, but from 85 short-period comets,

examined. On the other hand, for the mass supplied from long-period comets, such depression plays a key factor.

The dust production rate of the comets is assumed to be approximated as $10^5 r^{-2}$ g/s for short-period comets and $10^6 r^{-2}$ g/s for long-period comets, where r is a sun-comet distance in unit of AU(see Ney 1982). After that, the total mass of dust supplied from the comets are computed(see figure 4). The long-period comet with larger semimajor axis can supply less amount of the dust because the survival probability P for such a comet is, in general, low in the wide range of the grain radius. On the other hand, mass supply rate from short-period comet increases as the perihelion distance q of the comet decreases. This result is easily understood since the comet with smaller value of q can produce much dust near the sun.

Adding these computed values of the mass supply rates from 85 short-period comets and 101 long-period comets(the orbital data, respectively, came from Marsden 1986 and Marsden et al. 1978), we can predict that the total mass of dust supplied from short-period comets becomes 8×10^5 g/s, and that from long-period comets is 3×10^4 g/s. These results seem to be consistent with the values previously estimated by other authors based on the different way, e.g. Röser(1976) derived 2.5×10^5 g/s from short-period comets.

The uncertainties in the above estimation arise from the data of the dust production rate of the comets. The observed dust production rates scatter within the scale of one order of magnitude from comet to comet even at the same sun-comet distance r(see e.g. Hanner 1984). The dependence of dust production rate on r should be examined in detail in the future works.

5. Conclusion

Referring to the similarity of the shape of depression in the size spectrum of the observed interplanetary dust and that of the survived dust released from short-period comets, the origin of the interplanetary dust was discussed. It is found that short-period comets with high eccentricities play an important role for dust supplier, especially those with s< 10 μm. In addition, the contribution of the mass of dust supplied from long-period comets is roughly 30 times smaller than that from short-period comets.

Recently, IRAS found the dust bands associated with the orbits of the asteroids. As discussed above, the size spectrum of the survived dust released from the asteroids is supposed to have a different shape in the region of the grain radius less than 10 μm, compared with the observed shape of the size spectrum of the interplanetary dust. However, the thermal emission detected by IRAS mainly came from the grains larger than s=10 μm. This implies that the contribution of the dust with larger radii from the asteroids may be important in the interplanetary space. Further quantitative discussions about the dust supplied from the asteroids, such as the size spectrum of the survived asteroidal dust and the total mass of dust supplied from the asteroids, are strongly needed.

312

ACKNOWLEDGEMENT

This work was partially supported by the Grant-in-Aid for Scientific
Research on Priority Areas(Origin of the Solar System) of Japanese
Ministry of Education, Science, and Culture(No.63611007).

REFERENCES

Divine,N., and Newburn,RL,Jr.(1987). Modeling P/Halley before and after
    the encounters. **Astron. Astrophys.** 187, 867-872.
Grün,E., Zook,H.A., Fechtig,H., and Giese,R.H.(1985). Collisional
    balance of the meteoritic complex. **Icarus,** 62, 244-272.
Hanner,M.S.(1984). A comparison of the dust properties in recent
    periodic comets. **Adv. Space Res.** 4, 189-196.
Jessberger,E.K., Kissel,J., Fechtig,H., and Krueger,F.R.(1986). On the
    average chemical composition of cometary dust. **ESA EP-249**, 27-30.
Lamy,P.L., Grün,E., and Perrin,J.M.(1987). Comet P/Halley:implication of
    the mass distribution function for the photopolarimetric properties
    of the dust coma. **Astron. Astrophys.** 187, 767-773.
LeSergeant d'Hendrecourt,L.B., and Lamy,P.L.(1980). On the size
    distribution and physical properties of interplanetary dust grains.
    Icarus, 43, 350-372.
Marsden,B.G.(1986). **Catalogue of Cometary Orbits**(5th ed.)(Smithsonian
    Astrophysical Observatory, Cambridge, MA).
Marsden,B.G., Sekanina,Z., and Everhart,E.(1978). New osculating orbits
    for 101 comets and analysis of original orbits for 200 comets.
    Astron. J., 83, 64-71.
Mazets,E.P., Sagdeev,R.Z., Aptekar,R.L., Golenetskii,S.V., Guryan,Yu.A.,
    Dyachkov,A.V., Ilyinskii,V.N., Panov,V.N., Petrov,G.G.,
    Savvin,A.V., Sokolov,I.A., Frederiks,D.D., Khavenson,N.G.,
    Shapiro,V.D., and Shevchenko,V.I.(1987). Dust in comet P/Halley
    from Vega observations. **Astron. Astrophys.** 187, 699-706.
McDonnell,J.A.M., Alexander,W.M., Burton,W.M., Bussoletti,E.,
    Evans,G.C., Evans,S.T., Firth,J.G., Grard,R.J.L., Green,S.F.,
    Grün,E., Hanner,M.S., Hughes,D.W., Igenberg,E., Kissel,J.,
    Kuczera,H., Lindblad,B.A., Langevin,Y., Mandeville,J.-C.,
    Nappo,S., Pankiewicz,G.S.A., Perry,C.H., Schwehm,G.H.,
    Sekanina,Z., Stevenson,T.J., Turner,R.F., Weishaupt,V.,
    Wallis,M.K., and Zarnecki,J.C.(1987). The dust distribution
    within the inner coma of comet P/Halley 1982i:encounter by
    Giotto's impact detectors. **Astron. Astrophys.** 187, 719-741.
Mukai,T.(1985). Small grains from comets. **Astron. Astrophys.** 153,
    213-217.
Mukai,T., Mukai,S., Fechtig, H., Grün,E., and Giese,R.H.(1985).
    Adv. Space Res. 5, 339-342.
Ney,E.P.(1982). Optical and infrared observations of bright comets
    in the range of 0.5 um to 20 um. In **Comets**(L.L.Wilkening,
    Ed.). pp. 323-340, Univ. of Arizona Press, Tucson.
Röser,S.(1976). Can short period comets maintain the zodiacal cloud?
    Lecture Notes in Physics 48, 319-322.

THE ORIGIN AND PHYSICAL CHARACTERISTICS OF METEOROIDS

DUNCAN OLSSON-STEEL
Department of Physics and Mathematical Physics
University of Adelaide
G.P.O. Box 498
Adelaide, SA 5001
Australia

ABSTRACT. Recent investigations of the terrestrial influx of small meteoroids (sizes 100μm–1 cm) are reviewed, and it is shown that (i) Previous radar measurements have underestimated the influx of these bodies by at least an order of magnitude; (ii) The true height distribution of meteors in the atmosphere indicates a very low bulk density, in general of order 0.01–0.10 gm cm^{-3}; (iii) Several Apollo asteroids have associated meteoroid streams, implying a genetic relationship and hence the possibility that these asteroids are in fact extinct or dormant cometary nuclei; and (iv) Along with the other three retrograde intermediate-period comets, P/Halley most likely originated in the Kuiper Cloud of comets just beyond the planetary region, and not in the distant Oort Cloud as is usually assumed.

1. Introduction

The smaller particles in the inner solar system (sizes below a few centimetres) have generally been assumed to be derived directly or indirectly from comets. Such a conjecture is supported by recent observations of the heights of meteors ablating in the atmosphere, reported in section 2, which indicate that these objects are mostly of very low bulk density, and hence rather fluffy (i.e. comet-like) rather than compacted, rocky objects.

The origin of the Apollo-type asteroids has been a problem for some years, since it was not clear that these bodies could be supplied from the asteroid belt on a short-enough time-scale (Wisdom, 1983). The discovery in 1983 of 3200 Phaethon, which is apparently the parent of the Geminid meteoroid stream but nevertheless appears asteroidal in nature, strongly pointed towards the evolution of comets into asteroids, and recent work reported in section 3 has identified meteoroid streams associated with several other Apollo asteroids. Thus it seems that both meteoroids and asteroids (at least the planet-crossers) are products of comets, and these bodies are implied to be the primary sources of all inner solar system objects from micron-sized dust to kilometre-sized minor planets.

The question thus becomes one of the origin of comets. Previous models have assumed an origin in the Oort Cloud, at a heliocentric distance of 10^4–10^5 AU (Weissman, 1985; Bailey *et al.*, 1986), but for some time it has been realized that there is a problem in

313

D. McNally (ed.), Highlights of Astronomy, Vol. 8, 313–319.
© 1989 by the IAU.

explaining the number of captures from such large orbits to produce short-period comets (Everhart, 1973). Recent modelling has suggested a source rather closer to the planetary region, in the so-called Kuiper Cloud (Duncan *et al.*, 1988). Herein it is proposed that the four retrograde intermediate-period (RIP) comets, including P/Halley, may be objects which have been directly captured from the Kuiper Cloud in close encounters with the outer two planets. The RIP comets would thus be important diagnostics of this cloud and the origin of short-period comets in general, and hence the entire interplanetary system of solid bodies.

2. The heights and densities of meteors

For many years it has been realized that there is a lack of agreement between the fluxes of solid particles at the Earth as measured by radar meteor techniques (mass range 10^{-6}–10^{-2} gm), when compared to the satellite impact data at the low end of this mass range, and to the visual/photographic meteor data at the high-mass end (Hughes, 1978). The radar meteor influx was found to underlie the trend expected from an interpolation between the satellite and optical meteor data by about a factor of 20 to 30. This discrepancy has recently been removed as a result of observations with Australian HF radars (Olsson-Steel and Elford, 1987a,b; Thomas *et al.*, 1986, 1988), which have shown that 'conventional' VHF meteor radars miss the majority of the influx since meteors generally ablate above the so-called 'echo-ceiling' of VHF radars, but are detectable at lower frequencies. For example, in Fig.1a is shown a height distribution gained using a typical VHF meteor radar frequency (54 MHz), demonstrating a peak at around 93–95km and few meteors above 100 km, the 'echo-ceiling' for such a radar. In contrast, using a HF (2 MHz) radar, the resultant height distribution peaks at around 105–110 km, as shown in Fig.1b. Even at such a low frequency there is still a height-selection effect, and the 'true' peak of the distribution probably occurs near 120 km. These results imply that to date VHF radars have detected only the lowermost few percent of the meteoric influx.

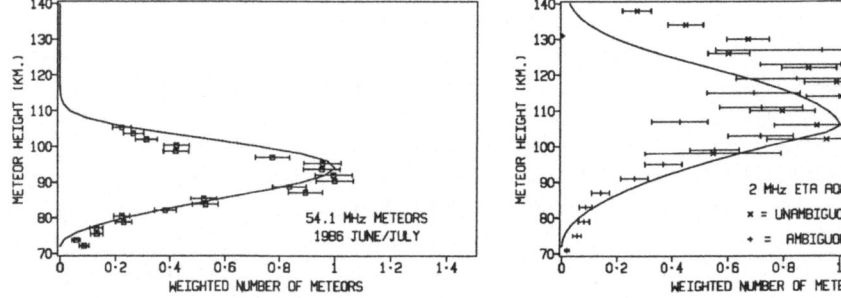

Figure 1: (a) The height distribution of radar meteors as observed at 54 MHz; the solid line shows a model which incorporates the various echo attenuation effects at this frequency. (b) As for (a) but for a 2 MHz radar: the model clearly underlies the data at high altitude, implying even more meteors at such heights than the model predicts.

Table 1: Inferred meteoroid densities for ionization peaks at the heights indicated under the classical ablation theory (Bronshten, 1983) or the dustball theory (Hawkes and Jones, 1975).

Height (km)	Bulk Density (gm cm^{-3})	
	Classical Ablation	Dustball Theory
85	3.5	5.9
90	1.6	2.6
100	0.3	0.5
110	0.06	0.10
120	0.012	0.020
125	0.005	0.009

Under the classical meteor ablation theory (e.g. Bronshten, 1983) the atmospheric density at the point of maximum ionization is:

$$d = \frac{\rho s \cos z}{A \Gamma H \sigma v_o{}^2} \tag{1}$$

where ρ is the meteoroid density, z the zenith angle of the radiant, H is the atmospheric scale height, s is the meteoroid 'size', A is a form factor such that the cross-sectional area is As^2, Γ is the drag coefficient, σ is the ablation coefficient, and v_o the original velocity. Putting in typical values ($\sigma v_o{}^2 = 40; \Gamma = 1.1; A = 1.5; H = 6$km; s = 1mm; $z = 45°$) one derives the meteoroid density as a function of height as given in Table 1.

Alternatively, using the dustball model of Hawkes and Jones (1975) one has:

$$d = \frac{8 \rho s X \cos z}{C H v_o{}^2} \tag{2}$$

where X is the energy required to fragment one kilogram of the meteoroidal material and C (= 1 to 3) is a factor depending upon the thermal conductivity. With $X/C = 10^6$ and $v_o = 30$ km sec^{-1} the values for ρ derived are again given in Table 1. Clearly our height determinations indicate a meteoroid density which is rather lower than that previously found from deceleration measurements of (relatively) low-altitude meteors (cf. Hughes, 1978; Bronshten, 1983); heights in the range 110–120 km are indicative of densities in the range 0.01–0.10 gm cm^{-3}. Such densities imply a very loose structure, and support a cometary origin for the particles (cf. the low density for P/Halley found by Rickman, 1986).

3. The asteroid-meteoroid stream link

To date the conventional wisdom on meteoroid streams, observed annually as meteor showers, has said that these originate as the larger particles in the dust tails of comets. However, using a new and powerful analysis technique Olsson-Steel (1988a) has been able to show that at least some, and perhaps many, of the Apollo-type asteroids have associated streams; in particular 1566 Icarus, 2101 Adonis, 2201 Oljato, 2212 Hephaistos, 3200 Phaethon, 1937 UB (Hermes), 5025 P-L, 1982 TA and 1984 KB are found to be linked with streams on the basis of the Adelaide meteor orbit data. This may be interpreted either (i) in terms of the asteroid being a remnant core after the de-volatilization of a cometary nucleus, or a nucleus which has formed a (temporary ?) insulating mantle; or (ii) collisional debris from boulder-sized impacts upon the asteroid.

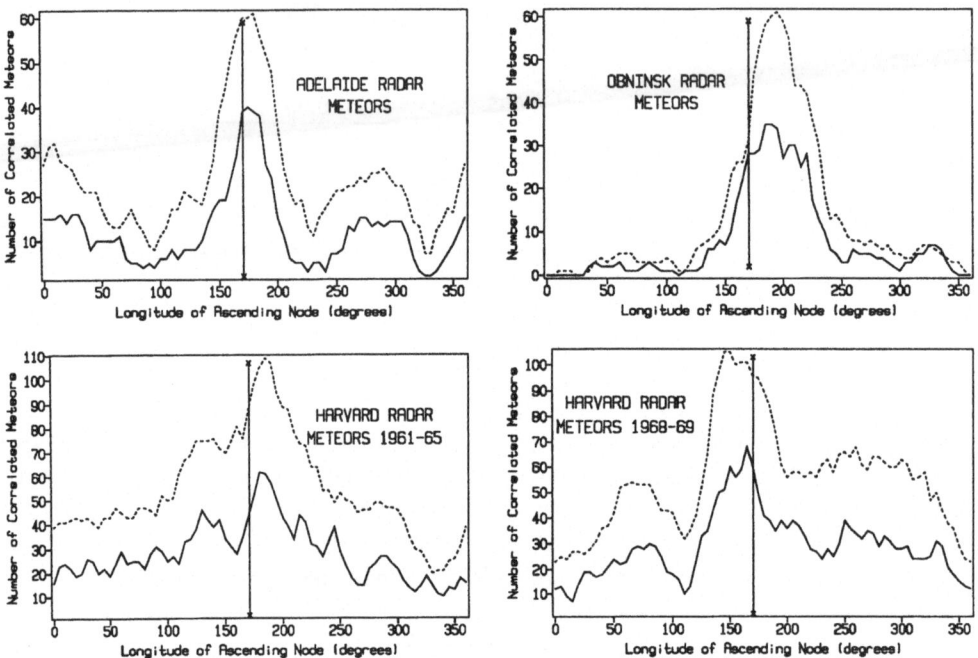

Figure 2: The number of correlated meteor orbits from various surveys as a function of the assumed Ω for 1984 KB; the vertical line at 170° is the actual Ω of this asteroid. The solid line is for $D < 0.20$ (Southworth and Hawkins, 1963) and the dashed line for $D' < 0.125$ (Drummond, 1981).

In order to confirm these links the data from other meteor orbit surveys are also being analysed, and the results so far back up the Adelaide data. As an example, in Fig.2 is shown the number of meteor orbits from the Harvard (Sekanina, 1973, 1976) and Obninsk (Lebedinets *et al.*, 1981, 1982) surveys which are correlated with asteroid 1984 KB as a function of nodal longitude (Ω). The concentration of this number at a particular value of Ω indicates the presence of a stream, and the fact that this occurs near to the nodal longitude of 1984 KB is indicative of a genetic relationship: for more details, see Olsson-Steel (1988a). It is noticeable that the plots for the two Harvard surveys show a rather broader peak than for the Adelaide and Obninsk data: this is apparently due to the fact that the Harvard data relate to rather fainter meteors (smaller meteoroids), for which the perturbational forces in space (radiative effects, larger ejection velocities from the parent) lead to such particles becoming dispersed from their original orbits much more quickly than the larger meteoroids detected from Adelaide and Obninsk, and hence forming a broader stream.

If, as is suggested by the above results, the Apollo asteroids are a major source of meteoroids, then this helps to solve one of the outstanding problems in the ecology of interplanetary objects: the apparent shortage of parents sufficient to explain the population of these bodies (e.g. see Whipple, 1967; Fulle, 1987). The objects observed in the Earth's atmosphere as meteors (i.e. sizes from $100\mu m$ to several centimetres) are known to power the zodiacal dust cloud (particle sizes mostly in the range 10–$100\mu m$) through catastrophic collisions (Grün et al., 1985; Olsson-Steel, 1986), but, as Grün et al. have shown, there seems to be a surfeit of meteoroids, by about an order of magnitude. This argues for a non-steady-state, with the present epoch reflecting a phase of decay after a recent large enhancement. The key to this situation may be the Taurid complex of interplanetary objects, which includes comet P/Encke [previously suggested as the major source of the zodiacal dust cloud: see Whipple (1967) and Gustafson et al. (1987)], four of the Apollo asteroids with associated meteoroid streams, and also possibly comet 1967 II Rudnicki (Olsson-Steel, 1987). The question of the origin of the whole interplanetary complex and the relationship between different object-types is thus opened up.

4. The origin of comet P/Halley

Of all the known comets with period $P < 200$ years, only five have $i > 90°$ and one of these (P/Hartley-IRAS) is an oddity in that it is on the borderline between the Jupiter-family/intermediate-period division, having $P = 21.5$ years, and is only just retrograde ($i = 95°.7$). The other four have larger inclinations and aphelia in the outer solar system, and seem to form a distinct group. Some of the orbital elements of these are listed in Table 2. It is notable that P/Tempel-Tuttle has aphelion very close to the orbit of Uranus, and P/Pons-Gambart close to that of Neptune. Similarly P/Halley has aphelion just beyond Neptune, and overall this suggests the possibility that these comets were originally in prograde orbits with perihelia in the outer planetary region (i.e. they were Kuiper Cloud objects), and were injected into their present orbits by close encounters with Uranus or Neptune, such encounters resulting in aphelion-perihelion exchanges and a switch from prograde to retrograde motion.

Table 2: The orbital elements of the retrograde intermediate-period (RIP) comets (data from Marsden, 1986).

	P(yr)	a(AU)	e	i	q(AU)	Q(AU)
P/Tempel-Tuttle	32.9	10.2	0.904	162°.7	0.982	19.5
P/Halley	76.0	17.8	0.967	162°.2	0.587	35.3
P/Pons-Gambart	57.5	14.9	0.946	136°.5	0.807	29.0
P/Swift-Tuttle	120	24.1	0.960	113°.6	0.963	47.7

This scenario has been considered in more detail by Olsson-Steel (1988b), who finds that a capture of P/Halley by Jupiter into an orbit like that presently followed by the comet from either (i) an original retrograde orbit with q near the present value, or (ii) an original retrograde orbit with q near Jupiter, is much less likely than a capture by Neptune from an original orbit with $q = 30$ AU and $i = 60°$. In addition, the required flux of long-period comets like (i) or (ii) in order to explain a steady-state population of four or more RIP comets is much higher than the observed flux, for sensible physical lifetimes. Thus it appears likely that P/Halley (and the other three RIP comets) has spent much of its lifetime at 50–100 AU, this having implications as regards the interpretation of abundance data in terms of how P/Halley relates to the origin and evolution of comets in general, and thus the other planet-crossing objects.

References

Bailey, M.E., Clube, S.V.M. and Napier, W.M. (1986) 'The origin of comets', *Vistas Astron.* **29**, 53–112.

Bronshten, V.A. (1983) *Physics of Meteoric Phenomena*, Reidel, Dordrecht.

Drummond, J.D. (1981) 'A test of comet and meteor shower associations', *Icarus* **45**, 545-553.

Duncan, M., Quinn, T. and Tremaine, S. (1988) 'The origin of short-period comets', *Astrophys. J.* **328**, L69–L73.

Everhart, E. (1973) 'The origin of short-period comets', *Astrophys. Lett.* **10**, 131–135.

Fulle, M. (1987) 'Meteoroids from Comet Bennett 1970 II', *Astron. Astrophys.* **183**, 392–396.

Grün, E., Zook, H.A., Fechtig, H. and Giese, R.H. (1985) 'Collisional Balance of the Meteoritic Complex', *Icarus* **62**, 244–272.

Gustafson, B.Aa.S., Misconi, N.Y. and Rusk, E.T. (1987). 'Interplanetary dust dynamics. III. Dust released from P/Encke: distribution with respect to the zodiacal cloud', *Icarus* **72**, 582–592.

Hawkes, R.L. and Jones, J. (1975) 'A quantitative model for the ablation of dustball meteors', *Mon. Not. Roy. Astron. Soc.* **173**, 339–356.

Hughes, D.W. (1978) 'Meteors', in J.A.M. McDonnell (ed.), *Cosmic Dust*, Wiley, Chichester, pp. 123–185.

Lebedinets, V.N., Korpusov, V.N. and Manokhina, A.V. (1981, 1982) *Radio Meteor Investigations in Obninsk: Catalogue of Orbits, Volumes 1 and 2*, Soviet Geophysical Committee of the Academy of Sciences of the U.S.S.R., Materials of the World Data Center B, Moscow.

Marsden, B.G. (1986) *Catalogue of Cometary Orbits*, I.A.U., Minor Planet Center, Cambridge, Mass.

Olsson-Steel, D. (1986) 'The origin of the sporadic meteoroid component', *Mon. Not. Roy. Astron. Soc.* **219**, 47–73.

Olsson-Steel, D. (1987) 'Asteroid 5025 P-L, comet 1967 II Rudnicki, and the Taurid meteoroid complex', *The Observatory* **107**, 157–160.

Olsson-Steel, D. (1988a) 'Identification of Meteoroid Streams from Apollo Asteroids in the Adelaide Radar Orbit Surveys', *Icarus* **75**, 64–96.

Olsson-Steel, D. (1988b) 'The inner Oort Cloud and the source of comet Halley', *Mon. Not. Roy. Astron. Soc.* **234**, 389–399.

Olsson-Steel, D. and Elford, W.G. (1987a) 'The height distribution of radar meteors: Observations at 2 MHz', *J. Atmos. Terr. Phys.* **49**, 243–258.

Olsson-Steel, D. and Elford, W.G. (1987b) 'The true height distribution and flux of radar meteors', *Publ. Astron. Inst. Czechoslov. Acad. Sci.*, No. 67, vol. 2, 193–197.

Rickman, H. (1986) 'Masses and densities of comets Halley and Kopff', *Proc. Comet Nucleus Sample Return Workshop*, **ESA SP–249**, 195–205.

Sekanina, Z. (1973) 'Statistical model of meteor streams. III. Stream search among 19303 radio meteors', *Icarus* **18**, 253–284.

Sekanina, Z. (1976) 'Statistical model of meteor streams. IV. A study of radio streams from the synoptic year', *Icarus* **27**, 265–321.

Southworth, R.B. and Hawkins, G.S. (1963) 'Statistics of meteor streams', *Smithson. Contrib. Astrophys.* **7**, 261-285.

Thomas, R.M., Whitham, P.S. and Elford, W.G. (1986) 'Frequency dependence of radio meteor echo rates', *Proc. Astron. Soc. Aust.* **6**, 303–306.

Thomas, R.M., Whitham, P.S. and Elford, W.G. (1988) 'Response of High Frequency Radar to Meteor Backscatter', *J. Atmos. Terr. Phys.* (in press).

Weissman, P.R. (1985) 'Cometary Dynamics', *Space Sci. Rev.* **41**, 299–349.

Whipple, F.L. (1967) 'On maintaining the meteoritic complex', in J.L.Weinberg (ed.), *The Zodiacal Light and the Interplanetary Medium*, **NASA SP–150**, Washington, DC, pp.409–426.

Wisdom, J. (1983) 'Chaotic behavior and the origin of the 3/1 Kirkwood gap', *Icarus* **56**, 51–74.

5. ATOMIC AND MOLECULAR DATA FOR ASTROCHEMISTRY

Scientific Organizing Committee

P.L. Smith (Chairman)

C. Arpigny, D. Flower

Supporting Commissions:

14, 15, 29, 34, 40, 44

ATOMIC AND MOLECULAR DATA FOR DIFFUSE CLOUD CHEMISTRY

Ewine F. van Dishoeck
Div. of Geological and Planetary Sciences
California Institute of Technology 170–25
Pasadena, CA 91125.

ABSTRACT. A brief overview of the procedure for modeling diffuse interstellar clouds is given, and specific needs for atomic and molecular data are pointed out.

1. INTRODUCTION

Diffuse cloud chemistry is the chemistry of the simplest interstellar molecules. This results from the fact that diffuse clouds are very tenuous concentrations of the interstellar gas with total visual extinctions A_V <2 mag. The interstellar ultraviolet radiation can easily penetrate these diffuse clouds and prevent the build–up of large molecules through rapid photodestruction processes. Most elements therefore exist primarily in atomic form in diffuse clouds, and any atom with an ionization potential less than 13.6 eV is mostly ionized. Thus carbon is present primarily as C^+, oxygen as O and nitrogen as N, with the most abundant molecule, CO, containing at most 1% of the total carbon. The only exception is formed by hydrogen, where H_2 may be comparable in abundance to H. Because of their simplicity, diffuse clouds are thought to be the best environment in which to test the basic concepts of interstellar chemistry.

The chemistry in diffuse clouds has recently been reviewed in great detail by van Dishoeck and Black (1988a), van Dishoeck (1988a), Dalgarno (1988), Black (1987), van Dishoeck and Black (1986) and Crutcher and Watson (1985). Because of space limitations, only a brief summary is given here with limited references to the original papers. Emphasis is placed on specific needs for atomic and molecular data. The reader is referred to the above papers for more details.

2. OBSERVATIONS

Diffuse clouds are studied mostly by the absorption lines of the atoms and molecules superposed on the spectra of bright background stars. The classic example is the line of sight toward ζ Oph. Although many atomic species have been identified, only few molecules have so far been found in diffuse clouds: H_2, HD, CH, CH^+, C_2, CO, CN and OH. The list of non–detections is considerably longer and includes such interesting species as H_2O, C_3, MgH, ... (see Table 1 of van Dishoeck and Black 1988a for a summary). The first need for atomic and molecular data in diffuse cloud studies lies in the availability

323

of high–resolution visible and ultraviolet laboratory spectra to identify the species. In addition, accurate oscillator strengths for the electronic transitions are required to convert the measured equivalent widths into column densities. Specific needs are discussed in the paper by J.H. Black in this volume.

There are two other classes of interstellar clouds that are chemically very similar to the classical diffuse clouds, but which are studied by different techniques. Translucent clouds with $A_V \approx 2 - 5$ mag are "thicker" versions of the classical diffuse clouds (Crutcher 1985; van Dishoeck and Black 1988b). The molecular clouds detected at high galactic latitudes have $A_V = 1 - 2$ mag, similar to the diffuse clouds (Magnani, Blitz and Mundy 1985). Both types of clouds have large enough molecular column densities to allow millimeter observations of CO and species such as CS, HCN, H_2CO, NH_3 and H_2CO. On the other hand, these clouds are still thin enough to permit absorption line observations, provided that there is a bright star located fortuitously behind the cloud. The interpretation of the millimeter data requires accurate rotational excitation cross sections, both by H_2 and by electrons. Only very limited information is currently available on electron–molecule excitation rates at low temperatures.

3. CHEMICAL MODELS

The basic outline of the chemistry responsible for the formation of the simplest molecules was established already more than 15 years ago. Detailed models for comparison with observations were first made by Black and Dalgarno (1977), whereas more recent studies include those of van Dishoeck and Black (1986) and Viala, Roueff and Abgrall (1988). With the well-known exception of CH^+, the models initially were able to reproduce the observed abundances quite well using reasonable parameters for unknown reaction rate coefficients. However, now that more accurate reaction rates are becoming available through detailed laboratory or theoretical work, it becomes increasingly difficult to develop models that are in harmony with the *complete* set of observational data for a cloud.

The modeling of a specific interstellar cloud consists of two steps:

3.1. Physical conditions

Since the reaction rates depend on temperature, density and the intensity of the radiation field, the physical conditions need to be constrained as closely as possible before the chemistry can be studied. In diffuse clouds, the observed excitations of various species are used as diagnostics, and accurate data on the atomic and molecular processes that enter the excitation calculations are needed.

The primary diagnostic on which most diffuse cloud models are based is H_2. The low rotational levels $J < 3$ of this molecule are populated primarily by collisional processes and are thus sensitive to temperature; the higher levels $J > 5$ are thought to be populated mostly by an ultraviolet pumping process, so that their populations are sensitive to the strength of the interstellar radiation field. The relative abundance of atomic to molecular hydrogen depends on the density in the cloud. Although the rates for collisional excitation of H_2 by H_2 are fairly well determined (Schaefer 1985; Danby, Flower and Monteiro 1987), the H_2–H cross sections are still uncertain by an order of magnitude (Green and Truhlar 1979; Allison and Dalgarno 1967). Better information on this system is badly needed.

Other important diagnostics in diffuse clouds include:
- C_2: the rotational levels of this homonuclear molecule are also populated by a combination of collisional and optical pumping processes, followed by infrared cascade (van

Dishoeck and Black 1982). The population distribution over the lower levels is again sensitive to temperature, whereas that over the higher levels depends on the ratio of the density and the strength of the radiation field in the red. Better information on the rotational excitation cross sections of C_2 by H_2 is needed, although theoretical calculations for this system have recently been reported by Chambaud et al. (1988). Note also that in diffuse clouds, the abundance of ortho-H_2 $(J=1)$ is comparable to that of para-H_2 $(J=0)$, so that excitation rates by both species are required. Another uncertain molecular parameter in the C_2 excitation calculation is the C_2 $a^3\Pi_u$-$X^1\Sigma_g^+$ intercombination transition moment.

- CO: its rotational levels are populated primarily by collisional processes so that the excitation is sensitive to both density and temperature. Collisional rates of CO with para-H_2 $(J=0)$ are now well determined (Schinke et al. 1985; Flower and Launay 1985), but rates with ortho-H_2 $(J=1)$ are still uncertain.
- Atomic species: the excitation of the fine structure levels of the C^+, C and O atoms occurs mostly by collisions with atomic and molecular hydrogen, and with electrons. Although theoretical calculations of rates for collisions with H have been available for some time, the rates for collisions with H_2 $(J=0,1)$ are based mostly on educated guesses. Monteiro and Flower (1987) have recently pointed out that $J=0\rightarrow1$ excitation of C and O by H_2 $(J=0)$ is in first order forbidden. Accurate calculations for the C^+-e cross sections exist (Hayes and Nussbaumer 1984), but not for the C-e or O-e collisions. Experimental information on all of these systems would be most welcome.

In diffuse clouds, the various diagnostics give physical conditions that are consistent within factors of a few, although it is often not possible to construct a unique model. The temperature in the models typically varies from >100 K at the edge to 30 K in the center, and the density from 100-200 cm^{-3} at the edge to 500-1000 cm^{-3} in the center. The largest uncertainty currently lies in the strength of the interstellar radiation field, which is constrained only by the H_2 high-J population under the assumption that ultraviolet pumping is the dominant excitation mechanism. Typically, the radiation field is found to be enhanced over the average field by factors of 3-5. As discussed in the paper by Hartquist in this volume, the levels may also be populated by collisions in shock-heated gas, in which case the inferred radiation fields are only upper limits.

3.2. Chemical network

The second step in the modeling consists of solving the steady-state chemical network at each depth. Integration of the concentrations over depth then yields the column densities. Overviews of the chemistry in diffuse clouds can be found e.g. in van Dishoeck and Black (1988a), van Dishoeck (1988a) and Black and Dalgarno (1977). Detailed information on the rate coefficients for the reactions that enter the networks is needed at this stage of the modeling procedure. It should be realized, however, that not all reactions are equally important, so that not all rate coefficients need to be known with equally high precision. It is the task of the astrochemists to identify the crucial reactions for which accurate rate coefficients are required. In the following, some of these crucial reactions with poorly-determined rate coefficients will be mentioned.

3.2.1. Carbon chemistry

The formation of the carbon-bearing molecules is thought to be initiated by the radiative association reaction

$$C^+ + H_2 \rightarrow CH_2^+ + h\nu. \tag{1}$$

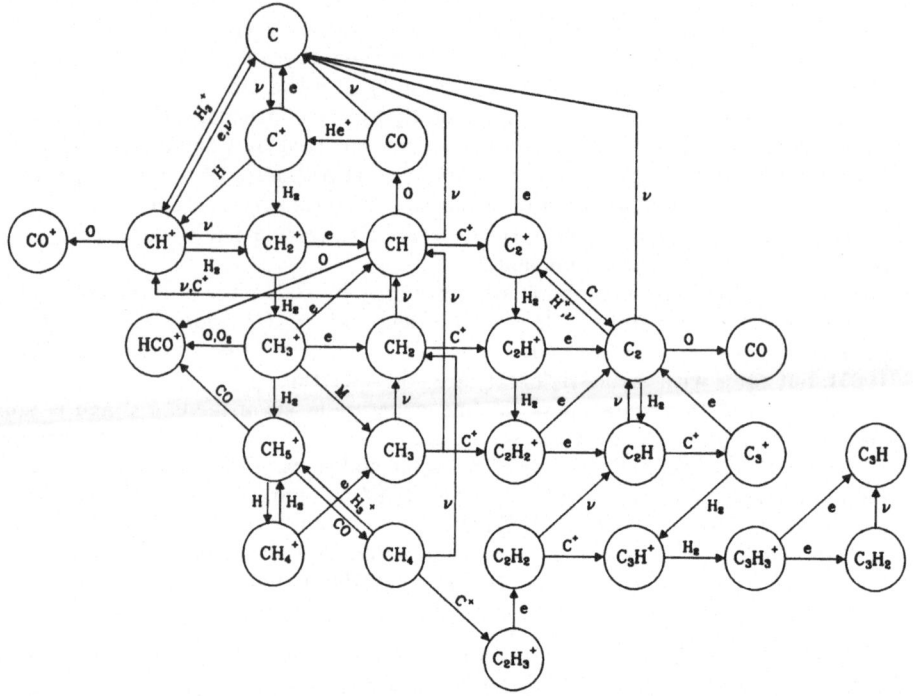

Figure 1. *The most important chemical reactions involving carbon–bearing molecules. M stands for metal.*

Although this reaction was postulated more than 15 years ago (Black and Dalgarno 1973), it has still not been measured in the laboratory, nor is there any theoretical calculation accurate to better than an order of magnitude. The approach of the astrochemists has therefore been to use the observed CH abundance to infer a value for this rate. The models for the ς Per and ς Oph diffuse clouds require $k_{ra} \approx 5 \times 10^{-16}$ cm³ s⁻¹, a value which is consistent with the theoretial estimate of $k_{ra} = 10^{-16} - 10^{-15}$ cm³ s⁻¹ (Herbst 1982), but which is close to the experimental upper limit $k_{ra} < 1.5 \times 10^{-15}$ cm³ s⁻¹ at 13 K (Luine and Dunn 1985). Note, however, that this experimental limit pertains to the reaction of C⁺ with normal hydrogen having an ortho:para ratio of 3:1, whereas the interstellar models require the rate for an ortho:para ratio of about 1:2. Radiative association with ortho–H₂ is expected to be slower than that with para–H₂ on theoretical grounds. An experimental verification of the astronomically determined rate is urgently needed. Even an upper limit lowered by a factor of four compared with that of Luine and Dunn (1985) would provide significant tests of the carbon chemistry

Further down in the carbon–chemistry network (see Figure 1), the dissociative recombination reaction

$$CH_3^+ + e \rightarrow CH_2 + H \tag{2a}$$

$$\rightarrow CH + H_2 \text{ or } H + H \tag{2b}$$

occurs. In this case, the reaction rate coefficient itself is not very important (as long as the reaction is rapid), but the branching ratio to form CH₂ and CH significantly affects the chemistry (see Black and van Dishoeck 1988 for an overview). If the branching ratio favors CH_2, the abundance of CH_2 is increased significantly in the models, and consequently also the abundances of subsequent molecules in the chain, such as C_2 and C_2H. Similarly, the branching ratios of the dissociative recombination of $C_3H_3^+$ to form cyclic or linear C_3H_2 and C_3H are important. Virtually no theoretical or experimental information on these branching ratios is currently available from experiment or theory.

The principal destruction mechanism of all neutral molecules in diffuse clouds is photodissociation. Many of the photo rates adopted in the early diffuse cloud models were based on educated guesses. Over the last ten years, more accurate rates have been determined for a number of species through detailed calculations or laboratory experiments. In most cases, the new rates are significantly larger than previous estimates, making it more difficult to reproduce the measured abundances. The photodissociation and photoionization rates for interstellar molecules have recently been summarized by van Dishoeck (1988b), and specific needs have been pointed out. In the carbon chemistry, the photo rates of the CH_2, C_2H and C_3H_2 molecules are particularly uncertain. In addition, the branching ratios to dissociation and ionization following absorption of photons with energies greater than the ionization potential of the molecule are not well known.

3.2.2. Oxygen chemistry

The formation of oxygen–bearing molecules is driven either by the near–resonant charge–transfer reaction

$$O + H^+ \rightarrow O^+ + H \tag{3}$$

or

$$O + H_3^+ \rightarrow OH^+ + H_2. \tag{4}$$

Reaction (4) proceeds rapidly at low temperature, but the rate coefficient for reaction (3) is more uncertain, and only theoretical estimates are available (Chambaud et al. 1980). Whether reaction (3) or (4) dominates the formation route depends on the abundance of H_3^+ in diffuse clouds. This abundance, in turn, depends on the rate at which H_3^+ is removed by dissociative recombination:

$$H_3^+ + e \rightarrow H_2 + H \text{ or } 3H. \tag{5}$$

The rate coefficient of reaction (5) for H_3^+ in its lowest vibrational level $v=0$ is currently very controversial. Although there was significant evidence from both theoretical and experimental sides that this process may be very slow under interstellar conditions (Michels and Hobbs 1984; Adams and Smith 1987), more recent experiments suggest that it remains rapid at low temperatures (Amano 1988; Hus et al. 1988). A discussion of the possible origin for these discrepancies has been given by Smith and Adams (1988). If $k_5 < 10^{-9}$ cm^3 s^{-1}, reaction (4) dominates the oxygen chemistry.

Other important molecular data needed in the oxygen chemistry include the branching ratios for dissociative recombination, in particular for H_3O^+ to form H_2O; the rate at which C^+ reacts with the polar molecule OH to form CO at low temperatures; and the photodissociation rates of the neutral species. Of particular importance is the rate at which CO photodissociates. Recent laboratory work on the vacuum ultraviolet absorption spectrum of ^{12}CO by Letzelter et al. (1987) and Yoshino et al. (1988) has clarified significantly the nature of its photodissociation processes, but crucial information on the isotopic species ^{13}CO, $C^{18}O$ and $^{13}C^{18}O$ is still lacking. A summary of the spectroscopic information needed for CO has been given by van Dishoeck and Black (1988c).

3.2.3. Nitrogen chemistry

The formation routes for nitrogen–bearing molecules in diffuse clouds are particularly uncertain. Previous models assumed that the principal initiating route is the exothermic reaction

$$N + H_3^+ \rightarrow NH_2^+ + H. \tag{6}$$

Theoretical calculations (Herbst et al. 1987), however, indicate that there are large barriers along the reaction path so that the rate is likely to be slow at low temperatures. An experimental upper limit would be very useful. The other initiating reaction

$$N^+ + H_2 \rightarrow NH^+ + H \tag{7}$$

is also likely to be slow under interstellar conditions (Rowe 1988). Alternative gas–phase reactions through which nitrogen–containing molecules can be formed include neutral–neutral reactions of atomic nitrogen with small oxygen– and carbon–containing molecules such as OH, CH and C_2 to form CN and NO, and ion–molecule reactions of atomic nitrogen with small hydrocarbons to form e.g. H_2CN^+, which can then recombine to form CN and HCN. Experimental information on the rate coefficients of the neutral–neutral reactions at low temperatures is lacking.

Neutral molecules such as CN are rapidly destroyed by photodissociation and by reactions with atomic oxygen. Experimental determinations of the reaction rate coefficients for the latter processes would be useful.

4. SUMMARY

Atomic and molecular data are needed in the modeling of diffuse clouds at three different stages:

- accurate transition wavelengths and oscillator strengths are required to identify the species and to convert the measured equivalent widths into column densities.
- accurate collisional excitation rates of various diagnostic species are needed to constrain the physical conditions.
- accurate rate coefficients for crucial reactions are needed to test the chemistry.

Although significant progress has been made in various areas, there are still a number of processes, such as dissociative recombination and radiative association, for which reliable experimental or theoretical results are only just beginning to appear. The models are therefore not yet conclusive as to whether the adopted chemical network is complete and whether steady–state gas–phase models can explain the abundances of most molecules in diffuse clouds. The current diffuse cloud models can reproduce the abundances of a number of the observed molecules such as CH, C_2, OH and HD with reasonable, although ad hoc values for rates of crucial reactions such as (1) and (5). On the other hand, the models tend to produce too little CO and CN compared with observations by factors of a few. These discrepancies most likely result from an overestimate of the strength of the interstellar radiation field incident on the model clouds. However, until more accurate data on the crucial atomic and molecular processes are available, we cannot be confident that we fully understand the chemistry of the simplest diatomic molecules in diffuse interstellar clouds.

REFERENCES

Adams, N.G. and Smith, D. 1987, in IAU Symposium **120**, *Astrochemistry*, eds. M.S. Vardya and S.P. Tarafdar (Reidel, Dordrecht), p. 1.

Allison, A.C. and Dalgarno, A. 1967, *Proc. Phys. Soc. London*, **90**, 609.

Amano, T. 1988, *Ap. J. (Letters)*, **329**, L121.

Black, J.H. 1987, in IAU Symposium **120**, *Astrochemistry*, eds. M.S. Vardya and S.P. Tarafdar (Reidel, Dordrecht), p. 217.

Black, J.H. and Dalgarno, A. 1973, *Astrophys. Letters*, **15**, 79.

Black, J.H. and Dalgarno, A. 1977, *Ap. J. Suppl.*, **34**, 405.

Black, J.H. and van Dishoeck, E.F. 1988, to appear in *Dissociative Recombination: theory, experiments and applications*, eds. J.B.A. Mitchell and S.L. Guberman (World Scientific Publishing, Singapore).

Chambaud, G., Launay, J.M., Levy, B., Millie, P., Roueff, E., and Tran Minh, F. 1980, *J. Phys. B*, **13**, 4205.

Chambaud, G., Lavendy, H., Levy, B., Robbe, J.M., and Roueff, E. 1988, *Chem. Phys.*, submitted.

Crutcher, R.M. 1985, *Ap. J.*, **288**, 604.

Crutcher, R.M. and Watson, W.D. 1985, in *Molecular Astrophysics*, eds. G.H.F. Diercksen *et al.*, NATO ASI Series **157** (Reidel, Dordrecht), p. 255.

Dalgarno, A. 1988, *Astro. Lett. and Communications*, **26**, 153.

Danby, G., Flower, D.R., and Monteiro, T.S. 1987, *M. N. R. A. S.*, **226**, 739.

Flower, D.R. and Launay, J.M. 1985, *M. N. R. A. S.*, **214**, 271.

Green, S. and Truhlar, D.G. 1979, *Ap. J. (Letters)*, **231**, L101.

Hayes, M.A. and Nussbaumer, H. 1984, *Astr. Ap.*, **134**, 193.

Herbst, E. 1982, *Ap. J.*, **252**, 810.

Herbst, E., DeFrees, D.J., and McLean, A.D. 1987, *Ap. J.*, **321**, 898.

Hus, H., Yousseif, F., Sin, A., and Mitchell, J.B.A. 1988, *Phys. Rev. A*, in press.

Letzelter, C., Eidelsberg, M., Rostas, F., Breton, J., and Thieblemont, B. 1987, *Chem. Phys.*, **114**, 273.

Luine, J.A. and Dunn, G.H. 1985, *Ap. J. (Letters)*, **299**, L67.

Magnani, L., Blitz, L., and Mundy, L. 1985, *Ap. J.*, **295**, 402.

Michels, H.H. and Hobbs, R.H. 1984, *Ap. J. (Letters)*, **286**, L27.

Monteiro, T.S. and Flower, D.R. 1987, *M. N. R. A. S.*, **228**, 101.

Rowe, B.R. 1988, in *Rate Coefficients in Astrochemistry*, eds. T.J. Millar and D. Williams (Kluwer, Dordrecht), p. 135.

Schaefer, J. 1985, in *Molecular Astrophysics*, eds. G.H.F. Diercksen *et al.*, NATO ASI Series **157** (Reidel, Dordrecht), p. 497.

Schinke, R., Engel, V., Buck, U., Meyer, H., and Diercksen, G.H.F. 1985, *Ap. J.*, **299**, 939.

Smith, D. and Adams, N.G. 1988, to appear in *Dissociative Recombination: theory, experiments and applications*, eds. J.B.A. Mitchell and S.L. Guberman (World Scientific Publishing, Singapore).

van Dishoeck, E.F. 1988a, to appear in *Molecular Astrophysics — A volume honoring Alexander Dalgarno*, ed. T.W. Hartquist (Cambridge University).

van Dishoeck, E.F. 1988b, in *Rate Coefficients in Astrochemistry*, eds. T.J. Millar and D. Williams (Kluwer, Dordrecht), p. 49.

van Dishoeck, E.F. and Black, J.H. 1982, *Ap. J.*, **258**, 533.

van Dishoeck, E.F. and Black, J.H. 1986, *Ap. J. Suppl.*, **62**, 109.

van Dishoeck, E.F. and Black, J.H. 1988a, in *Rate Coefficients in Astrochemistry*, eds.

T.J. Millar and D. Williams (Kluwer, Dordrecht), p. 209.

van Dishoeck, E.F. and Black, J.H. 1988b, *Ap. J.*, submitted.

van Dishoeck, E.F. and Black, J.H. 1988c, *Ap. J.*, Nov. 15 issue.

Viala, Y.P., Roueff, E., and Abgrall, H. 1988, *Astr. Ap.*, **190**, 215.

Yoshino, K., Stark, G., Smith, P.L., Parkinson, W.H., and Ito, K. 1988, in preparation.

ULTRAVIOLET, VISIBLE, AND INFRARED SPECTROSCOPY OF INTERSTELLAR MOLECULES

John H. Black
Steward Observatory
University of Arizona
Tucson, AZ 85721 USA

ABSTRACT. Absorption line spectroscopy in the ultraviolet, visible, and infrared regions can provide important probes of interstellar chemistry. Significant recent developments include the application of optical absorption line techniques to the study of thick molecular clouds and the improvements in infrared detectors that will eventually lead to effective interstellar spectroscopy in the infrared. Demands for basic molecular data will grow in scope and in level of precision.

1. INTRODUCTION

Owing to improvements in detectors and in astronomical spectrographs, optical observations can now be applied to very thick interstellar clouds and to the investigation of finer details of the internal structure of diffuse clouds. As a result, the demand for basic spectroscopic data on simple molecules is expanding in scope and is becoming more severe with respect to precision. In some cases, exploratory work is needed to identify the electronic and vibrational spectra of important interstellar molecules like C_3H and C_3H_2. In other instances, there are continuing needs for accurate oscillator strengths and for the resolution of discrepancies between experimental and theoretical determinations of them. Even in cases of well studied molecular spectra like those of CH and CN, the measurement of line positions needs to be improved in accuracy by an order of magnitude or more to satisfy existing needs in the interpretation of astronomical observations.

The number of interstellar molecules detected by optical methods has not increased much since earlier reviews of the subject were completed (Snow 1980; Black 1985; Federman 1987). In the meantime, however, upper limits have been obtained for additional species and the variety of interstellar clouds investigated through known molecules has grown. A recent list of molecules sought by optical methods in diffuse clouds can be found in van Dishoeck and Black (1988a). In particular, no polyatomic molecule has been detected in the interstellar gas by optical absorption measurements. The impending launch of the *Hubble Space Telescope (HST)* promises more sensitive, new searches for molecules in the ultraviolet. Improvements in infrared spectroscopy should open up a new realm of molecular studies in the next few years. In this brief review, recent astronomical results that illustrate needs for molecular data are considered. Some recent developments in molecular studies relevant to astrochemistry are summarized. Finally, an attempt is made to anticipate some paths that the subject might profitably follow in the future.

D. McNally (ed.), Highlights of Astronomy, Vol. 8, 331–338.
© *1989 by the IAU.*

2. BASIC SPECTROSCOPY

Many of the molecules of astrochemical interest are radicals and ions that lack detailed spectroscopic analyses. Moreover, there exist some stable, unreactive species for which no high–resolution spectroscopy has been performed in the ultraviolet. Although most of the previously unidentified interstellar absorption lines in the ultraviolet spectrum of ς Ophiuchi have since been assigned to H_2, HD, and atomic carbon, there remain as many as 19 unidentified interstellar lines at $\lambda = 1023 - 1317$ Å, some of which might be molecular (Morton 1978). We can expect that the number of unidentified ultraviolet lines will increase when the *Hubble Space Telescope* finally reaches orbit.

Interstellar absorption lines of CH were first identified more than 50 years ago (Swings and Rosenfeld 1937), yet increasing demands on the spectroscopy of CH continue to the present day. The widely observed interstellar CH feature $A^2\Delta-X^2\Pi$ (0,0) $R_f(1/2)$ is a closely spaced Λ–doublet (wavelengths 4300.3030 and 4300.3235 Å in standard air) which has never been resolved in astronomical spectra. With the recognition that some diffuse interstellar clouds exhibit complex Doppler velocity structure in multiple components on scales of 0.3–1.0 km s^{-1} (corresponding to wavelength shifts of 0.004–0.014 Å at 4300 Å), the accurate description of the intrinsic line shape (i.e. doublet structure in this example) becomes essential to the interpretation of the observations (Black and van Dishoeck 1988a). Laboratory measurements adequate for this purpose have only recently become available (Brazier and Brown 1984; Bernath 1987). Furthermore, the mean wavelength of the 4300 Å line blend differs by 0.008 Å (i.e. 0.54 km s^{-1} in Doppler velocity) from the value previously adopted in the astronomical literature. The wavelengths of the interstellar lines of the $B^2\Sigma^--X^2\Pi$ (0,0) band of CH have also been determined much more accurately (P. F. Bernath, preliminary results, as quoted by Black and van Dishoeck 1988a).

Interstellar absorption lines of CN are used not only to study its abundance and chemistry, but also as an indirect radiometric sensor of the cosmic background radiation, which governs the rotational excitation of CN in interstellar clouds of low density. In denser regions, the measurement of excess excitation can be used to estimate the electron density. The most recent observations of CN in diffuse clouds (Meyer and Jura 1985; Crane *et al.* 1986) determine the relative populations in the $N = 0$ and $N = 1$ rotational levels (or the excitation temperature) to approximately $\pm 1\%$ accuracy. At this level, the extraction of the true values of the column densities and the excitation temperature from the measurements can be quite sensitive to very small amounts of line saturation. The unresolved spin–doublets of the CN $B^2\Sigma^+-X^2\Sigma^+$ (violet) system have traditionally been treated as single, unresolvable lines. In fact, the B–X (0,0) $R(0)$ feature is a blend of $R_1(0)$ and $^RQ_{21}(0)$ lines with respective relative strengths of 2/3 and 1/3 and with a separation estimated to be 0.0035 Å (0.27 km s^{-1} in Doppler velocity) on the basis of published spectroscopic constants (Engleman 1974). This small splitting is uncertain and needs to be determined more accurately. There exist observed interstellar CN lines (Meyer and Hawkins 1987) for which the neglect of the unresolved structure would cause a 40% overestimate of the ground–state column density if the Doppler line broadening parameter were as small as $b = 0.5$ km s^{-1} (van Dishoeck and Black 1989).

Spectroscopic studies of small, carbon–bearing molecules of astrophysical interest have produced interesting results recently. The $B^4\Sigma_u^--X^4\Sigma_g^-$ electronic transition of C_2^+ has been identified (O'Keefe *et al.* 1984; Forney *et al.* 1987) and subsequently given a rotational anaylsis (Rösslein *et al.* 1987). The (0,0) band is calculated to have an oscillator strength $f_{00} = 0.0156$ (Rosmus *et al.* 1986), and potential interstellar absorption lines are at convenient wavelengths: (0,0) $R(1)$, 5065.19 Å; $R(3)$, 5063.29 Å. However, C_2^+ reacts rapidly with H_2; therefore, its interstellar abundance is predicted to be quite low. Abundances of the related ions C_2H^+ and $C_2H_2^+$ are likely to be higher and similar spectroscopic analyses of them would be valuable. Photoelectron spectroscopy of $C_2H_2^+$ indicates that the origin of the $\tilde{A}^2A_g$–$\tilde{X}^2\Pi_u$ system should lie near $\lambda = 2533$ Å (Reutt et al. 1986). Moreover, the ν_3 fundamental band of this ion has been observed directly at $\tilde{\nu} = 3136$ cm^{-1} (Crofton et al. 1987). Further exploratory spectroscopy is also needed for the ion C_2N^+, whose interstellar abundance may be high (Hartquist and Dalgarno 1980) and whose rotational spectrum has been investigated through theoretical calculations (Kraemer et al. 1984; Jensen and Kraemer 1988). The electronic spectrum of the neutral CCN is fairly well known (see Brazier et al. 1987 and references therein). It would also be of great interest to know about the electronic transitions of the first interstellar ring molecule, cyclopropenylidene (C_3H_2), because it is observed to be widespread and abundant even in diffuse clouds (Cox et al. 1988) through its pure rotational transitions. Electronic and vibrational spectra of C_3H, which has been observed in both cyclic and linear forms, would be of interest.

The C_3 molecule is observed in comets and carbon stars and has been sought unsuccessfully in interstellar clouds through its $\tilde{A}^1\Pi_u$–$\tilde{X}^1\Sigma_g^+$ transition at 4050 Å (Clegg and Lambert 1982; Snow et al. 1988). C_3 is expected to have an ultraviolet transition $^1\Sigma_u^+$–$\tilde{X}^1\Sigma_g^+$ with a very large oscillator strength, $f \approx 0.9$ (Römelt et al. 1978). Although ultraviolet absorption at 1580 Å has been attributed to C_3 (Shinn 1982) and matrix isolation studies indicate that the (0,0,0)–(0,0,0) band occurs near 1893 Å (Chang and Graham 1982), a rotational analysis based on a gas–phase spectrum is still needed. More recently, the ν_3 fundamental vibration–rotation band of C_3 has been identified and analyzed in the spectrum of the carbon star IRC+10216 (Hinkle et al. 1988) and in laboratory spectra (Matsumura et al. 1988). The oscillator strength of this band is large for an infrared transition (Kraemer et al. 1984), and interstellar C_3 is potentially observable in thick clouds through absorption lines near 4.9 μm wavelength. Positions of the interstellar absorption lines of the Phillips system of C_2 can now be determined much more precisely as a result of the work of Douay et al. (1988). The C_2H molecule, which is widely observed through its rotational transitions at millimeter wavelengths, has an electronic transition in the infrared (Curl et al. 1985) for which transition moments have been calculated (Reimers et al. 1985).

As high–resolution infrared spectrometers with more sensitive array detectors come into operation, infrared spectroscopy of interstellar molecules will begin to play a significant role in astrochemistry. The most abundant interstellar molecule, H_2, has been essentially unobservable in thick, quiescent molecular clouds, although it can be observed by means of ultraviolet absorption lines in thin, diffuse clouds and through infrared emission lines from molecular gas that is highly disturbed by shock waves or exposed to intense ultraviolet radiation. The fact that its infrared lines are weak dipole–forbidden, electric quadrupole transitions would seem to preclude absorption line studies; however, column densities of H_2 of the order of 10^{23} cm^{-2} or less are in principle detectable in absorption in the (1,0) $S(0)$ and $S(1)$ vibration–rotation lines toward suitably located infrared sources. In this instance, "detectable" refers to an equivalent width $W_\nu \leq 0.01$ cm^{-1} at $\lambda = 2.1 - 2.2$ μm, which requires only a signal/noise ratio of 10 or more at a resolving power $\lambda/\Delta\lambda \approx 10^5$. Moreover, it should be possible to observe infrared absorption lines of CO and H_2 simultaneously and thus to make *direct* determinations of the CO/H_2 abundance ratio, which is important for quantitative studies of the distribution of molecular material and the efficiency of star formation in galaxies (Williams 1985; van Dishoeck and Black 1987). The infrared absorption lines of interstellar CO have already been detected, both in actively star–forming regions (Hall et al. 1978; Scoville et al. 1983; Geballe and Wade 1985) and in a quiescent cloud (Black and Willner 1984). Solid CO that has frozen onto interstellar grains has been identified through its distinctive infrared absorption spectrum (Hagen et al. 1980; Lacy et al. 1984; Larson et al. 1985; Whittet et al. 1985; Geballe et al. 1985; Geballe 1986). Recent laboratory investigations have also made it possible to determine the column density of solid CO and to infer something about the surface chemistry (Sandford et al. 1988). Laboratory spectroscopy of other astrophysical ices has also been pursued (e.g. d'Hendecourt and Allamandola 1986, and references therein).

Infrared absorption line spectroscopy can provide significant tests of theories of interstellar chemistry, especially with respect to the abundances of important non–polar species like H_3^+, CH_4, and C_2H_2, which are otherwise not readily observable in interstellar space. The infrared absorption lines of H_3^+ at $\lambda = 3.7 - 4.1$ μm are expected to be detectable both in dark molecular clouds (Oka 1981) and in classical diffuse clouds (van Dishoeck and Black 1986; Black and van Dishoeck 1988b). Knacke et al. (1985) have reported a recent search for interstellar CH_4. Table 1 lists a selection of infrared transitions of actual or potential interstellar molecules that will be important for absorption line studies. The table presents band or line frequencies in cm^{-1} and, where information on oscillator strengths is available, a characteristic column density required to produce an absorption feature of equivalent width $W_\nu = 0.01$ cm^{-1}.

Metal atoms and ions may play a major role in the ionization balance of interstellar clouds, but very little is known about their participation in interstellar chemistry. Many metal–containing molecules have strong, well known transitions in the visible region. However, no molecules containing Mg, Ca, Na, Fe, or Ti have been found in the interstellar medium, with the possible exception of a detection of MgO at millimeter wavelengths. No sensitive optical searches for molecules like MgO and TiO have ever been reported in the literature. In view of the possibility that TiO^+ forms by an exothermic chemi–ionization process in interstellar clouds, $Ti + O \rightarrow TiO^+ + e$, it would be interesting to know something about the spectrum of this ion (Oppenheimer and Dalgarno 1977, Black 1988).

The subject of very large molecules (e.g. polycyclic aromatic hydrocarbons = PAHs) in interstellar space is very fashionable (see, e.g., Omont 1986, Lepp and Dalgarno 1988, and references therein) and much spectroscopic work remains to be done. Although PAHs have joined the long list of possible carriers of the diffuse interstellar bands (Josafatsson and Snow 1987 and references therein), no specific and convincing identification has been made. The large molecule buckminsterfullerene, C_{60}, has recently been shown to have an isolated, narrow absorption band at 3860 Å (Heath, Curl, and Smalley 1987).

3. QUANTITATIVE SPECTROSCOPY

Molecular abundances can be determined and quantitative tests of chemical models can be made only if oscillator strengths of molecular transitions are known. Only a few recent developments can be summarized here; for a recent review, see Smith (1987). Although detections or upper limits have been reported for at least 34 molecules in diffuse clouds (see Table 1 of van Dishoeck and Black 1988a), several of the limits are not very useful owing to the lack of information on oscillator strengths. Specifically, oscillator strengths are needed for ultraviolet transitions of MgH^+, SH, CaH, and SiO. Naturally, oscillator strengths will also be needed for various species whose electronic spectra have not yet been identified and analyzed (see §2 above).

The red systems of CN $A^2\Pi-X^2\Sigma^+$ and C_2 $A^1\Pi_u-X^1\Sigma_g^+$ are now being used for absorption line studies of diffuse and translucent clouds (van Dishoeck and Black 1989). Although much effort has been devoted to the determination of the oscillator strengths of these systems, there remain some unresolved questions. Recent experiments are in harmony for the C_2 A–X Phillips system (Erman et al. 1982; Bauer et al. 1985, 1986; Davis et al. 1984) with $f_{00} \approx 1.4 \times 10^{-3}$ and $f_{20} \approx 1.0 \times 10^{-3}$; however, oscillator strengths determined

Table 1. SELECTED INFRARED TRANSITIONS OF
INTERSTELLAR MOLECULES

Species	Transition	$\tilde{\nu}$ (cm^{-1})	N_0 (cm^{-2})
H_2	$v = 1 - 0$ S(0)	4497.8391	1.2(23)
H_3^+	$\nu_2 = 1 - 0$ $^R R_1(1)$	2691.444	1.3(14)
H_2D^+	$\nu_1 = 1 - 0$ $1_{01} - 0_{00}$	3038.177	2.3(15)
	$\nu_3 = 1 - 0$	2329.	4.9(14)
HeH$^+$	$v = 1 - 0$ R(0)	2972.91	7.6(13)
H^{35}Cl	$v = 1 - 0$ R(0)	2906.2474	1.8(15)
CO	$v = 1 - 0$ R(0)	2147.081	1.1(15)
	$v = 2 - 0$ R(0)	4263.83734	1.5(17)
CN	A$^2\Pi$–X$^2\Sigma^+$ (0,0) R$_1$(0)	9094.3192	6.6(12)
	$v = 1 - 0$	2046.13	2.6(15)
	$v = 2 - 0$	4062.26	8.0(15)
CH$^+$	$v = 1 - 0$ R(0)	2766.5341	3.4(16)
CH	$v = 1 - 0$ R$_2$(1/2)	2796.883,2797.008	6.0(14)
CH$_2$	$\nu_1 = 1 - 0$	3020.7	
	$\nu_3 = 1 - 0$	3152.5	
CH$_3$	$\nu_3 = 1 - 0$ $1_1 - 0_0$	3174.2935	6.0(15)
CH$_4$	$\nu_3 = 1 - 0$	3019.	9.4(14)
HC$_3$N	$\nu_1 = 1 - 0$ R(0)	3327.680	
C$_2$	A$^1\Pi_u$–X$^1\Sigma_g^+$ (0,0) Q(2)	8267.1336	1.6(13)
	A$^1\Pi_u$–X$^1\Sigma_g^+$ (1,0) Q(2)	9851.1169	1.4(13)
	A$^1\Pi_u$–X$^1\Sigma_g^+$ (2,0) Q(2)	11410.84115	2.3(13)
C$_3$	$\nu_3 = 1 - 0$	2042.665	9.7(13)
C$_2$H	$\tilde{A}^2\Pi$(0,0,0)–$\tilde{X}^2\Sigma^+$(0,0,0)	3772–4108	1.0(14)
C$_2$H$_2$	$010(11)^0_+$ R(1)	3286.585	2.4(15)
	0010^00^0 R(1)	3299.521	2.9(15)
	1000^01^1 Q(1)	4091.170	6.2(16)
	0100^03^1 Q(1)	4138.883	
HCO$^+$	$\nu_1 = 1 - 0$ R(0)	3091.679	
HCN	$\nu_3 = 1 - 0$	3311.	1.0(15)
DCN	$\nu_3 = 1 - 0$	2630.	1.8(15)
C^{32}S	$v = 1 - 0$ R(0)	1273.7898	
OH	$v = 1 - 0$ Q$_1$(3/2)	3568.47	9.1(15)
NH$_2$	$\nu_1 = 1 - 0$ $1_{11} - 0_{00}$	3250.54	
	$\nu_3 = 1 - 0$ $1_{01} - 0_{00}$	3322.50	
SH	$v = 1 - 0$	2598.03	
HF	$v = 1 - 0$ R(0)	4000.9894	6.4(14)

Note: The characteristic column density, N_0, is that required to produce a feature with an equivalent width of 0.01 cm^{-1} in the weak–line limit. Each transition is designated either as a band or as an individual line. In the former case, the characteristic column density refers to a sum over all lower rotational levels; in the latter case, the column density is that in the lower level of the line.

from ab initio theoretical calculations remain factors of 1.5 – 2 larger (van Dishoeck 1983; Chabalowski et al. 1983; Pouilly et al. 1983; ONeil et al. 1987; Klotz 1987). Theory and experiment are similarly discrepant for the A–X system of CN (Cartwright and Hay 1982; Larsson *et al.* 1983; Davis *et al.* 1986; Taherian and Slanger 1984; Sneden and Lambert 1982), although very recent calculations (Bauschlicher *et al.* 1988) come closer to experiment. Since the oscillator strength of the B–X system is well established, consistency tests can be carried out by astronomical spectroscopy for a few regions where both the violet and red system lines of interstellar CN are observed.

There remains an unresolved disagreement between two lifetime measurements for the $\tilde{A}$–$\tilde{X}$ system of H_2O^+ (Möhlmann *et al.* 1978; Curtis and Erman 1977), which is significant in view of the very low limits that have been placed on the strengths of its interstellar absorption lines in several clouds (see Federman 1987). This is important for the study of cometary spectra as well (see Lutz 1987).

In the Lyman and Werner systems in the ultraviolet spectrum of H_2, non–adiabatic rotational perturbations affect the oscillator strengths of individual rotational lines: the calculated line strengths of Ford (1975) have recently been extended to a larger number of transitions by Abgrall *et al.* (1987). Accurate transition probabilities in the ν_2 vibration-rotation band of H_3^+ have been calculated by Miller and Tennyson (1988).

Another important aspect of quantitative spectroscopy concerns information that is required for the full description of processes like photodissociation. One example of much current interest is the photodissociation of CO, which evidently occurs principally through pre–dissociating lines rather than by continuous absorption in the interstellar medium. Information on line positions, isotopic shifts, oscillator strengths, and pre–dissociation widths is becoming available as a result of diligent efforts by Eidelsberg *et al.* (1984), Letzelter *et al.* (1987), Stark *et al.* (1987), and Yoshino *et al.* (1988). The new spectroscopic data have already been applied to theoretical models of the CO chemistry in circumstellar envelopes (Mamon *et al.* 1988) and in interstellar clouds (Viala *et al.* 1988; van Dishoeck and Black 1988b). The rate of photodissociation of CO depends upon the intensity of ultraviolet starlight and on the oscillator strengths of the dissociating lines, while the depth–dependence of the absorption is governed partly by the line broadening which has a signicant contribution from the pre–dissociation itself. Observationally, the translucent clouds may provide the most important tests of our understanding of the CO photochemistry in that they occupy the realm of parameter space in which CO is just beginning to account for a large fraction of the gas–phase carbon and oxygen. Ultraviolet absorption line observations with the High Resolution Spectrograph on *HST* promise to be of some value in this regard since there are many strong bands of the $A^1\Pi$–$X^1\Sigma^+$ 4th–positive system at wavelengths as long as 1544 Å. On the other hand, these absorption features will be strongly saturated in translucent clouds, and the interpretation will be quite complicated owing to the dense rotational structure of these bands, which will at best be barely resolved at the highest resolution of the High Resolution Spectrograph. The fundamental and first overtone vibration–rotation bands of CO at 4.6 and 2.3 μm will eventually provide an important alternative source of information on the abundance and excitation of this molecule. Although atomic spectroscopy is outside the purview of this review, the chemistry of simple carbon–bearing molecules cannot be fully understood without knowledge of the abundances of C and C^+. *HST* is expected to make possible sensitive interstellar absorption measurements of the C II] intersystem multiplet near 2325 Å; therefore, it is important to determine the small oscillator strengths of these lines as accurately as possible (Nussbaumer *et al.* 1981; Stencel *et al.* 1981; Hobbs *et al.* 1982; Cowan *et al.* 1982; Lennon *et al.* 1985).

Preparation of this review has been supported in part by NASA through grant NAGW-763 to the University of Arizona. E. F. van Dishoeck provided very helpful comments.

REFERENCES

Abgrall, H., Launay, F., Roueff, E., and Roncin, J.-Y. 1987, J. Chem. Phys., 87, 2036.
Bauer, W., Becker, K. H., Hubrich, C., Meuser, R., and Wildt, J. 1985, Ap. J., 296, 758.
Bauer, W., Becker, K. H., Bielefeld, M., and Meuser, R. 1986, Chem. Phys. Lett., 123, 33.
Bauschlicher, C. W., Langhoff, S. R., and Taylor, P. R. 1988, Ap. J., 332, 531
Bernath, P. F. 1987, J. Chem. Phys., 86, 4838.
Black, J. H. 1985, in Molecular Astrophysics: State of the Art and Future Directions, G. W. F.
    Diercksen, W. F. Huebner, and P. W. Langhoff, editors, (Dordrecht: Reidel), p. 215.
Black, J. H. 1988, Adv. Atom. Mol. Phys., 25, in press.
Black, J. H., and van Dishoeck, E. F. 1988a, Ap. J., 331, 986.
Black, J. H., and van Dishoeck, E. F. 1988b, to appear in Dissociative Recombination, J. B. A.
    Mitchell and S. Guberman, editors, (Singapore: World Scientific), in press.
Black, J. H., and Willner, S. P. 1984, Ap. J., 279, 673.
Brasier, C. R., and Brown, J. M. 1984, Canadian J. Phys., 62, 1563.
Brasier, C. R., O'Brien, L. C., and Bernath, P. F. 1987, J. Chem. Phys., 86, 3078.
Cartwright, D. C., and Hay, P. J. 1982, Ap. J., 257, 383.
Chabalowski, C. F., Peyerimhoff, S. D., and Buenker, R. J. 1983, Chem. Phys., 81, 57.
Chang, K. W., and Graham, W. R. M. 1982, J. Chem. Phys., 77, 4300.
Clegg, R. E. S., and Lambert, D. L. 1982, M. N. R. A. S., 201, 723.
Cox, P., Güsten, R., and Henkel, C. 1988, Astr. Ap., in press.
Cowan, R. D., Hobbs, L. M., and York, D. G. 1982, Ap. J., 257, 373; 265, 582.
Crane, P., Hegyi, D. J., Mandolesi, N., and Danks, A. C. 1986, Ap. J., 309, 822.
Crofton, M. W., Jagod, M.-F., Rehfuss, B. D., and Oka, T. 1987, J. Chem. Phys., 86, 3755.
Curl, R. F., Carrick, P. G., and Merer, A. J. 1985 J. Chem. Phys., 82, 3479.
Curtis, L. J., and Erman, P. 1977, J. Opt. Soc. Am., 67, 1218.
Davis, S. P., Smith, W. H., Brault, J. W., Pecyner, R., and Wagner, J. 1984, Ap. J., 287, 455.
Davis, S. P., Shortenhaus, D., Stark, G., Engleman, R., Phillips, J. G., and Hubbard, R. P. 1986,
    Ap. J., 303, 892.
d'Hendecourt, L. B., and Allamandola, L. J. 1986, Astr. Ap. Suppl., 64, 453.
Douay, M., Nietmann, R., and Bernath, P. F. 1988, J. Mol. Spectrosc., in press.
Eidelsberg, M., Launay, F., Rostas, F., Le Floch, A., Breton, J., and Thieblemont, B. 1984, Ann.
    Isr. Phys. Soc. (Israel), 6, 240.
Engleman, R. 1974, J. Mol. Spectrosc., 49, 106.
Erman, P., Lambert, D. L., Larsson, M., and Mannfors, B. 1982, Ap. J., 253, 983.
Federman, S. R. 1987, in IAU Symposium 120, Astrochemistry, eds. M.S. Vardya and S.P. Tarafdar
    (Reidel, Dordrecht), p. 123.
Ford, A. L. 1975, J. Mol. Spectrosc., 56, 251.
Forney, D., Althaus, H., and Maier, J. P. 1987, J. Phys. Chem., 91, 6458.
Geballe, T. R. 1986, Astr. Ap., 162, 248.
Geballe, T. R., Baas, F., Greenberg, J. M., and Schutte, W. 1985, Astr. Ap., 146, L6.
Geballe, T. R., and Wade, R. 1985, Ap. J. (Letters), 291, L55.
Hagen, W., Allamandola, L. J., and Greenberg, J. M. 1980, Astr. Ap., 86, L3.
Hall, D. N. B., Kleinmann, S. G., Ridgway, S. T., and Gillett, F. C. 1978, Ap. J. (Letters), 223,
    L47.
Hartquist, T. W., and Dalgarno, A. 1980, in Giant Molecular Clouds in the Galaxy, P. M. Solomon
    and M. G. Edmunds, editors, (Oxford: Pergamon), p. 315.
Heath, J. R., Curl, R. F., and Smalley, R. E. 1987, J. Chem. Phys., 87, 4236.
Hinkle, K. H., Keady, J. J., and Bernath, P. F. 1988, Science, 241, 1319.
Hobbs, L. M., York, D. G., and Oegerle, W. 1982, Ap. J. (Letters), 252, L21.
Jensen, P., and Kraemer, W. P. 1988, J. Mol. Spectrosc., 129, 216.
Josafatsson, K., and Snow, T. P. 1987, Ap. J., 319, 436.
Klots, R. 1987, private communication.
Knacke, R. F., Geballe, T. R., Noll, K. S., and Tokunaga, A. T. 1985, Ap. J. (Letters), 298, L67.
Kraemer, W. P., Bunker, P. R., and Yoshimine, M. 1984, J. Mol. Spectrosc., 107, 191.

338

Lacy, J. H., Baas, F., Allamandola, L. J., Persson, S. E., McGregor, P. J., Lonsdale, C. J., Geballe, T. R., and van de Bult, C. E. P. 1984, *Ap. J.*, **276**, 533.
Larson, H. P., Davis, D. S., Black, J. H., and Fink, U. 1985, *Ap. J.*, **299**, 873.
Larsson, M., Siegbahn, P. E. M., and Ågren, H. 1983, *Ap. J.*, **272**, 369.
Lennon, D. J., Dufton, P. L., Hibbert, A., and Kingston, A. E. 1985, *Ap. J.*, **294**, 200.
Lepp, S., and Dalgarno, A. 1988 *Ap. J.*, **324**, 553.
Letzelter, C., Edielsberg, M., Rostas, F., Breton, J., and Thieblemont, B. 1987, *Chem. Phys.*, **114**, 273.
Lutz, B. L. 1987, *Ap. J. (Letters)*, **315**, L147.
Mamon, G. A., Glassgold, A. E., and Huggins, P. J. 1988, *Ap. J.*, **328**, 797.
Matsumura, K., Kanamori, H., Kawaguchi, K., and Hirota, E. 1988, *J. Chem. Phys.*, **89**, 3491.
Meyer, D. M., and Hawkins, I. 1987, *Bull. Am. Astr. Soc.*, **19**, 1054.
Meyer, D. M., and Jura, M. 1985, *Ap. J.*, **297**, 119.
Miller, S., and Tennyson, J. 1988, *Ap. J.*, in press.
Möhlmann, G. R., Bhutani, K. K., de Heer, F. J., and Tsurubuchi, S. 1978, *Chem. Phys.*, **31**, 273.
Morton, D. C. 1978, *M. N. R. A. S.*, **184**, 713.
Nussbaumer, H., and Storey, P. J. 1981, *Astr. Ap.*, **96**, 91.
Oka, T. 1981, *Phil. Trans. R. Soc. London A*, **303**, 543.
O'Keefe, A., Derai, R., and Bowers, M. T. 1984 *Chem. Phys.*, **91**, 161.
Omont, A. 1986, *Astr. Ap.*, **164**, 159.
ONeil, S. V., Rosmus, P., and Werner, H.-J. 1987, *J. Chem. Phys.*, **87**, 2847.
Oppenheimer, M., and Dalgarno, A. 1977, *Ap. J.*, **212**, 683.
Pouilly, B., Robbe, J. M., Schamps, J., and Roueff, E. 1983, *J. Phys. B*, **16**, 437.
Reimers, J. R., Wilson, K. R., Heller, E. J., and Langhoff, S. R. 1985, *J. Chem. Phys.*, **82**, 5064.
Reutt, J. E., Wang, L. S., Pollard, J. E., Trevor, D. J., Lee, Y. T., and Shirley, D. A. 1986, *J. Chem. Phys.*, **84**, 3022.
Römelt, J., Peyerimhoff, S. D., and Buenker, R. J. 1978, *Chem. Phys. Lett.*, **58**, 1.
Rosmus, P., Werner, H.-J., Reinsch, E.-A., and Larsson, M. 1986, *J. Elec. Spectrosc. Related Phen.*, **41**, 289.
Rösslein, M., Wyttenbach, M., and Maier, J. P. 1987, *J. Chem. Phys.*, **87**, 6770.
Sandford, S. A., Allamandola, L. J., Tielens, A. G. G. M., and Valero, G. J. 1988, *Ap. J.*, **329**, 498.
Scoville, N. Z., Kleinmann, S. G. Hall, D. N. B., and Ridgway, S. T. 1983, *Ap. J.*, **275**, 201.
Shinn, J. L. 1982, in "Thermophysics of Atmospheric Entry", T. E. Horton, editor, *Prog. Astronaut. Aeronaut.*, **82**, 68.
Smith, P. L. 1987, in IAU Symposium **120**, *Astrochemistry*, eds. M.S. Vardya and S.P. Tarafdar (Reidel, Dordrecht), p. 95.
Sneden, C., and Lambert, D. L. 1982, *Ap. J.*, **259**, 381.
Snow, T. P. 1980, in IAU Symposium **87**, *Interstellar Molecules*, ed. B.H. Andrew (Reidel, Dordrecht), p. 247.
Snow, T. P., Seab, C. G., and Joseph, C. L. 1988, *Ap. J.*, submitted.
Stark, G., Smith, P. L., Yoshino, K., and Parkinson, W. H. 1987, private communication, and in preparation.
Stencel, R. E., Linsky, J. L., Brown, A., Jordan, C., Carpenter, K. G., Wing, R. F., and Czyzak, S. 1981, *M. N. R. A. S.*, **196**, 4P.
Swings, P., and Rosenfeld, L. 1937, *Ap. J.*, **86**, 483.
Taherian, M. R., and Slanger, T. G. 1984, *J. Chem. Phys.*, **81**, 3814.
van Dishoeck, E. F. 1983, *Chem. Phys.*, **77**, 277.
van Dishoeck, E. F., and Black, J. H. 1986, *Ap. J. Suppl.*, **62**, 109.
van Dishoeck, E. F., and Black, J. H. 1987, in *Physical Processes in Interstellar Clouds*, eds. G. Morfill and M.S. Scholer (Reidel, Dordrecht), p. 241.
van Dishoeck, E. F., and Black, J. H. 1988a, in *Rate Coefficients in Astrochemistry*, T. J. Millar and D. A. Williams, editors, (Dordrecht: Kluwer), p. 209.
van Dishoeck, E. F., and Black, J. H. 1988b, *Ap. J.*, **334**, in press.
van Dishoeck, E. F., and Black, J. H. 1989, *Ap. J.*, **340**, in press.
Viala, Y. P., Letzelter, C., Eidelsberg, M., and Rostas, M. 1988, *Astr. Ap.*, **193**, 265.
Whittet, D. C. B., Longmore, A. J., and McFadzean, A. D. 1985, *M. N. R. A. S.*, **216**, 45P.
Williams, D. A. 1985, *Quart. J. R. Astr. Soc.*, **26**, 463.
Yoshino, K., Stark, G., Smith, P. L., Parkinson, W. H., and Ito, K. 1988, *J. de Phys.*, in press.

MICROWAVE SPECTROSCOPY OF ASTROPHYSICAL MOLECULES

WILLIAM M. IRVINE
Five College Radio Astronomy Observatory
University of Massachusetts
Amherst, Massachusetts 01003
USA

ABSTRACT. Recent detections of new molecules in dense interstellar clouds, first detections of certain chemical elements in interstellar molecules, and new information on isotopic fractionation of hydrogen in the interstellar medium are discussed in the context of the need for new laboratory data on transition rest frequencies, reaction rates, and branching ratios.

1. Introduction

This brief review will concentrate on some recent results in the chemistry of molecular clouds, in order to provide some examples of areas where new laboratory data are needed. I will not discuss laboratory microwave measurements in themselves. For a more complete description of recent progress, see reviews by Irvine (1988), Irvine and Knacke (1988), Turner and Ziurys (1988), Hjalmarson and Friberg (1988), and Guélin (1985).
    The discovery of new molecular constituents in these regions continues apace, with some 13 or 14 new species having been identified in the past couple of years. This brings to about 80 the number of molecular species known in dense clouds and in the envelopes expelled by evolved stars, including molecules with up to 13 atoms and weights up to about twice that of the simplest amino acid, glycine (cf. Irvine, 1988). The striking deviations from chemical equilibrium which characterize the interstellar medium are evident in the considerable number of highly unsaturated species (even in this very "reducing" environment) and in the large number of ions and radicals. The richness of the spectral material with which molecular astrophysicists now have to deal is exemplified by the detection of more than 800 individual transitions in the spectral survey of the Orion molecular cloud carried out at the Owens Valley Radio Observatory (Blake et al., 1987). Interestingly enough, the vast majority of these lines have been assigned to known molecular species. In fact, the proliferation of rotational transitions from asymmetric rotors and their isotopic variants threatens to create a confusion-limited spectrum at short millimeter and submillimeter wavelengths, for the richest molecular sources.

339

D. McNally (ed.), Highlights of Astronomy, Vol. 8, 339–344.
© 1989 by the IAU.

2. New Interstellar Molecules

At the time of the last IAU General Assembly, interstellar molecules (excluding near-photospheric species) were known which contained the elements H, C, N, O, S, Si, and probably Cl. Since that time the new 30m IRAM telescope has been used to discover several halides in the envelope surrounding the carbon star IRC+10216, a topic to be discussed later today by Professor Omont (see also Cernicharo and Guélin, 1987); and the first interstellar molecule containing phosphorus has been detected. Although phosphorus is essentially undepleted in diffuse interstellar clouds, it has heretofore been searched for unsuccessfully in denser regions. It is an element of personal interest to all of us, since biological material contains an anomalously high amount of phosphorus relative to terrestrial and cosmic abundances. The detection of PN in several molecular clouds by Ziurys (1987) and by Turner and Bally (1987) is thus an important development. It is important to note, however, that the bulk of the cosmic complement of phosphorus remains undetected in these regions and that the failure to observe PO is puzzling (cf. Thorne et al., 1984).

Abundance comparisons between chemical isomers may provide particularly useful tests of astrochemical models. The classical example is HCN and HNC, where the large HNC abundance in cold clouds has traditionally been one of the main arguments in favor of the importance of ion-molecule chemical schemes (Irvine et al., 1985; Goldsmith et al., 1986). Although the situation is more complicated, the pair CH_3CN/CH_3NC is somewhat analogous. The recent detection of the latter species by Cernicharo et al. (1988) in Sgr B2 with an abundance relative to CH_3CN in good agreement with theory is thus an important result.

The rotational spectra for the molecules discussed above were well known from laboratory work. In contrast, there continue to be a number of new molecular species for which the first observations of rotational transitions have been made with the radio telescope, and the detected unidentified lines have only subsequently been assigned on the basis of new laboratory measurements. For example, a number of initially unidentified emission lines shared the property of being weak in the rich molecular sources Sgr B2 and IRC+10216, apparently absent in Orion A, but strong in the cold, dark cloud TMC-1. Identification finally came on the basis of collaborative observational work at the Nobeyama Radio Observatory and laboratory spectroscopy at Nagoya University, and resulted in the detection of a new class of molecules, sulfur-containing carbon chains. Two new species were identified, C_2S and C_3S (Saito et al., 1987; Yamamoto et al., 1987). Both cases are unusual relative to previous sulfur-containing organic molecules in molecular clouds in that both are considerably more abundant than the oxygen-containing analogs (C_2O and C_3O). It has been suggested that this reflects the failure of the S+ ion to react with H_2, unlike the situation for O+ and H_2, leaving S+ free to insert itself in acetylene and polyacetylene species to produce C_2S and C_3S precursors (Smith et al., 1988). As with much of interstellar sulfur chemistry, however, there is a need for additional laboratory measurements of relevant reaction rates.

Hydrocarbon radicals of increasing length continue to be found in dense clouds: all of the species CnH for n=1-6 have now been observed. Interestingly enough, all of these molecules were observed astronomically before the corresponding spectra were studied in the laboratory. The newest addition to the group is C_6H, identified independently by Suzuki et al. (1986) at the Nobeyama Radio Observatory and Cernicharo et al. (1986) at IRAM and Bonn. Unlike the cyanopolyynes, which show a smooth fall-off in abundance with increasing length, the abundances of these hydrocarbon radicals show large variations which may be related to the ground electronic state. To verify this, laboratory measurements of the appropriate reaction rates are needed.

The first known interstellar hydrocarbon ring, C_3H_2, has proven to be widespread in the galaxy and promises to be a useful probe of both chemistry and physics in dense clouds (e.g., Madden, Irvine, and Matthews, 1986). It is hypothesized that C_3H_2 is produced by dissociative electron recombination with the precursor ion H_3C_3+. As is so often the case, branching ratios among possible exit channels for such a recombination are not known and are extremely important. It is hypothesized that one such channel would lead to C_3H, the linear form of which has been established in the interstellar medium (Thaddeus et al., 1985). Further information on the stereochemical processes involved are now provided by the identification of cyclic C_3H in TMC-1, the third ring molecule to be detected astronomically (Yamamoto et al., 1987). In this case laboratory work preceded the astronomical search and detection.

A classic (and highly unusual) example of the interaction of theory, laboratory work, and astronomical observation led a few years ago to the detection of interstellar tricarbon monoxide (C_3O); quantum chemical calculations established the structure and stability of this species, enabling accurate laboratory measurements of the rotational spectrum, which in turn allowed the astronomical detection and an accurate abundance determination. The research is also an illuminating example of international cooperation in astronomy, as it involved scientists in Australia, the United States, Canada, Japan, and Sweden (Brown et al., 1985). C_3O is thought to be produced by a set of reactions which culminate in the radiative association of CO and hydrocarbon ions followed by dissociative recombination (for example, of H_3C_3O+) to produce C_3O. No direct laboratory measurements of the relevant radiative associations are available, although some inferences on rates can be made from related three-body reactions. Laboratory data are also lacking on the branching ratios for the final dissociative recombination, but some theories would predict that the primary product would be more hydrogenated than C_3O itself. This led to a search for two isomers of H_2C_3O, propadienone ($H_2C=C=C=O$) and propynal ($HC\equiv CCHO$). Propynal was successfully detected in the cold cloud TMC-1, which is known as a particularly rich source of acetylene derivatives (Irvine et al., 1988a). The relative abundances of C_3O, propadienone, and propynal place important constraints on the detailed chemical reactions involved and suggest relevant laboratory measurements.

Another recent "detective story" leading to the identification of unassigned interstellar emission lines began with the observation by Thaddeus et al. (1985) that two lines in the Bell Laboratories spectral survey of the galactic center molecular cloud Sgr B2 were harmonically related to each other and to a third line observed in TMC-1 by L. Avery. Subsequent observations at FCRAO, NRAO Green Bank, Onsala, and Nobeyama, and laboratory spectroscopy at Nagoya University, have resulted in the identification of the heaviest non-linear interstellar radical found to date, CH_2CN (Irvine et al., 1988b; Saito et al., 1988). The presence of the unpaired electron as well as the spins of the nitrogen and hydrogen nuclei combine to produce a very complicated spectrum for the lowest rotational transitions. Interestingly enough, the abundance of CH_2CN appears to be significantly greater than that of the closed-shell species CH_3CN, at least in TMC-1. This seems to be consistent with separate routes for the production of these two superficially related species, including perhaps a route to CH_2CN that is analogous to that suggested for the production of the cyanopolyynes (reaction of atomic nitrogen with hydrocarbon ions; e.g., Millar and Freeman, 1984).

3. Isotopic Fractionation and High Temperature Chemistry

One of the most characteristic features of interstellar chemistry is the high degree of isotopic fractionation observed for several elements in certain regions. The effect is particularly pronounced for hydrogen, and is reasonably well understood in cold clouds (e.g., Wootten, 1987). Processes that can lead to significant deuterium enhancement in warmer clouds are discussed today by Millar. The effect may provide a link between the formation of the solar system and the molecular cloud where that event occurred, since significant deuterium enhancement is found in the organic components in carbonaceous chondrites (e.g., Epstein et al., 1987).

Although deuterium enhancement is found in virtually every hydrogen-containing interstellar molecule for which data is available (Irvine and Knacke, 1988), the largest effect measured accurately to date may be that in the cyclic species C_3H_2. The C_3HD/C_3H_2 ratio appears to be as much as 10,000 times the cosmic D/H ratio in cold, dark clouds (Bell et al., 1988; Gerin et al., 1987). It has been suggested that the effect may be enhanced by the preferential loss of H relative to D upon the recombination of the precursor ion C_3H_2D+, but laboratory measurements are needed.

Important clues to interstellar chemical processes are provided by the differences in chemical abundances observed under differing physical conditions. Ion-molecule processes are the paradigm for cold, dark clouds. On the other hand, in active star-formation regions a variety of other processes may occur including evaporation of volatile grain mantles and perhaps sputtering of grains themselves (e.g., Blake et al., 1987). Various models of "shock" or high-temperature chemistry have been proposed, and an interesting test appears to be provided by the abundance of the SiO molecule. Unlike the situation for a variety of other molecular species studied, it is unequivocal that SiO is enhanced

by at least two orders of magnitude in dense clumps associated with the
supernova remnant IC443, as well as in active regions of massive star
formation (Ziurys, Snell, and Dickman, 1988).

4. Conclusion

The observations described above and their interpretations point out the
need for laboratory data in a number of areas, including:
-Rotational frequencies for interstellar molecular species on which no
data currently exist. This tends to be the case for radicals, ions, and
other species which are "unstable" by terrestrial laboratory standards,
as well as for heavier molecules which are either of intrinsic interest
or whose presence might reasonably be extrapolated from known
constituents of molecular clouds.
-In order to determine abundances from observational spectra, dipole
moments are needed. These can normally be obtained with contemporary
quantum-mechanical computer codes, even for the exotic molecular species
characteristic of interstellar chemistry (e.g., recent calculations for
C_2S and cyclic C_3H; Saito et al., 1987; Yamamoto et al., 1987).
-For a number of important reactions branching ratios among possible
exit channels are needed. This is particularly the case for the
dissociative recombinations which are the final step in the production
of many neutral interstellar molecules. Such measurements would help to
resolve discrepancies between theoretical calculations (cf. discussion
in Herbst, 1988).
-Reaction rates are currently lacking for many potentially significant
reactions, including particularly processes of radiative association,
but also for many neutral-neutral reactions. Moreover, measurements at
the low temperatures characteristic of molecular clouds are needed in a
large number of cases for which only room temperature data are
available.

5. References

Bell, M.B. et al. (1988) Astrophys. J. 326, 924.
Blake, G.A., Sutton, E.C., Massen, C.R., and Phillips, T.G. (1987)
  Astrophys. J. 315, 621.
Brown, R.D. et al. (1985), Astrophys. J. 247, 302.
Cernicharo, J., Guélin, M., Menten, K.M., and Walmsley, C.M. (1986)
  Astron. Astrophys. 172, L5.
Cernicharo, J. and Guélin, M. (1987) Astron. Astrophys. 183, L10.
Cernicharo, J., Kahane, C., Guélin, M., Gomez-Gonzalez, J. (1988)
  Astron. Astrophys. 189, L1.
Epstein, S., Krishnamurthy, R.V., Cronin, J.R., Pizzarello, S., and
  Yuen, G.U. (1987) Nature 326, 477.
Gerin, M. et al. (1987) Astron. Astrophys. 173, L1.
Coldsmith, P.F., Irvine, W.M., Hjalmarson, Å., and Elldér, J. (1986)

Astrophys. J. 310, 383.
Guélin, M. (1985) in G. Diercksen, W. Huebner, and P. Langhoff (eds.), Molecular Astrophysics, D. Reidel, Dordrecht, pp. 23-44.
Herbst, E. (1988) in T.J. Millar and D.A. Williams (eds.), Reactive Rate Coefficients in Astrophysics, Kluwer, Dordrecht, in press.
Hjalmarson, Å. and Friberg, P. (1988) in A. Dupree and Lago (eds.), Formation and Evolution of Low Mass Stars, D. Reidel, Dordrecht, in press.
Irvine, V.M., Schloerb, F.P., Hjalmarson, Å., and Herbst, E. (1985) in D.C. Black and M.S. Matthews (eds.), Protostars and Planets II, Univ. Arizona Press, Tucson, pp. 579-620.
Irvine, V.M. (1988) Adv. Space Res., in press.
Irvine, V.M. and Knacke, R.F. (1988) in S.K. Atreya, J.B. Pollack and M.S. Matthews (eds.), Origin and Evolution of Planetary and Satellite Astrophysics, Univ. Arizona Press, Tucson, in press.
Irvine, V.M. et al. (1988a) Astrophys. J. (Lett.), submitted.
Irvine, V.M. et al. (1988b) Astrophys. J. (Lett.), in press.
Madden, S.C., Irvine, V.M., and Matthews, H.E. (1986) Astrophys. J. 311, L27.
Millar, T.J. and Freeman, A. (1984) M.N.R.A.S. 207, 405.
Saito, S., Kawaguchi, K., Yamamoto, S., Ohishi, M., Suzuki, H., and Kaifu, N. (1987) Astrophys. J. (Lett.) 317, L115.
Saito, S., Yamamoto, S., Irvine, V.M., Ziurys, L.M., Suzuki, H., Ohishi, M., and Kaifu, N. (1988) Astrophys. J. (Lett.), in press.
Smith, D., Adams, N.G., Giles, K., and Herbst, E. (1988) Astron. Astrophys. 200, 191.
Suzuki, H., Ohishi, M., Kaifu, N., Ishikawa, S., and Kasuga, T. (1986) Publ. Astron. Soc. Japan 38, 911.
Thaddeus, P., Vrtilek, J.M., and Gottlieb, C.A. (1985) Astrophys. J. (Lett.) 299, L63.
Thaddeus, P., Gottlieb, C.A., Hjalmarson, Å., Johansson, L.E.B., Irvine, V.M., Friberg, P., and Linke, R.A. (1985) Astrophys. J. (Lett.) 294, L49.
Thorne, L.R., Anicich, V.G., Prasad, S.S., and Huntress, W.T. (1984) Astrophys. J. 280, 139.
Turner, B.E. and Bally, J. (1987) Astrophys. J. (Lett.) 321, L75.
Turner, B.E. and Ziurys, L.M. (1988) in G.L. Verschuur and K.I. Kellermann (eds.), Galactic and Extragalactic Radio Astronomy, Springer-Verlag, Berlin, pp. 200-254.
Wootten, A. (1987) in M.S. Vardya and S.P. Tarafdar (eds.), Astrochemistry, D. Reidel, Dordrecht, pp. 311-320.
Yamamoto, S., Saito, S., Kawaguchi, K., Kaifu, N., Suzuki, H., and Ohishi, M., (1987) Astrophys. J. (Lett.) 317, L119.
Yamamoto, S., Saito, S., Ohishi, M., Suzuki, H., Ishikawa, S.I., Kaifu, N., and Murakami, A. (1987) Astrophys. J. (Lett.) 322, L55.
Ziurys, L.M. (1987) Astrophys. J. (Lett.) 321, L81.
Ziurys, L.M., Snell, R.L., and Dickman, R.L. (1988) Astrophys. J., submitted.

SOME SALIENT FEATURES OF EVOLVING MODELS OF INTERSTELLAR CLOUDS

S. P. TARAFDAR AND S. K. GHOSH
Tata Institute of Fundamental Research
Homi Bhabha Road
Colaba, Bombay 400005
INDIA

K. R. HEERE
Science Applications International Corporation
Los Altos, California
USA

S. S. PRASAD
Space Science Center
University of Southern California
University Park
Los Angeles, CA
USA

ABSTRACT.   Difficulties faced by various models of interstellar clouds
have been discussed. A new evolutionary model which uses energy equation instead of empirical temperature–density relation used in earlier
models has been presented. This calculation shows that for a given
initial density, the collapsing cloud has   a minimum mass which is
significantly smaller than the Jean's mass. The clouds with larger
mass than the critical mass continue collapsing and physical and
chemical evolution remain similar to earlier evolving models. Clouds
with mass smaller than the critical mass initially collapse but
ultimately bounce back, producing physically similar clouds in
collapsing and expanding phases. The chemical evolution in these two
physically similar clouds is different mainly due to differences in
their lifetime. The lifetime of this oscillating cloud is also longer
than the collapsing cloud.

1. INTRODUCTION

This review paper addresses recent attempts to find a common evolutionary thread through the vast diversity of interstellar clouds. We
will see how evolutionary models try to use the observed chemical composition of interstellar clouds for inferring their dynamical destiny.
As we go through this review it will become apparent that the the the
classical constant density–constant temperature models of interstellar
cloud chemistry may be inadequate because dynamical motion, both

345

quiescent and impulsive, prevent chemistry from attending equilibrium
at constant density and temperature.

   To put this review in proper perspective, let us start by noting
that our understanding of the interstellar medium in general and of
interstellar clouds in particular has taken a big step forward in the
last two decades because of observations from space platforms like COP-
ERNICUS and IUE in ultraviolet and discoveries of a large number of molecules from millimetre wave observations from the ground. The observations of molecules containing as much as 13 atoms in a low density
($<10^7$ cm^{-3}) and low temperature ($\sim10°$K) interstellar clouds have thrown
a challenge to astrochemists to provide an adequate theory of formation
of these molecules. Astrochemists met this challenge by inventing an
ion-molecular reaction scheme which is fast enough even at low
densities to provide a workable scenario of formation of molecules in
interstellar clouds. To determine the molecular abundances for comparison with observation it is necessary to know physical parameters
like density and temperature for the cloud. To start with it was simply
assumed to be of constant density and constant temperature (homogeneous and isothermal). The cloud was also assumed to be in steady
state which was relaxed later to allow time dependence. Since these
early models, models with varying complexity have been built to represent an interstellar cloud as realistically as possible. Most of
the models now present can broadly be classified into four categories:
   (A) constant density temperature (i.e. homogeneous isothermal)
       steady state models.
   (B) constant density constant temperature (homogeneous isothermal
       pseudo-time dependent models
   (C) hydrostatic (turbulence supported) models
   (D) evolutionary models.
Details of models (A)-(C) can be found in recent papers by van Dishoeck
and Black (1986), Viala (1986), Viala, Roueff and Abgrall (1987)
for diffuse clouds, and in papers by Prasad and Huntress (1980), de
Jong, Dalgarno and Boland (1980), Boland and de Jong (1982), Millar and
Freeman (1984), Millar and Nejud (1985), Herbst and Leung (1986a,b), Graedel,
Langer and Frerking (1982), Watt (1983), Suzuki (1983) for dense clouds. Model
(D) can be found in Gerola and Glasgold (1978), and . Tarafdar et al. (1985).
Different models have been reviewed recently by Langer and Graedel
(1987), Black (1987), Hartquist (1987), Wootten (1987), Dalgarno
(1987a,b), Herbst (1987), Winnewisser and Herbst (1987), Prasad
(1987), Prasad et al (1987) and van Dishoeck (1988a). Further progress
in models (A)-(C) and the need of atomic and molecular data in
these models will be presented elsewhere in this volume by van Dishoeck
(1988b), Black (1988) and Millar (1988). Our aim is to present a
few salient features including advantages and difficulties of model
(D). But before doing this, let us present some difficulties faced
by models (A)-(C), as some of these will be present in model (D) also,
and subsequently we shall point out how these difficulties are either
removed or alleviated and some others turn out to be advantages in
model (D).

2. SOME DIFFICULTIES OF MODELS (A) TO (C).

A common difficulty in models (A) and (C) is the problem of depletion
of gas phase molecules onto grains. The time scale of this depletion
can be written as $t_d = (s\sigma_g n_g v)^{-1}$, where s is the sticking coefficient,
σ_g and n_g are respectively grain cross section and density and v is the
thermal velocity of molecules. With $\sigma_g n_g = 10^{-21} n_H$ where n_H is the
density of hydrogen nuclei and $v = 10^4$ cm^2 for interstellar molecules
heavier than H_2, $t_d = 3 \times 10^9 / (s n_H)$ yrs. As $s \approx 1$ for molecules heavier
than H_2 (Hollenbach and Salpeter 1970), $t_d \leq 10^7$ yrs for n_H 300cm^{-3}. The
lifetime of clouds is 10^7-10^8 yrs. Thus clouds with $n_H > 300$cm^{-3} will
be devoid any gas phase molecule unless some efficient desorption
mechanism is found to bring back molecules from grain surfaces.
Several desorption mechanisms (cf. Boland and de Jong 1982, d'Hendecourt et al. 1982, Williams and Hartquist 1984, Leger, Jura and Omont
1985) have been proposed, but it is not yet clear whether any one of
these or together can work in real clouds to avoid the depletion
problem in models (A)-(C). We shall see later how the depletion
problem is alleviated in evolving models (D).
A common problem present in constant density constant temperature
models (A) and (B) is the energy balance in the cloud. In the absence
of any internal stellar source in an interstellar cloud (i.e. when the
cloud is not circumstellar) the source of energy of the cloud is from
outside as cosmic rays and average ultraviolet radiations from 0- and
B-stars. The ultraviolet radiation intensity decreases from the cloud
surface inward due to grain attenuation. Thus the heating rate (Γ)
at an optical depth point Av can be written as

$$\Gamma = [\Gamma_c + \Gamma_{uv} \exp(-\alpha A_v)] n_H \text{ ergs cm}^{-3} \text{ s}^{-1} \tag{1}$$

where α is a constant, n_H is the density of hydrogen nuclei and Γ_e and
Γ_{uv} are respectively the heating rates due to cosmic rays and ultraviolet radiation at the cloud surface. The cooling of the interstellar
cloud is by line emission as a result of spontaneous decay of an upper
level excited collisionally. Hence the cooling rate (Λ) can be represented approximately as

$$\Lambda = \Lambda_0 n_H n \exp(-E/kT) \text{ ergs cm}^{-3} \text{ s}^{-1} \quad , \tag{2}$$

where k is the Boltzman constant, T is the temperature, E is the excitation energy, n is the density of the coolant and Λ_0 is a constant of
dimension ergs cm^3 s^{-1}. Equating the heating rate to the cooling rate
and putting $\Gamma_c = 0$ for simplicity, the temperature T at optical depth A_v
can be expressed as

$$T = \frac{E/R}{\alpha A_v + \ln \dfrac{\Lambda n}{\Gamma_u}} \tag{3}$$

Eq. (3) shows that T decreases as A_v increases. Thus for a homogeneous
cloud (i.e., n=constant), the temperature should decrease inward for a

realistic interstellar cloud. Therefore, isothermal assumption of models (A) and (B) cannot be right for an interstellar cloud.

The difficulty of maintaining constant temperature in homogeneous clouds becomes an advantage for evolving clouds, because for a homogeneous cloud the inward decrease of temperature gives an inward force which helps the gravity to collapse the cloud to a higher density. As density increases the inside temperature drops further (eq. 3) adding the gravity to collapse the cloud further. Thus the inward decrease of temperature in interstellar clouds is the key point of evolutionary models and helps in collapsing a cloud with much lower mass for given density or with much lower density for a given mass than that which is possible under Jean's criteria.

3. AIMS OF EVOLVING MODEL

The basic aims of evolving model is to seek answers to questions like how different variety of interstellar clouds came to the state where they are, do they change with time and if so in what way this change takes place, how does the cloud evolution fit with the evolutionary sequence of other set of objects like stars. For example we would like to find out whether a link exists between interstellar clouds from diffuse to dense phase. A link in the sense that one type of clouds evolves to another in time will be very satisfying as it will establish an order among variety of clouds from diffuse to dense and avoid arbitrary nature of their presence.

4. EVOLVING MODELS

The basic equations of evolving models are:

conservation of mass, $\dfrac{\partial m(r)}{\partial r} = 4\pi r^2 \rho(r)$ (4)

conservation of momentum, $\dfrac{\partial^2 r}{\partial t^2} + \dfrac{1}{\rho} \dfrac{\partial P}{\partial r} + \dfrac{Gm(r)}{r^2} = 0$ (5)

conservation of energy, $\dfrac{\rho}{\mu} \dfrac{d}{dt} (\tfrac{3}{2}kT) - \dfrac{kT}{\mu} \dfrac{d\rho}{dt} = \Gamma - \Lambda$ (6)

radiative transfer equation, $\dfrac{dI_\nu}{d\hat{c}_\nu} = -I_\nu + S$ (7)

and

rate equations (i-1,..n) $\dfrac{dn_i}{dt}$ = formation-destruction rates. (8)

These equations are supplemented by

equation of state, $P = R\rho T/\mu$ (9)

heating rate, $\Gamma = \Gamma(n_H, T)$ (10)

and

$$\text{Cooling rate, } \Lambda = \Lambda(n_H, n_i, T) \quad . \tag{11}$$

Ideally, equations (4)-(11) need to be solved simultaneously to obtain physical variables like density ρ, temperature, T, pressure, P, radiation intensity I_v and chemical composition n_i (i=1...n) as a function of space variable r and time t. However, such a solution of equations (4)-(11) will be very time-consuming, as chemical reate equation (8) has to be for a large number of species, even if we want to cover a limited number of observed molecules. Therefore, we need to make some drastic simplification. In order to do this without loosing the basic physics of interstellar clouds, in Tarafdar et al. (1985) the energy equation which connects the dynamical equations with chemical equations through the cooling rate has been replaced by an empirical relation between temperature, density and optical depth of the form:

$$T = 163/[2.5+\ln n_H - \ln\{+1+500 \exp(-1.8A_v)\}] \quad . \tag{12}$$

The different constants are so chosen to fit the dependence of T on A_v and n_H given by de Jong et al.(1980). Note that relation (12) maintains the physical behavior of T in interstellar clouds pointed out earlier that T decreases as A_v increases. Further, the radiative transfer equation was replaced by a simple solution of the form $I_v=I_v^{\circ} \exp(-\alpha_v A_v)$, where α_v was determined from interstellar extinction laws, I_v° being the unattenuated interstellar background field. The details of this calculation and result have been presented in Tarafdar et al. (1985).

Three conservation equations (4)-(6) have now been solved simultaneously instead of the assumption of empirical temperature relation (12), but assuming as before that $I_v = I_v^{\circ}$ $(\exp(-\alpha_v A_v)$ and cooling is only by CII and CO. Further individual cooling rates of CII and CO have been replaced by a common cooling which is the same as CII cooling, assuming $n(CII)/n_H = 7.5\times10^{-5}$ throughout the cloud. The solution of equations (4)-(6) then gives $\rho(r,t)$ and $T(r,t)$ which have been used to solve rate equations(9) including a large number of species involving about 2000 reactions starting with the initial condition that all hydrogen are in the form of HI, all carbon in the form of C^+, oxygen in the form of OI, and all nitrogen are in the form of NI. Note that the subsequent chemical results do not depend on these initial values which are appropriate for diffuse interstellar clouds, as chemical equilibrium sets in soon because of longer dynamical time scale of the diffuse phase.

5. SOME RESULTS OF EVOLVING MODELS

5.1 Lowering of Jean's Mass

Now we present a couple of significant results. The first task confronting evolutionary models is the determination of the range of their applicability. For this purpose we must examine whether lower limit of

350

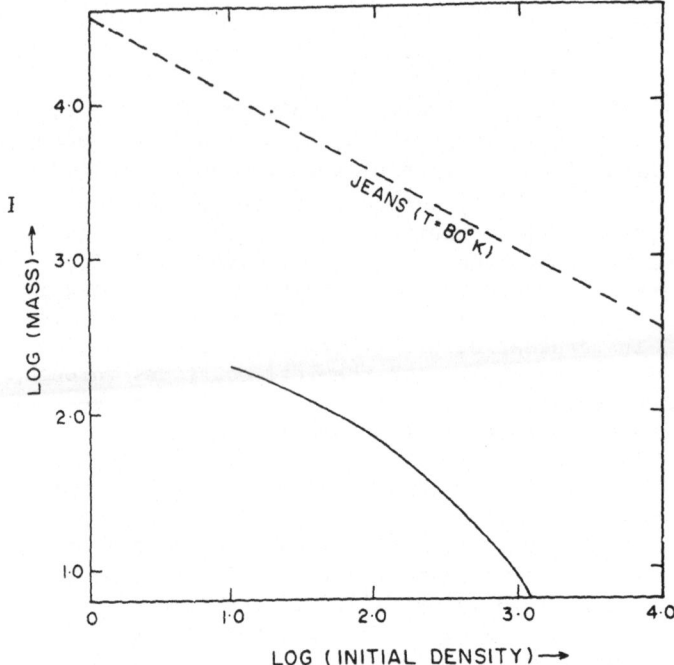

FIG. 1: VARIATION OF MIMIMUM MASS OF COL-LAPSING CLOUD AS A FUNCTION OF DENSITY: CONTINUOUS LINE-PRESENT CALCULATION AND DASHED CURVE-JEANS MASS AT 80°K.

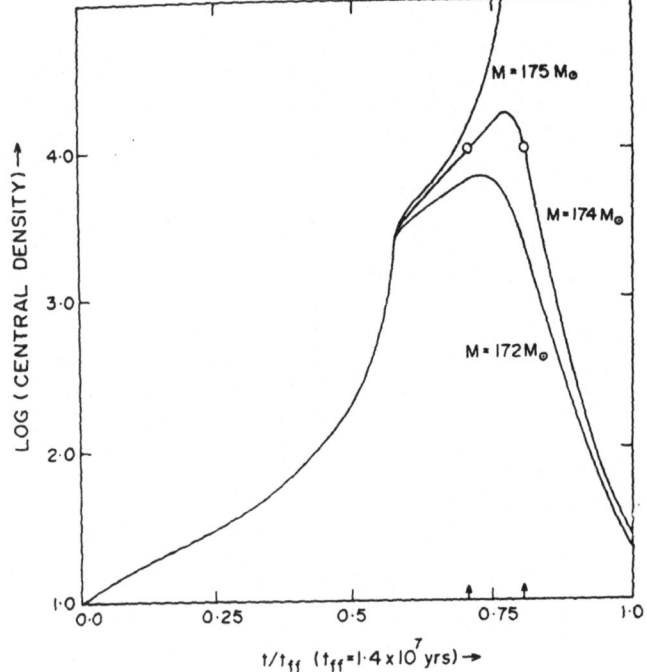

FIG. 2: VARIATION OF CENTRAL DENSITIES OF EVOLVING CLOUDS OF MASSES 175M_0, 174M_0, 172 M_0 WITH TIME. CENTRAL DENSITY IN-CREASES WITH TIME FOR 175M_0 CLOUD BUT PASSES THROUGH A MAXIMUM FOR CLOUD OF MASS 174M_0 PRODUCING TWO PHYSICAL-LY SIMILAR CLOUDS AT DIFFERENT PHASES (MARKED WITH DOTS) OF ITS EVOLUTION.

mass or density of a cloud, for collapse, can be lowered from those given by Jean's criteria. Starting the collapse at density as low as $10 cm^{-3}$ has the advantage of having a long-lived collapsing cloud for observation bringing the star formation under the realm of direct observation.

Fig. 1 shows the variation of minimum mass of a cloud which can collapse as a function of initial density. For comparison the variation of minimum mass from Jean's criteria with T=80°K has also been shown (dashed line). The figure shows that the minimum mass for a given density is significantly lower than the corresponding Jean's mass. Thus for an initial density of 10^3 cm^{-3}, a cloud with as low a mass as 10 M_0 can now be collapsed, whereas according to Jean's criteria, the necessary cloud mass is over 10^3 M_0. The figure also shows that for a given mass the density necessary for the cloud to collapse has also been lowered significantly. As for example, for a cloud of 200 M_0 mass to collapse, the necessary density is now of the order of 10 cm^{-3}, but according to Jean's criteria a density over 10^4 cm^{-3} is required. The lowering of mass and density for a cloud to collapse is the direct consequence of the proper treatment of the energy equation which gives an inward force due to inward temperature decrease in an interstellar cloud. Note that an isolated mass of stellar size can now collapse starting from realistic interstellar density and all stars need not form in clusters due to fragmentation of a large cloud. Moreover, as the collapse can start at a density of 10 cm^{-3}, the cloud lifetime is about 10^7 years, which is long enough for observation.

5.2 A Common Link between Diffuse and Dense Clouds

The variation of density and temperature with radius and time for a cloud of given mass larger than the minimum mass remains similar to those obtained earlier with empirical temperature-optical depth relation (Tarafdar et al. 1985). Therefore, the variation of column densities of various molecules with A_v remains the same as in previous models (Tarafdar et al. 1985), as long as the involved reaction rates are kept the same. Thus the agreement noted between the theoretical and the observed variation of column densities of CI and CO with A_v holds. The agreement suggests that the interstellar clouds from diffuse to dense are linked by evolution in the sense that a diffuse cloud evolves to a dense cloud after sufficient time has elapsed. This brings our aim into successful completion.

5.3 Oscillating Clouds and Their Properties

In order to examine the dynamical and chemical evolution of a cloud with mass smaller than the critical mass given in Fig. 1, we have evolved clouds with three masses of 175 M_0, 174 M_0 and 173 M_0. The initial homogeneous density for all clouds has been assumed to be 10 cm^{-3}. Fig. 2 shows the variation of central density with evolution time for these models. The figure shows that the central density of 175 M_0 cloud increases indefinitely, whereas it reaches a maximum for 174 M_0 and 173 M_0 at around t=0.75 t_{ff} and decreases thereafter. This rebounce back is due to competing effects of temperature and density gradient forces

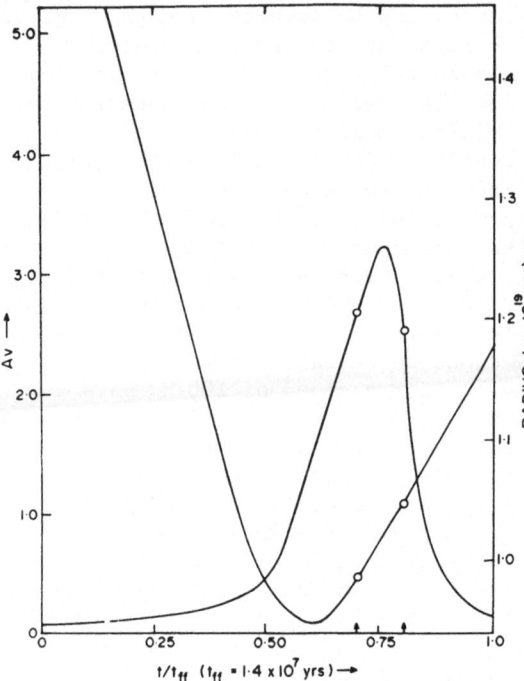

FIG. 3: VARIATION OF CENTRAL A_V AND RADIUS OF THE CLOUD OF $174M_O$ AS A FUNCTION OF TIME. DOTS SHOW THE TWO PHASES WHERE CENTRAL DENSITIES ARE SAME. FIGURE SHOWS THAT A_V AND RADIUS AT THESE TWO PHASES OF EVOLU--TION ARE ALSO SAME IMPLYING PHYSICAL SIMILARITY BETWEEN CLOUDS AT THESE TWO PHASES.

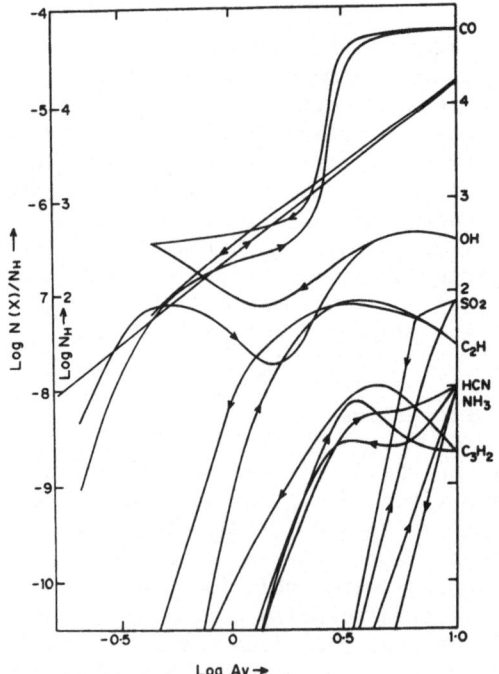

FIG. 4: VARIATION OF FRACTIONAL ABUNDANCES OF VARIOUS MOLECULES WITH A_V IN COLLAPSING (MARKED WITH ARROW TOWARDS INCREASING A_V) AND EXPANDING (MARKED WITH REVERSE ARROW) PHASES. NOTE THE ABUNDANCE DIFFERENCE IN TWO PHASES WITH SAME A_V.

which act in opposite sense. Thus the clouds, with mass smaller than the critical mass for collapse, initially contract and expand after attaining a maximum central density. This oscillation has significant physical and chemical consequences. Note that the central density has same value (marked by dots in 174 M_O curve) of 10^4 cm^{-3} at two different times.

Fig. 3 shows the variation A_V and cloud radius R of 174 M_O cloud as a function of evolving times. The dots in A_V and R curves correspond to the phases where the central density was the same as marked in Fig. 2. Note that the values of A_V and R are observationally indistinguishable in two phases of contraction and expansion of the cloud. Thus evolution of clouds with mass smaller than the critical mass will produce two physically similar clouds at two different times of its evolution. As chemical evolution is time dependent, the contracting and expanding clouds will have different chemical compositions. This is clear from Fig. 4 which shows the variation of abundances of a few molecules relative to that of hydrogen nuclei (i.e. $N(x)/(N_{HI} + 2N_{H_2})$) as a function of A_V. Note that A_V increases in the contracting phase and decreases in expanding phase with time. The curves giving molecular abundances in the contracting phase have been marked with arrows pointing toward large A_V. An arrow in the reverse direction indicates the abundances in the expanding phase. The figure shows that as expected, the molecular abundances may be different depending on whether the cloud is in contracting or expanding phase. The figure also shows that abundances of some molecules may be the same but others could be different in two phases. Whether an abundance of a particular species will be different or the same depends on A_V (i.e. on the phase of evolution). The possibility of having two physically similar but chemically different clouds has observational counterpart in TMC-1 and L134N which are physically similar but chemically different (Irvine, Goldsmith and Hjalmarson 1987).

6. DIFFICULTIES OF EVOLVING MODELS AND THEIR POSSIBLE REMEDIES

One of the difficulties of evolving models is its short lifetime compared to the generally accepted interstellar cloud lifetime of 10^7-10^8 yrs. The model reported by Tarafdar et al. (1985) has a lifetime of about 4×10^6 yrs which is a factor 2.5 smaller than the accepted cloud lifetime. The short lifetime of evolving models resulted as they started with an initial density of about 150cm^{-3}. The present model starts with initial density of 10cm^{-3} and hence increases the lifetime to 10^7 yrs. (Fig. 2) The lifetime of the cloud will further increase, if the effect of turbulence which has been ignored in these models, is included in the models. Moreover, if some of the interstellar clouds are oscillating, as shown above is possible, if their mass is appropriate, the problem of short lifetime of evolving clouds is further alleviated.

Lastly the depletion of gas phase molecules is also a problem in evolving clouds, though it is not as severe as it is in steady state models. This is because an evolving cloud spends most of its lifetime in the diffuse phase and a time $t_e = 4.1 \times 10^7/n_H^{\frac{1}{2}}$ yrs in the phase having density of n_H. Thus depletion becomes important in evolving models only at density higher than 5.0×10^3 cm^{-3}. This density can further be increased by proper treatment of cooling agents in the dense phase of its evolution. However, it is highly unlikely that depletion will be non-existent in these models at densities of 10^5-10^6 cm^{-3} unless some desorption mechanisms operate at that stage of evolution.

7. CONCLUSION

As can be seen from the above presentation and from Tarafdar et al. (1985), the evolving models even in their approximate form have been able to satisfy the goal of unifying the variety of interstellar clouds into an evolutionary sequence in the sense that a diffuse cloud evolves to a dense cloud. The apparent difficulties like short lifetime of the cloud and large star formation rates due to cloud evolution can be overcome by improved treatment of physical processes or by realizing that only the core of the cloud is to evolve into a star. The real test of evolving models and their usefulness in explaining and using observed molecular abundances to infer about the physical state of the cloud will come when better models and better determination of molecular abundances are available.

8. REFERENCES

Black, J. H. (1987) in M. S. Vardya and S. P. Tarafdar (eds.), IAU
    Symposium 120, Astrochemistry, Reidel, Dordrecht, p. 217.
Black, J. H. (1988) Highlights of Astronomy, vol. 8, p.
Boland, W. and de Jong, T. (1982) Ap. J. 261, 110.
Boland, W. and de Jong, T. (1984) Astr. Ap. 134, 87.
Dalgarno, A. (1987a) in G. Morfill and M. S. Scholer (eds.), Physical
    Processes in Interstellar Clouds, Reidel, Dordrecht, p. 219.
Dalgarno, A. (1987b) in A. E. Kingston (ed.), Recent Studies in
    Atomic and Molecular Processes, Plenum Press London, p. 51.
De Jong, T, Dalgarno, A. and Boland, W. (1980) Astr. Ap. 91, 68.
D'Hendecourt, L. B., Allamandola, L. J., Baas, F. and Greenberg, J. M.
    (1982) Astr. Ap. 109, L12.
Gerola, H. and Gllassgold, A. E. (1978) Ap. J. Suppl. 37, 1.
Graedel, T. E., Langer, W. D. and Frerking, M. A. (1982) Ap. J. Suppl.
    48, p. 321.
Hallenback, D. J. and Salpeter E. E. 1970, J. Chem. Phys. 53, p. 79.
Hartquist, T. W. (1987), in M. S. Vardya and S. P. Tarafdar (eds.),
    IAU Symposium 120, Astrochemistry, Reidel, Dordrecht, p. 297.
Herbst, E. (1987) in M. S. Vardya and S. P. Tarafdar (eds.), IAU
    Symposium 120, Astrochemistry, Reidel, Dordrecht, p. 235.
Herbst, E. and Leung, C. M. (1986a) M.N.R.A.S. 222, 689.
Herbst, E. and Leung, C. M. (1986b), Ap. J. 310, 378.
Irvine, W. M., Goldsmith, P. F. and Hjalmarson, A. (1987) in D. J.
    Hollenbach and H. A. Thronson, Jr. (eds). Interstellar Processes,
    Reidel, Dordrecht), p. 561.
Langer, W. D. and Graedel, T. E. (1987), in M. S. Vardya and S. P.
    Tarafdar (eds.),   IAU Symposium 120, Astrochemistry, Reidel,

Dordrecht), p. 305.

Leger, A., Jura, M. and Omont, A. (1985) Astr Ap. 144, 147.

Millar, T. J. (1988) Highlights of Astronomy, Vol. 8, p.

Millar, T. J. and Freeman, A. (1984), M.N.R.A.S. 207, 405; 425.

Millar, T. J. and Nejad, L. A. M. (1985) M.N.R.A.S. 217, 507.

Prasad, S. S. (1987) in M. S. Vardya and S. P. Tarafdar (eds). IAU
  Symposium 120, Astrochemistry, Reidel, Dordrecht), p. 259.

Prasad, S. S. and Huntress, W. T. (1980) Ap. J. Suppl. 43, 1.

Prasad, S. S., Tarafdar, S. P. Villere, K. R. and Huntress, W. T., Jr.
  (1987), in D. J. Hollenbach and H. A. Thronson, Jr. (eds.), Interstellar Processes, Reidel, Dordrecht), p. 631.

Suzuki, H. (1983) Ap. J. 272, 579.

Tarafdar, S. P., Prasad, S. S., Huntress, W. T. Villere, K. R. and
  Black, D. C. (1985) Ap. J. 289, 220.

van Dishoeck, E. F. (1988a), in T. J. Millar and D. A. Williams (eds.),
  Reaction Rate Coefficients in Astrophysics, (in press).

van Dishoeck (1988b) Highlights of Astronomy, vol. 8, p.

van Dishoeck, E. F. and Black, J. H. (1986) Ap. J. Suppl. 62, 109.

Viala, Y. P. (1986) Astr. Ap. Suppl. 64, 391

Viala, Y. P., Roueff, E. and Abgrall, H. 1987, Astr. Ap.

Watt, G. D. (1983) M.N.R.A.S. 205, 321.

Williams, D.A. and Hartquist, T. W. (1984) M.N.R.A.S. 210, 141.

Winnewisser, G. and Herbst, E. (1987) Topics in Current Chemistry 139,
  121.

Wootten, A. (1987) in M. S. Vardya and S. P. Tarafdar (eds.), IAU
  Symposium 120, Astrochemistry, Reidel, Dordrecht, p. 311.

MOLECULES IN CIRCUMSTELLAR ENVELOPES

Alain OMONT
Observatoire de Grenoble.
Universite Joseph Fourier.
BP 53X, F-38041 Grenoble-Cedex.

ABSTRACT. The chemistry of circumstellar envelopes around evolved stars is extremely rich, especially for C-rich ones. 34 molecules have been identified in IRC+10216. Recent observations and modeling prove that most of the unstable species are formed by photochemistry in the outer layers of the envelope.

I. INTRODUCTION.

Massive circumstellar envelopes (CSE) around late-type stars are amongst the most prominent objects for astrochemistry. Two well known points, out of many, examplify this importance: the most complex individual molecules ever observed outside of the Solar System were firstly detected there; CSE are believed to be the main location of formation of interstellar grains. Such molecular envelopes are generally found around late type stars in the final stages of their evolution: mostly when they are on (or just after) the asymptotic branch of the red giants (AGB), where they are loosing mass profusively at rates 10^{-7}-10^{-4} $M_{\odot}$/yr, with velocities in the range 10-30 km/s. The enormous amount of matter thus lost by the star has essential consequences both on its evolution, and on the composition of the interstellar gas. It allows stars as massive as ~ 7 $M_{\odot}$ to finish as white dwarfs, with masses $<\sim 1$ $M_{\odot}$, after passing through the stage of planetary nebula where the envelope expelled at the end of the AGB phase is photodissociated and ionized. The hot burning of H and He which occurs alternatively in AGB stars, enriches the star in elements such as ^{13}C, ^{14}N, ^{12}C and s-isotopes. At least some of this freshly synthesized nuclear material is very rapidly brought to the surface by convective dredge-up. The

357

D. McNally (ed.), Highlights of Astronomy, Vol. 8, 357–364.
© 1989 by the IAU.

relative abundances of the different elements and isotopes in CSE can thus be
very various. The most important consequence for the nature of the molecules
which can be found there, is certainly the possibility that the abundance of
carbon exceeds that of oxygen, at variance with the situation in most of
astronomical objects. In such conditions, a large amount of carbon is
available outside of CO to form carbon-rich grains and complex molecules.
AGB stars and especially their massive circumstellar envelopes are thought to
be the main contributors to the return of mass to the interstellar medium. A
good understanding of the chemistry of their circumstellar envelopes is also
necessary to infer the exact amount of the various elements and isotopes that
they inject into the interstellar medium, and the consequences on the
chemical evolution of galaxies.

A central constituent of CSE is dust. It has a profound influence on
their energetics, dynamics and chemistry. We will recall below the extreme
difficulty to properly model the chemical processes involving circumstellar
(as well as interstellar) dust. Its formation itself, which determines the
chemical composition of the remaining gas, is still poorly understood.
However, the information on circumstellar dust improves very rapidly with the
development of infra-red astronomy. The IRAS mission in particular has had a
tremendous impact on the studies of CSE: the Point Source Catalog has
identified tens of thousands of them with their far infra-red colors; and the
Low Resolution Spectrometer has taken thousands of their spectra, with a very
rich information on the composition of their dust.

As in other media, the discussion of the chemical processes in CSE
requires a good knowledge of the physical, geometrical and kinematical
conditions there. With respect to interstellar clouds, CSE are characterized
by extremely rapid evolution, on time scales $\sim 10^2 - 10^4$ years; and they span a
whole range of conditions from the photospheres to the external layers
similar to interstellar clouds. Descriptions of these conditions as well as
more details on the general features of these CSE and on their chemistry can
be found in various previous reviews, e.g.(1-10). These conditions are
extremely uncertain in the layers comprised between the photosphere and the
point where the acceleration is reasonably complete. It is believed that the
gas just above the photosphere is relatively stationary, levitated by
pulsations and cold enough to allow dust formation. However, it is clear
that these layers are basically complex and unstable: pulsations induce
complicated motions including infall and shocks; molecular instabilities
similar to the ones known to be occuring in the photosphere, are likely to be
present, in particular with dust formation. Therefore, even with simplifying
assumptions, the modeling of the temperature is particularly difficult there.

obviously follow the molecular composition of a gas sample in its transport from the photosphere up to its complete photodissociation and its merging into the interstellar medium. In expanding shells, the basic time constant is $t_{dyn} = r/V_{exp}$. Significant chemical processes must have rates not too small compared to t_{dyn}^{-1}. Accordingly, they are active for radii not much larger than a characteristic radius r_c. For $M' \sim 10^{-5} M_{\odot}/yr$ typical values of r_{c16} (in units of 10^{16} cm) are: 0.25, $6\ 10^5\ X^+$ and $2\ 10^4\ X_R$ for collisions with grains, reactions with ions of abundance X^+, and reactions without activation energy with radical of abundance X_R respectively.

A detailed modeling of the chemistry of the internal layers is extremely difficult because of the lack of knowledge of the physical and dynamical conditions, and of the complexity of grain processes. A qualitative discussion is developed elsewhere (see e.g. 24,6,25). The initial condition is thermodynamical chemical equilibrium in the photosphere (see e.g. 30,24 and references therein). Uncertainties on thermodynamical data of diatomic radicals still plague the determination of element abundances there (see e.g. 26). Anyway, these radicals are then destroyed in the cooling of the external layers of the photosphere (24). On the other hand, one still debates (27,24) the abundance of cyanopolyynes there, which could be transported to the external regions of the envelope.

Thermal equilibrium is also the first approach to grain formation. However, it is sure that it is far from being obeyed, as attested in particular by the large amount of condensable gaseous species in the external envelope. Detailed modeling of grain formation is a formidable task because of: i) the complexity of the chemical composition, and the uncertainties on thermodynamical data on possible intermediate species (clusters of various composition) in the nucleation theory, and on their reaction rates with important gaseous compounds such as H_2, H, C_2H_2, HCN, etc. ii) The non equilibrium of the various temperatures: kinetic, dust, internal excitation of molecules, radiation etc. iii) The complexity of the structure of these layers and of their dynamical evolution.

However, grain formation is essential because it completely determines the abundances in the gas of many elements in the external regions of CSE: carbon in C-rich envelopes, oxygen in O-rich ones; sulfur, silicon, etc. in all envelopes. Reaction on grain surfaces can proceed up to relatively external layers ($r \sim 10^{16}$ cm). Accretion of ice has been discussed by Jura and Morris (28). Ice is present in all massive and very cold OH/IR CSE. It exhibits spectacular 40-70μm emission bands in the special object IRAS

Outside of this region, the density is simply proportional to r^{-2} in the outer layers. However, the determination of the temperature is difficult there also: there are few and uneasy possible observational diagnostics, and one mainly relies on modeling such as worked out by Kwan and Linke (11 and references therein) for IRC+10216. However, it is frequently unrealized how different can the temperature be for stars with smaller or larger mass-loss rates (12).

II. MOLECULAR ABUNDANCES AND SPATIAL DISTRIBUTION.

The molecules presently detected in IRC+10216 are: CO, C_2H_2, HCN, NH_3, CH_4, C_2H_4, HNC, CH_3CN; HC_3N, HC_5N, HC_7N, HC_9N, $HC_{11}N$; CN, C_3, C_2H, C_4H, C_3N, C_6H, C_3H_2, C_3, C_5H; CS, C_2S, C_3S, H_2S; SiS, SiC_2, SiO, SiH_4; $ClAl$, FAl, $ClNa$, ClK. Their approximate abundances are given in ref.(4-9). Compared to IRC+10216, the other massive carbon rich stars appear to have a molecular content basically similar, but with significant differences in the abundances (see e.g. 13,14,15). Very few molecules, outside of CO and H_2 can survive for a long time in planetary nebulae. However, HCO^+, which is only marginally detected in IRC+10216, is very strong in very young planetary nebulae (16). The presence of polycyclic aromatic compounds ("PAH") with ∿20 to hundreds of atoms has been conjectured in CSE. The IR features attributed to these species are prominent in very young planetary nebulae, such as NGO 7027, direct progeny of CSE (17). However, there is no clear evidence yet of PAH in CSE, even in transition objects such as CRL 2688 and CRL 618 (46). The chemistry of O-rich CSE is of course poorer. However, the number of detected molecules there has significantly increased recently (18).

A knowledge of the spatial distribution of the different molecules is obviously fundamental in order to identify the dominant chemical processes. There is some indication that molecules such as Si-bearing molecules, metal halids (22), and possibly H_2S (23) are more concentrated in the internal regions where the gas is not yet fully accelerated. A more direct information is derived from observations with higher angular resolution, either with large millimeter wave single dishes, or with interferometers. Very important results of Bieging and Nguyen-Q-Rieu (45) have shown that C_2H, HNC and HC_3N are distributed in a basically hollow shell in IRC+10216, and are thus mostly formed in the external layers.

III. CHEMICAL PROCESSES. OVERVIEW AND INTERNAL PROCESSES.

Circumstellar chemistry is basically time dependent. Its modeling must

9371+1212 (29).

Radical reactions are still active in these internal layers. A key
uestion is the abundance of atomic hydrogen. Three-body reactions are
nefficient to transform all photospheric H into H_2 (25). However, the
bundance of H is probably reduced to very low values by grain processes, as
ttested by the presence of NH_3 and H_2S which are destroyed by H. Atomic
ydrogen is known to generate CH_4 in reaction with graphite. CO is believed
o be little affected after the gas leaves the photosphere, where CO contains
early all the oxygen in C-rich stars, and all the carbon in O-rich ones.
owever, the significant amounts of other C-bearing molecules observed in the
uter regions of O-rich CSE (Section III) prove that some processes extract C
rom CO in the intermediate layers: probably shocks, possibly chromospheric
V.

V. PHOTOCHEMISTRY IN THE OUTER LAYERS.
1. Photodestruction of molecules.

Photodestruction generally determines the outer boundary of molecular
SE. Photo-rates in the standard UV field are available (see e.g. 32), with
ajor uncertainties for some of them and the possibility of large
luctuations in the local UV intensity. An additional difficulty is the
valuation of the shielding by circumstellar dust. Photodissociation of CO
s peculiar because of self-shielding in the photodissociating lines. It has
ecently been remodeled by Mamon et al. (33) using the laboratory data of
etzelter et al (34). They have shown that no significant fractionation of
^3CO is expected. However, selective photodissociation could reduce the
atios $C^{18}O/C^{16}O$ and $C^{17}O/C^{16}O$. In CSE, photoionization of molecules is
lways small compared to photodissociation. However, it is very important
or species such as C_2H_2 and H_2O in initiating chains of ion reactions
4,35,36,37). A better knowledge of the corresponding rates for abundant
olecules is important. C^+ is also a major ion for initiating ion chemistry
43,35-37).
2. Radical formation and reactions.

Radical are obvious products of photodissociation of polyatomic
olecules. This is certainly the main source of prominent observed radicals:
H first, and C_2H and CN (possibly C_4H and C_3N). The observed extension of
H masers is reasonably well understood (38,25,39). However, a complete
etailed modeling of C_2H and CN in IRC+10216 still has some problems (see
.g.40). Reactions between abundant radicals or atoms without activation
nergy can form significant amount of new products in the photodissociation
egion. The main problem as usual in modeling these processes is to know the
eaction rates, and in particular the activation energies with an accuracy

comparable to the energy at very low temperature. SO_2 and SO are probably formed in this way in reaction of S and SO with OH (41,42). The observations of millimeter lines of SO_2 and SO (23) appear to be in agreement with this scheme. It has been proposed by Nejad and Millar (42, see also 20) that HCN is formed in a similar way in O-rich CSE, in reaction between N and CH_x produced from the photodissociation of N_2 and CH_4 respectively. However, this implies an extremely large abundance of CH_4 ($\sim 10^{-5}$) in order to account for the observed abundance of HCN (20); and it is also possible that HCN is directly formed in the internal layers.

3. Ion chemistry.

To the difference of the interstellar clouds, molecular ions such as HCO^+ are hardly observable in normal CSE. This is due partly to the relatively small masses of CSE, and partly to the slightly smaller ior abundances, as evaluated by Glassgold et al. (35-37). Ions without dipole moments such as $C_2H_2^+$ are probably more abundant. The ions generated by photoionization can generate a very efficient photochemistry, which has been modeled by several authors using more or less developed networks of reactions (see e.g. (10) and references therein). Some significant processes are the following:

$$H_2O \ --(h\nu)-->\ H_2O^+ \ --(H_2)-->\ H_3O^+ \qquad (37)$$
$$CO, \ C_2H_2 \ --(h\nu)-->\ C^+ \ --(C_2H_2)-->\ C_3H_3^+ \ -->\ C_3H, \ C_3H_2, \ C_3 \qquad (43)$$
$$C_2H_2 \ --(h\nu)-->\ C_2H_2^+ \ --(C_2H_2)-->\ C_4H_3^+, \ C_4H_2^+ \ -->\ C_4H_2, \ C_4H \qquad (35)$$
$$C_2H_2^+ + HCN \ -->\ H_2CN^+ \ -->\ HCN, \ HNC \qquad (36)$$
$$C_2H_2^+ + HCN \ -->\ H_2C_3N^+ \ -->\ HC_3N \qquad (44)$$

The agreement with observed abundances of HNC, HC_3N, C_3, C_3H, C_3H_2 and C_4H in IRC+10216 is reasonable (better than an order of magnitude). However, a detailed comparison with observations is still impeded by many uncertainties on rates and branching ratios of ionic reactions, photodissociations, dissociative recombinations, etc. (as well as in source parameters: mass-loss rate, distance, dust shielding, initial abundances of C_2H_2, HCN, etc.). The recent observational proof (45) that HNC and HC_3N are at least mainly formed in the external shells of IRC+10216, shows that ionio chemistry is as important in CSE as in the interstellar medium, for the synthesis of complex molecules. It is very probably responsible for most of those observed in IRC+10216 and in other C-rich CSE. Its modelisation still needs better rates, as in the interstellar medium, with a peculiar emphasis on those implying carbon chains and rings.

REFERENCES

1. B. Zuckerman: Ann. Rev. Astron. Astrophys. 18, 263 (1980)
2. H. Olofsson: In (Sub)millimeter Astronomy, ed. by P.A. Schaver and K.

ar (ESO Publications, Garching 1985)

3. H. Olofsson: Space Sci. Rev., in press (1988)
   H. Olofsson: In Evolution of Peculiar Red Giants, IAU Coll. N. 106, eds
R. Johnson and B. Zuckerman. Cambridge University Press 1988

4. A.E. Glassgold and P.J. Huggins: In M, S and C Stars, ed. by H.R.
hnson and F. Querci (NSF-CNRS 1988)

5. A. Omont: In Mass Loss in Red Giants, ed. by M. Morris and B. Zuckerman
eidel, Dordrecht 1985) p.269

6. A. Omont: In Astrochemistry, ed. by M.S. Vardya and S.P. Tarafdar
eidel, Dordrecht 1987) p.357

7. A. Omont: in Cosmic Chemistry, AG Meeting Koln, ed. by G. Klare
pringer-Verlag, Heidelberg 1988)

8. A. Omont: in Modeling the Stellar Environment, ed. by C.Magnan (1989)

9. A. Omont: in Physics and Chemistry of Interstellar Molecular Clouds,
. by G. Winnewisser (Springer-Verlag, Heidelberg 1989)

10. T.J. Millar: In Rate Coefficients in Astrochemistry, ed. by T.J.
llar and D.A. Williams (Kluwer Academic Pub. 1988) p.287

11. J. Kwan, R.A. Linke: Astrophys. J. 254, 587 (1982)

12. M. Jura, C. Kahane, A. Omont: Astron.Astrophys. 201, 80 (1988)

13. R.J. Sopka, H. Olofsson, L.E.B. Johansson, Nguyen-Q-Rieu, B.
ckerman: Preprint (1988)

14. Nguyen-Q-Rieu, S. Deguchi, H. Izumiura, N. Kaifu, M. Oshishi, H.
zuki. N. Ukita: Astrophys. J. 330, 374 (1988)

15. S. Guilloteau, R. Lucas, Nguyen-Q-Rieu, A. Omont: in preparation

16. S. Guilloteau, A. Omont, R. Lucas: in preparation

17. M. Cohen et al.: Astrophys. J. 302, 737

18. M. Morris, S. Guilloteau, R. Lucas, A. Omont: Astrophys. J. 321, 888
987)

19. S. Deguchi, M.J. Claussen, P.F. Goldsmith: Astrophys. J. 303, 810
986)

20. E. Nercessian, S. Guilloteau, A. Omont, J.J. Benayoun: Astron.
trophys. in press (1988)

21. M. Lindqvist, L.-A. Nyman, H. Olofsson, A. Winnberg: Astron.
trophys. in press (1988)

22. J. Cernicharo, M. Guelin: Astron. Astrophys. 183, L10 (1987)

23. A. Omont, R. Lucas, S. Guilloteau, M. Morris: in preparation

24. S. Lafont, R. Lucas, A. Omont: Astron. Astrophys. 106, 201 (1982)

25. A.E. Glassgold and P.J. Huggins: Mon. Not. Roy. Astr. Soc. 203, 517
983)

26. D.L. Lambert: In Astrochemistry, see [6] p.583

27. I.K. Shmeld. V.S. Strelnitskii, L.V. Gurvich: Sov. Astron. Lett. 11,
4 (1985)

28. M. Jura, M. Morris: Astroph. J. 292, 487 (1985)

29. A. Omont, H.M. Moseley, T. Forveille, P.H. Harvey et al.: in preparation

30. M.S. Vardya: In Astrochemistry, see [6] p.395

31. P.J. Huggins, A.E. Glassgold: Astrophys. J. 252, 201 (1982)

32. E.F. Van Dishoeck: In Astrochemistry, see [6] p.51
    E.F. Van Dishoeck: In Rate Coefficients in Astrochemistry, see [10]

33. G.A. Mamon, A.E. Glassgold, P.J. Huggins: Astrophys. J. 323, 306 (1987)

34. C. Letzelter, M. Eidelsberg, F. Rostas, J. Breton, B. Thieblemont: Chem. Phys. 114,273 (1987)

35. A.E. Glassgold, R. Lucas, A. Omont: Astron. Astrophys. 157, 35 (1986)

36. A.E. Glassgold, G.A Mamon, A. Omont, R. Lucas: Astron. Astrophys. 180 183 (1987)

37. G.A Mamon, A.E. Glassgold, A. Omont: Astrophys. J. 323, 306 (1987)

38. P. Goldreich, N.Z. Scoville: Astrophys. J. 205, 144 (1976)

39. N. Netzer, G.R. Knapp: Astrophys. J. 323, 734 (1987)

40. Truong-Bach, Nguyen-Q-Rieu, A. Omont, H. Olofsson. L.E.B. Johansson: Astron. Astrophys. 176, 285 (1987)

41. J.M. Scalo, D.B. Slavsky: Astrophys. J. Lett. 239, L73 (1980)
    D.B. Slavsky, J.M. Scalo: Preprint (1986)

42. L.A.M. Nejad, T.J. Millar: Mon. Not. Roy. Astr. Soc. 230, 79 (1988)

43. L.A.M. Nejad, T.J. Millar: Astron. Astrophys. 183, 279 (1987)

44. A.E. Glassgold, G.A. Mamon: In Second Haystack Conference on Interstellar Matter, ed. J.M. Moran (1987)

45. J.H. Bieging, Nguyen-Q-Rieu: Astrophys. J. Lett. 329, L107 (1988)

46. M. de Muizon, P. Cox, J. Lequeux: In "Interstellar Dust". IAU Symp. N. 135. X. Tielens and L. Allamandola eds. (Reidel, Dordrecht 1988)

Atomic and Molecular Data for Stellar Physics

R. A. Bell
Astronomy Program
University of Maryland

ABSTRACT. This paper discusses a number of problems in the field of
stellar astronomy which are caused by inadequate kinowledge of atomic
and molecular data.

1. INTRODUCTION

When I was asked to give this talk just a few days ago, I decided to
increase the size of the topic. The reason for which I did this was
that I felt it was necessary to talk about the atomic and molecular data
for stellar interiors.
    There are three fundamental points to be made.
(1) In stellar interiors, we need some checking of the existing
        calculations.
(2) In stellar atmospheres, we need a certain amount of data, obtained
        using very high precision.
(3) Again in stellar atmospheres, we need a lot of data, but we can
        accept a rather lower degree of accuracy.

2. STELLAR INTERIORS

In the field of stellar interiors, there is a rather disconcerting
disagreement between some work carried out at Livermore and the standard
Los Alamos results. Fortunately, the previous disagreement between
Carson and the Los Alamos group appears to have been resolved. However,
Iglesias, Rogers and Wilson (1987) have compared their results for a
number of species with the Los Alamos Astrophysical Opacity Library and
find that, at T=20ev, they calculate an iron opacity which is six times
larger. If such an increase occurs at higher temperatures (it does not
occur at T=10ev) than at least some evolutionary tracks, such as those
for low mass stars, will be affected. This will mean that globular
cluster isochrones will be altered thereby altering the ages found for
globular clusters. Simon (1982) has also suggested that stellar
interior opacities are underestimated, using as his data the various
results for the period ratios of Cepheids.

D. McNally (ed.), Highlights of Astronomy, Vol. 8, 365–368.
© 1989 by the IAU.

3. STELLAR ATMOSPHERES

In stellar atmospheres, whether you need a small amount of very good data or a lot of less good data depends on the' problem that you want to study.

The most accurate work on the determination of oscillator strength that I am aware of is that of Blackwell and his collaborators at Oxford, (e.g. Blackwell, et al. 1979, Andrews et al. 1979, Blackwell, Petford and Shallis, 1979, Blackwell and Shallis, 1979a) who claim an accuracy of 0.5 per cent. They have used this result to produce the very precise solar curve of growth shown by Blackwell and Shallis 1979b. This curve of growth is very well defined indeed. However, while very well defined, this curve is not ideal for determining the solar iron abundance--it simply does not extend to weak enough iron lines. The weakest line seen in it has $\log W/\lambda = -5.4$ or in addition, in a star like Arcturus, these lines would be much stronger and therefore less suitable for abundance work.

However, there is a problem once one starts to analyze very weak lines. It is necessary to check very carefully that the identifications are secure and that the features are not blended. I recall a personal story here. One of my students had analyzed the spectrum of CH and, as an extension, had computed the wavelengths of $C^{13}H$. He wanted to search for these lines and found one whose strength was greater than expected. After a while he discovered that the reason for this was because of blending with a line of, I think, ruthenium. This line had been included in our line data list, but we might have been a little cavalier in the value we used for the abundance of ruthenium in the sun. So, while we do need the weak line data for the analysis of stellar spectra, we also need to know a lot of data about all other possible blending contributors.

One further point about this weak line problem is that, as far as I know, every line which has been seen in the laboratory spectrum of neutral iron has been seen in the solar spectrum. This means that when you start looking at sufficiently weak lines in the solar spectrum, or other stellar spectra, you really cannot be sure of your identifications.

Two other problems where we need lots of data are (a) the actual calculation of stellar spectra and (b) the calculation of models of stellar atmospheres.

Why should anyone want to compute an accurate spectrum of a star? One basic reason is that we want to understand stellar photometry, while another is the understanding of spectra. We want to be able to convert stellar colors into T_{eff}, log g and metal abundance. And we even want to understand the influence of different metals on stellar colors--if you have observed a spectral region which contains a lot of CH lines for example, than your magnitudes are affected by the abundance of CH and, consequently, C. Now some colors we seem to calculate quite well e.g. B-V for F dwarfs, but we need help elsewhere. For example, our model spectra are too bright in the region 3,000-4,000 A. This is very unfortunate, because it makes our interpretation of the colors much less

certain. In order to do this job in this critical wavelength region, we
need to have more atomic data--more, not just more accurate values for
existing data.

What do we need for stellar model calculations? The point here is
that we have to take account of the absorption lines when we are
deriving the temperature structure of a model atmosphere. The problem
is made tractable by the use of "opacity distribution functions" or
ODFs. The idea here is that, instead of computing the opacity in great
detail at every depth in a model i.e. using millions of wavelength
points, we argue that it doesn't really matter where in a suitable
wavelength interval each line is. We determine a histogram for each 100
angstrom interval, say, giving the fractions of that interval where the
absorption coefficient exceeds the series of values. It is really
essential to use these ODFs instead of simply calculating mean
opacities.

What data is lacking? It depends upon the type of star. In the F
stars like Procyon, it must be the weak atomic lines. This is also true
in the G stars and, probably, the K stars. However, as we go to cooler
stars, more and more of the flux comes out at longer wavelengths. In
the M stars, we need the data for water vapor. This is still not
available in a suitable form for the calculation of ODFs.

While Bengt Gustafsson was unable to give this talk, he did
describe to me some of the problems which he and his colleagues had
worked on and gave a very interesting example of the importance of this
work. This concerns work on carbon stars. As you can imagine, in the
hotter carbon stars the line opacity is mainly due to CO and CN and
extensive data is available for these molecules. This data is adequate
for the hotter carbon stars. However, in the cooler ones, the molecular
equilibria change with the polyatomic molecules increasing in abundance
at the expense of the diatomics. With only CO and CN opacities, the
atmospheres become quite transparent and relatively thin. These
atmospheres have caused problems in the analysis of observations. They
have relatively high pressures. These high pressures lead to a lot of
the hydrogen being in the present in the form of H_2. However, since the
H_2 lines around 2 microns are observed to be relatively weak in the
spectral carbon stars, the early models (Johnson 1982) led to the belief
that cool carbon stars were relatively deficient in hydrogen.

Numerical experiments with a fudge factor i.e. a veil of weak metal
lines, show that even the inclusion of relatively weak lines would have
an important effect, reducing the pressures in the models and thereby
reducing the H_2 abundance and the predicted strength of the H_2 molecular
lines. Eriksson et al. (1984) were the first to consider the opacity of
HCN and C_2H_2 and found that their model atmospheres expanded
significantly, with a pressure reduction of two orders of magnitude
compared to models calculated without the HCN and C_2H_2 opacity.
Eriksson et al used semiclassical estimates for the absorption--later
Jorgensen et al (1985, 1988) performed ab initio CASSCF calculations for
HCN and C_3 and verified and strengthened these results.

References

Andrews, J.W., Coates, P.B., Blackwell, D.E., Petford, A.D. and
Shallis, M.J. 1979, M.N.R.A.S., 186, 651.
Blackwell, D.E., Ibbetson, P.A., Petford, A.D., and Shallis, M.J.
1979, M.N.R.A.S., 186, 633.
Blackwell, D.E., Petford, A.D., and Shallis, M.J. 1979, M.N.R.A.S.,
186, 657.
Blackwell, D.E., and Shallis, M.J. 1979a, M.N.R.A.S., 186, 669.
Blackwell, D.E., and Shallis, M.J. 1979b, M.n.R.A.S., 186, 673.
Eriksson, K., Gustafsson, B., Jorgensen, U.G. and Nordlund, A,
1984, Astr. Ap. 132, 37.
Iglesias, C.A., Rogers, F.J. and Wilson, B.G. 1987, Ap.J. Lett.,
322, L45.
Jorgensen, U.G., Almlof, J., Gustafsson, B., Larsson, M. and
Siegbalm, P. 1985, J. Chem. Phys. 83, 3034.
Jorgensen, U.G., Almlof, J., and Siegbalm, P. 1988, J. Chem. Phys.
(in press).
Simon, N.R. 1982, Ap.J. Lett., 260, 87.

CHEMISTRY IN DENSE INTERSTELLAR CLOUDS - DATA REQUIREMENTS

T. J. Millar
Mathematics Department
UMIST
PO Box 88
Manchester M60 1QD

ABSTRACT. Chemical models of dense interstellar clouds are reviewed with particular emphasis on recent results. The need for theoretical and experimental data on rate coefficients is pointed out and some observational studies are suggested.

1. INTRODUCTION

In this brief review I shall discuss some current research on the chemistry of dark, dense interstellar clouds. In particular, I shall restrict myself to the description of cold, quiescent clouds such as TMC-1 and L134N which have kinetic temperatures ~ 10K. Chemistry on grain surfaces and in evolving clouds is discussed elsewhere. Recent reviews of interstellar chemistry include those given by Herbst (1988) and Millar (1988). In the following section, I shall describe microscopic processes which may play important roles in the chemistry. Numerical results quoted in this section are taken from Millar (1988). In Section 3, I summarise recent chemical kinetic models of dark clouds, and finally in Section 4, give a brief list of data required.

2. MICROSCOPIC PROCESSES

2.1 Dissociative Recombination Branching Ratios

In models of interstellar chemistry, it has become standard practice to assume that when a molecular ion containing one or more hydrogen atoms, recombines with an electron, one or two hydrogen atoms, or possibly H_2, is released. This viewpoint is supported to some extent by the phase space calculations of Herbst (1978) and Green and Herbst (1979). Thus, for example, the reaction H_3O^+ + e results in the formation of OH and H_2O with a branching ratio, f, to OH generally assumed to ~ 0.5. More recently, Bates (1986) has argued that dissociation occurs

369

through excitation of a <u>particular</u> valence bond and hence results in the disruption of <u>one</u> bond, unless polarisation bonds are present in the molecular ion. For example, he argues that $H_3O^+ + e \to H_2O$ only (i.e. $f = 0$) and CH_5^+ ($\equiv CH_3^+.H_2$) $+ e \to CH_2 + H + H_2$, but not $CH_4 + H$. Bates and Herbst (1988) have given a review of dissociative recombination. Millar et al (1988) have applied Bates' ideas to a variety of complex molecular ions using *ab initio* techniques to determine the position of the charge on the ions. In general, the electron is delocalised, so that several product channels, rather than one, can result from the recombination process. Millar et al. have used their calculated branching ratios in a detailed chemical kinetic model of a dense interstellar cloud, and find that the most important branching ratio is that value assumed for f. If $f = 0$, then H_2O is as abundant as CO, H_3O^+ is the most abundant molecular ion and OH and molecules such as O_2, NO, SO, SO_2, N_2, N_2H^+ which form, directly or indirectly, from it, become much less abundant than their observed, or inferred, interstellar values. However, if $f = 0.1$, OH and its descendents are formed efficiently. The extra channels included for the dissociation of hydrocarbon ions tend to decrease the abundances of the hydrocarbons, though only at steady state ($\gtrsim 5\ 10^6$ yr), while the CI abundance at steady state can reach $\sim 6\ 10^{-8}$ for $f = 0$, since CI is destroyed mainly by O_2 and OH which have low abundances in this case.

2.2 Radiative Association via Electronic Stabilisation

Bates (1987) re-analysed the low temperature experimental results of Barlow et al. (1984) on the $CH_3^+ - H_2$ radiative association reaction. In particular, Bates argued that the experiment was carried out at an effective temperature of ~ 50K rather that 13K and hence that the CH_5^{+*} collision complex must stabilise via electronic, rather than vibrational, stabilisation in order to account for the measured rate coefficient. Subsequently, Herbst and Bates (1988) described two methods by which electronic stabilisation might occur and estimated expected enhancements to the radiative association rate coefficients. It is not known, as yet, whether electronic stabilisation applies to any other systems of astrophysical importance.

2.3 Cosmic Ray Induced Ultraviolet Photons

Prasad and Tarafdar (1983) recognised that the cosmic ray ionisation of H_2 in dense clouds would result in the generation of a low flux of UV photons within the clouds. Sternberg et al. (1987) and Gredel et al. (1987) have calculated the resulting spectrum of photons and used them to determine photodissociation rates for CO and other molecules (Gredel et al. 1988). These rates depend upon a number of parameters, most importantly the grain albedo, ω, and, for CO which is destroyed by line photons, on the cloud temperature, T, and on the CO line widths. Typical photorates are $\sim (10 - 10^3)\varsigma$ where ς is the primary cosmic ray ionisation rate of H_2. For $T = 10$K, $\omega = 0.5$ and $\varsigma = 10^{-17}$ s^{-1}, the inclusion of such photoprocesses affects abundances slightly, and then only at steady state. One important exception is CI whose abundance can increase by an order of magnitude or more as a result of CO photodissociation. Complex hydrocarbons have abundances reduced by at most a factor of five at steady state.

2.4 Large Molecules

Lepp and Dalgarno (1988) have explored the consequences of including a component of large molecules (LMs) - possibly the polycyclic aromatic hydrocarbons - in dense cloud chemistry. They find that the LMs take up almost all the electrons in such clouds with the result that [LM⁻] > [e] (here square brackets refer to abundance). Collisions of atomic and molecular ions with LM⁻ rather than with electrons become the dominant means of neutralisation. The reaction $C^+ + LM^- \rightarrow C + LM$ leads to a large abundance of atomic carbon, although it is unclear as to whether this reaction could lead to a [CI]/[CO] ratio as large as 0.01 (Lepp and Dalgarno 1988). The recombination of complex ions and LM⁻ may occur by mutual neutralisation, that is a reaction in which dissociation of the ion does not take place (Bates and Herbst 1988), which, together with the increased CI abundance, enhances the formation of complex species. The major uncertainty in this model is the LM abundance which Lepp and Dalgarno treated as a free parameter.

3. RECENT CHEMICAL KINETIC MODELS

3.1 Complex Molecule Formation

Herbst and Leung (1988) have extended the hydrocarbon chemistry to include species such as C_mH_n (m ≤ 9, n ≤ 2), the cyanopolyynes up to HC_9N, CH_3C_5N and so on. Although laboratory data on many of the reactions in their scheme are lacking, they have based their chemistry on the processes thought to important, and well determined, in the production of less complex species. They find reasonable agreement between their calculated abundances and those observed in TMC-1 given the large uncertainties in the chemical network used. It appears that a relatively simple chemistry involving carbon-chain growth via C^+ and CI reactions can explain, in a general fashion, the type of species observed in TMC-1.

3.2 Deuterium Chemistry

Brown and Rice (1981, 1986) performed pioneering studies on detailed models of deuterium chemistry in dark clouds. Recently, Millar, Bennett and Herbst (1988) have updated and extended their model to include the effects of deuterium fractionation in more complex molecules and to study the sensitivity of fractionation to the choice of dissociative recombination branching ratios. Their detailed results showed that (i) for hydrocarbon molecules, which are deuterated principally by CH_2D^+ and C_2HD^+ for T >20K, the degree of fractionation actually <u>increases</u> with temperature in the range 30 – 70 K. This behaviour is opposite to that of species such as H_2O and NH_3 which are deuterated by H_2D^+. Herbst et al. (1987) have shown that this behaviour is due to the competition between H_2 and HD in reactions with CH_3^+ and $C_2H_2^+$. These two ions react with H_2 in radiative association reactions whose rate coefficients therefore decrease with increasing temperature, hence allowing HD reactions, whose rate coefficients are independent of temperature, to be more effective. At T > 70 K, however, the degree of fractionation decreases because reactions of H_2 with the CH_2D^+ and C_2HD^+ ions dominate the loss of these species. (ii) As noted by Brown and Rice (1986), the fractionation of HCO^+ is strongly time dependent at T = 70 K. At early times fractionation results from the reactions of deuterated hydrocarbon ions with O and CO whereas at steady state, the hydrocarbon ions have been processed into CO and deuteration, which is now via H_2D^+, is inefficient. Observations of DCO^+ and HCO^+ in Orion indicate that this cloud has not yet reached a chemical steady state. (iii) Dalgarno and Lepp (1984) showed that the reaction D + OH → OD + H could lead to a large [OD]/[OH] ratio at low temperatures. Millar, Bennett and Herbst (1988) find this ratio to be ~ 0.5, although its value is uncertain due to some poorly known rate coefficients.

3.3 Carbon Isotope Chemistry

Langer and Graedel (1988) have included the chemistry of ^{13}C in dense cloud models and find some important differences from their earlier results (Langer et al. 1984). In particular, while ^{13}C enhancement occurs in CO and HCO^+, trace carbon-bearing molecules are enhanced in ^{12}C but to a degree less than in their earlier work, though still larger than those observed. If cosmic-ray produced UV photons dominate the loss of CO, then isotope selective photodissociation of ^{13}CO will occur and may change the ratios calculated by Langer and Graedel.

4. DATA REQUIREMENTS

4.1 Theoretical and Experimental Determination of Rate Coefficients.

As discussed in Section 2.1, branching ratios for dissociative recombination reactions are highly uncertain and experimental work in this area is essential for a proper understanding of interstellar chemistry. There has been some dispute recently (Amano 1988) over the rate constant for the H_3^+ + e reaction, but this reaction plays no important role in dense clouds because the H_3^+ ion is lost predominantly through reactions with species such as O, C, N, CO, H_2O, etc. Its value does have important implications for chemistry in diffuse clouds (van Dishoeck and Black 1988) and in Jupiter's atmosphere (Dalgarno 1988).

The most important recombination reaction in dense clouds is that of H_3O^+ + e, which can have significant effects on cloud chemistry if only H_2O is formed.

Low temperature data on neutral-neutral reactions are also needed. The reactions $O + OH \rightarrow O_2 + H$ and $C + O_2 \rightarrow CO + O$ are important in determining the O_2 and CI abundances in dense clouds. Smith (1988) has given some general rules which can be used to indicate the presence of barriers at low temperatures while Wagner and Graff (1987) and Leen and Graff (1988) have estimated rate coefficients for several systems of astrophysical importance.

Structural information on ions is extremely important as Adams and Smith (1987) have pointed out in their study of the linear and cyclic forms of $C_3H_3^+$. Such information is important in determining the HCN/HNC and linear - C_3H/cyclic - C_3H abundance ratios. It may also help us to understand the mechanisms by which the cyanopolyynes form. Lee and Amano (1987) have shown that protonated cyanoacetylene is linear, HC_3NH^+; hence, one expects the reaction $HC_3N + H_2D^+$ to produce the ion HC_3ND^+ upon deuteron transfer. In this case, it is difficult to see how DC_3N can result from dissociative recombination. The $[DC_3N]/[HC_3N]$ ratio should, therefore, not reflect the $[H_2D^+]/[H_3^+]$ or $[DCO^+]/[HCO^+]$ ratios but may be closely tied to ratios such as [CCD]/[CCH] or $[C_3HD]/[C_3H_2]$, depending upon its exact mode of formation. Such isomeric effects, which may also be important in the DCN/DNC system, have not been included in any study of deuterium fractionation.

4.2 Observational Data Requirements

It has been the case that observational studies have given impetus to the development of particular chemical models and, of course, observations provide a test for such models. Perhaps the most important measurement to be carried out is a study of CI in cold, dark clouds. To date, there has been only one source detected, L134N, and one other source, TMC-1, searched (Phillips and Huggins 1981). The chemical models predict that the $C_3H_3^+$ ion is the precursor to HC_3N and $c-C_3H_2$ indicating that these species should be correlated to some degree, although the exact correlation should depend on the [N]/[e] abundance ratio which can vary from source to source.

Detailed studies of particular molecular clouds are also of great value. Much observational data has been gathered for TMC-1, while Swade (1987) has studied L134N in great detail and presented evidence for a C/O abundance gradient within this cloud. Other studies such as this would be useful.

Finally there is evidence that, at least in hot core regions, the formation of complex molecules such as methanol, ethanol and dimethyl ether, is related to grain surface processes (Millar et al. 1988b). It is not clear, as yet, whether the grains play any significant role in the production of molecules in cold clouds but observations of large molecules in these regions may eventually provide information on this point.

REFERENCES

Adams, N. G. & Smith, D., 1987, Ap. J., 317, L25.
Adams, N. G., Smith, D. & Clary, D. C., 1985. Ap. J., 296, L31.
Amano, T., 1988, Ap. J., 329, L121.
Barlow, S. E., Dunn, G. H. & Schauer, M., 1984, Phys. Rev. Letts., 52, 902.
Bates, D. R., 1986, Ap. J., 306, L45.
Bates, D. R., 1987, Ap. J., 312, 363.
Bates, D. R. & Herbst, E., 1988, In: "Rate Coefficients in Astrochemistry"
    eds. T. J. Millar & D. A. Williams (Kluwer), p. 41.
Brown, R. D. & Rice, E. H. N., 1981, Phil. Trans. Roy. Soc., A303, 523.
Brown, R. D. & Rice, E. H. N., 1986, MNRAS, 223, 429.
Dalgarno, A. 1988, In: "Rate Coefficients in Astrochemistry"
    eds. T. J. Millar & D. A. Williams (Kluwer), p. 321.
Dalgarno, A. & Lepp, S., 1984, Ap. J., 287, L47.
Gredel, R., Lepp, S. & Dalgarno, A., 1987, Ap. J., 323, L137.
Gredel, R., Lepp, S., Dalgarno, A. & Herbst, E. 1988, Ap. J., in press.
Green, S. & Herbst, E., 1979, Ap. J., 229, 121.
Herbst, E., 1978, Ap. J., 222, 508.
Herbst, E., 1988, In: "Rate Coefficients in Astrochemistry"
    eds. T. J. Millar & D. A. Williams (Kluwer), p. 239.
Herbst, E., Adams, N. G., Smith, D. & DeFrees, D. J., 1987, Ap. J., 312, 351.
Herbst, E. & Bates, D. R., 1988, Ap, J., 329, 410.
Herbst, E. & Leung, C. M., 1986, Ap. J., 310, 378.
Herbst, E. & Leung, C. M., 1988, Ap. J. Suppl., in press.
Langer, W. D. & Graedel, T. E., 1988, Ap. J., in press.
Langer, W. D., Graedel, T. E., Frerking, M. A. & Armentrout, P. B., 1984
    Ap. J., 277, 581.
Lee, S. K. & Amano, T., 1987, Ap. J., 323, L145.
Leen, T. M. & Graff, M. M., 1988, Ap. J., 325, 411.
Lepp, S. & Dalgarno, A., 1988, Ap. J., 324, 553.
Millar, T. J., 1988, In: "Molecular Astrophysics" ed. T. W. Harquist
    (Cambridge University Press), in press.
Millar, T. J., Bennett, A. & Herbst, E., 1988, Ap. J., in press.
Millar, T. J., DeFrees, D. J., McLean, A. D. & Herbst, E. 1988a. A. & A.,
    194, 250.
Millar, T. J., Olofsson, H., Hjalmarson, A. & Brown, P. D., 1988b., A. & A.,
    in press
Phillips, T. G. & Huggins, P. J., 1981, Ap. J., 251, 533.
Prasad, S. S. & Tarafdar, S. P., 1983, Ap. J., 267, 603.
Smith, I. W. M., 1988. In: "Rate Coefficients in Astrochemistry" eds.
    T. J. Millar & D. A. Williams (Kluwer), p. 103.
Sternberg, A., Dalgarno, A. & Lepp, S., 1987, Ap. J., 320, 676.
Swade, D. A., 1987. Ph. D. Thesis, U. Mass. at Amherst.
van Dishoeck, E. F. & Black, J. H., 1988. In: "Rate Coefficients in
    Astrochemistry" eds. T. J. Millar & D. A. Williams (Kluwer), p. 209.
Wagner, A. F. & Graff, M. M., 1987. Ap. J., 317, 423.

CHEMISTRY IN SHOCKS

T.W. HARTQUIST
Max-Planck-Institute for Physics and Astrophysics
Institute for Extraterrestrial Physics
8046 Garching
Federal Republic of Germany

ABSTRACT. The column densities of interstellar CH^+, first detected about fifty years ago, cannot be explained with models of the chemistry in low temperature gas. The resolution of this classic problem is necessary for us to have confidence in our understanding of interstellar chemistry and its role in determining the physical conditions in interstellar clouds and in the utility of molecular abundance measurements as diagnostics. The possibility that the observed CH^+ is formed primarily in shocks in diffuse clouds is addressed. The way in which the chemistry affects the structure of such a diffuse cloud shock is also discussed. The analogous chemical influence on the structures of shocks in dense molecular clouds is also considered as is the possibility that gas in some dense molecular clouds passes repeatedly through dynamical cycles and is shocked frequently enough to influence the global chemical structures in those clouds. Some atomic and molecular data needs are mentioned.

1. Why Shocks?

The high temperatures ($1000 \text{ K} \lesssim T \lesssim 4000 \text{ K}$) in shock heated gas drive many endothermic reactions and reactions with activation barriers which are of no importance in the low temperature interstellar gas. In some types of shocks in weakly ionized magnetized media the relative motion of ions and neutrals is substantial and drives reactions which are normally slow at low temperature. Hence, the chemistry of shock heated gas is potentially rich, and when chemical models of low temperature clouds are incompatible with measured column densities of species, theorists have commonly considered shock chemistry models for the production of those species.

There exists the danger that shocks and dynamics will be invoked to solve outstanding chemical problems which arise not because dynamics have been ignored but rather because fundamental elements are missing in the chemical models. Hence, detailed observational studies (e.g. see the review by Langer (1989)) of the kinematics of the regions in which, for chemical reasons, shocks are supposed to exist are necessary.

Shocks are prevalent in the interstellar medium. On average, each diffuse cloud is engulfed by a supernova remnant once every $.10^6$ years

375

D. McNally (ed.), Highlights of Astronomy, Vol. 8, 375–382.
© *1989 by the IAU.*

(McKee and Ostriker (1977)) leading to the propagation of a shock in it. Dense molecular clouds are regions of star formation, and the young stars have winds which act back on the clouds and drive shocks. The well known shock in Orion and those shocks which may be associated with water maser formation are examples.

2. Shock Structure

The discussions of shock structure in standard texts on fluid mechanics are limited to considerations of one-fluid models. When they are appropriate, the flow is essentially discontinuous. Gas is compressed, heated, and accelerated in a negligibly thin transition zone and is further compressed, at roughly constant pressure, and accelerated in a cooling layer.

Mullan (1971), Draine (1980), and Draine, Roberge, and Dalgarno (1983) have noted that shocks in interstellar clouds often will not have such simple structures and multifluid hydromagnetic models must be used. For diffuse cloud shocks, three fluid models are often constructed. In them, the neutrals, ions, and electrons are treated as separate fluids, and grains are ignored (though grains may contribute significantly to the ion inertia). We will restrict attention to steady plane-parallel models of shocks propagating perpendicularly to the magnetic field. The requirement that no large charge separation induced electric field exists implies that the components of the electron and ion velocities perpendicular to the shock front are equal. However, if the medium is weakly ionized and the shock speed is not too high, the ion and neutral velocities can differ substantially.

Consider a magnetized medium with a nearly vanishing fractional ionization. Assume that a piston acts only on the ions and electrons and moves through the medium at a speed less than $v_{Ai} \equiv B_0/(4\pi\rho_{io})^{1/2}$ where B_0 and ρ_{io} are the upstream magnetic field strength and ion mass density. Then a magnetosonic wave will propagate in the upstream ion and electron fluids as the piston moves; behind its front the pressure of the ion and electron fluids will be increased. The neutrals will be unaffected.

Now, consider a medium which has a small but finite fractional ionization. Assume that a piston which acts on all fluids moves through it at a speed less than v_{Ai}. Because over some range, the ion and electron fluids are effectively decoupled from the neutral fluid, the signal speed in them is about v_{Ai}. Because of the large signal speed in the ion fluid, it responds much further upstream than the neutral fluid does. Hence, an element of ionized and electron fluids is accelerated to a substantial fraction of the shock velocity while the coresponding element of neutral fluid is still little affected by the shock. Eventually, ion-neutral friction will accelerate the neutral fluid to nearly the same velocity as the ions. Often, the lengthscale over which the neutral acceleration occurs is roughly

$$\Delta \equiv B_0^2/4\pi \, \alpha_{in} \, n_i \, \rho_0 \, v_s \tag{1}$$

where α_{in} is the ion-neutral collision rate coefficient, n_i is the ion number density in the neutral acceleration zone and ρ_0 and v_s are the upstream mass density and shock speed. This expression for Δ is obtained by equating the magnetic tension ($\sim B_0^2/4\pi\Delta$) to the ion-neutral friction.

In the acceleration zone, ion-neutral friction leads to heating. Henceforth, the term "dissipation zone" rather than "acceleration zone" will be used in this review; the term "magnetic precursor" is often employed in other papers. We will restrict attention to shocks for which the speed is low enough that the neutral gas does not get so hot that a discontinuous neutral subshock occurs; in other words, we consider only the so-called "C-type shocks" in which all flow variables are continuous. Because the dissipation zones are long for the C-type shocks (The expression given above for Δ is appropriate for them.), they generally have the largest column densities of those chemicals which form only under high temperature conditions.

3. CH⁺ in Diffuse Cloud Shocks

The column density of CH⁺ in the diffuse cloud (The visual extinction of a diffuse cloud is of order unity or less.) towards ζOph is about $3.\times10^{13}$ cm⁻². The reactions included in low temperature chemistry models produce CH⁺ column densities close to two orders of magnitude less than this. The reason that the cold chemistry model abundances are so low is that the reaction

$$0.4 \text{ eV} + C^+ + H_2 \rightarrow CH^+ + H \tag{2}$$

is so endothermic.

As a consequence, in cold diffuse molecular clouds the ionization balance is determined primarily by the photoionization of C and the radiative recombination of C^+ and nearly all carbon is contained in C^+. H^+ is also an abundant ion with a number density which is about an order of magnitude less than that of C^+.

Elitzur and Watson (1978) suggested that diffuse cloud CH⁺ is produced in shocks. They considered one fluid models only. The primary production mechanism for CH⁺ is reaction (2) while the primary removal mechanism is the reaction

$$CH^+ + H_2 \rightarrow CH_2^+ + H \tag{3}$$

which initiates the formation of other molecular species.

In all of the discussions in this section we consider regions in which H_2 contains all but a small fraction of the hydrogen. For the moment, ignore photodissociation and photoionization reactions other than the photoionization of C. Reaction (2) is rapid enough that all C^+ passing through the shock will be removed by it before the shocked gas cools. Reaction (3) is rapid in an analogous sense. Hence, the column density of CH⁺ in the shocked gas will be approximately

$$\frac{n_0(C^+)v_s}{\alpha_3 n_0(H_2)} \approx 3 \times 10^{11} \text{ cm}^{-2}$$

for $v_s = 10$ km^{-1}. α_3 is the rate coefficient for reaction (3), and $n_0(C^+)$ and $n_0(H_2)$ are the upstream number densities of C^+ and H_2.

Continue to make the same assumptions as those made for the discussion in the preceding paragraph, except now assume that photodissociation and photoionization reactions break CH_2^+ and the larger species formed from it back down to C^+ and CH^+ on a timescale, τ_{ph}. The timescale characterizing cooling in the one fluid shock model is taken to be τ_{cool}. Typically, $\tau_{ph} < \tau_{cool}$. Then the resulting column density of CH^+ in the shock heated gas is roughly

$$\frac{n_0(C^+)v_s}{\alpha_3 n_0(H_2)} \frac{\tau_{cool}}{\tau_{ph}}$$

Shock models for CH^+ formation are constrained by observations of other molecules. Higher lying ($3 \leq J \leq 6$) H_2 rotational levels are collisionally excited in the shock heated gas. OH is produced and removed by the reactions:

$$O + H_2 \rightarrow OH + H; \quad OH + H_2 \rightarrow H_2O + H; \quad H_2O + h\nu \rightarrow OH + H;$$

$OH + h\nu \rightarrow O + H$. The two neutral-neutral reactions are rapid in shock heated gas but slow in cold gas, and the shock models must be compatible with the OH data. Pineau des Forêts et al. (1986) have found that for reasonable parameters, the one fluid model column densities of CH^+ are at most only about 10^{12} cm^{-2} for shocks which have H_2 rotational level populations that are compatible with the data. The OH column densities do not constrain the one fluid models so severely.

Now consider multifluid models of shocks. Typically, $\Delta > v_s \tau_{cool}$. This implies that the maximum temperature for a multifluid shock model is not as high as that for a single fluid model of a shock with the same speed. However, for normal shock parameters, the multifluid model temperatures are high enough (The neutral temperature is usually about 1000 K.) to drive reaction (2) at a high rate. In addition, the ion-neutral streaming speed is comparable to the shock speed and can help drive the reaction. Hence, the column density of CH^+ for a multifluid model is roughly

$$\frac{n_0(C^+)v_s}{\alpha_3 n_0(H_2)} \frac{\Delta}{v_s \tau_{ph}}$$

Equation (1) shows that Δ depends on the ion number density in the dissipation zone. In the preshock medium, the most abundant ion is C^+, but in the dissipation zone C^+ is converted to molecular ions, which recombine four or five orders of magnitude more rapidly than C^+ does. Flower, Pineau des Forêts, and Hartquist (1985) pointed out that the subsequent drop in the flux of carbon bearing ions through the dissipation zone results in it being an order of magnitude longer than one would approximate by using the preshock ion density in equation (1). The drop in the flux of carbon bearing ions results in H$^+$-neutral scattering being an important source of ion-neutral coupling.

Two groups have independently investigated the production of CH^+ in multifluid models.

Papers by Pineau des Forêts et al. (1986) and Hartquist, Flower, and Pineau des Forêts (1989) and articles referenced in the latter summarize the results of one group. This group concluded that shocks which would produce column densities of CH^+ comparable to that measured towards ζOph would contain more H_2 in highly excited rotational levels than observed in that direction.

In contrast, Draine (1986) presented a shock model with CH^+ and excited H_2 rotational level populations which are comparable to those observed towards ζOph. Draine adopted higher photoionization and photodissociation cross sections for many carbon bearing species and a higher, but acceptable, value for the strength of the radiation field near the ζOph cloud. Hence, the value of τ_{ph} for Draine's model is smaller. Other differences include Draine's adoption of C^+ and H as the primary products of the photodissociation of CH^+ and of an incorrect endothermicity for the reaction $H^+ + O \rightarrow O^+ + H$ which dominates the removal of H^+ in the preshock gas. In reality, the dominant CH^+ photodissociation products are C and H^+ (Kirby (1980)). When Draine's "best fit" ζOph model was run again with these two discrepancies removed, the CH^+ column density dropped by a factor of three (Draine, private communication).

At this date, no shock model which is compatible with the CH^+ and H_2 observations towards ζOph exists.

4. Chemistry Affecting Dissipation in C-Type Shocks in Dense Clouds

If grains were of uniform radius, a, their fractional abundance in a cloud would be roughly 4×10^{-12} $(a/10^{-5}$ cm$)^{-3}$. In low temperature dark clouds, a grain with a = 10^{-5} cm carries on average a charge of -e, as a result of the collisions of electrons with neutral grains being much more frequent than collisions of the more slowly moving ions with neutral grains (Spitzer (1978)). The average number of negative charges on grains increases as the electron temperature rises, and a = 10^{-5} cm grains carry of the order of 10^2 negative charges in shocked gas. Because they are charged the grains may be well-coupled to the magnetic field and, hence, may move with the ions through the neutrals in the dissipation zones of C-type shocks in dense clouds. If so, the grain-neutral friction is roughly

$$10^{-9} \text{ g cm}^{-2} \text{ s}^{-2} \left(\frac{a}{10^{-5}\text{cm}}\right)^{-1} \left(\frac{v_s}{10\text{km s}^{-1}}\right)^2 \left(\frac{\rho_n}{1.4\times10^{-12}\text{g cm}^{-3}}\right)^2$$

In cold dense molecular clouds, the fractional ionization is probably roughly 3×10^{-6} $n^{-1/2}$ where n is the hydrogen nucleon number density in cm^{-3} (e.g. Elmegreen (1979)). The most abundant ions are metallic ions, such as Mg^+, which do not react with H_2 (Oppenheimer and Dalgarno (1974)). In low temperature gas, Mg^+ is formed primarily by the charge transfer of Mg with molecular ions and is removed most rapidly in collisions with grains. Gas phase electrons in low temperature gas are lost mostly by the dissociative recombination of the abundant molecular ions such as HCO^+. The ion-neutral drag in the dissipation zone of a dense cloud C-type shock is roughly

$$10^{-11} \text{g cm}^{-2} \text{ s}^{-2} \left(\frac{x_i}{10^{-8}}\right) \left(\frac{v_s}{10 \text{km s}^{-1}}\right) \left(\frac{\rho_n}{1.4 \times 10^{-12} \text{g cm}^{-3}}\right)^2$$

where x_i is the fractional ionization.

In shocks, x_i drops below its preshock value. The reactions of species like Mg^+ with H_2 are endothermic by about 2 eV but when they are driven by the high temperatures in shocked gas and ion-neutral streaming, they are followed by rapid dissociative recombination. When the temperatures are high, the grains become very charged and the recombination rate of ions onto them increases due to the enhanced Coulomb attraction as well as the higher ion thermal speeds.

Clearly, if grains remain well-coupled to the magnetic field, grain-neutral drag dominates ion-neutral drag in dense cloud shocks.

When the grains carry an insignificant fraction of the negative charge, they will be decoupled from the magnetic field when

$$\omega_c \, \tau_{STOP} = \frac{|\bar{Z}_g| eB}{m_g c} \frac{m_g}{(\pi a^2)\rho v_s} < 1$$

where ω_c, τ_{STOP}, $\bar{Z}_g e$, and m_g are the grain cyclotron frequency, stopping time, average charge, and mass. In all work by Draine and his collaborators on dense cloud shocks, the grain-neutral drag has been evaluated with an expression depending on $\omega_c \, \tau_{STOP}$. The expression is appropriate when the grains do not carry a substantial fraction of the negative charge and is a good approximation for the shock parameters which have been considered.

However, in shocks in denser media ($n \gtrsim 10^{7.5}$ cm^{-3}) such as those which may be associated with interstellar H_2O masers (e.g. Chernoff and McKee (1989)), the grains probably carry most of the negative charge. In such cases, a more complete treatment of the grain-neutral drag is necessary. Work following the suggestions made by Havnes, Hartquist, and Pilipp (1987) is in progress. Preliminary results suggest that in some cases when the grains acquire most of the negative charge, the ExB ion drift velocity, resulting from the electric field which is induced by charge separation between ions and grains subjected to large grain-neutral drag, can become great. The drift velocity may be limited by the onset of ion impact induced ionization of H_2.

5. Shocks and Global Chemical Structure in B5

Shocks may be prevalent enough in some regions in dense clouds to affect their chemical structures (Williams and Hartquist (1984)). Sputtering of grains in such shocks would constitute a mechanism for returning heavy elements to the gas phase; such elements are depleted onto grains on time scales of 10^6 yrs (n/10^3 cm^{-3})$^{-1}$ which are shorter than average cloud lifetimes.

Goldsmith, Langer, and Wilson (1986) have mapped B5 in optically thin $C^{18}O$ emission, and have delineated its clumpy structure. They argued that the cloud morphology and the presence of four T-Tauri stars are consistent with a dynamical cycling model which is reminiscent of the one proposed by Norman and Silk (1980) for dense molecular regions in general.

Charnley et al. (1988) have studied the time dependent chemical structure of gas which passes repeatedly through dynamical cycles which might be appropriate for B5. One such cycle consists of phases during which interclump material collapses to form clumps in which stars are born, the stars move from the centers of the clumps due to the fact that they are subject to gravitational forces only whereas the gas responds in addition to viscous forces, the stellar winds drive bubbles into the clumpy medium, the winds ablate the clumps and ionized material is mixed with clump molecular material in the wind-clump interfaces, the well mixed gas passes through a terminal bow shock where sputtering occurs, and the shell of shocked wind and ablated material and swept-up interclump gas propagates slowly outwardly and decreases in pressure until it merges with the ambient interclump medium. The mixing of H^+ and He^+ from the wind with clump material would drive the chemistry in the direction of atoms. The chemical model results show that approximate limit cycles obtain, and, hence, point by point mapping of a number of chemical species including C^0, C^+, CO, and others can, in principle, be used to diagnose the dynamical cycle in B5.

6. Important Data Needs

The CH^+ column density in diffuse cloud shocks depends on the rates at which photoionization and photodissociation of the species CH_n ($n \gtrsim 1$) and the photodissociation of the species CH_n^+ ($n \gtrsim 1$) occur. Most of the relevant rates are unknown.

The rotational level population distribution of H_2 constrains shock models for the production of CH^+. Danby, Flower, and Monteiro (1987) have considered H_2-H_2 collisional excitation. Further such studies are desirable but not as pressing as they once were.

In shocks in dense molecular clouds collisions of heavy ions with H_2 may constitute an important source of ionization. Knowledge of the cross sections near threshold is desirable.

7. References

Charnley, S.B., Dyson, J.E., Hartquist, T.W., and Williams, D.A. (1988) 'Chemical limit cycles for a region of low mass star formation', M.N.R.A.S., in press.

Chernoff, D.F. and McKee, C.F. (1989) 'Shocks in dense molecular clouds', in T.W. Hartquist (ed.), Molecular Astrophysics - A Volume Honouring Alexander Dalgarno, Cambridge University Press, Cambridge, in press.

Danby, G., Flower D.R., and Monteiro, T.S. (1987) 'Rotationally inelastic collisions between H_2 molecules in interstellar magnetohydrodynamical shocks', M.N.R.A.S. 226, 739-745.

Draine, B.T. (1980) 'Interstellar shock waves with magnetic precursors', Ap.J. 241, 1021-1038.

Draine, B.T. (1986) 'Magnetohydrodynamic shocks in diffuse clouds - III. The line of sight towards ζOph', Ap.J. 310, 408-418.

Draine, B.T., Roberge, W.G., and Dalgarno, A. (1983) 'Magnetohydrodynamic shocks in molecular clouds', Ap.J. 264, 485-507.

Elitzur, M. and Watson, W.D. (1978) 'Formation of molecular CH^+ in Interstellar Shocks', Ap.J. (Letters) 222, 141-144.

Elmegreen, B.G. (1979) 'Magnetic diffusion and ionization fractions in dense molecular clouds: the role of charged grains', Ap.J. 232, 729-739.

Flower, D.R., Pineau des Forêts, G., and Hartquist, T.W. (1985) 'Theoretical studies of interstellar molecular shocks - I. General formulation and effects of the ion-molecule chemistry', M.N.R.A.S. 216, 775-794.

Goldsmith, P.F., Langer, W.D., and Wilson, R.W. (1986) 'Molecular outflows, gas density distribution, and the effects of star formation in the dark cloud Barnard 5', Ap.J. (Letters) 303, 11-15.

Hartquist, T.W., Flower, D.R., and Pineau des Forêts, G. (1989) 'Shock chemistry in diffuse clouds', in T.W. Hartquist (ed.), Molecular Astrophysics - A Volume Honouring Alexander Dalgarno, Cambridge University Press, Cambridge, in press.

Havnes, O., Hartquist, T.W., and Pilipp, W. (1987) 'The effects of dust on the ionization structures and dynamics in magnetized clouds' in G.E. Morfill and M. Scholer (eds.), Physical Processes in Interstellar Clouds, D. Reidel Publishing Company, Dordrecht, pp. 389-412.

Kirby, K. (1980) 'The photodissociation of interstellar CH^+' in B.H. Andrew (ed.), Interstellar Molecules -IAU Symposium No. 87, D. Reidel Publishing Company, Dordrecht, pp. 283-287.

Langer, W.D. (1989), 'Observations of velocity and density structure in diffuse clouds', in T.W. Hartquist (ed.), Molecular Astrophysics - A Volume Honouring Alexander Dalgarno, Cambridge University Press, Cambridge, in press.

McKee, C.F. and Ostriker, J.P. 'A theory of the interstellar medium: Three components regulated by supernova explosions in an inhomogeneous substrate' Ap.J. 218, 148-169.

Mullan, D.J. (1971) 'The structure of transverse hydromagnetic shocks in regions of low ionization', M.N.R.A.S. 153, 145-170.

Oppenheimer, M. and Dalgarno, A. (1974) 'The fractional ionization in dense interstellar clouds', Ap.J. 192, 29-32.

Pineau des Forêts, G., Flower, D.R., Hartquist, T.W., and Dalgarno, A. (1986) 'Theoretical studies of interstellar molecular shocks - III. The formation of CH^+ in diffuse clouds' M.N.R.A.S. 220, 801-824.

Spitzer, L. (1978) Physical Processes in the Interstellar Medium, John Wiley & Sons, New York.

Williams, D.A. and Hartquist, T.W. (1984) 'On C^o and CO in dense interstellar clouds - evidence that cloud material is frequently shocked', M.N.R.A.S. 210, 141-145.

CHEMICAL EFFECTS OF INTERSTELLAR GRAINS

David A. Williams
Department of Mathematics
UMIST
Manchester   M60 1QD
United Kingdom

ABSTRACT The chemical effects of interstellar grains are briefly reviewed. Their dominant chemical role is to catalyze the formation of H_2 which is the seminal molecule for efficient gas phase chemistry. In regions of at least moderate extinction grains accumulate molecular mantles of CO, H_2O, etc. Solid state chemistry in such mantles may produce molecules of a type or in an abundance not achievable in the interstellar gas. Return of mantle material to the gas can - at least transiently - dominate gas phase chemistry. It is argued that the freeze-out of heavy atomic and molecular species on to grain surfaces limits the time available for chemistry, restricts molecular cloud chemistry to a "young" character, and suggests that chemical models of molecular clouds must have cyclic dynamics. Such models are briefly described.

2. Introduction

Some of the chemical effects of interstellar grains are listed in Table 1. Obviously, atoms locked in grains are unavailable in the interstellar gas and, thus, the presence of grains immediately affects possible chemistry. Table 2 indicates the approximate fraction of some elements available in the gas. The sensitivity of the chemistry of molecular clouds to, e.g., the C:O ratio has been explored by Watt (1984).

Table 1.   Chemical Effects of Interstellar Grains

Lock up certain elements (Si, Fe, Mg) in refractory cores
Catalyze H_2 formation; formation of other molecules?
Extinguish starlight, shield cloud interiors
Heat diffuse gas by photoelectric effect
Provide sites of mantle formation
Mantle processing in solid state
Storage of molecular mantles, release to gas
Role of grains and mantles in star formation

D. McNally (ed.), Highlights of Astronomy, Vol. 8, 383–386.

Table 2.   Abundances of elements in gas
and dust (by number)

Element	Fraction in	
	gas	dust
C	0.5	0.5
N	0.6	0.4
O	0.6	0.4
Mg	0.2	0.8
Si	0.05	0.95
S	0.6	0.4
Fe	0.02	0.98
Others	0	1.0

Perhaps the most significant effect of grains on interstellar chemistry is the efficient catalysis of H_2 in surface reactions. The observations require that in diffuse clouds almost every H atom striking a grain surface must leave as part of an H_2 molecule. Modern views of the nature of grains in terms of silicaceous and carbonaceous materials allow a more detailed description of the surface reaction, and the state of the nascent H_2 molecule can be predicted (Duley and Williams 1986). It appears likely that H_2 is kinetically and vibrationally excited on ejection from grain surfaces. However, radiative and collisional relaxation will generally ensure that the molecule ends up in an excited rotational state from which radiative relaxation is relatively slow.

The formation of molecules other than H_2 may also occur on grains. The observation of H_2O ice mantles in moderate extinction interstellar clouds implies that $O \rightarrow H_2O$ conversions occur *in situ* (Jones and Williams 1984). Ion reactions at charged defect sites on grains are sufficiently exothermic that products are ejected (e.g. Jones and Williams 1986 and references therein). The significance of surface chemistry to diffuse cloud chemistry is at present uncertain (Mann and Williams 1984, 1985). However, there is no doubt that the molecular mantles play an important part in the chemistry of dark clouds.

3.   Mantle Growth and Gas Phase Chemistry

Mantle composition reflects the current chemical composition of the gas, responds to chemical processing in the solid state, and is affected by any possible mantle limitation processes. Much laboratory and theoretical work has been performed describing the nature of mantle material, and comprehensive models of cloud chemistry, including both gas and solid components have been constructed (cf. d'Hendecourt et al. 1982). Models of specific molecular clouds such as TMC1, on simple assumptions of freeze-out and hydrogenation reactions, are in reasonable agreement with observations and expectations (Brown and Charnley 1988). In such models an "age" of the cloud can be inferred: for TMC1 it appears to be about 2×10^6 yr.

Such simple static models, however, suffer from a number of defects. For example: can freeze-out of heavy species on to grains occur without limit? Another class of objects seems better described by clouds which are undergoing collapse, in which gas phase chemistry and freeze-out of heavy species on to grains are simultaneously occurring (cf. Tarafdar, this volume). The Hot Core in Orion has a number of interesting features which make it distinct from cold molecular clouds: e.g., NH_3, CH_3OH and HDO seem strongly enhanced, and the HCN/HNC ratio is anomalously high. An explanation of this behaviour has been given by Pauls et al. (1983) in terms of the evaporation of molecular mantles (formed during a collapse phase) following star formation. This suggestion has been explored in detail by Brown et al. (1988) who conclude that such a model is capable of accounting for molecular abundances observed in the Hot Core. Some molecules (e.g. HCN, H_2CO) in the Hot Core were apparently formed in the cold gas, stored in the mantle and released in the hot phase, while others (H_2O, NH_3) arise from hydrogenation of atoms in the mantles. In particular, it appears that NH_3 must be deuterated (to NH_2D) on grain surfaces during the cold phase.

4. Dynamics

While it is encouraging that models like that of Brown et al. (1988) incorporating both gas and solid phase chemistry, cloud collapse and star formation have met with some success, it is clear that the actual situation in star formation regions will be considerably more complicated. The association of dark clouds and young stars has been well studied (cf. Fuller and Myers 1987) and the prevalence of high velocity flows from such stars is well established (Lada 1985). Charnley et al. (1988a) have explored the chemistry in a model of molecular clouds in which a sequence of events is supposed to occur: first, the wind of a newly formed star erodes dense cloud material; this mass-loaded wind is brought to local rest by a shock, accumulates and forms a new dense clump in which chemistry and mantle formation occur and from which a new generation star is formed. In such models chemistry is never in steady state, and chemical abundances reflect to some extent the star forming activity. The chemistry is generally "young" in the sense that not all the available carbon is processed into CO.

Such models imply that chemical and dynamical states in molecular clouds are closely linked and cyclic. Charnley et al. (1988b) have addressed the question: does the chemical state of the gas follow the dynamical cycle or does chemical chaos ensue? In two illustrative models, involving both long and short cycle times, their results indicate that chemical limit cycles exist and, therefore, that chemistry is a proper tracer of the physical conditions in molecular clouds. At one point in each cycle mantled grains are exposed to fast stellar winds: thus, the limitation of mantle growth has a natural explanation, yet mantle growth has its restricting effect on the chemistry. In the repeated cycling of the material, all memory of the initial conditions is lost. Such models predict the abundances of molecules point-by-point in clouds. Thus, the predictions of the model may be tested in relatively nearby clouds. Of these B5 (Goldsmith et al. 1986) appears to be an ideal example. It contains IRAS sources, dense clumps of gas, and molecular flows. Point-by–point molecular studies of B5 should indicate the validity of the dynamical/chemical model for molecular clouds.

5. Conclusion

Dust grains have a profound effect on interstellar chemistry in diffuse clouds by shielding cloud interiors, by photoelectric heating, and - primarily - by catalyzing the formation of H_2 (without which the chemistry is strongly suppressed). In dark clouds, mantles accummulate on grains and solid state reactions occur. This further loss of material from the gas influences chemical developments there. The return of mantle material, possibly stimulated by star formation, can lead to important observational consequences. The timescale for freeze-out is relatively short, and implies the need for efficient mantle removal. In the models described here, this mantle removal is associated with star formation.

REFERENCES

Brown, P.D., Charnley, S.B. & Millar, T.J. 1988 Mon. Not. R. astr. Soc. 231, 409.
Brown, P.D. & Charnley, S.B. 1988 in preparation.
Charnley, S.B., Dyson, J.E., Harquist, T.W. & Williams, D.A. 1988a. Mon. Not. R. astr. Soc. 231, 269.
Charnley, S.B., Dyson, J.E., Harquist, T.W, & Williams, D.A. 1988. Mon. Not. R. astr. Soc. in press.
Duley, W.W. & Williams, D.A. 1986. Mon. Not. R. astr. Soc. 223, 177.
Fuller. G.A. & Myers. P.C. 1987, Physical Processes in Interstellar Clouds, G.E. Morfill & M. Scholer (eds), D. Reidel Pub. Co., p.137
Goldsmith, P.F., Langer, W.D. & Wilson, R.W. 1986. Astrophys. J. Lett. 303, L11.
d'Hendecourt, L.B., Allamandola, L.J., Baas, F. & Greenberg, J.M. 1982. Astron. Astrophys. 109, L12.
Jones, A.P. & Williams, D.A. 1984, Mon. Not. R. astr. Soc. 209, 955.
Jones, A.P. & Williams, D.A. 1986 Mon. Not. R. astr. Soc. 219, 441.
Lada, C. 1985. Ann. Rev. Astron. Astrophys. 23, 267.
Mann, A.P.C. & Williams, D.A. 1984 Mon. Not. R. astr. Soc. 209, 33.
Mann, A.P.C & Williams, D.A. 1986 Mon. Not. R. astr. Soc. 213, 295.
Pauls, T.A., Wilson, T.L., Bieging, J.H. & Martin, R.N. 1984. Astron. Astrophys. 124, 33.
Tarafdar, S. 1988, this volume.
Watt, G.D. 1984. Mon. Not. R. astr. Soc. 212, 93.

THE VOLATILE COMPOSITION OF COMETS

H. A. WEAVER
Space Telescope Science Institute
3700 San Martin Drive
Baltimore, MD 21218
U.S.A.

ABSTRACT. Comets may be our best probes of the physical and chemical conditions in the outer regions of the solar nebula during that crucial period when the planets formed. The volatile composition of cometary nuclei, in particular, can be used to decide whether comets are the product of a condensation sequence similar to that invoked to explain the compositions of the planets and asteroids, or if comets are simply agglomerations of interstellar grains which have been insignificantly modified by the events that shaped the other bodies in the solar system. Although cometary nuclei are not generally accessible to observation, observations of cometary comae can illuminate at least some of the mysteries of the nuclei provided one has a detailed knowledge of the excitation conditions in the coma and also has access to basic atomic and molecular data on the many species present in comets. This paper examines the status of our knowledge of the volatile composition of cometary nuclei and discusses how these data are obtained.

1. Introduction

The volatile composition of a comet is a sensitive indicator of its formation environment. In particular, the relative abundances of molecules in cometary nuclei can be used to infer the physical and chemical state of either the solar nebula or the interstellar cloud from which the nebula condensed. However, measuring the volatile inventory in comets is a demanding task that involves combining *in situ* spacecraft data with observations from ground-based, airborne, rocket, and Earth-orbiting observatories.

2. Molecules in Comet Nuclei

2.1 THE DOMINANT VOLATILE: H_2O

During the past two decades cometary observers built a strong circumstantial case for H_2O

D. McNally (ed.), Highlights of Astronomy, Vol. 8, 387–393.
© *1989 by the IAU.*

ice as the dominant volatile in nuclei. However, the apparition of comet Halley provided the first opportunity to observe the H_2O molecule *directly*. Infrared (IR) observations of Halley in December 1985 from the Kuiper Airborne Observatory (KAO) [1] revealed intense, solar-pumped fluorescent emission in the ν_3 band of H_2O near $\lambda \sim 2.65\mu m$. This same band was subsequently observed in Halley post-perihelion by the IR spectrometer on Vega (the IKS experiment) [2] and from the KAO [3,4]. Measurements by the neutral mass spectrometer (NMS) on Giotto demonstrated that H_2O comprised $\geq 80\%$ of the total volatile inventory of comet Halley [5].

Water is not particularly well-suited for constraining the formation conditions of comets because its sublimation temperature is relatively high ($T_{sub} \sim 150$ K), and H_2O is expected to be a relatively abundant molecule in the gas phase throughout much of the solar nebula. Thus, measuring the relative abundances of the more volatile *trace* species in cometary nuclei is required for establishing the connection between comets and interstellar grains and between comets and possible solar nebula condensates.

2.2 MAJOR CARBON-BEARING SPECIES: CO, CH_4, CO_2, H_2CO

Carbon monoxide is ubiquitous in the interstellar medium and is also predicted to be the dominant carbon-bearing gas in non-equilibrium models of the solar nebula [6]. As illustrated in Table 1, CO may also be the second-most abundant volatile in cometary nuclei. Prior to Halley, CO had been observed definitely in only two comets: the CO fourth positive group at $\lambda \sim 1500$ Å was detected during sounding rocket observations of comet West (1976 VI) [7] and during IUE observations of comet Bradfield (1979 X) [8]. The derived CO abundance in comet West was huge ($\sim 30\%$ of the H_2O abundance). Similar observations were conducted on comet Halley [9], also showing a high CO/H_2O abundance ($\sim 17\%$ in this case). On the other hand, the UV observations of comet Bradfield yielded $CO/H_2O \sim 0.02$ [10]. Does the relative abundance of CO vary from comet to comet? The Giotto NMS measurements of CO in the coma of comet Halley provided some interesting insight into this question [11]. The NMS observations showed that only $\sim 7\%$ (or less) of the CO in the coma of Halley could be derived by sublimation from the nucleus. The rest of the CO in the coma is apparently produced from a distributed source, possibly due to the evaporation of organic mantles on cometary grains (the so-called CHON particles). Viewed from this new perspective, both the high CO abundances derived from UV observations of comets West and Halley, and the low CO abundance derived from the UV observation of comet Bradfield, are consistent. Whenever a large field-of-view (FOV) is used *and* the comet has a large dust production rate (the UV rocket observations of both West and Halley satisfied these conditions), a high abundance of CO will be observed in the coma due to extra CO production from grain evaporation. Whenever the FOV is small and/or the dust production rate is low, then the coma CO observation should be sampling something close to the "true" abundance in the nucleus (we are making a distinction here between the CO stored in the ice versus the presence of C-O bonds in complex hydrocarbons that may be coating cometary grains).

Equilibrium condensation models of the solar nebula indicate that CH_4 should be the dominant carbon-bearing gas in the outer regions of the nebula. Indeed, for many years people often referred to comets as being composed of "water, methane, and ammonia" although the latter two molecules had never been observed in comets and the indirect evidence for their presence indicated that they were only trace constituents. Recent non-equilibrium chemical models of the solar nebula demonstrate that CH_4 should be extremely depleted relative to CO in the nebula, except near the sub-nebulae associated with the Giant planets [6].

Spectroscopic observations of the ν_3 band of methane near $\lambda \sim 3.3 \mu m$ have yielded the first direct evidence for the presence of CH_4 in comets. The R(1) line of this band was marginally detected in comet Halley [12] while an average over six R-branch lines in this band gave a tentative detection during KAO observations of comet Wilson (1986l) [13]. The interpretation of these results in terms of the CH_4 abundance is somewhat problematical. The derived abundance can vary significantly depending on the choice of the "effective" excitation temperature of the CH_4 molecules in the coma, especially if only one line is used. Some of this uncertainty is reflected in the range of values given in Table 1. We simply note that the above-referenced KAO observations of H_2O indicate that the appropriate excitation temperature of CH_4 is probably ~ 50 K implying that the CH_4 abundance is closer to 1% than 5%.

Strong evidence for the presence of CH_4 in the coma of comet Halley was also obtained by analysis of data taken by the ion mass spectrometer (IMS) on Giotto [21]. Using the CH_3^+ ion as a probe of the CH_4 abundance in the coma, and by comparing the observed ion spectrum to a model spectrum containing all known important ion-molecule reactions, the IMS data indicate that $CH_4/H_2O \sim 0.02$. The uncertainty in this number is difficult to establish owing to the complexity of the modeling, but the nominal value reported is consistent with the results from the spectroscopic investigations.

Both the NMS experiment on Giotto and the IR spectrometer on Vega detected CO_2 in the coma of comet Halley and both measurements yield a CO_2 relative abundance of $\sim 3\%$ [5,15]. Assuming that the CO_2 observed in these experiments originated in the nucleus, the CO_2 abundance was approximately one-half of the CO abundance in Halley. UV observations of several comets indicate that the CO_2/H_2O abundance does not vary significantly from comet-to-comet (Feldman, private communication). If this is true, then CO_2 and CO may have comparable abundances in some comets (e.g., comet Bradfield [1979X]).

There was a tentative detection of H_2CO in Halley via IR fluorescence in a vibrational band near $\lambda \sim 3.5 \mu m$ by the IKS experiment [15]. The derived abundance from the IKS is given in Table 1 as an upper limit. If the detection stands, it indicates that H_2CO is an important trace constituent in comets and may contribute significantly to the "extended" source of CO in the coma.

2.3 MAJOR NITROGEN-BEARING SPECIES: NH_3, N_2, HCN

There has only been one direct observation of NH_3 in a comet: a radio line was marginally detected in comet IRAS-Araki-Alcock (1983d) [16]. The derivation of an NH_3 abundance from these data is extremely model-dependent; the radio observers reported $NH_3/H_2O \sim 0.06$ while I have derived a ratio of ~ 0.003 from the same data.

Since $\geq 90\%$ of the photodissociation of NH_3 by solar UV radiation results in the production of NH_2, visible observations of the NH_2 molecule should provide reasonably accurate measurements of the NH_3 abundance. Observations of the NH_2 band at $\lambda \sim 5700$ Å in comet Halley gave $NH_3/H_2O=0.003$ [17].

Both NH_3 and CH_4 are required to explain the Giotto IMS data [14] (see the earlier discussion on CH_4). An NH_3 relative abundance of $\sim$1-2% gives the best fit of the model to the data. However, the accuracy of this result has not yet been determined due to the complexity of the modeling, and the uniqueness of the interpretation has also been challenged [18].

The abundance of N_2 in comets is very difficult to establish with confidence. There are no known spectroscopic signatures of the neutral molecule and since its mass is coincident with that of CO, even *in situ* observations with mass spectrometers suffer from a confusion problem. Using the small difference in ionization potential between N_2 and CO, the Giotto NMS data were used to obtain $N_2/H_2O \leq 0.1$ [11].

The N_2^+ ion is observed in the visible spectra of comets near $\lambda \sim 3914$ Å. Since the ionization properties of N_2 and CO are similar, and since the CO/H_2O abundance can be determined by other means, the ratio of the intensities of the visible bands of N_2 and CO^+ can be used to infer the ratio of $N2/H_2O$. Using this technique for the Halley observations yields $N_2/H_2O \leq 0.001$ (Feldman and Wyckoff, private communication).

Fluorescent emission from CN is usually the strongest feature in the visual spectra of comets. Most of the CN in the coma is probably produced by the photodissociation of HCN that has sublimated from the nucleus, although some may have been produced as a result of evaporation of organic-coated grains [19] or from photodissociation of more complex molecules [20]. The HCN molecule is detected via a rotational transition at radio frequencies, and the apparition of Halley witnessed the first systematic investigation of HCN in comets [21]. The derived abundance from these observations is $HCN/H_2O \sim 0.001$.

2.4 MAJOR SULFUR-BEARING SPECIES: CS_2, S_2

Fluorescent emissions from CS and S are a prominent features in UV spectra of comets [22]. The most plausible parent for CS is the CS_2 molecule [23], although CS_2 has never been observed directly. Assuming that all of the CS observed in the coma comes from CS_2, then $CS_2/H_2O \sim 0.001$, i.e., CS_2 is a trace constituent of the nucleus.

UV observations of the "Earth-grazing" comet IRAS-Araki-Alcock (1983d) dramatically revealed the presence of S_2, a molecule which had never been observed previously in any astrophysical environment [24]. Since the S_2 discovery coincided with an "outburst" in cometary activity, and since the signal virtually disappeared during later observations, the presence of S_2 in IRAS-Araki-Alcock may not be typical. The abundance derived from the UV observations during the outburst is $S_2/H_2O \sim 0.001$.

The presence of the OCS molecule has been invoked to explain the spatial distribution of S atoms in the coma [25]. Although this hypothesis is speculative since OCS has never been detected in comets, it is also clear that if OCS is present it cannot be more than a trace constituent of the nucleus.

3. Conclusion

Table 1 summarizes the abundance data on cometary volatiles discussed in this paper. All values are expressed as volume abundances relative to H_2O. Although the ability to compile a table like this with reasonable confidence represents a significant improvement in our knowledge of the nature of cometary matter, several cautionary remarks must be made. First, the abundances of trace species have been measured in very few comets. Table 1 is heavily weighted toward results from Halley since the best data generally come from this comet. Systematic observations of trace species in both "new" and "old" comets are desperately needed to establish whether the results from Halley, and the handful of other comets mentioned here, are "typical" of comets as a class. Second, even when identifications of species are secure, the conversion from raw data into nucleus abundances frequently introduces significant uncertainties (factors of two or more are common). Finally, comets often exhibit dramatic temporal variability which may suggest that the observed composition is not constant in time (e.g., the case of S_2). Despite these present shortcomings, we are well on our way toward a more quantitative understanding of the link between comet compositions and cometary formation environments.

4. Acknowledgements

I would like to express thanks to Bruce Fegley, Paul Feldman, Jonathan Lunine, Hal Larson, and Mike Mumma for many stimulating discussions on the topic of this paper.

TABLE 1. Volatile abundances in comet nuclei

Molecule	Abundance
H_2O	1
CO	0.02 – 0.07
CH_4	0.01 – 0.05
CO_2	~0.03
H_2CO	≤0.05
NH_3	0.003 – 0.02
N_2	≤0.001
CS_2	≤0.001
S_2	≤0.001
HCN	~0.001

5. References

1. Mumma, M.J., Weaver, H.A. , Larson, H.P., Davis, D.S. and Williams, M. (1986) 'Detection of water vapor in Halley's comet', Science, 232, 1523-1528.

2. Combes, M., et al. (1986) 'Infrared sounding of comet Halley from Vega 1', Nature, 321, 266-268.

3. Larson, H.P., Mumma, M.J., Weaver, H.A. and Davis, D.S. (1986) 'Velocity-Resolved observations of water in comet Halley', Ap. J. Lett., 309, L95-L99.

4. Weaver, H.A., Mumma, M.J., Larson, H.P. and Davis, D.S. (1986) 'Post-Perihelion observations of water in comet Halley', Nature, 324, 441-444.

5. Krankowsky, D., et al. (1986) 'In situ gas and ion measurements at comet Halley', Nature, 320, 326-329.

6. Prinn, R.G. and Fegley, B. (1988) 'Solar nebula chemistry: origin of planetary, satellite, and cometary volatiles', in S. Atreya, J. Pollack, and M. Matthews (eds.), Origin and Evolution of Planetary and Satellite Atmospheres, University of Arizona Press, in press.

7. Feldman, P.D. and Brune, W.H. (1977) 'Carbon production in comet West 1975n', Ap. J. Lett., 209, L45-L48.

8. Feldman, P.D., et al. (1981) 'IUE Observations of the UV spectrum of comet Bradfield, Nature, 286, 132-135.

9. Woods, T.N., Feldman, P.D., Dymond, K.F. and Sahnow, D.J. (1986) 'Rocket ultraviolet spectroscopy of comet Halley and abundance of carbon monoxide and carbon', Nature, 324, 436-438.

10. A'Hearn, M.F. and Feldman, P.D. (1980) 'Carbon in comet Bradfield (1979l)', Ap. J. Lett., 242, L187-L190.

11. Eberhardt, P., et al. (1987) 'The CO and N_2 abundance in comet p/Halley', Astron. Ap., 187, 481-484.

12. Kawara, K., Gregory, B., Yamamoto, T. and Shibai, H. (1988) 'Infrared spectroscopic observation of methane in comet Halley', Astron. Ap., in press.

13. Larson, H.P., Weaver, H.A., Mumma, M.J. and Drapatz, S. (1988) 'Airborne infrared spectroscopy of comet Wilson (1986l) and comparisons with comet Halley', Ap. J., in press.

14. Allen, M., et al. (1987) 'Evidence for methane and ammonia in the coma of comet p/Halley', Astron. Ap., 187, 502-512.

15. Moroz, V.I., et al. (1987) 'Detection of parent molecules in comet p/Halley from the IKS-Vega experiment', Astron. Ap., 187, 513-518.

16. Altenhoff, W.J., et al. (1983) 'Radio observations of comet 1983d', Astron. Ap., 125, L19-L22.

17. Wyckoff, S., Tegler, S., Wehinger, P., Spinrad, H. and Belton, M.J.S. (1988) 'Abundances in comet Halley at the time of the spacecraft encounters', Ap. J., 325, 927-938.

18. Marconi, M.L. and Mendis, D.A. (1988) 'On the ammonia abundance in the coma of Halley's comet', Ap. J., 330, 513-517.

19. A'Hearn, M.F., et al. (1986) 'Cyanogen jets in comet Halley', Nature, 324, 649-651.

20. Bockelée-Morvan, D. and Crovisier, J. (1986) 'Possible parents for the cometary CN radical: photochemistry and excitation conditions', Astron. Ap., 151, 90-100.

21. Schloerb, F.P., Kinzel, W.M., Swade, D.A. and Irvine, W.M. (1987) 'Observations of HCN in comet p/Halley', Astron. Ap., 187, 475-480.

22. Weaver, H.A., et al. (1982) 'IUE observations of faint comets', Icarus, 47, 449-463.

23. Jackson, W.M., Halpern, J.B., Feldman, P.D. and Rahe, J. (1982) 'Production of CS and S in comet Bradfield (1979X)', Astron. Ap., 107, 385-389.

24. A'Hearn, M.F., Feldman, P.D., and Schleicher, D.G. (1983) 'The Discovery of S_2 in comet IRAS-Araki-Alcock 1983d', Ap. J. Lett., 274, L99-103.

25. Azoulay, G. and Festou, M.C. (1985) 'The abundance of sulfur in comets', in C.-I. Lagerkvist, B.A. Lindblad, H. Lundstedt and G. Rickman (eds.), Comets, Asteroids, and Meteors II, pp. 273-276.

ATMOSPHERES OF PLANETS AND THEIR SATELLITES

DARRELL F. STROBEL
Departments of Earth and Planetary Sciences &
Physics and Astronomy
The Johns Hopkins University
Baltimore, MD 21218   USA

ABSTRACT. A general review of the chemistry of planetary atmospheres was given with the emphasis on the outer planets and their satellites. The impact of the scientific findings of the Voyager Mission on our understanding of the chemistry and composition of the Jovian, Saturnian, and Uranian atmospheres was highlighted. Hydrocarbon photochemistry will be treated in the context of comparative planetary atmospheres and illustrated by important similarities and differences. The interesting chemistry of satellites with significant atmospheres, Titan, Triton, and Io was explored in the evolutionary context and the impact on the magnetospheres of their parent planet.
This subject matter is available in written form and may be found in the following references:

Strobel, D. F. (1983) 'Photochemistry of the Reducing Atmospheres of Jupiter, Saturn, and Titan', Intl. Rev. Phys. Rev. 3, 145.
Strobel, D. F. (1985) 'The Photochemistry of the Atmospheres of the Outer Planets and Their Satellites', in J. S. Levine (ed.), The Photochemistry of Atmospheres, Academic Press, New York, pp. 393-434.

D. McNally (ed.), Highlights of Astronomy, Vol. 8, 395.
© *1989 by the IAU.*

6. DISKS AND JETS ON VARIOUS SCALES IN THE UNIVERSE

Scientific Organizing Committee

J. Dyson (Chairman)

G.V. Bicknell, G. Miley, C. Norman, M. Rees

Supporting Commissions

28, 34, 40, 47, 48

THE FAR-INFRARED (IRAS) EXCESS IN ROBERTS 22 AND RELATED OBJECTS

M.PARTHASARATHY
Indian Institute of Astrophysics
Bangalore-560034,India

ABSTRACT: From an analysis of the IRAS data of Roberts 22,M1-92,M2-9,OH 231.8+4.2,M1-91,MWC 922,Hen 401,Mz-3,OH 19.2-1.0 and OH 26.5+0.6 it is found that the characteristics of the dust shells or disks around these objects are similar to that observed in planetary nebulae. These ten objects may be described as transition objects evolving from the tip of AGB towards left in the HR diagram. The bipolar and disk geometry of the dust envelopes around these objects may be the result of large angular momentum of the progenitor star or the central objects may be evolved binary systems embedded in thick disks formed from severe mass loss.

1.INTRODUCTION

Roberts 22 is a bipolar nebula and a strong OH maser source (OH 284.18-0.79) showing intense emission on both the 1612 MHz and 1665 MHz transitions (Allen et.al.,1980). The other objects similar to Roberts 22 are M1-92,CRL 618,CRL 2688,Red Rectangle (HD 44179),Hen 401,IRC +10420,M2-9,M1-91,Mz-3, and OH 231.8+4.2. Most of these objects show bipolar structure,infrared excess, and a large ratio of the fluxes f_{IR}/f_{vis}.

These objects are not associated with young stellar objects,and star forming regions. Calvet and Cohen (1978) and Morris (1981) suggested that these bipolar nebulae like M1-92,Hen 401,M2-9,M1-91,Mz-3 and also other objects mentioned above are in the early stages of the formation of planetary nebulae. Morris (1981) proposed that these bipolar nebulae resulted from non - corotating binary systems in which severe mass loss from the evolved component forms a disk in the equatorial plane around the system. Recent OH observations show that OH 19.2-1.0 and OH 26.5+0.6 are bipolar nebulae with characteristics similar to the objects described above. In this paper an analysis of the IRAS observations of ten bipolar nebulae described above is reported.

2. IRAS OBSERVATIONS

The IRAS fluxes of the bipolar nebulae mentioned above are obtained from the IRAS point source catalogue (Beichmann et.al.,1985). The IRAS fluxes for ten objects at 12μm,25μm,60μm, and 100μm are listed in Table 1. The LRS 8 - 21μm spectra of these sources (Olnon and Raimond,1986) are also used in the present analysis.

399

D. McNally (ed.), Highlights of Astronomy, Vol. 8, 399–401.
© *1989 by the IAU.*

3.ANALYSIS

The LRS 8-21μm spectra of the objects listed in table 1 shows only continuum radiation by the dust. The flux maximum is around 25μm or 60μm. The 12μm to 100μm flux distribution clearly shows the large infrared emission from these objects. The 12μm to 100μm fluxes are integrated and are listed in Table 2. The temperatures T_d of the dust envelopes are derived from the far infrared flux distributions and colours. The dust temperatures (Table 2) are found to be in the range of 100 to 140K. For MWC 922 and OH 19.2-1.0 the dust temperature is 200K. The distances of these objects are taken from the literature and are listed in Table 2. The far infrared luminosities L_{IR} are derived from the integrated fluxes and using the distances. The L_{IR} values are found to range from 4.5x10^4 L$\odot$ for Roberts 22 to 9.5x10^2 L$\odot$ for OH 26.5+0.6. The mass M_d of the dust shells are estimated from the equation M_d = 4/3 (a d^2 f$_\nu$ f /Q$_\nu$B$_\nu$(T$_d$)), for details see Hilderbrand(1983). The masses of the dust envelopes for the ten objects are listed in Table 2, and are found to be in the same range as that found in planetary nebulae (Pottasch et.al.,1984). If the ratio of gas to dust mass is about 100 as it is in the interstellar medium then the total shell mass around Roberts 22 is 0.12M$\odot$. The characteristics of the dust shells of these ten bipolar nebulae (Table 2) are found to be similar to the characteristics of dust shells around evolved stars and planetary nebulae. These ten objects are not associated with star forming regions and young stellar objects. The most likely explanation for the presence of dust shells around these ten objects with properties similar to that found in planetary nebulae is that they experienced severe mass loss in the recent past on their AGB stage of evolution. The mass loss rate from the central stars in these objects is of the order of 10^{-4} to 10^{-5} M$\odot$ per year. The central stars in these objects appears to be evolving from the tip of AGB towards left in the HR diagram. These ten objects may be described as transition objects. Peimbert and Peimbert (1983) listed 29 planetary nebulae of Type I,most of them are bipolar. The far infrared (IRAS) characteristics of these Type I planetary nebulae are found to be similar to the objects described here (Table 2). The bipolar Type I planetary nebulae are found to be He and N rich. Some of the objects discussed here (M2-9,CRL 618, Mz-3) are also found to be N rich similar to Type I planetary nebulae. The ten objects (Tables 1 and 2) discussed here appear to be related to bipolar planetary nebulae. The bipolar disk geometry of Roberts 22 and other objects may be similar to that of M2-9. Aspin et al.,(1988) obtained high spatial resolution IR images of M2-9. They found a predominant disk-like structure 20$''$ in size stretching across the core region of M2-9. Peimbert and Peimbert (1983) suggest that the bipolarity is due to the high angular momentum of their progenitor stars and they estimate the progenitors masses to be in the range of 2<M/M$\odot$<6. However Morris (1981) proposed that these are young planetary nebulae in which the central objects are binaries and the bipolar structure and disk in the equatorial plane is the result of mass loss from the evolved components. The detection of central stars and their radial velocity variation study and also CNO abundance analysis may enable us to understand the evolutionary stage of these objects.

Table 1.Bipolar PN type objects similar to Roberts 22

	mag	Sp	IRAS fluxes (Jansky)			
			12μm	25μm	60μm	100μm
Roberts 22	16	A2Ie	200.0	1091.4	588.0	<271.6
M1-92	12	B1V	17.6	59.7	118.0	67.0
M2-9	13.7	B1e	50.6	110.2	123.1	74.8
OH231.8+4.2		M6I	19.0	226.3	548.3	292.1
M1-91	11.8	B0	3.9	8.3	12.0	9.4
MWC 922	12.5	Be	336.0	595.0	252.8	
Hen 401	13.5	Be	4.2	38.3	75.9	41.1
Mz-3	14.1	B0	89.1	340.5	276.4	111.2
OH19.2-1.0			14.8	32.7	20.0	
OH26.5+0.6			359.8	633.7	463.1	309.3

Table   2.Luminosities,Temperatures, and Masses of the   Dust
Envelopes

	d(kpc)	$F_{IR}(\mathrm{Wm}^{-2})$	$L_{IR}(L\odot)$	$T_d(K)$	$M_d(M\odot)$
Roberts 22	2.5	2.3×10^{-10}	4.5×10^{4}	140	1.2×10^{-3}
M1-92	3.0	14.8×10^{-12}	4.2×10^{3}	100	4.3×10^{-4}
M2-9	3.0	22.8×10^{-12}	6.4×10^{3}	140	1.7×10^{-4}
OH231.8+4.2	1.3	58.6×10^{-12}	3.1×10^{2}	80	7.8×10^{-4}
M1-91	3.1	2.0×10^{-12}	6.0×10^{4}	130	0.2×10^{-4}
MWC 922	2.0	1.7×10^{-10}	2.1×10^{3}	200	1.4×10^{-4}
Hen 401	4.0	7.6×10^{-12}	3.8×10^{3}	100	3.9×10^{-4}
Mz-3	2.0	57.0×10^{-12}	7.1×10^{3}	140	1.9×10^{-4}
OH19.2-1.0	4.5	5.5×10^{-12}	3.5×10^{2}	200	2.2×10^{-5}
OH26.5+0.6	0.5	121.0×10^{-12}	9.5×10^{2}	150	0.2×10^{-4}

References
Allen,D.A.,Hyland,A.R.,Caswell,J.L.1980,Mon.Not.R.Astr.Soc.192,505
Aspin,C.,McLean,I.S.,Smith,M.G.1988,Astron.Astrophys.196,227
Beichman,C.A.,et.al.,1985,IRAS point source catalogue,JPL
Chapman,J.M.1988,Mon.not.R.Astr.Soc.230,415
Calvet,N.,Cohen,M.1978,Mon.Not.R.Astr.Soc.182,687
Hilderbrand,R.H.1983,Quart.J.Roy.Astron.Soc.24,267
Morris,M.1981,Astrophys.J.249,572
Olnon,F.M.,Raimond,E.1986,Astron.Astrophys.Suppl.65,607
Peimbert,M.,Peimbert,S.T.1983,IAU Symposium 103 p.233
Pottasch,S.R.,et.al.,1984,Astron.Astrophys.138,10

RECENT OBSERVATIONS OF THE BEAMS IN SS 433

R.C. VERMEULEN
Leiden Observatory
P.O. Box 9513
2300 RA Leiden
The Netherlands

ABSTRACT. Preliminary results are presented of an intensive multifrequency observing campaign on SS 433 conducted in 1987. There were participants from all over the world; they brought all the observations together at the 1988 Dwingeloo workshop on SS 433. With a rapid sequence of VLBI maps, we have been able to follow the onset and evolution of a series of radio flares. We have also obtained a wealth of optical spectra allowing studies of the variability of the Doppler shifted lines on timescales of hours. There is little correlation between the activity in the radio regime, and that seen in the Doppler shifted lines.

1. Introduction

A decade has now elapsed since the discovery of the unusual nature of the binary system SS 433. An excellent review of the knowledge gathered during the first five years has been given by Margon (1984). The single most peculiar phenomenon in SS 433, the "moving lines" in its optical spectrum, was by that time thoroughly described in terms of the "kinematic model", which was first given in nearly its present form by Abell and Margon (1979). The kinematic model assumes only that matter is ejected from a component of SS 433 in two oppositely oriented collimated beams, and that the beam axis precesses. The physical conditions governing the ejection, collimation and propagation of the beams are not specified.

The kinematic model parameters were first determined from the optical moving lines, which show the radial velocity component of the beams. Later, radio maps, which show the shape and motion of the beams projected onto the plane of the sky, were made with the VLA by Hjellming and Johnston (*e.g.*, 1985) and with VLBI (*e.g.*, Vermeulen *et al.* (1987) and Romney *et al.* (1987)). These maps confirmed the kinematic model parameters and unambiguously fixed the direction of the precession axis in space; they also allowed an independent distance estimate to SS 433.

Much of the research into SS 433 was prompted by the hope that this system would prove to be a nearby, and therefore relatively easy to observe, version of the extragalactic jet sources associated with quasars. Indeed, SS 433 seems to have more in common with those extragalactic sources, discussed earlier in this Joint Discussion, than with the more typical galactic outflow sources treated in this session. For, just like the well-collimated jets in those large systems, the beams of SS 433 have an opening angle of only a few degrees (see Margon (1984)). The relativistic bulk speed, 0.26 c, too, likens SS 433 more to quasars

403

than to, say, Herbig-Haro objects. Again, just like in classical double radio sources, one can trace the beams of SS 433 over at least eight orders of magnitude in size. On the one hand, the beams in SS 433 are accelerated to terminal speed within 10^{12} cm, as was deduced from EXOSAT spectroscopic observations by Watson *et al.* (1986). On the other hand, the radio shell of W 50, which has a diameter of about 10^{20} cm, has clearly undergone the influence of the beams of SS 433 (see for instance the radio map by Downes *et al.* (1981)).

2. Overview of the 1987 Campaign

In order to exploit the advantageous scales offered by SS 433 in both time variability and angular size, compared to active galaxies, we have organised an intensive monitoring campaign in the spring of 1987. We concentrated on obtaining very frequent measurements in a limited (fourteen day) period at the end of May (JD 2446933 to JD 2446947), but the first measurements were made on JD 2446920 and the last on JD 2446957.

All participants in the campaign were invited to a workshop held on June 21 and 22 1988 in Dwingeloo, where the observations were presented and discussed. We will try to publish all our data as a series of articles in a single issue of a scientific journal.

The campaign was scheduled as close as possible to a time when the Doppler shifted lines reach their maximum separation; that is, when the angle between the beams of SS 433 and the line of sight is at its minimum. Differential effects between the approaching and the receding beam are then as large as possible. We wished to compare our observations to those made in an earlier, less extensive campaign in 1985, when the beams were at right angles to the line of sight (see Vermeulen *et al.* (1987) and Vermeulen *et al.* (1989a)).

We aimed to obtain a sequence of high dynamic range VLBI maps, spaced so closely in time that any changes could be properly followed at the available resolution. We expected that some features in the beams of SS 433 would expand and fade. Therefore, MERLIN observed SS 433 exactly simultaneously with the VLBI network, thus enabling visibilities from both networks to be combined in one map, which results in an improved sensitivity for extended low surface-brightness features. Despite scheduling and logistic constraints, we have made six eleven-hour observations at two-day intervals.

Simultaneously, we wanted to track the behaviour of the Doppler shifted lines in the optical spectrum of SS 433, so as to be able to combine the radial and tangential velocity components of the beams, and in order to study correlations between the thermal and non-thermal emission from SS 433. However, the exposure time needed to record the optical spectrum of SS 433 is typically only fifteen minutes, and we could therefore aim to study the variability on timescales much shorter than one day. In order to obtain proper twenty-four hour coverage, and to reduce the observing time needed on any particular instrument, we used a total of ten telescopes around the globe. This also minimised the risks of bad weather; for instance, at La Silla, Chile, it was cloudy on ten out of fifteen nights, but our overall program was not substantially damaged.

In support of the spectroscopic observations, multicolour photometry was done. Unfortunately, adverse weather conditions led to a somewhat incomplete coverage.

There were also daily radio flux density measurements at numerous frequencies. This effort started well before the radio mapping, since we wanted to investigate a possible correlation between the occurrence of flares and the generation of bright patches in the beams of SS 433.

Lastly, the Ginga X-ray satellite observed SS 433 on five consecutive days, at the start of the multifrequency campaign.

3. Discussion of the First Results from the 1987 Campaign

3.1 RADIO FLUX DENSITY MONITORING

Measurements at a total of sixteen frequencies, many of them at daily intervals, were obtained by Vermeulen et al. (1989b). They find that SS 433 was in a quiescent state in the last few weeks before the VLBI mapping started, and during the first few days of that central part of the campaign. Then, there was a sequence of flares, in which at least five different events, each lasting several days, can be distinguished. The flaring had not yet ended when monitoring stopped. The authors have found a dip in the total flux density of SS 433 just before the onset of the flaring. They speculate that when flaring starts there is first of all an ejection of some material with large optical depth, such that it blocks some of the steady quiescent emission from SS 433.

After subtracting the quiescent radio emission of SS 433, Vermeulen et al. (1989b) analysed the time and frequency dependence of each of the flares. They conclude that the first event is clearly different from the others. Also, it is clear that these flares are not due to single instantaneous injections of relativistic electrons; rather, processes of sustained generation of relativistic particles are indicated.

3.2 RADIO SYNTHESIS MAPPING

Vermeulen et al. (1989c) have obtained a series of six maps at 5 GHz, at two-day intervals using Mark III mode A recording with the five-station European VLBI Network (EVN, consisting of Effelsberg, Jodrell Bank, Medicina, Onsala, and the Westerbork tied array). Six hundred and ninety tapes were used, making this the largest experiment to date on the EVN. Additionally, the NRAO 140 ft telescope at Greenbank joined the network for the first, fifth, and sixth of the observations. Also, as was mentioned before, simultaneous MERLIN measurements improved the sensitivity for extended low surface brightness features. However, there were no common stations with the VLBI network, which complicates the relative calibration of the arrays.

The variability of SS 433 leads to difficulties in interpreting the obtained visibilities. As mentioned above, SS 433 was flaring during the measurements; the data of Vermeulen et al. (1989b) indicate flux density changes of up to ten percent during a single eleven-hour synthesis observation. Moreover, the kinematic model predicts that matter in the beams of SS 433 moves by about half a resolution element (4.5 mas) during each observation. Work by Vermeulen and Rugers (1989d) is in progress to develop a hybrid mapping algorithm that takes account of such variability. Despite these problems, the currently available EVN maps already have a higher image fidelity than any previous ones, and they confirm and extend very nicely some of the conclusions of Vermeulen et al. (1987) and Romney et al. (1987).

The radio beams of SS 433 are not continous (jet-like), but rather composed of a series of discrete "blobs" emanating from a core. Note that, by using the difference in proper motion between the approaching (eastern) and receding beam, one can easily identify the central source of the system in a sequence of maps, in contrast to the usual situation for VLBI maps. The locus and proper motion of the blobs are completely consistent with the kinematic model parameters derived from optical observations. In the 1987 series of VLBI maps there is even evidence for the six-day nodding period. Evidently, the motion of these blobs, which emit synchrotron radiation, is the same as that of the thermal matter, which emits Doppler shifted lines.

Vermeulen *et al.* (1987) and Fejes *et al.* (1988) have suggested that there is a radio "brightening zone" at a relatively large distance from SS 433. The flux density evolution of the blobs in the 1987 series of maps is in good agreement with that idea. They brighten for several days after ejection, reaching their maximum flux density at a distance of about 200 AU from the central object, after which they fade rapidly. Work is in progress to verify whether the flux density evolution of the blobs agrees with the suggestion by Vermeulen *et al.* (1987) that brightening occurs when a blob crosses the bowshock of its predecessor, and with the idea of these authors that the subsequent fading must be ascribed to adiabatic expansion in a region where the passage of previous blobs has created a lower matter density.

The combination of radio maps at one frequency and total flux density measurements at many frequencies yields a powerful diagnostic tool; however, the analysis of this database is as yet far from completed. It is clear that the first of the series of flares discussed in section 3.1 is related to events in or very close to the core of SS 433. Possibly, that first flare occurs when the matter is ejected that later gives rise to the prominent blobs in the VLBI maps. The second flare in the series, on the other hand, reflects the increase in flux density of the blobs in the brightening zone. Indeed, in the map where it is most prominent, the eastern blob alone emits about half of the total flux. This difference in origin between the flares may well explain why they evolve differently, as was discussed in section 3.1. It should be noted that the prominent blobs seen in 1987 are likely to be unusual; in all previous maps of SS 433 the core was the brightest feature. Interestingly, the peak flux density reached by the main "eastern" blob is about three times higher than that of the main "western" one, but this ratio comes closer to unity when account is taken of the expected differential Doppler boosting (Fejes, 1986).

3.3 OPTICAL SPECTROSCOPY

The optical spectra collected by Vermeulen *et al.* (1989e) were obtained on many different instruments; the resolution, spectral range, and integration time vary widely. Therefore, some spectral features can be studied in greater detail than others. The best covered portion of the spectrum contains the blueshifted Hα line; it was observed nearly a hundred times in twenty-one days.

The Doppler shifted lines exhibit variations on a large range of timescales, from several days down to half-an-hour. Episodes of a rather smooth change in wavelength are present, as well as instances of the well-known "bullet-like" behaviour, when a line at one particular wavelength diminishes in strength while a new, separate one rises to prominence at a different wavelength. Up to five such bullets have been seen simultaneously, quite apart from the structure which is often present in the profile of one line.

The wavelength of the shifted lines in general agrees quite well with the predictions of the kinematic model; in the few weeks of the 1987 campaign, the six-day nodding motion shows up especially nicely. When the line profile is complex, it is generally the brightest part which is closest to the predicted wavelength. In the case of multiple lines, the brightest is often the newest one, in agreement with the fast rise to prominence (usually a few hours) and the slower decay (one to several days). However, it should be stressed that the behaviour can vary enormously. During the measurements in 1987 there is a two-day and a three-day period when no Doppler shifted emission lines could be detected at all. Clearly, the beams of SS 433 are fragmented, but models must allow for wide ranges in size, emissivity, and radiative lifetime of the bullets.

Often, the red- and blueshifted lines behave like mirror images, for example during the switch-off episodes mentioned above. However, there are many instances in the dataset of Vermeulen *et al.* (1989e) when the redshifted lines change ahead of the blueshifted ones by a few hours or even as much as one day. Moreover, in this dataset the redshifted lines occasionally have a more complex profile than their blueshifted counterparts, but the reverse does not occur. The interpretation of these effects is unclear. Possibly, they find their origin near the compact object, due to some form of partially one-sided ejection. It is perhaps more likely that differences in the local conditions at a greater distance from the binary system are involved, which lead to a dissimilar fragmentation, or to unequal emission measures, or simply to a different obscuration of the two beams.

3.4 X-RAY MONITORING

Thanks to a collaborative effort by the Japanese Ginga team and workers at Leicester, X-ray data are available for five consecutive days at the start of the 1987 SS 433 monitoring campaign. A preliminary report was presented at the 1988 Dwingeloo workshop on SS 433 by H.C. Pan.

The 1987 observations confirm and extend the earlier EXOSAT work of Watson *et al.*, (1986) and Stewart *et al.* (1987). The X-ray spectrum of SS 433 can be described by a thermal bremsstrahlung continuum and an Fe XXVII line, at the position predicted by the kinematic model for the blueshifted system. The absence of the redshifted line constrains the length of the X-ray emitting portion of the beams; it also shows that the matter is already fully accelerated at 10^{12} cm from the compact object.

The Ginga data trace a primary eclipse, in which the compact object is obscured by its companion. The equivalent width of the Doppler shifted line remained constant, indicating that most of the underlying continuum originates in the same region as the emission line (*i.e.*, in the approaching beam). The spectral index softens in the eclipse, which is evidence for a temperature gradient along the X-ray beam.

4. Summary; Correlations between Different Wavelengths

There is no clear evidence for a correlation between the evolution seen in the radio maps and the behaviour of the optical Doppler shifted lines. No specific event in the spectra can be linked to the occurrence of the radio flares, and it has thus far not been possible to identify a specific feature on the VLBI maps in which a certain Doppler shifted line originates. That lack of correlation between such widely different emission processes should perhaps not be too surprising. The 1987 campaign has shown that, whereas the beams of SS 433 are clumped on a wide range of scales (minutes to days) in the optical emission lines, the radio emission, in the form of flares or blobs, typically varies on a somewhat larger timescale (days). Thermal material and synchrotron emitting plasma evidently have a different distribution. It is remarkable that they nevertheless both move at 0.26 c according to the predictions of the same kinematic model parameters.

The series of photometric measurements by Aslanov *et al.* (1989) unfortunately has some gaps during the monitoring campaign. However, these workers do find that SS 433 underwent an optical flare just when the object changed from the radio quiet to the flaring state. Further examination of the shape of the lightcurve during the campaign may reveal whether this simultaneous flare is significant, and, if so, how it should be explained.

More speculatively, there is a link between the onset of the radio flaring, and the drop in the X-ray flux seen at the very end of the Ginga measurements. This behaviour, if real, would somewhat resemble the suggestion by Band and Grindlay (1986) that when SS 433 is flaring, its X-ray spectrum is (mostly) non-thermal, with a low flux density, but that its spectrum is (partly) thermal, with a higher flux density, when SS 433 is quiescent at radio frequencies.

The 1987 multifrequency campaign has yielded an impressively large set of first class data, thanks to the efforts put in by all the authors of Aslanov et al. (1989) and Vermeulen et al. (1989a to 1989e), by the Japanese Ginga team and the group at Leicester, as well as by other people acknowledged in the references mentioned above. We were fortunate to catch SS 433 in a transition from its quiescent state to the radio flaring state. However, it is not easy to disentangle the emission coming from the central region, near the binary system, from what is being generated further out in the two beams. Perhaps we may conclude that future studies of the radio flaring behaviour of SS 433 would not necessarily have to be accompanied by optical spectroscopy, but that such radio studies would again greatly benefit from simultaneous optical photometry and X-ray measurements, given the tantalising hints of correlated behaviour seen in the 1987 campaign.

References

G.O. Abell, B. Margon; 1979: Nature **279**, 701–703
A.A. Aslanov, A.M. Cherepashchuk, V.P. Goranskij, V.Yu. Rahimov,
    R.C. Vermeulen; 1989: in preparation
D.L. Band, J.E. Grindlay; 1986: Astrophys. J. **311**, 595–606
A.J.B. Downes, T.Pauls, C.J. Salter; 1981: Astron. Astrophys. **103**, 277–287
I. Fejes; 1986: Astron. Astrophys. **166**, L23–L25
I. Fejes, R.T. Schilizzi, R.C. Vermeulen; 1988: Astron. Astrophys. **189**, 124–127
R.M. Hjellming, K.J. Johnston; 1985: in *Radio Stars, eds. R.M. Hjellming,*
    D.M. Gibson, 309–323
B. Margon; 1984: Ann. Rev. Astron. Astrophys. **22**, 507–536
J.D. Romney, R.T. Schilizzi, I. Fejes, R.E. Spencer; 1987: Astrophys. J. **321**, 822–831
G.C. Stewart, M.G. Watson, M. Matsuoka, W. Brinkmann, J. Jugaku, K. Takagishi,
    T. Omodaka, J.C. Kemp, G.D. Henson, D.J. Kraus, T. Mazeh, E.M. Leibowitz; 1987:
    M.N.R.A.S. **228**, 293–303
R.C. Vermeulen, R.T. Schilizzi, V. Icke, I. Fejes, R.E. Spencer; 1987: Nature **328**, 309–313
R.C. Vermeulen, P.G. Murdin, P. Angebault, S. D'Odorico, R.T.Schilizzi; 1989a:
    in preparation
R.C. Vermeulen, W.B. McAdam, S. Trushkin, S. Facondi, R. Fiedler, R. Hjellming,
    K. Johnston, J. Corban; 1989b: in preparation
R.C. Vermeulen, R.T. Schilizzi, R.E. Spencer, J.D. Romney, I. Fejes; 1989c: in preparation
R.C. Vermeulen, M. Rugers; 1989d: in preparation
R.C. Vermeulen, P.G. Murdin, E.P.J. van den Heuvel, S. Fabrika, R.M. Wagner,
    B.H. Margon, J.B. Hutchings, R.T. Schilizzi, L.B. van den Hoek, M. van Kerkwijk,
    E. Ott, S. D'Odorico; 1989e: in preparation
M.G. Watson, G.C. Stewart, W. Brinkmann, A.R. King; 1986: M.N.R.A.S. **222**, 261–271

LARGE SCALE JETS IN CLASS I AND CLASS II RADIO SOURCES AND QUASARS

G.V. BICKNELL
Mount Stromlo and Siding Spring Observatories
Institute of Advanced Studies
Australian National University

ABSTRACT. The physics of large scale jets in class I and class II extragalactic radio sources and quasars is discussed. Class I jets appear to be turbulent, transonic jets which entrain the interstellar medium. The related jet deceleration causes a slow surface brightness decline which is usually observed. Class II jets are supersonic and terminate in an advancing shock against the external medium. Both types of jet are initially light but the ratio of jet density to external density of class I jets increases owing to entrainment. It is quite plausible that quasar jets are hypersonic and light and this may solve problems of confinement. The velocities of class I jets are of the order of a few thousand kilometers per second. Class II and quasar jets may be at least mildly relativistic. However, it is not clear whether the velocities of large scale jets in powerful sources are close to the speed of light. Recent depolarization measurements provide an interesting focus for discussion of this question.

1. Introduction

Almost a decade and a half ago Fanaroff and Riley (1974) demonstrated a distinct morphological difference between radio sources whose total powers are respectively less than and greater than the critical value

$$P_{\text{tot}}^{1.4 \text{ GHz}} \approx 10^{24.5} \text{ W Hz}^{-1}.$$

They noted that the low luminosity (class I) sources are edge–darkened whereas the high luminosity (class II) sources have their brightest emitting regions towards their extremities. This morphological difference is illustrated in figure 1 which is a VLA radio image of the radio source in the cluster Abell 3744 (Cameron, 1988). Fortuitously, this cluster contains an example of both generic types (albeit with large scale structure which is influenced by the properties of the intracluster medium and the velocities of the galaxies with respect to it).

Fanaroff and Riley noted that the morphological difference between class I and class II sources probably reflects the different manner in which energy is transported in the two types of source. At around the same time, Blandford and Rees (1974) and Scheuer (1974)

D. McNally (ed.), Highlights of Astronomy, Vol. 8, 409–416.

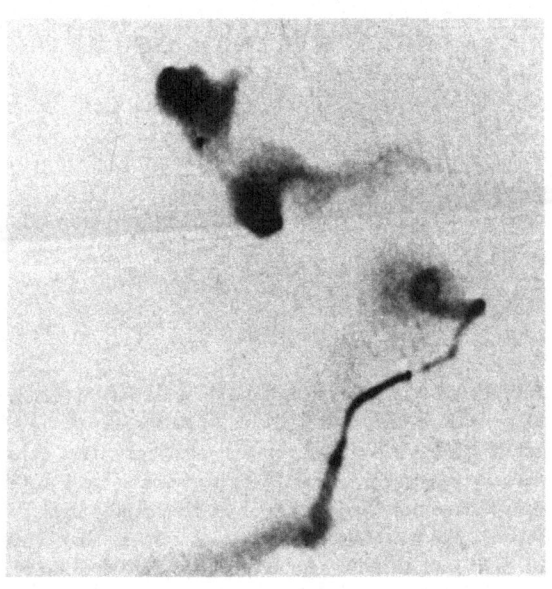

Figure 1. An N.R.A.O. V.L.A. image of the radio source PKS 2104-25 in the cluster Abell 3744. The class I source is the southern component and is associated with the elliptical galaxy NGC 7016; the class II source is the northern component and is associated with the dumbell galaxy NGC 7018. The N.R.A.O is operated by Associated Universities Inc. under contract to the National Science Foundation.

published papers proposing that "beams", consisting of energetic particles and magnetic field transported the energy to the lobes of radio sources over a timescale $\sim 10^8$ yrs. The subsequent detection of "jets" between the core and extended lobes of radio sources verfied this idea and it became apparent from later work that the major differences between class I and class II sources are strongly related to the physics of jet propagation, bearing out Fanaroff's and Riley's original proposition. For instance, Bridle (1984) showed that there are strong systematic differences between jets in class I and class II ssources. In particular, class I sources tend to have two jets of similar brightness whereas class II sources are generally one–sided with respect to their jets. Thus, over the last few years, the study of radio sources has been dominated by the study of the jets they contain. The reward, for what has been an intensive observational and theoretical effort which can not be covered adequately in this short review, is that the relevant physical differences between the two types of radio source are beginning to be understood. Moreover, the detailed study of jets in quasars (see, for example Barthel et al 1988a,b) has added a fascinating extra dimension to the studies of jets. However, despite the progress that has been made in this field, the question of the velocities of jets, in particular class II jets, remains a controversial one. The recent discovery of substantial depolarization variations in powerful radio sources has added significant fuel to the debate and I shall come to this topic towards the end of this paper. The first topics I shall discuss relate to the physics which is considered to be relevant for class I, class II and quasar jets and the models which have been constructed to explain them.

2. Class II Jets

The first models that were constructed for radio sources were directed towards understanding the properties of powerful radio sources. Thus it is appropriate to discuss these first.

The paper by Blandford and Rees (1974) essentially mapped out a great deal of the physics which is relevant to the large scale structure of class II radio sources. In a class II source a supersonic jet terminates at a "hotspot" via a shock against the intergalactic or interstellar medium. A bow shock precedes the jet as it propagates out into the external medium and a contact discontinuity between the jet and external plasma is also established. The hotspot progresses outwards with a velocity $v_h \approx \eta^{-1/2} v_{jet}$ where the important parameter η is the ratio of the jet density to the density of the surrounding medium. Most lobes in class II radio galaxies are observed to be large compared to the related jets, when they can be detected, and the parameter η has the largest effect on this. The expansion of the lobe is the greater the smaller η is and the axisymmetric simulations by Norman et al. (1984) demonstrate this effect quite clearly. This raises the possibility that η could be estimated from the relative size of the jet and the lobe. However, wiggling of the end of the jet can lead to a much larger lobe and the production of multiple hot spots (Scheuer 1974). This has been verified by the three dimensional simulations of Williams and Gull (1984) and recent higher resolution simulations of two dimensional slab jets by Norman and Hardee (1988) (see figure 2). Nevertheless, jet wiggling can not account for the entire size of a lobe and at present most numerical modeling of class II sources entails values of $\eta < 0.01 - 0.1$. This is also probably the correct physical range for this parameter; the lowest realistic limit to η is unknown at present.

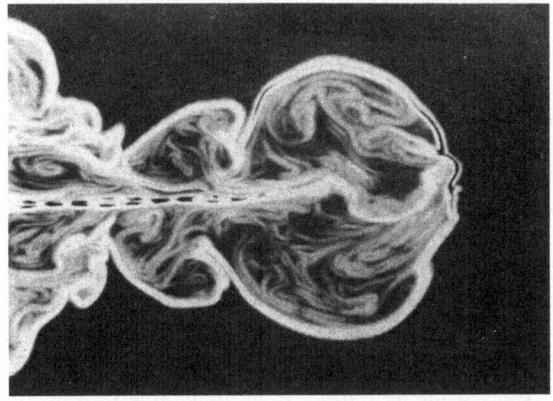

Figure 2. Simulation of a supersonic slab jet by Norman and Hardee (1988). Oscillation of the end of the jet causes it to generate a larger cavity than would be the case if the jet were perfectly straight.

With the high resolution provided by the V.L.A. and Merlin interferometers, radio astronomers have been able to observe more detail in distant sources. In particular, quasars have revealed intetesting jet structure (see, for example Potash and Wardle 1979, Barthel et al. 1988a,b) which poses some important questions. Jet confinement often proves to be a problem enunciated clearly by Pottash and Wardle. Moreover, while a number of quasars have typical class II morphology, many are bent or distorted indicating significant interaction with large ($\sim 10^7 - 10^8\ M_\odot$) clouds in the ambient medium. Some quasar jets, e.g. 3C273 (Davis et al. 1985), do not have the typical class II morphology, lacking the usual inflated lobe at the end of the jet. One possibility therefore, is that these jets are heavy but this would be surpising in view of what appears to be the case in powerful class II radio galaxies. Norman (1986) has proposed an attractive alternative and that is that these jets are light and hypersonic. In this flow regime the bow shock is highly swept back preventing the lateral expansion of shocked jet plasma. Moreover, the pressure behind the advancing bow shock is well above ambient and confines the jet. Thus Norman's model provides a neat solution for both the morphology and the apparent overpressure of quasar jets. The idea of a hypersonic jet also fits in qualitatively with the generally higher quasar power output.

One of the early objections to beam models of radio sources was that beams would be Kelvin–Helmholtz unstable. Supersonic, light jets tend to avoid destructive Kelvin–Helholtz effects through the "reflecting mode" which sets in when

$$v_{\rm jet} > c_{\rm s,int} + c_{\rm s,ext}$$

where $c_{\rm s,int}$ and $c_{\rm s,ext}$ are the internal and external jet sound speeds (Payne and Cohn 1985). This mode is related to internal waves which are reflected from the surface of the jet and which saturate in travelling biconical shocks (Norman et al. 1984). These shocks are less disruptive than the stationary "ordinary mode" shocks which can occur in lower velocity, higher density radio jets. Reflecting mode shocks may be responsible for at least some knots observed in class II jets.

Another aspect of jets which saves them from destruction by the Kelvin Helmholtz instability is the fact that they propagate through a pressure gradient which tends to reduce the level of turbulence. It has been proposed that this results in a critical Mach number of around 2 where jets make a transition from class I behaviour (see the following section) to class II behaviour (Bicknell 1986a).

3. Class I Jets

Whereas the physics of jets in the powerful class II radio sources is dominated by the physics of supersonic or hypersonic flow, the physics of low luminosity class I sources is dominated by considerations of turbulent transonic or subsonic flow. There are two main reasons for this: Near their respective cores, class I jets spread more rapidly than can usually be accounted for by adiabatic laminar flow in a normal galaxy atmosphere and they decline in surface brightness more slowly than expected for a constant velocity laminar jet. It is natural to interpret the spreading as being due to turbulent expansion and the second property as being due to the compression of matter and fields by the related entrainment (Fanti et al. 1982, Bicknell 1984, 1986b). Models of class I jets have generally focussed on possible explanations for the surface brightness behaviour since this was judged as the main feature of class I jets that was most difficult to understand. Nevertheless, the realistic calculation of turbulent flow models is fraught with a number of difficulties, not the least of which is the turbulent closure problem for light, transonic flow. This is highlighted by the following ensemble–averaged equation for the z–momentum of a jet:

$$\frac{\partial}{\partial z}\left(\bar{\rho}\tilde{v}_z^2\right) + \frac{1}{r}\frac{\partial}{\partial r}\left(r\,\bar{\rho}\tilde{v}_r\tilde{v}_z\right) = -\left(1 - \frac{\bar{\rho}}{\rho_{\text{ext}}}\right)\frac{dP_{\text{ext}}}{dz} - \frac{1}{r}\frac{\partial}{\partial r}\left(r\,\langle\rho v_r'\tilde{v}_z'\rangle\right)$$

In order to solve this, and the related hydrodynamic equations, it is necessary to relate the Reynolds stress, $\langle\rho v_r'\tilde{v}_z'\rangle$, to the variables describing the mean flow. Unfortunately, such a prescription will probably never become available for light transonic/supersonic flow and detailed direct modelling of such flows (necessarily three dimensional and of high resolution) requires at least the capabilities of modern supercomputers. It is possible to make useful progress however, using a semi–empirical approach (Bicknell 1986b) wherein data on the spreading rate of class I jets can be used to infer the rate of change of velocity and surface brightness.

As is evident from the above equation the local external pressure gradient is an important input to these models. A simple isothermal atmosphere is used whose pressure is described by:

$$\frac{P}{P_c} = \exp\left[-\beta\frac{(\phi - \phi_c)}{\sigma^2}\right]$$

where the c subscripts indicate central values and ϕ is a King potential with velocity dispersion σ. This has many of the features of a more realistic atmosphere model, namely a power law region outside the core followed by a flattening to a background pressure at large radii. The crucial parameter in these models is β, the ratio of the virial temperature to the ISM temperature.

The surface brightness of a jet described by such an adiabatic model goes as

$$I_\nu \propto \left(\tilde{v}_z a^2\right)^{-s/3} \times \left(\tilde{v}_z a\right)^{-(s-1)/2}$$

where the two separate factors indicate the effect of adiabatic compression of the relativistic electron distribution and the magnetic field respectively. (The parameter s is the exponent of the distribution function.)

This model has been fitted to the data on a number of sources. Most recently, Morganti et al. (1988) have fit the surface brightness data on 23 jets from the Bologna survey obtaining satisfactory fits in most cases. An important result is that the initial value of the parameter η, the ratio of the jet density to the interstellar medium density is less than unity, consistent with the inferred situation in class II sources. Entrainment causes this parameter to increase towards unity. The model fits also generally lead to an initial Mach number between 1 and 2, consistent with the notion that these jets are transonic. A sample model fit to the eastern jet in the radio galaxy B2 0034+25 is given in figure 3.

Perhaps the most satisfactory aspect of the model fits is the estimated values of the parameter β. The median value is 0.7 which is identical to the median β inferred by Thomas et al. (1985) from an independent sample of X-ray ellipticals.

414

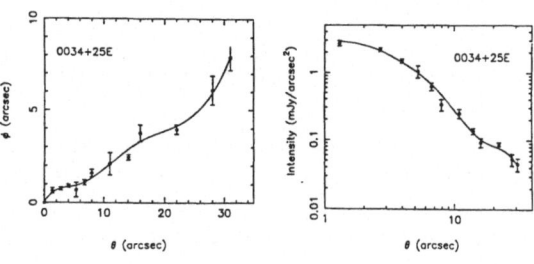

Figure 3. The FWHM data (left panel) and a light, transonic jet model fit (right panel) to the surface brightness of the eastern jet in the source B2 0034+25.

The notion of low Mach numbers for class I jets is supported by the work of O'Dea (1985) on the characteristics of bent jets in clusters of galaxies.

4. Jet Velocities and Depolarization

With the understanding that class I jets are low Mach number, confined jets with $\eta \sim .01-1$, the implied velocities

$$v_{\text{jet}} = 4.3 \times 10^7 \eta^{-1/2} M \left(\frac{T_{\text{ism}}}{10^7 \, \text{K}} \right)^{1/2} \text{km s}^{-1}$$

where M is the Mach number and T_{ism} is the temperature of the interstellar medium, are of the order of a few thousand kilometers per second. Velocities of this magnitude are consistent with the energy budget in class I sources (Bicknell 1986a). Allowing that class II jets have higher Mach numbers and possibly lower density ratios, at least mildly relativistic velocities are possible. This has also been argued for by Dreher (1984) on the basis of an energy budget calculation relevant to this type of source. Nevertheless, the question of whether jet velocities are relativistic on the large scale is a controversial one at present. Relativisitic velocities have long been invoked as an explanation for the sidedness of jets in powerful sources through doppler boosting/dimming of the radiation from the approaching/receding jet. Moreover, superluminal motion, generally attributed to the effect of viewing a relativistic jet which is moving at a small angle to the line of sight, has been observed on the VLBI scale in a number of quasars (such as 3C 273, Unwin et al. 1985) and one radio galaxy, 3C 120 (Walker et al. 1987).

Recently, there have been some very interesting observations that have injected new life into the velocity debate. Laing (1988) reported results on the depolarization properties of powerful radio sources all of which have one-sided jets. From a sample of ten sources, he concluded that the jetted side is less depolarized than the non-jetted side. This conclusion was supported by Garrington et al. (1988) in a follow-up sample made up of 25 sources. One possible interpretation of this is that both jets are observed through a uniformly distributed Faraday screen which causes depolarization through differential Faraday rotation. This implies that the jetted side is closer to us, supporting the relativistic interpretation.

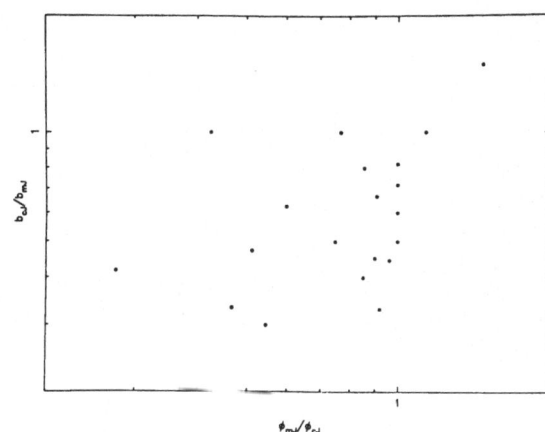

Figure 4. Plot of the ratio of counter jet surface brightness to main jet surface brightness versus the ratio of spreading rates for 20 sources from the Bologna survey. In all but two cases the brighter jet spreads the least rapidly.

On the other hand, as both Laing and Garrington et al. remark, the depolarization asymmetry could be intrinsic, in which case one is required to search for some mechanism relating the jet asymmetry to it.

The basis for such an explanation may come from the following (Bicknell, Parma and de Ruiter 1988). Suppose we attribute the depolarization asymmetry to an asymmetric distribution of matter within the galaxy. Then it is possible that the jet propagating through the denser material is "splattered" causing it to be dimmer, partially because of less collimation. If this is the case, then one may expect to see some evidence for the effect in class I sources. Although the jet brightness ratios in class I sources are not as extreme as they are in class II sources, brightness differences do exist within the same source. Thus Parma, de Ruiter and I examined the data on a number of jets from the Bologna survey Parma et al. 1986) in order to determine whether there is any correlation between the ratio of jet surface brightnesses and the ratio of spreading rates within the same source. The result of plotting such ratios is shown in figure 4 where the ratio of the brightnesses of the counter jet and main jet at 5 kpc is plotted against the ratio of the spreading rates. It is apparent that the fainter jet almost always (there are only two exceptions) spreads more rapidly than the brighter jet. This plot suggests that sidedness may not be related to relativistic beaming; however the final verdict is certainly not in. One phenomenon that is not immediately catered for by this splattering argument is that where a one–sided VLBI jet is observed it always points in the direction of the brighter large scale jet. If splattering is responsible for this as well, then it implies that the asymmetry in the interstellar medium extends from parsec to kiloparsec scales. Some find this unappealing. Whatever the case, the resolution of the jet sidedness question is not merely a matter of determining a number to attach to the velocity. It impacts upon the nature of jet formation and the interaction of the jet with the interstellar medium on all scales within the galaxy.

Acknowledgement

I would like to acknowledge the assistance of Paola Parma, Hans de Ruiter, Mike Norman and Phil Hardee in preparing material for this talk.

416

References

Barthel P.D. and Miley G.R., 1988a, *S.T.Sc.I preprint*
Barthel P.D., Miley G.K., Schilizzi R.T. and Lonsdale C.T., 1988b, *S.T.Sc.I preprint*
Bicknell G.V. 1984, *Ap.J.*, **239**, 433
Bicknell G.V., 1986a, *Ap.J.*, **286**, 68
Bicknell G.V., 1986b, *Ap.J.*, **305**, 109
Bicknell G.V., Parma P., and de Ruiter H., 1988, in preparation
Bridle A.H., 1984, *A.J.*, **89**, 979
Blandford R.D. and Rees M.J., 1974, *M.N.R.A.S.*, **169**, 395
Cameron R.A., 1988, *Ph.D. thesis*, Australian National University
Davis R.J., Muxlow T.W.B. and Conway R.G., 1985, *Nature*, **318**, 343
Dreher J., 1984, in *Physics of Energy Transport in Extragalactic Radio Sources*, ed. Bridle A.H. and Eilek J.A., N.R.A.O, Greenbank, 109
Fanaroff B.L. and Riley J.M., 1974, *M.N.R.A.S.*, **167**, 31P
Fanti R., Lari C., Parma P., Bridle A.H., Ekers R.D., and Fomalont E.B., 1982, *Astr. Ap.*, **110**, 69
Garrington S.T., Leahy J.P., Conway R.G., and Laing R.A. 1988, *Nature*, **331**, 147
Laing R.A., 1988, *Nature*, **331**, 149
Morganti R., Parma P., de Ruiter H.R., Bicknell G.V. and Fanti R., 1988, in preparation.
Norman M.L., 1986, *Bull. A.A.S.*
Norman M.L. and Hardee P.E., 1988, in preparation.
Norman M.L., Winkler K.-H.A., and Smarr L.L., 1984, in *Physics of Energy Transport in Extragalactic Radio Sources*, ed. Bridle A.H. and Eilek J.A., N.R.A.O, Greenbank, 150
O'Dea C., 1985, *Ap. J.*, **295**, 80
Parma P., de Ruiter H.R., Fanti C., Fanti R., 1986, *Astron. Astrophys. Suppl. Ser.*, **64**, 135
Payne D.G. and Cohn H. 1985, *Ap.J.*, **291**, 655
Pottash R.I. and Wardle J.F.C., 1979, *A.J.*, **84**, 707
Scheuer P.A.G., 1974, *M.N.R.A.S.*, **166**, 513
Thomas P.A., Fabian A.C., Arnaud K.A., Forman W., Jones C., 1985, *M.N.R.A.S.*, **222**, 655
Unwin S.C., Cohen M.H., Biretta J.A., Pearson T.J., Seielstad G.A., Walker R.C., Simon R.S. and Linfield R.P., 1985, *Ap.J.*, **289**, 109
Walker R.C., Benson J.M. and Unwin S.C., 1987, *Ap. J.*, **316**, 546
Williams A.G., and Gull S.F., 1984, *Nature*, **310**, 33

SYNCHROTRON THERMAL INSTABILITIES AND RADIO FILAMENTS IN THE LOBES OF CYGNUS A

G. Bodo[1], A. Ferrari[1],[2], S. Massaglia[3], E. Trussoni[4]

[1] Osservatorio Astronomico di Torino, Pino Torinese, Italy
[2] Istituto di Fisica Matematica, Università di Torino, Italy
[3] Istituto di Fisica Generale, Università di Torino, Italy
[4] Istituto di Cosmo-geofisica del C.N.R., Torino, Italy

I. INTRODUCTION

Recent VLA observations of the lobes of Cygnus A exhibit complex "filamentary" structures, with typical scale width ~ 1 arcsec (Dreher, Carilli and Perley, 1987, Perley, 1987). The filaments appear aligned with the magnetic field, as results from polarization measures, suggesting that the field may play a fundamental role in the process of their formation.

We propose a mechanism for the possible formation of these filaments based upon a thermal instability connected with synchrotron emission from relativistic electrons. This type of instability was studied by Simon and Axford (1967), who discussed it in connection with the Crab Nebula filaments, and by Eilek and Caroff (1979), who generalized the previous study for application to quasar atmospheres.

The treatment followed here assumes in addition that the energetic equilibrium is maintained by continuous replenishment of relativistic electrons streaming from the "hot spot" region. In section II we outline the general picture of the stability process, in section III we derive the conditions and the growth rate of the instability, and in section IV we discuss the application to the formation of filaments in the lobes of Cygnus A.

II. THE SYNCHROTRON THERMAL INSTABILITY

Following the approach of Simon and Axford (1967) we consider a two components magnetized plasma, where the inertia is provided by protons and the internal energy by relativistic electrons. Therefore the system can be described by classical MHD equations:

$$\frac{\partial \rho}{\partial t} + \nabla \cdot (\rho \mathbf{v}) = 0, \tag{1a}$$

$$\rho \frac{d\mathbf{v}}{dt} = -\nabla P + \frac{1}{c}\mathbf{J} \times \mathbf{B}, \tag{1b}$$

$$\frac{dP}{dt} - \gamma \frac{P}{\rho}\frac{d\rho}{dt} = (\gamma - 1)(\mathcal{Q} - \mathcal{L}), \tag{1c}$$

417

D. McNally (ed.), Highlights of Astronomy, Vol. 8, 417–422.
© 1989 by the IAU.

$$\frac{\partial \mathbf{B}}{\partial t} = \nabla \times (\mathbf{v} \times \mathbf{B}) \quad , \tag{1d}$$

where the pressure is given by $P = N\bar{\epsilon}/3$, N is the number density of relativistic electrons, $\bar{\epsilon}$ the average energy per electron and ρ is the mass density of the protons; consistently with our assumptions $\gamma = 4/3$). In addition $\mathbf{B}$ and J are the magnetic field and current density respectively, and Q and $\mathcal{L}$ the absorbed and emitted power per unit volume. If the zeroth-order configuration is chosen in energy balance, the process of instability can be studied by linearizing the MHD equations, and assuming for the perturbed quantities a form $\propto \exp(-ik_x - ik_z + \omega t)$.

In the present astrophysical application, the energy losses are due to the synchrotron emission from relativistic electrons, while the energy input is given by continuous supply of particles into the instability region ("lobe region"). If we assume that relativistic electrons are accelerated in the "hot spot" (which is overpressured with respect to the whole lobe), they diffuse across the lobe region streaming along pressure gradients. In this case the energy balance at each point can be maintained by continuous advection of energetic particles. While in previous works a maxwellian distribution was assumed for electrons (Simon and Axford 1967, Eilek and Caroff 1979), here we discuss a power law distribution as expected in radio lobes.

We model this situation considering a rectangular domain, with uniform magnetic field directed along the z axis, where energetic electrons are injected at one end, and diffuse along the magnetic field with velocity v_0 (of the order of the Alfvèn velocity V_A), while losing energy by synchrotron radiation. The system of equations (1) can be linearized; in agreement with the previous considerations, and assuming that the equilibrium variables vary only along the z direction, parallel to the magnetic field), one obtains:

$$\rho_0 v_0' = -v_0 \rho_0' , \qquad \rho_0 v_0 v_0' = -P_0' , \qquad v_0 P_0' - \gamma \frac{P_0}{\rho_0} v_0 \rho_0' = -(\gamma - 1)\mathcal{L}_0 , \tag{2}$$

where primes indicates d/dz derivatives, and where we set $Q_0 = 0$, as appropriate in the lobe region, and

$$\mathcal{L}_0 = \sigma_T c \frac{B_0^2}{8\pi} N_0 \bar{\epsilon}_0^2$$

For consistency the scale lengths of the variation of the equilibrium variables must be much larger than the perturbations wavelengths (local instability criterion), i.e. $k_z l_p \gg 1$, where $l_p = -P_0/P_0'$ and

$$l_p = l_{sync} \left(1 - \frac{c_s^2}{v_0^2} \right) ,$$

with $c_s^2 = \gamma P_0/\rho_0$, and $l_{sync} = v_0 \tau_{sync}$ (τ_{sync} is the time scale of synchrotron losses). By introducing the nondimensional variables:

$$K_{x,z} = k_{x,z} c_S \tau_{sync}, \qquad V_{AS} = \frac{V_A}{c_S}, \qquad V_A^2 = \frac{B_0^2}{4\pi\rho_0},$$

$$L = k_z l_p, \qquad V_0 = \frac{v_0}{V_A}, \qquad V_0' = v_0' \tau_{sync} ,$$

we get the following conditions for the validity of our approximation:

$$V_0 = \frac{L}{V_{AS}K_x}(1 + \gamma V_0'), \quad V_0' = \frac{K_z}{LV_0V_{AS}}, \quad L \gg \frac{K_x}{K_z}, \tag{3}$$

The dispersion relation, obtained annulling the determinant of the linearized MHD equations (1), is a 6th order polynomial:

$$\sigma^6 + a_1\sigma^5 + a_2\sigma^4 + a_3\sigma^3 + a_4\sigma^2 + a_5\sigma + a_6 = 0, \tag{4}$$

with $\sigma = (\omega - ik_zv_0)\tau_{sync}$. The coefficients of the polynomial (see the Bodo et al., (1988) for their general expressions) depend upon the zeroth order configuration, and have the following functional form:

$$a_i(K_x, K_z, V_{AS}, \alpha_N, \alpha_\epsilon, \alpha_B, L)$$

where we have defined

$$\alpha_{N,\epsilon,B} \propto \left[\frac{(N, \epsilon, B)}{\mathcal{L}}\frac{\partial \mathcal{L}}{\partial(N, \epsilon, B)}\right]_o, \tag{5}$$

For a homogeneous plasma ($L \to \infty$) $a_6 = 0$, and Eq. (4) is reduced to a 5th degree polynomial, while, if $K_z = 0$ also, the dispersion relation becomes a third degree polynomial that, for the particular case of $K_x \gg 1$, leads to the following stability condition:

$$\alpha_N - \alpha_\epsilon\left(1 + \gamma V_{AS}^2\right) + \alpha_B > 0, \tag{6}$$

(Simon and Axford 1967).

In this case the unstable mode is a purely growing filamentary perturbation aligned with the magnetic field. In the following Section we shall discuss some numerical roots of Eq. (4) by choosing for the parameters values consistent with the physical conditions in the lobes of Cygnus A.

III.*STABILITY ANALYSIS*

The growth rate of unstable perturbations has been evaluated numerically by means of a NAG routine for the search of the roots of a high order polynomial, and are presented in function of the values of the physical parameters. Concerning the values of α, it is found for synchrotron losses $\alpha_N = 1$, and $\alpha_B = 2$, while the value of α_ϵ depends from the spectrum of relativistic electrons. For a maxwellian distribution with an average energy $\bar{\epsilon}$, is $\alpha_\epsilon = 2$. For a power law distribution $\bar{\epsilon}$ depends from the spectral parameters, namely the maximum (ϵ_{max}) and minimum (ϵ_{min}) energy cut-offs, and the spectral index Γ. The values of α are then evaluated form Eq.(5) by assuming that in the instability process the spectrum is varying only one of the three parameters ϵ_{max}, ϵ_{min}, and Γ. If we assume a typical spectral index $\Gamma = 2.5$, and $\epsilon_{max}/\epsilon_{min} = 10^3$, we have from Eq.(5) (see also Bodo et al. 1988) $\alpha_\epsilon \approx 4$ (varying Γ), $\alpha_\epsilon \approx 1.5$ (varying ϵ_{min}) and $\alpha_\epsilon \approx 0$ (varying ϵ_{max}).

The nondimensional wavenumbers K_x and K_z, can be expressed in terms of physical variables:

$$K_{x,z} \approx \frac{10^6}{\lambda_{pc}^{\perp,\parallel}} (\nu_{10} B_{-5}, n_{th_{-4}})^{-1/2},$$

where ν_{10} is the frequency of radio emitting electrons in units of 10 GHz, B_{-5} is the magnetic field in units of 10^{-5} gauss, $n_{th_{-4}}$ is the number density of thermal particles in units of 10^{-4} cm^{-3}, and $\lambda_{pc}^{\perp,\parallel}$ are the components of the perturbation wavelengths, in units of 1 parsec, perpendicular and parallel to the magnetic field direction. From the typical values of the physical parameters of the extended radio lobes, and for perturbations much smaller than the typical sizes of radio components, we must have in general $K_x, K_z \leq 10^{5-6}$. If furthermore is $V_{AS} \sim 1$ (as from equipartition conditions) conditions (3) provide a rough estimate of L: for the results presented here it will be assumed $L = 10^5$

The growth rate ($Re(\sigma)$) is plotted vs. K_z for a fixed vale of K_x ($= 10^5$), for some values of V_{AS}, and for three values of α_ϵ: $\alpha_\epsilon = 4$ (Fig. 1), $\alpha_\epsilon = 1.5$ (Fig. 2) and $\alpha_\epsilon = 0$ (Fig. 3). First of all we notice as a general feature for all unstable perturbations, that they are always travelling modes ($Im(\sigma) \neq 0$), as expected from the assumption of a travelling plasma, therefore the ratio K_x/K_z is related with the propagation direction of the perturbation with respect to the magnetic field direction. Three unstable modes are found: the perturbation analyzed by Simon and Axford (1967), modified by the streaming plasma (we call it the condensation mode), and two slow MHD modes (with opposite phase velocities). The former mode has large growth rate for small values of K_z, but it is sharply damped by increasing K_z; conversely the two slow MHD modes have a constant growth rate for a large range of values of K_z, and are damped for K_z approaching K_x. This stabilization is expected taking into account that for $K_x/K_z \to 0$ the magnetic field perturbation, which drives the instability, tends to vanish. In addition we see that increasing the magnetic field strength tends to stabilize the perturbations since it inhibits the pressure build-up associated with them, more precisely, in the present case, stability is found for $V_{AS} \approx 1$. Finally, the general behavior of the instability is not very sensitive to the spectral parameters, even though the exact value of the growth rate can vary.

III.SUMMARY

We can summarize our results as follows:
1) The thermal instability related to the synchrotron emission can be present in the lobes of Cygnus A under a quite wide range of conditions.
2) The typical time scale of development of instability is $\sim \tau_{sync}$.
3) The most unstable modes are found for small values of K_z, i.e. for modes propagating almost perpendicular to the magnetic field direction, which is in accord with the observational data.

REFERENCES

Bodo G., Ferrari, A., Massaglia, S., and Trussoni, E., 1988, Proc. Int. School *Plasma Astrophysics*, Varenna (Italy), September 1988, in press.

Dreher, J.W., Carilli, and Perley, R.A., 1987, *Astrophys. J.*, **316**, 611.

Eilek, J.A. and Caroff, L.J., 1979, *Astrophys. J.*, **233**, 463.

Perley, R.A., Cargese Workshop on "Magnetic Fields and Extragalactic Objects", June 1987, ed. E. Asseo and D. Gr"esillon, p. 255.

Simon, M. and Axford, W.I., 1967, *Astrophys. J.*, **150**, 105.

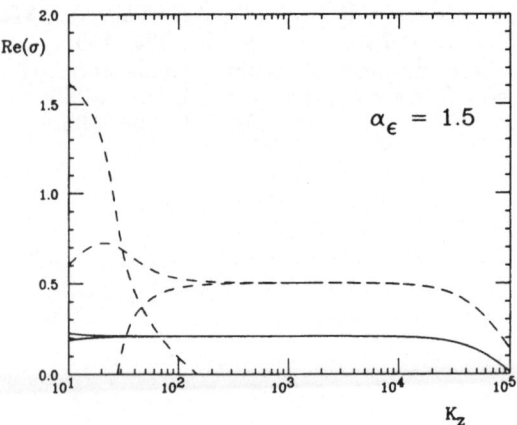

Figure 1: Growth rate $(Re(\sigma))$ as a function of K_z for $K_x = 10^5$, $V_{AS} = 0.1$ (dashed lines), and $V_{AS} = 0.5$ (solid lines), with $\alpha_\epsilon = 4$.

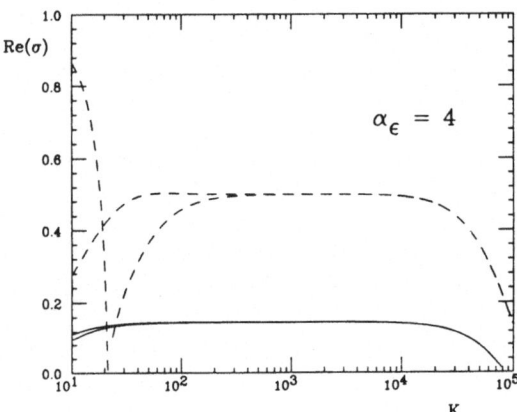

Figure 2: Same as in Fig. 1, with $\alpha_\epsilon = 1.5$.

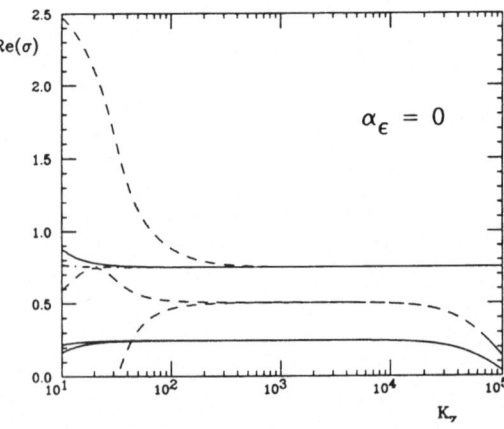

Figure 3: Same as in Fig. 1, the dot–dashed line is for $V_{AS} = 1$, and $\alpha_\epsilon = 0$.

GRAVITATION AND JET INDUCED VELOCITIES IN THE NARROW LINE REGION OF ACTIVE GALAXIES

MARK WHITTLE
Astronomy Department
University of Virginia
Charlottesville VA 22903
U.S.A.

INTRODUCTION

The basic question I want to address is : "What mechanism or mechanisms accelerate the gas in the narrow line region of active galaxies, yielding the observed profile shapes." At the present time there is no generally accepted answer to this question. Part of the problem dates back to the first few Seyferts that were discovered which, in retrospect, seem to have had anomalously broad lines. For example NGC 1068, MKN 3, and MKN 78 have [OIII] λ 5007 FWHM $\sim$ 1000 km s^{-1}, which is clearly greater than anything associated with a normal galaxy velocity fields. An obvious implication was that these high velocities were in some way related to the *activity*. This view was reinforced by the discovery of a correlation between [OIII] FWHM and non-thermal radio luminosity (Wilson and Willis 1981). When many radio sources were found to have linear double or triple morphology (e.g. Ulvestad and Wilson 1984a,b), a natural explanation seemed to be outflowing radio jets which stir up the narrow line region and thus generate the correlation between line width and radio luminosity.

On the other hand, some facts seemed to undermine this "active" acceleration picture. First, the [OIII] FWHM do not correlate with the overall level of activity. For example, the mean FWHM for Seyfert 2s, Seyfert 1s and QSOs are the same, despite an enormous range of non-thermal luminosity (Whittle 1985b). Second, as more [OIII] line widths were measured, it became clear that the mean FWIIM is only $\sim$ 350 km s^{-1}, much less than previously thought. Such a velocity is similar to a typical *stellar* velocity in a spiral bulge of $M_B \sim -21.5$. Although this seems large, it is in fact typical of Seyferts, which tend to inhabit relatively massive early type spirals. Thus, we have a second possibility that for the majority of Seyferts, the basic acceleration mechanism may be *gravitational* (Whittle 1985a,b).

To test this latter possibility it is important to look for explicit correlations between [OIII] linewidth and diagnostics of the bulge potential. Two such diagnostics are the rotation amplitude of the galaxy and the bulge absolute magnitude.

[OIII] FWHM AND GALAXY ROTATION

The observed rotation amplitude is simply the peak to peak velocity difference on the flat part of the rotation curve, measured either from long slit optical emission line

423

D. McNally (ed.), Highlights of Astronomy, Vol. 8, 423–427.
© 1989 by the IAU.

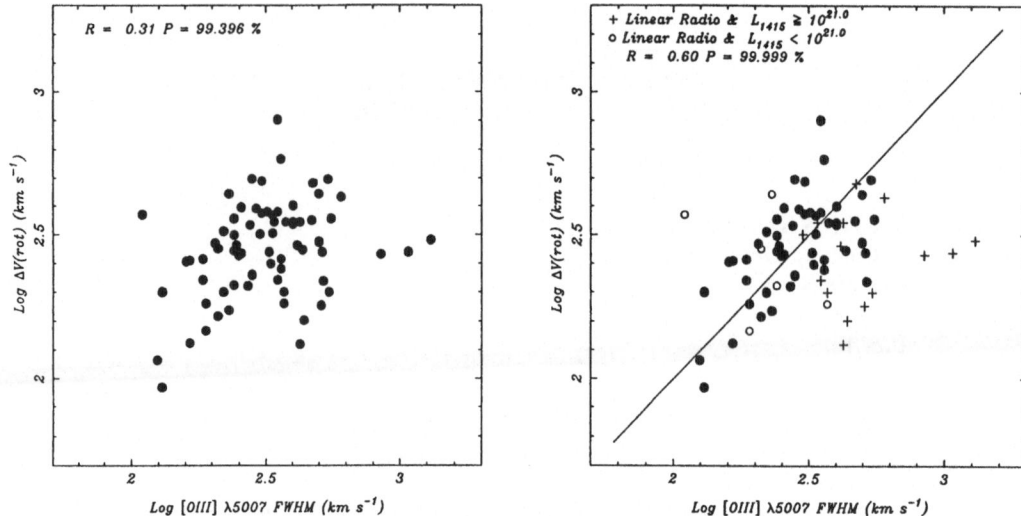

Figure 1 (left) : Plot of [OIII] FWHM against observed galaxy rotation amplitude for about 60 Seyferts.

Figure 2 (right) : Same data as Figure 1, objects flagged by radio properties.

studies or single beam HI profiles. Figure 1 shows a plot of Log [OIII] FWHM against Log observed rotation amplitude, ΔV, for about 60 Seyfert galaxies. There is a general trend with considerable scatter. The overall correlation strength is only moderate : 99.4% (two tailed). The diagram becomes much more interesting when the radio properties of the individual objects are considered. Figure 2 shows exactly the same data but those objects which have been classified by Wilson and Ulvestad (1984a,b) as having linear radio morphology on the Kpc scale have been flagged differently. Luminous linear radio sources are plotted as plusses, weaker linear radio sources as open circles, the rest as filled circles. There is a clear tendency for the luminous linear radio sources to have anomalously broad [OIII] lines and contribute most of the high velocity scatter. The rest of the sample falls on a tight correlation with linear correlation coefficient 0.6 and very high significance. The best fit straight line falls almost exactly on the line y = x. In other words, the [OIII] FWHM not only correlate with but are equal to the observed rotation amplitude.

We conclude from this that the basic acceleration mechanism is *gravitational*. For some objects there is an additional acceleration mechanism associated with the linear radio structure, and this almost certainly results from the interaction of bipolar jets with the surrounding medium.

JET INTERACTIONS

Let us first consider the radio jet interaction briefly, before returning to discuss the gravitational processes. A detailed description of 10 Seyferts with linear radio morphology is given by Whittle *et al.* 1988. Figure 3 shows the archetype MKN 78. The spectrograph slit was placed along the radio axis to yield high resolution [OIII] $\lambda5007$ line profiles running east-west across the nuclear regions. In addition to a rotating component, there is a redshifted component which is physically located close to the eastern radio lobe, and also a

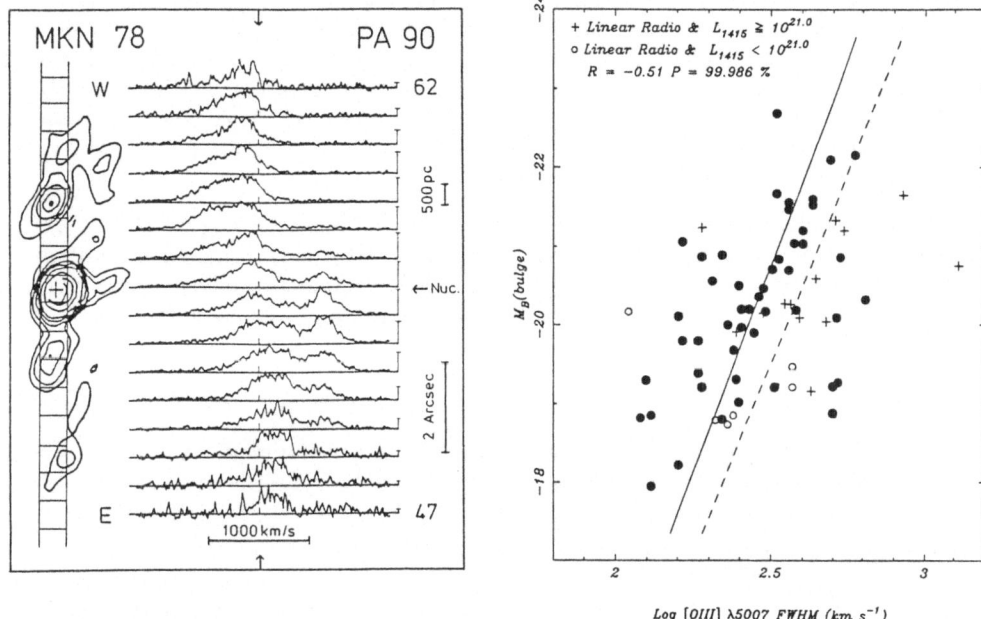

Figure 3 (left) : Long slit [OIII] λ5007 observation of MKN 78 (autoscaled) with radio continuum map plotted at the same scale (from Whittle *et al* 1988).
Figure 4 (right) : Plot of [OIII] FWHM against bulge absolute blue magnitude, with best fit line (solid) and relation between stellar velocities and bulge magnitudes for normal galaxies (dashed line)

blueshifted, somewhat blended, component which is located close to the western radio lobe. The simplest interpretation is that the two high velocity components are accelerated locally by the interaction of a jet with the interstellar medium. It isn't yet clear whether this is an entrainment process or accelaration by an advancing bow shock. Both these processes need to be modelled in more detail.

GRAVITATIONAL VELOCITIES

Returning now to the group of objects for which gravity seems to be the dominant acceleration mechanism. How can the rotation amplitude, which is measured on large scales in the disk of the galaxy, be related to a highly nuclear velocity measurement taken, typically, with a 2 – 3 arcsec aperture ? There are two scenarios to account for this, although both are probably unrealistic extremes. First, it might be that the rotation velocity remains high right into the nucleus, so that there is significant rotational smearing even within a small aperture. Observations of normal galaxy rotation curves suggest that this is not the explanation since the nuclear rotation gradients are too shallow. However, some galaxies have recently been found to have rapidly rotating nuclear stellar disks with very steep rotation curves (e.g. Franx and Illingworth 1988, Jedrzejewski and Schechter 1988). If Seyferts possess such disks and these have rotation amplitudes similar to the disk amplitude, then rotational smearing can indeed contribute to the correlation shown in Figure 2. Second, although the rotation curve may be low within a nuclear aperture, the stellar velocity *dispersion* remains high across the bulge. If the ionised gas moves anything like the stars, then a nuclear aperture will yield a broad profile. Now, Whitmore and Kirshner (1981) find that the central velocity dispersion is proportional to the rotation amplitude with only a weak Hubble type dependence. Hence the correlation between rotation amplitude and [OIII] FWHM. In fact, by using their relation to infer nuclear velocity dispersion from rotation amplitude, it can be shown that [OIII] FWHM is proportional to and approximately equal to the inferred stellar velocity dispersion.

An independent demonstration of this is shown in Figure 4. Log [OIII] FWHM is plotted against bulge absolute magnitude (in B). This bulge magnitude has been derived from the total absolute galaxy magnitude and a bulge/total ratio inferred from normal spiral galaxies of the same Hubble type. There is a moderately strong correlation with a best fit (solid) line which lies parallel to (and slightly displaced from) the classic $L \sim \sigma^4$ relation between luminosity and stellar velocity dispersion for spheroidal systems (shown as the dashed line).

Thus, there are two independent demonstrations of gravitational origin to the [OIII] FWHM — correlations with rotation amplitude and correlations with bulge absolute magnitude.

ROTATION OR DISPERSION ?

It may be possible to distinguish between the two extremes described above (pure rotation and pure dispersion) by using the galaxy inclination. There are really *two* rotation measures, the *observed* rotation amplitude, ΔV, and the inclination corrected or *true* rotation amplitude, $\Delta V_c = \Delta V / \sin i$. If rotation in the plane of the galaxy dominates the velocity field of the NLR then there should be a tighter correlation between [OIII] FWHM and ΔV than between [OIII] FWHM and ΔV_c. Conversely if dispersion velocities dominate in the NLR then there should be a weaker correlation between [OIII] FWHM and ΔV than between [OIII] FWHM and ΔV_c, since [OIII] FWHM would not itself suffer projection effects. Comparing the two correlations for Seyfert galaxies with well measured inclinations in fact shows no significant differences in their correlation strengths. This result is consistent with an intermediate situation in which both rotation and dispersion velocity fields are present. The result also agrees with a direct analysis of the inclination dependence of [OIII] FWHM (Whittle 1985a,b).

[OIII] LINE WINGS

The previous discussion considers only the [OIII] line core width, as measured by the FWHM parameter. Do similar correlations exist for the base and wings of the [OIII] profile, which are presumably produced by gas residing closer to the nucleus ? The correlations with rotation amplitude and bulge magnitude are progressively weaker for base width, FW20, and wing width, IPV20, parameters. Unfortunately, the interpretation of this result is not clearcut. First, the results seem consistent with the plausible notion that non-gravitational acceleration processes may become more important closer to the active nucleus itself. While intuitively appealing, the weak point of this explanation is that the base and wing widths do not seem to depend on overall level of activity (i.e. they do not correlate with non-thermal luminosity or Seyfert type). Second, it is possible that the base and wing widths are still determined by the gravitational potential but that close to the nucleus the gravitational velocities are significantly different from the average bulge velocities measured on larger (arcsec) scales. For example, a dense star core possibly (though not necessarily) surrounding a black hole could be dynamically quite isolated form the rest of the bulge. Finally, the breakdown in correlation for base and wing width parameters may simply reflect the fact that while these parameters are well defined observationally they are not well defined physically. A particular problem with the wing width parameters is that they are quite sensitive to the intensity ratio of the core and wings. It may be that a component fitting parameterization scheme, similar to that employed by Pelat et al (1981) for example, would recover the correlation strength between wing widths and gravitational diagnostics.

SUMMARY

The main result of this study has been to show that the [OIII] FWHM in Seyfert galaxies correlates quite tightly with tracers of the bulge gravitational potential, implying that gravity is usually the dominant acceleration mechanism for the ionised gas in the NLR. For objects with luminous linear radio sources, there is usually an additional acceleration mechanism which is almost certainly associated with a jet interacting with the interstellar medium. Somewhat more tentative results come from an analysis of the inclination dependence of these correlations and suggest the NLR velocity field is only partly confined to the plane of the galaxy. Finally, the correlations between base and wing width and gravitational diagnostics are less good than the correlation with core width, although the interpretation of this is ambiguous.

REFERENCES

Franx, M., and Illingworth, G.D., 1988, *Astrophys. J. Lett.*, **327**, L55.

Jedrzejewski, R.I., and Schechter, P.L., 1988, *Astrophys. J. Lett.*, **330**, L87.

Pelat, D., Alloin, D., and Fosbury, R.A.E., 1981, *M.N.R.A.S.*, **195**, 787.

Ulvestad, J.S., and Wilson, A.S., 1984a, *Astrophys. J.*, **278**, 544.

Ulvestad, J.S., and Wilson, A.S., 1984b, *Astrophys. J.*, **285**, 439.

Whitmore, B.C., and Kirshner, R.P., 1981, *Astrophys. J.*, **250**, 43.

Whittle, M., 1985a, *M.N.R.A.S.*, **213**, 1.

Whittle, M., 1985b, *M.N.R.A.S.*, **213**, 33.

Whittle, M., Pedlar, A., Meurs, E.J.A., Unger, S.W., Axon, D.J., and Ward, M.J., 1988 *Astrophys. J.*, **326**, 125.

Wilson, A.S., and Willis, A.G., 1981, *Astrophys. J.*, **240**, 429.

TWO-FLOW MODEL FOR EXTRAGALACTIC RADIO JETS

H. SOL, D.A.R.C., Observatoire de Paris-Meudon, 92195 Meudon,
E. ASSEO, Ecole Polytechnique, 91128 Palaiseau, France,
and G. PELLETIER, Université de Grenoble, CERMO BP 68,
38402 Saint-Martin d'Hères, France.

Different models of extragalactic jet formation have been proposed in the litterature. They usually considered only one population of particles, ascribing one typical bulk velocity to the jet. However from the point of view of plasma physics it seems very likely that beams of particles with different velocity will be generated in the surroundings of the central engine by any kind of jet acceleration mechanism. This can lead to a typical beam-plasma configuration with a slow ambient plasma pervaded by energetic beams which constitute a population inversion and are a potential source of energy and of instability.

To investigate the role of the beam-plasma instability in the physics of astrophysicals jets, we considered the simplest scenario with a beam-plasma configuration but still consistent with the current knowledge on extragalactic jets. This includes three components :

(i) a relativistic beam of electrons and positrons, related to the VLBI jets, with a bulk Lorentz factor deduced from the observed superluminal motion,
(ii) a longitudinal magnetic field, as inferred from VLBI polarization maps and also from theoretical studies of black hole magnetosphere evolution,
(iii) and a slow ambient plasma of electrons and protons due to the accreted interstellar medium and to a wind from the accretion disk.

The presence of a relativistic beam and of a longitudinal magnetic field is directly grounded on observational facts. The presence of a slow ambient plasma has no direct observational signature yet. However it seems hard to avoid especially in the vicinity of the central engine. It might be responsible for a part of the X-ray emission of the active galactic nuclei and play a role in the confinement of broad and narrow emission line clouds. Evidence of this ambient plasma and estimates of its density can be expected from further Faraday depolarization data which up to now provide only upper

429

D. McNally (ed.), Highlights of Astronomy, Vol. 8, 429–432.
© *1989 by the IAU.*

limits.

Assuming given a relativistic beam, a wind from the accretion disk and a longitudinal magnetic field, we studied the strong Langmuir turbulence excited by the beam-plasma interaction and the development of the self-modulation instability (the quasilinear theory is not expected to apply for high values of γ, the Lorentz factor of the beam). Under simplifying assumptions such as small transverse wavenumber $k_\perp$ and small thermal velocity v_{th} in the ambient wind, we deduced the evolution of the Langmuir waves packets. Following the perturbation method developed by Karpman and Maslov (1978) and Kaup and Newell (1978) we found that they are stable against transverse perturbations for high values of the magnetic field, namely when the cyclotron frequency $\omega_c = eB/m_e c$ is greater than the plasma frequency

$$\omega_p = \left(\frac{4\pi \, n_p \, e^2}{m_e}\right)^{1/2}$$. Conversely, for weak magnetic field, when $\omega_c < \omega_p$,

we found that the Langmuir waves packets are unstable and collapse towards small scales (Pelletier et al, 1988). Then qualitative arguments about the detuning of the beam-plasma interaction due to large density fluctuations in the strong field zone lead to the idea that the beam itself is stable and can propagate as long as the magnetic field is strong enough ($\omega_c > \omega_p$). However when the beam reaches the weak magnetic field zone it starts losing its energy. The beam-plasma interaction generates Langmuir waves in the ambient wind which evolue through the self-modulation instability and collapse towards the Debye's sphere where they transmit their energy to the ambient wind through Landau effect. Energy is then transferred from the beam to the ambient wind, which induces the destruction of the beam after some relaxation length.

Using a self-similar solution for the collapse of the Langmuir waves packets and under the assumption that all the energy of the beam goes through such a mechanism we obtain an estimate of the beam relaxation length

$$z_R = \gamma^{14/3} \left(\frac{m_e \, c^2}{T}\right)^{5/3} \left(\frac{n_b}{n_p}\right)^{1/9} \frac{c}{\omega_p}$$

The value of the relaxation length is thus very dependent on the Lorentz factor of the beam. We can adopt for γ a conservative value of the order of 10 or less, as deduced from the standard model of superluminal sources. The beam to ambient plasma density ratio n_b/n_p does not have a great influence on the relaxation length. We assume it of the order of 0.01. Up to now, observational data do not provide direct information on the density and temperature of the ambient plasma. All we can say is that in a realistic range of density and temperature, for instance n_p between 10^{-3} cm^{-3} and 10^2 cm^{-3} and T between 10^4K and 10^8K, the relaxation length is always much smaller than typical lengths of VLBI jets. It is also smaller than the relaxation length due to other dissipation effects such as relativistic bremsstrahlung, diffusion on turbulent Alfven waves, ionization losses, electron-positron annihilation, synchrotron losses or inverse Compton effect at some parsecs from the central engine (Sol et al, 1988). The dynamics of the beam seem therefore controlled by the magnetic field strength. The beam propagates for a magnetic field larger than a critical value $B_c = 3.2 \times 10^{-3} \sqrt{n_p}$ and is rapidly destroyed as soon as $B < B_c$. The problem is then to have an estimate of the relative importance of the magnetic field.

For the fiducial values given by Reos (1984), we obtain a ratio $B/B_c \simeq 40$ close to the active galactic nuclei which is consistent with the beam extraction and propagation from the vicinity of the central engine. A simple flux conservation model leads to a ratio B/B_c inversely proportional to $\sqrt{S}$, (S is the cross-sectional area of the jet). This corresponds to a decrease of B/B_c with the distance from the central engine, if the jet widens out. When B/B_c reaches unity the beam enters its relaxation zone and is destroyed, with a subsequent heating and entrainment of the wind. The heating likely induces a local enhancement of the radio emission. Acceleration of the wind can be described by the equations of hydrodynamics for two coupled fluids. We showed the possibility to have regular solutions with no shocks for the flow and to reach wind asymptotic velocities larger than $2c/5$ (Sol et al, 1988). If the wind is identified with the jets observed at the kiloparsec scale, such midly relativistic values for the speed are high enough to explain the one-sidedness of large scale jets by a Doppler beaming effect.

Combining VLBI and VLA observations, Walker and cowokers (1987) obtained very detailed data on the radio galaxy 3C120. In particular, they derived under the usual minimum-energy assumption the variation of the magnetic field along the jet $B = 4.1 \times 10^{-5} r^{-0.97}$ G, where r is the measured full width at half maximum (FWHM) of the jet in arcseconds. The absence of any significant Faraday rotation gives an upper limit on the ambient plasma density inside the emitting jet $n_p \lesssim 9.9 \times 10^{-4} r^{-0.03}$ cm^{-3}, assuming a uniform magnetic field mainly oriented along the line of sight. With these values the ratio B/B_c reaches unity at $r \gtrsim 180$ pc which corresponds to a minimal distance of about 1.4 kpc from the central engine. This is amazingly close to the location of the 4"-radio knot, a "rather curious structure" described by Walker et al (1987). We propose to interpret this peculiar 4"-knot as the beam relaxation zone. The relativistic beam then survive for the first 4" from the nucleus and starts to be dissipated when it reaches the 4"-knot where $B/B_c = 1$. This picture is quite coherent with recent results of Benson et al (1988) and Walker et al (1988) which show some evidence for superluminal motion on kiloparsec scales in 3C120. Several properties of this curious 4"-knot can find a natural explanation in this context. Especially the morphology of the jet can suggest a change in the flow regime at the location of the 4"-knot, which would correspond here to the beam (respectively the wind) being mainly responsible of the radio emission before (respectively beyond) the 4"-knot.

Our two-flow model is not drastically different from the usual scenario for extragalactic jets. It introduces a new realistic latitude with allows qualitative explanations of the different types of extragalactic radio sources, according to the relative importance of the relativistic beam and of the slower ambient wind. It also allows high bulk Lorentz factor in the first part of the jets where superluminal motion are detected and midly relativistic bulk velocity on larger scales, without inducing any strong discontinuity in the jet structure. Moreover only a small fraction of the jet particles needs to reach very high bulk Lorentz factor, which can help to solve the problem of jet formation. Different steps in our study rely on simplifying assumptions or qualitative arguments required to progress further. This brings about uncertainties in the scenario. However there are several possible observational tests such as other estimates of the magnetic field and ambient plasma density in 3C120 and other superluminal radio sources showing peculiar knots somewhat reminiscent of the 3C120 4"-knot (NRAO 140, 4C39.25, 3C279, 3C345, 3C454.3, 3C245, 1642+690, 1928+738). Generally speaking, any direct evidence of a slow wind from the active galactic nuclei, for instance from positive Faraday depolarization measures or from detection of very faint non superluminal radio emission at VLBI scale in superluminal sources, would be an important clue in favour of two-flow models.

REFERENCES

- Benson, J.M., Walker, R.C., Unwin, S.C., Muxlow, T.W.B., Wilkinson, P.N., Booth, R.S., Pilbratt, G., and Simon, R.S., 1988, ApJ, in press.
- Karpman, V.I., and Maslov, E.M., 1978, Sov. Phys. JETP, 46, 281.
- Kaup, D.J., and Newell, A.C., 1978, Proc. R. Soc. London A, 361, 413.
- Pelletier, G., Sol, H., and Asséo, E., 1988, Phys. Rev. A, 38.
- Rees, M.J., 1984, ARAA, 22, 471.
- Sol, H., Pelletier, G., and Asséo, E., 1988, MNRAS, in press.
- Walker, R.C., Benson, J.M., and Unwin, S.C., 1987, ApJ, 316, 546.
- Walker, R.C., Walker, M.A., and Benson, J.M., 1988, ApJ, dec. 15.

7. THE HUBBLE SPACE TELESCOPE - STATUS AND PERSPECTIVES

Scientific Organizing Committee

G. Miley (Chairman)

J. Bergeron, R. Giacconi, C. Humphries, Y. Kondo,
P. van der Kruit, H. van der Laan, M. Rees, G. Setti, R.A. Sunyaev, S.G. Wang

Supporting Commissions:

9, 28, 44, 47

THE SCIENCE PROGRAM OF THE HUBBLE SPACE TELESCOPE

Neta A. Bahcall
Space Telescope Science Institute
3700 San Martin Drive
Baltimore, MD 21218

ABSTRACT. Highlights of the science program to be carried out by the Hubble Space Telescope are reviewed. Some of the main scientific projects to be carried out by the General Observers (GOs) and the Guaranteed Time Observers (GTOs) during the early phases of the Space Telescope are discussed. The method and criteria for selecting science projects are also discussed.

1. SCIENCE WITH THE HUBBLE SPACE TELESCOPE

The Hubble Space Telescope (HST) will be the first long-lived international optical observatory in space. Its location above the obscuring and distorting effects of the earth's atmosphere will provide HST with unique capabilities that will yield a major improvement in observational optical astronomy. The main unique capabilities of HST are:

      (i)   High angular resolution: ~0.1";
      (ii) Faint stellar limiting magnitude: ~28";
      (iii) UV observations: ≥ 1150Å.

The telescope will be equipped with a complement of six scientific instruments, including cameras and spectrographs that will take advantage of HST's unique capabilities. Only scientific projects that require at least one of HST's unique capabilities will be carried out by the telescope. Projects that can be done from ground-based or other telescopes will generally not be accepted for HST observations.

The high angular resolution and sensitivity of HST and its broad wavelength coverage will allow astronomers to perform scientific observations currently unfeasible with ground-based telescopes. These programs will be carried out by General Observers (GOs) as well as the Guaranteed Time Observers (GTOs). I summarize below some of the main science projects that will be investigated by the GOs and GTOs during the initial cycles of the HST. I divide the projects according to their scientific subdiscipline.

D. McNally (ed.), Highlights of Astronomy, Vol. 8, 435–439.
© *1989 by the IAU.*

Cosmology

- Determination of accurate distances to nearby galaxies and the Hubble constant. HST will provide an extension of a factor of ten in the distance over which properties of the standard candles can be measured.
- Determination of the cosmological deceleration parameter, q_o. This fundamental determination may be attempted by observing the dependence of standard candles (such as the size of galaxies) with redshift to $z \sim 1$ or by using supernovae as standard candles (if they prove to be standard).
- Confirmation of the expansion of the universe by studying the surface brightness dependence on the inverse of $(1 + z)^4$ to large redshifts.

Evolution

- The evolution of galaxies and clusters of galaxies with redshift to $z \sim 1$. Galaxy morphology determination at high redshifts will be a most exciting study with HST and will provide new insight into the nature of galactic evolution.
- Evolution in QSO absorption lines. This will allow an understanding of the evolution of matter in the universe from nearby to very large redshifts.
- Deep surveys with HST will allow the investigation of galaxy and quasar counts to faint limiting magnitudes.

QSOs and AGNs

- A study of the physics of the nuclear regions of QSOs and AGNs, including the central source and the line regions. High resolution imaging and spectroscopy will be uniquely suited to HST capabilities and they are expected to reveal exciting discoveries.
- A study of the host galaxies around QSOs, which can be seen with high resolution only with HST.
- QSO absorption line survey and high resolution observations in the UV.
- Gravitational lenses. The high resolution structure and properties of these systems will be investigated.

Galaxies and Clusters

- The evolution of galaxies and clusters (see above).
- Detailed properties of nearby galaxies using the unique high resolution

and sensitivity of HST.

- Detailed studies of galactic nuclei at high resolution.
- Stellar population studies in nearby galaxies.

Stars and Interstellar Medium

- Stellar populations in our and other galaxies.
- Detailed studies of globular clusters, their cores and stellar populations.
- Star formation regions
- Stellar binary systems
- High resolution studies of supernovae and supernovae remnants.

Planetary Astronomy

- Atmospheres, features, satellites, and rings of planets.
- Primitive bodies in the solar system.
- Companions of nearby stars.

Surveys

- Deep surveys of the sky will reveal faint stars, galaxies, supernovae, variable objects, and, most importantly, unknown types of objects.

Included within the above list are the three Key Projects identified by the community and listed in the HST Call for Proposals; these are:

(i)   Distances to galaxies and H_o;

(ii)   QSO absorption line survey;

(iii) Medium deep survey.

2. SELECTION OF SCIENCE PROGRAMS

Observing time on HST will be allocated to openly solicited, peer-reviewed, and competitively selected proposals from the international astronomical community.   During the first two and half years of HST operations, the scientists associated with the development of HST and its instruments (GTOs) are guaranteed an average of 30% of observing time.   Scientists from ESA member states will receive, on the average, at least 15% of the available time over the lifetime of the observatory.

All proposals will be submitted to the Space Telescope Science Institute (ST ScI). A two-phase selection process will be carried out:

Phase I: • Proposal submission
• Time Allocation Committee (TAC) review and recommendation

Phase II: • Detailed program submission (for those who pass Phase I)
• Detailed feasibility and resource evaluation
• Final acceptance

The peer-review process will also be a two-stage procedure. First, proposals of the various science disciplines will be reviewed by scientific peer-review panels; the panels will include the following disciplines (which may change according to proposal volume):

• Quasars and AGNs
• Galaxies and Clusters
• Stellar Populations
• Stellar Astrophysics
• Interstellar Medium
• Solar System

The panels will review the proposals following the selection criteria listed below, and recommend time and resource allocations. The panels will then rank the proposals in their discipline in order of recommendation. A cross-discipline TAC will then follow; the TAC will recommend the final, cross-discipline program to be considered for observations with the HST.

The selection criteria for the review and selection of science proposals include the following:

• Scientific merit
• Need of unique HST capabilities
• Technical feasibility
• Observation plans and data analysis
• Capability of proposers to carry out project
• Demand on HST resources; program efficiency

In addition to the general GO program, a small amount of time will be reserved for Director's Discretionary Time. This time will be used, for instance, for special programs such as targets-of-opportunity observations, exploratory observations and other programs judged to be appropriate by the Director.

All data will be kept proprietary to the observers for a period of one year after the observations. At that time, the data will be placed in a public archive and will be available to interested scientists. It is expected that the HST public archive, with its enormous volume of astronomical data, will be a highly useful source of scientific investigations for many years to come.

HUBBLE SPACE TELESCOPE SECOND GENERATION INSTRUMENT SELECTION

E. J. WEILER
Hubble Space Telescope Program Scientist
Code EZ
NASA Headquarters
Washington, DC 20546
USA

ABSTRACT. The Hubble Space Telescope Second Generation Instrument Program is described. The original instrument selection process in 1985 is discussed as well as the NASA plan to make a final selection of an infrared instrument in late 1988.

An Announcement of Opportunity (AO) for Hubble Space Telescope (HST) Second Generation Instruments was issued by NASA in October, 1984. NASA received eight proposals as a result of this AO. A sixteen member scientific peer review team evaluated these proposals and provided NASA with a summary of each proposal's strengths and weaknesses. They also provided unsolicited advice on the relative priorities of various classes of instruments. Although the team was primarily composed of ultraviolet and visible astronomy experts, they rated an infrared camera/spectrometer as the number one priority. An ultraviolet/visible spectrograph was rated second priority.

NASA considered the peer panel's advice and selected three proposals for Phase A/B studies. This decision was announced in December, 1985. Since the peer panel could not provide clear discriminators between the two top infrared instruments, NASA decided to select both for definition studies with the intent of carrying only one into development. The selected infrared instruments were the Near-Infrared Camera and Multi-Object Spectrometer (NICMOS), proposed by Dr. Rodger Thompson of the Unversity of Arizona, and the Hubble Imaging Michelson Spectrometer (HIMS), proposed by Dr. Donald Hall of the University of Hawaii. The third instrument selected was the Space Telescope Imaging Spectrograph (STIS), proposed by Dr. Bruce Woodgate of the Goddard Space Flight Center.

Although NASA's original plan was to carry both infrared instruments through full Phase A and Phase B studies, it was decided in late 1987 to initiate an early selection process. This decision was made both to save money and because the respective Principal Investigators (P.I.'s) felt that enough progress had been made to enable an early selection. A plan to enable an early selection was developed by NASA and the P.I.'s during the early part of 1988. This plan calls for final proposals from both the NICMOS and HIMS teams by August 26, 1988. During September, 1988, two travelling peer review teams will visit contractor and home institution sites to conduct indepth technical

D. McNally (ed.), Highlights of Astronomy, Vol. 8, 441–442.
© *1989 by the IAU.*

reviews of detector technology and instrument cryogenics and mechanisms. In addition to these specific technical reviews, the Goddard Space Flight Center will conduct a broad technical review of both instruments. The overall proposals and the results of the aforementioned reviews will be evaluated by a summary peer review panel in early November, 1988. Selection criteria will be those listed in the original A.O. (e.g., overall scientific merit, technical merit, etc.). NASA will announce a final selection in early December, 1988. The selected instrument will then be funded to complete its Phase B study and then proceed into instrument development.

The current NASA plan calls for the development of the infrared instrument first. This is based on the fact that there is no infrared capability in the original complement of HST instruments and on the strong scientific recommendations of the original peer review team. Under this plan, the infrared instrument would be launch ready in August, 1994. The STIS would follow with a launch readiness about 1-2 years later. This plan could be altered in the event that there are early problems with one or both spectrographs in the original complement of HST instruments. In other words, the STIS could precede the infrared instrument if such a decision were made shortly after the launch of the HST.

Wide Field / Planetary Camera-II for the Hubble Space Telescope

John Trauger (Jet Propulsion Laboratory, California Institute
of Technology, Pasadena, California, USA)

Abstract
The Wide Field / Planetary Camera-II will be available to replace the
WF/PC at the first opportunity for in-orbit Shuttle maintenance and
refurbishment of the Hubble Space Telescope. It is fundamentally an
engineering "clone" of the original radial camera, which nevertheless
incorporates a number of science and operational enhancements based on
improved technologies. Generic WF/PC performance issues identified and
understood early in the WF/PC-II program provide the opportunity for
carefully considered, cost effective hardware refinements. The result
is improved science performance and operational efficiency for WF/PC-II
which will be highly visible to the GO astronomer, and will provide
savings in data reduction and calibration overhead.
Biased-platinum-gate CCD technology is introduced for the eight imaging
sensors, in order to provide long and short term QE stability while
obviating the solar UV light channel. Signal chain electronics have
been modified in several ways to improve science performance and
operational efficiency. The WF/PC set of 48 filters has been reviewed
and a number of modifications have been implemented. The WF/PC-II is
well into construction at JPL, scheduled for delivery in late 1991.

WF/PC-II CCD Technology
Sensors are drawn from an existing inventory of 800x800 Texas
Instruments CCDs, but differ from WF/PC sensors in new processing of the
(illuminated) backside surface and improved control of the packaging
procedures. WF/PC-I UV flooded CCD technology is adequate, but provides
little margin against QE instabilities and hysteresis, and has
substantial operational impact for HST. Our goal is to improve QE
stability, with as much margin as possible, in order to simplify
photometry procedures and improve the viability of a long-term image
calibration database. New processing steps demonstrate long and short
term QE stability beyond requirements for 1% photometry. These steps
include 1) selection of "thick p+" sensors from the existing inventory
of unpackaged chips, 2) use of lumogen rather than coronene as the UV
phosphor, 3) application of a platinum flashgate, and 4) electrically
biasing the flashgate to a fixed potential. Each step independently

443

adds margin against QE instabilities and hysteresis. Taken together, these steps are more than adequate for 1% photometry. Lumogen provides better quantum efficiencies in the FUV. Manufacturing procedures will be finalized this fall, based on extensive laboratory investigations now in progress with flight-packaged test devices. Packaging of flight sensors, following critical review of process qualification test results, will commence in December 1988. The UV light flood is obviated, hence the WF/PC solar UV light channel will be eliminated.

Signal Chain Electronics
Residual image removal circuitry has been added to the standard exposure sequence. This eliminates the potential difficulties associated with deeply overexposed star images, which would otherwise impose themselves in subsequent WF/PC exposures. The WF/PC "missing code" phenomenon has been eliminated. This recovers the otherwise ambiguous three least significant bits of the digital pixel values. Baseline stabilization circuitry has been added, adding margin against particle radiation upset events in the on-chip amplifier. "Partly-inverted" clock voltages are implemented for parallel phases, improving full well, cosmetics, and performance during overexposures.

FUV Performance
A FUV-attenuating condensate was identified in testing of the WF/PC, with substantial impact on operational efficiencies and FUV performance. Steps are being taken to eliminate sources of such contamination in WF/PC-II, through vacuum qualification of material samples and individual flight components prior to assembly. In addition, improved venting of the instrument to space will reduce the risk of condensation.

Filter Set
A baseline filter set has been defined, following an STScI workshop and HST Science Working Group concurrence. The ability of vendors to meet filter specifications has been demonstrated, and procurement of the filter set is in progress. Filters which were not requested in GTO proposals have been studied, most of these (six filters) have been modified or eliminated for WF/PC-II. WF/PC "UBVRI" is supported, providing continuity with the WF/PC system (avoids the Balmer jump). Stromgren photometry is now supported. The wide (30% FWHM) WF/PC filter sequence is extended with four new filters in the UV and FUV. Filters near the IS absorption feature have been adjusted for compatibility with other instruments. Narrow-band filter profiles are standardized in spectral shape and width to improve the determination of line ratios, especially in objects with significant velocity structure or red-shifts. "Quad" format filters are introduced for some red-shifted emission lines, polarizers, and solar system methane bands. A laboratory set of high fidelity witness filters will track the filter performance over time, as part of an integrated plan for fabrication, testing, and verification of on-orbit filter performance. The 48 baseline filters are fabricated with currently available technologies, however parallel and independent development of advanced technology broad-band UV and FUV filters is being pursued.

The Space Telescope Imaging Spectrograph

Bruce E Woodgate

NASA/Goddard Space Flight Center

Instrument

The Space Telescope Imaging Spectrograph (STIS) is a
second generation instrument to be installed into the
Hubble Space Telescope (HST) in-orbit 5-9 years after the
first launch. Together with the infra-red instrument, it
will provide a large increase in capability of the
observatory, and be able to replace a first generation
that had failed or degraded.

The STIS will be able to replace both first generation
instruments with an increase in angular and/or spectral
coverage of a factor 200-2000, and will operate in the
spectral range 1050-11000A. Also it will provide
broad-band camera capability over this range, and
high-speed photometry.

The capability is centered around two kinds of 2048x2048
pixel detectors, Multi-Anode Microchannel Arrays (MAMAs)
in the UV and charge-coupled devices (CCDs) in the
visible and near-IR. Interchangeable spectral formats
allow selection of spectral resolving powers between 80
and 140000 and a field-of-view up to 50 arc-sec. The MAMA
detectors allow UV spectrographic and camera observations
without obscuration due to scattering from the visible.
The operating modes (spectral range, resolving power,
effective area and angular range) are shown in the Table,
and the optical arrangement to accomplish them in the
Figure. The mode is selected by rotating the grating
wheel which inserts the appropriate optic (grating, prism
or mirror) and steers the beam to the next optic
(grating, mirror or detector). The slit wheel is also
rotated to match the mode. In this way the detector
(spectral band) and resolving power as shown in the Table
are selected. For each mode, the full angular resolution
of the HST of 0.05 arc sec is maintained with 2-pixel
sampling. Limited spectral selectivity in the camera
modes will be provided in each of the 4 spectral bands by
inserting 4 filters each covering a quadrant of the 50 x

445

D. McNally (ed.), Highlights of Astronomy, Vol. 8, 445–448.
© 1989 by the IAU.

50 arc sec field.

Scientific Objectives

The capabilities of STIS will allow astronomers from a broad range of disciplines to pursue programs that are not feasible or are very sharply limited in scope with the first generation of HST instruments. These investigations will be aimed at a deeper physical understanding of the origin and evolution of galaxies and their nuclear activity, of the nature and energy balance of the material comprising the interstellar and intergalactic medium, of the dynamical processes of star formation and subsequent variable stellar phenomena, and of the origin and evolution of the solar system.

These topics may be illustrated by examples of observations which could answer questions of great current interest, while recognizing that the HST first generation cameras and spectrographs will have discovered entirely new phenomena and questions for us to investigate.

Absorption lines in the spectra of quasars enable us to study the evolution of the intergalactic medium, intervening clusters of galaxies, and haloes and disks of galaxies as a function of cosmic time. The strong absorption lines fall in the UV at the absorbing object, and can be studied from the ground only at high redshift, where very few of the objects can be detected in emission. An evolutionary comparison with more nearby objects can only be made by systematic studies in the UV. By combining simultaneous broad spectral coverage and fairly high spectral resolving power (20000), STIS will allow sensitive searches for absorption features over a wide redshift range. For example, Lyman alpha systems may be searched for over the redshift range $0 < z < 1.5$ in two exposures. A resolving power above 10000 is required to find low density clouds in Lyman alpha, to find weak lines of low abundance species in the denser clouds, to separate weak components such as deuterium from hydrogen, and to separate velocity components. For particularly narrow features in galactic disks a resolving power of 140000 may be used. A current unresolved question is whether the low density clouds observed only in hydrogen are of primordial composition, or are enriched with heavy elements like the higher density absorbing systems. This can only be answered by a substantial survey with high sensitivity to faint lines, with high resolving power. Another question is the evolution with redshift in density and ionization of the intergalactic medium, as shown by the Lyman alpha clouds, which again requires a survey sensitive to weak lines over a large redshift range. The general topic of quasar absorption line spectroscopy was considered so important that it was

selected as a key project, worthy of hundreds of hours of
observing time with the first generation instruments.
Even if this is carried out, it will barely scratch the
surface of the topic, which STIS will be able to pursue
10-30 times faster.
Another topic selected as a key project, better suited to
STIS, is the investigation of the mass distribution at
the centers of active galaxies. For this, we need to know
the velocity of the stellar distribution very close to
the nuclei. STIS will perform long slit spectroscopy with
the full angular resolution of HST in the visible, with a
spectral resolving power of 20000, providing simultaneous
observations across the nuclei, with more adequate
angular and spectral resolution than previously
available. This will allow the determination of the mass
distribution, including measuring the mass and
determining the existence of black holes at the nuclei.
Similar observations in the UV will measure the gas
distribution and the ionization in the narrow-line region
for nearby AGNs.

   With its visible-blind UV area-array detectors, unique
within HST, STIS will also be able to study the angular
and spectral distribution of the diffuse UV background.
Estimating from the Shuttle payload results by the Johns
Hopkins and Berkeley groups, we see that by using a broad
slit and moderate resolution grating, STIS could measure
the UV continuum and spectral lines of the background.
Also, in UV camera mode, the small-scale arc sec
structure could be measured in individual long exposures,
and the large scale structure determined by comparing
observations as a function of galactic latitude. Clusters
of galaxies and quasar environs could be surveyed. Much
of this could be done as parallel observing, avoiding the
use of prime observing time.

Table   **Instrument Performance Summary**

Angular Resolution ~ 0.05 arcsec (FWHM) - HST Limit (2 pixel sampling)
Maximum Field-of-View = 50 arcsec
Approximate Spectral Resolutions are 10^5, 2×10^4, 1200, 100

Mode	Wavelength Range Å / Detector	1050-1700 MAMA/CsI	1700-3000 MAMA/CsTe	3000-6000 CCD	5500-11000 CCD	
High Resolution	Resolving Power*($\lambda/\Delta\lambda$)	1.4×10^5	1.0×10^5	---	---	
	Effective Area	370	560			cm^2
	Angular Range	0.1	0.1			arcsec
Medium Resolution	Resolving Power	2.1×10^4	2.0×10^4	1.9×10^4	1.5×10^4	
	Effective Area	370	560	3800	3500	cm^2
(a) Wide Spectral Coverage	Angular Range	1.4	1.0	0.6	0.6	arcsec
(b) Wide Angular Coverage	Resolving Power	1.4×10^4	1.6×10^4	1.5×10^4	1.2×10^4	
	Effective Area	860	1060	6400	6300	cm^2
	Instantaneous Spectral Coverage of Band	1/8	1/11	1/13	1/13	
	Angular Range	50	50	50	50	arcsec
Low Resolution	Resolving Power	1300-2200	1000-1800	800-1600	800-1600	
	Effective Area	1800	1900	8900	7800	cm^2
	Angular Range	50	50	50	50	arcsec
Very Low Resolution	Resolving Power	160-300	120-250	80-200	80-200	
	Effective Area	2200	2800	12600	11600	cm^2
-includes 50x50 arc s Objective Prism Mode	Angular Range	50	50	50	50	arcssec
Camera (Including Acquisition)	Angular Range Total Image	50x50	50x50	50x50	50x50	arcsec
	Effective Area	3100	2800	12600	11600	cm^2

*At least 2 pixel sampling for all resolution.

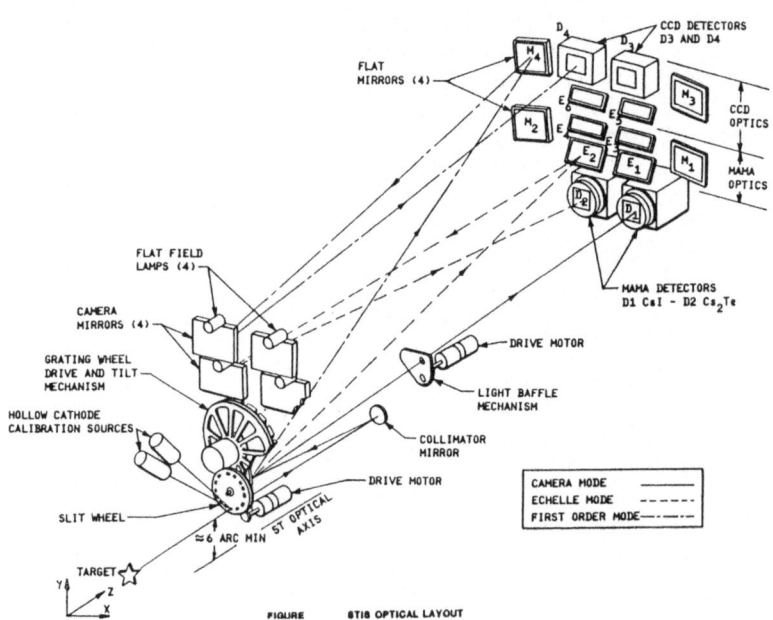

FIGURE   STIS OPTICAL LAYOUT

THE NEXT GENERATION: AN 8-16 M SPACE TELESCOPE

Garth D. Illingworth

Lick Observatory/Board of
Studies in Astronomy and Astrophysics

Introduction

The Hubble Space Telescope is to be a long-lived observatory with wide-ranging spectroscopic and imaging capability in the UV and the visible, and in the near-IR as the second-generation instruments are implemented. HST will have a dramatic impact on our view of the universe. However, it is by no means premature to address the issue of its successor, even though HST has not yet been launched. We must look ahead with the realistic view that HST will degrade and will need to be replaced. The question that we must then address is:

What is the UV-Visible-IR Observatory that will follow HST?
I will make the case for this being an 8-16 m class telescope.

In developing the case for such a successor to HST, one that we might think of as being the Son of HST (or in this day, the Daughter of HST), we need to first focus on certain broad questions. We need to be clear in our own minds as to what the role of such a successor will be, and to elucidate clearly its scientific potential so that we can make the strongest possible case to:

- our colleagues and to the science community at large;
- to NASA, and to ESA and other space agencies, since I think that it is desirable that such a facility be international;
- to those who must budget and appropriate the funds in Congress, in the Administration, and in other governments.

The questions that we have to address are fivefold and are:

- Why do we need a successor to HST?
- What are the outstanding scientific objectives?
- What are the capabilities that we need?
- What is the successor to HST?
- Why do we need to start now?

While the answers to these questions may appear obvious to us, they are not to all our colleagues, nor to scientists in other fields, and certainly not to those being asked to fund such a major enterprise.

The Case for a Successor

Why do we need a successor to HST? To my mind there are two major reasons:

a) Continuity. HST is an observatory of wide-ranging capability. The scientific case for such capability will remain as strong in the future as it was in the seventies, and is now.

b) Discoveries. New technology will allow us to make a major step in the power of the next generation telescope, allowing us to open up new "discovery space" by a significant amount.

The argument (a) for continuing capability to carry out major programs and to support other missions is a strong one. It is my view that the development of scientific understanding is a mixture of both *discovery* and *goal-oriented programs*, and that these two aspects of scientific research play comparable and complementary roles. There is no doubt that we can make a strong case for continuity of access to the wide-ranging observational capabilities exemplified

449

D. McNally (ed.), Highlights of Astronomy, Vol. 8, 449–453.
© 1989 by the IAU.

by HST. I am sure that any astronomer, particularly the observers, can imagine the impact on their research programs if access to their primary facility was cut off for 10-15 years. It would be devastating, yet this is what has happened in X-ray astronomy. The Einstein satellite had a dramatic impact in many areas. It offered us a tantalizing glimpse of data on many important problems – and then the door was shut, in most cases until AXAF (the Advanced X-ray Astrophysics Facility) flies in the mid-1990's. Clearly this is not the way to proceed. We need to work towards missions that, while not necessarily overlapping or contiguous, at least have hiatuses that are of the order of 5 years or less.

It is greatly to NASA's credit that they have recognized this. The development of the Great Observatories program is an excellent response to this clear need for continuity. Continuity is, however, not the only gain from such a program. There is a synergistic aspect here as well that has also been recognized. Real scientific gains come from having concurrent operation of observatories that operate over a wide range of wavelengths with broad spectroscopic and imaging capability.

I think that NASA's OSSA scientists should be commended for their development and support of the Great Observatories concept, as should those in ESA responsible for the development and support of the Horizon 2000 program. Both these programs reflect the more global thinking that is now seen as an integral part of the planning for Space Science missions. The Great Observatories program is in itself also a driver for consideration of a successor to HST. AXAF will be launched well into the life of HST – and SIRTF (the Space Infra-Red Telescope Facility) even further along. To continue the overlap as HST degrades in 10–15+ years, we need to plan now for its successor.

Beyond continuity we also have (b) the enhanced power and capabilities that the Son of HST will bring. We can, and must, make substantial gains in resolution, sensitivity and throughput. Scientifically, the case for doing so is extremely strong, as is the political necessity. We must open up the "discovery space". In my view, the way to accomplish these joint goals is with an 8-16 m passively-cooled telescope.

Most of us have thought about or can appreciate the gains to be made with 8-16 m class telescopes on the ground, and have recognized the power of HST and other space telescopes. To have the flux collecting area of an 8-16 m telescope with the gains in resolution, wavelength coverage, and low background that come from having a cool diffraction-limited telescope in space makes for a remarkable and astonishingly powerful observatory. Let me highlight the major gains:

- $\approx$ 0.01 arcsecond (FWHM) images in the visible (3-6× smaller than HST);
- a collecting area that is 10-40× larger;
- an IR background that is < 1/1000 that from the ground.

Scientific Potential

It is not difficult to think of programs on which such a telescope would have a dramatic effect. Let me highlight a few:

A. One of the most exciting and highest priority scientific objectives for such a telescope must be the detection and measurement of structure in distant galaxies, particularly those at redshifts $z > 1$. There are two characteristics of this telescope that would ensure that we would make dramatic advances in this area:

- Resolving power. A resolution of 10 milliarcseconds (mas) corresponds to $\approx$ 100 pc resolution in *galaxies at any redshift*, given the currently accepted cosmological parameters. This is remarkable. This is the resolution with which we see Virgo galaxies from the ground. Such resolution is of particular importance for understanding the evolution of galaxies, since much of the structure in galaxies relevant to this question occurs on length scales of 100 pc to 1 kpc. Of course, measurement will be difficult at high z because the galaxies will be faint and often of low surface brightness. But there is no doubt that we would make remarkable gains in our knowledge of galaxy structure and evolution at high redshift.

- Much of the light from galaxies at high z comes out in the near-IR, where the sky is darkest from space. This is in the 2-4 μm band between zodiacal scattering and emission.

The background in the near-IR with a passively-cooled space telescope will be less than 1/1000 that from the ground.

B. The same high resolution will be of great value for studying galactic nuclei and other compact structures (e.g., VLBI jets and superluminal clouds). By imaging in the near-UV (0.3-0.4 μm), an 8-16 m should be able to resolve $\approx$ 0.5 pc at Virgo. This resolution will be particularly important for studying nuclei with complex structure. Such structure is difficult to elucidate cleanly with interferometers. In fact, I think that this observatory will complement and enhance the power of the coming interferometric experiments.

C. The resolution will also prove decisive for astrometry, particularly of faint objects. With current detectors we can centroid routinely to 10^{-2} FWHM, and even to 10^{-3} FWHM (as has been demonstrated on the ground), given adequate S/N, good sampling and a well-characterized detector. With such capability we could measure the distance to stars directly throughout the galaxy to 10 kpc. Also, by measuring the proper motions and radial velocities of objects in circular orbits, we should be able to measure the distances to nearby galaxies *directly* over a 5-10 yr baseline.

D. Finally, but by no means last – since the list is long! – is the detection of earth-like planets around nearby stars. This would probably be the one of the most important discoveries ever made. But it does offer a huge technical challenge. Angel, Cheng and Woolf (1986) have discussed using an apodized 16 m telescope to find earth-like planets around the nearest stars. Such planets should be found $\approx$ 0.25'' from a star at a distance of a few parsecs. It seems just possible to detect such an object at $\approx$ 10 μm with a cooled, apodized 16 m where the first dark diffraction ring corresponds to the planet's orbit. Then the signature of ozone O_3 at 9.5 μm can be searched for spectroscopically. This is a tough, tough problem, and whether it is really feasible deserves more study, but it is a tremendous idea and a great challenge.

I have only touched the surface of what is a vast number of outstanding scientific problems that could be tackled with such an observatory.

Performance Goals

With these scientific goals in mind, let me summarise the desirable characteristics of an 8-16 m space telescope.

- Resolution: < 10 mas at 0.3-0.4 μm, which is > 50× better than that usually available from the ground, and > 6× that of HST.
- Collecting area: 10-40× HST, making it a particularly powerful spectroscopic system.
- Wavelength coverage: 0.1 μm to beyond 10 μm.
- High sensitivity: the "sky" background would be $\approx$ 33 mag *per resolution element*, a remarkable 10 mag or more fainter than that for ground-based telescopes, and some 4 mag less than that for HST.
- Low IR background: < 10^{-3} the background from the ground. Passive cooling to $\approx$ 100°K makes the telescope particularly sensitive in the 2-4 μm dark window.
- Multi-purpose Instrumentation: spectroscopic and imaging systems for the UV-Visible-IR.
- Detector mosaics: large, several-arcminute field with diffraction-limited imaging.
- Stable pointing and tracking to $\ll$ 10 mas.
- High observational efficiency.

The optical performance of this telescope is crucial. Fortunately, new optical fabrication and polishing techniques (below) promise to greatly improve the quality of optical surfaces. The development and demonstration of such techniques must be given particular attention during the early phases of this program.

We want to take advantage of the high optical performance of this telescope by implementing large area detectors in all wavelength regions. While the availability of large-format, solid-state detectors (CCDs) has been a significant concern during the last decade, there should be significant gains in this area in the coming decade.

The last two points bring us to the issue of the location, i.e., the choice of orbit, for the new generation of space observatories. The use of low-earth orbit has clearly added to the complexity of the HST system, and to the overall mission planning and operation. And complexity can be equated with cost. It is my view that we should take advantage of emerging technologies and capabilities, and plan to put the next generation of observatories into geosynchronous orbit. Let me just summarize the advantages of geosynchronous orbits for a mission such as this:

- The elimination of short period thermal cycling, thereby lessening the difficulty of achieving the pointing and tracking requirements.
- The attitude control requirements are minimized, since there is low to zero gravity gradient and aerotorque.
- Easier baffling and more effective passive cooling with the small subtended earth-angle.
- Direct data link, allowing high-speed data transfer and simplified control.
- Greatly simplified program optimization and scheduling.
- High observing efficiency.
- Long integrations.

The technologies that will allow us to consider boosting large payloads (8+ m?) to near-earth orbit, allowing some limited on-orbit assembly, if necessary, and transfer to geosynchronous orbit are the likely development of:

1. HLV - Heavy Lift Vehicle ("Space Truck");
2. Space Station;
3. OTV - the Orbital Transfer Vehicle.

The development of these capabilities will influence the type of observatory we launch, and so we should not let the current limited launch capability condition our thinking regarding orbits – or the size of the payloads.

The Telescope

Given these capabilities, and the performace goals for this HST successor, what is a reasonable system for conceptual development and technology evaluation? I think that we should consider an observatory based on:

- 16 m primary (four 8 m segments?);
- Wavelength coverage of 0.1 μm to > 10 μm;
- Diffraction-limited ($\approx$ 10 mas in the visible);
- Passively-cooled ($\approx$ 100°K);
- Wide-field (several arcminutes);
- Geosynchronous orbit.

Scientifically, the case is very strong for a 16 m telescope. A plausible approach might be to assemble $\approx$ 8 m-segments on the ground, launch them, and then bring them together to form a 16 m filled-aperture telescope out of 4 such segments. Such an approach is not beyond likely technology or launch capability developments over the next decade. Segmented mirror technology is being applied to major optical facilities, e.g., the 10 m Keck telescope (Nelson et al 1984) and the 4 m LAMP mirror (Aviation Week and Space Technology, Nov 23, 1987). For space we require higher precision in the segment location, but support is easier since there is no gravity and no wind!

Improved optical performance is likely within the next decade. An important development is the dynamic lap approach for polishing optical surfaces. Examples are the membrane approach being developed at Zeiss, and the stressed lap technology being developed by Roger Angel (Angel 1984) and collaborators at Arizona. The techniques should overcome many of the problems associated with CCP (Computer Controlled Polishing) where the small tools leave too much surface structure on scales critical for UV-Visible optics. The new approaches have the potential to give us diffraction-limited performance into the UV.

Passive cooling also seems plausible. The IRAS mirror has settled to $\approx$ 100°K, and calculations indicate that a large telescope in geosynchronous orbit could be configured to have its mirrors operate at $\approx$ 120°K (Bely et al 1987). The methods used to polish and test such mirrors do, however, need to be the focus of a significant study and development effort.

It is clear that a project of this magnitude is a costly enterprise. A critical, and integral, part of the technology development will be to evaluate new approaches and technologies whose goal will be to minimize cost. This is not an unrealistic goal. Just as the Keck 10 m and other large ground-based telescopes had to utilize new technology and novel approaches to break the cost scaling laws before they could be considered feasible projects, so we expect that similar advances will lead to a system cost much less than would be suggested by simple scaling from the cost of HST.

Moving Ahead

I think that a brief review of the history of the HST project will help put the last question "why start now?" into perspective. A workshop was held in 1962 under the auspices of the National Academy of Science. The workshop was titled "A Review of Space Science". It included a recommendation to build a large diffraction-limited space telescope. Yet it was nearly 10 years before the concept was taken up and detailed studies begun on LST, as it was then known. Another five years passed before funding was appropriated and construction started. With launch in 1990 that means that we have seen 25+ years from a major recommendation to 18 years from detailed design to 12 years for construction and testing.

Now it is realistic to expect that we could move a little faster in the future – but not a whole lot faster. The hurdles are large for any major project. The timescales have been very long, for example, for AXAF. If we are looking to the launch of an HST successor in 15-20 years, now is clearly the time to move.

Where are we today for an 8-16 m observatory? We are a long, long way from our goal. We are at the 1962 stage. Just recently a study was published that was performed under the auspices of the National Academies of Science and of Engineering, and of the Institute of Medicine. The study group was tasked with developing a program for Space Science in the 21st century. The reports of the study groups and an overview volume have recently been published. It is entitled:

Space Science in the Twenty-First Century:
Imperatives for the Decades 1995 to 2015.

The astronomy section contains a recommendation for an 8-16 m passively-cooled UV-Visible-IR telescope. The first step has been taken, and we need to move on. We have a project that, while ambitious, offers major scientific rewards, is a striking technical challenge, but is not out of the question. Let me quote the report's concluding remark (from the section discussing an 8-16 m telescope in space):

"........ Given a well-directed technology development program, the task group anticipates that an 8- to 16 m telescope will prove to be within closer reach than a simple extrapolation from HST would suggest."

It is clear that we must move now into a serious evaluation of the scientific, technical, and cost questions and issues, with the goal of defining an 8-16 m successor to HST within the near future.

Acknowledgements

I have enjoyed and learnt from stimulating conversations with many people about how to proceed with the next generation of space telescopes. I am particularly grateful to R. Angel, P. Bely, R. Giacconi, and P. Stockman.

References

Angel, J.R.P. 1984, in *Very Large Telescopes, their Instrumentation and Programs*, eds. M-H. Ulrich and K. Kjar, ESO, Garching, p.11.
Angel, J.R.P., Cheng, A.Y.S., and Woolf, N.J. 1986, *Nature*, **322**, 341.
Bely, P.Y., Bolton, J.F., Neeck, S., and Tulkoff, P. 1987, *Proc. SPIE*, **751**, 29.
Nelson, J.E., Budiansky, M.P., Gabor, G., and Mast, T.S. 1983, *Proc. SPIE*, **444**, 274.

SPACE ASTRONOMY - THE NEXT THIRTY YEARS

M.S. Longair
Royal Observatory,
Blackford Hill,
Edinburgh EH9 3HJ

1. Introduction

I have a strong feeling that I am not the right person to give this lecture. Many of you will have noticed that recently the National Academy of Sciences has published a report entitled "Space Science in the 21st Century – Imperatives for the Decades 1995 – 2015". This assessment was requested by NASA and was headed by the Space Science Board of the National Academy of Sciences. This is a very large and ambitious report and is essential reading for those who wish to have a broad vision of what might become possible in the coming decades. As a result, this brief review will make no attempt to be complete but will simply raise some points which may contribute to the discussion of future directions of space astronomy. I would emphasise that these are personal views.

In looking at the next thirty years of space astronomy, it is salutary to look at the state of astrophysics thirty years ago. In Table 1 I show a selected list of discoveries and achievements which occurred over that period along with a selection of space missions which have had a major impact upon astrophysics. It is salutary to remember that the space age itself is only just thirty years old. There is no question but that the last thirty years have been the golden age of astrophysics – there has never been a period when so many new disciplines were opened up or so many key discoveries made about the ways in which matter behaves in the extreme physical conditions found in the Universe. Space science has contributed in a central way to many of these achievements.

Looking to the future, it is interesting to look at these successful space astronomy missions and likely future space missions in terms of their size and cost. I display this information in Table 2 in which the missions are arbitrarily divided into small, medium and large programmes and presented in terms of the coverage of the electromagnetic spectrum. Missions already in orbit are shown in bold, roman letters, approved future programmes in underlined, bold letters and proposed but unapproved future missions in italics.

There are a several well known features of Table 2. The first is the fact that the exploratory phases of most of the wavebands which can only be observed from space have now been carried out by small to medium sized missions. There are still some regions of the spectrum which have yet to be explored. I would include among these the extreme ultraviolet waveband, the difficult γ–ray waveband about 1 MeV and higher energies and possibly the sub–millimetre wavebands where there may well still be surprises waiting for us. You would have to ask Martin Harwit

455

D. McNally (ed.), Highlights of Astronomy, Vol. 8, 455–460.
© *1989 by the IAU.*

TABLE 1 – Selected Astronomical Discoveries and Achievements of the Last Thirty Years and Selected Space Missions

	Radio Galaxies	Sputnik
1960	------------------------------------	----------
	Quasars	
	Microwave Background radiation Pulsars	SAS–2
1970	------------------------------------	----------
	γ–rays from the Galaxy Binary X–ray sources X–ray emission from Clusters Infrared "protostars"	UHURU IUE COS–B
	Binary pulsar	
	Superluminal radio sources	Einstein
1980	------------------------------------	----------
	Gravitational lenses	
	Galaxies with redshifts > 1	IRAS EXOSAT
1988	------------------------------------	----------

TABLE 2 – Selected Past, present and future space missions

Waveband Size of Mission	Radio	Infrared	Optical	UV	X-ray	γ-ray
Large		LDR *SIRTF* *FIRST*	HST		AXAF XMM	
Medium	*Quasat* *Radioastron*	ISO IRAS	Hipparcos IUE	Lyman EUVE	Einstein EXOSAT *XTE*	*Advanced* *GRO* GRO
Small		COBE			UHURU	SAS–2 COS–B

about the number of qualitatively new phenomena we might uncover in these wavebands but I would be surprised if there were none.

The next natural step in all the wavebands is the construction of larger scale missions aimed at developing real astrophysical understanding. It is true of all great discoveries in astronomy that, whilst the discovery phase has a unique excitement and intellectual stimulation for astrophysicists, it is the systematic follow-up phase by dedicated observatories which sets the new science in its real physical and astrophysical context. There is therefore a trend towards astrophysical observatories with a wide range of scientific capabilities. It is gratifying to see that what are referred to as the "great observatories" are planned for all the astronomical wavebands only accessible from space but it can also be seen that the tendency is for the missions to migrate towards the upper part of the Table, i.e. towards large programmes, to order of magnitude on the scale of the Hubble Space Telescope project. Indeed, there are already some very ambitious programmes in this Table, the Large Deployable Reflector (LDR) probably being the most ambitious of all these missions and I would not be surprised if it slipped right off the top of my diagram.

2. The New Projects

It has been remarked many times that, for the large space astronomy missions, there is a very long time between the initial concept and the successful deployment of the spacecraft in orbit. The Hubble Space Telescope is an example of a mission which needed about ten years to reach the final approval phase and then more than ten years to construct the telescope and place it in orbit. The logic of this experience is that we already know most of the possibilities for the first fifteen years of our 30-year forward look period. If a large project has not reached a certain degree of maturity by now, we know that it will be beyond 2000 before it will fly. This makes half of my task easy – we already know the likely programme of large missions for the next 15 years. Indeed, one might take the position that, if the programme laid out in Table 2 were achieved by 2000, that would indeed be a considerable achievement.

The remarks in the last paragraph refer to the large Observatory-class missions. The small and medium-class missions have, however, a key role in the future development of the space astronomy programme. There is a continuing need to undertake these classes of mission in order to carry out qualitatively new types of astronomy, where the exploratory phases have yet to be undertaken, and also to exploit new techniques and instruments for astronomical observations. There are many such possibilities. Some examples which spring to mind in the first category might include:

• **Studies of the Microwave Background Radiation** There will be a continuing need to define its intensity and fluctuation spectrum with greater and greater precision. As shown by the RELIKT experiments, a great deal can be achieved with small experiments.

• **"Simple" optical and infrared interferometry from space** Here I am not proposing an enormous experiment but one carried out with small mirrors but taking full advantage of the lack of phase fluctuations in the wavefronts of the signals as observed from space.

• **"Solar probe"** type experiments in which a heavily protected space vehicle passes close to the surface of the Sun.

• The **"Interstellar probe"** in which a small space vehicle is sent beyond the Solar System to sample the nearby interstellar medium.

This list is simply illustrative, designed to make the point that important new classes of experiments should be undertaken which need not fall in the Observatory–class category and which are likely to open up new ways of doing astronomy.

As an illustration of the second point, there is an excellent case to be made for a small to medium sized mission in γ–ray astronomy, what I have called *Advanced GRO* in Table 2, in which new technologies are exploited to make very large increases in scientific capability possible. To a certain extent, the same argument can be applied to the types of X–ray instruments to be flown on the USSR *Specrum–X* mission. There are many other examples in other space astronomy disciplines which could be listed.

3. The Large and Very Large Projects

The big question concerns the strategy for the large and very large missions which could be achieved in the future if the resources were made available. The natural tendency is to think in terms of large and very large observatory–type missions, the goals being the need in increase **angular resolution, sensitivity, field of view** and the **precision of astronomical observations.** The scientific case for such projects has to be outstanding and the proposed missions must exceed by at least one order of magnitude, and preferably by many, what has been achieved before.

I would make two points about these missions. First of all, I believe that it is not at all difficult to make a very convincing scientific case for constructing such facilities. The enormous scientific and technical advances made over the last 20 years and the need to understand the many new facets of physics and astrophysics opened up by them makes the writing of the proposals remarkably straightforward. Even in the case of optical astronomy where it might be thought that the prime initiatives might lie with the ground–based astronomers, the case for very large facilities in space can be made very compelling. To give a simple example of this in the case of cosmology, we can now undertake real astrophysical studies of the very brightest galaxies at redshifts greater than 1. However, we all know that what we really need is to study the common stuff of the Universe as it was when the Universe was very much younger than it is now. We would aim to read off directly by observation the evolutionary history of our Universe. I have no doubt but that this could be dressed up as a very compelling case for a 16–metre optical telescope in space but the essence of the case would in the end not be very different from these few words.

The second point concerns advanced technology. Unlike a few years ago, I believe that, although there are unquestionably technological problems to be solved in producing enormous space astronomy facilities, progress in the design of telescopes and their instrumentation in all wavebands has been so rapid that, provided the resources are made available to accomplish the programmes, the big future programmes are unlikely to be impossible on technological grounds. An excellent example is the 8 or 16-metre spcae telescope described by Dr. Illingworth. There are plainly many technical problems to be solved but I would be amazed if, by the year 2015, it were not entirely feasible to put together the segments of a mirror in space to produce a huge telescope which could be maintained in essentially perfect alignment. There is a clear requirement to be able to fabricate large facilities in orbit but one would imagine that this should be exactly the motivation for the construction of space stations. The natural extension of this single mirror concept is to arrays of optical-infrard interferometers which would have been preceded by the small types of mission discussed above. Again, I would be amazed if there were any real technical show-stoppers in this area by the year 2015. Similar arguments can be made for the other astronomical wavebands.

The really big problem is that I have not attached a price tag to any of the huge observatories. This is where the crystal-ball gazing really begins. The generation of space observatories after those shown in Table 2 are likely to be orders of magnitude more expensives than the large projects. It is not a scientific question to ask whether or not it is sensible to plan for such possibilities. The astronomers have no problem in comparing the costs of such huge observatories with, say, the cost of a manned flight to Mars or selected defense programmes and then showing that they are really quite modest and should be affordable by the advanced nations. The reality is of course that we are talking about different types of money. An optimist might argue that the great powers should come together to pool their resources for space astronomy and in this way bring the huge projects within the bounds of possibility.

Projects of such enormous scale and cost require major political initiatives and we must continue to press for the next generation of space astronomy missions stressing the great value of international collaboration without which many of the most ambitious and scientifically important projects are unlikely to be affordable. It is only realistic to note that there are real limits to the what can be afforded by agencies such as NASA and ESA. Everyone is aware of the fact that the likely cost of a 16-metre optical telescope in space exceeds by a large factor the typical cost of an ESA cornerstone mission.

In this situation, it is interesting to look at some of the lessons which have come out of the major astronomy missions so far. One striking feature is that they have all been relatively simple missions. For example, IUE was essentially a single purpose UV spectroscopic mission, IRAS was a fixed scanning photometric telescope with limited capability for pointed observations, Einstein contained more complex instruments but the pointing requirements were kept simple. This contrasts with the complexities of the Hubble Space Telescope project where there are six complex scientific instruments, a pointing accuracy of 0.007 arcsec and enormous numbers of modes of operation of the telescope, let alone the facility for on-orbit servicing and replacement of scientific instruments in low Earth orbit. It is an interesting question to ask to what extent the overall costs, and even more the operating costs, have been driven by the complexity factor. On the one hand, there is a certain minimum infrastructure needed to maintain a space vehicle in orbit and the question is really what the incremental cost is for each added layer of complexity in addition to the basic instrument. I believe the moral which will come out of such a study will be **keep it simple!**

A second concern is the ability of space astronomy to do a complete job in any particular area of astrophysical endeavour. It is interesting to note the time needed to complete any major programme of astronomical observations on ground-based telescopes with a generous allocation of observing time. My own experience is that the major programmes require a great deal of time to reach a successful result as opposed to one-off observations which often produce interesting results but which are difficult to set in their astrophysical context. I consider that it was an enlightened decision of the Space Telescope Project to accept the concept of **Key Projects**, the aim being to ensure that these programmes have adequate observing time to enable good results in major programmes to be obtained. It is important to note how few of these large projects can be accomodated even in a long-lived space observatory such as the Hubble Space Telescope. There is no question about the wealth of new science which will come out of project like the Hubble Space Telescope but it is important to recognise that even with a 15-year lifetime, only a limited number of major programmes can be accomplished.

These concerns lead me to wonder if we might not adopt a somewhat different approach to the large and very large missions which would be rather more astrophysically driven than facility driven. I have invented a game for astronomers in which the participants have to decide what would be the single most important space mission for the solution of a specific astrophysical problem. No more than 2 scientific instruments are permitted aboard the satellite observatory to solve the problem. The participants may throw up their hands and say "But I need information from all wavebands to solve my problem!" That answer is disallowed. Just to get the game going, I would suggest that the following astrophysical problems as candidates for these missions : the formation of planetary systems, the origin of stars, the origin and evolution of galaxies, the physics of black holes, the physics of relativistic and non-relativistic cosmic plasmas, the physics of dark matter and so on. The winner is the person who comes up with the cheapest, realistic space project which is likely to achieve its scientific goals as agreed by the other players. Tactical voting is not allowed.

I am not pretending that this is a solution to the problem of deciding the directions of future very large missions. Astronomers are always wary about putting all their eggs in one basket. However, in many areas, the exploratory phases are over and one needs dedicated facilities to make substantial progress. The only virtue of the above game is to suggest that, with the increasing maturity of the astrophysical sciences which can be carried out in space, we might with advantage plan missions with much more specific astrophysical goals and design the satellites and their scientific payloads specifically to address major physical and astrophysical problems. Of course, given such a facility, an enormous range of other science could be undertaken but I would allocate at least half the time to the solution of a small number of really major problems. The rest of the time would be available for all other types of science. It may well be that after this exercise is carried out, we would find ourselves back where we started with "Great Observatory" class missions but I believe the case for them would then be that much stronger.

JOINT COMMISSION MEETINGS

1. FOR MILLIARCSECOND OR BETTER ACCURACY

Chairman and Editor: P.K. Seidelmann

Supporting Commissions:

4, 7, 8, 19, 24, 31, 40

"FOR MILLIARCSECOND OR BETTER ACCURACY"
P. K. SEIDELMANN
U S NAVAL OBSERVATORY, WASHINGTON, D C 20392

INTRODUCTION

The accuracies being achieved in astrometry, celestial mechanics, Earth Orientation, ephemerides and time have been improving significantly in recent years.

The introduction of the improved astronomical constants, ephemerides, time scales and nutation as adopted from 1976 to 1984 has had the desired effect of permitting the investigation of systematic effects at precisions of an order of magnitude better than previously possible.

Therefore, there have been many developments in observational data, in theories, and in astronomical computations that have promised, or claimed, to deliver accuracies of a milliarcsecond or better.

Working Groups had been established with interrelationships in their scopes of activities. It did not appear that any of the working groups were prepared to present final recommendations that would be generally accepted.

As a result a number of commissions requested that there be a Joint Commission Meeting, or a Joint Discussion, on the general topic of milliarcsecond accuracy. The IAU Executive Committee asked me to organize a Joint Commission Meeting involving commissions 4, 7, 8, 19, 24, 31 and 40. That meeting, and these resulting proceedings, are designed by means of a series of invited papers to give the background and status of observational, theoretical, and computational efforts necessary to achieve milliarcsecond or better accuracy. It was hoped that these presentations would provide an overall background of the considerations and current status to be considered by the working groups on the use of the millisecond pulsars, nutation, astronomical constants and reference systems.

Achieving milliarcsecond accuracy, requires significant improvements in many different areas. The breadth of the disciplines involved in these accuracy improvements is indicated by the number of commissions sponsoring Joint Commission Meeting 1. Considerations that in the past were negligible now have to be included. This means that people working in one discipline are required to include the latest knowledge, techniques and constants from another discipline.

These are a number of issues which need to be considered as accuracies of a milliarcsecond or better are being sought by different techniques. The following issues were drafted as a basis for the papers.

1). Are we achieving accuracy or precision?

D. McNally (ed.), Highlights of Astronomy, Vol. 8, 465–499.
© 1989 by the IAU.

2). Are the theories and computational procedures consistent with observational accuracies?

3). Do the anticipated observational accuracies require new levels of accuracy in theories?

4). Do the anticipated accuracies of the future require fundamentally new methods of proceeding in astronomy?

5). How do we ensure that there are not systematic errors in a type of observation or computation?

6). Is our knowledge, understanding and application of relativistic theory sufficient for current observational and theoretical activities?

7). Are the old divisions between physics, astronomy, geodesy and mathematics causing barriers to achieving accuracy improvements?

8). Can we have a single definition of a reference system and practical realizations thereof for terrestrial and celestial coordinates that satisfy all requirements?

9). Is the accuracy of the reference system satisfactory for the accuracy of observations and theories?

10). Are radio and optical based reference systems consistent, compatible and equivalent?

11). Should there be IAU adopted constants, reference frames, and theories?

12). Should there be IAU adopted values of constants in some cases and best estimates in other cases?

13). Are the constants sufficiently accurate for the theories and observations?

14). Are inaccuracies of constants, theories, or observations in one area limiting the accuracies which can be achieved in another area?

15). Are we making approximations based on past accuracy requirements that are no longer satisfactory for the present?

16). How can we ensure documentation and consistency in constants, reference systems and computational procedures for different theories and observations?

The presentations were to provide an interdisciplinary overview of the considerations and current status on important observational, theoretical, and computational subjects involved. These presentations then provided an introduction to the Working Group reports which followed. The following list of presentations were given by the indicated speakers.

Observational accuracies

Radio Interferometry*   K. J. Johnston

Current and Potential Accuracies
of Optical Interferometry   M. Shao

Lunar Laser Ranging   P. Bender

Millisecond Pulsars   J. H. Taylor

Theoretical Developments

Relativistic Framework for
Precision Astrometry*   I. I. Shapiro

New Nutation Theory   H. Kinoshita

Earth Models   J. Wahr

Non-rotating Origin   N. Capitaine

Procedures for Accurate Origin   S. Aoki

Computational Considerations

The Mean Motion in Modern
Planetary Ephemerides   E. M. Standish

Determination of Earth Orientation   M. Feisel

Apparent Place Computations   B. Yallop

Galactic Coordinates   C. A. Murray

Review of Current/Future
Catalog Accuracies   H. Schwan

Working Group Reports

The Use of Millisecond Pulsars*   D. Allan

Nutation   R. L. Duncombe

Astronomical Constants   B. Morando

Reference Systems   J. A. Hughes

The same speakers provided the following written summaries of
their presentations and working Group Reports. The asterisks
indicate the cases where summaries have not been received.

Reports of the working groups reflect the results of the
discussion that took place during the IAU and to that extent,
are improved versions of the reports presented at the General
Assembly.

The recommendation of the Working Group on Nutation was judged
not acceptable for high precision requirements, unnecessary for
lower precision needs, and not consistent theoretically.
Therefore, it was not adopted by the Commissions.

During the discussion of the Working Group reports it became
evident that the working groups' reports were not ready for
adoption at this General Assembly. Rather, more effort and
discussion were required. The Scientific Director of the U S
Naval Observatory extended an invitation for a meeting to be
held in 1990 at the U S Naval Observatory on this general
subject.

The Resolution C2 as follows was drafted and adopted by the IAU
General Assembly:

Commissions 4, 7, 8, 19, 20, 24, 31, 33 and 40

noting   the proliferation of Working and Study
                        Groups which deal with various matters of
                        concern to these Commissions;

recognizing   the necessity of considering such matters
                        carefully along with the inevitability of
                        scientific interrelationships among them;

thanks   the Chairperson and Members of the Working
                        Groups on Nutation and Astronomical
                        Constants for their efforts; and

recommends   that the Working Group on Reference
                        Systems (WGRS) be continued as an
                        intercommission project and that it
                        concern itself with Nutation, Astronomical
                        Constants, Origins, Reference Frames and
                        time;

                        that appropriate Study Groups be formed as
                        required and that the current chairman
                        continue in office, and that Commissions
                        4, 7, 8, 19, 20, 24, 31, 33 and 40 and the
                        IAG be invited to appoint members;

                        that the International Astronomical Union
                        support the efforts of the Intercommission
                        Project by providing funds for travel of
                        members to attend the Working Group
                        meetings;

                        that the WGRS produce a draft report with
                        specific recommendations at least six

months before the General Assembly;

that close ties be maintained between the International Astronomical Union, as represented by the WGRS and the Geodetic Community, as represented by the IAG/IUGG;

that a close liaison with the IERS be continued.

Thereby, the Working Group on Reference Systems was continued and its purview broadened to specifically include the questions of nutation, astronomical constants, time and the origins.

Current Status of Optical Interferometric Astrometry M. Shao

Traditionally built for astrophysical research, such as stellar diameter measurements, the success of radio interferometry in making extremely accurate astrometric measurements has motivated the recent work in optical interferometric astrometry. Interferometric astrometry at optical wavelengths can be divided into 3 categories, very narrow field astrometry (double stars), narrow field astrometry (a few degrees), and wide angle astrometry (1 radian).

The technique most widely used in milliarcsecond (mas) double star astrometry is speckle interferometry where separations of stars can be measured with accuracies slightly better than 1 mas. The major source of error is in the calibration of the effective focal length of the telescope. Long baseline interferometers can also be used to measure the separation of double stars with the potential of much higher accuracy because of the higher resolution of the longer baseline and the ability to accurately measure the baseline vector. The Mark III Interferometer has measured the diameter of large stars (>0.010 arcsec) with accuracies of 0.1 mas for one night. Interferometric techniques work best for stars separated by less than 5-10 arcsec.

In narrow field astrometry, the Mark III is currently the only interferometer making such measurements. With the addition of several subsystems such as a laser metrology system to measure siderostat bearing errors, an internal white light metrology to measure the thermal drift of the delay offset, and the use of two color astrometry, a precision of 3-4 mas rms have been demonstrated for six stars in an 8 degree field. These preliminary results must be verified with a much larger data set. However, the precision is almost competitive with ground based photoelectric long focus astrometry with much smaller fields of view.

The Mark III was built as a prototype wide angle astrometric instrument. Instrumental systematic errors are most severe in wide angle measurements. Our measurements in 1986 showed repeatability of a night's observation to be 50 mas in declination using a 12 meter N-S baseline. In 1987, we added a second E-S baseline that had a 5 meter E-W component and the night to night repeatability of 50 mas in DEC and about 60 in RA. Averages of about a dozen nights of data would give a formal error of 20 mas for the 20-30 stars observed over a range of 45 deg in DEC and 10 hrs in RA. In 1988, we moved the E-S baseline so that it had a 10 meter E-W component. More important, we refined the software so that two color astrometry was now an operational system, using two colors for atmospheric correction and three colors for central fringe identification. In addition we installed a system of 12 laser interferometers for siderostat monitoring and the white light delay offset measurement system. Currently we have night to night repeatability of 20-25 mas in the relative position of stars that cover a DEC range of 55 deg and 12 hrs in RA. The formal error for an average of 6-10 nights of data will then be in the 10 mas range. It should be noted that the narrow and wide angle numbers represent precision, since we lack the large data sets needed to demonstrate accuracy. Optical interferometry compared to other astrometric techniques is in its infancy and we expect that progress towards higher accuracy will continue and accelerate as more people and resources migrate to the field.

Lunar Laser Ranging: P. Bender

The lunar laser ranging data through 1983 was obtained almost completely with the 2.7 m telescope at the McDonald Observatory. Some additional measurements were made with this telescope in 1984 and 1985, but most of the data from 1984 on has come from 3 other instruments. These are the 1.5 m telescope at the Claern-CERGA Observatory in France, the Multi-Lens Telescope at the Haleakala Observatory on Maui, and the 0.75 m. telescope at the McDonald Observatory. The most prolific producer of lunar range data has been the CERGA station, which also has done an excellent job of obtaining ranges to the Apollo 11, Apollo 14, and Lunakhod 2 reflector packages, as well as the Apollo 15 site.

The length of the observing period which goes into forming one range normal point varies considerably, but is roughly 20 minutes. From 1976 through 1983 both the precision and the accuracy of the normal points was typically 15 to 20 cm. The precision was improved to roughly 5 cm by 1987, with the accuracy being somewhat worse because of limitations in the calibration procedures. However, recent improvements have been dramatic. By the spring of 1988, all 3 stations had demonstrated repeatabilities for their best range normal points over periods of 4 hr or more of 2 cm. During the IAU General Assembly Dr. Christian Veillet reported that more recent analyses at CERGA and at the USNO gave better than 1 cm normal point repeatabilities. The accuracy is currently estimated to be 2 or 3 cm, but plans are to make major efforts to improve this to 1 cm in the next year or two.

Almost all of the scientific results so far from lunar ranging have been based on the earlier data with roughly 6 to 20 cm accuracy. Most of the analysis work has been carried out by the following organizations: JPL; MIT; the Center for Astrophysics; the USNO; the University of Texas; and CERGA. Major objectives of the work have been: the determination of the Earth's rotation and nutation, in conjunction with VLBI and

Satellite laser ranging measurements; the continual improvement
of the lunar ephemeris for use in planetary ephemeris
development work; and studies of lunar interior properties from
measurements of the lunar librations.

One important gravitational physics test from lunar
ranging is a negative result for what is called the Nordvedt
effect. This result shows that the gravitational self-energy
of the Earth behaves in the same way as other forms of energy
in determining the gravitational interaction of the Earth with
the Sun. However, if energy and momentum conservation are
assumed and preferred frame theories are not considered, the
lunar ranging results provide an accurate determination of the
parameter ß in the Robertson-Walker metric. The accuracy
currently is about 0.15%, which is roughly an order of
magnitude better than has been obtained from the precession of
perihelion for Mercury. Tests of the geodetic precession of
the lunar orbit predicted by De Sitter in 1916 based on general
relativity also have been reported recently by the Center for
Astrophysics and JPL groups, with 2% accuracy. And finally, a
combination of the secular acceleration of the moon from lunar
ranging with determinations of ocean tides on the Earth from
satellite laser ranging provides confirmation that the
Newtonian gravitational constant G is not changing with time.

Astrometry of Millisecond Pulsars   J. H. Taylor

Soon after the discovery of the first millisecond pulsar
it became clear that this class of objects would provide
unusual opportunities for high-precision astrometric
observations. Even for the longer-known class of "ordinary"
pulsars, timing observations had yielded celestial position
measurements with accuracies at the <0.1" level. Since the
uncertainty in pulsar time-of-arrival measurements tends to be
a fixed fraction (typically 10^{-4} to 10^{-3}) of a period,
millisecond pulsars obviously afford even better
possibilities. Consequently, it is not surprising that early
work on PSR 1937+21 yielded position measurements with
precision at the milliarcsecond level (Davis et al 1985), or
that more recent work (Rawley, Taylor, and Davis 1988, and
unpublished results) have improved these measurements by a
further factor of 8. Millisecond pulsar timing observations
are already accomplishing astrometry at the 0.1 milliarcsecond
level.

Precision is one thing, and absolute accuracy another.
The intrinsic reference frame underlying the analysis of pulsar
timing data is that of the planetary ephemerides. In current
practice, this means the reference frame of a model solar
system fitted to a large archival data base of optical and
radar observations, numerically integrated to construct the
tabular ephemerides. Celestial coordinates quoted in this
system have real meaning, and clear definition - but obviously
a different definition from, for example, FK5 positions. Some
of the ramifications and difficulties relating the reference
frames have been discussed by Becker et al. (1986).

About a year ago Rawley, Taylor & Davis (1988) were
surprised to discover that Arecibo Observatory timing data on

PSR 1937+21 revealed that this pulsar has remarkably small proper motion in the planetary reference frame. Galactic rotation should contribute a proper motion of -5mas/y in galactic longitude, but the observed rate was only -0.6 ±0.3 mas/y. If the planetary reference frame is truly inertial and thus non-rotating, as our ephemeris-oriented colleagues very reasonably insist, then the peculiar velocity of PSR 1937+21 must have just the right magnitude (about 85 km/s) and direction to cancel most of the contribution from galactic rotation. Further results on this and other millisecond pulsars are steadily accumulating, and it will be of considerable interest to see whether special values of peculiar velocity will be required to explain them. In any event, millisecond pulsar timing observations promise to provide some of the most accurate astrometric data available over the next few years.

References:

Backer, D.C., Fomalont, E.B., Goss, W.M., Taylor, J.H., & Weisberg, J.M., 1986, Astron J., 90, 228.
Davis, M.M., Taylor, J.H., Weisberg, J. M., & Backer, D.C., 1985, Nature 315, 547.
Rawley, L.A., Taylor, J.H., & Davis, M.M. 1988, Astrophys. J., 326, 947.

Preliminary Results of Reconstruction of Nutation Series of the Rigid Earth   H. Kinoshita & J. Souchay

At present for the orientation of the Earth in the space, we have an observational accuracy of a millaircsecond by VLBI. The accumulated 8 years of VLBI observations clearly show the systematic and periodic residuals, which indicate that the present IAU nutation series should be revised. Recently Kubo (1986) compared the IAU nutation series with numerical integration and found long periodic systematic deviations. The order of these systematic deviations from the theory is of a milliarcsecond. On the other hand we are going to achieve an accuracy of submilliarcsecond by VLBI and other high precision techniques in the near future. Consequently, a nutation theory with an internal precision better than 0.1 milliarcsecond is necessary to be compatible with the observational accuracy.

The present internationally adopted nutation series are based on Kinoshita's rigid theory (1977) and Wahr's non-rigid theory (1981) that uses the Earth model 1066A (Gilbert and Dziewonski, 1975). Wahr's theory gives a ratio of the nutation amplitude for the non-rigid Earth to that for a rigid Earth model. Therefore, we have to improve both a rigid theory and a non-rigid theory.

For a rigid theory, we have to: 1) Adopt more precise orbital theories of the Moon and the Sun for the computation and disturbing functions (for example ELP2000 for the Moon and VSOP82 for the Sun). The present rigid theory (Kinoshita 1977) adopts Brown's theory for the Moon and Newcomb's theory for the

Sun for the calculation of the disturbing function by these two bodies. 2) Take into account the direct torques from planets. 3) Take full account of second order effects such as the disturbing potential due to J3 and J4, interactions among nutation, coupling effects between the rotational motion of the Earth and the orbital motion of the Moon.

Among these three items the most important one is the second order coupling effect between the rotation of the Earth and the orbital motion of the Moon. Kinoshita's theory (1977) is not complete in this respect, which was pointed out by Kubo (1982). In order for a theory of nutation to be complete up to the second order, we have to treat the rotation of the Earth as a dynamical system with six degrees of freedom (3 for rotation and 3 for orbital motion) instead of a restricted problem with three degrees of freedom (the orbital motions of the Moon and the Sun are given and only the rotation of the Earth is to be solved.)

The preliminary corrections to the nutation amplitudes with argument Ω and 2Ω (Ω is the longitude of the node of the Moon) arising from the coupling effect mentioned above are:
nutation in longitude: 0.00007 $\sin\Omega$+0.00121 sin 2Ω (1)
nutation in obliquity: 0.00069 $\cos\Omega$ -0.00024 cos 2Ω.   (2)
where the unit is arcsecond. These numerical values of (1) and (2) may change after taking account of other effects mentioned above. We have already finished calculations related to items (1) and (2) above. The difference in the nutation amplitudes due to the change of planetary theories and Moon's theory is of order 0.1 milliarcsecond and the contribution from direct planetary torques is also of order 0.1 milliarcsecond.

References:

Kubo, Y. 1986 Proceedings of the Nineteenth Symposium on Celestial Mechanics, Kinoshita, H. and Nakai, H. editors, 78-81.
Kinoshita, H. 1977 Celest. Mech. 15, 277.
Wahr, J. 1981, Geophy. J. Royal Astr. Soc. 64, 705.
Gilbert, F. and Dziewonski A.M.: 1975 Phil. Trans. Soc. London A278, 187.
Kubo, Y. 1982 Celest. Mech. 26, 97-112.

The Effects of the Earth's Non-Rigidity on Nutation   J. Wahr

Recent VLBI nutation results (Herring, et al, 1986) disagree with the current IAU nutation series at the milliarcsecond level. Studies of this problem have demonstrated that there are certain properties of the Earth that geophysicists do not presently understand well enough to allow them to predict the nutation for a non-rigid Earth accurate to the sub-milliarcsecond level. This is good news for geophysicists, because it implies that the VLBI results can be used to constrain those properties. But, it is bad news for those astronomers who only want to be able to accurately remove the nutation from their data.

Some important features of the real Earth that are not

included in the current IAU model include:
(1)   The oceans. Wahr and Sasao (1981) find that the ocean's
       contributions to the nutation can be as large as 1
       milliarcsecond. These oceanic corrections require
       knowledge of the Y^1_2 spherical harmonic components of
       certain diurnal ocean tides. The Y^1_2 components from
       different ocean models agree reasonably well, suggesting
       that the estimated ocean corrections are probably pretty
       good. On the other hand, Wahr and Sasao did not consider
       the possible effects of global, tidal currents.
(2)   Mantle anelasticity. Wahr and Bergen (1986) and Dehant
       (1988) conclude that the effects of mantle anelasticity
       can be as large as .5 milliarcseconds. However, these
       corrections require knowledge of mantle Q at diurnal
       periods, and this is a poorly constrained parameter.
(3)   Non-hydrostatic core structure. The largest disagreement
       between the VLBI results and the IAU series is a 2
       milliarcsecond discrepancy for the annual nutation.
       Gwinn, et al (1986) use this discrepancy to constrain the
       Y^0_2 component of core/mantle boundary topography. Seismic
       results for this component are far from reliable enough
       to predict the annual nutation to within the accuracy of
       the VLBI results. In fact the Y^0_2 component inferred by
       Gwinn, et al, is proving to be a useful constraint for
       seismic (and other types of) core/mantle topography
       models. To complicate the issue, however, Wahr and de
       Vries (1989) find that if the core has a low density
       fluid layer just inside the core/mantle boundary, then
       the VLBI results could also be caused by Y^0_2 structure
       within that layer.

References:

Dehant, V. 1988. in Variations in Earth Rotation. AGU
Monograph Series, D. McCarthy, ed. (in press)
Gwinn, C. R., Herring, T.A., Shapiro, I. I. 1986 J. Geophys,
Res 91:4755-4766.
Herring, T.A., Gwinn, C.R., Shapiro, I.I. 1986. J. Geophys.
Res. 91:1745-4754.
Wahr, J.M. & Sasao, T., 1981. Geophys. J.R. Astr. Soc., 64,
747-766.
Wahr, J.M. and D. deVries, 1989. Geophysical Journal,
submitted.

The Non-Rotating Origin   N. Capitaine

INTRODUCTION;

       The non-rotating origin (NRO) has been proposed by Guinot
(1979) to replace the traditional equinox as a reference point
on the moving equator. Its principal purpose is to give an
exact and clear description of the Earth's rotation angle which
does not critically depend on theories, leading therefore to a
strict definition of UT1.

REASONS FOR A CHANGE IN THE TRADITIONAL REFERENCE POINT ON THE EQUATOR

Due to the expanding use of observational techniques, such as Very Long Base Interferometry (VLBI) and Satellite Laser Ranging (SLR), the definition of UT1 must be extended to Celestial Reference Systems (CRS) in which the equinox cannot be determined. On account of the improvement in observational accuracy, such a definition of UT1 must be given to a submilliarcsecond accuracy. A new reference point must therefore be chosen in order to measure strictly the specific Earth's rotation in space and take advantage of the quasi-ideal CRS (without residual rotation in space) as realized by the positions of quasars. Among the alternative choices, only the NRO has both conceptual and practical advantages with respect to the traditional equinox.

DEFINITION OF THE NRO AND CONCEPTUAL ADVANTAGES

The definition of the NRO (Guinot 1979, and 1981), is such that the instantaneous system formed by the celestial directions toward the pole of rotation and the NRO respectively, has no component of instantaneous rotation around the axis of rotation with respect to space when the pole of rotation moves in the CRS. Such a concept of "no-rotation" along the moving equator is unavoidable to describe any motion of rotation along the equator, and especially, the Earth rotation. It is the most natural concept as it only involves both the Earth rotational and orbital motions. An additional advantage is that a clear definition of the "instantaneous origin of the longitudes" can be given as being the "non-rotating origin" in the Terrestrial Reference System (TRS).

COMPARISON BETWEEN THE USES OF THE NRO AND THE EQUINOX

The hour angle of the NRO from the prime meridian, or "stellar angle", θ , (Guinot 1979, 1981) is such that its time derivative is exactly the Earth angular velocity around its axis of rotation, so that UT1 should be conceptually defined as an angle proportional to θ. The position of the NRO in the CRS can be easily obtained through a simple formula which only depends on the celestial motion of the pole. This leads to a realization of θ which is not sensitive to the representation of the pole trajectory up to 10^{-4} arcsecond (Capitaine, et al., 1986, Capitaine & Guinot, 1988). If no residual rotation is introduced, a definitive numerical relationship between θ and UT1 can therefore be given which should be considered as the primary conventional definition of UT1; secondary definitions would then be derived for particular types of observations.

On the other hand, as the position of the equinox in the CRS depends on both the equator and the ecliptic, Greenwich Sidereal Time is non-proportional to UT1 and its relationship with UT1 is largely sensitive to models for precession and nutation. Moreover, the use of the NRO in the transformation from the TRS to the CRS allows a clear separation of the precession and nutation motions from the Earth rotation, which is more adapted for deriving UT1, polar motion and the celestial pole coordinates than the classical transformation involving the equinox, in which the precession and nutation motions are coupled with the Earth rotation.

REFERENCES

Capitaine, N., Guinot, B., Souchay, J., 1986, Celest. Mech. <u>39</u>, 283.
Capitaine, N., Guinot B., 1988, in The Earth's Rotation and Reference Frames for Geodesy and Geodynamics, A.A. Babcock and G. A. Wilkins ed. D. Reidel Publishing Company, 33.
Guinot, B., 1979, in Time and the Earth's Rotation, D.D. McCarthy and J.D., Pilkington ed, D. Reidel Publishing Company, 7.
Guinot, B., 1981, in Reference Coordinate Systems for Earth Dynamics, E.M. Gaposchkin and B. Kolaczek ed, D. Reidel Publishing Company, 125.

Procedures for Accurate Origin   S. AOKI

RELATION BETWEEN CELESTIAL AND TERRESTRIAL REFERENCE SYSTEMS

In the previous paper (Aoki 1988b) we have solved the equations of rotational motion of a rigid Earth up to the second order. The results are summarized as follows: (i) the forced oscillation of the polar position in space is expressed with the celestial ephemeris pole (ii). The remaining part includes nutation and Oppolzer terms multiplied by wobble. (iii) The Greenwich apparent sidereal time (GAST) is expressed in the form, GAST-GMST + (q)p where (q)p called the equation of equinoxes in the wider sense, includes the additional terms, $0.00264 \sin\Omega + 0.000063 \sin2\Omega$, besides $\phi \cos e$. (iv) The relation between GMST and UT1 is given by eq. (13) of Aoki et al (1982).

DEMERITS OF NRO

The coordination using the Non-Rotating Origin (NRO) proposed by Guinot (1979) and restated by Capitaine et al. (1986) has following demerits; (i) the NRO is only locally inertial and moves with respect to space even to the right ascension direction, by $(\cos e-1)P_1 = 413''/$ century on average (Aoki, 1988a). Even for an object near the equator, NRO moves by 0.00386"/century, by the second order effect of nutation (Aoki 1988b). (iii) The nutation with respect to NRO looks as if it includes out-of-phase terms (Aoki and Kinoshita 1983). (iv) The position of NRO is given by an indefinite integral. If we want to give it definitely, it depends on the initial position as well as the adopted precession constant. This reveals that the position of NRO, depending on its hysterisis, cannot be corrected if the NRO is chosen continuously across a changeover date in future. In other words, the NRO is theory-dependent (Aoki 1988a), whereas the equinox is observable (Aoki, 1988b).

References:

Aoki S. 1988a. in Proc. Japanese Symp. on Earth Rotation, Astrometry and Geodesy, pp. 196-206.
Aoki, S.: 1988b. Celest. Mech. in press.
Aoki, S., Guinot, B., Kaplan, G.H., Kinoshita, H., McCarthy, D.D. and Seidelmann, P.K., 1982. Astron Astrophys 105, p. 359-361.
Aoki, S. and Kinoshita, H., 1983 Celest Mech 29, 335-360.
Guinot B., 1979. IAU Symp No. 82, 7 - 18.
Capitaine, N., Guinot, B., and Souchay, J. 1986. Celest. Mech. 39 283-307.

The Mean Motion in Modern Planetary Ephemerides   E. M. STANDISH, JR.

In 1984, Stumpff and Lieske illustrated an inconsistency of approximately 1"/cty which exists between three astronomical parameters: Fricke's (1972) correction to Newcomb's value of general precession in longitude, the mean motion of the Earth in modern planetary ephemerides and the rate of the apparent longitude of the Sun as given by Newcomb's Theory of the Sun (1898). Of course, at least one (if not more) of the three quantities could be in error. It is instructive to examine each.

Precession

It seems certain that Fricke's correction of 1.10"/cty to the Newcomb precession is correct, at least to within 20%. More modern determinations show 0.85"/cty from Lunar Laser Ranging, 0.90"/cty from VLBI at JPL and 0.80"/cty from VLBI at MIT.

Planetary mean Motion

Realistic estimates of the accuracy of the Earth's mean motion have been about 0.01"/cty - two orders smaller than the sought-for inconsistency mentioned above. These estimates have now been substantiated in two ways:

1) A comparison between the two independently created ephemerides, JPL's DE118 and MIT's PEP740, shows agreement for the Earth's mean motion of about 0.005"/cty.

2) An experiment was performed during which a change in mean motion of 1"/cty was artificially forced into the ephemeris for the Earth. A new adjustment was then made for all relevant parameters while keeping the Earth's mean motion artificially changed. The results show how badly distorted the solution becomes and are perhaps best illustrated by a comparison of the Viking Lander residuals in Figure 1 which shows a normal fit and the residuals in Figure 2 which shows the best possible fit with an artificially changed mean motion.

Newcomb's Theory

The apparent longitude is given as a function of time, but it isn't clear exactly which time to use. Further, the theory is based upon 19th century observations, about which Fricke had the following to say. "It may be mentioned that absolute observations carried out before 1890 show not only a large scatter but also clear indications of neglected instrumental errors..."

Conclusion

Newcomb's theory was a remarkable achievement and it is a credit that he was able to attain an accuracy of 1"/cty. However, one could hardly expect more.

Figure 1. Viking Lander residuals from a normal solution.

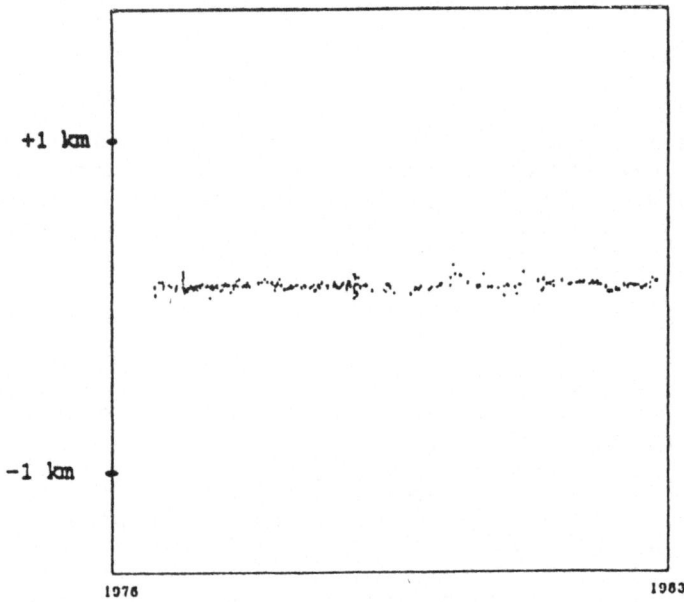

Figure 2. Viking Lander residuals when the earth's mean motion
           is forced to fit a 1"/cty change to its normal value.

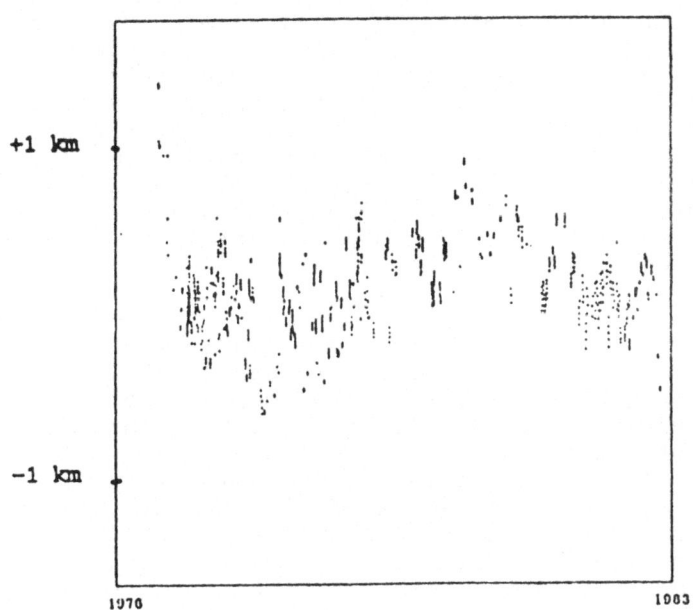

+1 km

-1 km

1970   1983

Determination of the Earth Orientation   M. FEISSEL

     The Earth orientation can be described by five parameters
which provide the transformation between a reference frame
attached to the Earth and a quasi-inertial reference frame, as
a function of time. The parameters measured give the direction
of the rotation axis in space (offsets in longitude and in
obliquity with respect to a modelled direction), and in the
Earth (x, y), and universal time (UT1-UTC). Due to the
incomplete theoretical modelling of the Earth rotation
irregularities, these parameters have to be monitored, for
practical applications (e.g. space navigation) and for
improvement of the theory. The main observation methods are
Very Long Baseline Interferometry (VLBI), laser ranging to the
Moon (LLR), and high altitude satellites (SLR), organized in
permanent programs involving worldwide networks. From 1970
through 1983 they have progressively replaced the less precise
optical astrometry method, which had been in use since the end
of the 19th century. Global analyses of these observations
over several years include, for each program, the computation
of a terrestrial and celestial frame, of time series of the
Earth orientation parameters, and other parameters pertaining
to the body of the Earth or to specific aspects of the
observing method. The celestial frame realized by VLBI is a
set of coordinates of extragalactic compact radio sources; in
the laser techniques, it is the ephemeris of the target, lunar
reflector or artificial satellite. The terrestrial frame is in
all methods a set of geodetic coordinates for the observing
sites; the VLBI terrestrial frames include about 15 sites and

the SLR about 55.

The limiting factors to precision and accuracy of the analyses are in general at the milliarcsecond level or lower, e.g. tectonic plate motions (VLBI, LLR, SLR) precession-nutation models (VLBI, LLR) radio source structure (VLBI), lunar ephemeris (LLR); orbit perturbations by the oceanic tides prevent SLR from accurately determining UT1 for frequencies lower than 1c/80d; due to a sparse network (3 stations), LLR cannot determine polar motion. For the participation of analysis centers in the International Earth Rotation Service (IERS), a part of the models and astronomical or geodetic constants are unified, for the models and constants which are not known accurately enough, their improvement is pursued on the basis of the observations themselves.

The precision of the Earth orientation measurements is estimated by the analysis centers, which associate formal uncertainties to their results. The accuracy, or consistency, of these determinations can also be estimated through comparisons of the global results obtained independently from the different analysis centers for the same technique. Taking advantage of several time series of similar quality over several years. the long term consistency of results can also be evaluated. Various evaluations of precision and consistency are summarized in Table 1.

Table 1 - Precision and consistency of Earth orientation determinations
Units : 0.001".

methods :	VLBI				LLR		SLR		
estimation (1984-1987)	sampl. time	celest. pole	terr. pole	UT	sampl. time	UT	sampl. time	terr. pole	UT
formal uncert.	1d	0.4	0.9	0.6	0.1d	4.5	3d	0.7	1.0
consistency of time series	5d	0.5	1.2	0.7	0.5d	4.2	5d	1.3	1.8
	1m		0.5	0.4			1m	0.5	1.0
	1y		0.2				1y	0.4	

The VLBI polar motion determinations have an error spectrum consistent with a white noise model. The VLBI series of UT1 as well as the SLR determinations of polar motion and UT1 have an error spectrum which is nearer to a flicker noise model, suggesting that some time dependent errors are present in the analyses at the submilliarcsecond level.

Other consistency tests can be performed with the help of closure equations which should be verified between the relative orientations of the individual reference frames and the biases between the corresponding series of Earth orientation parameters. The inconsistencies between two parallel solutions from the same method are at the level of 0.0002" for SLR and

0.002" for VLBI, probably due to the poor distribution of the
VLBI network; the inconsistencies between SLR and VLBI
independent solutions are at the level of 0.002".

In summary, while the long term inconsistency of the
series of Earth orientation parameters is well under the
milliarcsecond, the evaluation of their accuracy is presently
limited to about 0.002" by the coverage of terrestrial
networks.

References

BIH, 1988, Annual Report for 1987, Observatorie de Paris IERS,
1988, IERS Standards, Technical Note No. 1, (Ed. D.D.
McCarthy).
Moritz, H., and Mueller, I.I.: 1987, Earth Rotation, Theory
and Observation, Ungar.

Apparent Place Reduction   B. D. Yallop

A procedure based on that published in the Astronomical
Almanac, section B, is recommended. That procedure was
intended to produce apparent places to 0.01". In most cases it
may be used without modification to produce apparent places to
mas.

For stars and extragalactic sources the main inaccuracy
occurs at the start of the procedure which is to obtain a
barycentric position and velocity of the object on the FK5
system at J2000.0. The FK5 catalogue itself contains standard
errors of 0.02" in position and 0.7 mas per year in proper
motion.

In general it will be necessary to include observations
made in the FK4 system to produce final positions to mas
precision. The FK4 system is non-inertial due to known errors
in precession and the motion of the equinox. The
transformation recommended by Aoki et al from FK4 at B1950.0 to
FK5 at J2000.0 ignores second order effects of non-inertial
motion, which produces errors of up to 5 mas in position and up
to 13 mas per century in proper motion. Murray has taken fully
into account the non-inertial motion and has shown that it is
more logical to make the changeover from FK4 to FK5 at 1950 as
Standish suggested originally.

The effect of removing the E-terms of aberration may
still introduce inaccuracies in the positions and proper
motions of catalogue mean places. The effect of these terms
can only be properly eliminated from individual observations
when the original procedure is known. This demonstrates the
importance of using agreed standard procedures for apparent
place reductions.

At the next stage of forming the mean place of a star at
the epoch of date, Stumpff has pointed out that the calculation
of space motion in the Astronomical Almanac is not rigorous.
Observables are used instead of inertial quantities and the
effect of light time has been ignored. Fortunately these
effects are very small and are only important for nearby stars

whose apparent motions are changing rapidly across the line of sight. The errors increase progressively with time so that they may become significant at the mas level only after several decades.

Several general relativity effects up to order μ/c^2 have been accounted for in the reduction procedure, because they do not exceed 0.1 mas. While photons are crossing the solar system, the Sun moves relative to the barycentre. This effect, which is estimated not to exceed 0.1 mas, has been ignored in the algorithm for solar light deflections.

The Transformation Between the Coordinate Systems of FK4 at B1950.0 and FK5 at J2000.0
C. A. Murray

The essential changes in the fundamental coordinate reference system of FK5 relative to that of FK4 are (i) revision of the precession constants, (ii) correction for the zero point error in right ascension and its rate of change, (equinox correction and motion), and (iii) the adoption of J2000.0 as fundamental epoch instead of B1950.0. We are not concerned here with regional systematic errors.

Two matrix formulations of the transformation have been proposed, by Aoki et al. (1983) and by Standish (1982); these differ in the following two important respects. (a) The osculating epoch at which the coordinate axes in the two systems coincide is taken to be 1984 January 1 by Aoki et al. and B1950.0 by Standish. and (b) the equinox correction at a general epoch is applied in the reference frame of date by Aoki et al. and in the frame of B1950.0 by Standish.

We consider first point (a). The FK4 system is defined at B1950.0 by positions and proper motions in the catalogue, and it is these proper motions which were analyzed by Fricke (1967) in order to derive corrections to precession and equinox motion. It follows therefore that these corrections must be applied at this epoch. We assume, with both authors, that the proper motion system defined by FK5 is inertial whereas that of FK4 is not. This implies that a linear space velocity is represented by coordinates varying linearly with time in the FK5, but not in FK4. We can therefore calculate true coordinates at any epoch in the J2000.0 reference frame just from the positions and proper motions in FK5 (with radial velocity and parallax if necessary), but B1950.0 is the only epoch for which true coordinates are known in the FK4 system; it is therefore necessary to adopt B1950.0 as the osculating epoch.

On point (b) it is clear that the motion of the equinox obtained from proper motions is a rotation of coordinates about the pole of the B1950.0 frame. It is also evident that Fricke (1985) regarded the equinox corrections derived from observations of solar system bodies at successive epochs as being in the B1950.0 frame. Therefore, it is essential that the correction for equinox motion be applied in this frame. In the special case in which the precession is unchanged and the equinox correction does not vary with time, both coordinate systems are inertial and should differ only by a constant rotation; this follows from the formulation proposed by

Standish, but that of Aoki et al. depends on the epoch, which is absurd.

The main difference between the transformations formulated by Aoki et al. and by Standish are coordinate rotation of 0.005" at B1950.0 and a relative rotation rate of 0.013" per century, both about the direction to the vernal equinox. Since at least one of these transformations leads to a non-inertial coordinate system at the milliarcsecond level, it is important to arrive at a consensus as to which should be adopted. My own view is that Standish's formulation is correct with respect to both points.

References:

Aoki, S., Soma, M., Kinoshita, H., Inoue, K.: 1983, Astron. Astrophys. 128, 263.
Fricke, W.: 1967, Astron. J. 72, 1368
Fricke, W.: 1985 Veroff, Astron Rechen-Institut Heidelberg No. 31.
Standish, E.M.: 1982, Astron. Astrophys. 115, 20.

Review of Current/Future Catalog (Radio/Optical) Accuracies
H. Schwan

Considerable progress in the determination of the positions of celestial objects has been made in the recent years. In optical astrometry new fully automated and photoelectric meridian circles have come into operation (in La Palma, Bordeaux and Tokyo) which can perform about 100,000 observations per year for objects brighter than 13th magnitude with a precision of 0.15" to 0.20" for a single observation. Including such observations in addition to those made in the past with the aid of other meridian circles, astrolabes, vertical circles and transit instruments, we have achieved in the basic part of the FK5 (consisting of the classical 1535 fundamental stars) a systematic and individual accuracy of about 20 mas for the positions at the mean epoch (about 1950) and of 0.7 mas/year for the proper motions. Positions and proper motions of inferior precision will be derived for about 3000 new fundamental stars (the FK5 Extension) selected from the FK4 Sup and IRS list.

In the future the astrometry satellite HIPPARCOS will hopefully measure positions and proper motions of more than 100,000 stars brighter than 13th magnitude with a precision of 2 mas and 2 mas/year, respectively. The precision of these proper motions (which is comparatively low because of the short period of the mission) can be significantly improved by combining the HIPPARCOS measurements with existing ground based catalogues. A combination of HIPPARCOS with the Basic FK5, e.g., would provide proper motions with a precision of 0.4 mas/year. Optical Interferometry is a promising means to determine high precision positions. Based on experience at MIT, SAO and USNO one can expect an accuracy of 3 mas from one night for stars brighter than 12th magnitude.

The most precise positions have been obtained by means of VLBI measurements of compact extragalactic radio sources. The

precision given in existing radio catalogues is typically
better than 5 mas. Analyzing only objects with more than 100
observations, a precision of better than 1 mas had been
achieved; in that case corrections to the 1980 IAU nutation
series had to be included in the reduction procedure.

Future improvements in radio (and optical) astrometry
will arise from an increased number of the measured objects, in
particular in the southern sky, from improved techniques, from
more instruments and therefore more base-line orientations,
from more sophisticated reduction procedures and improved
theories, and from the accumulation of more observing time per
object.

Report of the IAU Working Group on the Theory of Nutation
R. L. Duncombe

The IAU Working Group on the Theory of Nutation
comprises; N. Capitaine, T. Herring, G. Kaplan, H. Kinoshita,
M. Rochester, J. Vondrak, J. Wahr, D. McCarthy with R. Duncombe
as Temporary Chairman during the beginning phases of the work.
The Working Group was formed in April 1987 and from that time
on it has not been possible to assemble this group as a whole.
Consequently all of the work has had to be done by
correspondence. Helpful comments and contributions to the task
of this Working Group have been received not only from within
the group but also from other scientists who are actively
engaged in problems concerning the Earth's nutation. This
report draws heavily on contributions by N. Capitaine, J.
Vondrak, V. Dehant, J. Dickey, T. Herring, and O. Sovers and C.
Edwards.

The IAU 1980 Nutation Series are based on Kinoshita's
(1977) rigid Earth Theory, using Newcomb's Theory for the
motion of the Earth, Brown's Theory for the motion of the Moon
and the IAU 1976 System of Astronomical Constants. These
theoretical coefficients are modified on the basis of Wahr's
theory (1981) in the ratio of the amplitudes of each circular
nutation relative to a realistic Earth model and to a rigid
Earth Model. This ratio is computed for an elliptical,
rotating, elastic and oceanless, hydrostatically pre-stressed
Earth with a fluid core. While this nutation series has proven
adequate for many astronomical reductions, the introduction of
high-precision VLBI, LLR, and SLR observing techniques has
revealed some significant inadequacies. VLBI observations at
the milliarcsecond level (or better) have indicated possible
amendments to a number of terms in the nutation series from the
18.6 year term on down. To enable the systematic reduction, on
a common basis, of VLBI, LLR and SLR observations, a new more
precise nutation series is required.

To this end, the consensus of the Working Group indicates
that it would be desirable to repeat Kinoshita's theory for a
rigid Earth, using modern theories for the motion of the Earth
and Moon, current values for the astronomical constants and
including the effect of the planetary perturbations. To match
the accuracy of the VLBI, LLR and SLR techniques this new rigid
Earth theory should attempt to incorporate all known

contributions at the one tenth milliarcsecond level.   In
addition the Wahr theory should be redone, again attempting to
incorporate all known contributions at the one tenth
milliarcsecond level. N. Capitaine has pointed out that the
amplitudes of the in-phase and out-of-phase components of
nutation should include the contribution of the non-hydrostatic
core flattening as well as the oceanic and anelastic effects
which have been shown to be significant. She adds, however,
that present models do not allow these effects to be derived at
the sub-milliarcsecond level.

The tasks outlined above will clearly occupy the
endeavors of this Working Group and others for the next
triennium and possibly beyond. In the interim, it becomes
necessary to take some action to provide standardized working
coefficients for those terms where observations have clearly
shown modifications to be desirable, to allow for systematic
nutation reduction of VLBI, LLR and SLR observations. It is
the consensus of this Working Group that these corrections
should not be empirical (that is taken directly from the
observations) but rather "adopted" values based on plausible
modifications to present theory which produce corrections
consistent with the observed results. It is a further
consensus of this Working Group, that of the several terms
shown by observation to need possible amendment, only the
annual and semiannual terms are sufficiently verified to
consider at this time. The working group consensus favors the
"adopted" values of corrections to these terms given by the
discussion of T. Herring (1987). These are for nutation in
longitude annual term, in-phase +5.23 mas, out-of-phase +0.61,
mas; semi-annual term, in phase +1.02 mas, out-of-phase -1.18
mas. For nutation in obliquity the corrections are: annual
term, in-phase +2.08 mas, out-of-phase -0.24 mas; semi-annual
term, in-phase -0.41 mas, out-of-phase -0.47 mas. Herring
states that these "adopted" corrections to the "coefficients in
the 1980 nutation series were obtained assuming that the semi-
annual nutation corrections are due solely to an error in the
prograde semi-annual circular nutation, and that the annual
correction is due to an error in the retrograde annual circular
nutation. These circular nutations are those which
geophysically should be most affected by the fluid core,
elasticity and the oceans. These assumptions are consistent
within the observational uncertainties."

It is the consensus of the Working Group that these
corrections be used, not as new constants, but as standardized
working numbers to facilitate the uniform nutation reduction of
VLBI, LLR, SLR and other observations, so that results from all
techniques may be intercompared.

In concluding this report, I would like to thank those
scientists, both within and without the Working Group, who have
contributed so significantly to our task.

References:

Herring, T. 1987, private communication.
Kinoshita, H. 1977 Celest. Mech. 15 277
Wahr, J. 1981 Geophy J. Royal Astr. Soc 64, 705.

Report of the IAU Working Group on Astronomical Constants B. Morando

Setting up of the Working Group

The Working Group was created following Resolution C1 of the International Astronomical Union adopted at the General Assembly in New Delhi in November 1985. This resolution reads as follows:

"Commission 4, (Ephemerides), 7 (Celestial Mechanics), 8 (Positional Astronomy), 19 (Rotation of the Earth), 31 (Time) recognizing the importance of ensuring that the IAU system of astronomical constants is rigorously defined and is well suited to current applications,
invite the presidents of IAU Commissions 4, 7, 8, 19, and 31 to form a Working Group to serve in collaboration with the appropriate special study group of the International Association of Geodesy which will; 1) review current determinations of astronomical and geodetic constants, 2) provide for informational purposes the current best estimates of the values, accuracies and sources of these constants, 3) propose appropriate changes in the relevant definitions and values of the constants of the IAU system, 4) urge all authors to specify completely the values and accuracies, as well as the sources, of the constants used in their work and 5) submit a preliminary report in 1987."
The Working Group is composed as follows: Chairman; B. Morando (Bureau des Longitudes, France), Members; V. A. Abalakin (Pulkova Observatory, USSR), W. E. Carter (National Geodetic Survey, USA), H. Kinoshita (Tokyo Astronomical Observatory, Japan), J. Lieske (Jet Propulsion Laboratory, USA), J. Schubart (Astronomisches Rechen Institut, GFR), H. Schwan (Astronomisches Rechen Institut, GFR), P. K. Seidelmann (U S Naval Observatory, USA), E. M. Standish (Jet Propulsion Laboratory, USA), J. M. Wahr (University of Colorado, USA), G. Wilkins (Royal Greenwich Observatory, UK), B. Yallop (Royal Greenwich Observatory, UK), Ya S. Yatskiv (Kiev Main Observatory, USSR).

Philosophy for a system of constants

The activity of astronomy has broadened to such an extent that the various activities of astronomy encompass a broad spectrum of different types of constants. These range from geodetic constants to constants for physics, so it is difficult to draw a line and say that these are the constants for astronomy. At the same time the increase in knowledge has progressed at such a rate that the constant adopted today may not be satisfactory a year later.
On the other hand one of the principal requirements for utilizing observations from the past or for understanding the observations of today in the future, will be a requirement to know what constants and procedures were used for the reduction of those observations.

     Thus if the need for the adoption of an apriori set of
constants has lessened, it has become important to publicize
the procedures that were used. The question of what procedures
were followed includes the reference system used, in what order
were the computations made, and what accuracy and weighing
procedures were used. The need is for a means of comparing
different results today and for documenting what was done. It
might even be possible to have computer readable, updated files
available almost immediately over some sort of telephone
network with a monitoring group responsible for the contents of
such a file. An archiving system would have to be associated
with this by which one could retrieve previous files in order
to determine how a given set of data has been processed.
     In the past the IAU has regarded as one of its duties to
provide conventional values of some astronomical constants
considered as being especially significant. Such systems of
constants were the 1964 IAU system of constants adopted at the
Twelfth General Assembly and introduced in the ephemerides in
1968 and the 1976 IAU System of Constants adopted at the
Sixteenth General Assembly and introduced in the ephemerides in
1984.
     The difficulties of an adopted system of constants are
exemplified by the system adopted in 1976. When ephemerides
were subsequently fit to the observations, it became apparent
that there was no way in the fitting process, to enforce upon
the observations given values for some constants. For example,
the value of the obliquity of the ecliptic determined by
Newcomb could not and should not be used to fit modern
ephemerides with the modern observations. Also between 1976
and 1984 a satellite of Pluto was discovered which indicated
that the mass of Pluto was very different from the value
adopted in 1976. To force such a mass of Pluto on the
ephemerides would be a serious mistake. The process of
fitting the ephemerides to the observations and solving for
unknowns, such as the obliquity and the masses of the planets,
results in a better fit and better ephemerides. So the apriori
adoption of constants is inconsistent with the process of
obtaining the best fit of the ephemerides to the observations.
The values of some constants are improved, sometimes
drastically, thanks to numerous and precise, mainly space
borne, observations. These observations so frequently alter
the values of the constants involved that it is impossible for
a system of constants intended to last ten years to keep up
with them. This is the case, for instance, for the radii of
the satellites of the planets and it may become very soon the
case for the masses of the minor planets.
     This situation will inevitably remain for the future.
The inadequacy of the current system of constants has, of
necessity, led to the practice of different groups defining and
adopting their own system of constants. Examples are the Merit
standards and the standards of the International Earth Rotation
System.

Special Issue: The astronomical units. The concept of the
unit of length.
     The 1976 IAU system gives a definition of the unit of

time in the relativistic framework, but the unit of length is
not clear, as in fact two different time scales exist each one
being related to a specific frame. It is suggested that the SI
second, and the SI meter in the 1976 IAU system should be
understood as the barycentric second sB and the barycenter
meter mB respectively. These quantities would then be related
to the SI units sL and mL by the following relations:

$$sB= \frac{1}{n} S_L \quad m_B = \frac{1}{n} m_L \quad \text{where } n=1-1.55051.10^{-8}$$

n is the mean value of the derivative of local time with
respect to barycentric time. Then the day would be 86 400
barycentric seconds and the astronomical unit would be defined
as the semi-major axis of a planet with a mass equal to zero
having a mean motion equal to the Gauss constant. As a derived
constant it would be equal to
$1.4959787014953416 \times 10^{11}$ barycentric meters. The velocity of
light would have the same value in the local frame or in the
barycentric frame. Some members of the Working Group, instead
of defining the SI second in terms of the barycentric second,
had rather use the local second which would have the advantage
of being more accessible to time-service people and laboratory
physicists. The definition of the unit of length remains so
far linked to the two body problem in the frame of newtonian
mechanics. It might be approached quite differently by
defining first, as above, a second of coordinate time, the day
being 86400 such seconds, and defining then the unit of length
(the astronomical unit) by adopting a given value for the
velocity of light expressed in astronomical units per day.
        Another strong feeling of the Working Group has to do
with the names given the time scales TDT and TDB which are
deemed confusing and misleading. Proposals are made to drop
the word "dynamical" from those names.

New approach to constants

        In order to comply with resolution C1 of the IAU there is
the need for communication between the IAU and the various
organizations specifying constants. This includes the
International Union of Geodesy, Codata and other such constant
defining groups. The constants could be divided into the
following groups:

1. Defining constants should be adopted and accepted as they
are unlikely to change over a long period of time. First the
unit of time would be clarified as explained in section 3.
Then there are two possible solutions:
        a - The unit of mass is the mass of the Sun and the unit
of length is defined given the Gaussian gravitational constant.
        b - The velocity of light expressed in astronomical units
per day is a defining constant which defines the unit of length
and the unit of mass is defined using the Gaussian
gravitational constant.
        The velocity of light in meters per second is now a
defining constant of the SI system of units. It is then
unnecessary to give it as part of an astronomical system of

units.
2.   All the other constants would only be considered as part of
a recommended set of values.   This set would include a list of
constants obtained as the result of the solution of fitting the
ephemerides to observations.   Such a system of constants would
be some of the values underlying, for instance DE200/LE200.
Also a list of best estimates of values needed for the
multitude of purposes in Astronomy would be given.   This list
would include the masses of the minor planets and satellites,
the equatorial radii of the planets, etc.   The list of these
constants could be revised on an annual or tri-annual basis.
        The defining constants and the list of recommended values
could be published annually in the national ephemerides.
3.   In addition a set of procedures and computational
algorithms should be suggested for use as applicable.   Thereby
astronomers could document their procedures and observational
reductions by consistently including in publications a
statement that the procedures and constants documented in a
reference are utilized in this work.   To provide a draft
example of such a system of astronomical constants and
procedures the following is put forth as an example.

- Suggested Procedures and Computational Algorithms are:
    (1)   Planetary, Solar and Stellar Reduction as described
    in the national ephemerides.
    (2)   Conversion of standard epoch B 1950.0 to J2000, from
    Standish, E.M., (1982) Astron. Astrophys, 115, 20-22 and
    from Aoki, S., Soma M., Kinoshita, H., Inoue, K., (1983)
    Astron. Astrophys., 128, 263-267.
    (3)   IAU Theory of Nutation
    (4)   Radiation Pressure Model of Merit Standards 1983,
    Appendix A4.
    (5)   Ocean Tide Model of Merit Standards 1983, Appendix
    6.
    (6)   General Relativistic Terms for Propagation and Time,
    Merit Standards 1983, Appendix 12.
    (7)   Radio Source Positions of Merit Standard 1983,
    Appendix 12.
    (8)   FK5 Star Catalogue.

- Defining constants
    (1)   Gaussian gravitational constant k=0.01720209895, the
    unit mass being the mass of the Sun.
        Alterative
    (1)   Gaussian gravitational constant k=0.01720209895
    (2)   Velocity of light c=173.14463331 au/day

Report Of The Working Group On Reference Frames

J.A. Hughes, Chairman

The following is an abreviated and edited version of the report of the Working Group as given at the General Assembly in Baltimore. The six recommendations of the group which are given here are slightly edited in order to agree with the exact wording adopted by the IAU as Resolution C1. The original report contained two additional recommendations which are essentially subsumed in Resolution C2 as adopted at the IAU General Assembly (GA), and which therefore, are not given here. The essential text of the report is unchanged.

The Working Group on References Frames (WG), was founded by Resolution C2 of the XIX GA held in New Delhi in 1985. The members of the WG and the Commissions which they represent are: B. Morando (4), J. Kovalevsky (7), H. Schwan (8), N. Capitaine (19), E. Roemer (20), C.A. Murray (24), I. Mueller (31), R. Wielen (33), K. Johnston (40) and D. McCarthy representing the IAG.

At the outset, it was anticipated that the matters listed in the founding document of the WG could not be completely and definitively addressed prior to this GA. Such has proved to be the case. On the other hand however, the WG has been very effective in stimulating discussions and promoting a wider awareness and deeper understanding of the essential issues which are involved in the matters of concern to the WG. Open discussions were held by the WG on three occasions: IAU Symposium 128, *The Earth's Rotation and Reference Frames for Geodesy and Geodynamics,* Coolfont, West Virginia, October 1986; IAU Symposium 133, *Mapping the Sky,* Paris, June, 1987; IAU Colloquium 100, *Fundamentals of Astrometry,* Belgrade, September 1987. Discussions also took place during the XIX GA of the IUGG, Vancouver, August, 1987. A great deal of personal correspondence was also exchanged by those most interested in the matters with which the WG was charged, some relevant papers were given at various meetings and a few publications have appeared. As a result of this activity, a better informed membership of the IAU should be more capable of fully considering any resolutions regarding reference frames and time which may come before it.

The following report is divided into three sections, each dealing with a major area of concern to the WG. These sections are: Celestial Reference Frames; Terrestrial Reference Frames; and Time. The use of the word "Frames" rather than "Systems" appeared desirable to the Chairman of the WG. The recommendations of the WG appear in boldface type throughout the report.

Celestial Reference Frames

At the present time there exists only one generally acknowledged, global reference frame. This is the FK5 classical, optical system which, including the fainter stars added to the FK4 star list, contains a total of approximately 5,000 stars, (rather than 1535 stars as in the FK4).

The WG recommends that: (WGRF Recommendation No. 1)

In order to avoid a confusing proliferation of reference frames, the FK5 should be retained as the IAU reference frame at optical wavelengths for the present and immediate future.

As is generally known, progressions such as FK3 to FK4 and FK4 to FK5 were achieved by incorporating new observations into the predecessor system in order to improve it and thus generate a successor system. Efforts are now underway to refine this procedure by discussing all observations simultaneously and *ab initio*. These efforts will produce a successor to FK5 which, it is anticipated, will be superior to FK5 in both its random and systematic error characteristics.

The WG recommends that: (WGRF Recommendation No. 2)

In order to derive the maximum possible information from the accumulated, classical observations, and most especially from the fundamental observations, ab initio discussions of these latter observations should be encouraged and supported.

Current definitions of celestial reference frames at radio wavelengths make use of the extragalactic references provided by suitable objects, primarily quasars. Such references, when combined with interferometric astrometry, are *nonpareil* when applied to the determination of Earth Rotation Parameters (ERP) and Crustal Dynamics (CD) studies. Indeed, these programs have made the major contribution to the various radio reference catalogs currently available. It is now necessary to extend the use of such systems to astronomy in general. This must include extensions in wavelength, in applications, and in the classes of objects included in such systems. There are, of course, obstacles to be overcome if the full potential of this conceptually straightforward approach is to be realized. However, if the phenomena of precession, nutation and polar motion as well as the concepts of the ecliptic and the vernal equinox can be disconnected from the realization of a reference frame, and be regarded as simply describing various aspects of the Earth's complicated motions, then a great simplification will have been achieved. Of course all of the above phenomena and concepts are basic, and a knowledge of them is absolutely necessary. This knowledge will continue to be supplied by the classical, dynamical observations, radio astrometry and pulsar observations. However, it is now possible to consider these items in their proper context and to define a reference frame which is independent of them. Such independence will benefit not only the reference frame, but also aid in the study of the very phenomena from which the concept of a reference frame will have been freed. Essentially, observations will have been decoupled from the observing platform. As a result of this, the accuracy of the reference frame will become primarily dependent upon the precision and accuracy of the underlying measurements, and will have a minimal, non-critical dependence upon any companion theories.

The WG recommends that: (WGRF Recommendation No. 3)

The IAU should adopt a celestial reference frame based upon a consistent set of coordinates for a sufficient number of suitable extragalactic objects when the required observational data have been successfully obtained and appropriately analyzed. This reference frame should be based upon a common, simultaneous discussion of the observations using agreed upon conventions. This reference frame is likely to be

based, initially at least, exclusively upon radio astrometry, and transformations between this reference frame and the conventional celestial and terrestrial reference systems as well as the dynamical frame should be defined. The reference frame should be updated as required.

The wording of the above recommendation implies that a reference frame other than the Conventional Celestial Reference System should be adopted by the IAU. This is not necessarily the case, but could be the case. That is, the title, Conventional Celestial Reference System, is used by the ERP community in conjunction with a similar name for the adopted Conventional Terrestrial Reference System. The use of the modifier "Conventional", came into general use during the MERIT program, possibly due to the earlier use of the word in conjunction with the origin of the pole, i.e., the Conventional International Origin (CIO). The determination of ERP does not require a global distribution of sources nor a large number of them. The needs of the purely astronomical community are not adequately known at this time and therefore it is not possible to assert that the conventional system of the ERP/geodetic community will precisely fulfill the requirements of the IAU in general. On the other hand, ERP observations provide much of the data used to define the extragalactic radio frame and will perforce contribute greatly to whatever frame might be adopted. If the conventional system is ultimately adopted as it stands, then the transformation mentioned in the recommendation above becomes simply an identity.

In any event, the achievement of a general reference frame on a global basis, encompassing a range of both magnitudes and wavelengths, will not be easy, but the difficulties are primarily observational and not conceptual or theoretical. Indeed, great progress has already been made in the areas of ERP and CD and VLBI experiments in general. For example, an extragalactic reference frame which will serve as the initial system of the International Earth Rotation Service (IERS) was received by the Chairman of the WG as this report was being completed. This frame was compiled on the basis of four individual catalogs from the Goddard Space Flight Center, the Jet Propulsion Laboratory and the U.S. National Geodetic Survey. The compilation was carried out at the IERS (E.F. Arias, M. Feissel and J.F. Lestrade, *Bureau International de l'Heure Annual Report for 1987*, Observatoire de Paris), and includes 228 extragalactic, compact sources divided into primary, secondary and complementary sources depending upon geometrical and physical considerations as well as observational histories. Unfortunately, this reference frame contains no sources south of -45°, and of the 23 primary sources which define the directions of the axes, only 8 are in the southern hemisphere (between the equator and -29°). This points up the fact that even with the excellent ERP results, the distribution of well observed radio sources and radio interferometry baselines is far from ideal for the purposes of a global reference frame. Nevertheless great improvements are taking place. Indeed, since the formation of the WG, dozens of new sources have been observed, most recently many in the southern hemisphere. Thus the problem of sufficient coverage on a global basis, whatever coverage that may ultimately prove to be, is being addressed. At present the density and the distribution of radio sources necessary to provide an acceptable transformation between radio and optical systems depend primarily upon the homogeneity or isotropy of the optical system. If, for example, one had an optical catalog with relative coordinates of the stars at some epoch with the same accuracy as in radio catalogs, then merely applying a correction to the zero points could serve as the transformation for that epoch. The forthcoming HIPPARCOS catalog is intended to approach this ideal and will provide an excellent example regarding the matter of radio/optical transformations. If the HIPPARCOS system is successfully referred to an extragalactic frame, then the extension of this frame to magnitudes intermediate to HIPPARCOS stars and the quasars will follow through the use of astrographs and Schmidt telescopes. Imperfect proper motions complicate the situation of course, but the whole point of improving the reference frame is to provide a better standard coordinate system within which improved stellar motions, for example, can be determined. It is important to note that given an accessible extragalactic reference frame, optical reference frame positional observations would be freed of the burden of simultaneously determining the zero points of a dynamical system, the improvements to the assumed planetary orbits and the individual star positions. The emphasis could then be upon achieving isotropy and observing to fainter magnitudes. Questions involving source structure and any evolution thereof can only be resolved by repeated and carefully programmed observations.

The WG recommends that: (WGRF Recommendation No. 4)

The determination of the positions of radio sources at all possible wavelengths should be continued and accelerated so as to achieve the best possible all sky coverage and overall accuracy, while testing the suitability of candidate sources. The International Astronomical Union should encourage institutions to provide adequate time on appropriate instruments to ensure that the necessary astrometric observations are obtained.

The accessibility of an extragalactic reference frame to astronomers dealing with brighter, optical objects is of great concern to many and this question must be satisfactorily addressed. At the present time, as discussed above, the matter of accessibility really concerns the transformations, or links, between radio and optical reference frames. The identification and observation of galactic radio stars at both optical and radio wavelengths and in both optical and radio reference frames plays a crucial role in this area. Action is necessary if the full potential of the radio reference frame is to be realized.

The WG recommends that: (WGRF Recommendation No. 5)

The detection of radio stars and the determination of their positions and proper motions should be a major goal of astrometry.

The determination of optical positons and proper motions of stars with respect to extragalactic objects should be encouraged.

All applicable methods, particularly astrometry on large reflectors, should be used.

In the longer term, further progress will be possible only when an optical/IR reference frame is available which is comparable to the radio reference frame in accuracy and which is also based directly upon extragalactic objects. This implies a need for improvements in existing methods and the use of new techniques, especially interferometry and space astrometry, the latter perhaps also using interferometric techniques. Having independent radio and optical/IR reference systems of comparable accuracy will permit much more physically significant astrometric comparisons of objects. At present, with but few exceptions, transformations have been derived whose essential purpose is to improve optical systems by using the accuracy of the radio system. Even so it should be noted that significant research is being carried out, for example regarding stellar maser activity, using available astrometric data.

The WG recommends that: (WGRF Recommendation No. 6)

Optical and infrared astrometric interferometry should be developed vigorously for use on the ground and possibly later in space. The related efforts in imaging interferometry have astrometric implications and these developments should also be supported. In all cases the direct determination of the positions of extragalactic objects at optical/IR wavelengths must be a major goal.

Interferometric observations have provided absolute declinations and so-called "relative right ascensions." That is, observations contributing to reference frames based upon extragalactic objects do not automatically define a zero point for right ascensions as do observations leading to a dynamically based frame with its vernal equinox. As a matter of fact the former observations do not, strictly speaking, measure right ascensions at all. For this reason it is necessary to devise a procedure which can uniquely define an origin for this measured coordinate. In addition, it must be decided how such a coordinate is to be distinguished from right ascension. The problem is solvable and is just as much a matter of convention and protocol as of scientific principles. The important points are that a common origin, whose basis and construction are understood and agreed upon, must be defined and utilized by all, and that a similarly agreed upon nomenclature regarding coordinates must be adopted.

The vernal equinox is the origin of right ascension, and its definition involves both the rotational and orbital motion of the earth. As perceived by some, there are intrinsic difficulties with such involvements since problems with the definition of the orbital plane of the Earth immediately lead to related problems with the definition of the dynamical equinox. For example, assuming a continuosly moving equator versus using instantaneous orientations of the equator can lead to a difference as large as $0\overset{''}{.}1$ in the location of the equinox. Similarly, the definition of a "mean" orbital plane involves various assumptions. For these reasons a proposal has been made which is intended to remove the dependence of the origin on the orbital motion of the Earth. Reference is made to the *Non-Rotating Origin* (NRO) first proposed by Guinot and which has been described in various places, but perhaps most fully in the article, *A Non-Rotating Origin on the Instantaneous Equator: Definition, Properties and Use,* N. Capitaine, B. Guinot and J. Souchay, Celestial Mechanics 39 (1986), 283-307. Although proposed primarily for use in defining the sidereal rotation of the Earth and the definition of Universal Time, such an origin could be used for celestial positions. Of course a distinction between right ascension and the "corresponding" coordinate would be required. The name, *instantaneous ascension* has been suggested. The NRO requires a celestial reference frame based upon extragalactic objects with respect to which the motion of the Earth's pole is specified. This is, of course, exactly what the IERS provides. Reaction to the concept of the NRO has been mixed, but a thorough evaluation of the concept as it applies to celestial positions and motions is definitely called for.

Terrestrial Reference Frames

Various terrestrial reference frames exist around the world. Some are local systems while others may be used for global applications. In general it is the latter which are of most interest to the IAU. The name, Conventional Terrestrial Reference System, is used to delineate that system which is defined by the most precise geodetic techniques. Currently the BIH Terrestrial System (BTS), adopted for use in the determination of Earth orientation parameters by the IERS, makes use of these techniques to provide the most suitable terrestrial reference system. This system is described most recently in the *Bureau International de l'Heure Annual Report for 1987*, Observatoire de Paris. For general information consult, *Realization of the BIH Terrestrial System*, Boucher, C. and M. Feissel, Proc. Internat. Symp. on Space Techniques for Geodynamics, 1984, Sopron, Hungary.

Briefly, the BTS consists of a reference frame defined by the station coordinates of the observatories contributing observations of Earth orientation to the IERS plus a model describing the motions of the tectonic plates on which the observatories are located. Transformation parameters relating the terrestrial systems used in the determination of Earth orientation data are also given. The epoch of the coordinates is 1984.0 The plate motion model, AMO-2, of Minster and Jordan is used, (*Present-Day Plate Motions*, J. Geophys. Res., **83**, 1978, pp. 5331- 5354).

The reference frame of the BTS is Earth-centered, with the pole designated as the BIH pole and longitude origin as the BIH Origin of Longitudes. See, *Comments on the terrestrial pole of reference, the origin of longitudes, and on the definition of UT1*, Guinot, B., Proc. IAU Coll. No. 56, 1981, D. Reidel Pub. Co. The pole is offset from the Conventional International Origin of the International Latitude Service (no longer in existence).

In practice, the system is accessed through the use of the Earth orientation parameters which are published routinely by the IERS. By employing these data in transforming from the Conventional Celestial System, the user obtains coordinates or directions in the BTS. Uncertainties in the reference frame of the BTS are: a few centimeters in the origin, 0.002 parts per million in scale, and up to $0''004$ in orientation.

The WG has no explicit recommendations to make regarding terrestrial reference frames, but it should be noted that the IAU must maintain an on-going liaison with the geodetic community regarding terrestrial systems, most especially regarding matters of Earth orientation and the reference frame to which the orientation is referred. Of course such a liaison occurs naturally in the work of Commission 19 and also with Commission 31, the WG merely reaffirms the necessity of supporting close collaboration together with the timely exchange of information.

Time

The current names, definitions and underlying resolutions which, when taken together, represent the official position of the IAU regarding time, have been found to be unsatisfactory by many. This dissatisfaction has arisen for both practical and theoretical reasons, and therefore the consideration of possible changes in the present posture was made a part of the charter of the WG.

The specific complaints involve the facts that:

1) There are perceived differences between the viewpoints of the IAU and those of the time keeping and physics communities.

2) There are questions regarding the units of Terrestrial Dynamical Time (TDT) and Barycentric Dynamical Time (TDB), and regarding the definitions of these times and their relationship to International Atomic Time (TAI).

3) There are many reservations about the use of the word "Dynamical" in the naming of TDT and TDB.

4) There are outstanding disagreements regarding the characterization of TAI as a proper and/or coordinate time.

5) There is a need to clarify the relationship between the IAU and the International Radio Consultative Committee (CCIR) and the Comité Internationale des Poids et Mesures (CIPM).

After the discussions held during IAU Symposium 128 in Coolfont, B. Guinot and P.K. Seidelmann indicated an interest in pursuing these questions. The Chairman of the WG encouraged this effort, and the result was the publication of an article, *Time Scales: their history, definition and interpretation,* [Astron. Astrophys. **194**, (1988), 304-308]. As the title indicates, the events leading up to the adoption of TDT and TDB are described and the reasons for the subsequent ambiguities and disagreements are explained. This historical section may be considered a part of the report of the WG, and therefore the WG wishes to express its appreciation for the work undertaken by Guinot and Seidelmann. Following the historical background, their article culminates with a recommendation regarding the time reference for the ephemerides together with the grounds upon which the proposal is based. Since the publication of the article, Guinot and Seidelmann in conjunction with D. Allan, S. Aoki, M. Fujimoto and T. Fukushima, have communicated a revised recommendation. The revised recommendation, which is a replacement for Section 6.2 as printed in the article, but which is not at present a recommendation of the WG, follows.

"It is recommended that:

(a) the time reference, or the independent variable, of the apparent geocentric ephemerides be Terrestrial Time, TT,

(b) TT be the proper time for the geocenter,

(c) the time unit of TT, the terrestrial day, being chosen so that the reading of TT agrees with that of a proper time on the rotating geoid whose time unit is 86400 SI seconds,

(d) at the instant of 1977 January $01^d00^h00^m00^s$ TAI, TT be synchronized with TAI plus 0.0003725 day exactly.

(e) the time reference for ephemerides referred to the barycenter of the solar system be Barycentric Time, TB,

(f) the time unit of Barycentric Time, the barycentric day, be chosen so that there are only periodic variations between the readings of TT and TB.

Notes

1. The recommendation refers to the definition of the SI second adopted by the 13th CGPM, 1967 (atomic second).

2. In practice, realizations of the ideal TT are needed. Such realized time scales are designated by TT(xxx), where xxx is an identifier.

3. A realization of TT is that it be synchronized, according to the conventions of the CCIR and the CCDS, with TAI plus 0.0003725 day. The synchronization can be improved by analysis of atomic time data at future times.

4. The designation of a time as being "proper" or "coordinate" time is a cause of much confusion. Perhaps the following statement by Fukushima helps clarify the problem. "Usually in the general relativistic theories, a coordinate time is defined as the proper time of a standard clock which rests at the space origin of a chosen coordinate system. For example, the coordinate time of a geocentric coordinate system is defined as the proper time of a clock which rests at the geocenter while the gravitational effect of the Earth itself is ignored in computing the proper time." Thus, TT should be the proper time of an ideal clock moving with the geocenter while suffering the gravitational effects due to the solar system bodies except for the Earth. Similarly, the proper time for the barycenter of the solar system in the universe means the proper time of an ideal clock comoving with the barycenter of the solar system while suffering the gravitational effects due to the universe except for the solar system. TB is used as a time coordinate in a frame referred to the barycenter of the solar system.

5. Differences in gravitational potentials and velocities may introduce secular terms in the conversion formula between two time scales if the same definition of the unit of time is used for each of them. Therefore, in order to satisfy the requirement that TT and TB differ only by periodic terms, their units must be different; the difference amounts to 1.5×10^{-8}. In a like manner, the unit of TT differs from the unit of TAI by 0.7×10^{-9}. This might be understood if one considers an ideal atomic clock on the geoid keeping TAI. If that ideal clock could be slowly moved to the geocenter, the change in the gravitational potential and the velocity, due to the Earth's rotation, would cause a change in the rate of the clock. Thus, the clock would not keep TT without a rate adjustment. Similarly, if the clock were moved to the barycenter of the solar system, there would be a change in the rate of the clock and clock would not be keeping TB unless it were adjusted in rate.

6. The statement that there are only periodic variations between TB and TT may be considered equivalent to stating that there is no secular

term in the relationship between TB and TT. The number of terms and the length of the periods included in the relationship depends on the accuracy and the time period of concern. Further clarification of the relationship between TB and TT may be desired in the future.

7. Prior to the existence of atomic time, the accuracy achieved by time scales was not as good, so that the practical time scale was not sufficiently precise to distinguish between TB and TT. Thus, prior to 1955 the realization of either TT or TB can be assumed to be Ephemeris Time (ET); but another theoretical relationship might be required in the future."

The recommendation quoted above represents the viewpoint of a number of people, some of whom have contributed directly or indirectly to the discussions which led to the formulation as given. However, it would be incorrect to imply that the recommendation is acceptable to all, or indeed at present, even to a majority. Nevertheless, the recommendation can and should serve admirably as a starting point for thoughtful discussion and indeed possibly for final adoption. In addition, it can help to define and refine any contrasting positions held by others. Most importantly, it can be a catalyst in the process of informing the great majority of IAU members, many of whom are very unfamiliar with the issues and scientific ramifications involved even though their work may be affected in one way or another.

It is impracticable to attempt to list here all of the different points of view or explicit objections to the present or other proposed definitions. However, the following points should be noted.

1. Although no one has come forward to the WG and expressed a desire to retain the word "dynamical" in perpetuity for either TDT or TBT, some have stated that, for the present, the names should remain unchanged. There is little doubt that a majority favors the eventual elimination of this word.

2. In conjunction with No.1, immediately above, it is held that the names Terrestrial Time and Barycentric Time should be reserved until clearly defined and understood transformations of both time and space are agreed upon.

3. The objection has been raised that the IAU appears to have at least three classes of clocks which, if colocated and comoving, would run at different rates and have different associated meter rods.

4. In conjunction with No.3, immediately above, it is held that anyone familiar with relativistic concepts would not understand why there should exist several standard clocks which do not run at the same rate when colocated and comoving. This apparent contradiction arises from the implied rate adjustments which would be made to hypothetical, physical TT and TB clocks.

5. There appears to be general (although not universal) agreement that

TAI may be either a proper or a coordinate time. Indeed, depending upon the application, TAI is considered to be one or the other by various users and their distinct characterizations are entirely appropriate.

6. The position has been taken that ephemerides should be published using TAI or UTC as the time argument, the choice depending upon the period which a particular ephemeris covers. The use of TB as a coordinate time by the celestial mechanician, for example, is a part of the calculation of the ephemerides, but the user need not be forced to deal with concepts such as TT and TB.

7. In conjunction with No. 6, immediately above, it is held that there exist conceptual problems with TB. If TB is a coordinate time in the solar system, it has then been interpreted as being the proper time of a clock at the barycenter with all solar system mass removed, (see Note 4 in the quoted, revised recommendation above). However, removing this mass also removes the cause of the periodic differences between TT and TB. But, if the mass is not removed, then serious difficulties occur when the barycenter moves into the Sun or moves while inside the Sun. Considerations such as these are used as examples of the conceptual difficulties.

8. The position has been taken that there must exist periodic differences between TT and TAI due to the changes in the potentials at the geocenter and on the rotating geoid due to solar system bodies other than the Earth. The change in this potential is not identical for clocks at the geocenter and on the rotating geoid.

The above comments while not exhaustive, do show that valid issues remain to be settled and that alternative viewpoints exist which deserve a comprehensive examination. Although a final disposition of this matter at the XX General Assembly would be highly desirable, the interests of the IAU are more likely to be served in the long term by a thorough airing of contrasting viewpoints, even if such a course of action requires an additional effort with a concomitant delay.

General Considerations

Passing on to a matter related to the responsibilities of the WG, the following is presented for general consideration.

If one examines the history of reference frames within the IAU, it becomes evident that there has existed a symbiotic relationship among the various commissions which have contributed to reference frame work. For example, the FK5 represents a combination of: observations and compilation by Commissions 8 and 24, of ephemerides and the underlying celestial mechanics by Commissions 4 and 7, of polar motion and time determinations by Commissions 19 and 31, and of binary star orbits from Commission 26. Although each of these contributors have unique attitudes and interests regarding reference frames, the combination did work well in the past. However, it is not clear that such a distributed, cooperative effort would function satisfactorily if the basis for reference frames were to be changed as recommended. Hence the question now arises; How should the matter of responsibility for reference frames be addressed in the future? The WG has not been charged with answering this question and can make no official recommendations regarding the matter. However, since many, varied, *ad hoc* suggestions have been made to the WG, thus indicating a great deal of general interest in the matter, it appears appropriate for the WG to offer general comments upon the situation for consideration by the members of the IAU and by the Executive Committee if it chooses to do so.

There are various options: 1) do nothing, 2) assign the WG the responsibility for the next reference frame, 3) name an existing commission to bear the responsibility for reference frames, 4) institute a new commission for reference frames, 5) restructure the existing commissions.

If the potential for difficulties is acknowledged to exist, then the first option is not a realistic choice. Furthermore, a very good case can be made for the necessity of having a dedicated forum and meeting place for those interested in and working on reference frames in general. Such a meeting place would help meld individual, specialized approaches to reference frames into a cohesive, more unified conception, reminiscent of the symbiosis described earlier. If one considers the many tasks which will have to be undertaken to unify and develop the various approaches to reference frames, then such a common meeting ground does indeed become a necessity. Given this assessment, it then becomes necessary to take action along the lines of one, or possibly of a combination, of the options listed above.

These comments are offered in order to stimulate discussion and thought based upon the perception that something should be done. However, with the realization that careful consideration is required, hasty actions at this GA are not envisioned nor encouraged.

2. <u>SOLAR AND STELLAR CORONAE</u>

A Joint Commission Meeting in Honour of

GORDON NEWKIRK Jr

Chairmen and Editors: E.R. Priest, R. Falciani

<u>Supporting Commissions:</u>

10, 12, 29, 35, 36, 44

GORDON NEWKIRK'S CONTRIBUTIONS TO CORONAL STUDIES

J. A. EDDY
Office for Interdisciplinary Earth Studies
University Corporation for Atmospheric Research
Box 3000
Boulder, Colorado 80307
USA

I.   Brief Biography

Gordon Allen Newkirk, Jr. was born in West Orange, New Jersey June 12, 1928 and died in Boulder, Colorado December 21, 1985 at age 57. He was graduated from Harvard University in 1950 and in 1953 earned a Ph.D. in astrophysics from the University of Michigan. In 1955, after service in the Signal Corps of the U.S. Army he took a position at the High Altitude Observatory in Boulder where he worked the remaining thirty years of his life. For 11 of those years (1968-1979) he was director of the observatory and associate director of the National Center for Atmospheric Research. He was also active as a teacher and from 1965 through 1985 was an adjoint professor at the University of Colorado. From 1972 through 1975 he served as Chairman of the Solar Physics Division of the American Astronomical Society and from 1976 through 1979 as President of Commission 10 (Solar Activity) of the International Astronomical Union.

On April 11, 1956 Dr. Newkirk was married to Nancy Buck in Boulder and together they built a home in Sunshine Canyon and reared their family of three daughters. Gordon loved the mountains and knew them well. He was an active outdoorsman and naturalist for whom backpacking and skiing were important family endeavors. His scientific life, as his private one, was characterized by a lifelong fascination with nature and a skilled and patient approach in learning its secrets and mastering its ways.

2.   The Nature of Newkirk's Research

Although a scientist of broad interests Newkirk's particular interest was the corona of the sun and fully half of the papers he published dealt with coronal physics. Through his work and his influence on others the High Altitude Observatory became known during his tenure there as the world leader in optical studies of the extended corona. He was by nature a methodical and deliberate scientist whose career was focused on a few fundamental problems, which he followed, patiently, hand over hand to their conclusion, until he became the master of every detail that was involved. This was the pattern of his coronal research, and it led to pioneering achievements.

D. McNally (ed.), Highlights of Astronomy, Vol. 8, 503–512.
© *1989 by the IAU.*

Newkirk's research in coronal physics was directed at four fundamental problems:

o   the development of a model of the electron corona;
o   the perfection of the coronagraph;
o   studies of the structure of the corona; and
o   an uncompleted study of the propagation of cosmic rays in the corona.

2.1   A MODEL OF THE ELECTRON CORONA.

In a paper published in 1959 (1) Newkirk developed a coronal model that is still used to describe the electron density of the corona. To do this he took on the task of interpreting measurements that had been made routinely since 1956 at the Climax station of HAO with the K-coronameter --- a photoelectric coronagraph that had been developed there by Gérard Wlérick and James Axtell (2). The instrument was designed to distinguish the outer electron corona on the basis of the linear polarization of the K-component, and it provided useful data, under the best conditions, to a distance of about 1 radius above the solar limb.

Newkirk's model, fitted empirically to the Climax data, was of especial value in solar radio astronomy, which in the 1950's was a particularly active field. To this end he employed the model in conjunction with models of the chromosphere and of a plage to develop intensity profiles of the solar disk as they should be seen in various radio frequencies, and to demonstrate that the "slowly varying" component of coronal radio emission was due solely to density enhancements (3).

2.2   PERFECTING THE CORONAGRAPH

The optical coronagraph, developed by Lyot in 1929 (4), had by 1955 become an instrument of routine --- fundamentally limited, in terms of spectral data, to bright coronal features within about $0.5 R_o$ above the limb of the sun, and to white-light polarimetric observations extending somewhat higher but of much coarser resolution. Moreover, the K-coronameter observations of the intermediate corona were secured at the expense of long integration times. Images of the corona with detail sufficient to identify changes in fine scale structure were restricted to a narrow range of the brightest inner corona, extending but a few arc minutes above the limb. The dimmer intermediate and outer corona --- seen so clearly in fleeting moments of natural eclipse --- lay just beyond the reach of ground-based, Lyot coronagraphs (5). It is fair to say that images from coronagraphs were, to any who had seen the extended beauty of the corona at eclipse, at best a tantalizing substitute for the real thing.

From the day that Newkirk joined HAO he bent his efforts to perfect the 25-year old instrument to the point where the full extent of the corona might be captured outside of eclipse. It was a challenge that would test his patience and tax his skill as an experimentalist for almost 20 years.

He found success, as had Lyot, by returning to basic principles: first in the physics of scattering and sky brightness, then in optics and optical design, and last in adapting the instrument for remote operation above the atmosphere.

Newkirk had begun the quest during his military service, when on special assignment he measured the brightness of the high altitude sky near the sun from altitudes of 5000 to 13000 ft., from Colorado mountaintops that he sometimes reached by ski-lift (6). For this, he used a portable photographic sky photometer that had been designed earlier at HAO by Jack Evans (7). The Evans photometer was a portable Lyot f/100 coronagraph, modified to be externally occulted, and used to make broad-band measurements of the brightness of the sky near the limb of the sun. A 1.6 cm external occulting disk at the solar end of a meter-long stem blocked direct sunlight from reaching the 1.5 cm objective lens, with a considerable reduction in instrumentally-scattered light. Evans had developed the instrument to evaluate potential coronagraph sites for a U.S. Air Force solar observatory that was eventually built on Sacramento Peak in New Mexico. He had also attempted sky measurements with the instrument from the open hatch of a propeller-driven military aircraft in an early attempt to measure the brightness of the near-sun sky above the ground.

In 1959 Newkirk tried the Evans sky photometer as a true coronagraph, in a manned balloon flight to an altitude of about 40,000 ft, launched from the "Stratobowl" --- a natural depression in the Black Hills of South Dakota from which earlier manned balloon flights had been made. On board as pilot was Cdr. Malcolm Ross, USN, who held the altitude record for manned ascent, and Robert Cooper, a technician at HAO who endeavored to point the occulting photometer/coronagraph at the sun from the balloon gondola in the cold air of the tropopause, in the hope of capturing photographs of the outer corona in the field of view of the photometer. It was not a successful experiment: Cooper's heroic measurements were severely contaminated by the radiance introduced by the scattering of sunlight from everpresent, local clouds of water particles from his and Ross' exhalations.

Newkirk learned from the experience (8). The manned balloon measurements were followed by a series of unmanned flights to explore sky brightness as a function of angle and of wavelength. For this, he employed the "Stratoscope" gondola that had been developed at Princeton by Martin Schwarzschild in the late 1950's to secure high-resolution, white-light photographs of sunspots and solar granulation (9). This necessitated the design of a new and larger version of the external-occulting photometer and the perfection of skills and auxiliary equipment for remote operation. In the autumn of 1960, Newkirk's balloon-borne "Coronascope", launched from an airfield near Minneapolis, succeeded in securing measurements on two flights into the stratosphere, to an altitude of about 80,000 ft. These were followed by papers that interpreted these pioneering measurements of the solar aureole to derive the particle size distribution as a function of height in the upper atmosphere, using the theory of Rayleigh and Mie scattering (10-15). These studies also established for the first time the near-sun sky brightness that fixed the limits of observation of the corona outside of eclipse, as a function of wavelength and altitude above the ground.

This done, Newkirk turned to the next step in the challenge: that of developing a coronagraph to operate in the stratosphere that was sufficiently improved over Lyot's original design to take full advantage of the much-reduced sky brightness, and thus be capable of detecting the intermediate and outer corona. He had found that at an altitude of 80,000 ft. the near-sun sky brightness had fallen enough (in the photographic infrared) to equal the brightness of the outer corona, or about $10^{-8}B_o$, in units of the brightness of the solar disk. To photograph the corona to at least $6R_o$ required that the instrumentally scattered light in the coronagraph be at least an order of magnitude lower than this, or about $10^{-9} B_o$. The level of scattered light in Lyot's coronagraph --- in which instrumentally scattered light had been reduced by at least a factor of 100 below that in the best simple telescopes --- was about $5 \times 10^{-5} B_o$. The challenge Newkirk faced was to improve the Lyot design by a further factor of 1000!

And he did it. Newkirk first considered a reflecting coronagraph as a possible solution (16), but returned instead to a more conventional instrument that was externally occulted in the manner of the Evans sky photometer. Newkirk found, upon measurements, that the limit of detection in the Lyot coronagraph was set principally by body-scattering in the singlet objective lens: a fundamental property of the glass itself. The way around this was to shade the lens from direct solar light, by employing an external occulter at the cost, however, of the innermost corona. The height above the limb at which the corona could be imaged with an external occulting coronagraph was fixed by the aperture of the objective and the distance between the occulter and the lens that it shaded. Newkirk adopted $0.5R_o$ above the solar limb as a workable inner limit, to overlap with the range available in conventional Lyot coronagraphs. This required a separation of 2.3 m between the occulting disk and the 3.3 cm objective lens. He achieved further reductions in scattered light by suppressing diffraction at the edge of the occulting disk through principles of "apodization" --- in this case the use of multiple, co-aligned external occulting disks (17,18).

Newkirk launched his new coronagraph in the old Stratoscope gondola from Palestine, Texas, in 1964 (19) and succeeded in recording the form of the intermediate and outer corona on infrared film, on the first try, from an altitude of about 80,000 ft. (20). Following these successes, he turned immediately to the challenge of adapting the external occulting coronagraph for use on orbital platforms, where sky brightness was no longer a factor. This, too, he passed on as observatory director to other hands (21). These efforts, guided by Robert MacQueen, reached fruition in 1973 with the launch and successful operation of the Skylab manned spacecraft, which carried, as one of its primary observatory instruments, a three-meter Newkirk coronagraph. The instrument recorded the intermediate and outer corona on photographic film that was returned to earth with each of the three teams of Skylab astronauts (22). On these film were the first high-resolution records of temporal changes in the intermediate and outer corona, including the first clear looks at the birth, development and expulsion of a new class of solar features that came to be known as "coronal transients".

2.3   THE STRUCTURE OF THE SOLAR CORONA

In the years that Newkirk was perfecting the coronagraph for orbital use he worked as well on the general question of coronal structure: a problem that led him to develop a new camera for securing photographs of unprecedented quality of the corona at natural eclipses.

2.3.1   A CORONAL ECLIPSE CAMERA

Since 1851, astronomers had made broad-band photographs of the corona at
eclipse --- but almost never bringing back on plates or film the wonders they
had seen in fleeting moments by eye. The obstacle, Newkirk recognized, was
the steep brightness gradient of the white-light corona --- falling 5 orders of
magnitude from the innermost corona to visible limits of coronal streamers.
The great dynamic range of the dark-adapted eye was able to accommodate
the full range; photographic emulsions, with a dynamic range of typically two
and at most three orders of magnitude could capture but a part of the corona
in a single exposure. The best photographs of the corona made at eclipse
showed in exquisite detail either the innermost corona and imbedded
prominences, or an equally restricted glimpse of some part of the more
ethereal outer corona, extending above an indistinguishable, over-exposed
glare of the inner and brighter region. The problem had come home to
Newkirk in 1963, when for a laboratory exercise he had dispatched graduate
students with simple coronal cameras to a line of stations along the path of
totality of the Alaskan-Canadian eclipse of that year. The goal was to search
for temporal changes in the corona, by post-eclipse comparison of photographs
taken at different stations and hence at separated intervals of time. The
experiment failed. Key stations were beclouded and pictures from successful
sites were inadequate for the task.
        Newkirk attacked the problem by designing and having built a radially-
graded neutral filter, of deposited metal on glass, to be placed immediately
before the focal plane of the telescope. The circularly-symmetric density
gradient of the filter was designed to compensate for the brightness gradient
of the average corona. The film at the focal plane saw a flattened gradient of
coronal brightness that was well within its range. To ensure that the
remainder of the system would not further degrade the result, Newkirk
worked with Lee Lacey of the HAO engineering staff to design and build a
new, single-purpose eclipse telescope whose focal plane was driven, west to
east, to compensate for the opposite motion of the sky. The telescope itself
(f/15 with 11 cm aperture) was mounted rigidly in a fixed position and hence
extremely stable.
        In the autumn of 1966, Newkirk and Lacey took their new coronal
camera to Pulacayo on the altiplano of Bolivia to try it at the November 11
total eclipse of the sun --- at a site not far from where Schaeberle had gone,
73 years before, to test his 40-foot camera at the 1893 eclipse in Chile (23).
The result, as for Schaeberle, was the finest photograph ever taken of the
solar corona, by anyone, anywhere. Far more, the 1966 photograph had
captured for the first time on film, the essential structure of the entire
corona. Additional exposures were made through polarizing filters for
quantitative analyses of the corona. But the unpolarized, radially-graded
picture was worth a thousand words, and more, for it revealed associations
that had never before been seen; oft reprinted, it guided a fresh understanding
of the solar corona as the product of magnetic forces and the outward flow of
the solar wind. It made the field lines that hold the corona together at once
visible throughout the coronal form, giving to coronal physics what x-rays had
once given to medicine. It was bad news, however, for romantics, for it
bettered what could be seen by eye or through any simple telescope. After

105 years of repeated attempts, photography had finally gained the upper hand in recording the coronal form.

Newkirk took his coronal camera, with results equally fine, to Oaxaca in Mexico for the 1970 eclipse (24), and, thwarted by clouds, to the Gaspe Peninsula for the eclipse of 1972. It was his last expedition, for by then, as observatory director he felt that he could go no more. Since that time, the Newkirk camera has been taken, with always good results, to total eclipses in Africa, Australia, India, Siberia, Java, and in 1988, to the Philippines, compiling an unequalled record of the changing corona through a complete Hale cycle of 22 years.

2.3.2   INTERPRETIVE STUDIES

Newkirk used the 1966 eclipse photograph as an aid for a seminal review of coronal structure that was published the following year (25). This led in turn to papers with Altschuler, Harvey, Howard and others on the relationship between magnetic fields and the structure of the corona (26-35). These demonstrated a remarkable correspondence between calculated force-free fieldlines derived from photospheric magnetograms and the white-light structure of the corona revealed in Newkirk's remarkable eclipse photographs.

In intervening years, while he was almost fully occupied as director of HAO Newkirk become heavily involved in studies of solar activity (36,37)--- coincident with and following his term as president of Commission 10 of the IAU. This included directing a planning project for NASA in 1980 on the study of the solar cycle from space (38), a study in the same year of the faint early sun paradox and of solar activity on time scales of 10^5 yrs and longer (39), an exhaustive review in 1983 of solar luminosity variations (40), and a venture in 1984 into the use of ice-borne ^{10}Be as a tracer of solar activity in the past (41), in connection with a sabbatical year in Zürich.

The last of these brought home to Newkirk that much more needed to be known about the mechanism of the solar modulation of the incoming galactic rays that produced ^{10}Be and ^{14}C in the atmosphere. This led him, in turn, to a fourth fundamental problem regarding the extended corona of the sun.

2.4   PROPAGATION OF GALACTIC COSMIC RAYS IN THE CORONA

Newkirk had made earlier, preliminary studies of this problem with Wentzel (41), with Hundhausen and Pizzo (42), and with Lockwood (43). These inspired him to tackle, alone, the fundaments of the problem, in his typical, methodical manner --- first collecting all relevant data, as he had done 30 years before when he took up the backlog of early K-coronameter measurements. This time he would harness the power of computers to trace the paths of incoming particles through the corona. As his health began to fail, he attempted to work the data from a remote terminal in his home.

Life did not give Gordon time to complete the study, though we should wish that it had --- for without a doubt he would have followed this tangled, still-unresolved problem through to its very end, as was his way with all he ever touched.

3.   References

1.   Newkirk, G.A. Jr. (1959) 'A model of the electron corona with reference to radio observations', in R.N. Bracewell, ed., Paris Symposium on Radio Astronomy, IAU Symposium No.9 and URSI Symposium No.1, Stanford Univ.Press, Palo Alto, pp.149-158.

2.   Wlerick, G. and Axtell, J. (1957) 'A new instrument for observing the electron corona', Ap.J., 126, 253-258.

3.   Newkirk, G. Jr. (1961) 'The solar corona in active regions and the thermal origin of the slowly varying component of solar radio radiation', Ap.J., 133, 983-1013.

4.   Lyot, B. (1930) 'La couronne solaire etudiee en dehors des eclipses', C.R.Acad.Sci.Paris, 191, 834-837.

5.   Lyot, B. (1939) 'A study of the solar corona and prominences without eclipses', M.N.Roy.Astron.Soc., 99, 580-595.

6.   Newkirk, G.A. Jr. (1956) 'Photometry of the solar aureole', J.Opt.Soc.Am.,46, 1028-1037.

7.   Evans, J.W. (1948) 'A photometer for measurement of sky brightness near the sun', J.Opt.Soc.Am., 38, 1083-1085.

8.   Newkirk, G. Jr. (1967) 'The optical environment of manned spacecraft', Planet.Space Sci., 15, 1267-1285.

9.   Schwarzschild, M. (1959) 'Photographs of solar granulation taken from the stratosphere', Ap.J., 130, 345-363.

10.   Newkirk, G. A. and Eddy, J.A. (1962) 'Daytime sky radiance from forty to eighty thousand feet', Nature, 194, 638-641.

11.   Newkirk, G. Jr. and Eddy, J.A. (1962) 'A coronagraph above the atmosphere', Sky and Telescope, 24, 77-81.

12.   Newkirk, G.A. Jr. and Eddy, J.A. (1963) 'Influx of meteoric particles in the upper atmosphere of earth as determined from stratospheric coronagraph observations', in W. Priester, ed., Space Research III, North Holland Pub. Co., Amsterdam, pp.143-154.

13.   Newkirk, G. Jr. and Eddy, J.A. (1964) 'Light scattering by particles in the upper atmosphere', J.Atmos.Sci., 21, 35-60.

14.   Newkirk, G. Jr. and Kroening, J. (1965) 'Aerosols in the stratosphere: A comparison of techniques of estimating their concentration', J.Atm.Sci., 22, 567-570.

510

15. Newkirk, G. Jr. (1967) 'Meteoric dust in the stratosphere as determined by optical scattering techniques', Smithsonian Contributions to Astrophysics, 11, The Proceedings of a Symposium on Meteor Orbits and Dust, NASA, Washington D.C. and the Smithsonian Institution, 349-358.

16. Newkirk, G. Jr. and Zirin, H. (1963) 'Feasibility of a reflecting coronagraph', Applied Optics, 2, 977-978.

17. Newkirk, G. Jr. and Bohlin, D. (1963) 'Reduction of scattered light in the coronagraph', Applied Optics, 2, 131-140.

18. Newkirk, G. Jr. and Bohlin, D. (1964) 'Scattered light in an externally occulted coronagraph', Applied Optics, 3, 543-544.

19. Newkirk, G. Jr. and Bohlin, J.D. (1964) 'The first flight of Coronascope II', Sky and Telescope, 28, 16-19.

20. Newkirk, G.A. Jr. and Bohlin, D. (1965) 'Coronascope II observation of the white light corona from a stratospheric balloon', Ann.d'Astrophys., 28, 234-238.

21. MacQueen, R.M, Gosling, J.T., Hildner, E., Munro, R.H., Newkirk, G. Jr., Poland, A,I., and Ross, C.L. (1974) 'The High Altitude Observatory white light coronagraph', Proceedings of the Society of Photo-Optical Instrumentation Engineers, Tucson, 44.

22. MacQueen, R.M, Eddy, J.A., Gosling, J.T., Hildner, E., Munro, R.H., Newkirk, G. Jr., Poland, A.I, and Ross, C.L. (1974) 'Outer solar corona as observed from Skylab', Ap.J., 187, L85-88.

23. Eddy, J.A. (1971) 'The Schaeberle 40-ft eclipse camera of Lick Observatory', Journal for the History of Astronomy, 2, 1-22.

24. Newkirk, G. Jr. and Lacey, L. (1970) 'The corona during the March 7 1970, eclipse,' Nature, 226, 1098 only.

25. Newkirk, G. Jr. (1967) 'Structure of the solar corona,' Ann.Rev.Astron.Ap., 5, 213-266.

26. Newkirk, G., Altschuler, M.D. and Harvey, J. (1968) 'Influence of magnetic fields on the structure of the solar corona', in K. O. Kiepenheuer, ed., Structure and Development of Solar Active Regions, IAU Symposium 35, D. Reidel, Dordrecht, pp.379-383.

27. Altschuler, M.D. and Newkirk, G., Jr. (1969) 'Magnetic fields and the structure of the solar corona I: Methods of calculating coronal fields', Solar Phys., 9, 131-149.

28. Newkirk, G. Jr., Dupree, R.G. and Schmahl, E.J. (1970) 'Magnetic fields and the structure of the solar corona II. Observations of the 12 November 1966 solar corona', Solar Phys., 15, 15-39.

511

29. Newkirk, G. Jr. (1971) 'Large scale magnetic fields and their consequences', in R.Howard, ed., Solar Magnetic Fields, IAU Symposium 43, D. Reidel, Dordrecht, pp.547-568.

30. Newkirk, G. Jr. (1971) 'Coronal magnetic fields', in C.J. Macris, ed., Physics of the Solar Corona, D. Reidel, Dordrecht, 66-87.

31. Trotter, D.E. and Newkirk, G. Jr. (1971) 'Coronal magnetic field of the sun on 7 January 1969', Solar Phys., 20, 372-374.

32. Altschuler, M.D., Newkirk, G. Jr., Trotter, D.E. and Howard, R. (1971) 'Time evolution of the large-scale solar magnetic field', in R.Howard, ed., Solar Magnetic Fields, IAU Symposium 43, D. Reidel, Dordrecht, pp.588-594.

33. Newkirk, G. Jr. (1972) 'Coronal magnetic fields and the solar wind', in C.P. Sonnet, ed., Solar Wind, Proceedings of the Asilomar Conference on the Solar Wind, , NASA SP-308, Washington, pp.11-29.

34. Altschuler, M.D., Trotter, D.E. and Newkirk, G. Jr. (1974) 'The large-scale solar magnetic field', Solar Phys., 39, 3-17.

35. Altschuler, M.D., Trotter, D.E. and Newkirk, G. Jr. (1975) 'Tabulation of the harmonic coefficients of the solar magnetic fields', Solar Phys., 41, 225-226.

36. Newkirk, G. Jr. and Frazier, K. (1982) 'The solar cycle', Phys.Today, 35, 25-34.

37. Newkirk, G. Jr. (1982) 'The nature of solar variability', in J. A. Eddy, ed., Sun, Weather and Climate, National Research Council, National Academy of Sciences, pp.33-47.

38. Newkirk, G. Jr. (1980) 'Study of the Solar Cycle from Space: Solar Cycle and Dynamics Mission', Final Report, NASA SCADM #3, Washington.

39. Newkirk, G. Jr. (1980) 'Solar variability on time scales of 10^5 years to $10^{9.6}$ years', in R. O. Pepin, J. A. Eddy, and R. B. Merrill, eds., The Ancient Sun, Pergamon, pp.293-320.

40. Newkirk, G. Jr. (1983) 'Variations in solar luminosity', Ann.Rev.Astron. and Ap., 21, 429-467.

41. Newkirk, G. Jr. (1985) 'What accelerator mass spectroscopy can do for solar physics', in Nuclear Instrumentation and Methods, Third Internatl. Symposium on Accelerator Mass Spectroscopy, Zurich.

42. Newkirk, G. Jr. and Wentzel, D. (1978) 'Rigidity-independent propagation of cosmic rays in the solar corona', J.Geophys.Res., 83, 2009-2015.

512

43.   Newkirk, G. Jr., Hundhausen, A.J. and Pizzo, V. (1981) 'Solar cycle
       modulation of galactic cosmic rays: speculation on the role of coronal
       transients', J.Geophys.Res., 86, 5387-5396.

44.   Newkirk, G. Jr. and Lockwood, J.A. (1981) 'Cosmic ray gradients in the
       heliosphere and particle drifts', Geophys.Res.Lett., 8, 619-622.

Structure of the Solar Corona

Takashi Sakurai and Eijiro Hiei
Solar Physics Division
National Astronomical Observatory
Mitaka, Tokyo 181
Japan

ABSTRACT. Our recent understanding of the structure of the solar corona
is reviewed, with an emphasis on the ground-based observations.

1. INTRODUCTION

Our knowledge on the structure of the solar corona has dramatically
increased with the advent of X-ray observations from space, especially
by the Skylab experiment. The solar corona seen in X-rays clearly demonstrates the magnetic structuring. The magnetic field provides the heating mechanisms, and the magnetic confinement of the heated plasma
creates the closed magnetic structures (loops or arcades) in the corona.
The open magnetic field lines which are connected to the interplanetary
space lead to the coronal streamers and the coronal holes.

Unfortunately, X-ray observations have been carried out only inter-
mittently. On the other hand, ground-based observations provide more
continuous coverage of data. For example the intensities of the K-corona
and the coronal emission lines have been monitored by using coronagraphs
for many years. Spectroheliograms of He I 10830Å line and radio maps
also give us information on the coronal structures. This review mostly
discusses the recent results on the structure of the solar corona by
means of ground-based observations.

2. SOLAR CYCLE VARIATIONS

A 'Butterfly Diagram' can be constructed based on the coronal green
line intensities monitored for several solar cycles (Hiei and Okamoto,
1988). In a broad sense the coronal Butterfly diagram resembles the
sunspot Butterfly diagram. As was pointed out by Waldmeier(1964), however, coronal Butterfly diagram contains poleward migrating branches in
the high latitude zones. This feature is also inferred from the migration of polar crown prominences. Leroy and Noens (1983) plotted the
standard deviation of yearly green line intensities and revealed the
existence of high-latitude branches unambiguously. Coronal emission in

513

D. McNally (ed.), Highlights of Astronomy, Vol. 8, 513–516.
© *1989 by the IAU.*

the high latitude zones will come from the streamers which extend above
the polar-crown prominence belts. Therefore the poleward migration of
these branches reflects the dynamo process of reversing the polar magnetic field of the Sun.

Another important aspect of the solar cycle variation is the population of X-ray bright points (XBPs), which may correspond to small
concentrations of magnetic fields. Davis (1983) summarizes the XBP
number count available to date. Strikingly, there are more XBPs near
activity minimum than near activity maximum. Based on this, Golub et al.
(1979) conjectured that the decrease in the sunspot magnetic flux near
the activity minimum might be compensated for by the XBP-associated
magnetic flux. However this argument does not hold quantitatively. Total
magnetic flux as measured at Mt. Wilson (Howard and LaBonte, 1981) shows
that there is more magnetic flux on the Sun in the activity maximum.

3. MAGNETIC FIELD CONFIGURATION IN THE CORONA

The magnetic field in the corona is hard to observe directly. Therefore
the magnetic field configuration in the corona is usually inferred by
extrapolating the magnetic field measured at the photospheric level. The
simplest assumption adopted is that there is no electric current in the
corona (viz. current-free modeling). The correspondence between the
calculated current-free field lines and the coronal loop structures is
generally satisfactory (Sakurai and Uchida, 1977).

The rotation of the Sun makes the coronal structures to be seen
from various directions. Therefore the three-dimensional shape of stable
coronal structures can be reconstructed stereoscopically. Berton and
Sakurai (1985) determined the three-dimensional shape of coronal loops
by using this technique.

For more global magnetic structures extending beyond 1 R☉ or so,
the effect of the solar wind on the magnetic field becomes important.
Such effect is approximately taken into account in the so-called source-
surface modeling. This model revealed the correspondence between the
open magnetic fields and the coronal hole regions (Levine et al.,1977).

4. ROTATION OF CORONAL STRUCTURES

Daily record of coronal intensities observed by coronagraphs can be used
to deduce the rotation speed of coronal structures. The differential
rotation thus obtained is flatter (i.e. close to rigid rotation) than
the sunspot differential rotation law (Antonucci and Dodero, 1979;
Parker et al., 1982; Fisher and Sime, 1984). Coronal holes show almost
rigid rotation (Bohlin, 1977).

Hoeksema and Scherrer (1987) showed that the coronal magnetic field
expected from the observed photospheric magnetic field by means of the
source-surface modeling exhibits nearly rigid rotation. Wang et al.
(1988), following the evolutional modeling of surface magnetic field
initiated by Sheeley et al. (1985), also found the same results. That
is, global configuration of the corona reflects the large-scale

components of the surface magnetic field, which rotate nearly rigidly. The reason why the large scale magnetic fields rotate rigidly remains yet to be explained.

5. CORONAL HOLES

In considering the influence of the solar wind on the environment of the earth's magnetosphere, it is important to constantly monitor the coronal holes. This has been done by using the infrared line of Helium at 10830Å. The energy levels involved in this transition will populate according to the XUV radiation from the corona. Therefore active regions show stronger absorption and coronal holes show less absorption. X-ray images and He 10830Å images are positive and negative photographs of the corona, so to speak.

The other method is to make use of radio observations. Generally the coronal holes appear as depressions in the radio maps. Kundu et al. (1987) found that the position of a coronal hole shifts eastward as they observe it in lower frequencies (i.e. higher in the corona). This might suggest the Archimedian spiral of magnetic field lines bending toward east, but the magnitude of the shift is too large as the authors pointed out. Kosugi et al.(1986) found that the polar coronal holes observed at millimeter wavelengths are brighter than the surroundings. This may be interpreted as due to the existence of polar faculae.

6. X-RAY BRIGHT POINTS, MAGNETIC BIPOLES, AND HELIUM DARK POINTS

XBPs are visible in He 10830Å spectroheliograms as dark points (HDPs). As is expected from the mechanism of providing He 10830Å absorption by XUV radiation, the correspondence between XBPs and HDPs is generally good. Golub et al.(1977) claimed that XBPs are formed above ephemeral active regions (ERs), viz. small and short-lived bipolar magnetic fields. However ERs are more numerous near the activity maximum (Harvey, 1985), which contradicts the solar cycle variation of XBPs.

Harvey (1985) showed that about one-third of HDPs are formed above ERs(=small bipolar regions), but the rest of dark points are found where opposite magnetic polarities encounter by chance. Therefore two-thirds of HDPs do not correspond to 'genuine' magnetic bipoles. This latter component may be more numerous near the activity minimum in which a more mixed appearance of magnetic fields is seen.

Kundu et al. (1988) observed bright points in the microwave range (MBPs) and found that some bright points are found above networks where the magnetic field is presumably unipolar. Such bright points may correspond to clumps of network magnetic fields. Note that microwave observations see the chromosphere and the transition region. Network magnetic fields will become diffuse in the corona so that no X-ray conterpart of unipolar microwave bright points will be seen. On the other hand HDPs may form due to XUV radiation from the corona as well as from the transition region. Therefore the possibility is that not all HDPs are XBPs and network bright points also show up as HDPs. The expected relation

516

among several features could be as in Figure 1.

7. OUTLOOK

Because of the lack of persistent X-ray observations from space, recent observational progresses concering the solar coronal phenomena have mostly been brought by the ground-based observations. We hope that the soft X-ray observations in the 1990's (Solar-A project of Japan-USA-UK, SXI project of NOAA/USA) will give us a big step toward better understanding of coronal physics.

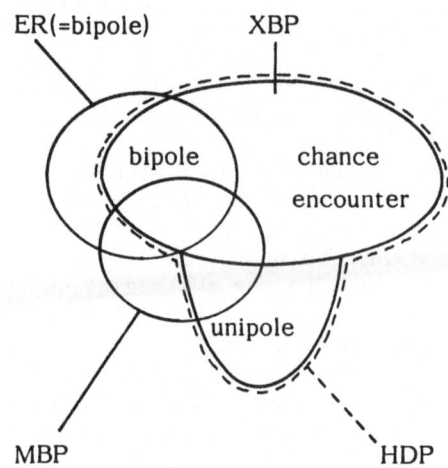

Fig.1. XBP, HDP, MBP and ER.

REFERENCES

Antonucci, E. and Dodero, M.A.:1979, Solar Phys., **62**, 107.
Berton, R. and Sakurai, T.:1985, Solar Phys., **96**, 93.
Bohlin, J.D.:1977, Solar Phys., **51**, 373.
Davis, J.M.:1983, Solar Phys., **88**, 337.
Fisher, R. and Sime, D.G.:1984, Astrophys. J., **287**, 959.
Golub, L., Krieger, A.S., Harvey, J.W. and Vaiana, G.S.:1977, Solar Phys., **53**, 111.
Golub, L., Davis, J.M. and Krieger, A.S.:1979, Astrophys. J., **229**, L145.
Harvey, K.L.:1985, Aust. J. Phys., **38**, 875.
Hiei, E. and Okamoto, T.:1988, in 'Solar and Stellar Coronal Structure and Dynamics', ed. R.C.Altrock.
Hoeksema, J.T. and Scherrer, P.H.:1987, Astrophys. J., **318**, 428.
Howard, R. and LaBonte, B.J.:1981, Solar Phys., **74**, 131.
Kosugi, T., Ishiguro, M. and Shibasaki, K.:1986, Publ. Astron. Soc. Japan, **38**, 1.
Kundu, M.R., Gergely, T.E., Schmahl, E.J., Szabo, A., Loiacono, R., Wang, Z. and Howard, R.A.:1987, Solar Phys., **108**, 113.
Kundu, M.R., Schmahl, E.J. and Fu, Q.-J.:1988, Astrophys. J., **325**, 905.
Leroy, J.-L. and Noens, J.-C.:1983, Astron. Astrophys., **120**, L1.
Levine, R.H., Altschuler, M.D. and Harvey, J.W.:1977, J. Geophys. Res., **82**, 1061.
Parker, G.D, Hansen, R.T. and Hansen, S.F.:1982, Solar Phys., **80**, 185.
Sakurai, T. and Uchida, Y.:1977, Solar Phys., **52**, 397.
Sheeley, N.R., Jr., DeVore, C.R. and Boris, J.P.:1985, Solar Phys., **98**, 219.
Waldmeier, M.:1964, Z. Astrophys., **59**, 205.
Wang, Y.-M., Sheeley, N.R., Jr., Nash, A.G. and Shampine, L.R.:1988, Astrophys. J., **327**, 427.

CORONAL HEATING: THEORETICAL IDEAS

Joseph V. Hollweg
Space Science Center
Institute for the Study of Earth, Oceans, and Space
 and Department of Physics
University of New Hampshire
Durham, NH 03824 USA

Introduction

After four decades of study, the mechanisms for heating the
corona are not understood. However, the development of the field is
vigorous. A variety of new ideas have been proposed, and these new
ideas have generated lively debate. The goal of this short review is
to present a broad overview of the ideas currently under consideration.

It is now accepted that the solar magnetic field is the key
ingredient. It is not known whether waves are important or not, but
it is agreed that the magnetic (and possibly associated kinetic)
energy must eventually appear in thin structures so that the (generally weak) dissipative processes can heat the plasma. This poses a
problem of how the heat gets distributed throughout the corona.

The coronal heating requirements range from about 3×10^5 erg
$cm^{-2} s^{-1}$ in the quiet corona to some 10^7 in coronal active region
loops. From observations of the Doppler widths of spectral lines we
have some knowledge of the "turbulent" kinetic energy in the corona
(rms line-of-sight turbulent velocities are about 20-40 km s^{-1}). If
this kinetic energy were associated with sound waves, the wave energy
flux density would be much too small. This is one reason for
invoking the magnetic field. The other reason is from inspection of
the remarkable x-ray images of the corona, which clearly show loops
of plasma confined by the field. However, the corona outside of the
loops also requires substantial energy. It is probably a mistake to
isolate the coronal loop heating problem as many workers have tended
to do.

D. McNally (ed.), Highlights of Astronomy, Vol. 8, 517–520.
© *1989 by the IAU.*

Observational constraints are very limited. There is general agreement[1] that the heating is positively correlated with the magnetic field strength, B. And Parker[2] has emphasized that the surface brightness of the x-ray loops is roughly independent of loop size, implying an inverse correlation of heating with loop length, L.

Stressed Magnetic Fields

There are difficulties with wave theories. One is that the waves must damp efficiently in the corona, e-folding in a few periods. This is a difficult (but not impossible) demand. On short loops, the transit time for an Alfvén wave can be less than 10s. Unless there is substantial wave power at such short periods, it makes more sense to think of the field lines as simply being slowly displaced by the footpoint motions. If one footpoint moves with velocity V_h, the magnetic energy along the loop grows as $d(ME)/dt = B^2 V_h^2 t/4\pi L$ ergs cm^{-2} s^{-1}. The inverse dependence on L is qualitatively desired. Parker has argued that random motions will inevitably produce coronal current sheets which lead to reconnection. He further argues that a steady-state will be achieved when t = (flux tube diameter)/(velocity of reconnection). Thus $d(ME)/dt = (B^2/4\pi) (V_h^2/v_{rec}) (d/L)$. If we require $d(ME)/dt = 10^7$ erg cm^{-2} s^{-1}, and take $d/L = 0.1$, B = 50G, and $V_h = 0.4$ km s^{-1}, we find a steady-state after 90 hrs, when $V_h t = 0.2 r_o$. We also require $(v_{rec}/v_{Alfvén}) = 10^{-5}$. Observationally, it is not known if these numbers are reasonable.

Parker's ideas have stirred up a controversy. He, B.C. Low, and others have argued that current sheets are inevitable. A. van Ballegooijen[4], and S. Antiochos[5] have argued that current sheets are not inevitable[4], but[5] are associated with surfaces of discontinuous magnetic connectivity or magnetic null points.

Van Ballegooijen[6] has produced a statistical cascade model in which random walks at the footpoints produce a power spectrum for the current density which increases exponentially with time at large wave numbers. Eventually, joule dissipation at high wavenumbers gives a steady-state. He then obtains $d(ME)/dt \sim 10^5$ erg cm^{-2} s^{-1} which is too small by two orders of magnitude.

Sturrock and Uchida[7] considered twists produced on magnetic flux tubes by random walks of the footpoints; observational evidence for or against twists is needed. They obtain $d(ME)/dt = (B B^*/4\pi) V_h^2 \tau_{corr}/L$, where B^* (~ 1500 G) is the photospheric field strength. If $B = 50G$, $L = 10^5$ km, $V_h = 1$ km s^{-1} and $\tau_{corr} = 300$s, then $d(ME)/dt = 1.8 \times 10^6$ erg cm^{-2} s^{-1}. This is closer to the required value of 10^7, but still too small. A peculiar feature of this model is that there is no need to specify the dissipation mechanism.

Heyvaerts and Priest[8] have produced a sophisticated model in which magnetic stresses repeatedly release energy subject to the conservation of magnetic helicity, leading to a constant-α force-free state after each relaxation. In a poster at this meeting, P.K. Browning applied these ideas to a cylindrical loop of radius R being twisted at a velocity v. She obtained a steady-state heating $d(ME)/dt = (8/75)(B^2 v/4\pi)(R/L)(\tau_{rel}/\tau_v)$ where τ_{rel} is a relaxation time and τ_v is a time scale for the twists; generally $\tau_{rel}/\tau_v \lesssim 1$. This expression is basically a measure of the Poynting flux into a loop with $B_\theta/B \sim \tau_{rel}/\tau_v$. If $B_6 = 50G$, $v_2 = 30$ km s^{-1}, R/L = 0.1, and $\tau_{rel}/\tau_v = 1$, we obtain 6.4×10^6 erg cm^{-2} s^{-1}, which is close to the observed values.

These models have their successes and failures. Except for the Browning calculation, the heating rates tend to be rather low. It is not clear what role they play in the chromosphere. It is generally agreed that these models fail on open magnetic field lines (e.g. coronal holes) where the stresses propagate away as Alfven waves.

These models do not include dynamics, although reconnection must certainly bring dynamics into play[9]. Numerical simulations will probably help in this regard. We should also mention a recent simulation by Mikic, Schnack, and van Hoven[10] which showed the rms current density increasing exponentially with time, in qualitative agreement with van Ballegooijen's model. However, the simulation did not critically test other aspects of the model.

A strong plus for these models is evidence that some heating occurs impulsively, as might be expected if the reconnection leads to "microflares". Porter and Moore[11] observed microflares in CIV occuring in low-lying (2-4000 km high) loops. They estimate that at any time there are 10^4 events of 10^{26} erg on the Sun. If each lasts 30 sec the global average is 5×10^5 erg cm^{-2} s^{-1}. It is not yet clear if this number would approach 10^7 if only the active regions were considered. It is also not known how the energy gets to higher heights; perhaps the impulses launch waves. (We also caution the reader that these events are not guaranteed to be due to reconnection; short-lived dynamics events could also be waves or shocks.)

Wave Theories

Wave theories may seem old-fashioned, but there are still reasons for considering them. The most compelling reason is that Alfven waves have been observed in the solar wind. There are other reasons as well. It is generally agreed that waves are needed to heat the open coronal field lines. We know that there are "turbulent" motions in the corona; if the observed motions are assumed to propagate at the Alfven speed, then the energy fluxes are adequate to heat the corona. There are some hints of wave-excited resonances on active region loops[12]. There are other dissipation mechanisms besides electrical resistivity and reconnection. And

520

because the waves propagate through the photospheric flux tubes and chromosphere, they can be expected to heat those regions as well.

One wave damping mechanism of current interest is resonance absorption either of surface waves[13,14] or of waves propagating toward a smooth boundary. For a coronal surface wave, the damping is[14] $\gamma/\omega = \pi k a (\rho_2 - \rho_1)/8(\rho_2 + \rho_1)$, where 'a' is the surface thickness. For typical numbers, the wave can e-fold in a few periods as required. It has also been pointed out[14] that resonance absorption can give a surface brightness independent of loop length. Resonance absorption dumps the surface wave energy into a thin layer, but that in turn could lead to turbulence which could actually distribute the energy over a substantial fraction of a loop's diameter[14].

Turbulence is also a possible damping mechanism for waves. This is poorly understood theoretically, but it does seem to be happening in the solar wind, where we see waves but with a Kolmogorov turbulent $(k^{-5/3})$ power spectrum. The observed wave damping corresponds to the turbulent heating rate $\rho < \delta v^2 >^{3/2}/\ell_{corr}$, where ℓ_{corr} is the correlation length. It turns out that this heating rate "works" almost everywhere in the solar atmosphere, but it fails in coronal holes where the turbulent heating produces a large proton temperature peak near $3r_o$, if it is assumed that the energy eventually heats the protons.

References

1. Golub, L., Maxson, C., Rosner, R., Serio, S., and Viana, G.S.: 1980, Ap.J., 238, pp. 343-348.
2. Parker, E.N.: 1986, in Coronal and Prominence Plasmas, NASA-CP-2442, pp. 9-17.
3. Parker, E.N.: 1983, Ap.J., 264, pp. 642-647.
4. van Ballegooijen, A.A.: 1985, Ap.J., 298, pp. 421-430.
5. Antiochos, S.K.: 1988, Bull. Amer.Astron. Soc., 20, p. 681.
6. van Ballegooijen, A.A.: 1986, Ap. J., 311, pp. 1001-1014.
7. Sturrock, P.A., and Uchida, Y.: 1981, Ap.J., 246, pp. 331-336.
8. Heyvaerts, J., and Priest, E.R.: 1984, Astron. Astrophys, 137, p. 63.
9. Dahlburg, R.B., Dahlburg, J.P., and Mariska, J.T.: 1988, Astron. Astrophys., in press.
10. Mikic, Z., Schnack, D.D., and van Hoven, G.: 1988, Ap.J., submitted.
11. Porter, J.G., and Moore, R.L.: 1987, NASA-Marshall preprint 87-146.
12. Koutchmy, S., Zugzda, Y.D., and Locans, V.: 1983, Astron. Astrophys., 120, p. 185.
13. Davila, J.M.: 1987, Ap.J., 317, pp 514-521.
14. Hollweg, J.V., and Yang, G.: 1988, J. Geophys. Res., 93, pp. 5423-5436.
15. Hollweg, J.V., and Johnson, W.: 1988, J. Geophys. Res., in press.

AN UPDATE ON X-RAY EMISSION FROM STARS

R. ROSNER
Enrico Fermi Institute and Dept. of Astronomy and Astrophysics
The University of Chicago
5640 South Ellis Avenue
Chicago, IL 60637

ABSTRACT. With the closing of the *Einstein* and *EXOSAT* eras, the characteristics of stellar x-ray emission are now fairly well understood. In this very brief review, I will focus on two specific topics central to the physics underlying stellar activity: the rotation-activity connection and the "decay" of activity at the low mass end of the main sequence.

1. Introduction

Given the brevity of space for this report, I have chosen to focus on two topics which I think well-illustrate both the breadth of results now available for stellar activity studies and the depth of analysis now possible: the nature of the activity-rotation connection, and the nature of activity at the low-mass end of the main sequence. Both topics were recognized early on to be central to our understanding of how stellar activity comes about; and the recent results which I shall discuss provide significant information on this central problem of stellar activity.

Of necessity, a substantial number of interesting topics will not be discussed here: the novel high time resolution radio observations of stellar activity reported by J. Bookbinder, G. Dulk, and collaborators; the new correlation studies of x-ray luminosity and UV flux reported by K. Schrijver and collaborators; the novel theoretical modeling of hot and cool "loops" by S. Antiochos and G. Noci; and the remarkable "stellar"-like flare observed by J. Schmitt, F.R. Harnden, and H. Fink by using *Einstein* data of florescent emission from the solar flare-illuminated terrestrial atmosphere. This omission is regrettable, but unavoidable.

2. Is Rotation *the* Determinant of Stellar X-ray Emission?

In a recent preprint, Fleming, Gioia, and Maccacaro (1988) have used the Extended Medium Sensitivity Survey (EMSS) from *Einstein* to construct an x-ray-selected sample of 128 F-M dwarf stars, and have combined this sample with optical observations (including spectroscopy) to establish the relations between spectral type, rotation rate, x-ray luminosity, etc. among these sample stars.

D. McNally (ed.), Highlights of Astronomy, Vol. 8, 521–524.

The key result is shown in their Figure 1, which seems to indicate that unlike the classical result of Pallavicini et al. (1981), stellar x-ray luminosity L_x for this sample seems to scale like $(v_e \sin i)^1$, rather than $(v_e \sin i)^2$, where v_e is the equatorial stellar rotation rate and i is the inclination angle of the stellar rotation axis to the line-of-sight. The question of exactly how L_x scales with the rotation rate Ω has of course been a rather hotly debated topic since the early results of Pallavicini et al. and Walter (1982); and it is therefore crucial to understand the present result, which emerges from the first x-ray-selected sample constructed to date for this purpose.

To begin with, we must note that since an x-ray-selected sample of the kind examined here by its very nature preferentially picks out the x-ray-bright sources, one way of reconciling the apparently conflicting results is if the intrisically brighter sources (e.g., the ones in the present sample) are simply showing the effect of stellar activity "saturation"; that is, it may be that at the upper end of the stellar luminosity function, the dependence of activity on rotation rate is weaker than it is for less-active stars (cf. Vilhu and Walter 1987). Can we determine whether this simple interpretation is correct?

To answer this question, Fleming et al. considered the correlations between L_x, Ω, and stellar radius R; and found that (1) L_x is uncorrelated with $\Omega \sin i$, while (2) the x-ray luminosity instead scales with R^2. These latter results seem to deepen the mystery: it appears that for this well-determined sample of late-type dwarf stars, it is the stellar radius (or surface area) which sets the x-ray emission level, not the stellar rotation rate. Is there a simple explanation for the discrepancy between this result and earlier studies?

The following argument seems to me to resolve the difficulty. Since the sample we are considering is x-ray-selected, and indeed represents the upper tail of the L_x distribution, one might expect that the stars in this sample have mean surface fluxes which are comparable to those in solar active regions; this is in fact borne out by comparing L_x of stars in this sample with the expected L_x if these stars were fully covered by solar-like active regions (cf. Vaiana and Rosner 1978). In that case, we expect that the x-ray luminosity of stars in this sample can be "predicted" simply by determining the stellar surface area for each star, and multiplying this area by the mean solar active region x-ray flux; that is, we would expect a strong correlation between L_x and the square of the stellar radius for stars in this sample. This is exactly what is found by Fleming et al.

However, recall that $v_e = \Omega R$, and that Fleming et al. found L_x to be uncorrelated with $\Omega \sin i$ for this sample; the implication is then immediate that $L_x \approx v_e^2$, contrary to what is in fact found — Fleming et al. find instead a linear dependence of L_x on v_e. Is there then an internal inconsistency in the data analysis? I believe not. Careful reading of Fleming et al.'s analysis shows that the power law fit connecting L_x to $v_e \sin i$ is obtained by simply assuming that the upper limits on rotation rates $v_e \sin i < 10$ km/s can be regarded as "true detections", with fractionally-large errors, and applying a least-squares fit to all of the data. The effect is to overweight (with respect to the upper bounds) the few "detections" at large rotation rates and relatively low luminosities, and thus to underestimate the exponent in the power law. This can be avoided by instead applying the by-now standard detection-and-bounds correlation analysis (cf. Schmitt et al. 1985). Indeed, if one uses the data presented in Table 1 of Fleming et al. to compute the power law index, one finds that within the statistical errors of the data, the power law index connecting v_e to L_x for this sample cannot be distinguished from 2! This is remarkable: the exponent is 2 despite the fact that L_x for this sample is demonstrably *uncorrelated* with stellar rotation. I suspect that if the $L_x \approx v_e^2$ relation had been found first, then there would have been little reason to probe further, since this result would have confirmed the "well-known" result that activity and rotation are correlated in precisely this manner; it is to Fleming *et al.*'s credit that they did in fact probe further, and now provide the first good evidence for saturation of stellar coronal activity.

A rather different perspective on the activity-rotation problem is provided by the recent completion of the *Einstein* survey of late-type giants and supergiants (Maggio *et al.* 1988). The total sample encompassed 380 stars, and was used in the first instance to resolve the problem of locating the "dividing line" separating coronally-active from inactive evolved stars in the H-R diagram. Having confirmed earlier work placing this boundary at ≈ spectral type K, Maggio *et al.* proceeded to consider the x-ray luminosity-rotation rate correlation for this stellar sample; it is of some considerable surprise that the sought-for correlation is remarkable weak.

The reasons for this surprise are of course the fact that for late-type dwarf stars, x-ray luminosity and rotation are well-correlated; and that there is a rather commonly-accepted and intuitively-appealing picture of how this correlation might come about. However, the data for giants serve to remind us that in some sense the data for dwarf stars are too "neat". That is, the common explanation for the activity-rotation connection is based on the idea that in standard $\alpha - \omega$ dynamo theory, the rate of magnetic flux production depends on the rate of differential rotation, and hence (for fixed convection zone depth, for example) should scale with the rotation rate. What is commonly forgotten is this latter caveat: namely, in order for this correlation to be evident, it must be the case that the depth of the region over which the differential rotation field acts either does not vary with stellar type, or varies randomly with stellar type. Thus, the x-ray data from evolved stars tell us quite directly that, given that we are wedded to standard $\alpha - \omega$ dynamo theory, the depth over which the ω-dynamo must act within the convection zone of these stars must be correlated with stellar type. This would seem to be a rather stringent constraint for dynamos for evolved stars; and is confirmed by Maggio *et al.*'s success in correlating L_x with the Rossby number (at least for F and early G giants).

3. The "Decay" of Stellar Activity at the Low-mass End of the Main Sequence

The possibility that the nature of stellar activity may change in some dramatic way for very low-mass stars has been the focus of detailed studies for a number of years, and plays an important role in subjects ranging from the galactic contribution to the diffuse soft x-ray background and stellar dynamo theory to the genesis of the "period gap" in cataclysmic variables (cf. Rosner, Golub and Vaiana 1985). Very recently, the *Einstein* Extended Medium Sensitivity Survey (EMSS), already alluded to above, has been used by Fleming, Liebert, Gioia, and Maccacaro (1988) to attack this problem as well.

Table 1. Number of predicted and observed stars as a function of detection threshold

	Volume [pc^3]	Predicted [3] # [Sp>M5]	Observed # [Sp>M5]
$\log <L_x>^{[1]} = 28.73$	3,870	91	0
$\log <L_x>^{[2]} = 27.98$	265	6	0

[1] Mean L_x for young disk population stars (Bookbinder 1985), dM only.
[2] Mean L_x for old disk population stars (Bookbinder 1985), dM only.
[3] Using stellar mass function of Reid (1987).

Fleming *et al.* obtain two basic results from their study of the EMSS x-ray selected sample (consisting of some 31 dM stars, all optically identified). First, they show that the sample as a whole has a tendency for $<L_x/L_{bol}>$ to be roughly constant; second, they pointed out that no dM stars later than M5 were either detected or identified. The above table (extracted from Fleming *et al.*) summarizes the second result quantitatively, and shows quite explicitly that the absence of dM stars later than M5 is not likely to be a statistical fluke. Since this sample is x-ray-selected, it is clearly the most explicit demonstration of an effect first noticed in the dM star analysis of Bookbinder (1985), namely that there simply are no x-ray-bright stars later than M5 (there are certainly x-ray-emitting stars later than M5, but *all* of these stars detected to date show very modest (solar-like or less) x-ray emission levels.

The fundamental reason for this deficit of x-ray-bright stars remains a mystery. Rosner and Vaiana (1980) some time ago pointed out that for stars later than roughly M5, the presumptive change in stellar structure as a result of convection dominating energy transport everywhere within the star ought to be reflected in the efficiency with which such stars produce magnetic fields, and hence in how active they were; this would be especially the case if the dominant dynamo process for main sequence stars later than M5 were a "shell" dynamo located at the interface between the outer convection zone and the radiative core. This expectation is apparently borne out by the x-ray data; but unfortunately, we are no closer to understanding exactly in what way this change in stellar structure has brought about a *decrease* in stellar magnetic field production efficiency. That is, magnetic field production has certainly not ceased for the very low mass stars (since we still observe coronal emission), so that it cannot be simply a matter of "shutting down" field production. Instead, one presumes that the basic dynamo process has been modified; this remains a problem for the future.

4. References

Bookbinder, J. (1985) PhD Thesis, Harvard University.

Fleming, T.A., Gioia, I.M., and Maccacaro, T. (1988) 'The Relation Between X-ray Emission and Rotation in Late-type Stars from the Perspective of X-ray Selection', *Astrophys. J.*, in press.

Fleming, T.A., Liebert, J., Gioia, I.M., and Maccacaro, T. (1988) 'M Dwarfs From the Einstein Extended Medium Sensitivity Survey', *Astrophys. J.*, in press.

Maggio, A., Vaiana, G.S., Haisch, B., Stern, R.A., Bookbinder, J., Harnden, Jr., F.R., and Rosner, R. (1988) *The Einstein Observatory Magnitude-limited X-ray Survey of Late-type Giant and Supergiant Stars'*, *Astrophys. J.*, in press.

Pallavicini, R., *et al.* (1981) 'Relations among Stellar X-ray Emission Observed from *Einstein*, Stellar Rotation and Bolometric Luminosity', *Astrophys. J.*, **248**, 279.

Reid, I.N. (1987) *M.N.R.A.S.*, **225**, 873.

Rosner, R., Golub., L., and Vaiana, G.S. (1985) 'On Stellar X-ray Emission', *Ann. Rev. Astron. Ap.*, **23**, 413-52.

Rosner, R. and Vaiana, G.S. (1980) 'Stellar Coronae from *Einstein*: Observations and Theory', in R. Giacconi and G. Setti (eds.), *X-ray Astronomy*, D. Reidel Publ., Dordrecht, pp. 129-51.

Schmitt, J.H.M.M., *et al.* (1985) *Astrophys. J.*, **290**, 307.

Walter, F.M. (1982) *Astrophys. J.*, **253**, 745.

Vaiana, G.S. and Rosner, R. (1978) 'Recent Advances in Coronal Physics', *Ann. Rev. Astron. Ap.*, **16**, 393-428.

Vilhu, O. and Walter, F.M. (1987) *Astrophys. J.*, **321**, 958.

SOLAR AND STELLAR WINDS

G. L. WITHBROE
Harvard-Smithsonian Center for Astrophysics
60 Garden Street
Cambridge, MA 02138 U.S.A.

ABSTRACT. This review discusses winds from stars with hot outer atmospheres, stars with coronae similar to the sun. It illustrates how solar observations can be used to test a theoretical model for solar-stellar winds, and thereby provide some insights concerning applications to studying winds from stars other than the sun.

1. Introduction

Because the sun is so close, its outer atmosphere can be studied by a wide variety of techniques not applicable to other stars. The coronal source region of the solar wind can be probed with optical and radio remote sensing techniques and the properties of the asymptotic solar wind can be sampled directly via in situ techniques. Because the solar wind outflow is not spherically symmetric, but originates from magnetically open regions of the solar surface which cover only a fraction of the surface, one can observe solar wind streams which originate in regions with different conditions. Data from different regions provide multiple constraints for testing theoretical wind models for stars with hot coronae.

A type of model which appears to be particularly useful for such stars is the radiative energy balance model, whose properties have been discussed in some detail by Hammer (1982a, 1982b). These models assume that the corona is heated by a mechanism in which the mechanical energy flux F_m responsible for the heating is dissipated with a characteristic dissipation length H_m. The model also assumes that the flux of energy carried by thermal conduction inward toward the stellar surface from the corona is dissipated by radiation in the low corona and chromospheric-coronal transition region. This assumption, along with the solar wind equations for conservation of mass, momentum, and energy, yield a model with two adjustable parameters for a star of known mass and radius. These parameters are F_m and H_m. For solar observations, it necessary to include the effects of non-spherical expansion and the effects of acceleration provided by Alfvén waves (see Holzer 1988, Leer 1988, Withbroe 1988a).

D. McNally (ed.), Highlights of Astronomy, Vol. 8, 525–528.
© 1989 by the IAU.

2. Solar Observations

In situ measurements of the solar wind have been acquired near the ecliptic from 0.3 AU to the outer limits of the solar system. They provide information on electron, proton, and alpha particle temperatures and densities, flow velocities, magnetic field strength and direction, and charge state distributions. There are two classes of steady-state wind, high speed and low speed solar wind, with typical speeds of, respectively, 700 km/s and 350 km/s. High speed wind comes from coronal holes, low density coronal regions with unipolar magnetic fields. Low speed wind appears to come from denser, magnetically complex regions. Optical observations provide measurements of coronal electron densities throughout the corona (broad-band white light measurements made at eclipses and with ground-based and satellite coronagraphs) and information on temperatures, densities, abundances, nonthermal random velocities, and flow velocities near the coronal base (EUV and x-ray instruments), and temperatures, densities, nonthermal random velocities, and flow velocities out to several solar radii (UV coronagraphs). Radio observations provide information on flow velocities by monitoring interplanetary scintillations of natural and artificial sources located beyond the sun. Some optical and radio measurements relevant to solar wind studies are summarized by Withbroe (1988a).

3. Application of Solar Observations to Theoretical Models

Data from different solar regions can be used to evaluate theoretical solar-stellar wind models. Figure 1 shows the radial variation of the electron density N_e, temperature T, and flow velocity V for an equatorial coronal hole. The points are the measurements and the curves the results of model calculations. The solid curve is for the best-fitting thermally driven wind. It predicts flow speeds at large distances from the sun which are too low, a common failing of solar wind models in which the wind is driven solely by thermal pressure gradients. If one includes additional acceleration by waves, most often assumed to be Alfvén waves (which can pass through the dense low corona and deposit their momentum and energy in the supersonic region of the wind), then it is possible to account for the high asymptotic speeds observed. The required Alfvén wave flux is consistent with amplitudes of nonthermal velocities measured in the low corona via the widths of spectral lines.

One can make similar comparisons between parameters measured in other solar regions (see Withbroe 1988a, 1988b and references cited therein). The results indicate that radiative energy balance models provide a good fit to solar data. For coronal holes it appears that the mechanical energy flux at the coronal base $F_m \approx 4 \times 10^5$ erg cm^{-2} s^{-1}, the Alfvén wave flux $F_A \approx 2 \times 10^5$ erg cm^{-2} s^{-1}, and the characteristic dissipation length for the mechanical energy flux is 0.4 to 0.8 $R_\odot$ (and about 0.2 $R_\odot$ in the more dense quiet regions). The mechanical energy flux heating the corona in coronal holes (large scale magnetically open regions) is approximately equal to that in large scale magnetically closed regions, which appear to cover most of the solar disk outside of active regions (which are smaller scale closed structures with much stronger fields).

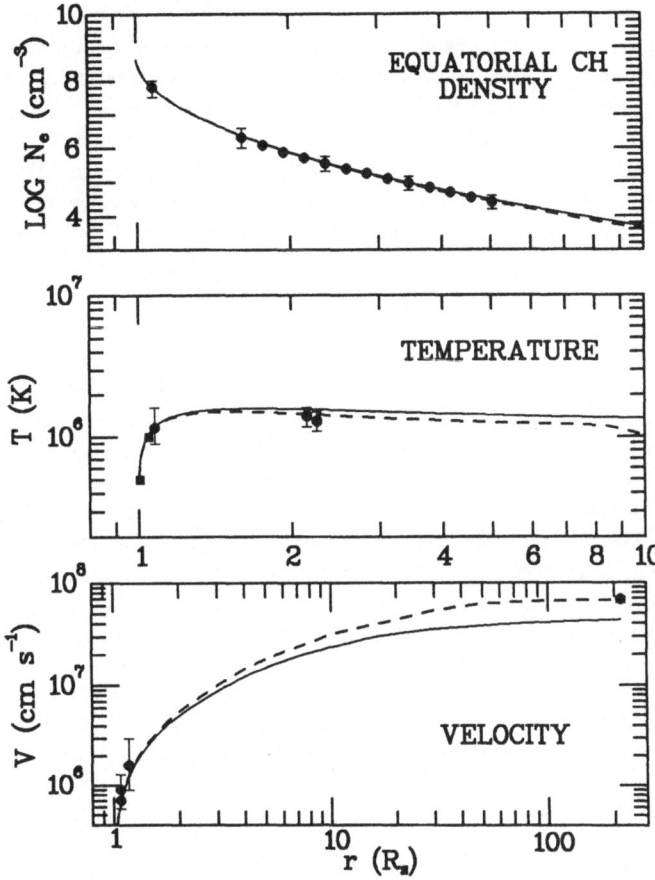

Figure 1. Comparison of empirical densities, temperatures, and flow velocities (points) with those calculated with a radiative energy balance model (curves). The solid curves are for a purely thermally driven wind; the dashed curves show the effect of adding sufficient energy in the form of Alfvén waves to raise the asymptotic flow speed to that observed.

4. Implications for Mass Loss From Stars with Hot Coronae

The solar mass loss is determined by: (1) the fraction of the solar surface which is magnetically open, typically about 20%, and (2) the heating of the coronal plasma which gives rise to a solar wind which appears to be primarily thermally driven in the subsonic region of the wind (where the mass loss is determined, see review by Leer, Holzer, and Fla 1982). To first order, it appears that the effects of Alfvén waves can be ignored in calculating mass losses for stars similar to the sun. Calculations with radiative energy balance models indicate that the stellar mass loss rate (for a spherically symmetric outflow) depends primarily on the mechanical energy input and is insensitive to its dissipation length for dissipation lengths comparable to a stellar radius, as is the case for the sun (see Hammer 1982b, Withbroe 1988a).

These results suggest that for a star with know mass and radius and a thermally driven wind, two parameters are needed to estimate the mass loss rate, the mechanical energy flux F_m and fractional area of the star covered by magnetically open regions. If only F_m is know, an upper limit to the mass flux can be estimated. Determining F_m is difficult due to the fact that coronal fluxes are likely to be dominated by emissions from closed magnetic regions on the star, as is the case for the sun. However, solar observations suggest the F_m has the same magnitude in large scale magnetically open and closed regions (see Withbroe 1988b), hence an estimate of F_m for closed regions (obtained from, for example, observations of the C IV $\lambda 1548$ or O VI $\lambda 1032$ flux) should provide a good first order estimate for F_m and the mass loss rate for the open regions. The resulting mass loss rate will be an upper limit (for a thermally driven wind) because (1) of the assumption of spherical symmetry (overestimate by a factor of 5 for the sun) and (2) the possibility that some of the spectral line flux may come from active regions. For some stars it may be possible to determine the contributions of active regions from information on rotational and stellar cycle variations.

For some solar-type stars, deposition of energy/momentum by waves may influence the mass flux (see Holzer 1988, Leer 1988). Measurements of line widths of stellar transition or coronal lines could be used to place an upper limit on the magnitude of the Alfvén wave velocity amplitude, and thereby place constraints on the relative importance of thermal and Alfvén contributions to the energy driving the mass flux.

To summarize, the presently available results suggest that radiative energy balance models provide a good first order solution to the problem of calculating wind models and estimating mass losses for stars with hot solar-type coronae.

References

Hammer, R. (1982a), 'Energy balance of stellar coronae. I. Methods and examples', Astrophys. J., 259, 767-778.

Hammer, R. (1982b), 'Energy balance of stellar coronae. II. Effect of coronal heating', Astrophys. J., 259, 779-791.

Holzer, T. E. (1988), 'Acceleration of stellar winds', in V. J. Pizzo, T. E. Holzer, and D. G. Sime (eds.), Proceedings of the Sixth International Solar Wind Conference, NCAR/TN-306+Proc, Nat. Cent. Atmos. Res., pp. 3-22.

Leer, E. (1988), 'Wave acceleration mechanisms for the solar wind', in V. J. Pizzo, T. E. Holzer, and D. G. Sime (eds.), Proceedings of the Sixth International Solar Wind Conference, NCAR/TN-306+Proc, Nat. Cent. Atmos. Res., pp. 89-106.

Leer, E., Holzer, T. E., and Fla (1982), 'Acceleration of the solar wind', Space Sci. Rev., 33, 161.

Withbroe, G. L. (1988a), 'The temperature structure, mass, and energy flow in the corona and inner solar wind', Astrophys. J., 325, 442-467.

Withbroe, G. L. (1988b), 'Acceleration of the solar wind as inferred from observations', in V. J. Pizzo, T. E. Holzer, and D. G. Sime (eds.), Proceedings of the Sixth International Solar Wind Conference, NCAR/TN-306+Proc, Nat. Cent. Atmos. Res., pp. 23-48.

CORONAL INSTABILITIES

G. Einaudi
Istituto di Astronomia
Universita' di Firenze
50125 Firenze, Italy

ABSTRACT. The interest in the stability of coronal structures derives from their observed lifetime (much longer than the relevant hydromagnetic timescale) coupled with their active behavior. This fact implies that these structures must be globally stable with respect to fast and destructive instabilities and, at the same time, must allow some local, non-disrupting, dissipative process to take place. In highly magnetized media as solar and stellar coronae a large number of plasma instabilities can occur. The present review will concentrate on those governed by the magnetohydrodynamic (MHD) equations with the inclusion of the effects of finite resistivity and viscosity and the use of an energy equation where radiation, mechanical heating and thermal conduction are considered.

1. INTRODUCTION.

The stability of the magnetic structures observed in the solar and stellar coronae has received a great deal of attention in the last decade, since observations have clearly shown that the magnetic field plays a fundamental role. In fact the geometric arrangement of coronal plasma is directly related to the structure of the magnetic field, which, in active regions, confines and collimates the emissive plasma in a myriad of loops and arcades (Vaiana and Rosner, 1978). These structures maintain their identity for a long time (in some examples more than a day) (Foukal, 1976, Krieger, 1977) and on the other hand appear to be the site of high energy active phenomena , like the presence of extended heating, the recurrence of flares, the appearance and disappearance of prominences, the occurence of coronal mass ejections, etc.
The dynamical behavior of coronal structures is not surprising, the tendency of a magnetized plasma to be unstable being well known (Bateman, 1978). The nature of the instability depends on the features of the equilibrium configuration and on the boundary conditions the perturbations must satisfy. The challenge in the last years has been to understand precisely the link between the nature, and therefore the effects and the timescales, of the instability and the physical conditions of the structure in which the instability evolves. In this paper we will concentrate on phenomena governed by the fluid equations neglecting the influence both of the microinstabilities and of most kinetic effects on the macroinstabilities. Of course this influence is important and can be invoked to explain a number of observed features; however within the MHD approximation we have to deal with an already difficult problem and, at the same time, we are able to describe the properties of the most dangerous and energetic instabilities, such as the ideal magnetic instability, the resistive instability, the Kelvin-Helmholtz instability and the thermal instability.
We want to stress that each type of instability is driven by a different physical mechanism and therefore we can study separately the properties of each instability by considering conditions in which only one mechanism at the time is dominant. In reality this separation is quite artificial and the nature of the growing perturbation is governed by the timescales on which each driving mechanism acts. When the

D. McNally (ed.), Highlights of Astronomy, Vol. 8, 529–533.
© *1989 by the IAU.*

timescales are comparable, the coupling between different types of instability produces the appearance of new features in the resulting growing perturbation.

We believe it is worth to outline here that the instabilities which will be reviewed in the present paper have been extensively studied for laboratory purposes with the conclusion that they are quite sensitive to the particular regime in which the plasma is found. The operation of transferring results obtained for one particular regime to different ones is, generally speaking, unsafe. This is the reason why for coronal applications many original calculations have been made and only the properties of these "coronal instabilities" will be discussed.

In the next section we will briefly summarize what is known on coronal instabilities and in the last section we will discuss the lines for future work.

2. NATURE OF THE INSTABILITIES.

We assume that the collisional magnetohydrodynamic theory is applicable, that the plasma radiates and is mechanically heated according to some known laws and that the only non-ideal effects which are important are finite constant and isotropic resistivity and shear viscosity and finite parallel thermal conduction. The dynamics of such a plasma is then described by the following set of equations:

$$\partial\rho/\partial t + \text{div} (\rho v) = 0 \tag{1}$$

$$\rho (\partial/\partial t + v. \text{grad}) v = -\text{grad} \, p + (\text{curl} B) x B/4\pi - \rho g + \mathcal{V} \, \text{div} \, \text{grad} \, v \tag{2}$$

$$(\partial/\partial t + v.\text{grad})(p/\rho^{\gamma}) = (\gamma\text{-}1) \, \rho^{1\text{-}\gamma} (E_H - E_R + \text{div} (b\kappa b \, \text{grad} T)) \tag{3}$$

$$\partial B/\partial t = \text{curl} (vxB) - (c^2/4\pi)\eta \, \text{div} \, \text{grad} \, B \tag{4}$$

where ρ is the plasma density, p the pressure, v and B the velocity and magnetic field respectively, T the temperature, γ the specific heat ratio. E_R represents the losses per unit mass due to radiation, E_H the heating and $b\kappa b$ the parallel to the magnetic field thermal conduction coefficient. Finally η is the resistivity and $\mathcal{V}$ the shear viscosity.

It is easy to identify in Eqs 1-4) the terms which can be responsible for the the onset of an instability.

The magnetic term in Eq 2) can drive an ideal magnetic instability due to the curvature of the magnetic lines. Generally speaking, all the cylindrical or toroidal pinches have the tendency to disrupt on the very short Alfven time scale. In the laboratory, stabilization is achieved by a combination of strong axial applied fields and close-fitting concentric conductors. Such conditions are unrealistic in modelling coronal plasmas, and it has been necessary to perform completely new calculations to study under which conditions realistic equilibrium configurations, when the influence of the photospheric boundary is taken into account, can be stable.

The resistive term in Eq. 4) is generally very small with respect to the ideal convective term and therefore negligible everywhere except in the regions close to the zeroes of curl (vxB). When these zeroes are present in the structure and the ideal instabilities either are stabilized or evolve on time scales longer than the Alfven time, the resistive term leads to reconnection of the magnetic lines and consequent conversion of the magnetic energy into other forms of energy.

The v.gradv term can originate a Kelvin-Helmholtz instability whose source lies in the energy stored in the kinetic energy of relative motion of the different layers when a sheared velocity field is present. The magnetic field line tension can in some cases play the role of the surface tension as stabilizing effect. On the other hand, in the presence of a magnetic field the possible interplay of magnetic and K.-H. instabilities can considerably change the nature of the growing perturbation with respect to the static or non-magnetized case.

The gravitational term can be responsible for the onset of Rayleigh-Taylor instability and Kruskal-Schwarzschild instability. We will not be concerned with these kinds of processes in this paper, since, at least in the solar case, the gravitational scale length is bigger than the typical height of the coronal structures.

The balance between gains (E_H) and losses (E_R) can also be destabilizing if its temperature and density dependence are appropriate, namely when E_H- E_R is a monotonically increasing funtion of the temperature and therefore, for example, to a local decrease of the temperature corresponds a local decrease of E_H- E_R. A thermal instability develops when the thermal conduction, which tends to stabilize by smoothing the perturbation temperature gradient, is not effective in balancing the destabilizing term.

Let us now review some of the most interesting results obtained recently in the study of the instabilities mentioned above.

2a. Magnetic instabilities. As far as the ideal magnetic instabilities are concerned one has to choose a realistic equilibrium configuration and proper photospheric boundary conditions to mimic the effect of the much denser photosphere on the perturbations arising in corona. The coronal magnetic field is embedded in the high-inertia photospheric plasma, which effectively anchors the magnetic lines, producing the so-called line tying. The relatively long photospheric Alfven time is taken to mean that the medium below the boundary, which models the transition region between photosphere and corona, cannot move on the relevant coronal time scale. This circumstance, in turn, ties the magnetic field in the photosphere against perpendicular motions and maintains the continuity of the component of the magnetic field perpendicular to the boundary at the ends (Van Hoven et al. 1981, Velli et al. 1988a). As far as the equilibrium is concerned both magnetic arcades and loops are usually described in cylindrical geometry with the z-axis parallel to the photospheric surface (assumed planar), when modelling an arcade, and perpendicular to it in the case of a loop. Various authors have shown that ideal stability can be achieved for these cylindrical equilibria when line-tying is taken into account.

In the case of an arcade stability is found when the field is force-free and all the magenetic lines reach the photosphere, that is if the axis of the cylinder is in the photosphere (Cargill et al. 1986). When the field is not force free, radial pressure gradients can drive an interchange-type instability which has been found using a ballooning transformation with a large wavenumber perpendicular to the magnetic field (Hood 1986).

A loop has been described as a cylindrical structure where currents are flowing, embedded in a much more extended potential field (Chiuderi et al. 1980). Radial pressure gradients can be present (Chiuderi and Einaudi 1981) as suggested by some of the Skylab EUV observations (Foukal 1978). The potential blanket and the pressure gradients improve the stability, which can be achieved only when line-tying is considered (Einaudi and Van Hoven 1981,1983). Raadu (1972) was the first to suggest the importance of the line-tying, while Hood and Priest (1979,1980) studied the constant shear field of Gold and Hoyle (1960). All these works have shown, using an energy principle, that a line-tied coronal loop goes unstable once a critical twist (or ratio of average poloidal to longitudinal flux) is exceeded. The critical twist depends on the length of the loop and on the kind of magnetic equilibrium adopted. Recently Velli et al. (1988b) have realized the importance of studying the growth rate and the spatial profile of the growing perturbations in order to determine the capacity of the small plasma resistivity to enhance instability producing reconnection of magnetic lines. As discussed above, resistive field reconnection involves a localized break-down of the frozen-in-field constraint of infinite conductivity MHD. In the infinite length or periodic cases this break-down occurs in narrow layers around the surfaces where $mB_\theta/r+kB_z=0$ (Coppi et al. 1966), where a θ and z dependence $\sim\exp(i(m\theta+kz))$ of the perturbed quantities has been assumed. Line-tying however excludes a simple harmonic dependence in the axial direction. It has been suggested (Mok and Van Hoven 1982) that in this case only θ-independent (m=0) perturbations are resistively unstable in configurations where the field component B_z vanishes at some point. Mok and Van Hoven (1982) and Velli and Hood (1988) have found that the instability is localized in vicinity of such points and that line-tying does not influence its properties with respect to the infinite or periodic case. However also θ-dependent perturbations can be resistively unstable, as Velli et al. (1988b) have shown for the resistive kink (m=1) by studying the behaviour of the operator $\mathbf{B}.\mathrm{grad}$ on the ideal eigenfunction at marginal stability. They have found that the derivative along the magnetic field vanishes on the same magnetic surface as in the infinite case but only at the center of the cylinder and when an inversion of the axial component of the field is present.

The properties of the m=1 resistive instability have been extensively studied for laboratory purposes (Coppi et al. 1976, Ara et al. 1978, Finn and Manheimer 1982). We want to point out two results which have important coronal implications. The first is that close to the ideal marginal stability the interplay of ideal and resistive effects markedly modifies the plasma behaviour (Batistoni *et al.* 1985) producing an instability which grows on a faster time scale than the well-known tearing mode. The second one concerns the effects of finite pressure which, even at very small values of the plasma β, can either tranform the tearing mode into a more robust and faster growing interchange-type instability or can stabilize it, depending on the geometry of the equilibrium configuration (Valdettaro *et al.* 1988).

2b. Kelvin-Helmholtz instability. As already remarked, this instability arises when the different layers of a stratified heterogeneous fluid are in relative horizontal motion. Its properties have been widely studied in the hydrodynamic framework (Chandrasekhar 1961), whereas the magnetic effects have been scarcely considered, only in the ideal limit and adopting elementary magnetic configurations. There have been some efforts to study analitically the properties of resistive instabilities in presence of a velocity field aligned withe the magnetic field (Hoffman 1975, Dobrowolny *et al.* 1983, Paris and Sy 1983). Recently Einaudi and Rubini (1986,1988), using a numerical approach, have found that the nature of the instability in presence of velocity and magnetic sheared fields is determined by the ratio R between the magnetic and the velocity shear scales and the ratio V between the amplitude of the velocity field and the Alfven velocity. Depending on the values of R and V an interplay between magnetic and fluid effects arises and dissipative processes as resistivity and viscosity can play different roles. In particular, resistivity can drive reconnection of magnetic lines even when the fluid effects are dominant on a time scale shorter than in the static case. Viscosity can enhance the growth rate of the perturbation, in contrast to its stabilizing effect on all other instabilities, when the sheared velocity field is subject to the so-called "cat's-eye" instability (Drazin and Reid 1981). In this case both reconnection of magnetic and stream lines occurs, leading to the formation of magnetic islands and small scale vortices. This instability has been used by Carbone *et al.* (1987) to explain some observed coronal features associated to solar surges.

2c. Thermal instability. The possible occurence of thermal instabilities in coronal structures has been investigated by many authors since the pioneering paper by Field (1965), who found that isobaric perturbations grow provided the length of the isothermal structure $L \geq 3 \times 10^{-11} T^{13/4}/P$ with P the pressure and T the temperature. The driving mechanism of such instability is the form of the solar radiation function at high temperatures. The non-linear evolution leads to the formation of a thin prominence-like condensation (Oran *et al.* 1982). The results obtained in the isothermal one-dimensional case cannot be applied to the corona because the presence of the transition region has a dramatic influence on the stability for three reasons. First of all L is not a free parameter as in the isothermal case but is related to the pressure and the apex temperature. Secondly, the typical radiative time scales in the transition region are much shorter than those in the corona. Finally, the static and stationary solution of Eq. 3) reproduces the observed temperature profile provided the heating dominates the radiation in the corona while it is negligible in the transition region (Chiuderi *et al* 1981). As a result, the radiation drives an instability which mainly influences the transition region leading to the formation of low-lying cold loops, rather than prominence-like structures (Klimchuk *et al* 1987 and references therein). The quasi-isothermal coronal part can be destabilized only if the heating has an appropriate pressure and temperature dependence and in this case a linear growing perturbation similar to the isothermal one has been found by Demoulin and Einaudi (1988).

The only magnetic effect considered in all these one-dimensional studies is the channelling of the heat flux along the magnetic lines. In a three-dimensional magnetic structure the thermal conduction can vanish in the same points as curl(**v**x**B**), namely where the perturbation propagation vector is perpendicular to the field. When this is the case, the growing perturbation is localized in the regions around these points (Chiuderi and Van Hoven 1979, Van Hoven *et al.* 1987), and a local decrease of the temperature can lead to an increase of the value of the resistivity and to a consequent accaleration of the reconnection process (Steinolfson and Van Hoven 1984).

3. CONCLUSIONS.

The above discussion on the nature of the various instabilities which can arise in a coronal plasma has clearly shown that the understanding of the evolution of the coronal structures is far from being satisfactory. The knowledge of which physical mechanisms can influence the growth of an instability, and therefore can determine the dynamical behaviour of the coronal plasmas, has improved consistently in the last few years. The actual non-linear development of the relevant instability, however, is in many cases unclear because only highly idealized computations have been performed. In particular the choice of the initial equilibrium configurations, in which curvature, flows and localized currents effects should be included, has been determined by computational convenience rather than by realistic considerations. Moreover the boundary conditions adopted in very few cases have properly modelled the photospheric line-tying or the presence of nearby regions. Finally the thermal properties of coronal plasmas should be considered more in detail.

The problem is that the observations are unable at the moment to provide direct details on the small scale structure of coronal features. Only a joint observational and theoretical effort can lead to a better insight on the dynamics of coronal structures through observations triggered by theoretical results on one hand and theoretical work establishing connections between unknown and observable quantities.

REFERENCES

Ara, G., Basu, B., Coppi, B., Laval, G., Rosenbluth, M.N., Waddell, B.V., Ann. Phys.,**112**,443, 1978
Bateman, G. , *MHD Instabilities*, M.I.T., Cambridge Mass., 1978
Batistoni, P., Einaudi, G., Chiuderi, C., Solar Phys.,**97**,309, 1985
Carbone, V., Einaudi, G., Veltri, P., Solar Phys.,**111**,31, 1987
Cargill, P.J., Hood, A., Migliuolo, S., Ap.J., **309**,402, 1986
Chandrasekhar, S., *Hydrodynamic and Hydromagnetic Stability*, Oxford U.P., New York, 1961
Chiuderi, C., Van Hoven, G., Ap. J.,**232L**,69, 1979
Chiuderi, C., Einaudi, G., Ma, S.S., Van Hoven, G., J. Plasma Phys., **24**,39, 1980
Chiuderi, C., Einaudi, G., Solar Phys.,**73**,89, 1981
Chiuderi, C., Einaudi, G., Torricelli-Ciamponi, G., Astron. Astrophys.,**97**,27, 1981
Coppi, B., Greene, J.M., Johnson, J.L., Nucl. Fusion,**6**,101, 1966
Coppi, B., Galvao, R., Pellat, R., Rosenbluth, M.N., Rutherford, P.H., Fiz. Plazmy,**6**,961, 1976
Demoulin, P., Einaudi, G., Ap.J., submitted, 1988
Dobrowolny, M., Veltri, P., Mangeney, A., J.Plasma Phys.,**29**,393, 1983
Drazin, P., Reid, W., *Hydrodynamic Stability*, Cambridge U.P., New York, 1981
Einaudi, G., Van Hoven, G., Phys. Fluids,**24**,1092, 1981; Solar Phys., **88**,163, 1983
Einaudi, G., Rubini, F., Phys. Fluids,**29**,2563, 1986; Phys. Fluids, submitted, 1988
Field, G.B., Ap. J.,**142**,531, 1965
Finn, J.M., Manheimer, W.M., Phys. Fluids,**25**,697, 1982
Foukal, P.V., Ap.J.,**210**,575, 1976; Ap.J.,**223**,1046, 1978
Gold, T., Hoyle, F., M.N.R.A.S.,**120**,89, 1960
Hoffman, I.,Plasma Phys.,**17**,143, 1975
Klimchuk, J.A., Antiochos, S.K., Mariska, J.T., Ap.J.,**320**,409, 1987
Krieger, A.S., 'X-Ray Observations of Solar Structural Features', Proc. OSO-8 Workshop, Univ. Colorado, Boulder, 1977
Hood, A.W., Priest, E.R., Solar Phys.,**64**,303, 1979 ; Solar Phys.,**66**,113, 1980
Hood, A.W., Solar Phys.,**105**,308, 1986
Mok, Y. and Van Hoven, G., Phys. Fluids,**25**,636, 1982
Oran, E.S., Mariska, J.T., Boris, J.P., Ap.J,**254**,349, 1982
Raadu, M.A., Solar Phys.,**22**,425, 1972
Steinolfson, R.S., Van Hoven, G., Ap. J.,**276**,391, 1984
Valdettaro, L., Einaudi, G., Pegoraro, F., J. Plasma Phys., submitted, 1988
Van Hoven, G. Ma, S.S., Einaudi, G., Astron. Astrophys.,**97**,232, 1981
Van Hoven, G., Sparks, L., Schnack, D.D., Ap. J.,**317**,l91, 1987
Velli, M, Hood, A.W., Solar Phys.,**106**,354, 1986
Velli, M., Einaudi, G., Hood, A.W., Ap.J., submitted, 1988a ; 1988b

ACCRETION DISK CORONAE

Max Kuperus
Sterrekundig Instituut
P.O. Box 80 000
3508 TA Utrecht
The Netherlands

SUMMARY. Accretion disk coronae around compact objects are the result of strong magnetic activity in the inner regions of accretion disks. Part of the accreting energy is dissipated in te corona and can be observed as hard X-ray emission with a time variability caused by the coronal structures. The interaction of disk coronae with neutron stars and black holes may cause quasiperiodic oscillations respectively flare type emission.

INTRODUCTION. Many accretion driven X-ray sources show a two component spectrum characterized by a soft X-ray component and a hard X-ray power law tail. The spectrum originates from an optically thick accretion disk radiating as a blackbody with a radial temperature distribution and a hot corona comptonizing the soft black body photons, thus forming a hard tail. Burm (1988) calculated the resulting comptonization spectrum using a Monte Carlo method applied to a disk-corona model consisting of magnetic coronal loops with a length ℓ for which a scaling law

$$T_{cor}^{5/2} = 3.6 \times 10^{5} \ p_{cor} \ell \tag{1}$$

has been derived (Burm; 1986).
It is assumed that all accreting matter is assembled in a thin disk drifting inwards with a velocity that is slow compared to the orbital velocity and where the vertical structure is determined by hydrostatic equilibrium. The structure of such a thin steady disk is described by the standard model, based on the conservation equations. The energy sink is the black body radiation of the optically thick disk and the energy source is produced by the viscous dissipation in the differentially rotating disk represented by the turbulent and magnetic stress tensor $t_{R\phi}$. The key assumption in the standard model is that the stress tensor is assumed to be proportional to the pressure ($t_{R\phi} = \alpha p$), where α is a constant independent of the radius, which can be made plausible on dimensional grounds.

D. McNally (ed.), Highlights of Astronomy, Vol. 8, 535–538.
© 1989 by the IAU.

MAGNETIC ACTIVITY IN ACCRETION DISKS

A small magnetic field in an accretion disk will be amplified by the differential rotation, while turbulence creates a fluctuating component. The combination of differential rotation and helical turbulence acts as a dynamo whose efficiency is described by the ratio of growth and decay time expressed by the dynamo number N

$$N = A\Omega h^3/\beta^2, \tag{2}$$

where the net turbulent helicity $A = tv^2/3\beta$ and $\beta = v^2 t$, where t is the turbulent correlation time. Following Pudritz (1981), we can esti-mate the dynamo number assuming $t = \Omega^{-1}$ and the turbulent correlation length $\ell = h$ (the disk thickness). Then with $v = Mv_s$, M being the turbulent Mach number, and $h/r = v_s/v_\phi$ for a thin disk it follows that $N = 0.3 \ M^{-2}$. For subsonic turbulence we expect an efficient dynamo action resulting in strong and localized field fluctuations. The growth of magnetic fields is limited by magnetic buoyancy caused by the gravity acting on magnetic flux tubes. Burm (1985) estimated the Maxwell stress limited by buoyancy: $t_{R\phi} = 0.23 \ M_t^3 \ p_{gas}$. Consequently buoyancy puts severe limits to the field strengths in the disk causing large scale fields to emerge from the disk, thus forming magnetic layers on both sides of the disk consisting of loops and streamer type structures similar as observed in the Solar Corona.

However, small scale magnetic fields may be confined in the disk, thus forming magnetic cells. Equipartition type fields or even much stronger fields may be present on small scales, where line tying prevents them from escaping. As a first approximation the disk field may be considered to exist of turbulent magnetic cells stretched by the differential rotation until the magnetic stresses oppose the shear. Eardley and Lightman (1975) and Coroniti (1981) made such an approach, focussing on magnetic reconnection as the dominant field limiting mechanism for small scale fields in the disk. After the cell stretching is stopped by the magnetic stress the magnetic pressure gradients pinch the plasma in the center of the cell. The magnetic field lines are thus forced to reconnect, which ultimately results in the fission of the elongated flux cell into two smaller cells. This fission process occurs on an typical reconnection timescale. Hence the distortion stores shear motion into magnetic energy which is released by the reconnection process primarily into radial motion. The two remaining cells do not follow the Kepler rotation since they are slowed down. The two cells move apart radially in order to adjust their angular momentum.

FORMATION OF A DISK CORONA

The presence of a corona coupled to the disk exerts a torque on the disk whose effect is hidden in the parameter α describing the total stresses. Ionson and Kuperus (1985) separated the internal stresses α_i from the external stresses α_e where of course $\alpha = \alpha_i + \alpha_e$. Then the disk luminosity, which is related to the soft X-ray luminosity is $L_{disk} = 3\ \alpha_i p\ h\ \Omega\ /\ 4$, while the total luminosity is $L_{tot} = L_{disk} + L_{cor} = 3\ \alpha\ p\ h\ \Omega\ /\ 4$. Hence $\alpha = \alpha_i(1 + L_{cor}/\ L_{disk})$; the external stresses are directly related to the coronal emission. Only a fraction of the accreted gravitational energy is emitted by the disk; the remainder is radiatively vented by the overlying corona. It is important to note that coronal dissipation has a stabilizing effect on the disk since it reduces the ratio of radiation pressure to gas pressure in proportion to the rate at which the disk couples non-thermal energy into the corona. Kuperus and Ionson (1985) estimated the upper limit of L_{cor} by considering a coronal magnetic loop as the load of a resonant electric circuit. If one assumes that all the energy that is delivered to the corona is used to heat the corona which consists of loops of length $\ell = h$ filling a fraction f of the disk a good approximation of the coronal luminosity is given by:

$$L_{cor} = 8\ f\ \frac{B_{cor}}{B_{disk}}\ (v_A^d)^2\ \alpha\ \rho\ v_s. \tag{3}$$

Using the hydrostatic equilibrium condition $h/r = v_s/v_\phi$, the ratio of coronal to disk luminosity is given by

$$\frac{L_{cor}}{L_{disk}} = 10\ f\ (\ \frac{B_{cor}}{B_{disk}})\ (\frac{v_A^d}{v_s^d})^2. \tag{4}$$

This expression is remarkable in the sense that it neither depends on the disk thickness nor on the parameter α. The last statement is only correct if $\alpha \ll 1$. Let us assume that the fields fan out into the corona in the same way as they do in the solar corona. Then $f = 1$ and eq. (4) shows that the emission ratio is directly proportional to the ratio of magnetic pressure and gas + radiation pressure in the disk, which may be locally much larger than one but averaged over the disk of the order of unity. It therefore seems that the observed large ratios of hard to soft X-ray emission of black hole accretion disks can be understood by the emission of magnetically structured and electrodynamically coupled disk coronae.

The observable ratio of hard to soft X-ray emission is a new way to search the internal magnetic structure of an accretion disk. Not only does it reveal the magnetic energy content but the important turbulent nature of the disk magnetic field may be studied by careful Fourier analyses of the time fluctuations on a wide variety of scales such as observed in Cygnus X-1. Note that in the Sun the ratio $L_{cor}/L_{phot} = 10^{-6}$, which is indicative for the small magnetic energy content of the solar photosphere. Actually $\langle B^2\rangle/8\pi p(phot) \approx 10^{-5}$.

INTERACTION OF DISK CORONAE WITH NEUTRON STARS AND BLACK HOLES

In order to explain the spectral characteristics of quasiperiodic oscillating X-ray sources (QPO), Stollmann and Kuperus (1988) developed a model where the magnetic field of the central neutron star interacts with disk coronal loops located around the Alfvén radius. It can be shown that in this case energy is transferred from the neutron star to the disk corona. Consider a loop with an obliquely rotating neutron star magnetic field. The loop is anchored in the disk rotating with the Kepler velocity, while the neutron star magnetic field inside the Alfvén radius is supposed to rotate with the neutron star angular velocity $\Omega*$. In a coordinate system fixed to the loop, the loop experiences an oscillating stellar field with a frequency equal to the beat frequency $\Omega_B = \Omega_K - \Omega*$. The magnetic periodicity at radius R is given by

$$B_S(R) = B_0(R) + B_1(R) \cos \Omega_B t \tag{5}$$

In order to estimate the power consumption the loop is considered as a forced oscillator with a forcing term $F = B_0 B_1 \ell \cos \Omega_B t$. The maximum amplitude is reached when $\Omega = \omega_0$, the characteristic frequency of the loop. Assuming that the damping of the oscillation is due to the emission of magnetoacoustic waves generated by the oscillating surface of the filament, Stollman and Kuperus found that one loop may contribute up to a few percent of the Eddington luminosity. This shows that the rotating neutron star may be an important energy source to heat the disk corona.

The magnetic field in the accretion disk corona around a black hole becomes detached from the accreting matter once the matter approaches the horizon in the sense that closed loops will dissipate at the stretched horizon, while open field lines accumulate near the horizon (Thorne et al., 1986).

This decoupling of field and matter near the horizon has profound consequences for the activity around a black hole. The open field lines of one polarity generate a magnetic pressure that may eventually prevent further accretion and divert the matter along the field funnel, thus forming jets. However, there is no reason why only one polarity will be present. Hence the accumulation of one polarity is likely to be followed by the accumulation of the opposite polarity, thereby generating flare type disturbances. The peculiar behaviour of the accreting coronal magnetic fields around the horizon might be another source for time variability of hard X-ray emission in black hole accreting X-ray sources.

REFERENCES

Burm, H.   (1985), Astron. Astrophys. 143, 389
Burm, H.   (1986), Astron. Astrophys. 165, 120
Burm, H.   (1988), Astron. Astrophys., (submitted)
Burm, H. and Kuperus, M.   (1988), Astron. Astrophys. 192, 165
Coroniti, F.V.   (1981), Astrophys. J. 244, 587
Eardley, D.M. and Lightman, A.P.   (1975), Astrophys. J. 200, 187
Ionson, J.A. and Kuperus, M.   (1984), Astrophys. J. 284, 389
Kuperus, M. and Ionson, J.A.   (1985) Astron. Astrophys. 148, 309
Pudritz, R.E.   (1981), Mon. Not. Roy. Astron. Soc. 195, 897
Stollman, G.M. and Kuperus, M.   (1988), Astron. Astrophys. (in press)
Thorne, K.S., Price, R.H. and Macdonald, D.A.   (1986), Black Holes:
   The Membrane Paradigm, Yale University Press

SOLAR AND STELLAR FLARES

A.O.BENZ
Institute of Astronomy, ETH
8092 Zurich
Switzerland

ABSTRACT. This review concentrates on some selected topics concerning the release of magnetic energy and associated phenomena in flares. Emphasis is on microflares, recent studies of different phases of flares, and propagation and trapping of flare accelerated electrons. The ongoing analysis of the observations of the previous solar cycle reaches a state where quantitative models become possible. The subject of solar flares can be broken up into several, now well defined physical problems.

1. Introduction

Flares are of great interest for the understanding of solar and stellar atmospheres. because they make fundamental processes visible. They occur in a large variety of stars, particularly

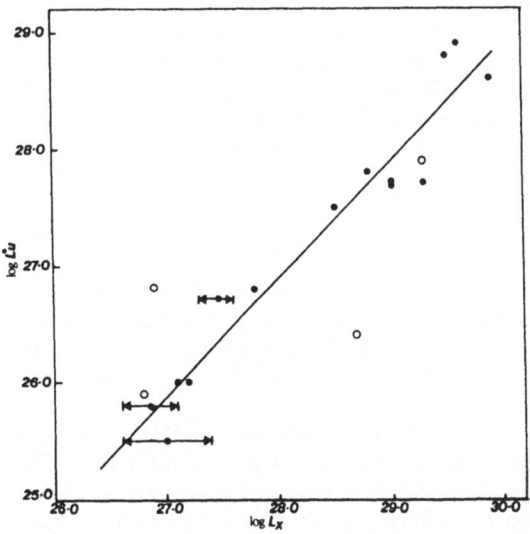

Figure 1. Time-averaged flare energy in the U band, L_u vs. quiescent soft X-ray flux, L_x (both in erg/s) of dMe stars. Open circles represent data points based on $< 15^h$ of observation and thus are probably uncertain (from Doyle and Butler, 1985).

539

D. McNally (ed.), Highlights of Astronomy, Vol. 8, 539–542.
© 1989 by the IAU.

in their early phase and in main sequence types later than F. They clearly are related to other important processes in the atmospheres such as the heating mechanism of the corona. This has been made obvious by the correlation of the average flare power in UV and the quiescent soft X-ray fluence in dwarf M stars (Figure 1). The origin of both is generally believed to be the subphotospheric dynamo driving the coronal magnetic field.

2. Microflares

A more direct relation between flares and coronal heating is postulated in the hypothesis that a large number of very small flares provide the energy. The best evidence for such microflares occuring all the time in the solar case even in quiet regions comes from UV observations (Brueckner and Bartoe, 1983; Porter et al., 1987). Lin et al. (1984) found small amplitude band X-ray (> 22 keV) bursts during a balloon flight of a high-sensitivity detector. These events occured at a rate of about one in five minutes. Some of them, however, were shown to be associated with strong metric and decimetric type III bursts (Benz, 1983) known to be generally associated with active regions. Thus the observed HXR microflares cannot be the cause of heating in quiet regions. However, much weaker type III bursts at meter-dekameter wavelengths have been reported by Kundu et al. (1986) which occur several times per hour during quiet-Sun periods. There maybe associated HXR flares, but they are below the sensitivity of current HXR telescopes.

Small events or fluctuations have long been reported for solar radio observations. Correlation studies between Arecibo and Effelsberg (100m telescope) have shown that most radio fluctuations recorded by single dishes are instrumental (Benz and Fürst, 1987). Only in a weakly active region correlated fluctuations of typically 500Jy amplitude and 90s correlation length have been found. About 2 orders of magnitude smaller peaks were observed in quiet regions at the same wavelength of 6cm with the VLA by Fu et al. (1987). Such brightenings seem to be related to coronal X-ray bright points.

SXR microflares of dMe stars were reported by Butler et al. (1986). However, there seem to be continuous periods of several hours with no statistically significant variability (Pallavicini, 1987). The SXR emission of dMe stars is not a superposition of low level flares.

It is concluded that microflare heating of coronae may be theoretically attractive, but the observational evidence from both Sun and stars is still weak.

3. Flare Phases

Considerable effort has been made to disentangle the different phases of energy release in flares. Klein et al. (1987) have studied the onset of HXR emission in solar flares. In the pre-flash phase HXR at increasingly higher energies are observed. The onset of relativistic electrons and proton lines coincides with the flash phase. A qualitative difference seems to distinguish the two phases.

The spatial evolution of solar flares was investigated in HXR and Hα by Martin and Svestka (1988). The two emissions are not cospatial. First the HXR loop fills starting from the two footpoints. This is compatible with the idea of explosive evaporation.

Beams of down-going electrons have recently been observed in microwaves as type III bursts (Stähli and Benz, 1987). Most important is the fact that the corona is transparent to plasma emission down to densities of at least 10^{11}cm. This is where energy release in flares and other interesting phenomena occur. Since radiation in a homogeneous atmosphere could not escape from this level, emission and/or propagation must take place in a fibrous plasma. Recent works on this problem include Roelof and Pick (1989) and Poquérusse et al. (1989).

4. Trapped Particles

Millisecond radio spikes are one of the most fascinating topics in current flare research. They are very closely related to HXR emission and typ III bursts, thus seem to have an intimate relation to the energy release. An example is shown in Figure 2.

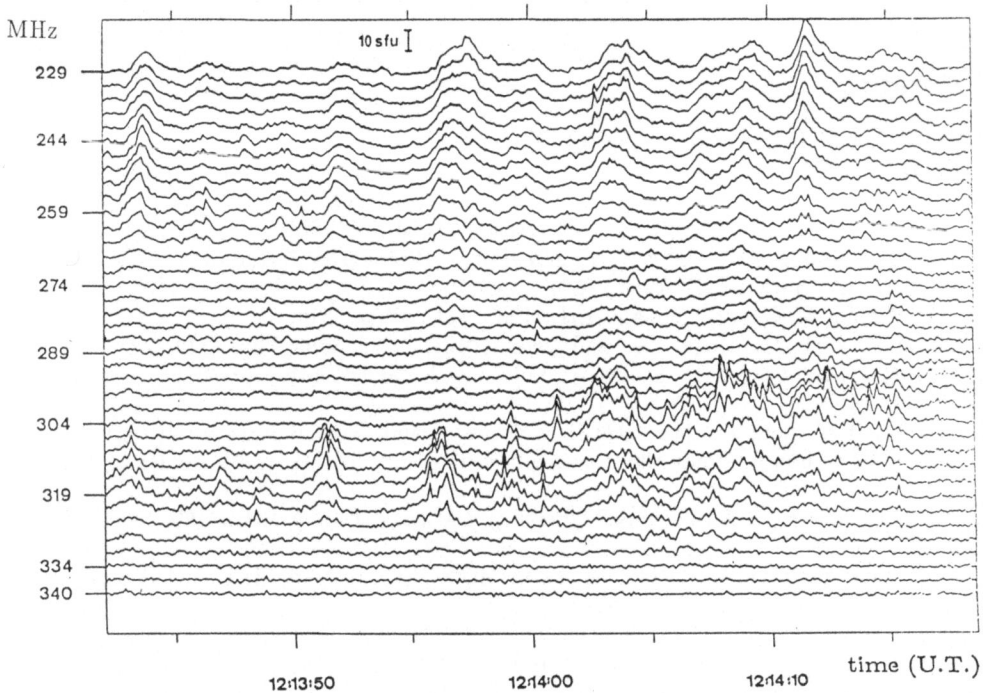

Figure 2. Radio flux in 38 channels observed by Zurich spectrometer. In the 229 – 280 MHz band (upper half) type III bursts are visible. Spikes appear loosely correlated to them in the 280 – 340 MHz band.

A possible emission mechanism is maser action of trapped electrons. Winglee et al. (1988) have numerically simulated the process and found spiky emission but with nanosecond timescale. An alternative model are thousands of electron beams injected at skew angles to the magnetic field (Li et al., 1984).

Quantitative models of synchrotron and HXR emissions of trapped and precipitating particles have been fitted to observations by several people (most recently e.g. Lu and Petrosian, 1989). A new development has started with quantitative comparison of maser emission of decimetric pulsations with HXR (Aschwanden, 1987). Combining observations at various wavelengths eliminates free parameters and reduces the range of possible models. Further progress can be expected from spatially resolved observations.

Quantitative investigations of the flare particles, plasma and magnetic field open the path for the solution of the major flare problems:
  - global MHD configuration for energy release
  - current sheet physics
  - acceleration (or heating) of electrons

542

- role of protons
- distribution of flare energy in the corona.

Flare physics is a large field and includes many interesting problems that are of general relevance. The major problems remain unsolved, but they have come into clearer focus. Considerable progress has been made in the observation of temporal and spatial evolution, as well as in the understanding of some restricted problems such as particle trapping in loops.

References

Aschwanden, M.J.: 1987, Ph.D. thesis, ETH Zurich.

Benz, A.O.: 1981, Proc.21st ESLAB Symposium, Bolkesjø, Norway, ESA SP-275, 105.

Benz, A.O. and Fürst, E.: 1987, Astron.Astrophys. **175**, 282.

Brueckner, G.E. and Bartoe, J.D.F.: 1983, Astrophys.J. **272**, 329.

Butler, C.J., Rodonò, M., Foing, B.H., and Haisch, B.M.: 1986, Nature **321**, 679

Doyle, J.G. and Butler, C.J.: 1985, Nature **313**, 378.

Fu, Q., Kundu, M.R., and Schmahl, E.J.: 1987, Solar Phys. **108**, 99.

Klein, K.-L., Pick, M., Magun, A., and Dennis, B.R.: 1987, Solar Phys. **111**, 225.

Kundu, M.R., Gergeley, T.E., Szabo, A., Loiacono, R., and White, S.M.: 1986, Astrophys.J. **308**, 636.

Li, C., Li, H.W., and Fu, Q.: 1984 in C.de Jager and Chen Biao (eds.) Proceedings of Kunming Workshop on Solar Physics and Interplanetary Travelling Phenomena, Vol.2. p. 542.

Lin, R.P., Schwartz, R.A., Kane S.R., Pelling, R.M., and Hurley, K.C.: 1984, Astrophys.J. **283**, 421.

Lu, E.T. and Petrosian, V.: 1989, The ratio of microwaves to X-rays in solar flares: The case of the thick target model, submitted.

Martin, S.F. and Svestka, Z.F.: 1988, Solar Phys. **116**, 215.

Pallavicini, R.: 1987, Lecture Notes in Physics **292**, 98.

Poquérusse, M. Steinberg, J.L., Caroubalos, C., Dulk, G.A., and MacQueen, R.M.: 1988, Astron.Astrophys. **192**, 323.

Porter, J.G., Moore, R.L., Reichmann, E.J., and Harvey, K.L.: 1987, Astrophys.J. **323**, 380.

Roelof, F.C. and Pick, M.: 1989, Astron.Astrophys., submitted.

Stähli, M. and Benz, A.O.: 1987, Astron.Astrophys. **175**, 271.

Winglee R.M., Dulk, G.A., and Pritchett, P.L.: 1988, Astrophys.J. **328**, 809.

3. HIGH ANGULAR RESOLUTION IMAGING FROM THE GROUND

Chairmen: J. Baldwin and J. Davis; Editor: J. Davis

Supporting Commissions:

9, 40

INTRODUCTION TO THE JOINT COMMISSION MEETING ON HIGH RESOLUTION IMAGING FROM THE GROUND

JOHN DAVIS
Chatterton Astronomy Department, School of Physics
University of Sydney,
N.S.W. 2006, Australia.

1. INTRODUCTION

As a result of advances in instrumentation and techniques, from radio through to optical wavelengths, we have before us the prospect of producing very high resolution images of a wide range of objects across this entire spectral range. This prospect, and the new knowledge and discoveries that may be anticipated from it, lie behind an upsurge in interest in high resolution imaging from the ground. Several new high angular resolution instruments for radio, infrared, and optical wavelengths are expected to come into operation before the 1991 IAU General Assembly.

Although several meetings and workshops have been held in recent times in the general area of high resolution imaging they have catered primarily for those already active in the field. The majority of astronomers are probably unaware of the status of imaging techniques - what can and what cannot be done. With this in mind, the aim of this Joint Commission Meeting is to inform the astronomical community of the status of high angular resolution imaging from the ground and of the capabilities which will soon be available for astrophysical studies. To meet this aim, the program consists of a number of invited reviews covering the techniques, the instruments, and the astrophysical potential of high resolution imaging for the different regions of the spectrum. Imaging from space is not included since the emphasis of the meeting is on opportunities in the near future.

2. SOME BACKGROUND MATERIAL

Radio astronomers are producing remarkable maps of the sky which have revealed details of a wide range of objects such as supernova remnants, radio galaxies with complex structures including jets and tails, and apparent superluminal motion infrared from images obtained at different epochs. Infrared and optical astronomers have lagged behind their radio colleagues in high resolution imaging but work is under way at a number of centres to adapt radio techniques to the shorter wavelengths.

Objects for which images are already being obtained, or soon will be, include asteroids, planets and satellites, protostars and envelopes, stellar surfaces and circumstellar shells, binary and multiple systems, pulsating stars (Cepheids, Miras etc.), planetary nebulae, supernova remnants, central regions of globular clusters, galactic nuclei and quasars. The list is not intended to be exhaustive but simply to indicate the wide range of objects which can be studied with the aid of high resolution imaging.

Unfortunately, while the imaging capabilities of optical telescopes are being improved

D. McNally (ed.), Highlights of Astronomy, Vol. 8, 545–546.

by speckle interferometry and adaptive optics, the angular resolution required to achieve images of the objects listed above is generally well beyond the capability of existing and, in many cases, of any conceivable single aperture instrument at any wavelength. It is necessary to use an array which may range from a simple two aperture interferometer to multi-element arrays like the Very Large Array (VLA).

The task of producing images from interferometric arrays is non-trivial and later speakers will address some aspects of the problems posed by the atmosphere, by instrumental factors, by the dilute nature of the aperture and the limited u,v plane cover. Nevertheless, there has been a revolution in imaging capabilities and some of the major advances in instrumentation and techniques that have brought this about include the development of aperture synthesis, large radio telescope arrays (e.g. Westerbork, VLA, Merlin), very long baseline interferometry (VLBI), electronics and computers, image processing algorithms such as CLEAN and MEM, self-calibration using closure phases and amplitudes, speckle interferometry, measurement of accurate optical and infrared visibilities, optical fringe tracking and the demonstration of phase closure at optical wavelengths.

Some of the major instruments currently under development (with expected completion dates where known) which make this meeting timely include:

- The Australia Telescope (AT) (radio) (1989)
- The Berkeley Infrared Spatial Interferometer (ISI) (USA) (1989)
- The Sydney University Stellar Interferometer (SUSI) (optical) (Australia) (1990)
- The Very Long Baseline Array (VLBA) (radio) (USA) (1992)
- The Giant Meterwave Radio Telescope (GMRT) (India) (1992)
- The Cambridge Optical Aperture Synthesis Telescope (COAST) (ir) (UK)
- The Institut de Radio Astronomie Millimétrique Interferometer (France & W. Germany)
- Development of the CERGA I2T into an INT with N telescopes (ir and optical) (France)
- The United States Naval Observatory Astrometric Optical Interferometer (USA)
- The Very Large Telescope (VLT) (ir and optical) (ESO) (1994-9)

In addition there are several optical and infrared projects in the early planning or prototype stage.

In summary, techniques which have been developed and used successfully at radio wavelengths are now being tried in the infrared and optical spectral regions. Several major high resolution instruments covering the spectrum from radio to optical wavelengths are under construction. The field is at an exciting stage and it is hoped that this meeting will stimulate further interest in the use of high resolution imaging in astrophysical studies.

PRINCIPLES OF IMAGING USING ARRAYS

T.J. CORNWELL
The National Radio Astronomy Observatory[1]
Socorro, NM 87801, U.S.A.

Since diffraction-limited imaging with a single aperture yields angular resolution $\sim \lambda/D$, the attainment of high angular resolution with single apertures requires the construction of correspondingly large monolithic apertures, the whole surface of which must be figured to much less than a wavelength. At the longer wavelengths, it is impossible to build a sufficiently large single aperture: for example, at $\lambda 21$ cm, arcsecond resolution requires an aperture of diameter ~ 50 km. At the shorter wavelengths, the atmosphere imposes a natural limit in resolution of about one arcsecond. However, another route is possible: that of using synthetic apertures to image the sky. The problem of figuring synthetic apertures is considerably simpler, and can be implemented in a computer. Synthetic apertures are now in use in many fields, e.g. radio-interferometry, radar imaging, magnetic resonance imaging. Radio-interferometric techniques developed in radio-astronomy over the past forty years are now being applied to optical and IR astronomical imaging by a number of groups.

At the begining of this century, Michelson investigated the use of interferometry for high resolution measurements of stellar diameters. This relies upon the van Cittert-Zernike theorem, which states that for an incoherent object, the coherence of the electric field far from the object is the Fourier transform of the sky brightness function. In 1960, Ryle pointed out that one could synthesize a large aperture by collecting coherence samples with an interferometer using many different spacings of the elements, and then Fourier-transforming the resulting sampled coherence function in a computer to make an image. Ryle's concept of a synthetic aperture holds for all wavelengths but the technology required for the measurements differs considerably. These differences will be addressed in the following talks.

Sampling of the coherence function over the synthesized aperture can be accomplished either by physically moving the interferometer elements or by allowing the rotation of the earth to do so or by a combination of both approaches. As long as the light can be interfered coherently, the elements may be an arbitrarily large distance apart.

Two generic problems afflict the measured coherence function: first, the sampling is often incomplete, and second,the calibration of the coherences may be uncertain because of the effects of the Earth's atmosphere or uncertainties in the geometry of the interferometer. The first problem may be addressed using deconvolution algorithms which can use a priori information about the sky brightness to interpolate missing values of the coherence

---

[1]The National Radio Astronomy Observatory (NRAO) is operated by Associated Universities, Inc., under contract with the National Science Foundation.

D. McNally (ed.), Highlights of Astronomy, Vol. 8, 547–548.
© *1989 by the IAU.*

function. Examples of such algorithms are CLEAN, the Maximum Entropy Method, the Gerchberg-Saxton-Papoulis algorithm, and the Lucy algorithm. The second problem is of varying importance in different applications. A good rule of thumb is that for wavelengths shorter than about 30 cm (including IR and optical), imaging at better than arcsecond resolution requires some countermeasures to the neutral atmosphere. In other regimes, countermeasures are necessary for high quality imaging. The geometric uncertainties are worst for long baselines (note the similarity to the problem of figuring a single aperture). Most of the effective techniques are related to the concept of closure phase introduced by Jennison about 30 years ago. Since calibration errors are predominantly associated with the interferometer elements, rather than pair of elements, a sum of the observed coherence phase around any closed loop of interferometers will be invariant to those errors. A similar observable can be derived for the coherence amplitudes. High-resolution imaging therefore uses these closure quantitites rather than the observed coherences as constraints on the final image.

Pearson and Readhead (1984) have discussed these techniques in more detail for radio-astronomy; as discussed above, the same procedures apply equally well at other wavelengths.

References

Pearson, T.J. and Readhead, A.C.S., 1984, *Ann. Rev. Astron. Astrophys.* **22**, 97-130.

IS THE IMAGING PROBLEM IDENTICAL IN ALL WAVE BANDS?

J.E. BALDWIN
Cavendish Laboratory,
Cambridge, U.K.

One aim of this meeting is to tempt us to study methods of imaging in wavebands outside our own. This might seem perverse since the imaging problems in the radio and optical/IR regions appear at first sight to differ profoundly in several ways: the rapid evolution of radio imaging over the last forty years has been largely ignored in optical imaging whilst the well developed theory and measurements of atmospheric fluctuations for the optical regime have been disregarded by radio astronomers; there have been long and exacting searches for the best sites for optical telescopes, but rarely for radio telescopes; the instrumental techniques appear to differ in many and complicated ways.

But such a view is quite mistaken. The problem is not only the same one but is much more nearly the same than could be expected a priori. The apparently complicated nature of the comparison arises mainly from the wide range of wavelengths in each regime, from 0.4 - 20 μm in the optical/IR and 0.4 mm - 10 m in the radio.

In space no one doubts that the problems are virtually identical. Only the photon fluxes differ, which introduces several interesting issues which there is no time to pursue here. From the ground the main problem is the atmosphere, in particular its irregular variations in refractive index. At optical/IR wavelengths this arises from turbulence in dry air giving fluctuations in density. At cm wavelengths the fluctuations in water vapour content predominate, whilst at wavelengths greater than $\sim$10 cm the effects of the ionosphere become overwhelming. In each case there is a spectrum of scales in the turbulence. Qualitatively there are three scales which are distinct in their effects on an image from an optical telescope. The very smallest scales give scattering over wide angles but the depth of the phase fluctuations is so shallow that little power is scattered from the incident wave and their effects can be ignored. On intermediate scales, but still smaller than the size of the telescope, speckles arise in the image if the phase fluctuations are significant ($\sim$1 rad). The largest scales produce phase tilts across the whole telescope and corresponding shifts in the image. The overall behaviour of the image is determined by the ratio of these last two effects, which depends mainly on the relative size of the telescope and the scale at which the phase variations reach $\sim$1 rad. The slope of the power spectrum of the turbulence is a second but usually somewhat less important factor.

The theoretical expectation for the turbulence in the dry atmosphere is for a Kolmogoroff spectrum. This is well confirmed experimentally, the chief arguments centering on the inner and outer scales of the turbulence. Somewhat surprisingly, the observed spectra of the phase fluctuations at cm and m wavelengths, corresponding to the water vapour and ionospheric components, also have very similar slopes. Whilst there is no theoretical explanation for

549

D. McNally (ed.), Highlights of Astronomy, Vol. 8, 549–550.
© *1989 by the IAU.*

550

this, it simplifies the description of the atmospheric turbulence to a single quantity, Fried's parameter r_o, the separation of two points for which the rms phase difference is ~2.6 rad, or for which the long-exposure seeing disk has an angular size $\theta \sim \lambda/r_o$.

Fig. 1 shows the variation of r_o/λ across the whole wavelength range from 0.1 μm to 10 m. The three sets of curves correspond to excellent and rare seeing, good seeing and very poor seeing (e.g. 0.5, 1 and 3 arcsec seeing for λ = 500 mm). They delimit the range of seeing for > 90% of the time on good sites, or rather on the sites where telescopes have been built; no radio array has yet been constructed on a site which has been adequately tested for radio seeing. The straight lines show telescopes of different sizes. This picture illustrates the familiar fact that a 1 m telescope is larger than r_o in the visual, giving speckled images, but smaller than r_o in the infra-red, giving diffraction limited images. It also shows results which are unfamiliar in this notation:

1. all radio dishes built so far are smaller than r_o.

2. many radio arrays are larger than r_o. Some may be larger than r_o at short cm wavelengths and metre wavelengths but smaller than r_o at intermediate wavelengths.

3. the numerical values of r_o/λ at optical and short cm wavelengths are very similar.

The second of these points clarifies the apparent complexity of comparisons between the atmospheric problems at optical and radio wavelengths. The problem is the same one; only r_o/λ changes rapidly with λ.

The third point suggests something more interesting. A 1 m telescope at λ500 μm is very similar to a 40 km radio telescope at λ2 cm. The imaging properties of radio arrays of this size such as the VLA with 27 × 25 m dishes, using aperture synthesis and closure phase or self-calibrations techniques are excellent, giving diffraction limited images with high dynamic ranges. This tells us that if an array of small apertures over 1 m baselines is used in the optical, then diffraction limited images of equal quality must also be achievable. The prospects for resolutions in the optical and IR equivalent to those in radio VLBI are even more exciting.

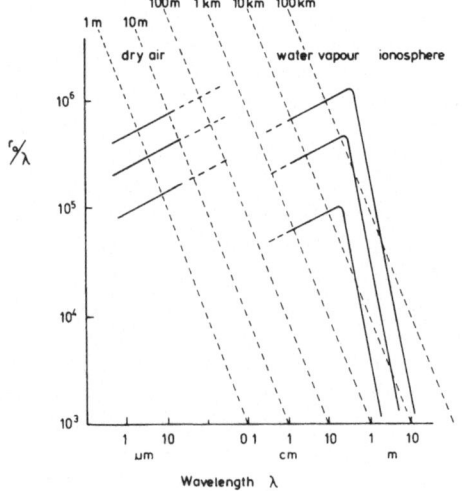

Figure 1. Atmospheric seeing as a function of wavelength.

REVIEW OF LINKED ARRAY INSTRUMENTS

R.D. EKERS
Australia Telescope
CSIRO Division of Radiophysics
Epping, N.S.W. 2121, Australia

At cm wavelengths aperture synthesis radio-telescopes (arrays of linked antennas which synthesize an image of the sky with high angular resolution) are now becoming the dominant astronomical research tool. Major new facilities such as the VLA are in full operation, others such as the Australia Telescope are nearing completion and a number of telescopes designed to form images in real time have been converted to operate in the aperture synthesis mode (e.g. MOST, Bologna Cross). See Napier et al. (1983) for a review of modern synthesis telescopes. The high resolution, sensitivity and freedom from confusion have led the aperture synthesis telescopes into very diverse astronomical applications.

In addition to the obvious advantages of high angular resolution for increased detail in an image and avoiding confusion, there are a number of other less obvious advantages. Deconvolution algorithms take advantage of the precision with which the beam of a linked array is known to correct for sidelobes. New algorithms, such as those described in a previous lecture by Cornwell, which can factor out antenna dependant errors, now enable imaging with extremely high dynamic range. The highest dynamic range images which have been made have a ratio of more than a million between the brightest feature and the faintest believable feature. There may also be dynamic range advantages for spectral line observations. Errors which modulate the gain of an antenna, and hence its system temperature, as a function of frequency do not normally affect the output of the interferometer since the receiver noise is uncorrelated. This has been particularly advantageous in imaging spectral lines which are a very small fraction of the continuum signal.

The ability of an array to filter out extended emission, corresponding to Fourier components with spacing less than the minimum spacing in the array, can be both an advantage and disadvantage. It is an obvious advantage to filter out atmospheric emission and its variability with time. In more specialised cases, for example to measure the Crab Pulsar in the presence of the extended emission from the Crab Nebula, the signal to noise can be improved with an array which resolves out the extended emission causing an increase in the system temperature. However it is a disadvantage when this filter distorts the structure (causing a bowl) and decreases the sensitivity to low brightness objects.

The computations required to form an image using observations from different spacings and different days must be done with considerable precision to avoid degradation of the beam quality. As a result astrometric accuracy is an automatic by-product of any synthesis observation.

In many cases where it is necessary to map an area of the sky the synthesis telescope has a considerable advantage because of the number of simultaneous beams which can be

D. McNally (ed.), Highlights of Astronomy, Vol. 8, 551–552.
© *1989 by the IAU.*

formed. In general, the speed of a synthesis telescope to map an area of sky is only matched by a single dish of the same total collecting area if it has an array of feeds equal to the number of elements in the synthesis telescope array.

A synthesis telescope measures the low spatial frequencies with less weight (and for very low spatial frequencies zero weight) so it is always relatively less sensitive than a single dish of the same total area for large diffuse objects.

Because of the large number of receivers required it is far more expensive to obtain frequency agility with an array than with a single dish.

An array generates its synthesized beam by manipulating the phase of the received signals instead of the delay. As a result it suffers from severe chromatic aberration and wide-bandwidth imaging is only possible when the band is divided into many small pieces. This adds to the complexity of the back end and the imaging process and will limit the total bandwidth to that available in the correlator. An instrument which forms its image in the focal plane is not affected by this form of aberration and much wider bandwidths may be employed.

To optimize the quality and quantity of science done with a linked array radio-telescope a balance must be found between two opposing factors. On one hand it is necessary to make the instrument readily available to users with interesting astrophysical problems but who are not "black belt interferometerists". On the other hand the most demanding observations and the unexpected discoveries are most likely to be made by astronomers with a clear understanding of the property of the instrument they are using (eg Harwit 1981). Both of these goals can be satisfied by providing interface between the user and the instrument which is as far as possible free from the use of technical jargon and other factors which obscure the underlying principles of operation. As systems become more complex and flexible the use of expert systems may be the best way to provide a suitable user interface.

The development of low noise solid state receivers has greatly simplified and improved reliability of arrays requiring large numbers of low noise front-ends. The limiting factor in the design of arrays with a large number of elements is now becoming the correlator since its complexity scales as the square of the number of elements. Here also large scale integration is making it possible to design even larger correlators. The use of fibre optic cables for IF transmission pioneered in the Australia Telescope will make wider bandwidths available and again the limitation will depend on the maximum bandwidth achievable in the next generation of correlator.

Some of the most dramatic changes have come from the developments of new algorithms of the type discussed in a previous lecture by Cornwell. Deconvolution algorithms are becoming more robust and the implementations more efficient. The cost for given computing capacity continues to decrease making increasingly complex imaging problems accessible. Forefront research in this area includes the deconvolution of data combining multiple instruments or multiple fields on the sky (mosaicing). Another development enables us to take advantage of the better spatial frequency coverage obtained through bandwidth synthesis by allowing for variations in both intensity and spectral index across the image.

In the area of self calibration current research involves questions of uniqueness, application of additional constraints based on physical models of the atmosphere and, in the case of low frequencies, solving the self calibration problem in the non-isoplanatic case.

References

Harwit, M. "Cosmic Discovery", Basic Books Inc., New York, 1987.
Napier, P.J., Thompson, A.R. and Ekers,R.D., Proc. IEEE 71, pp.1295-1322, 1983.

VERY LONG BASELINE INTERFEROMETRY

J.M. MORAN
Harvard-Smithsonian Center for Astrophysics
Cambridge, MA 02138, U.S.A.

The technique of very long baseline interferometry (VLBI) has undergone two decades of steady growth and refinement since its inception in 1967. In the beginning, only crude measurements of visibility on single baselines were possible. Now 18-station arrays have been used to produce images with dynamic ranges exceeding 2000:1; relative motions of cosmic masers have been tracked at the microarcsecond level of accuracy; and angular size measurements have been made with baseline lengths up to 2 two earth diameters with an orbiting satellite as a receiving element.

VLBI has two characteristics that usually differentiate it from linked-element interferometry: (1) coherence is maintained by independent atomic frequency standards at each station that provide stable local oscillator signals for heterodyne conversion and time tags for the received signal, and (2) the received signals are recorded on magnetic tape at the Nyquist rate (e.g., 10^8 bits s^{-1} for a bandwidth of 50 MHz) and processed at a later time. These characteristics are not fundamental but dictated by economics: the feasibility of distributing the local oscillator signal by satellite and the real time transmission of data via satellite have been demonstrated. Most linked-element radio interferometers maintain phase coherence by frequently observing calibration sources near the source to be imaged. Few such calibration sources are available to VLBI, so that maintaining coherence is a problem.

The first step in the postcorrelation data analysis is to find the interference fringes in the two-dimensional space of delay and fringe frequency, which is proportional to the rate of change of delay. This search can be carried out on each baseline. Alternatively, a global analysis of all the data from an array of telescopes can be carried out and clock (delay) and clock rate (fringe rate) parameters determined for each station (Global Fringe Fitting). This is closely related to the phase closure technique and is particularly effective in facilitating the detection of fringes on baselines formed with stations of low sensitivity. Once the fringe fitting analysis is complete, the data can be passed to image formation procedures such as self calibration or hybrid-mapping.

Currently, VLBI observations are conducted with telescopes scheduled on an ad hoc basis or with telescopes in networks that are organized for routine VLBI observations. The oldest network is the US VLBI Network, which was formed in 1976 and consists of six continental US stations and the MPI 100 m telescope. Observing sessions are scheduled for about 2 weeks every 3 months at wavelengths from 1.3 to 90 cm. The European VLBI Network (EVN) consists of nine telescopes. The operations of the EVN and US networks are closely coordinated. The IRIS (International Radio Interferometric Surveying) Network

D. McNally (ed.), Highlights of Astronomy, Vol. 8, 553–554.

consists of five telescopes, and, although it is dedicated to studying geodetic and geophysical phenomena, it provides useful astrophysical images. Other networks are under development in Russia (six elements), China (five elements), US (ten elements: VLBA), and Australia (five elements: Australia Telescope). The number of telescopes that have participated in VLBI experiments at one time or another is more than 50.

The VLBI images with the greatest dynamic range and detail have been produced at 18 cm wavelength with "world array" observations involving 18 telescopes. Maps of 3C120, 3C236 and M87 have unprecedented dynamic range. The high quality of these images is due to good (u,v) plane coverage, careful editing and calibration of the data, and the use of self-calibration programs.

The resolution of VLBI is ultimately limited by scintillation caused by irregularities in the ionized component of the interstellar medium. The seeing limit is approximately

$$\theta_s \sim \frac{1.5\lambda^2}{\sqrt{|sinb|}}\mu as,$$

where b is the galactic latitude, and λ is the wavelength in cm.

Imaging at mm wavelengths is an important new frontier in VLBI research. The image quality currently available resembles that achieved at cm wavelengths about 15 years ago. VLBI at mm wavelengths is difficult because of the lack of sensitivity, the small number of telescopes currently available, and the problems of phase stability at the shorter wavelengths. Nevertheless, images of 3C84 at 3 and 7 mm wavelengths have been produced with 4- and 6-station arrays at resolutions of 150 and 100 μas, respectively.

The Very Long Baseline Array (VLBA), now under construction by NRAO, consists of ten 25 m diameter antennas, located in Hawaii, in mainland U.S.A. and in the Virgin Islands. The array will operate at 9 wavelengths from 0.7 to 90 cm, providing a maximum resolution of about 200 μas at 0.7 cm. The VLBA correlator will be able to process data from up to 20 stations simultaneously and will provide 256,000 frequency-baseline or delay-baseline channels. It can be organized in many ways to accommodate different requirements (e.g., 20 stations [190 baselines], 8 IF bands, 128 spectral channels each). The VLBA is expected to improve greatly the quality and diversity of VLBI images because (1) the station locations provide excellent (u,v) plane coverage, (2) the electronics are well matched to reduce closure errors, (3) the array can operate well at short wavelengths, and (4) the large correlator makes ambitious spectral line projects possible.

The worldwide capability for VLBI work will continue to improve as more telescopes are equipped with terminals and frequency standards. The EVN has a plan to build a correlator of capacity similar to the one for the VLBA.

Interferometers with baselines longer than an earth diameter require stations in earth orbit. The feasibility of VLBI with satellite-borne stations has recently been demonstrated with the TDRS satellite. There are currently three viable projects for large antenna ($\sim$ 15 m) VLBI stations in earth orbit. The Russian project, RADIOASTRON, is well advanced, and the station could be deployed by 1991 in an orbit with apogee radius of 75,000 km. The QUASAT project is in a phase A study at ESA. Its initial orbit has an apogee radius of 42,000 km. Finally, plans for the Japanese project VSOP call for a station with apogee radius of 16,000 km. All these projects require a high degree of international cooperation to make effective use of ground stations and data transmission facilities. The relative timing of these projects will undoubtedly emerge as an important issue. Having two satellites available simultaneously would increase their (u,v) plane coverage greatly because of the different orbits and make satellite-to-satellite VLBI possible.

MILLIMETER WAVE INTERFEROMETRY

D. DOWNES
Institut de Radio Astronomie Millimétrique
Grenoble, France

Millimeter interferometers give high-resolution (few arc sec) information on cool (100-1000 K) matter, which cannot be easily studied by other techniques. The objects include regions of high extinction (A$_V$ ~ 10-200 mag), planetary atmospheres, outflows from forming stars, mass loss from stellar envelopes and molecular clouds in galaxies. The advantages which mm arrays have over cm arrays are the greater number of molecular spectral lines and the larger linewidths, which allow a higher sensitivity for the same angular resolution and system performance. In the continuum mm arrays can observe solar system objects, stellar winds, compact HII regions, galaxies and quasars in thermal and synchrotron emission, and thermal radiation from dust in dense clouds. The main technical challenges are (1) sensitivity, which limits the longest usable antenna spacing, (2) small field of view, and (3) atmospheric effects, which introduce phase noise. For the next 5-10 years, the main arrays will be those of the Berkeley-Illinois-Maryland collaboration, Cal Tech, Nobeyama and IRAM. There are proposals to add more antennas to existing arrays, to build new ones, and to extend their cover to higher frequencies.

The arguments for extending the cover to higher frequencies are that the Einstein A-coefficient for molecular transitions varies as ν^3 so, for optically thin lines, the brightness temperature, peak flux and frequency-integrated flux vary as ν^2, ν^4, and ν^5, respectively. For optically thick lines, the peak and frequency-integrated fluxes vary as ν^2 and ν^3. For dust, the continuum flux density goes as ν^3 or ν^4, depending on grain emissivity. With these strong dependencies a high frequency array may detect many sources, even if the system is noisier than for lower frequency arrays.

Important considerations for sub-mm arrays are the greater atmospheric extinction, which limits u,v plane cover with a 1-D array, even for high-Dec. sources. The observing cut-off will be ≤ 2 air masses, so a 2-D array is essential. The short wavelengths imply narrow fields and poor coverage of low spatial frequencies, unless the dishes are small (≤ 6 m). The periods of excellent sub-mm observing conditions are short, so there should be ≥ 6 antennas, giving ≥ 15 simultaneous baselines, for a reasonable map in one night.

There are two ratios which are lower for mm arrays than for cm arrays, maximum-to-minimum baseline, and field of view-to-source size. In the first of these ratios, the minimum baseline is about the dish size, while the maximum useable baseline is limited by sensitivity. Unlike most cm arrays, mm arrays have had smaller and fewer antennas, higher system temperatures, more atmospheric problems, and weaker calibrator sources. For mm arrays until now, the minimum detectable brightness temperature in the synthesised beam is a few K at a 100 m baseline. However, the temperatures of thermal line sources are also

555

D. McNally (ed.), Highlights of Astronomy, Vol. 8, 555-556.

a few K, so it has been hard to get useful signal-to-noise ratios at baselines > 100 m.

The synthesised beams for typical baselines used at 3 mm are 4″ 11″, close to the sizes for which information is missing from the maps because of the minimum baseline. At 2.6 mm, a minimum spacing of 25 m reduces the visibility of 9″ structures by half, and structures > 20″ to zero. In principle, missing zero-spacing information can be recovered from maps by single dishes like the Onsala 20 m, IRAM 30 m and Nobeyama 45 m telescopes. Other ways to get the short-spacing information are scanning the antennas during the synthesis and sub-aperture illumination of the dishes.

The second ratio which is less favourable for mm arrays than for cm arrays, is the field of view (∼ 1′ at 3 mm) relative to the size of typical sources, namely several arc min for molecular clouds and nearby galaxies. To overcome this problem, a mosaic of adjacent fields could be observed by multi-beaming of several fields simultaneously, or by observing the fields sequentially, with only one receiver per antenna. Another possibility might be to combine data from different arrays, e.g., the Hat Creek array could synthesise a large field at short baselines, and the IRAM array could observe the field hot spots at longer baselines.

Another effect with stronger consequences for mm arrays than for cm arrays, because of the smaller primary beams, is anomalous refraction. In these events sources appear to move by ≤ 20″ for ≤ 30 sec of time. Anomalous refraction is caused by variations in the "wet" component of the refractive index and in mm and sub-mm interferometry it will increase phase noise, broadening the synthesized beam, and will reduce fringe amplitude as sources move out of the narrow primary beams.

As the number of antennas in mm arrays increases, it may be possible to use optical interferometry methods to overcome atmospheric phase errors, such as the bi-spectrum method in which triple products of the complex visibility are formed on short time scales, then ensemble averaged to improve sensitivity, and then used to derive phases. This method may improve phase closure or self-calibration methods by 3-10 times in sensitivity.

Table 1 lists current and proposed mm arrays, representing different approaches to solving these problems. Probably both types of array will be useful, those with wide field of view, and those with high sensitivity on longer baselines. Given the active interest in enlarging existing arrays, in constructing new and more powerful ones, and the rewards to be obtained at higher frequencies, mm interferometry obviously has a very promising future.

TABLE I Existing and proposed millimeter and submillimeter arrays

Array/Status	Number of Antennas	Instantaneous Area $= n\pi D^2/4$ (m^2)	Shortest Usable λ (mm)	Relative Speed for full synthesis
Hat Creek (in operation)	3 × 6 m	85	1.3	1.0
Hat Creek (funded)	6 × 6 m	170	1.3	5.0
Owens Valley (in operation)	3 × 10.4 m	255	1.3	1.0
Nobeyama (in operation)	5 × 10 m	393	1.3	5.0
Nobeyama (future)	+45 m	1983	1.3	5.0
Plateau de Bure (construction)	4 × 15 m	530	0.8	1.0
Plateau de Bure (funded)	4 × 15 m	707	2.6	2.0
Australia T'scope (later phase)	5 × 15 m	884	2.6	3.3
Smithsonian (proposed)	6 × 6 m	170	0.35	5.0
NRAO (current idea)	40 × 7.5 m	1740	∼ 1.	63.3

METER WAVE INTERFEROMETRY

G. SWARUP
Tata Institute of Fundamental Research
P.B. 1234
Bangalore - 560012, India

There are many outstanding astrophysical problems which are best studied at meter wavelengths. However this part of the spectrum has not been fully exploited for astronomical observations for several reasons. Firstly, the arrays at cm wavelengths provide higher resolution. Secondly, the level of man-made radio interference is very high at the longer wavelengths in industrialised countries, although it is not so high in countries such as India. Finally, the phase variations caused by the ionospheric irregularities are quite severe at meter waves, particularly for a large array ($\sim \lambda/35$ rad.km^{-1}).

Recent developments of self-calibration and closure-phase techniques have provided a revolutionary breakthrough for interferometric data processing in the presence of atmospheric phase variations (Cornwell and Wilkinson, 1981). However, the large field of view of each element of the interferometric array poses two very severe constraints for data processing: (a) the need for a three-dimensional rather than the usual 2-D Fourier transform due to the "w" term for a large non-EW array such as the VLA at 327 MHz and (b) the wavefront is not isoplanatic, i.e. phase variations are not only functions of time but also of the position of each source in the field of view. Schwab (1983) proposed the division of the field of each antenna into m cells, but this increases data processing nearly m times. Using the MX task for AIPS, maps are being made for 327 MHz observations at the VLA following a Hybrid approach in which maps are being analyzed over only a square degree in the first instance and a larger mosaic is patched up subsequently. Subrahmanya (1988) has recently proposed that the ionospheric region corresponding to the field of view of the array be divided into m cells, which minimizes the number of unknown variables. It seems that the new generation of relatively low cost parallel/distributed processing/vector machines and also new algorithms may allow achievement of high dynamic range over a wide field of view of a meter wave interferometer.

A Giant Meterwave Radio Telescope (GMRT) is being set up at Khodad, about 80 km north of Pune in India, for high resolution mapping of galactic and extragalactic radio sources. One of the primary aims of the telescope is to search for the highly redshifted 21 cm spectral line emitted by neutral hydrogen clouds to determine the epoch of formation of galaxies.

GMRT will consist of 34 fully-steerable parabolic dishes of 45 m diameter each. A novel design has been adopted for the dishes. Sixteen of these would be placed in a compact central array of about 1 km in size, including nine forming a 3×3 grating to facilitate deep searches for new pulsars. The other 18 antennas will be placed along the arms of a Y-shaped array with each arm extending to about 14 km. GMRT is being designed to operate at

557

six different frequency bands between 38 and 1420 MHz. The reflecting surface of the dish will be made of a mesh of stainless steel wires of 0.55 mm diameter. The size of the mesh in the central part will be 10 ×10 mm, in the middle 15 ×15 mm and in the outer part it will be 20 ×20 mm. The surface accuracy of the reflecting surface will allow operation up to about 1420 MHz. The multi-frequency operation is planned by mounting the different fe and low noise eds RF amplifiers on a rotatable turret placed near the focal point of the dish. Optical fibre links will be used to distribute coherent local oscillator signals to all the antennas and to bring received antenna signals to a central laboratory. A 256- channel FFT unit with a bandwidth of 16 MHz placed after each antenna (with > 35 dB sideband rejection) followed by a complex correlator (FX system) is planned to cross-correlate outputs of all the 34 antennas for both polarizations, providing about 560,000 complex outputs. Of these, a maximum of 140,000 channels will be available for recording after rejection of any channels in which radio interference is present.

GMRT will be a versatile instrument for radio astronomy research. It is to be completed by 1992.

References

Cornwell, T.J. and Wilkinson, P.N., 1981, *Mon. Not. R. astr. Soc.,* **195**, 1067.

Schwab, F.R., 1983, VLA Scientific Mem. No. 151; *Proc. SPIF*, 1984.

Subrahmanya, C.R., 1988, in preparation.

LONG BASELINE OPTICAL INTERFEROMETRY

S.T. RIDGWAY
Kitt Peak National Observatory
National Optical Astronomy Observatories[1]
Box 26732, Tucson AZ 85726-6732,U.S.A.

During the last three years significant results have been obtained from several operational, long baseline optical interferometers. Precision stellar angular diameters (accuracy of order 2% and better) have been reported in the infrared (DiBenedetto and Rabbia, 1987) and in the visible (Davis and Tango, 1986). Astrometric precision of order 20 milliarcsec has been demonstrated over large angles (Shao et al., in press). Spectro-spatial resolution of the disk of a Be star in the hydrogen emission line (Thom et al., 1986) suggests spectacular imaging science to come with many-telescope coherent and phased optical arrays.

Stimulated by such results, by the success of radio telescope arrays, and by rapid progress in control, detector, and optical fiber technology, a large number of interferometer projects have been initiated. The accompanying table summarizes the information currently available to me concerning these projects. It seems quite remarkable that by the year 1995 we may have more than 15 operational optical interferometers.

The first instruments coming on-line during the next five years provide a capability for the measurement of accurate stellar diameters for samples of most spectral types, will provide direct observations of stellar pulsations in many variable types, will provide elementary imagery of circumstellar material in a variety of early and late type stars, especially including disks around YSO's, will yield direct orbit determination for large numbers of spectroscopic binaries, will provide precision astrometry of large numbers of bright stars, and many other applications will surely be demonstrated.

The capital investment in long baseline optical interferometry is significant. Nevertheless, most of the long baseline plans involve small numbers of apertures, hence are not well suited for imaging, and the few many-aperture instruments are limited by cost constraints to very small telescopes, hence bright limiting magnitudes. Thus these projects, ambitious though they may appear, are primarily a proving ground, where astronomers will master the techniques for the ground and space observatories of the next century.

References

DiBenedetto, G.P. and Rabbia, Y., 1987, *Astron. Astrophys.*
Davis, J. and Tango, W.J., 1986, *Nature* **323** 234-235.
Shao, M.*et al. Astron. Astrophys.* (in press).
Thom, C. *et al. Astron. Astrophys.* **165**, L13- 15.

---

[1]National Optical Astronomy Observatories, operated by the Association of Universities for Research in Astronomy under contract to the National Science Foundation

D. McNally (ed.), Highlights of Astronomy, Vol. 8, 559–560.
© *1989 by the IAU.*

TELESCOPE ARRAY PROJECTS[1]

Project Name (Center or Location)	Telescopes No.	Diameter (cm)	Baseline (m)	Array Shape	Status Notes
(Interferometric Operation Demonstrated)					
I2T (CERGA)	2	25	144	N-S	1
G12T (GERGA)	2	150	70	N-S	2
MMT (Mt. Hopkins)	6	180	6.8	Hexagon	3
Mark III (Mt. Wilson)	2	7.5	12	N-S	4
SUSI-Prototype (U. Sydney)	2	10	11.4	N-S	5
SOIRDETE (CERGA)	2	100	15	E-W	6
(Funded for Construction)					
SUSI (U. Sydney)	2	14	640	N-S	5
Optical IF (U. Erlangen)	16	11	7.5	Line	
Distributed Array (NOAO)	5	60	100	Star	7
IRMA (U. Wyoming)	2	10	15	N-S	8
IOTA (CFA)	2	45	50	N-S	9
COAST (Cambridge)	4	40	100	Y	10
VLT (ESO)	4	800	150	Linear	11
Astrometric IF (SAO)	3	100	20		12
Imaging IF (SAO)	6	60	100		13
(Planned)					
Columbus (U. Arizona)	2	800	22	Rot. Line	14
CHARA array (Georgia State U.)	7	100	400	Cobweb	15
VISIR (France)	3	150	300	Cross	16
OVLA (CERGA)	27	150		Platform	17
NNTT (NOAO)	4	800	22	Square	18

1. I2T - Baseline extended, new lab. telescopes being refurbished for remote operation.
2. GI2T - Operation interrupted for drive mods., new computer and correlation system.
3. MMT - All reflective beam combiner; coherent speckle; phased operation planned.
4. Mark III - install baselines up to 30 m, improvements in limiting mag.
5. SUSI - Prototype closed after successful tests; first light expected early for SUSI.
6. SOIRDETE - A recent, experimental modification of an existing 2-telescope facility.
7. NOAO - Construction of first 2 telescopes suspended due to NOAO budget reductions.
8. IRMA - First light expected in August 1988.
9. IOTA - Contracts for telescope fabrication to be committed soon; assembly in 1989.
10. COAST - two telescopes funded and under construction.
11. VLT - funded detailed interferometric beam combination plan not yet developed.
12. Astrometric IF - Hope to reach quasars; telescope diam. cost factors under study.
13. Imaging IF - 6 t'scope for phase and amp. closure; t'scope diam/cost under study.
14. Columbus - Detailed design study in progress, some funding already committed.
15. CHARA - Detailed design study in progress, some funding already committed.
16. VISIR - Engineering study at IRAM of telescope designs for interferometry.
17. OVLA - One of several concepts with many innovative features.
18. NNTT - Suspended after years of design, due to slim prospects for early funding.

---

[1]A collection of project descriptions like this is certain to contain some errors, if only because plans change. I apologise in advance for any inaccuracies which I have introduced.

SPECKLE INTERFEROMETRY

J.C. CHRISTOU
Kitt Peak National Observatory
National Optical Astronomy Observatories[1]
Box 26732, Tucson AZ 85726-6732, U.S.A.

Speckle interferometry is a technique which utilizes the full diffraction-limited imaging potential of ground-based telescopes. Short exposure images, or specklegrams, with an exposure time less than that of the atmospheric correlation time ($\sim$5- 50 ms) preserve the high-spatial frequency information lost in long exposure imaging. In 1970, Labeyrie computed the power spectrum of a set of specklegrams and showed that they contained diffraction-limited information. Since then the field has grown with improvements in both instrumentation and the phase recovery algorithms necessary for imaging. It has been applied at both visible and near-infrared wavelengths although, until recently, the latter has used slit-scanning techniques with single pixel detectors because of the lack of array detectors. The current state of speckle interferometry has been well covered in the proceedings of two recent joint National Optical Astronomy Observatories - European Southern Observatory workshops on Interferometric Imaging in Astronomy (Oracle, 1987 & Garching, 1988).

A number of phase recovery algorithms have been investigated and of these the bispectrum algorithm developed by the Erlangen group is the most attractive as it represents a phase closure technique. However, it is cumbersome in that the bispectrum of a 2-D image is four dimensional. Recent research by a number of groups (Garching, 1988) suggest that it is not necessary to compute the whole bispectrum, especially for visible data, and the advent of faster and larger processors make it attractive. The widely used Knox-Thompson algorithm is a subset of the bispectrum algorithm. Other phase recovery techniques being used include variants on "Shift-and-add" and Speckle Holography. Included in the latter is a technique which simultaneously measures the wavefront and a specklegram. The former can be used to measure an instantaneous point spread function, thus eliminating the requirement of a point source.

Calibration is a major problem in speckle interferometry. Accurate Fourier modulus recovery requires a point source observed under essentially identical seeing conditions. In addition, noise biases contaminate the measured Fourier spectra. For infrared data these biases are detector induced whereas for visible data they come about because of photon statistics. Their removal is not always straightforward with faint object data being affected more severely than bright object data. In addition, detectors contaminate the signal from convolution with a detector response function affected by image tube "pincushion". A

[1]Operated by the Association of Universities for Research in Astronomy Inc., under contract with the National Science Foundation

D. McNally (ed.), Highlights of Astronomy, Vol. 8, 561–562.
© *1989 by the IAU.*

description of these calibration problems is given by Christou (Garching, 1988).

Two types of detectors are currently in use for visible speckle interferometry. These are (i) intensified television systems which record photon events over more than one pixel within a standard 30 Hz frame, e.g. the Steward Observatory and Erlangen systems, and (ii) the newer generation of individual photon counting systems (IPCS) such as the PAPA and MAMA detectors which also time-tag the events. The latter tend to be more stable and calibratable. At near-infrared wavelengths 2-D detectors are now coming into use, i.e. the NOAO IR speckle camera (Bekers et $al.$, Garching, 1988). The NOAO system uses a 58×62 SBRC InSb array. It is currently limited by readout noise, ≤ 400 e^{-}, but as the technology improves larger format detectors with reduced noise will soon be available.

Speckle interferometry has been applied to a broad variety of astronomical objects, from bright objects such as α Orionis to Pluto/Charon (15th magnitude) and the 17th magnitude triple QSO, PG 1115+08. Astrometric measurements of binary and multiple stars represent the majority of the results. In addition there have been measurements of stellar diameters, solar granulation, circumstellar environments and Active Galactic Nuclei. However, until recently, most of these results were obtained through either power spectrum or autocorrelation analysis. With the introduction of improved algorithms, more systematic error free detectors and careful calibration, reliable images have been produced.

A large number of speckle interferometry papers are technical in nature as is evidenced from the NOAO/ESO workshops. However, a perusal through the recent astronomical journals show that there have been ~ 30 papers utilizing speckle interferometry results since January 1987. Of these, the majority ($\sim 50\%$) report multiplicity. However, a number of these, most notably the one-dimensional infrared results, have not only produced astrometric measurements but also magnitude differences and images. Approximately one half of the other papers report stellar diameters, circumstellar shells etc. Although power spectral analysis is the most common, a number of these papers also show reconstructed images, in particular the MMT and 4m measurements of the atmosphere of α Orionis. Most of the other papers have dealt with improved application of the technique. In addition, images of the asteroid 4 Vesta have been reported, the size of SN1987A has been measured at a number of wavelengths and an apparent transient companion has also been detected. The application of speckle interferometry to highly extended objects is also represented with power spectral analysis of solar granulation. The improved understanding of the technique apparent from the workshop proceedings will lead to more imaging papers in the astronomical literature. The use of large aperture instruments such as the family of 4m class telescopes and the co-phased MMT (6.86m) is already producing images with the same resolution as 2cm VLA maps.

Future developments will include (i) the improved resolution of the larger telescopes currently in development, (ii) the larger format detectors especially for near-infrared imaging which will also have lower readout noise and (iii) improved processors which will allow for real-time image reconstructions giving the observer a preliminary idea of the object's morphology. In addition there is improvement of the technique either by the use of wavefront sensing or by the use of partially- or non-redundant apertures, obtained by masking the pupil. Recent results from pupil masking observations suggest improvement in the signal-to-noise ratio.

References

Proceedings of Joint ESO-NOAO Workshop on Interferometric Imaging in Astronomy, Oracle, January 1987, Ed. J.W. Goad, NOAO, 1987.

Proceedings of Joint NOAO-ESO Conference on High Resolution Imaging by Interferometry, Garching, March 1988, Ed. F. Merkle, ESO, 1988.

INFRARED LONG BASELINE INTERFEROMETRY

W.C. DANCHI, M. BESTER, P.R. McCULLOUGH and C.H. TOWNES
Space Sciences Laboratory, Physics and Astronomy Departments
University of California at Berkeley
Berkeley, CA 94720, U.S.A.

During the last few years, two new instruments using long baseline interferometry have been constructed for high angular resolution astronomy in the mid-infrared spectral region (8-12 μm). One called SOIRDETE–Synthese d'Overture en InfraRouge a DEux TElescopes– was built by J. Gay and his collaborators at CERGA. SOIRDETE has a fixed E-W 15 m baseline and two 1 m diameter telescopes of conventional design. This instrument obtains interference fringes by adjusting an optical-precision delay line in discrete steps to compensate for the geometrical delay of the projected baseline. The interference fringe from the source is detected using HgCdTe photodiodes. Because the instrumental delay has discrete steps a time-domain interferogram is created. This interferogram, upon Fourier transformation to the frequency domain, yields information about the spectral characteristics of the source. First fringes have recently been obtained with this instrument (Gay, 1988).

The other instrument, which uses heterodyne detection of the infrared signal, was built by our group and is called ISI - Infrared Spatial Interferometer. It is constructed from moveable telescopes of a novel design, which allows a number of baselines to be sampled depending on site geometry. ISI has recently obtained its first interference fringes. Because ISI is constructed of telescopes of unusual design, and since its first fringes have not been reported elsewhere, this report gives an overview of its design and initial results.

The ISI telescopes are of Pfund-design and each consists of a 2.03 m diameter flat mirror that reflects star light to a stationary 1.65 m diameter parabolic mirror. The flat mirrors sit on alt-az mounts and track the source. Each parabolic mirror brings light to a focus behind the front surface of its flat mirror through a hole in the mirror. The choice of geometry allows a range of $\pm55°$ in azimuthal rotation angle from where the flat mirror points directly at the parabola, and $-2°$ to $+55°$ in altitude. Unlike conventional alt-az mounts, the Pfund-design has no singularity at the zenith. The flat and parabolic mirrors are mounted in a special semi-trailer. On site, the flat and parabolic mirror mounts rest on kinematic supports sitting on firm, reinforced concrete bases. The trailer bed, which is mechanically decoupled from the mounts, contains an optics room and control room behind the flat mirror. The current site allows for E-W baselines ranging from approximately 4 to 28 m and N-S baselines up to about 15 m, as well as baselines of intermediate angles (cf. Townes, 1984; Townes et al.,, 1986; Danchi et al.,, 1986).

Infrared and visible light are separated by a dichroic mirror near the prime focus of the parabola. The visible light is imaged on a solid-state MOS camera for acquisition and guiding. The infrared radiation propagates onto an optical table, where it is combined at a beamsplitter with CO_2 laser radiation and detected on a HgCdTe photodiode. This

563

D. McNally (ed.), Highlights of Astronomy, Vol. 8, 563–564.
© *1989 by the IAU.*

heterodyne detection scheme is conceptually similar to the technique used in centimeter and millimeter wavelength interferometry. The theoretical sensitivity of the system is sufficient to allow observation of infrared sources about six magnitudes weaker than the brightest sources. A detailed discussion of the detection system and its sensitivity limits has been reported by Danchi *et al.*, (1988).

After a design and construction phase lasting from October 1983 to December 1987, ISI was moved to Mt. Wilson, in January 1988 and installed with a 4 m E-W baseline. ISI obtained its first fringes on IRC +10216, displayed in Fig. 1, on June 29, 1988. To interpret Fig. 1 it is important to know some details about the detection system. The two CO_2 laser local oscillators are offset in frequency to take out the normal time variation of the fringe frequency, which results from the changing projected baseline due to the rotation of the earth. As a result, the fringe frequency at the output of the correlator is fixed and, for convenience, we have chosen a value of 10 Hz. Fig. 1 shows a narrow peak at 10 Hz as expected. The width of the peak is limited by the finite length of the scan to about 0.01 Hz. This peak sits on a broader, shallower peak of approximately 1 Hz width, which is likely due to atmospheric fluctuations. Further work is necessary to obtain the visibilities of this source and others as well as the spectrum of phase fluctuations due to the atmosphere.

To summarize, the Infrared Spatial Interferometer has just recently obtained its first fringes and should soon begin producing results on many late-type stars, infrared sources embedded in dark molecular clouds, and other interesting astronomical objects.

References

Danchi, W.C., Arthur, A., Fulton, R., Peck, M., Sutton, E.C., Townes, C.H. and Weitzmann, R.H., 1986, *Proc. of the SPIE Conference on Advanced Technology Optical Telescopes III*, **628**, 422.

Danchi, W.C., Bester, M. and Townes, C.H., 1988, *Proc. of the NOAO/ESO Conference on High Resolution Imaging by Interferometry*, Ed. F. Merkle, Garching, FRG.

Gay, J., 1988, private communication.

Townes, C.H., 1984, *Astrophys. and Astron.*, **5**, 111.

Townes, C.H., Danchi, W.C., Sadoulet, B. and Sutton, E.C., 1986, *Proc. of the SPIE Conference on Advanced Technology Optical Telescopes III*, **628**, 281.

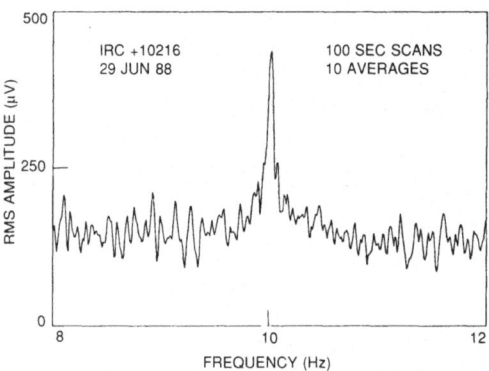

Figure 1. RMS amplitude of an interference fringe as a function of frequency for IRC +10216. This is the first fringe from an astronomical source obtained by ISI. Note the narrow peak at 10 Hz, which sits on a shallow, broad feature of width about 1 Hz, also centered at 10 Hz.

ACTIVE CONTROL AND ADAPTIVE OPTICS FOR OPTICAL INTERFEROMETERS

FRITZ MERKLE
European Southern Observatory
Karl-Schwarzschold-Str. 2, D-8046 Garching
Federal Republic of Germany

Long baseline interferometry requires the full phasing of a telescope array. Especially for future arrays with large unit telescopes active control systems are mandatory. Adaptive optics can be applied for real-time phase compensation of the individual pupils due to atmospheric distortions. Additional to phasing of the individual pupils of independently mounted telescopes, the whole array has to be phased, including pupil position corrections due to pupil foreshortening and shift effects in order to reach a reasonable phased field-of-view.

From the beginning it can be excluded that a linear array and its large unit telescopes offer passive stability sufficient for coherent beam combination in interferometry. The finally tolerable vibrational amplitudes are less than a tenth of a wavelength, which is not achievable, even in the 10 to 20 μm wavelength ranges. Therefore, it is important to develop techniques for active stabilization of the light beams, after they have left the unit telescopes. Nevertheless, the unit telescopes should be as stable as possible, in order to limit these stabilization efforts. For pathlength control of the telescope and combining optics, laser interferometers with closed loop active stabilization of the opto-mechanical system are proposed.

In order to guarantee phasing of the desired field-of-view, off-axis as well as on-axis for an array of telescopes, the Lagrange invariants of individual telescopes including their combining trains must be equal, and additionally the overall Lagrange invariant of the array must be conserved. The co-phasing of an array of independent telescopes differs significantly from the requirement for telescopes arranged in the same mechanical mount (like multi-mirror telescope type interferometers). This has a major impact on the optical design of the final combination optics. In order to maintain the necessary geometrical scaling of the lateral pupil geometry, the separation of the pupil images at the combining optics has to be adjusted according to the zenith angle. Additionally, the longitudinal pupil position needs a correction due to the shift of the relative locations of the optical elements in the combining train with respect to the wavefronts. Besides these pupil corrections, which determine the off-axis phasing, the overall on-axis phase difference between the telescopes to be combined has to be compensated. In order to perform the necessary pathlength compensation an optical delay line is proposed for each telescope. These pathlength and pupil corrections are mandatory for interferometers composed of large telecopes.

Obtaining diffraction limited images or spectra from the ground, either with a single large pupil or with two or more smaller pupils, suffers basically from the same limitation: the fluctuations of the index of refraction of the atmosphere due to turbulence. For long baseline

D. McNally (ed.), Highlights of Astronomy, Vol. 8, 565–566.
© *1989 by the IAU.*

interferometry the gain of large apertures is only given in combination with adaptive optics. Otherwise the signal-to-noise ratio will not be improved compared with interferometers with smaller apertures. Phasing the individual pupils, at least partially, is a fundamental requirement for the efficient use of a large aperture interferometer. Large scale atmospheric wavefront aberrations have to be corrected by a measurement of the relative phase between the telescopes to be combined. The same correction system which is used to compensate phase fluctuations due to mechanical vibrations could be applied. Fringe tracking will allow long integration times resulting in a considerable sensitivity gain.

GALACTIC AND EXTRAGALACTIC APPLICATIONS

G.B. FIELD
Harvard-Smithsonian Center for Astrophysics
Cambridge, MA 02138, U.S.A.

For the purposes of this talk I will take "high resolution" to mean 1 milliarcsecond (mas). Only objects of high surface brightness when averaged over 1 mas are considered; as H. McAlister will discuss stellar applications, I will speak only about nonstellar applications.

Interesting nonstellar objects in the Galaxy include disks in binary systems, in protostellar systems, and in the galactic center. Unfortunately, accretion disks in X-ray binaries are probably too small to be resolved, about 0.03 mas in CygX-1. A protostellar disk in Orion at 400 pc could be resolved if its diameter exceeds 0.4 AU; as the temperature is low, this requires infrared or radio observations. At the distance of the galactic center, 1 mas corresponds to 10 AU; if there is a black hole of $10^6 M_\odot$, the material in its accretion disk would have a velocity at that distance of 10,000 km s^{-1} and a dynamical time scale of the order of a day. VLBI observations have already shown that the radio source is about 10 AU across. Infrared or longer wavelengths are necessary to penetrate the extinction.

An angular resolution of 1 mas at optical wavelengths would permit new studies of stellar kinematics, as a proper motion of 1 mas/y corresponds to 150 km s^{-1} at 30 kpc. The proper motions of selected stars in globular clusters would make it possible to derive the space motions of globular clusters and hence to constrain models of the galactic gravitational field at substantial galactocentric distances. Imaging at the 1 mas level would reveal stars in the cores of globular clusters without confusion.

At the distance of M31, 1 mas corresponds to 0.004 pc, sufficient to probe concentrations of stars at the center of M31 or its companions. At the Virgo cluster 1 mas corresponds to 0.05 pc, about 1/10 of the upper limit (from VLBI) for the sizes of hot spots in the jet of M87. As the hot spots could be sources of energetic electrons (we know from lifetime arguments that electrons don't propagate all the way from the central engine), it would be interesting to study them both in the optical and with the improved dynamic range to be expected in the radio with VLBA. High-resolution optical emission line studies might reveal whether the gas in a cooling flow at the center of M87 forms stars or falls further into the nucleus, perhaps feeding the black hole believed to be there.

Supernovae occur often in the Virgo cluster, and a light echo from a supernova travels 0.3 pc = 1 mas per year. Study of such echoes would provide a way to estimate the distance of Virgo.

Seyferts, about 1% of all galaxies, contain active nuclei that are fainter than quasars. At the 10 Mpc distance of the Seyfert galaxy NGC 4151, 1 mas corresponds to two light months. As its optical light varies on a time scale of months, observations could help to clarify the mechanism responsible. At the quasar 3C273, whose redshift is 0.15, 1 mas

567

D. McNally (ed.), Highlights of Astronomy, Vol. 8, 567–568.
© *1989 by the IAU.*

corresponds to 10 ly in a $q_o = 1/2$ universe with $H_o = 50$. Light variations occur on the scale of years, and it would be interesting to find out whether the broad emission lines originate at this radius or less in 3C273.

Radio observations of quasars reveal jets which carry energy from a central engine probably located in the nucleus of a galaxy. At a VLBI resolution of ~ 1 mas, jets appear to originate in a high-brightness core which is self-absorbed at centimeter wavelengths. It is important to penetrate into the core with the highest possible angular resolution and at the highest possible frequency. The VLBA, operating at its shortest wavelength, 7 mm, will be able to do this. A preliminary study, carried out with a few of the antennas of the VLBA, has recently achieved a resolution of 0.1 mas on 3C84 at a wavelength of 7 mm.

Improved dynamical range at high resolution will also bear on the problem of super-luminal motion in quasar jets. The model that assumes that superluminal motion is an optical illusion due to nearly luminal motion of a quasar jet directed nearly at the observer and seen more readily because of Doppler boosting, runs into the difficulty that such alignments are too improbable to account for the high incidence of the phenomenon. Perhaps the alternative suggestion, that jets themselves are rarely pointed at the observer, but that energetic acceleration towards the observer occurs here and there within the jet, is more likely. If so, the greater dynamic range of VLBA as compared with previous VLBI experiments will be useful in identifying the features not directed toward the observer, but much fainter because of the lack of Doppler boosting.

As we consider objects of large redshift, such as Lyman-α galaxies with redshifts from 1.6 to 3.8 and quasars with redshifts up to 4.1, we recall that in relativistic models of the universe, the normalized angular diameter $c\delta/H_0$ subtended by an object of size D is D times a function $f(z)$ of redshift z that is z^{-1} for $z \ll 1$ and that flattens out for $z \gtrsim 0.5$. For $q_0 = 0, f(z)$ decreases from 3.8 to 2.0 as z increases from 0.5 to 4, while for $q_0 = 1/2, f(z)$ decreases from 4.3 to a minimum value of 3.4 at $z = 1.25$, then increases again to 4.5 at $z = 4$. If we adopt $c\delta/H_0 \sim 3D, \delta = 1$ mas corresponds to $D \sim 10$pc in these objects. The most distant objects are of great interest because they are being observed at the greatest look-back time. If there is a critical density of cold dark matter in the universe, galaxies are expected to form over the range $z = 0$ to 5, with a peak of activity at $z = 3$; this theory can be tested by studying star formation at various redshifts. The Lyman-α galaxy 0902+34 with a redshift of 3.4 has an exponential dependence of brightness on radius, with characteristics scale 2200 mas = 22 kpc, much larger than the high resolution element under discussion. If Lyman-α line is due to many HII regions ionized by hot young stars, this could be tested with a high-resolution image.

Finally, one always hopes that a "standard rod", expected to have the same length in galaxies of all redshifts, will be discovered, as observations of such an object would reveal its angular size, and, when z is known, $f(z)$ as explained above. Comparison with theory yields q_0. Although no such objects are known now, the enormous increase in resolution in going to 1 mas improves the chances they will be discovered.

THE APPLICATION OF OPTICAL ARRAYS TO SOLAR SYSTEM AND STELLAR PROBLEMS

H.A. McALISTER
Center for High Angular Resolution Astronomy
Georgia State University
Atlanta, Georgia 30303, U.S.A.

Unprecedented activity exists in very high angular resolution astronomy at optical wavelengths. Applications of telescope arrays to modern problems have been described in publications too numerous to list here. For illustrative purposes, we shall here consider an array with a maximum baseline of 350 m corresponding to a limiting resolution at the first visibility null of about 0.3 milli-arcsec at 550 nm wavelength.

For solar system applications, such an interferometer would be capable of resolving structures as small as 200 m from a distance of 1 a.u. The imaging of minor planets could be carried out for thousands of these objects pending instrumental sensitivity limits. The Galilean satellites would be resolvable with more than 20 million resolution elements showing details less than 1 km across. Pluto and Charon could be inspected with 9 km resolution. These exciting possibilities may be somewhat remote from present analysis capabilities, but the feasibility of imaging highly extended objects with low surface brightnesses must also have seemed remote to the pioneers of radio aperture synthesis imaging.

Multiple telescope arrays will be the premier instruments for stellar astrophysics during the coming decades, their applications to extragalactic astronomy being somewhat more problematic due to sensitivity considerations. The angular diameter of a star when combined with the measured integrated flux from the entire photosphere yields the emergent flux at the stellar surface. Emergent flux, or its equivalent, effective temperature, is the basic parameter describing stellar radiation, required in many fundamental theoretical and observational applications. These data, when obtained in large samples as functions of spectral type, luminosity class, metallicity, rotation rate, etc., will provide a powerful means of testing model atmospheres. Measurement of angular diameters provides the only means of directly determining effective temperatures, except for the relatively small number of eclipsing binaries with accurately measured parallaxes and well-determined linear radii. Equally important is the calibration of temperature sensitive photometric indices, thus effectively extending the method to distances as great as the particular index can be measured. Davis has shown that diameter accuracies are required at the $\pm 2\%$ level if a solid calibration is to be applied to indices such as the $(B-V)_0$ scale.

Along the main sequence, the interferometer could resolve arbitrarily large samples of stars for all spectral types except for the very hottest and coolest stars. The rarity of O-type stars is the limitation although there are some 15 resolvable O stars, primarily in the southern sky. Success with M-type dwarfs is a question of sensitivity. For main sequence stars with types ranging from B to K, hundreds of stars of each class are readily accessible. Thousands of cool giants and supergiants are resolvable out to kpc-scale distances.

569

D. McNally (ed.), Highlights of Astronomy, Vol. 8, 569–570.
© *1989 by the IAU.*

Arrays will resolve the photospheres of variable stars exhibiting radial and non-radial pulsations. Scores of Cepheids and Mira variables and hot stars such as the Beta Cephei stars will be resolved. The direct calibration of the Cepheid period-luminosity relation follows from the careful combination of interferometric and spectroscopic results. The extended atmospheric phenomena associated with Be stars and with Wolf-Rayet and P Cygni stars will be directly observable. For these complex structures, or for detecting rotationally induced oblateness, it is important that arrays offer good two-dimensional u-v coverage.

Binary stars play a fundamental role in observational astrophysics by providing the only direct means of measuring stellar masses, quantities that, with initial chemical composition, determine the entire course of evolution for a star. In spite of the fact that tens of thousands of binaries have been discovered, measures are accurately known for only some fifty systems. This situation is due to the largely non-overlapping selection effects among the various techniques for observing binaries, each technique being individually incapable of extracting the masses. By basing the empirical mass-luminosity and mass-radius relations upon masses with errors of $\pm 15\%$, as we are now compelled to do, sensitivity to fine structure due to differing metallicity, core rotation and main sequence evolution is largely lost. There exists essentially no information on the masses of extreme metal-poor stars and only the barest of data pertaining to highly evolved, disk population stars.

Telescope arrays will revolutionize binary star astronomy by eliminating the gap now separating the visual and spectroscopic binaries. At 100 pc, a 350 m array could resolve a 2 $M_\odot$ binary if the period exceeds 12 hours, corresponding to a semi-major axis of only 0.016 a.u. This same system could be resolved at 1 kpc with a period as short as 16 days. Most promising will be the resolution of double lined spectroscopic binaries for which individual masses and the distance to the system will be derivable. This offers the possibility for directly measuring distances well beyond the practical cut-off of trigonometric parallax. Full two-dimensional u-v coverage is necessary to extract the orbital inclination, the crucial information that is lacking from spectroscopy. Of the 978 systems in the spectroscopic orbit catalogue of Batten, Fletcher, and Mann, 683 binaries, or 70% of the entire catalogue, would be resolvable by the array. For many systems individual diameters will also be resolved, providing cases in which all the basic physical parameters for a star will be measurable. Applications to binary star studies are almost boundless. Surveying with new degrees of completeness, imaging the rich variety of phenomena associated with close binaries, measurement of apsidal motions, and the astrometric detection of very low mass companions come to mind.

These brief examples show but a portion of the almost limitless possibilities for modern high resolution optical astronomy.

OPTICAL INTERFEROMETRY: SUMMARY AND PERSPECTIVES

P.J.LÉNA
Observatoire de Paris
F-92195 Meudon, France.

The outstanding progress of radio astronomical interferometric imaging - a gain in angular resolution of 10^{10} in 50 years - demonstrates that once basic physical principles are established and technology developed, the field can develop with immense benefits for astronomy. Today, physics and technology give solid foundations for rapid growth of optical interferometry.

In a graph of angular resolution vs. wavelength there is a large empty domain. This "ecological niche" (Ekers) contains fundamental astrophysical phenomena where the combination of physical sizes, distances and temperatures places objects such as forming stars and planets, galactic nuclei and observable motions on the scale of years (e.g. masers or jets from AGNs). Optical interferometry in the visible and infrared will fill most of this "niche", leaving only the 300 nm and the 20-500 μm ranges for future space ventures.

Optical interferometry sometimes appears strange to classical optical astronomers, while it is increasingly achieving a common language with radioastronomers. This allows the new field to draw from the considerable expertise of the mature one. But the hope of considering an optical interferometer as a "black box" producing images is still in the distant future.

The phase perturbations due to the Earth's atmosphere are largely understood, although the value of the outer scale of turbulence seems to vary from meters to km depending on site, while millimetric results indicate H_2O fluctuations extending to km scale. Optical interferometry is heading towards complete internal phase control of the instrument (Berkeley, Mt. Wilson) using servo-loops, and for tracking of external atmospheric errors (CERGA). On single dishes, adaptive optics (ESO, NOAO) may fully control the phases and make speckle interferometry obsolete on the new generation of optical telescopes, at least in the infrared. Conversely, large dishes operated at mm wavelengths encounter multispeckle images, a nuisance only bound to grow with the advent of multi pixel mm detectors.

Isoplanatic limitations are encountered through ionospheric phase drifts on metric telescopes (Swarup) and tropospheric index fluctuations in the optical: this deserves systematic studies for the future of adaptive optics.

Speckle interferometry, where no real time phase correction may be applied, continues to develop in the infrared and the visible (Christou), despite the intrinsic limitations of: signal-to-noise degradation, large computing time requirements, and systematic errors. Outstanding results continue to come, such as evidence for structures in circumstellar envelopes or the envelopes of SN1987a.

Optical interferometry is currently limited to the combination of two apertures, very much as the early Cambridge array (Baldwin), but instruments under construction, design

D. McNally (ed.), Highlights of Astronomy, Vol. 8, 571–572.

or consideration extend to as many as 20 telescopes. Fig. 1 illustrates the tendencies: simple, low cost small telescopes, $\sim$10 cm=r_0(500nm) in size, built in large numbers; average size telescopes, dedicated to interferometry, in the 1 to 1.5 m class, well suited for a phased aperture in the IR; finally, high sensitivity large new generation telescopes, where adaptive aperture phasing of the aperture becomes possible in the IR, may be interferometrically coupled for part of the time. Radio VLBI shows that high quality (pixel number and dynamic range) images may be obtained with a small number of telescopes. Interesting convergences appear in imaging strategies: the Indian metric telescope combines an array (20 km scale) with a "wide field camera" (1 km scale), fairly comparable to the European VLT with its single dish base (8 m) and its interferometric base (ca. 120 m). New methods such as hypersynthesis (Vivekanand) emerge as specific to optical where u-v frequency sampling may be made with short time integrations (seconds).

Coherent detection extends to IR interferometry (Townes), while incoherent imaging detectors rapidly progress in the IR, with formats 64 × 64. First interferometric applications (Mariotti) are promising, although the detectors do not yet reach the signal photon counting mode of visible CCDs. Other convergences reveal themselves: multi-spectral imaging in radio is paralleled with the proposed and demonstrated double Fourier (spatio-spectral) technique in the IR (Ridgway & Mariotti).

Technological progress in optical interferometry is rapid. The MultiMirror Telescope cophasing (Beckers & Hege) has proven the possibility to align large steel structures at the required accuracy, while recent mechanical measurements at ESO and CERGA telescopes (Bourlon) show that classically designed telescopes almost achieve interferometric stability. To paraphrase the "steel-silicon issue" (Ekers) as alternate routes for radio imaging, active and adaptive optics stabilization of large telescope/interferometers is a "glass-silicon issue" of similar nature. The Mt. Wilson small optical interferometer (Shao) indeed demonstrates, by routinely observing in a phase stable mode tens of stars per night, that an optical interferometer may be operated similarly to conventional telescopes.

Many questions remain open. For example, what will be the ultimate sensitivity of an optical interferometer? What dynamic range can be achieved? What is the optimum number of telescopes for imaging? However, the impressive number of projects in this field, the strength and dedication of the community and the way shown by radio interferometry, all combine to predict an interesting future.

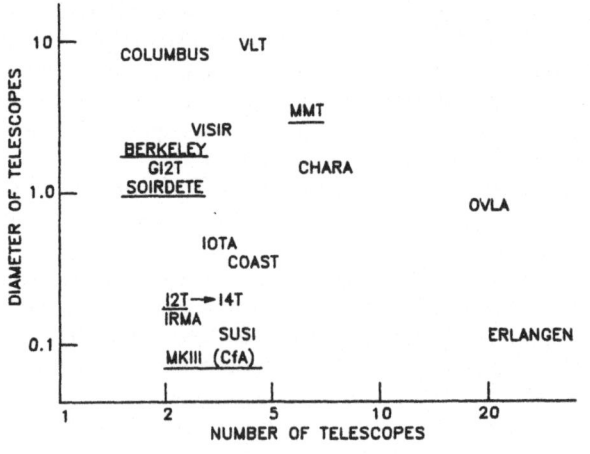

Figure 1. Existing (____) or planned optical interferometers.

4. <u>MOLECULES IN EXTERNAL GALAXIES</u>

Chairmen: F. Combes, N.Z. Scoville, J. Young
Editor: F. Combes

<u>Supporting Commissions:</u>

28, 34, 40

THE MOLECULAR SPIRAL STRUCTURE IN M51 DERIVED FROM CO(J=2-1) LINE OBSERVATIONS

M. GUÉLIN, S. GARCIA-BURILLO, R. BLUNDELL, J. CERNICHARO,
D. DESPOIS and H. STEPPE
*IRAM, Domaine Universitaire, 38406 Saint Martin d'Hères, France
and Avenida Divina Pastora 7, 18012 Granada, Spain*

ABSTRACT. We present preliminary results of a high angular resolution-high sensitivity survey of $CO(J = 2 - 1)$ line emission in M51 made with the IRAM 30 m telescope.

1 Introduction

Since our Galaxy's spiral pattern is difficult to trace, the relation between molecular clouds and spiral structure must be studied in external systems. M51, with its "grand design" spiral pattern and favorable inclination, is well suited for this purpose. It is strong in CO, the most sensitive molecular probe, and close enough for its arms be resolved by the largest millimetre wave telescopes. Several surveys of the $J = 1 - 0$ emission in M51 have been reported in the literature (e.g. Rydbeck *et al.* 1985, Lo *et al.* 1987, Vogel *et al.* 1988, see also this Commission Meeting). They lack, however, angular resolution or sensitivity to resolve the interarm emission from the arm emission. We present here CO $J = 2 - 1$ line observations of the western half of M51, made with the IRAM 30-m telescope. This is the first survey of an external galaxy combining high angular resolution (HPBW = 12") with high sensitivity, thus allowing a thorough study of the interarm molecular gas.

2 Observations

The $J = 2 - 1$ ^{12}CO line emission was mapped over a 2'x3.5' area covering the western inner part of M51. Most of this area was fully sampled (6" spacings in **r.a.** and **dec.**) using the "basket-weaving" technique; the outer edges and southernmost part were half-sampled in declination. Good pointing and accurate calibration were ensured through frequent observation of nearby quasars and reference positions.

The IRAM 1.3 mm SIS receiver had an SSB noise temperature of 200-250 K over its 600 MHz-wide IF band. A 512x1MHz channel filterbank provided a velocity resolution of 1.3 kms^{-1}. Fig. 1 diplays the spectra observed in the central 100"x100" region, smoothed to a 13 kms^{-1} (10 MHz) resolution. Except for particular positions re-observed with longer integration times, they have a r.m.s. noise of $\simeq$15 mK. The spectra outside this central region have a r.m.s. of 20-30 mK. The velocity-integrated antenna temperature contours are presented in Fig. 2.

D. McNally (ed.), Highlights of Astronomy, Vol. 8, 575–577.

576

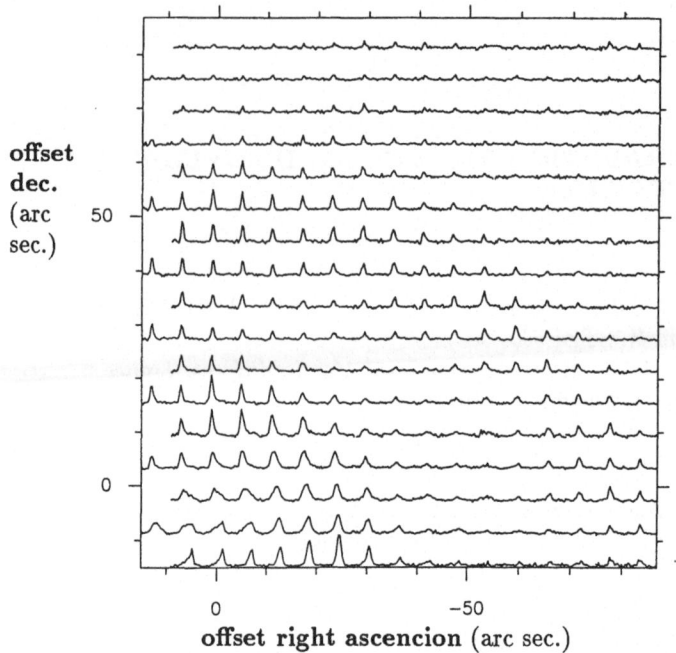

offset dec. (arc sec.)

50

0

0   −50

offset right ascencion (arc sec.)

Figure 1: ^{12}CO (J=2-1) line profiles plotted against the offset from the nuclear continuum source (1950.0) right asc. = $13^h27^m46.1^s$ dec. = $47°27'14$". The velocity span for each spectrum is 300 kms^{-1}. The maximum antenna temperature, corrected for atmospheric absorption, is 0.7 K.

Less extensive observations, aimed at determining the $(J = 2 − 1)/(J = 1 − 0)$ and ^{12}CO/^{13}CO line intensity ratios, have also been made in the ^{13}CO $(J = 2 − 1)$ line and in the ^{12}CO and ^{13}CO $(J = 1 − 0)$ lines.

3 Results

The main results, from Fig. 1 and 2, are:

i) The CO arm-interarm contrast is large and highly variable. The arm-interarm velocity-integrated intensity ratio, observed with our 12" beam, is typically 3-6 for the arm at the centre of Fig. 1. It is only 2 at 90" west, 45" south of the galaxy's centre, and reaches 10-17 for the inner southwest arm (e.g. 10" west, 50" south of the centre);

ii) The CO arms are thick. Figs. 1 and 2 show that half of the CO emission in the western arm arises from a broad component (HPW$\simeq$20") which was missed by the interferometric study of Vogel et al. (1988). Actually, the $J = 2 − 1$ line brightness contours of Fig. 2 correlate closely with the 6 cm continuum emission contours observed with the VLA (van der Hulst 1988);

iii) CO is detected everywhere between the arms. Although weak, the interarm emission is not uniform and shows cloud complexes of small velocity dispersion (HPW$\simeq$6-20 kms^{-1}, vs 30-40 kms^{-1} in the arms);

iv) contrary to the finding of previous studies, there is plenty of CO at the centre of M51. The smaller peak temperature at the centre is compensated by a broader linewidth. The CO-derived rotation curve rises so steeply that the full span of velocities is almost reached in the central 12"-radius region;

v) the ^{12}CO(2−1)/(1−0) line brightness ratio, calculated after smoothing the $(J = 2−1)$ data to 21" and correcting for the different beam efficiencies, is found to decrease from the centre (aver. ratio: 1.2), to the interarm region ($\simeq$0.9) and to the arms ($\simeq$0.6).

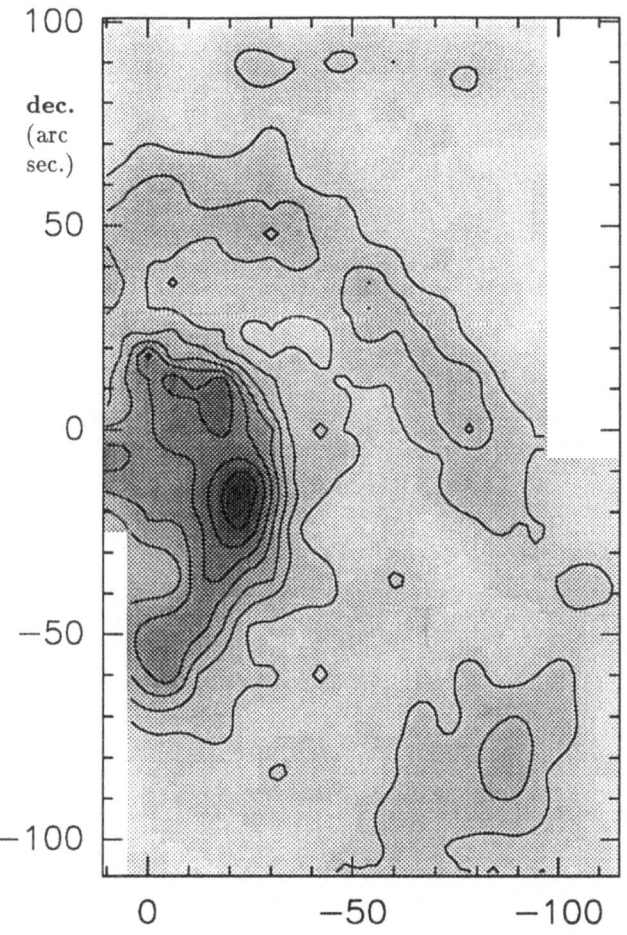

Figure 2: velocity-integrated antenna temperature contours of the $^{12}CO(J=2-1)$ line. First contour and contour step are 5 K·kms^{-1}. Abscissa and ordinate are offsets in **r.a.** and **dec.** from the nuclear continuum source.

right ascencion (arc sec.)

A detailed analysis of the $^{12}CO\,(2-1)$ and $(1-0)$ line profiles, as well as of complementary ^{13}CO data, is under way. It should help to understand the fate of the molecular clouds streaming across spiral arms.

References

Lo, K.Y., Ball, R., Masson, C.R., Phillips, T.G., Scott, S., and Woody, P.D. 1987, *Astrophys. J.* **317**, L63

Rydbeck, G., Hjalmarson, A., and Rydbeck, O.E.H. 1985, *Astron. Astrophys.* **144**, 282

van der Hulst, Kennicutt, R.C., Crane, P.C., and Rots, A.H. 1988. *Astron. Astrophys.* **195**, 38

Vogel, N.S., Kulkarni, S.R., and Scoville, N.Z. 1988, *Nature* **334**, 402

MOLECULAR CLOUD SPIRAL ARMS AND
RESULTS FROM TIDAL INTERACTION MODELING

Åke Hjalmarson

Onsala Space Observatory, S-439 00 Onsala, Sweden

I wish to report on some results from mapping of molecular cloud distributions in galaxies and from tidal interaction modeling - work performed at Onsala Space Observatory and in the Astrophysics Group of Institute of Theoretical Physics, Chalmers University of Technology/University of Göteborg.

There are a number of important questions in this context:

- Do molecular cloud spiral arms exist (and what is their location, width, structure)?
- Do interarm molecular clouds exist and at which mass fraction?
- What is the nature of the arm/interarm clouds (temperature, density, size, mass, velocity dispersion)?
- Where and how do dense clouds form?
- Do density waves , tidal interactions or bars play a role in the cloud/star formation processes, and how?
- Do molecular "superclouds" exist?

Detailed mapping of the CO(1-0) emissions of M51 and IC342 have become major projects for the Onsala 20 m telescope (beam width 33"), beginning with the first M51 observations in 1981.

<u>M51</u>. Spiral arm features, traced already in the raw 16" spacing data, became very conspicuous when the average radial brightness distribution was subtracted, as discussed already in Tokyo 1985 (Rydbeck etal. 1987). Radial streaming across arms - in the sense predicted by density wave theory - is clearly observed. The estimated velocity shifts are surprisingly large, at least 50 km s^{-1} (Rydbeck etal. 1985, 1987). A maximum entropy type deconvolution now has been applied to a more extensive data set (Rydbeck etal. 1988). While the antenna beam size is 33" the true resolution approaches 10" (0.5 kpc) where the signal is strong. The spiral arm distribution of molecular clouds then becomes very clear. The patchy structure of the arms suggests that the molecular clouds are assembled into giant complexes of mass up to 10^8 M$_\odot$. The average arm-interarm contrast is about 4 (actually this is a lower limit since it is hard to know how much of the signal left between the arms is due to error - and sidelobe contributions). With the higher angular resolution resulting from the MEM deconvolution large and abrupt tangential as well as radial velocity shift across arms are "observed". Spatial overlap of pre and post arm velocities is also evident. This suggests that cloud- cloud collisions must occur in the arms.

D. McNally (ed.), Highlights of Astronomy, Vol. 8, 579–580.
© *1989 by the IAU.*

<u>IC342</u>. The center of our CO map is dominated by the "bar" previously seen with the OVRO interferometer. However, connected to this bar spiral-arm-like features can also be traced (Wiklind etal. 1988). Arm-interarm contrasts of about 4 have been estimated for a few arm cuts. Like in M51 spatially overlapping pre and post arm velocity componenets are observed. Although of much smaller magnitude than in M51 this velocity structure suggests cloud- cloud collisions in the spiral arms of IC342.

We note that for 25% of a disk in spiral phase an arm-interarm contrast of 3:1 (9:1) is expected if 50% (75%) of the H_2 mass is contained in the arms. From these observations we may conclude that:

- Massive molecular cloud spiral arms do exist.
- Interarm molecular clouds may still contain a sizable mass fraction.
- Density waves (tidal interactions, bars) do play a role in organizing (existing) clouds into arms ("orbit crowding") and very massive cloud complexes. In M31 support has been found for formation of molecular clouds inside HI "superclouds" in a spiral arm region (Lada etal. 1988).
- Density waves (tidal interactions, bars) may also indirectly enhance high-mass star formation in the arms via observed cloud- cloud collisions. This suggestion is supported by the observed quadratic dependence of the number density of giant HII regions on the local H_2 density in our Galaxy (Scoville etal. 1986).

Numerical simulations of tidal triggering of spiral arm structures in disk galaxies is a parallel development (cf Sundelius etal. 1987). Large, abrupt velocity shifts across arms and also "spatially overlapping pre and post arm velocity components" - like those observed in M51 - do indeed appear in simulations where impact parameters and perturber mass are similar to those of the M51 (NGC5194/5195) system. An ongoing detailed study of disk evolution vs time shows how tidal, material, arms first develop very suddenly - subsequently triggering density wave arms. This rapid organization of the gas into arms during the early phase of the tidal interaction presumably would lead to rather violent cloud- cloud collisions - and hence may be useful for our understanding of the IR/CO luminous "star-burst" galaxies.

References:

Lada, C.J., Margulis, M., Sofue, Y., Nakai, N. and Handa, T. 1988, Astrophys.J. <u>328</u>, 143.
Rydbeck, G. Hjalmarson, Å., and Rydbeck, O.E.H. 1985, Astron.Astrophys. <u>144</u>, 282.
Rydbeck, G. Hjalmarson, Å., Johansson, L.E.B., and Rydbeck, O.E.H. 1987, in <u>Star Forming Regions</u>, eds. Peimbert and Jugaku, Reidel, p.535.
Rydbeck, G., Hjalmarson, Å., Wiklind, T., and Rydbeck, O.E.H. 1988, in <u>Molecular Clouds in the Milky Way and Nearby Galaxies</u>, eds. Dickman, Snell and Young, Springer.
Scoville, N.Z., Sanders, D.B., and Clemens, D.P. 1986, Astrophys.J. <u>310</u>, L77.
Sundelius, B., Thomasson, M., Valtonen, M.J., and Byrd, G.G. 1987, Astron.Astrophys. <u>174</u>, 67.
Wiklind, T., Rydbeck, G., Hjalmarson, Å., and Rydbeck, O.E.H. 1988, in <u>Molecular clouds in the Milky Way and Nearby Galaxies</u>, eds. Dickman, Snell and Young, Springer.

CO IN NGC4438 AND TIDAL STRIPPING IN THE VIRGO CLUSTER

F. COMBES [1,2], C. DUPRAZ [2,1], F. CASOLI [2,1], L. PAGANI[1]
[1] Observatoire de Paris, Section de Meudon, F-92 195 MEUDON
[2] Ecole Normale Supérieure, 24 Rue Lhomond, F-75 231 PARIS

1. Background

It is now well established that the environment plays an essential role in the morphology and evolution of galaxies: in particular, the HI gaseous content is often deficient for galaxies in clusters, and the deficiency increases towards the cluster center (cf for the Virgo cluster: Chamaraux et al 1980, Cayatte et al 1988, in prep). The gas is mostly deficient in the outer parts of galaxies, which considerably reduces the size of HI disks (van Gorkom & Kotanyi 1985). However the central gaseous content, usually under the form of molecular hydrogen traced by CO emission, seems normal, at least in the Virgo cluster, the only one surveyed at millimetric wavelengths (Kenney & Young 1986).

Two main mechanisms have been proposed to explain the HI deficiency: either interaction with the hot intracluster medium (ICM), or tidal stripping during collisions with other cluster galaxies. The former process, which can be ram-pressure sweeping (Gunn and Gott 1972), thermal evaporation (Cowie and Songaila 1977) or viscous stripping (Nulsen 1982) has been favoured in Virgo and the other X-ray emitting clusters. In particular NGC4438, one of the closest galaxies to the Virgo center M87, has been considered as the archetype of galaxies swept out by the ICM (Kotanyi 1981): radio-continuum, X-ray and HI emissions reveal NW extensions, in the opposite direction to M87 (Kotanyi et al 1983).

The determination of the molecular gas distribution provides a crucial test for these mechanisms: indeed the molecular material is too dense to be easily swept out, and large perturbations in the CO emission are likely to be of tidal origin.

2. CO observations

We mapped the galaxy at 23" resolution in CO(1-0) with the IRAM 30m telescope at Pico Veleta (Spain). The contours of CO integrated emission are superimposed onto the Arp photograph in fig.1.

A strong, highly-concentrated component of diametre 25" (2.5 kpc at the adopted distance of 20Mpc) at half maximum is centered on the continuum source (Hummel et al 1983). Roughly elongated along the major axis, it appears associated to the unperturbed inner galaxian disk, and coincides with regions of non-thermal radio-continuum and X-ray emissions that are likely due to star

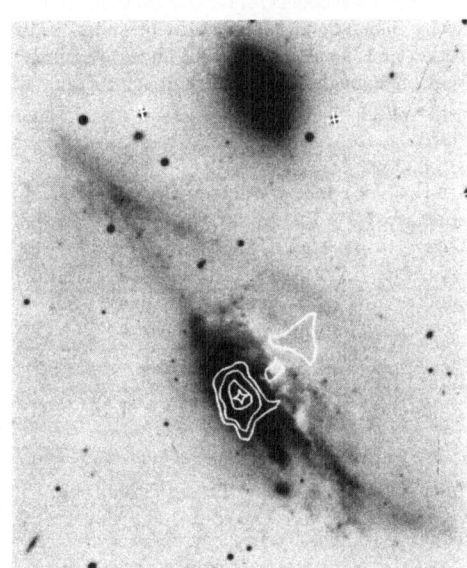

Fig.1: *Photograph of NFC4438/4435 (Arp 1966). Solid lines: contours of $\int T_R{}^* CO(1\text{-}0)\ dV$; levels 8, 20, 50 Kkm/s. The star is the radio continuum position.*

D. McNally (ed.), Highlights of Astronomy, Vol. 8, 581–582.

582

formation. We derive a total H_2 mass of $2.9\ 10^9\ M_\odot$ for this central component, assuming a standard NH_2/I_{CO} of $4\ 10^{20}$. The CO velocity field corresponds to a very regular rotation, in agreement with the $H\alpha$ rotation curve (Chincarini and de Souza 1985).

We also mapped the NW-displaced component of CO emission up to more than 1' from the centre. This component reveals very unusual features, such as very large linewidths (300km/s at zero level), and almost no central velocity gradient. The total H_2 mass in this component is $1.2\ 10^9\ M_\odot$.

3. Theoretical interpretation

The discovery of molecular gas displaced in the same direction as HI, X-ray and radio emissions does not support the hypothesis of the ICM sweeping. Indeed, the average surface density of the molecular gas σ_{gas} = 200 $M_\odot/pc^2$ is too high for the ram-pressure to be efficient. Assuming a mean ICM density of $2\ 10^{-4}$ cm^{-3} (Fabricant and Gorenstein 1983) and a velocity through the ICM of 1200 km/s for NGC4438, a crude estimate of the ram-pressure exerted by the ICM on the galaxy is $p_{ram} = 6\ 10^{-12}$ dyne/cm^2. The ratio R of the gravitational restoring force (GM/r^2) to the ram-pressure force can then be estimated for r=10kpc at R= 100 >> 1. Also, the presence of two stellar tails (NE and SW), strongly suggests a tidal interaction with a nearby companion, which is only 4.'5 = 26kpc away in projection on the sky.

We therefore simulated the encounter with a test-particle code similar to that used by Toomre and Toomre (1972), the details of which are described in Combes (1978). We take a mass ratio of 1/2 between NGC4435 and NGC4438, according to their luminosities. Since their relative line of sight velocity is high (690km/s), we choose an hyperbolic orbit, of eccentricity e =4. The minimal distance of approach is 6kpc, since perturbations begin only beyond that distance in the CO emission. Only a retrograde passage can fit the observations.

Fig. 2 shows that the tidal-encounter model successfully accounts for the overall shape of NGC4438 and its companion. Contrary to previous arguments, even a highly hyperbolic retrograde collision can inflict severe damages to a disk, provided that the impact parameter is small. Such small impact parameters are frequent enough in rich clusters to account for the HI-gas stripping observed.

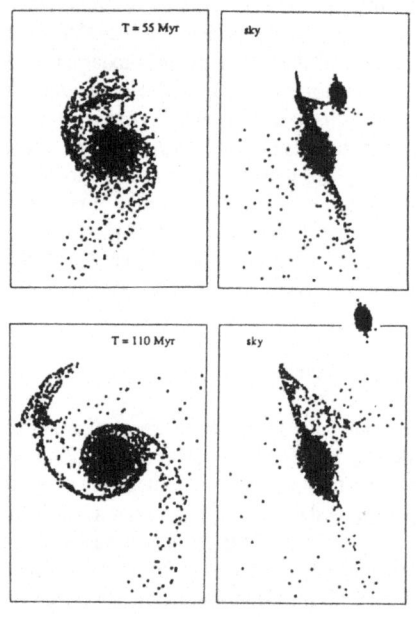

References

Arp H. (1966) *Atlas of Peculiar Galaxies* (C.I.T.)
Chamaraux P., Balkowski C., Gérard E. (1980) *Astron. & Astrophys.*, **83**, 38
Chincarini G., de Souza R. (1985) *A&A.*,**153**, 218
Combes F. (1978) *Astron. & Astrophys.*, **65**, 47
Cowie L.L., Songaila A. (1977) *Nature*, **266**, 501
Fabricant D,Gorenstein P. (1983) *Ap.J.*, **267**, 535
Gunn J.E., Gott J.R. (1972) *Astrophys.J.*, **276**, 1
Hummel E., van Gorkom J.H., Kotanyi C.G. (1983) *Astrophys.J.*, **267**, L5
Kenney J.D., Young J.S. (1986) *Ap.J.*, **301**, L13
Kotanyi C.G. (1981) PhD Thesis, Groningen
Kotanyi C.G., van Gorkom J.H., Ekers R.D., (1983) *Astrophys. J.*, **273**, L7
Nulsen P.E.J. (1982) *M.N.R.A.S.*, **198**, 1007
Toomre A., Toomre J., (1972) *Ap.J.* **178**, 623
van Gorkom J.H., Kotanyi C.G. (1985) in*Virgo Cluster of Galaxies*, eds Richter and Binggeli, p 61

Fig.2: *Two snapshots of the simulation for t = 55 Myr and t = 110 Myr. Right panels: projection on the sky plane. Left panels: left-hand-side projection, for which the galaxies are almost seen face-on.*

CO OBSERVATIONS OF THE CENTRAL REGION OF NGC4258

Yoshiaki SOFUE

Institute of Astronomy
The University of Tokyo, Mitaka, Tokyo 181, Japan

ABSTRACT. $^{12}CO(J = 1 - 0)$ observations of the central region of NGC4258 at a resolution of 17" revealed a high-density concentration of molecular gas toward the center. The molecular clump lies on a central depression in the HI bar. No clear interaction of the molecular gas with the continuum anomalous arms has been found.

The SBb type spiral galaxy NGC4258 is known by its anomalous radio continuum arms (van der Kruit et al 1972; van Albada 1980). The anomalous arms may be some ejection from the galactic center. The central activity responsible for such an energetic ejection may be occuring in a dense, compact gaseous disk with high accretion rate. Dynamics of the gas in such a very inner region is better investigated in the molecular line than in the HI, and we have undertaken $^{12}CO(J = 1 - 0)$ line observations of the central region using the Nobeyama 45-m telescope.

The ob servations covered the central $1' \times 2'$ approximately along the HI bar. The CO intensity of about $T_A^* \sim 0.1$ K has been detected near the center. Fig.1 shows the integrated CO intensity superposed on the HI emission distribution obtained by van Albada (1980). The CO-bright region is confined in a small area of 20" × 30", while the intensity decreases steeply outside this area, composing a high-density clump of molecular gas. This clump is elongated in the direction of the major axis of the HI bar. The clump has the H_2 mass of about $10^{10} M_\odot$, which makes up a few percent of the total dynamical mass involved in the same region. Although the CO clump is located near the center of the HI bar, the CO intensity distribution along the bar seems to be anti-correlated with the HI intensity (Fig. 2).

Fig.3 shows a comparison of the CO intensity distribution with the radio continuum emission (van Albada and van der Hulst 1982). The CO clump is bout 5" off set from the continuum center. The eccentric CO distribution seems to be correlated with the asymmetric intensity distribution of the anomalous arms: the brighter arm appears to extend toward the north-west where the CO intensity is weaker compared to the south-east side of the center. However, there is no indication that the molecular gas is interacting with the anomalous arms. This is not inconsistent with the suggestion that the arms are not located in the galactic plane but the ejection is into the halo out of the nuclear disk (Sofue 1980; Sanders 1980).

D. McNally (ed.), Highlights of Astronomy, Vol. 8, 583–584.

584

NGC 4258   CO+HI

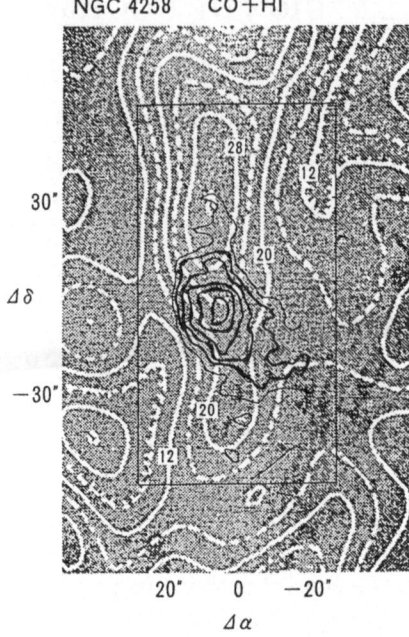

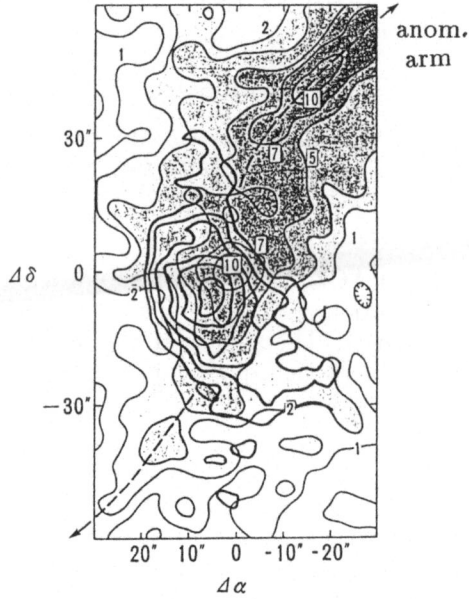

anom.
arm

Fig.1: Distributions of the integrated CO (black contours) and HI (white cotours: van Albada 1980) intensities of the central region of NGC4258. The contour interval of the CO intensity is 5 K km s^{-1} and the lowest contour is 10 K km s^{-1}.

Fig.2: Cross sections of the CO and HI intensities along the HI bar. Note the anticorrelation of the two emissions, or the depression of HI near the peak of the CO emission.

Fig.3: CO intensity distribution superposed on the radio continuum emission distribution. Note the eccentric distribution. No clear interaction between the CO and continuum features is found.

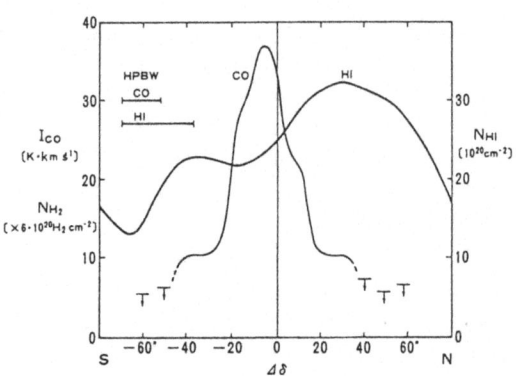

References

van Albada,G.D. 1980, *A. Ap.*, **90**, 123
van Albada,G.D., van der Hulst,J.M. 1982, *A. Ap.*, **115**, 375
van der Kruit,P.C., Oort,J.H., Mathewson,D.S. 1972, *A. Ap.*, **21**, 169
Sanders,R.H. 1982, in *Extragal. Radio Sources, IAU Symp. 97*, eds. D.S.Heeschen and C.M.Wade (D.Reidel, Dordrecht), p.145
Sofue,Y. 1980, *Pub. Astr. Soc. Jpn.*, **32**, 79
Sofue,Y., Krause,M., Doi,M., Nakai,N., Handa,T. 1988, *Pub. Astr. Soc. Jpn.*, **40**, in press

THE CORRELATION OF CO AND IR EMISSION FROM GALAXIES: WHAT DOES IT TELL US?

Frances Verter
NRC Research Associate,
NASA Goddard Space Flight Center

Almost all galaxy properties correlate with each other to some degree; in this bounty of agreement, how are we to extract essentials? Fortunately, there is a framework in which to interpret CO correlations. Empirical studies have indicated that the fundamental galaxy properties are scale and form (Whitmore 1984), and theoretical work has implied that this follows from the physics of galaxy formation (see Lin and Pringle 1987, and references therein). It is known that the CO luminosity of galaxies is a function of both their scale (Young and Scoville 1982) and their form (Verter 1987). But the interpretation of any individual correlation is shrouded by uncertainties in the conversion from observables to source population properties.

As an example, consider the correlation between CO luminosity, $L(CO)$, and far-infrared luminosity, $L(FIR)$, of entire galaxies. If we lump together all galaxies from IR-luminous mergers, through normal spirals, to dwarf irregulars, this correlation spans 7 orders of magnitude (Tacconi and Young 1987). Among normal spirals there is an order of magnitude scatter, but still a very strong correlation (Kenney 1987, Verter 1988). Neither axis of this plot can be interpreted with equanimity.
The 3 main uncertainties in converting CO emission to molecular mass are:
(1) Cloud Ensemble – Is the ISM dominated by GMC's (eg. M51) or cirrus (eg. M31)?
(2) Metallicity – The non-linearity of CO self-shielding is such that spirals are largely insensitive to this, but in irregulars cloud emissivity is drastically reduced.
(3) Gas Temperature – Optically thick CO emission is proportional to source temperature.
These issues are summarized in Verter (1987) and Maloney and Black (1988).
The 3 main uncertainties in converting far-IR emission to either total gas mass or number of young stars are:

D. McNally (ed.), Highlights of Astronomy, Vol. 8, 585–586.
© 1989 by the IAU.

(1) Cloud Ensemble – What fraction of the IR arises from dust heated by OB associations, or from dust heated by the general interstellar radiation field?

(2) Metallicity – Optically thin dust emission is proportional to the dust/gas ratio.

(3) Dust Temperature – Optically thin dust emission is proportional to the 5th or 6th power of T, depending on grain emissivity.

These issues are discussed by Maloney (1987). The operation and importance of all 6 issues above are still under debate. Since L(FIR) is a convolution of gas density with stellar heating, much of the heating occuring at large distances from the energy source (Boulanger and Perault 1988), the L(CO)–L(FIR) correlation cannot easily reveal the nature of either the cloud or stellar population.

An alternate approach to placing CO luminosities in the framework of galaxy properties is taken by Verter (1988). The CO luminosities of a representative sample of Sa–Sc galaxies was tested for correlation with 10 galaxy properties that express a mixture of galaxy scale and form. The relative dependence of CO emssion on these two axes was assessed from the relative strengths of the correlations. Very few previous correlation studies, summarized in Verter (1988), were compatible for intercomparison. Verter (1988) found that L(CO) correlates strongly with tracers of galaxy scale, weakly with tracers of galaxy form, and best with L(FIR). Not surprisingly, L(CO) is closely tied to the reservoir of gas and young stars; apparently this reservoir is not strictly a measure of galaxy scale or form. More comparative studies of mapped galaxies are needed to bridge the gap between the triggering of local star formation processes, and their dependence on global galaxy parameters set at the time of formation.

REFERENCES

Boulanger, F., and Perault, M. 1988, Ap.J., 330, 964
Kenney, J. D. 1987, Ph.D. thesis, U. of Massachusetts
Lin, D., and Pringle, J. E. 1987, Ap.J. Lett., 320, L87
Maloney, P. R. 1987, Ph.D. thesis, U. of Arizona
Maloney, P., and Black, J. H. 1988, Ap.J., 325, 389
Tacconi, L. J., and Young, J. S. 1987, Ap.J., 322, 681
Verter, F. 1987, Ap.J. Suppl., 65, 555
Verter, F. 1988, Ap.J. Suppl., to appear Oct.
Whitmore, B. C. 1984, Ap.J., 278, 61
Young, J., and Scoville, N. 1982, Ap.J. Lett., 260, L11

Can Galactic GMCs be Identified from l-v Diagrams?

David S. Adler and William W. Roberts, Jr.
Departments of Astronomy and Applied Mathematics
University of Virginia
Charlottesville, VA 22903

Identifying the spiral nature of the distribution of gas in the Galaxy has been a subject of much research in the past thirty years. The position of the sun in the disk of the Galaxy presents us with a problem of perspective: how does one identify the cloud system from within the system? Longitude-velocity (l-v) diagrams have been used to try to determine the distribution of interstellar gas, but problems inherent in the methods have been pointed out previously (Burton 1971). Recent Galactic CO surveys have been used in attempts to map the distribution of molecular cloud complexes in the disk of the Galaxy (Dame, $et\ al.$ 1986). Here, we use numerical simulations of the molecular cloud system in a spiral galaxy to consider the following question: to what extent can concentrations of emission in the l-v diagram (LVCs) be considered complexes of gas in the disk of the Galaxy (GMCs)?

The galactic disk used here is from the models of Roberts and Hausman (1984), with 20,000 particles representing gas clouds orbiting in a spiral perturbed, galactic gravitational potential. The observer's reference point is placed at a galactocentric radius of 10 kpc between major spiral arms. An l-v diagram is then created for all clouds in the first quadrant of the model galaxy. Clouds are binned in the l-v diagram with a cell size of Δl_{cell}=0.25° and Δv_{cell}=1.25 $km\ sec^{-1}$.

A clipping method (Dame, $et\ al.$ 1986) is used to reduce the background emission of clouds not belonging to complexes. For a clipping level of c=1, each cell with a density of one cloud per cell (or less) is considered background and set to a density of zero clouds/cell. Thus every cell with a density of 2 clouds/cell and above is considered to be part of an LVC. Neighboring cells that meet the clipping criterion are considered to be part of the same LVC.

The fifteen largest LVCs (labelled A-O) for a clipping of c=1 are displayed in Figure 1. When the clouds belonging to these LVCs are plotted spatially (Figure 2), we see that what appear as complexes in the l-v plane are not necessarily complexes in the disk. Investigation of any one LVC shows that it is made up of clouds that are spread out along the entire line of sight. The same effect can be seen for different clipping levels.

The net result is that apparent clumping in the l-v plane can be misinterpreted

587

D. McNally (ed.), Highlights of Astronomy, Vol. 8, 587–588.
© *1989 by the IAU.*

588

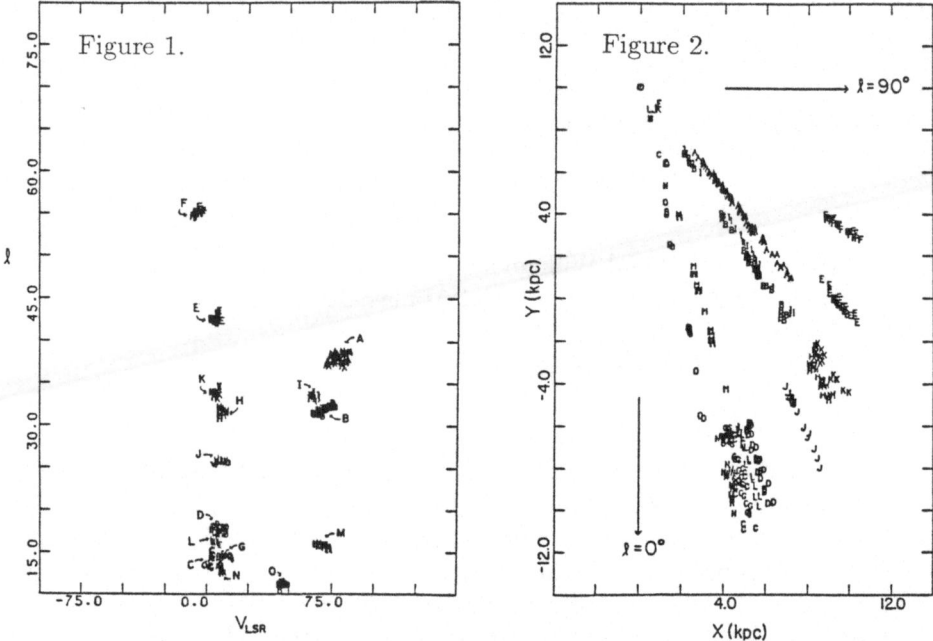

Figure 1. The fifteen largest complexes for the c=1 case. Each complex is represented by a letter, A-O.

Figure 2. The spatial distribution for the LVCs in the c=1 case. The lettering scheme for the separate complexes is consistent with Figure 1. The observer's reference point is marked with a circle; lines of $l = 0°$ and $l = 90°$ are marked.

unless care is taken. This was first pointed out by Burton (1971), who demonstrated that certain geometrical effects appear in l-v digrams no matter what parameters are used. He also pointed out the importance of including streaming motions in rotation curves when determining kinematic distances of clouds.

We conclude that the current method of determining molecular cloud complexes from the analysis of the l-v diagram is not reliable. The unfortunate circumstance of being located in the disk of the Galaxy makes it exceedingly difficult, if not downright impossible, to determine the distribution of molecular cloud complexes with the data currently available.

This work was supported in part by the National Science Foundation under Grants AST-82-04256 and AST-87-12084, and the National Aeronautics and Space Administration under Grant NASA-NAGW-929.

References

Burton, W.B. 1971, Astr. Ap., **10**, 76.

Dame, T.M., Elmegreen, B.G., Cohen, R.S., and Thaddeus, P. 1986, Astrophys. J., **305**, 82.

Roberts, W.W., and Hausman, M.A. 1984, Astrophys. J., **277**, 744.

Warm Gas and Spatial Variations of Molecular Excitation in the
Nuclear Region of IC342

A. Eckart(1), D. Downes(2), R. Genzel(1), A.I. Harris(1),
D.T. Jaffe(3) and W. Wild(1)

(1) Max-Planck-Institut für extraterrestrische Physik,
    München, West-Germany
(2) Institut de Radioastronomie Millimetrique (IRAM),
    Grenoble, France
(3) Department of Astronomy, University of Texas at
    Austin, USA

Using the IRAM 30m millimeter radio telescope we mapped the line
emission of the J=1-0 and J=2-1 transitions of ^{12}CO, ^{13}CO, and $C^{18}O$
in the nuclear region of the spiral galaxy IC342. This study demonstrates the value of multi line studies to investigate the neutral
interstellar medium in extragalactic sources. Our observations as
well as calculations of simple models of CO excitation and radiative
transport show that the molecular gas in the nucleus is warm and that
physical conditions vary with position in the galaxy. The molecular
gas in the central kiloparsec of IC342 has a kinetic temperature of
at least 30K and a molecular hydrogen density of about 3×10^3 cm^{-3}. At
distances more than 500pc north and south of the center the kinetic
temperature is significantly less (≥ 13K). About 500 pc north east of
the center we find evidence for optically thin CO emission originating in a component of warm gas with a temperature of at least 40 K.
Our model calculations result in conversion factors between the H_2
column density and the ^{12}CO (1-0) line intensity close to the value
of $3-4 \times 10^{20}$ cm^{-2}/K km s^{-1} derived for molecular clouds in the Galaxy.
The molecular mass contained in the central two kpc of IC342 is of
the order of 2×10^8 M$_\odot$.
The CO millimeter emission comes from a compact structure elongated
at a position angle of zero degrees. After deconvolution with the
beam profile the intrinsic source size is 30"x$\leq$13" (FWHM). The
emission is centrally peaked and shows an extension to the northeast.
We also find underlying extended emission over more than 90" (2 kpc)
at a position angle of 25°.
Our J=1-0 (resolution 21") and J=2-1 (resolution 14") maps are
consistent with the ^{12}CO (1-0) interferometer map at 7" resolution
(Lo et al. 1984) and the ^{12}CO(1-0) Nobeyama map (Hayashi et al. 1986)
at 13" resolution. Minor differences among the maps can be explained
by the different beam sizes or are due to the fact that components on
size scales larger than 30" are resolved out by the interferometer.
The extended emission in all three maps is consistent with the 65"

D. McNally (ed.), Highlights of Astronomy, Vol. 8, 589–590.

resolution ^{12}CO (1-0) measurement by Rickard et al. (1981). In this map, the dominant central component is extended to the north and south.

The observed [CII] 158 μm brightness toward the center of IC342 is 4.5×10^{-4} ergs^{-1} cm^{-2} ster^{-1} (Crawford et al. 1985). We measure a [OI] 63 μm/[CII] 158 μm line ratio of 1.5 ± 0.7. In collisionally excited neutral gas (ne$\leq 10^{-3}$n(H+H$_2$)) with a O/C fractional abundance near the solar neighborhood value ($\approx$2) this line ratio corresponds to a gas pressure n(H)T $\approx 3 \times 10^5$ cm^{-3} K. A comparison of the HI emission measured and predicted from the [CII] 158 μm line flux indicates that a significant fraction of the 21 cm HI emission may originate in warm, dense photodissociation regions at the surfaces of molecular condensations. This is in contrast to the standard interpretation (c.f. Kulkarni and Heiles 1987) that most of the HI column density originates in $\approx$80 K "HI clouds" of volume density 10 to 100 cm^{-3}.

Lo, K.Y., Berge, G.L., Claussen, M.J., Heiligman, G.M., Leigthon, R.B., Masson, C.R., Moffet, A.T., Phillips, T.G., Sargent, A.I., Scott, S.L., Wannier, P.G., Woody, D.P. 1984, Ap.J. 282, L59

Hayashi, M., Handa, T., Sofue, Y., Nakai, N., Hasegawa, T., Lord, S., Young, J. 1986, NRO Contribution to the IAU Symp. No. 115 on "Star Forming Regions", Tokyo, 11.-15.Nov. 1985

Rickard, L.J., Palmer, P. 1981, Astron.Astrophys. 102, L13 Crawford, M.K.,Genzel, R., Townes, C.H. Watson, D.M. 1985, Ap.J. 291, 755

Kulkarni, S.R., Heiles, C. 1987, in Proc. of the Symposium on Interstellar Processes, D.J. Hollenbach and H.A. Thronson (ed.), D. Reidel, Astrophys. and Space Science Library Vol. 134

RECENT CO(2-1) OBSERVATIONS OF GALAXIES WITH THE CSO

A. I. SARGENT, T. G. PHILLIPS, D. B. SANDERS, N. Z. SCOVILLE
California Institute of Technology
Pasadena, CA 91125

ABSTRACT CO(2→1) observations of a number of galaxies have recently been obtained with the CSO. Examples are presented here to illustrate the capabilities of the instrument.

1. Introduction

The newly-constructed Caltech Submillimeter Observatory (CSO) on Mauna Kea, Hawaii, has been used to acquire observations of several galaxies in the CO(2→1) transition at 230 GHz. Currently, the 10.4 m diameter Leighton telescope is equipped with an SIS receiver with double-sideband noise temperature $\sim$ 200 K, and an acousto-optic spectrometer of bandwidth 500 MHz. The high surface accuracy dish, 30 μm, and the excellent atmospheric transmission make it well-suited for sensitive measurements of galaxies. Observations presented here demonstrate the versatility of the system and its potential for studying distant galaxies.

2. Results

New CO(2→1) observations of Centaurus A (NGC 5128) have been made to compliment the initial 7-point map of Phillips *et al.* (1987). These confirm that the molecular gas is rotating at velocities of a few hundred km s^{-1} in a direction perpendicular to the ellipsoidal stellar component (*c.f.* Wilkinson *et al.* 1986), reflecting the separate dynamical systems in this highly disturbed merger. They also allow an examination of the kinematics of the central core, where HI VLA observations (van Gorkom 1987) are handicapped by self-absorption. The molecular gas and far infrared emission distributions (Joy *et al.* 1988) are remarkably similar.

NGC 3256, whose characteristic "tails" and chaotic nuclear appearance at optical wavelengths clearly identify it as a merging system (Toomre 1977; Schweizer 1986), has also been studied. Molecular gas extends over 7 – 10 kpc around the nucleus, as does 10 μm emission (Graham *et al.* 1987), indicating that the enhanced luminosity seen by IRAS, 3.2×10^{11} L$_\odot$, is plausibly accounted for by a burst of star formation. Although the total mass of gas, 2×10^{10} M$_\odot$, is a factor of two higher than that associated with Arp 220, the order of magnitude lower luminosity suggests that in NGC 3256 the merger is less advanced (*c.f.* Joseph and Wright 1985). Indeed, ESO photographic plates, kindly provided by J. Bergeron, show two central condensations separated by only 4″, which could reflect the presence of independent nuclei.

CO(2→1) observations of four more distant galaxies, NGC 1614 (cz $\approx$ 4800 km s^{-1}), NGC 2623 (cz $\approx$ 5500 km s^{-1}), NGC 6090 (cz $\approx$ 6090 km s^{-1}), and Markarian 231 (cz $\approx$ 12700 km

591

D. McNally (ed.), Highlights of Astronomy, Vol. 8, 591–592.

s^{-1}), have also been made. For the nearer objects, typical integration times were about 25 minutes, while the spectrum of Mrk 231 shown in Figure 1 was acquired in an hour.

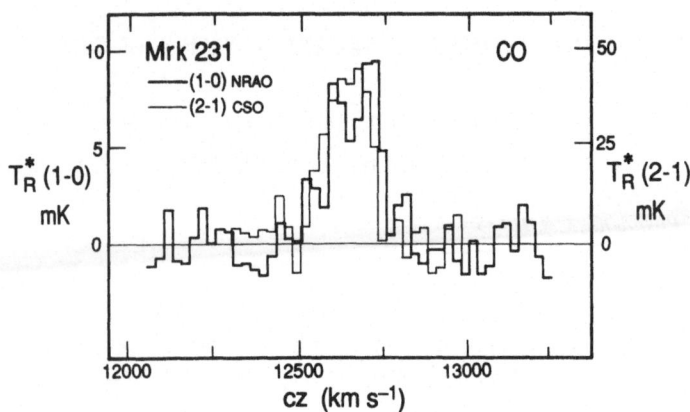

Figure 1. A spectrum of the CO(2→1) emission from Markarian 231 acquired at the CSO, superimposed on the CO(1→0) spectrum from NRAO

These CSO CO(2→ 1) spectra have been compared with CO(1→0) profiles obtained with the NRAO 12-meter telescope. The results for Mrk 231, displayed in Figure 1, are typical of galaxies unresolved by both instruments; the line shapes are very similar and the substantially higher values of T_R^* for the CO(2→1) line probably reflect only the ratio of telescope beam size. Thus we have as yet no evidence that the molecular gas, even in these distant galaxies, differs in its properties from molecular clouds in our Milky Way.

References

van Gorkom, J. 1987, in I. A. U. Symposium No. 127, *Structure and Dynamics of Elliptical Galaxies*, ed. T. de Zeeuw (Dordrecht:Reidel), p. 421.

Graham, J. R., Wright, G. S., Joseph, R. D., Frogel, J. A., Phillips, M. M., and Meikle, W. P. S. 1987, in *Star Formation in Galaxies*, ed. C. J. Lonsdale (U. S. Government Printing Office), p. 517.

Joseph, R. D., and Wright, G. S. 1985, *M. N. R. A. S.*, **214**, 87.

Joy, M., Lester, D. F., Harvey, P. M., and Ellis, H. B. 1988, *Ap. J.*, **326**, 662.

Phillips, T. G. *et al.* 1987, *322*, L73.

Schweizer, F. 1986, *Science*, **231**, 227.

Toomre, A. 1977, in *The Evolution of Galaxies and Stellar Populations*, (Yale University Observatory, New Haven), p. 401.

Wilkinson, A., Sharples, R. M., Fosbury, R. A. E., and Wallace, P. T. 1986, *M. N. R. A. S.*, **218**, 297.

MOLECULES IN GALAXIES: RESULTS FROM BELL LABORATORIES

A. A. Stark
AT&T Bell Laboratories
Crawford Hill Laboratory; Holmdel, NJ 07733

ABSTRACT. A decade of galaxy observations at the Crawford Hill 7 m antenna are summarized. Significant results include the mapping of spiral arms in CO, the demonstration that gas in many spiral galaxies is roughly half atomic and half molecular, the detection of extragalactic HCO^+, CS, ^{13}CO and $C^{18}O$, and the observation of molecular material in the halo of M82.

The 7 m diameter millimeter-wave antenna at AT&T Bell Laboratories, Crawford Hill, has been used for a variety of extragalactic observational projects. This paper briefly summarizes the results.

Early observations concentrated on the feasibility of detections. HCN and HCO^+ were detected in M82 and NGC253 (Stark and Wolff 1979). CS was detected by Henkel and Bally (1985). ^{13}CO was detected in several galaxies (Encrenaz *et al.* 1979), showing that the $^{12}CO/^{13}CO$ ratio varied from galaxy to galaxy, but that typical values were higher than GMCs in the Milky Way. Stark and Carlson (1984) detected $C^{18}O$ in M82, at a level consistent with terrestrial isotopic abundances. Stark and Carlson detected CO emission from the filaments above and below the disk of M82. This molecular gas is in an unusual state, probably a result of galaxy collisions. Seyfert galaxies were detected in ^{12}CO (Bieging *et al.* 1981), but subsequent work (Blitz *et al.* 1986) showed that most Seyferts are not extraordinary in their CO brightness.

Many of the observations relate to the nature of spiral structure, and its appearance in the CO line. Stark *et al.* (1979), Stark (1979, 1984a), Boulanger *et al.* (1981), Ryden and Stark (1986), and Casoli *et al.* (1987) mapped spiral arms in M31, where the arm/interarm contrast in CO brightness exceeds 10. Ryden and Stark observed a sudden shift in velocity across the arm, interpreted by them as a dynamical effect of the spiral density wave; however, Casoli *et al.* argue that this is in fact an effect of the warp which is known from HI line observations. Stark *et al.* (1987) observed total CO emissivity in a sample of 29 galaxies, some with well-defined spiral structure, and some without. They found that spiral structure does not greatly affect a galaxy's CO brightness, and

593

D. McNally (ed.), Highlights of Astronomy, Vol. 8, 593–594.
© *1989 by the IAU.*

interpreted this in a model where GMC formation is very efficient: if a spiral density wave is present, it organizes the formation of GMCs along the arms; if a spiral density wave is not present, then GMCs form throughout the disk anyway.

Recent work concentrates on the overall CO emissivity of galaxies. Verter (1983) showed that CO emissivity increases with luminosity class, and that Sb spirals seem to have the highest CO emissivity per unit mass, although this quantity has considerable unexplained scatter. Polk *et al.* (1988) modeled the overall CO emission from galaxies as the result of two populations of molecular clouds: giant molecular clouds and diffuse clouds. They showed that diffuse clouds can make a disproportionate contribution to the total emissivity. A sample of Virgo cluster spiral galaxies (Stark *et al.* 1986) has been analysed (Knapp *et al.* 1987) to show that the CO emissivity per H_2 molecule is about the same in these galaxies as in the Milky Way, and that this sample of "normal" spiral galaxies contains some galaxies which are mostly molecular, and some galaxies which are mostly atomic. Richmond and Knapp (1986) found that NGC 4565 is approximately half atomic and half molecular, but the molecular component is more centrally concentrated. Total CO emission is well-correlated with far-IR emission, but Stark (1984b) showed that there is little molecular gas in the center of M31, even though IRAS shows more far-IR radiation from the center than from the spiral arms, which are CO bright.

REFERENCES

Bieging, J. H., Blitz, L., Lada, C. J., and Stark, A. A. 1981 *Ap. J.*, **247**, 443.

Blitz, L., Mathieu, R. D., and Bally, J. 1986 *Ap. J.*, **311**, 142.

Boulanger, F., Stark, A. A., and Combes, F. 1981 *Astr. Ap.*, **93**, L1.

Casoli, F., Combes, F., and Stark, A. A. 1987 *Astr. Ap.*, **173**, 43.

Encrenaz, P. J., Stark, A. A., Combes, F., and Wilson, R. W. 1979 *Astr. Ap.*, **78**, L1.

Henkel, C., and Bally, J. 1985 *Astr. Ap.*, **150**, L25.

Knapp, G. R., Helou, G. and Stark, A. A. 1987 *A. J.*, **94**, 54.

Polk, K. S., Knapp, G. R., Stark, A. A., and Wilson, R. W. 1988 *Ap. J.*, **332**, 432.

Richmond, M. W., and Knapp, G. R. 1986 *A. J.*, **91**, 517.

Ryden, B. S., and Stark, A. A. 1986 *Ap. J.*, **305**, 823.

Stark, A. A. 1979 *Ph. D. thesis, Princeton University*.

Stark, A. A. 1984a in *The Milky Way Galaxy* ed. H. Van Woerden *et al.*, p.p. 445.

Stark, A. A. 1984b *B.A.A.S* **16**, 538.

Stark, A. A., and Carlson, E. R. 1984 *Ap. J.*, **279**, 122.

Stark, A. A., Elmegreen, B. G., and Chance, D. 1987 *Ap. J.*, **322**, 64.

Stark, A. A., Frerking, M. A., and Linke, R. A. 1979 *B.A.A.S.* **11**, 415.

Stark, A. A., Knapp, G. R., Bally, J., Wilson, R. W., Penzias, A. A., and Rowe, H. E. 1986 *Ap. J.*, **310**, 660.

Stark, A. A. and Wolff, R. S. 1979 *Ap. J.*, **229**, 118.

Verter, F. 1983 *Ph. D. Thesis, Princeton University*.

CO IN M82 AND OTHER MILDLY ACTIVE GALAXIES

R. Wielebinski
Max-Planck-Institut für Radioastronomie
Auf dem Hügel 69
D-5300 Bonn 1
West Germany

A large project of studying various transitions and isotopic species of CO in M82 has been carried out with the 30-m IRAM radio telescope. A 1'6×1'0 area was mapped in $^{12}CO(2{\to}1)$ with 13" resolution (Loiseau et al., 1988a). The purpose of these observations was to obtain high signal-to-noise, high resolution data for comparison with the $^{12}CO(1{\to}0)$ observations of Nakai et al. (1987). A $^{13}CO(2{\to}1)$ observation covered the central 1'0×0'7 region (Loiseau et al., 1988b) and allowed to establish the velocity field of CO in the inner nuclear region. The position angle of the rotation axis of ^{13}CO is aligned with the optical rotation and at some 30° relative to the ^{12}CO velocity field. The ^{13}CO map also shows the existence of clumped emission in the central region. All data show the rotating

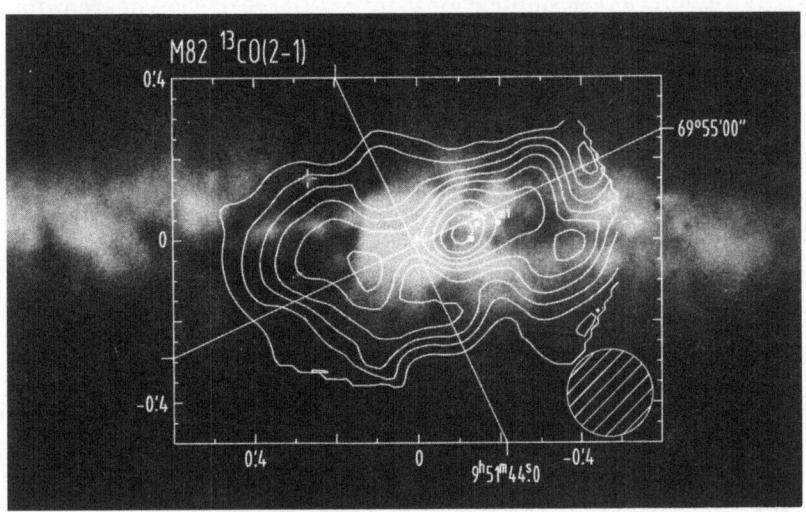

molecular ring. At some selected points (nucleus, major axis) $^{12}CO/^{13}CO$ ratios in both the (1→0) and (2→1) transition have been determined. The earlier suggestion that optical depth plays an important role in M82 has been confirmed. Variations of optical depth have been determined in the

D. McNally (ed.), Highlights of Astronomy, Vol. 8, 595–596.

area where common data is available. More recently, $C^{18}O$ studies of M82 have started (H.P. Reuter, private communication) with the aim of understanding both the optical depth and isotopic ratios in M82. The optical filaments in the nuclear area of M82 suggest a magnetic field along the minor axis. Radio continuum studies of M82 (Klein et al., 1988) indicate that this galaxy has the highest magnetic field of any galaxies. The casual connection between the rotating torus (CO, HI, etc.) and a poloidal magnetic field has been investigated (Lesch et al., 1988).

From this consideration we selected a number of mildly active galaxies and began investigations in CO and radio continuum. The galaxy M104 has a VLBI source in the nucleus, a radio disk and polarised emission which is aligned in the poloidal direction at the nucleus (Bajaja et al., 1988). A detection of CO in this early-type galaxy has been made (H.P. Reuter, M. Krause), but at a level of only ~50 mK so that the question of a rotating torus could not be definitely answered. Another candidate galaxy which has been investigated is NGC 1808. The optical filaments are clearly along the minor axis and an active nucleus is known to be present (Sersic and Pastoriza, 1965). Recent VLA observations (M. Dahlem, private communication) showed extended jets originating in the nucleus. Also a system of HI absorption was detected. SEST observations (R.S. Booth et al., private communication) again indicate a rotating ring of CO. The southern galaxy NGC 4945 is known to be a very active object. It is a source of dense HI, H_2CO, OH and CN emission. It is also a megamaser galaxy. A recent radio continuum survey of NGC 4945 (Harnett et al., 1988) has shown the existence of two polarisation maxima above the disk, symmetrically disposed about the nucleus. This galaxy has been recently extensively mapped with the SEST telescope (J.B. Whiteoak, M. Dahlem et al., private communication) in $^{12}CO(1\rightarrow0)$. The search for ring structure will be continued in the $(2\rightarrow1)$ transition.

The study of a number of mildly active galaxies indicate that there may be a casual relation between poloidal magnetic fields, radio continuums jets along the minor axis and rotating torus ring structures (CO, HI, OH, etc.) in the nuclei of such galaxies.

References

Bajaja, E., Dettmar, R.-J., Hummel, E., Wielebinski, R.: 1988, Astron. Astrophys. **202**, 35

Harnett, J.I., Haynes, R.F., Klein, U., Wielebinski, R.: 1988, Astron. Astrophys., submitted

Klein, U., Wielebinski, R., Morsi, H.W.: 1988, Astron. Astrophys. **190**, 41

Lesch, H., Crusius, A., Wielebinski, R., Schlickeiser, R.: 1988, Astron. Astrophys., submitted

Loiseau, N., Nakai, N., Wielebinski, R., Sofue, Y., Klein, U.: 1988a, in *Molecular Clouds in the Milky Way and External Galaxies*, eds. R. Dickman, R. Snell and J. Young, Springer-Verlag, in press

Loiseau, N., Reuter, H.P., Wielebinski, R., Klein, U.: 1988b, Astron. Astrophys. **200**, L1

Nakai, N., Hayashi, M., Handa, T., Sofue, Y., Hasegawa, T., Sasaki, M.: 1987, Publ. Astron. Soc. Japan **39**, 685

Sersic, J.L., Pastoriza, M.G.: 1965, Publ. Astron. Soc. Pacific **77**, 287

MOLECULAR CLOUDS IN DWARF IRREGULAR GALAXIES

C. Henkel
Max-Planck-Institut für Radioastronomie,
Auf dem Hügel 69, 5300 Bonn 1, F.R.G.

The study of molecular clouds in dwarf galaxies was, until recently, an arduous task and provided detections in only a handful of sources (see Thronson and Bally 1986; Wiklind et al. 1986; Tacconi and Young 1987; Arnoult et al. 1988). Line strengths observed in CO were small ($<$100mK) and measurements were confined to one position per source, thus inhibiting any detailed work. However, dwarf galaxies appear to be much more abundant than either spirals or large ellipticals and play an important role in our understanding of star and galaxy formation (see e.g. Gerola et al. 1980; Gallagher and Hunter 1984; Silk et al. 1987). With small rotational velocities (typically 50-100kms^{-1}), the absence of density waves, and low metallicities, they provide, by "local" standards, an interstellar environment with even more extreme conditions than the outer Galaxy. Detailed molecular studies are hence important.

In small galaxies, star formation is believed to occur in bursts which are followed by long inactive periods during which new molecular material may be built up (e.g. Gerola et al. 1980). The observation of blue compact galaxies proves that star formation can proceed, during the burst phase, as prodiguously as in spirals and a detailed investigation of molecular clouds in this as in the quiescent phase would be worthwhile. What are the properties of the individual clouds? And is the distribution of clouds, thermal and nonthermal radio continuum, and infrared sources consistent with stochastic self-propagating star formation as suggested by Gerola et al. (1980)?

At present, only the star burst phase itself is beginning to be investigated. CO emission in NGC4214 and probably also in DDO 126 was detected with the 30-m IRAM telescope. More important, molecular clouds in NGC3077 and IC10 (see Becker et al. 1988, 1989) could be mapped. The cloud in NGC3077, located on the line of sight toward the central region of this galaxy, has a mass in excess of $10^7 M_0$ and an FWHP size of 280pc. Toward the Magellanic irregular IC10, two clouds are found to be associated with the central bar: one with a size of $180d_{Mpc}$pc, the other with a size of $<40d_{Mpc}$pc (d=1-3Mpc). Both are associated with H_2O masers (see Henkel et al. 1986). Also Ohta et al. (1988) reported the detection of clouds in the extended HI envelope of IC10. These observations have been confirmed. More detailed maps reveal in at least one of these

D. McNally (ed.), Highlights of Astronomy, Vol. 8, 597–598.
© *1989 by the IAU.*

sources that there are several extended (>100pc) molecular complexes.

While both NGC3077 and IC10 may, for a variety of reasons, not be "normal" dwarf irregulars, we nevertheless can draw the following very preliminary conclusions:

(1) There are indeed CO rich dwarf galaxies.

(2) The detected GMC's are associated with regions of enhanced HI column density ($\gtrsim 10^{21} cm^{-2}$), even in the extended HI shell of IC10.

(3) The detected GMC's are associated with HI "superclouds" (see Elmegreen and Elmegreeen 1987) with typical masses of $10^7 M_\odot$, whose existence (as that of the GMC's) appears not to depend on density waves, speed of rotation, degree of shear or metallicity.

(4) The GMC's in the central regions of their parent galaxies are associated with IRAS point sources. This is, however, not the case for the clouds in the HI shell of IC10.

(5) The GMC's can be larger than those observed in the galactic disk, in spite of the small size of the parent galaxies and the relatively small number of large clouds. Such very large clouds are also seen toward the 30 Dor cloud in the LMC (Cohen et al. 1988), and toward the lenticular galaxy NGC404 (Wiklind and Henkel 1989). All these galaxies have one property in common: massive star formation is not periodically triggered by density waves. Hence more time might be available to build up extended clouds before they are dispersed by newly formed massive stellar objects.

For the future, we expect detailed studies from a large number of dwarf irregulars, located both in the northern and in southern hemisphere, where NGC55 was recently detected with the 15-m SEST telescope (Heithausen, private communication). In addition, investigating the extended HI envelopes of both dwarf and giant irregulars will provide important clues with respect to the unknown origin of this material. Assuming that the CO clouds are in virial equilibrium, a determination of the $I_{CO}-N(H_2)$ conversion ratio will provide a first estimate of the metallicity, which can then be compared with theoretical predictions, as e.g. with the models of Dekel and Silk (1986) and Silk et al. (1987).

References

Becker, R., Appenzeller, I., Wiklind, T., Wouterloot, J.G.A., Henkel, C., Wilson, T.L., Diamond, P.: 1988, Astron. Ges. Abstract Ser. 1, 42
Becker, R., Schilke, P., Henkel, C.: 1989, Astron. Astrophys., in press
Cohen, R.S. et al.: 1988, Astrophys. J. **331**, L95
Dekel, A., Silk, J.: 1986, Astrophys. J. **303**, 39
Elmegreen, B.G., Elmegreen, D.M.: 1987, Astrophys. J. **320**, 182
Gallagher, J.S., Hunter, D.A.: 1984, Ann. Rev. Astron. Astrophys. 22, 37
Gerola, H., Seiden, P.E., Schulman, L.S.: 1980, Astrophys. J. **242**, 517
Henkel, C., Wouterloot, J.G.A., Bally, J.: 1986, Astron. Astrophys. **155**, 193
Ohta, K., Sasaki, M., Saito, M.: 1988, Pub. Astron. Soc. Japan, in press
Silk, J., Wyse, R.F.G., Shields, G.A.: 1987, Astrophys. J. **322**, L59
Tacconi, L.J., Young, J.S.: 1987, Astrophys. J. **322**, 681
Thronson, H.A., Bally, J.: 1986, in "Star Formation in Galaxies", NASA Conf.Pub. **2466**, ed. C.J. Persson, p. 267
Wiklind, T., Henkel, C.: 1989, Astron. Astrophys., submitted
Wiklind, T., Rydbeck, G.: 1986, Astron. Astrophys. **164**, L22

MOLECULAR CLOUDS IN THE LARGE AND SMALL MAGELLANIC CLOUDS

Mónica Rubio
Departamento de Astronomía, Universidad de Chile

To determine the distribution of molecular gas in the Large and Small
Magellanic Clouds (LMC and SMC), we made complete surveys in the J=1-0
transition of CO, of the central 6°x6° and 3°x2° areas of the LMC and
SMC, respectively. The observations were made with the 1.2m Columbia
Telescope at Cerro Tololo, which provides and angular resolution of 8.8
arcmin at 115 GHz. The spectral resolution of the surveys was 1.3 km s^{-1}
and the sensitivity in antenna temperature was 0.05 K for the LMC and
0.02 K for the SMC.

In the LMC we detected 40 CO complexes. Positions, masses and other
physical characteristics of the identified molecular clouds are given in
Cohen et al. (1988). In the SMC we detected CO emission from two large
complexes located in the southwest, near N19, and northeast, near N78,
regions of the SMC bar (Rubio et al. 1989).

The general correspondence of molecular complexes in the Magellanic
Clouds to other Pop I objects is close. In the LMC, nearly all the CO
sources appear projected toward peaks in the HI emission having HI
column densities $>2\times10^{21}$ cm^{-2}. In the SMC, both complexes are projected
toward regions of atomic gas having the largest HI column densities
$(N(HI)\sim 10^{22}$cm$^{-2})$. The CO sources in the Magellanic Clouds are found
projected near either optical HII regions, radio continuum sources, and
peaks in the IRAS 100μ emission, suggesting that in these galaxies the
presence of molecular clouds is a prerequisite for star formation.

The intensity of the CO emission from molecular clouds in the LMC and
SMC is weak compared with that from Galactic giant molecular clouds (see
Table 1). Molecular clouds in the LMC and SMC have antenna temperatures,
T_a, 10 and 50 times smaller, respectively, than those of Galactic clouds.
The average CO luminosity, L_{CO}, of the complexes in the LMC and SMC is 4
and 6 times smaller than in the Galaxy. The observed size of molecular
clouds are 3 to 4 times greater in the Magellanic Clouds than in the
Galaxy, but we note that they do not correspond to the cloud physical
dimensions but to correlated sizes of CO emission.

The most likely explanation for the striking low antenna temperature
of the ^{12}CO emission from molecular complexes in the Magellanic Clouds is
beam dilution already suggested by Rubio and Garay (1988). They conclude
that the characteristic physical length of individual CO clumps, making

599

D. McNally (ed.), Highlights of Astronomy, Vol. 8, 599–600.
© *1989 by the IAU.*

Table 1. Observed parameters of the ^{12}CO emission

Object	T_A K	DV km s^{-1}	R pc	L_{CO} K km s^{-1} pc^2
SMC	0.04+0.01	15+5	220+50	0.6 10^5
LMC	0.2+0.1	11+4	180+65	1.0 10^5
Galaxy	(2-3)	9+4	63+35	3.6 10^5

up a molecular complex, in the LMC and SMC is 3 and 10 times smaller than that in the Galaxy, respectively. In addition, they propose that molecular clouds in the LMC and SMC are the UV shielded cores of large diffuse HI clouds. The small dimensions of the CO clouds are probably a consequence of the low dust content and the low metallicity of the interstellar gas of the Magellanic Clouds. Further support to this suggestion are given by the good agreement between the theoretical critical HI column density required to shield the cloud interior from the external UV radiation field (Franco and Cox 1986) and the observed column densities of HI toward the CO sources, and by the similar radial velocities between the HI and CO gas.

To determine the molecular mass of the Magellanic Clouds we assume that the velocity integrated CO emission (W_{CO}) is a tracer of the molecular hydrogen column density $N(H_2)$. However, due to differences in the physical conditions in the LMC and SMC and the Galaxy it is not wise to adopt the Galactic value for the ratio $N(H_2)/W_{CO}$. From a comparison of the empirical CO luminosity versus linewidth relationship of molecular clouds in the Magellanic Clouds and our Galaxy we propose a conversion factor, to derive molecular masses from the CO luminosities, that is six and twenty five times larger than the Galactic value for the LMC and SMC, respectively. The total mass of molecular hydrogen is 1.5×10^8 $M_\odot$ in the LMC and 3.5×10^7 $M_\odot$ in the SMC. The ratio of molecular to atomic mass is ~ 0.3 and ~ 0.09 for the LMC and SMC, respectively, about three and ten times smaller than the ratio derived in our Galaxy.

Preliminary results of observations of the molecular clouds in the Magellanic Clouds with the 15m SEST telescope at La Silla (Booth 1988) have confirmed the suggestion that the size of the CO clumps making up a molecular complex in the LMC and SMC is smaller than the typical size of Galactic giant molecular clouds.

We acknowledge support from FONDECYT grant 486/88 and Universidad de Chile grant E2604-8824 (DTI).

REFERENCES

Booth, R. 1988, these proceedings.
Cohen, R.S., Dame, T.M., Garay, G., Montani, J., Rubio, M., and Thaddeus, P. 1988, Ap. J. (Letters), 311, L95.
Franco, J., and Cox, P.C. 1986, Pub. A.S.P., 98, 1076.
Rubio, M., and Garay, G. 1988, Molecular Clouds in the Milky Way and External Galaxies, (Springer-Verlag), in press.
Rubio, M. et al., 1989, In preparation.

CO in Early Type Galaxies

Tommy Wiklind[1] and Christian Henkel[2]

1. Onsala Space Observatory, S-43900 Onsala, Sweden
2. Max-Planck-Institut für Radioastronomie, Auf dem Hügel 69, D-5900 Bonn, FRG

During the last two years, observations of the molecular cloud content of early type galaxies have drastically changed our view of these systems as inert gas-poor galaxies with little or no star formation activity. Systematic surveys in the CO (J=1-0) line (Wiklind and Henkel, 1988a,b; Thronson, private communication) have shown that an IR selected sample of early type galaxies contains typically 10^7-10^8 $M_\odot$ of molecular hydrogen gas. This is similar to the typical HI masses found in these galaxies (Knapp et al., 1985; Wardle and Knapp, 1986).

The majority of the detections comes from a survey that we are conducting, using the IRAM 30-m telescope and the newly constructed SEST telescope in Chile. We have also incoporated results from Thronson (private communication), as well as NGC 185 (Wiklind and Rydbeck, 1986) and NGC 4472 (Huchtmeier et al., 1988). The H_2 masses have been computed from the CO integrated intensity, using a $N(H_2)/I_{CO}$ conversion ratio of $2 \cdot 10^{20}$ cm^{-2} (K km s^{-1})$^{-1}$. The distances have been derived from our measurements of the radial velocities, corrected for the Solar motion relative to the center of the Local Group and for the Virgocentric flow (Aaronson et al., 1982). We have adopted a Hubble constant of 100 km s^{-1} Mpc^{-1}, with an assumed distance of 13.5 Mpc to the Virgo cluster. Since both the conversion ratio and the assumed Hubble constant are "conservative", this means that the derived H_2 masses are lower limits, unless the molecular cloud properties of the early type galaxies are greatly different from those of the Milky Way.

For comparision with the molecular cloud properties of our sample of early type galaxies, we have from the literature compiled a list of the molecular cloud properties, as well as FIR and blue luminosities, for 123 spiral galaxies. In Figure 1 we show the H_2 mass distributions for the two samples (light bars for the spirals, dark bars for the early type galaxies). It is evident that the spiral sample, on the average, has about an order of magnitude more H_2 gas than the early type galaxies. A similar difference can be seen in Figure 2, which shows the distributions of the log(SFR). The star formation rate (SFR) can be derived directly from the FIR emission, as estimated by the IRAS, in a similar manner as Thronson and Telesco (1986). The SFR is here assumed to be directly proportional to the FIR luminosity: SFR=$3.2 \cdot 10^{-10} L_\odot$ $M_\odot$ yr^{-1}. The typical star formation rates for the early type galaxies are around 0.1 - 1 $M_\odot$ yr^{-1}. In Figure 3 we have plotted the distributions of the ratio of F_{60}/F_{100}, as measured by the IRAS. Although the L_{FIR} (SFR) distributions are different by about an order of magnitude, the distributions of the dust temperatures appear to be the same. Since we do not belive that the dust properties are significantly different in the early type galaxies than in spiral galaxies, this result means that the heating mechanism of the dust is as efficient in the early type sample as in the spiral sample. This indicates that the efficiency of star formation in early type galaxies might be higher than for the spiral sample. In Figure 4 we have plotted the distributions of log(L_{IR}/L_{CO}) for the two samples. The ratio of L_{IR}/L_{CO}, which is proportional to the ratio SFR/M(H_2), is usually taken as a measure of the star formation efficiency (SFE). From this figure it is evident that *the SFE for the early type galaxies is, on the average, as high or possibly higher than that of the spiral galaxies.* This result may indicate that *spiral density waves are not necessary for efficient formation of massive stars.*

601

D. McNally (ed.), Highlights of Astronomy, Vol. 8, 601–602.
© 1989 *by the IAU.*

602

Acknowledgement. We are grateful to Dr. H. Thronson for communicating results prior to publication.

References:

Aaronson, M., Huchra, J., Schechter, P.L. and Tully, R.B., 1982, Astrophys. J., **258**, 64
Huchtmeier, W.K., Bregman, J.N., Hogg, D.E. and Roberts, M.S., 1988, Astron. Astrophys., **198**, L17
Knapp, G.R., Turner, E.L. and Cunniffe, P.E., 1985, Astron. J., **90**, 454
Thronson, H.A. and Telesco, C.M., 1986, Astrophys. J., **311**, 98
Wardle, M. and Knapp, G.R., 1986, Astron. J., **91**, 23
Wiklind, T. and Rydbeck, G., 1986, Astron. Astrophys., **164**, L22
Wiklind, T. and Henkel, C., 1988a, submitted to Astron. Astrophys.
Wiklind, T. and Henkel, C., 1988b, in preparation

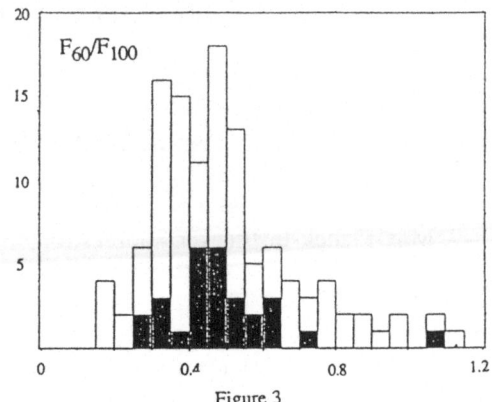

Figure 3

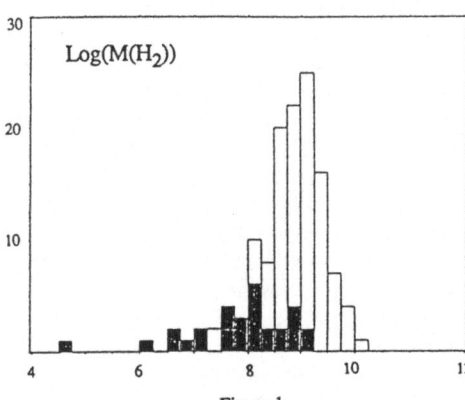

Figure 1

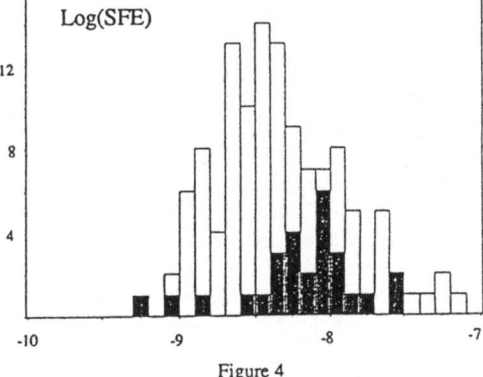

Figure 4

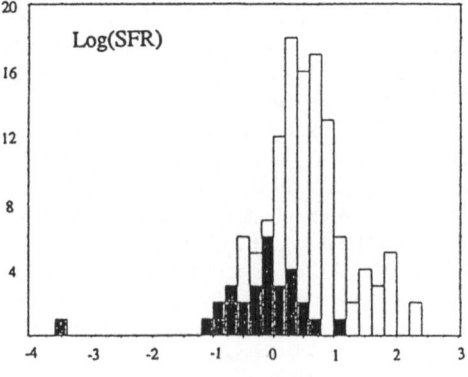

Figure 2

THE RATIO OF H_2 TO HI GAS IN INFRARED LUMINOUS GALAXIES

I. F. MIRABEL, and D. B. SANDERS
Division of Physics, Mathematics and Astronomy
California Institute of Technology
Pasadena, CA 91125

ABSTRACT. The majority of infrared active galaxies ($f_{fir}/f_b \geq 2$) have molecular to atomic mass fractions in the range of 0.5 to 2.0. Among the galaxies with the higher infrared excesses there are spectacular cases of HI deficient systems, where less than 15% of the total mass of interstellar gas is in atomic form. The optical morphology of luminous infrared galaxies suggest that the overall mass fractions of molecular to atomic gas, and the infrared luminosities per nucleon of interstellar gas are enhanced during galaxy-galaxy interactions.

1. Results

The relations between the masses of molecular and atomic gas with the far-infrared and blue luminosities were studied in an unbiased sample of galaxies with far-infrared luminosities in the range of 2×10^{10} $L_\odot$ - 2×10^{12} $L_\odot$. The galaxies were selected from the IRAS Bright Galaxy Sample of Soifer et al. (1987). The atomic and molecular components were determined from two large and homogeneous surveys of that sample: one in HI conducted at Arecibo by Mirabel and Sanders (1988), the other conducted in CO(1-0) by Sanders et al. (in progress) with the 12-m NRAO and 14-m FCRAO antennas. All the galaxies considered for the present study are unresolved by the telescopes used for the observations.

Figure 1a shows that the majority of the galaxies with infrared excesses in the range $2 \leq f_{fir}/f_b \leq 40$ have $M(H_2)/M(HI)$ mass ratios between 0.5 and 2. However, among galaxies with $f_{fir}/f_b \geq 50$ there are striking HI deficient systems, with less than 15% of the gas in atomic form. Despite the large scatter, Figure 1a shows an enhancement of the overall fraction of gas in molecular form with increasing far-infrared excess.

Figure 1b shows that there is an enhancement of the infrared luminosity per nucleon of interstellar gas with increasing far infrared excess. Galaxies with the highest infrared excess may have infrared luminosities per nucleon that are as high as ~ 20 times that found for the Galaxy.

Figure 2 summarizes the morphologies of the sample galaxies as revealed by optical images obtained with the 1.5-m and 5-m Palomar telescopes. A striking result is that most, if not all infrared luminous galaxies appear to be interacting systems. Furthermore, advanced mergers (M) and strongly damaged disks (D) predominate among galaxies with the highest far-infrared excesses, while less strongly interacting systems (C and P) predominate among galaxies with more moderate far-infrared excess. This suggests that the far-infrared excess in these galaxies is related to the intensity of the tidal interactions.

References
Mirabel, I. F., and Sanders, D. B. 1988, *Ap. J.*, **335**, in press.
Soifer, B. T., Sanders, D. B., Madore, B. F., Neugebauer, G., Persson, C. J., Persson,
    S. E., and Rice, W. L. 1987, *Ap. J.*, **320**, 238.

603

D. McNally (ed.), Highlights of Astronomy, Vol. 8, 603–604.
© 1989 by the IAU.

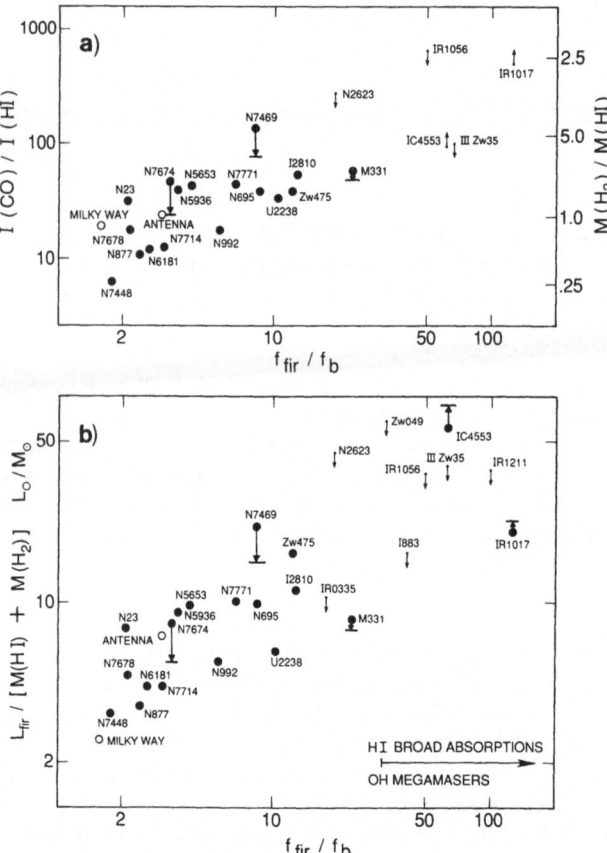

Figure 1. a) CO(1→0)/HI flux ratios measured at λ2.6 mm and λ21 cm, as a function of the far-infrared-to-blue flux ratio, f_{fir}/f_b, for 28 luminous infrared galaxies. The arrows are due to lower and upper limits in the measurements of the HI emission. The verticle scale on the right was computed using the relation $M(H_2) = 5.82 \, L_{CO}$. b) The far-infrared luminosity per unit mass of HI+H_2 vs. f_{fir}/f_b. Most of the galaxies with $f_{fir}/f_b \geq 40$ exhibit HI absorption and OH maser emission with velocity widths of several hundred km s^{-1}.

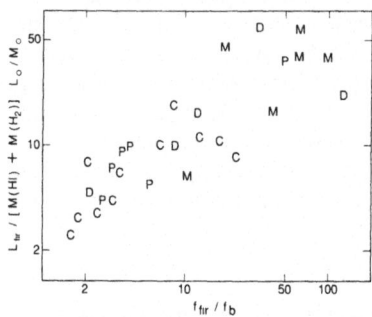

Figure 2. Morphology of the galaxies in Figure 1b. C = companion 0.5 - 2.0 diameters away, P = companion ≤ 0.5 diameter, D = single tidally distorted disk, M = advanced merger.

MOLECULAR GAS IN GALACTIC NUCLEI

Nick Scoville
Owens Valley Radio Observatory
California Institute of Technology

Recent high resolution interferometric observations of the molecular gas in luminous IR galaxies reveal extraordinary concentrations of star forming material in the central few kpc. In several of the nearest IR bright galaxies, the molecular gas in the central regions is concentrated in a bar-like distribution (IC342, NGC 6946, and NGC 253) and in NGC 1068, approximately 40% of the molecular gas is confined to two arms or a ring at approximately 1.6 kpc radius. Interferometry on the most luminous galaxies ($L_{IR} \geq 10^{11} L_\odot$) reveals that approximately half of the total interstellar matter is contained in the central kpc with mean densities of several hundred H_2 cm^{-3}. Such gas concentrations should result in the very rapid formation of stars, i.e. a central star burst yielding a massive central star cluster.

1. LUMINOUS GALAXIES

In the nearby bright IR galaxies, high resolution millimeter-wave interferometry and single dish observations have revealed a variety of morpohologies in the neutral gas. Three of the galaxies first mapped with the Owens Valley millimeter-wave inteferometer showed elongated bar-like distributions for the molecular gas in the central kpc. The results for IC 342 and NGC 6946 have been published by Lo *et al.* (1984) and Ball *et al.* (1985). The more recent CO interferometry for NGC 253 consisting of a mosaic of seven 1' fields (Canzian, Mundy, and Scoville 1988) shows a massive bar of molecular gas aligned with the stellar bar seen in optical and near infrared maps (Scoville *et al.* 1985).

A rather different morphology is found in the nearby Seyfert II galaxy NGC 1068. In this case, approximately $4 \times 10^9 M_\odot$ of molecular gas resides in a ring at the outer edge of the bright optical disk (Myers and Scoville 1986). This ring of neutral gas situated just outside the stellar bar recently discovered in the near infrared (Scoville *et al.* 1988) is somewhat surprising in view of the abundant evidence for a high rate of star formation in the interior optical disk. On the other hand, the kinematics of the molecular gas indicate a substantial component of radial motion suggesting that at times in the past, there has been an abundance of star forming material within the central disk.

Perhaps most dramatic in terms of star burst activity are the high luminosity and ultraluminous galaxies discovered as a result of the IRAS survey. At the higher

605

D. McNally (ed.), Highlights of Astronomy, Vol. 8, 605–607.

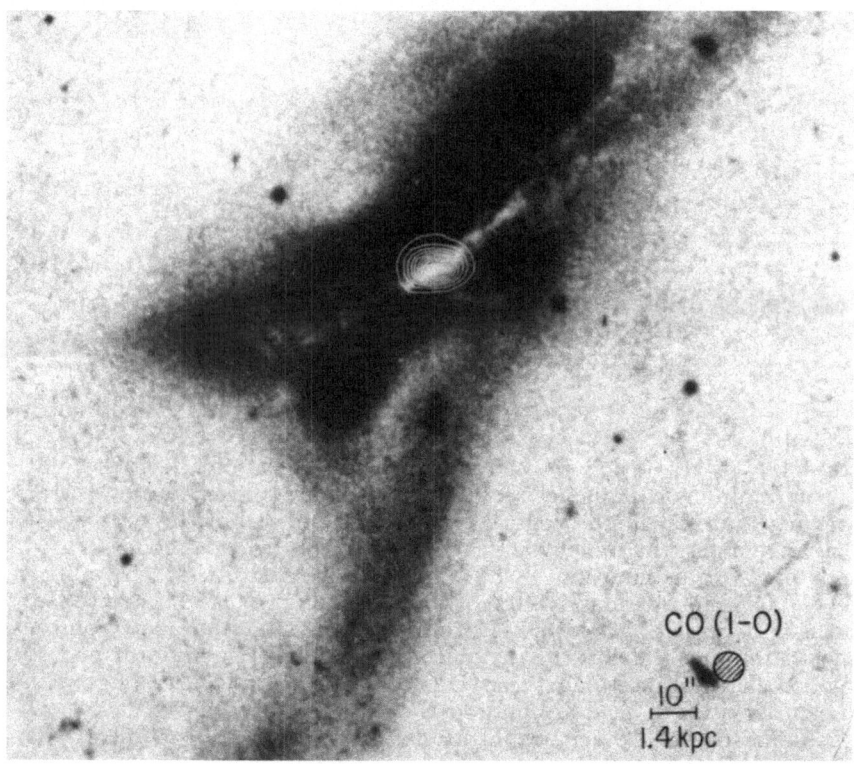

Figure 1: Integrated intensity map of CO emission in NGC 520 superimposed on the optical photograph from the Arp atlas. Contour levels are 10% of peak. The hatched beam symbol indicates both the position angle and size of the synthesized beam (Sanders *et al.* 1988).

luminosities, one sees a high preponderance of double nuclei and/or extended tidal tails indicative of strong galactic interactions or the merging of two galaxies. It is also evident that the optical spectra of the ultraluminous galaxies are dominated by non-thermal emission characteristic of a narrow line AGN or Seyfert nucleus rather than thermal HII region-type spectra seen in the lower luminosity galaxies. This qualitative assessment of the optical data strongly suggests that *the highest luminosities are initiated by galactic collisions, and the dominant energy source may in fact be a non-thermal AGN*. Virtually all the luminous IRAS galaxies have also been shown to be extremely rich in interstellar gas, predominantly molecular hydrogen (e.g. Sanders, Scoville, and Soifer 1987).

The molecular gas gas is also highly concentrated in their nuclei. Over the last two years, the millimeter wave interferometer at Owens Valley Radio Observatory has been used for aperture synthesis mapping of the CO emission in eight of the luminous galaxies (Scoville *et al.* 1986, Sargent *et al.* 1987, Sanders *et al.* 1988, Scoville *et al.* 1989). In each case, 30-70% of the total CO emission is confined to a region $\leq 10''$ in size, centered on one of the galactic nuclei. Most spectacular is Arp 220 ($L_{IR}=1.3\text{x}10^{12}$ $L_\odot$) where $1.3\text{x}10^{10}$ $M_\odot$ of H_2 is contained within the central R≤750 pc. The mean molecular gas density *averaged* over a spherical volume of this size is approximately 200 H_2 cm^{-3}. The ratio of far infrared luminosity to H_2 mass is 100 $L_\odot/M_\odot$. For comparison, the mean value of this luminosity-to-mass ratio in Galactic GMCs is 3 $L_\odot/M_\odot$ and the maximum value, obtained in localized areas immediately adjacent to galactic HII regions, is 40 $L_\odot/M_\odot$. Thus, in Arp 220 the overall "star formation" efficiency is 30 times the average in the Milky Way.

We have recently undertaken a theoretical investigation of the evolution of interacting gas-rich galaxies (Norman and Scoville 1988, Scoville and Norman 1988). Our model starts with the formation of a single coeval stellar cluster of total mass $4\text{x}10^9$ $M_\odot$. This mass is distributed among stars with a Salpeter initial mass function ($\alpha=2.35$) over the range 1-50 $M_\odot$. The total number of stars in the cluster is $1.4\text{x}10^9$ and evolution of the stellar population is followed. The small radius of the cluster (10-50 pc) implies an extremely high escape velocity ($>10^3$ km s^{-1}) ensuring that all mass-loss occurring during the late stellar evolution phases (red giant mass-loss and supernovae) will be trapped in the cluster and eventually sink, dissipatively, to the center. Thus, based on standard stellar evolution and an assumed initial mass function, it is possible to predict the growth rate of the central black hole and its accretion luminosity.

2. REFERENCES

Ball, R., Sargent, A.I., Scoville, N.Z., Lo, K.Y., and Scott, S.L. 1985, *Ap.J. (Letters)*, **298**, L21

Canzian, B.J., Mundy, L.G., and Scoville, N.Z. 1988, in preparation

Lo, K.Y. *et al.* 1984, *Ap.J. (Letters)*, **282**, L59

Myers, S.T. and Scoville, N.Z. 1986, *Ap.J. (Letters)*, **312**, L39

Norman, C.A. and Scoville, N..Z. 1988, *Ap.J.* **332**, 163

Sanders, D.B., Scoville, N.Z., Sargent, A.I., and Soifer, B.T. 1988, *Ap.J. (Letters)*, **324**, L55

Sargent, A.I., Sanders, D.B., Scoville, N.Z., and Soifer, B.T. 1987, *Ap.J. (Letters)*, **312**, 235

Scoville, N.Z., Matthews, K., Carico, D., and Sanders, D.B. 1988, *Ap.J. (Letters)* (submitted)

Scoville, N.Z. and Norman, C.A. 1988, *Ap.J.*, **332**, 163

Scoville, N.Z., Sanders, D.B., Sargent, A.I., Soifer, B.T., Scott, S.L., and Lo, K.Y. 1986, *Ap. J. (Letters)*, **311**, L47

Scoville, N.Z., Soifer, T., Neugebauer, G., Young, J.S., Matthews, K., and Yerka, J. 1985, *Ap.J.*, **289**, 129

Soifer, B.T., *et al.*. 1984, *Ap. J.*, **283**, L1

Young, J.S., Kenney, J., Lord, S.D., and Schloerb, F.P. 1984, *Ap.J. (Letters)*, **287**, L65

5. SPECTROSCOPY OF INDIVIDULAR STARS IN GLOBULAR CLUSTERS

AND THE EARLY CHEMICAL EVOLUTION OF OUR GALAXY

Chairman: G. Cayrel de Strobel
Editors: G. Cayrel de Strobel & M. Spite

Supporting Commissions:

29, 35, 36, 37, 47

JOINT COMMISSION MEETING 5, JCM5:
SPECTROSCOPY OF INDIVIDUAL STARS IN GLOBULAR CLUSTERS AND THE EARLY
CHEMICAL EVOLUTION OF OUR GALAXY.

Chair: G. Cayrel de Strobel
Supporting Commissions : 29, 35, 36, 37, 47

COMMISSION MEETING CM37/3:
The abundance spread within globular clusters.

Chair: T. Lloyd-Evans

The Proceedings of these two meetings will be edited by G.. Cayrel de
Strobel, M. Spite and T. Lloyd-Evans and published by the printing office of
Paris Observatory.
    We present here an overview of what has been discussed during these
meetings.

I - INTRODUCTION

    The long lasting interest of astronomers in globular clusters and their
star population was once more proved during the scientific sessions at the IAU
XX General Assembly. Nothing but, three meetings on globular clusters have been
organized; one meeting (CM37/3) was chaired by T. Loyd Evans and was devoted to
"The abundance spread within globular clusters". Another meeting (JCM7),
directed by Pierre Demarque gave us an overview of : "Stars clusters in the
Magellanic Clouds". The third meeting, organized by G. Cayrel de Strobel, has
concentrated of what can be done nowadays in high signal/noise spectroscopy of
individual stars in globular clusters. The great progress in high resolution
spectroscopy of very faint objects has allowed to obtain in recent years a
wealth of new results on detailed abundances of cluster stars. These results
come principally from observations with CCD detector systems on 4m-class
telescopes, as well as from the explosive growth of VAX class computers for
both, data analysis and theoretical modeling.
    The interpretation of such a wealth of new and sometime contradictory
results is not a small affair. A great number of specialists in several fields
is needed to understand the spectroscopic messages which are sending the far
away cluster stars. This is perhaps the reason why, in spite of two recent IAU
Symposia (J.E. Grindlay and A.G.D. Philip 1988, and G. Cayrel de Strobel and M.
Spite 1988) dedicated entirely or in part to the better understanding of this
subject, three astronomers had the same idea : to make globular clusters the
focus of scientific sessions at the XXth General Assembly.

    The meetings, JCM5 and CM37/3 were focused on the chemical abundances of
globular clusters. Because of the shortness of time allocated to them, being both
one half day IAU scientific sessions, they were organized as panel discussions
with interventions from the floor.

II - SCIENTIFIC PROGRAM OF JCM5:

    Four panel discussions addressed the general subjects of: "The
spectroscopy of individual stars in globular clusters and the early chemical
evolution of the Galaxy".

    The first panel: "Spectroscopic metallicities in Galactic and Magellanic
globular clusters : individual chemical abundances in metal poor and metal rich
clusters" has had as leaders:

R. Gratton: New spectroscopic cluster to cluster abundance analyses.

D. McNally (ed.), Highlights of Astronomy, Vol. 8, 611–614.
© *1989 by the IAU.*

C. Pilachowski: Individual spectroscopic abundances in metal poor and metal rich clusters.

M. Spite: Results from detailed analyses in stars of LMC and SMC.

F. Spite: The lithium abundance in the Magellanic globular clusters: a step toward the primordial lithium abundance.

V. Castellani: Globular clusters, Spectroscopy and evolutionary theories.

The second panel has discussed on: "Metallicity of globular clusters versus metallicity of field halo stars. Do cluster and field halo stars have a common history?" Its leaders have been:

J.B. Laird: The Halo metallicity distribution function.

N. Suntzeff: Can we compare extreme metal poor field stars to very metal poor globular cluster stars ?

R. Peterson: Relative abundances of stars in several extremely-metal-poor systems and in the field Halo stars.

M. Bessell: Oxygen abundances in field halo and globular stars.

J. Truran: Abundance constraints on the dynamical and chemical evolution of the Milky Way globular clusters.

The third panel:"The chemical inhomogeneities within globular clusters : example ω Centauri" has discussed on :

G. Da Costa: Chemical inhomogeneities within clusters.

V. Smith: New spectroscopic results of stars in ω Centauri.

M. Spite: Chemical inhomogeneities in ω Centauri.

K. Freeman: CN inhomogeneities in ω Centauri and 47 Tucanae from individual stars correlated with their velocity distribution.

The last and fourth panel of JCM5 concerned only indirectly the spectroscopy of individual stars in globular clusters, but it seemed interesting to conclude JCM5 with the question: "Is there an age spread among the galactic globular clusters?"

Three specialists have been chosen to lead this discussion:

R.J. Dickens: The age of globulars?

B. Carney: Is there an age spread among the galactic globular clusters?

P. Demarque: Globular cluster ages from new data on turnoff-fitting and horizontal-branch modelling.

III - SCIENTIFIC PROGRAM OF CM37/3.

The subject of CM37/3: "The abundance spread within globular clusters" was quite similar to that of the third panel discussion in JCM5. However it has been discussed, reviewed and commented in a different way than in JCM5 by the panelists and speakers of CM37/3.

In CM37/3 five speakers presented results and theoretical constraints and five panelists commented on them.

The speakers have spoken on:

N. Suntzeff: The metal-poor clusters

G.H. Smith: The metal-rich clusters

R.J. Dickens: Omega Centauri

C.A. Pilachowski: High resolution spectroscopy: quantitative results and
                        critical tests

A.V. Sweigart: Theoretical considerations.

The panel discussion was animated by:
R.A. Bell, K.C. Freeman, R. Gratton, J.E. Hesser, R.P. Kraft.

IV - SCIENTIFIC HIGHLIGHTS OF THE MEETINGS, JCM5 and CM37/3.

Chemical abundances of individual stars in different galactic globular clusters have been widely discussed. Evidence of a number of fascinating variations among stars in different clusters have been displayed and the theoretical implications of these results have been analyzed.

Galactic globular clusters are "fossils" of the epoch of galaxy formation and samples of a very early but still present stellar generation. The situation is different in the Magellanic Clouds where globular clusters with different ages can be found. The question why young globulars, in other words "blue globular clusters", can be found in the Magellanic Clouds and not in our Galaxy is very important for the general understanding of galactic evolution. For two of these young clusters, one of the SMC and one of the LMC, preliminary abundance values could be derived for some stars. Surprisingly their [Fe/H] value was very low for such young stars, ($\approx$-1.3 dex).

The most striking features of the analyzed stars in the young globulars of the Magellanic Clouds are overabundances of the light metals, in particular sodium, and also an overabundance of the rare earth.

Another problem has been discussed concerning similarities and differences between halo field stars and stars in galactic globular clusters.

In recent years, a considerable effort has been made in the determination of accurate element-to-element abundance ratios in field halo stars, mainly thanks to the advent of high S/N linear detector. A short review of the abundances of the principal elements in the field halo stars is given here below:

O and other "even" light elements (Mg, Si, Ca and Ti) are overabundant in metal poor ([Fe/H]<-1) field halo stars by $\approx$+0.4 dex with respect to Fe.

C scales approximately as Fe in metal poor stars (it might be overabundant in the most extreme metal poor stars, ([Fe/H]<-2.5).

light odd elements (Na and Al)⁻ are underabundant with respect to Mg (the run with respect to Fe is still somewhat uncertain).

there is an enhanced odd-even effect in the Fe-group elements in metal poor stars.

Ba and some other neutron rich element are underabundant; however other neutron rich elements (probably originated mainly from r-processes) are not deficient (see the case of Eu).

N shows a very special behaviour not presently understood.

The field halo and globular cluster stars do not appear to come from the same parent population, although the mean metallicities are almost identical, the cluster metallicity function is narrower than that of the field. Does the difference in the metallicity distribution function between field halo and globular clusters stars arise only from a lack of an unbiased sample of field stars and its small number, or is this difference real? This is one question which remained unanswered.

The abundance variation of stars within clusters has also raised many questions. Omega Centauri has been taken as the most glamorous example of a system exhibiting wide metallicity variations among member stars. Various explanations have been invoked to explain this property, including primordial origin and enrichment from a previous generation of stars.

Another open question was of course the age of globular clusters, which has been discussed very intensively during the final panel.

Have the globular clusters been formed in a common short interval of time or has their formation spread over a period of several billion of years?

As we can see many interesting and important questions concerning the mode of formation and the early evolution of globular clusters remained unanswered. However we hope, that the panel discussions during JCM5 and CM37/3 injected new enthousiasm among the specialists of globular clusters and will attract new energies toward this very exciting domain.

6. <u>STELLAR PHOTOMETRY WITH MODERN ARRAY DETECTORS</u>

Chairman and Editor: F. Rufener

<u>Supporting Commissions:</u>

9, 25, 26, 28, 37

INTRODUCTION AND BASIC REFERENCES FOR STELLAR PHOTOMETRY WITH CCD

F. Rufener
Geneva Observatory
CH-1290 Sauverny / Switzerland

1 Project and aims of the meeting

An object of this meeting is to define and emphasize the best practices that allow stellar photometry done with array detectors to be as accurate as possible. We dream of applying simple and clearly justified principles; in reality, we find the equipment has physical and technical properties that interfere with each other. Photometry is a method, a metrology; so we first have to define the desired accuracy. Some will be happy with 0.1 magnitude; others need 0.01, or even 0.001 magnitude. The necessary precautions must be matched to these ambitions.

2 Conditions for stellar photometry

The first requirement is precision of measurements. Although there are some difficulties, there are already means of extracting the signal with high precision. This topic has been studied most thoroughly, but some difficulties remain, particularly when fields are crowded.

The second requirement is for known, stable passbands, which must match the standard ones. This point has frequently been underestimated. If we all want to measure the same thing, accurately as well as precisely, it is of first priority to fix and check the right passband. Later transformations by mathematical recipes cannot recover the physical information that has not been included in the measurements. When using CCD's, it is vital to take the greatest possible care about this point.

The reduction to outside the atmosphere, and the relation to a standard system is also a problem. It is necessary to observe enough standard stars, of the right distribution of colors and magnitudes. To have adequate sets of such stars remains a challenge.

The above, unavoidable requirements cannot easily and perfectly be fulfilled with an array detector.

However, if the following conditions are met:

a) The matching of the passbands is tested and good enough.

b) The knowledge and control of the stability of all the functions of the camera (filters and detector) are ascertained.

D. McNally (ed.), Highlights of Astronomy, Vol. 8, 617–621.
© *1989 by the IAU.*

c) The chromatic properties of the instrument (the colour equation) are methodically and periodically evaluated relatively to a standard having an extended range in colours and magnitudes.

d) The star's signals are precisely deconvolved and summed-up.

e) Some stars within each field are well known in the given system.

It should then be possible to reach or even improve the threshold of accuracy required.

3 Basic references for stellar photometry with CCD

In the following I have attempted to set up, with the help of several experts in the field, a number of references, of course non-exhaustive, which are listed as follows:

3.1 General overview

(1975)	**Sequin, C.H., Thompsett, M.F.** *Charge Transfer Devices*, Academic Press, N.Y.	
(1975)	**Barbe, D.F.** *Proc. IEEE* **63**, 38	
(1984)	**Djorgovski, G.** in Proc. of the workshop on improvements to photometry, NASA Conf. Pub. 2350	CCD's: Their cause and cure.
(1985)	**King, I.R.** in "Data analysis in astronomy" Proc. of the Erice workshop (40.012.096)	Cluster photometry: present state of the art and future developments.
(1985)	**Fort, B.** in "New aspects of galaxy photometry", 8th IAU European Regional Astr. Meeting Toulouse, p. 3	
(1986)	**Tyson, J.A.** JOSA. A. **3**, 2131	Low light-level CCD imaging in astronomy.
(1986)	**Walker, A.R.** in IAU symp. 118, p. 33. Ed. J.B. Hearnshaw, P.L. Cottrell, Reidel, Dordrecht	CCD photometry with small telescopes.
(1986)	An ESO-OHP workshop held in Saint-Michel, ed. Baluteau & D'Odorico	The optimization of the use of CCD detectors in astronomy.
(1988)	**Janesick, J.R.** as guest editor, Optical Engineering **26**, 685-1076	15 papers on characterization, modelling, manufacture, theory and application of CCD.
(1988/89)	**McLean, I.S.** under preparation	An introductory text to CCD's.

3.2 Hardware description and development report

(1980)	*Proceedings of the SPIE* **264** *(29.012.068)*	
(1980)	**Leach, R.W. et al.** PASP **92**, *233*	Description, performance and calibration of a CCD camera.
(1981)	*Proceedings of the SPIE* **290** *(32.012.098)*	
(1984)	*Proceedings of the SPIE* **445** *(38.012.098)*	
(1984)	*Proceedings of the SPIE* **501** *(40.012.047)*	
(1985)	**Gudehus, D., Hegyi, D.** *Astron. J.* **90**, *130*	The design and construction of a CCD image system.
(1986)	**Mackay, C.D.** *Ann. Rev. of A. & A.* **24**, *255*	"CCD in Astronomy".
(1986)	*Proceedings of the SPIE* **627** *(42.012.092)*	
(1987)	**Gunn, J.E. et al.** PASP **99**, *618*	Description of the Palomar Observatory CCD camera.

3.3 Software packages and descriptions

(1985)	**Stetson, P.** *Dominion Astrophysical Observatory, Victoria, B.C., Canada*	DAOPHOT User's Manual.
(1986)	**Tody, D.** *SPIE* **627**, *733*	The IRAF Data reduction and analysis system.
(1987)	**Stetson, P.** PASP **99**, *191*	DAOPHOT: A computer program for crowded-field stellar photometry.
(1988)	*MIDAS User's guide, vol. A & B*	Image processing group ESO, Garching.

3.4 Specific problems

3.4.1 PSF fitting

(1983)	**Buonanno, R., et al.** *Astron. Astrophys.* **126**, *278*	Automated photographic photometry of stars in globular clusters.
(1983)	**King, I.R.** PASP **95**, *163*	Accuracy of measurement of star images on a pixel array.
(1984)	**Walker, A.R.** *MNRAS* **209**, *83*	CCD observations of photoelectric standard stars.
(1985)	**Gudehus, D., Hegyi, D.** *Astron. J.* **90**, *130*	The design and construction of a CCD image system.
(1986)	**Smith, G.H. et al.** *Astron. J.* **91**, *842*	CCD photometry of the globular cluster Pal 5.
(1987)	**Bendinelli et al.** *Astron. J.* **94**, *1095*	The Newton-Gauss regularized methods.

620

| (1987) | **Mighell, K.J.**
The Messenger **47**, *24* | Crowded field photometry using EFOSC and Romafot. |

3.4.2   Flat fielding (see paper hereafter by P. Stetson)

3.4.3   Cosmic ray removal

| (1980) | **Goad, L.E.**
SPIE **264**, *136* | Statistical filtering of cosmic ray events from astronomical CCD images. |

3.4.4   Fringes removal

| (1987) | **Bica, E., Alloin, D.**
Astron. Astrophys. **186**, *49* | Near IR spectral properties of star clusters and galactic nuclei. |
| (1987) | **Yee, H., Green, R.**
Astron. J. **94**, *618* | The environment of the quasar PG 1613 +65: a close interacting pair. |

3.4.5   Artifacts

| (1980) | **Lorre, J., Gillespie, A.**
SPIE **264**, *123* | Artifacts in digital images. |

3.4.6   Sensitization to UV

(1985)	**Cullum, M. et al.** *Astron. Astrophys.* **153**, *L1*	Spectroscopy to the atmospheric transmission limit with a coated GEC CCD.
(1986)	**Robinson, L., Osborne, J.** *SPIE* **627**, *492*	CCD's at Lick Observatory.
(1988)	**Oke, J.G. et al.** *PASP* **100**, *116*	CCD testing at Palomar Observatory.

3.4.7   Pass-band matching

| (1986) | **Walker, A.R.**
in IAU symp. 118, p. 33, Ed. J.B. Hearnshaw, P.L. Cottrell, Reidel, Dordrecht. | CCD photometry with small telescopes. |
| (1987) | **Massey, P. et al.**
NOAO Newsletter **12**, *28* | Out with the Mould, in with the Blue. |

3.4.8   Linearity

3.4.9   Rapid sampling

(1985)	**Dunham, F.W. et al.** *PASP* **97**, *1196*	A high-speed dual-CCD imaging photometer.
(1986)	**Howell, S.B., Jacoby, G.H.** *PASP* **98**, *802*	Time-resolved photometry using a CCD.
(1986)	**Stover, R.J.** *SPIE* **627**, *195*	High-speed CCD imaging stellar photometer.

3.4.10 Standard stars and correlation to standard

(1984)	**Schild, R.**	CCD photometry of M67 stars use-
	PASP **95**, *1021*	ful as BVR1 standards.
(1985)	**Christian, C. et al.**	Video camera / CCD standard stars
	PASP **97**, *363*	(BVR1).
(1985)	**Olszewski, E.W., Aaronson, M.**	The Ursa minor dwarf galaxy: still
	Astron. J. **90**, *2221*	an old system.

3.5 Sample or reference applications

(1980)	**Leach, R.W. et al.**	Description, performance and cali-
	PASP **92**, *233*	bration of a CCD camera.
(1984)	**Walker, A.R.**	CCD observations of photoelectric
	MNRAS **209**, *83*	standard stars.
(1986)	**Murray, C.A., Dierker, S.B.**	Use of an unintensified CCD de-
	JOSA A **3**, *2151*	tector for low-light-level Raman
		spectroscopy.
(1987)	**Hesser, J.E. et al.**	47 Tuc color-magnitude diagram.
	PASP **99**, *739*	
	Stetson, P., Harris, W.	M92 Color-magnitude diagram.
	Astron. J., in press	

CCD IMAGERS FOR ASTRONOMY: PAST PROBLEMS AND FUTURE HOPES

John C. Geary
Smithsonian Astrophysical Observatory
60 Garden St.
Cambridge, MA 02138

1 Introduction: The long trek.

The search for high-quality, large-format CCD imagers during the 1980's reminds me of a journey across a large and forbidding desert. One starts off full of great expectations of a quick completion, but discouragement sets in around mid-journey as the barrenness of the landscape becomes apparent and the goal on the horizon is shown to be only a mirage. Finally, mountains resolve themselves from the mirages and we take heart in knowing that we seem finally to be making some progress. I think the analogy fits the CCD situation well, starting with the view from the time period 1979-80 and working toward the present.

Back at the beginning, spirits were high and everyone looked forward to the new vistas offered by the emerging technology, with every expectation that progress in CCD technology would quickly yield up near-perfect devices for our demanding astronomical applications. We had, for instance, the RCA 320 X 512 backside-illuminated chip, which, despite its mediocre noise and charge-transfer performance, remains virtually unexcelled in quantum efficiency to this day. GEC in Britain was in the process of developing a promising low-noise device, and of course, there were tremendous expectations for the Space Telescope 800 X 800 imagers from TI, although the secrecy surrounding their performance and availability was a most annoying feature for those of us not in the ST loop. With such a good start in both the commercial TV sphere (RCA, GEC, and later Thomson) and the non-commercial specialty product development (TI), we had every reason to suppose that our ideal imagers were close at hand. Then the landscape began drying up.

What happened to those of us involved in CCD instrument development during the period of 1981 - 87 can only be described as a descent into ever-increasing depression at the lack of progress in device availability, punctuated be brief moments of hope from time to time. Things started to go bad when we discovered how hard it was to interact with the major commercial producers to try to combine into one imager all the desireable features that had been demonstrated individually. What we wanted was the high quantum efficiency of the RCA backside- illuminated CCD combined with the low noise and good charge transfer of devices like those of GEC (later produced by EEV) or Thomson. What we got was indifference on the part of the manufacturers, who were locked in a mostly losing battle with the Japanese to develop the solid-state hand-held color TV camera. RCA made one largely unsuccessful attempt to produce a scientific grade device based on their commercial

623

D. McNally (ed.), Highlights of Astronomy, Vol. 8, 623–627.
© *1989 by the IAU.*

chip, but along the way a decision was made to terminate CCD development entirely, and they have sadly passed from the scene. EEV for years resisted pleas to develop a thinning process for their devices so that high blue quantum efficiency might be realized in a backside-illuminated mode. Their one excursion into large-area devices (1500 X 1500 pixels) for scientific applications has evidently not been very successful, as I have heard nothing of it for several years. Thomson has also been reluctant to try thinning until recently, but has remained at least a stable supplier of one useful imager over the past few years. Finally, to the great disappointment of all concerned, TI chose early on to terminate any further work on the technology developed for the Space Telescope imagers, leaving the astronomical community with only a few crumbs from what was hoped would be a great banquet.

We have occasionally had our spirits lifted by the prospect of real progress in putting the CCD puzzle together right. Thanks to the efforts of Jim Janesick and co-workers at JPL, the two great problems plaguing the TI imagers, namely deferred charge and unstable quantum efficiency (quantum efficiency hysteresis), have been largely solved, the first by a cleaver change in the clocking scheme and the latter by various surface treatments such as UV floods or thin platinum overcoats. At the same time, concerted efforts by a consortium of research institutions and the National Science Foundation resulted in at least a few of the devices rejected from the space program being made available for astronomical research. Alas, many of these imagers proved to be of the "overly thick" variety, having low blue quantum efficiency and not responding well to surface treatments. They have, however, whetted our appetites enormously, as their other characteristics (noise and charge transfer) are often superb.

The other event that has from time to time lead to great expectations has been the formation of a team at Tektronix to produce scientific CCD imagers with both outstanding performance and large (even huge) formats. Beginning in 1984 - 85, it was learned that this effort would indeed be aimed at producing low-noise thinned devices, initially in 512 X 512 format with quite substantial pixels (27 microns) and subsequently in a truly awesome 2048 X 2048 size. By mid-1986, the first experimental chips became available for initial testing, and it was then that a new problem, large numbers of traps called "pockets", became evident, rendering most devices useless for precision work. Once again our hopes for a speedy end to the CCD drought were dashed, as Tektronix was forced back into fundamental investigations of the chemistry and physics involved instead of device production.

1.1 The Recent Past: Hope on the horizon?

Beginning 18 months ago or so and continuing right up to the present, several developments have occurred that appear to have some substance to them. Compared to the preceding 6 or 7 years, one can certainly say that there is at last some progress to cheer. These developments arise from several independent efforts and involve a variety of players, giving us some reason to believe that we will at last have access to scientific grade imagers of moderate to large format, even if one or more of these projects ultimately fail.

The first new development to catch our attention was the use of a so-called "silicon foundry" by Photometrics Ltd. to produce for them under contract a 516 X 516 CCD, which has subsequently been shown to possess outstanding charge transfer and noise performance. By concentrating solely on the technical aspects of device production, not subsequent marketing considerations, the silicon foundry can provide its contracting customer with a product tailored to specific needs. Entry level costs for such best-effort projects run several hundred thousand dollars, too high for most institutions alone, but increasingly within reach of a consortia of institutions operating in a cooperative fashion. At least two silicon foundries advertise special competence in CCD technology and more may eventually be added to the list.

At about the same time, Photometrics announced the commercial availability of a phosphor coating which could be applied by vacuum sublimation. This offers the possibility of getting significant response in the blue / UV spectral regions from devices that otherwise are weak there, such as unthinned frontside-illuminated imagers and perhaps also the "overly-thick" TI devices.

The National Science Foundation has funded a grant through Lick Observatory to develop a large CCD imager at Reticon Corp. The format chosen is 400 X 1200, with large 27 micron pixels. While ideal for many spectroscopic applications, this device should also prove interesting for direct imaging programs as well. Some of the initial experimental devices have already been tested and shown to be very promising.

Also, a new push at Thomson CSF toward production of scientific grade CCD's is in evidence. A few preliminary data sheets on their 512 X 512 and 1024 X 1024 developmental devices have been circulated and there is now talk of progress in thinning, to be ultimately made available on these imagers.

Finally, the long ordeal at Tektronix may be coming to an end, as they are now shipping 512 X 512 CCD's to back-ordered customers. It seems that the infamous pocket problem may finally be solved, or at least reduced to a manageable level, and testing is now under way at several locations to verify this. If so, then we can expect to start seeing the first high-grade 2048 X 2048 devices in the not so distant future, perhaps even in a matter of a few months.

1.2   The Immediate Future and Beyond:

Those faced with making decisions on specific CCD's to buy in the next year or two will at least have the luxury of a nonzero number of options available. I have listed below some of the best bets for the near future, starting with those that are essentially "off the shelf" and working on to those still in development, which include all of the really large devices. The actual choice of one device over another will depend mostly on system requirements and availability of the requisite cosmetic quality, not so much on the cost of the particular CCD itself, except in the case of truly huge formats such as the large Tektronix device. This reflects the often unappreciated fact that, for the most part, the cost of the CCD itself (when finally available) is usually a small fraction of the total cost of the camera system. This is especially true when one takes into account the hardware and software burden necessary to support electronic imaging and subsequent reductions.

Photometrics 516 X 516:
20 micron pixels; 10.2 mm square format.
Unthinned, frontside-illumination; can be phosphor coated.
Noise less than 10 electrons possible.
Charge transfer is excellent.

Thomson 384 X 576:
23 micron pixels; 8.8 X 13.2 mm format.
Unthinned, frontside-illumination; phosphor for UV.
10 electron noise and good transfer efficiency.
Versions buttable on three sides are now being developed, allowing arrays of 2 X N to

be fashioned.

EEV 384 X 576:
22 micron pixels; frontside illuminated.
Recent work toward thinning has been undertaken, and a new series of larger format devices (approx. 1150 pixels square) is being developed.

Tektronix 512 X 512:
27 micron pixels; 13.8 mm square format.
Noise expected to be about 10 electrons.

Thomson 512 X 512 and 1024 X 1024:
Currently under development for scientific applications.
19 micron pixels; formats of 9.7 mm and 19.5 mm square.
Thinned version eventually is under active development.

Reticon 400 X 1200:
27 micron pixels; 10.8 X 32.4 mm format.
Initial unthinned versions in development production.
First samples exhibit good transfer and low noise.
Thinned version is planned as soon as development is completed.

Tektronix 2048 X 2048:
Available next year (?)
Thinned, backside-illuminated.
27 micron pixels; 55.3 mm square format !!!
Expensive – about 100K $!!!
Huge burden on data processing.

Formation of contractual efforts for custom devices, drawing on the services of imager manufacturers and silicon foundries. This last option for the procurement of just the right device for a specific class of instrumentation may ultimately be the most rewarding for the field. It does however require a level of cooperation between institutions at present, as the costs for a new start are at the moment beyond what most of us can budget.

2 New-technology imagers:

Looking toward the more distant future, things are of course not clear at all. Proposals for the next generation of CCD (or other solid-state imager) have not been extensively pursued because we have had such a devilishly hard time getting good devices from the present generation of manufacturing technology, or perhaps more correctly, from the past generation of device fabricators. With a shift in the offing, new ideas will now start to be investigated. The three listed below are certainly not comprehensive, but may serve as examples for the audience.

1. New output structures capable of resolving single electron signals. Even some of the older, less well optimized output designs have shown themselves capable of approaching this ideal, and there is no fundamental reason why this goal cannot be routinely achieved. The CCD would thus become at least statistically equal to a true photon counter down at the single photoelectron level, allowing work in even the lowest signal-to-noise regimes.

2. Devices designed to operate in electronic inversion of the electrode-side surface. This can reduce dark current a hundred fold or more, allowing much higher temperatures to be used, even for long integrations. This in turn can bring both increased quantum efficiency and increased charge transfer efficiency.

3. CCD structures designed for ease of manufacturing and increased yield of usable devices, while still retaining all the essential characteristics of precision imagers. Such techniques may ultimately make possible devices of very large format but without the very large price tags of present technology.

THE CCD MOSAIC PROJECT BY ESO[1] AND INSU[2] / TOULOUSE OBSERVATORY[3]

Sandro D'Odorico[1]
Jean-Louis Prieur[3]
[1] *European Southern Observatory, Garching (FRG)*
[2] *Institut national des Sciences de l'Univers, Paris (France)*
[3] *Observatoire de Toulouse (France)*

In the last few years, as the CCD cameras have been routinely used in most of the Observatories, there has been a growing concern among the scientific community about the major weakness of this very good detector: the format of the available chips is too small for many applications.

ESO has been particularly interested in acquiring and testing large format CCD's, mainly because of the enormous potential possibilities of the NTT and VLT projects. The instrumentation which was planned for these projects assumed the availability of the large CCD array Tektronix 2048 x 2048. In the beginning of 1987, as it seemed that this circuit would not be available in the near future, ESO and INSU decided to support industrial development of buttable CCD's with Thomson CSF (France).

This contract lead to the designing of a new chip (THX 31157), which is derived from the standard CCD THX 7882, but with a very thin ceramic baseplate smaller than the sensitive part of the chip, and the connectors on one side only, to allow for joining the CCD's together with a smaller gap as possible in between. It is a front illuminated array of 579 x 400 pixels, with square pixels of 23 x 23 μm. The general properties are similar to those of the THX 7882, but the charge transfer efficiency has been improved. The quantum efficiency reaches about 45% at 600 nm and the read-out noise can be as low as 5 e^-.

A special machine for assembling these chips, MAM (Mounting and Alignment Machine), has been designed and built in Toulouse Observatory, with financial support from ESO and INSU. The idea was to align the chips and stick them on a sapphire baseplate with a minimal gap in-between. In this way large matrices of two rows of CCD's can be mounted on the same baseplate. The expected accuracy for aligning the chips is about $\pm 3\,\mu$m, and the coplanarity requirement about $\pm 15\,\mu$m. Designed for moving the chip in all directions, with 6 degrees of freedom, the machine allows for assembling matrices of CCD's up to 2 x 12. The first mechanical tests were performed with 4 chips in Toulouse in spring 1988.

A multipurpose electronic control camera is being developed jointly by ESO / Toulouse Observatory for data acquisition and pre-processing. The core of the system is a microcode programmable sequencer (clock pattern generator) supplying all the digital signals necessary to run a CCD. It will be able to control up to 16 buttable 579 x 400 CCD's, or four 1024 x 1024, or two 2048 x 2048. The first images with a 4 x 4 matrix are expected before the end of the year.

D. McNally (ed.), Highlights of Astronomy, Vol. 8, 629–630.

ESO/INSU also participated to the development of large monolithic 1024 x 1024 CCD's with Thomson CSF. This led to the new chip THX 31156. The performances are still to be evaluated accurately, but from the first measurements made at Photometrics (Tucson, USA) it seems that they are similar to those of the buttable THX 31157, which is very encouraging. Further tests should be done in September-October 1988 in Toulouse Observatory.

In collaboration with the French CNES/CEA[1] Thomson has also started the development of thinned back-illuminated CCD's. Though the results are still preliminary and many problems remain to be solved, chips with good quantum efficiency in the blue could become available in the near future.

Despite its financial participation with Thomson, ESO is open to any other alternative solution to obtain large CCD arrays, and ready to test any other large format CCD which would be available on the market. The multi-purpose camera which is in development can be easily modified to control any other chip, only by programming a new RAM. In the next few years, the large 1024 x 1024 THX 31156 matrices could also be transformed for butting them together, as the THX 31157, to form large 2048 x 2048 matrices. We can therefore be rather confident for the future and think that, provided that the financial choices are made, large format CCD's will be available for use on the VLT instruments in the middle 1990's.

ACKNOWLEDGEMENTS

We cannot conclude without mentioning H. Bauer, M. Cullum, S. Deiries, H. Dekker, J.L. Lizon, R. Reiss at ESO, and B. Bertin, J.P. Dupin, B. Fort, G. Gallou, C. Lours at Toulouse Observatory, for their contribution to this project.

REFERENCES

Fort, B., 1988, ESO/NOAO proceedings "Instrumentation on Large Telescopes"

D'Odorico, S., Bauer, H., Deiries, S., Lizon, J.L., Reiss, R., Fort, B., Bertin, B., Dupin, J.P., Gallou, G., 1988, ESO/NOAO proceedings "Instrumentation on Large Telescopes"

Vigroux, L., 1988, Internal report on the thinned THX 7882, CEA Saclay/Gif sur Yvette, France

DISCUSSION

Bessel, M.S.: *What is the likely cost of the new buttable CCD's? Will ESO sell 2x2 or larger joined matrices? Will ESO sell the controller (camera)?*

Prieur, J.L.: *THX 31156 would cost about twice the price of current THX 7882, so about $ 3000. Though ESO is not a commercial company, special requests may be satisfied by discussing directly with ESO.*

---

[1] Centre National d'Etudes Spatiales, and Centre d'Etudes Atomiques

GROUND-BASED PHOTOMETRIC CALIBRATION OF THE SPACE TELESCOPE CCD CAMERA

D.A. Hunter[1], H.C. Harris[2], W.A. Baum[1], J.H. Jones[1], T.J. Kreidl[1]
[1] Lowell Observatory, 1400 West Mars Hill Rd., Flagstaff, AZ U.S.A.
[2] U.S. Naval Observatory, PO Box 1149, Flagstaff, AZ U.S.A.

1 Introduction

The Wide-Field and Planetary Camera (WF/PC) is a CCD imaging instrument which is part of the Hubble Space Telescope. Ground-based observations have been made with a CCD system similar to those of the WF/PC in order to establish the standard star sequence to be used for in-flight photometric calibration. Because the WF/PC passbands differ from those in previous photometric use, the filters and CCDs will define a new photometric system. We outline here the procedures used to establish the calibration fields to be used in flight (see Harris et al. 1988 for additional details).

Two globular cluster fields were chosen for use as calibrators: a position 13.6' southwest of the center in ω Cen and a position 10.2' northwest of the center in NGC 6752. Observations were made with copies of 15 of the 42 WF/PC filters, covering 3000 to 10000 Å (see Griffiths 1985 for filter curves). Five of these filters approximate the Johnson-Cousins $UBVRI$ system. However, the filters that are likely to be the most popular substitutes for V and I – F555W and F785LP – are broader than the Johnson-Cousins filters.

2 Observations

The observations were made using the Las Campanas 1-m telescope in 1985. Additional observations were made at Lowell Observatory. The ground-based instrument was a thinned TI 816 x 800-pixel CCD. The chip was coated with coronene so that ultraviolet photons below 3800 Å cause the chip to fluoresce in the green where the quantum efficiency of the chip is higher (Blouke et al. 1980). To improve the quantum efficiency and stability, the uncooled chip was flooded with ultraviolet light by illuminating the window with a mercury discharge lamp for 30-60 minutes during pumping prior to observing.

Most of the images were flat-fielded using exposures of a white spot projected on the dome. A color-balancing filter was used to approximate the color of the night sky, and for the widest passbands this made a difference of 2% in the flat fields. For F336W twilight sky flats were used because of the greater photon flux. Potential problems with using twilight sky as flats include the presence of emission lines, polarization, and the blue color. Tests indicated, however, that these problems contribute errors less than 1%. The instrument was sufficiently stable that an average flat could be constructed for a given observing run.

631

D. McNally (ed.), Highlights of Astronomy, Vol. 8, 631–634.
© 1989 by the IAU.

Approximately 30 Landolt (1973, 1983, 1987) standard stars were observed each night, and stars were observed several times at different locations on the detector and at different air masses. For a few bright red stars the telescope was defocussed to keep the exposure time over 3 seconds. The cluster exposures consisted of a short- and long-integration pair, typically 1 and 5 minutes, and were repeated on 2 – 5 nights. Approximately 2000 CCD frames were obtained.

3 Standard star reductions

Aperture photometry was simulated using DAOPHOT (Stetson 1987) to determine the magnitudes of the standard stars. Growth curves were determined, and the magnitudes were corrected to a common 25" aperture. The correction depended on the seeing and was typically 1-2%. An iterative least squares fit was used to place each night on a common internal system. A color term was unnecessary. The mean sigma in the fits was 0.003 - 0.018 magnitude. The filters F336W and F1042M gave worse photometric errors at all steps of the process. For F336W the sigma in the fit is 0.04 and for F1042M it is 0.03 magnitude.

Atmospheric extinction determinations were modelled using Tüg et al.'s (1977) extinction as a function of wavelength and extrapolated to 2900 Å (Allen 1973). This formulation was applied for air masses of 1.2 and 2.0 to the Gunn and Stryker (1983) stellar spectral atlas, WF/PC passband magnitudes were synthesized, and extinction coefficients were calculated. A color term was necessary and (F555W - F785LP) was used. For F336W a quadratic term in air mass was also required.

The final zero point of the system was defined so that an AOV star has colors of zero. To accomplish this a plot was made of the WF/PC magnitude minus the nearest $UBVRI$ magnitude versus the $UBVRI$ color index. The zero point is where the fit of these points intersects the origin. However, the curvature of the fit was determined not from the standard stars, which were too few, but from magnitudes synthesized from the stellar spectra of Gunn and Stryker (1983). A least squares fit was then applied to the standard stars with the coefficient of the quadratic color term being set by the Gunn and Stryker stars. Only normal giants and dwarfs were included. Sigmas of the fits were 0.007 – 0.037 with an average of 0.02 magnitude, excluding F336W and F1042M.

4 The cluster field reductions

Because the cluster fields are crowded, DAOPHOT was used to measure the stellar photometry. A master coordinate list was set up for each cluster using one of the deep images and was transformed to the coordinate system of subsequent images. Only a few frames were deeper than the one used to set up the master list of stars. Stars over most of the field of view of the original images were included: about 3100 stars in ω Cen and about 1300 stars in NGC 6752. The 91 cluster frames were reduced on the WF/PC-team Micro Vax II workstation and the SSC Cray X-MP/48 with time provided through NOAO and code provided by M. Mateo. Approximately 10% of the stars were deleted on the basis of quality indicators returned by DAOPHOT.

After the magnitudes were obtained with the NSTAR fitting routine, the aperture correction was determined using the short exposure of the short-long pair. The long exposure was forced to agree with the short exposure as determined by stars that were well measured on both frames. The cluster magnitudes were placed on the same aperture system as the standard stars using the growth curves of standard stars bracketting the cluster in time.

The cluster magnitudes were corrected for extinction in the same manner as for the standard stars. Observations made on different nights were compared. If there was any systematic offset, the difference was split and added or subtracted as a constant to each of the nights since there could be stars that were measured on one frame and not on another. Typical offsets were less than 0.02 magnitude. The results from different frames were combined and the zero point constant added. Magnitudes were weighted by sigma. A sigma was computed that is a combination of the sigmas of individual observations and one that is the dispersion around the mean. The higher of the two sigmas was adopted as the final measure of uncertainty. Figure 1 shows the final sigmas of all of the stars which were measured in ω Cen as a function of the magnitude for filter F555W. Figure 2 shows a F569W vs. F569W – F791W ($\sim V$ vs. $V - I$) color-magnitude diagram.

5 Selection of standard stars

Bright, relatively unblended stars of low sigma in the WF/PC fields were selected as stars to be used as in-flight standards. In addition some fainter stars with somewhat higher sigmas were included in the center of the fields in order to increase the number of stars available to the limited field of view of the Planetary Camera. In ω Cen 80 stars and in NGC 6752 67 stars were selected. The standard stars have sigmas which range from 0.005 to 0.03 with an average of about 0.01 (see Figure 3).

6 References

Allen, C.W. 1973, *Astrophysical Quantities*, (London: Athlone).

Blouke, M.M., Crowens, M.W., Hall, J.E., Westphal, J.A., and Christensen, A.B. 1980, *Applied Optics*, **19**, 3318.

Griffiths, R. 1985, *Wide Field and Planetary Camera Instrument Handbook* (STScI).

Gunn, J.E. and Stryker, L.L. 1983, *Ap J Suppl.* **52**, 121.

Harris, H.C., Hunter, D.A., Baum, W.A., Jones, J.H., and Kreidl, T.J., in preparation.

Landolt, A.U. 1973, *A J*, **78**, 959.

— 1983, *A J*, **88**, 439.

— 1987, Letter to STScI.

Stetson, P.B. 1987, *P A S P*, **99**, 191.

Tüg, H., White, N.W., Lockwood, G.W. 1977, *A A*, **61**, 679.

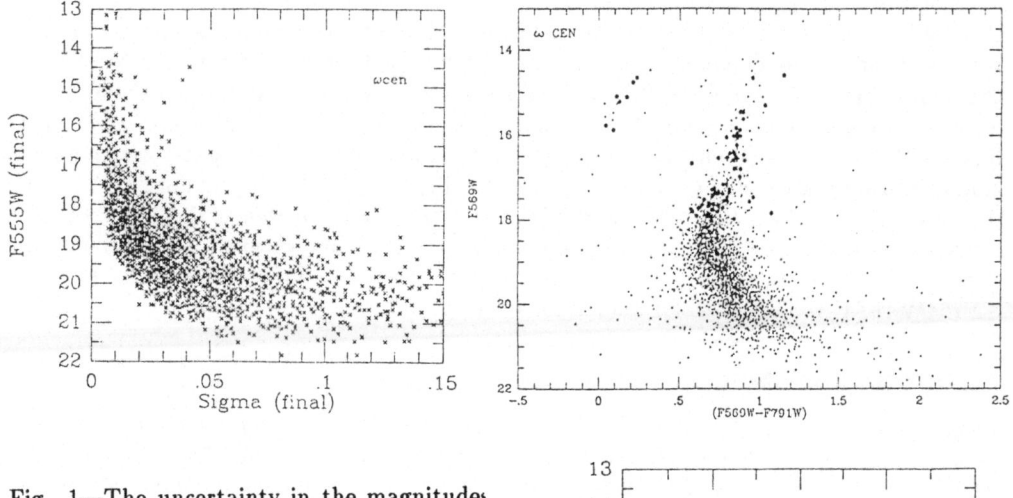

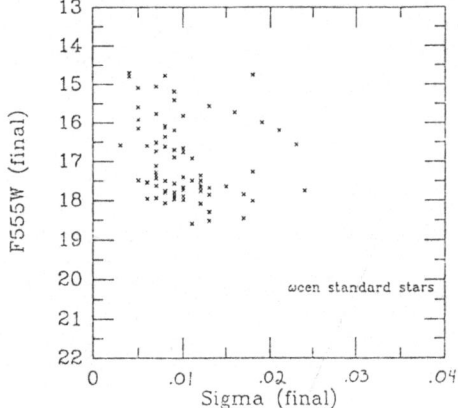

Fig. 1—The uncertainty in the magnitudes of F555W are shown as a function of magnitude for all of the stars measured in the ωCen field.

Fig. 2—A color-magnitude diagram (∼V-I vs. V) for the stars measured in the ωCen field. The enlarged dots are stars chosen for use as in-flight standards.

Fig. 3—The uncertainty in the F555W magnitudes are shown for the standard stars chosen in the ωCen field.

SOME FACTORS AFFECTING THE ACCURACY OF STELLAR PHOTOMETRY WITH CCDs (And some ways of dealing with them)

Peter B. Stetson
Dominion Astrophysical Observatory, Herzberg Institute of Astrophysics,
National Research Council
5071 West Saanich Road
Victoria, British Columbia V8X 4M6 / Canada

ABSTRACT. There are many factors which make it difficult to relate instrumental CCD photometry to a fundamental standard system with an accuracy much better than about 1%. Here I will address only three of them: (1) infrared leaks in the filters; (2) the finite opening and closing times of mechanical shutters; and (3) changes in the air mass for long integrations. I will be approaching these subjects from the point of view of a visiting astronomer at someone else's observatory, who gets three or four nights of observing time a year, and has only the afternoons preceding those nights to perform whatever tests can be carried out while the equipment sits on the telescope.

1 Infrared leaks and flat fields

For normalizing CCD photometry of bright stars (those that are *not* very much fainter than the night-time sky, especially standards and stars observed with narrow-band filters), two types of flat-field calibration frames are particularly useful: dome flats, which are observations of the inside of the dome or of a white diffusing screen, usually illuminated with an incandescent lamp; and twilight-sky flats, which are observations of the bright sky just after sunset or just before sunrise. These frames are used to map out the spatial variations in the sensitivity of the detector, so that the derived apparent magnitude of a star can be made independent of where on the detector the star's image falls.

It is commonly found, however, that dome flats and twilight-sky flats differ by significant amounts, making it rather uncertain which – if either – of the two techniques is correct. For instance, in data from an observing run which Jim Rose and I had in 1986 on the University of Hawaii 2.2-m telescope with the IfA/Galileo 500^2 TI CCD, I found that the twilight flats in the I, V, B, and U filters had structure at the 7%, 9%, 11%, and 13% levels (root-mean-square), respectively. The dome flats in exactly the same filters, on the other hand, showed structure at the 5%, 6%, 7%, and 8% levels. The U dome flats much more closely resembled the I sky flats in general appearance than they did the U sky flats, and even the B and V dome flats appeared to show significant dilution by an I-like response pattern. This led me to suspect that the dome flats were being contaminated by very-long-wavelength light (longer-wavelength than I, probably), which is much stronger in the tungsten lamp than in the light of the twilight sky.

It seemed to me that I could see the same effect in dome- and twilight-flats which Bill Harris and I had obtained in 1984 and 1985 at Kitt Peak, albeit at a somewhat lower level.

635

D. McNally (ed.), Highlights of Astronomy, Vol. 8, 635–644.
© *1989 by the IAU.*

Therefore I have carried out some crude numerical simulations based on published curves for "Mould "interference filters and RCA CCDs (which cover the principal peaks of the filter passbands, and the wavelength range 4000 – 9500 Å for the CCDs). Actual scans of the Kitt Peak "old Mould" filters out to 12.000 Å were kindly made for me by George Jacoby and Ed Carder, and I have taken rough curves for the reflectivity of aluminium and for the atmospheric opacity from Allen's *Astrophysical Quantities* (3rd edition, The Athlone Press, London). Fig. 1 shows one example of such a simulation: the effective spectral response for a dome flat obtained through a "Mould" B filter. I start with the Planck curve for a 2300 K black body which, according to data kindly provided by Bill Schoening, is the temperature of the Kitt Peak 1-0.9m telescope's flat-field lamp when operated at 7 volts – the voltage setting we used in 1984. This is multiplied by the throughput of the filter itself (shown in the second panel from the top) and by the response of the CCD, including the effects of two reflections off aluminium (which are unimportant). Note that the vertical axes in Fig. 1 represent the logarithm of the number of photons per unit wavelength interval.

In 1984 I had asked the Kitt Peak staff about the possibility of red leaks in the "Mould" filters, and was told that "the red leaks are negligible – less than 1% out to one micron." You can see from Fig. 1 that this statement was completely accurate: in fact the red leak is less than about 0.2% out to one micron. What they and I had both forgotten, however, is that when dealing with light sources as cold as the flat-field lamp, the relative flux at B is *also* less than 1%. The very red color of the lamp has two effects on the spectral response of the system. First, the short-wavelength side of the filter's principal bassband is eroded by the rapid falloff in the lamp's flux. Second, and much more important, a long red-leak region is raised up. True, the red leak never gets much above 2%, but on the other hand it may span as much as 5000 Å in wavelength. Numerical integration indicates that 10% of the light getting detected comes from wavelengths longer than the nominal cutoff of the filter at 5300 Å; this contaminating light has an effective wavelength near 10.000 Å – quite different from the 4500 Å light the filter is supposed to be letting through. Since the spatial character of the detector's sensitivity is a strong function of wavelength (*cf.* data given above for the IfA/Galileo CCD), this means that our B dome flat will have superimposed on it a quite different pattern corresponding to the longer wavelengths.

I should point out one particular source of uncertainty in this simulation: the published sensitivity curves for the Kitt Peak RCA CCDs only extend to 9500 Å, and I have been unable to locate a quantitative description of how the detector response falls off longward of that. Apparently, the infrared sensitivity differs strongly from chip to chip, and is also a strong function of temperature. For this simulation I have assumed that the curve drops exponentially toward longer wavelengths with the last known slope, but the real situation with this CCD or with any other could be better or, of course, worse.

In the interval between our observing runs in 1984 and 1985, Kitt Peak introduced the use of "color-balance" filters to modify the spectral characteristics of the lamps employed for dome flats. These are commercial filters designed to let photographers use ordinary daylight-balanced color film in situations where the light comes from incandescent lamps. A tracing of one of these filters, once again kindly provided by George Jacoby, Ed Carder, and Bill Schoening, is shown as Fig. 2. The filter progressively attenuates longer-wavelength light throughout the visible region of the spectrum – just what is needed to make orange artificial light look as yellow as sunlight. However, longward of about 7000 Å – a region that most photographers don't care about – the color-balance filter also has a red leak. In Fig. 3 I compare the spectral sensitivity of our 1984 dome flats (without color-balance filter;

denoted "u" for "unbalanced") to that of our 1985 dome flats (with balance filter; denoted "b"). I also show here comparable simulations of the response of the system to the light of a Solar-type star (a 5700 K black body, including the effects of terrestrial atmospheric extinction; denoted "s" for "star"), and to the light of the twilight sky (modelled by starting with a 5700 K black body, attenuating this by the atmospheric absorption and *multiplying* it by the wavelength-dependence of the atmospheric scattering; denoted "t" for "twilight").

Part of the improvement in going from the unbalanced dome lamp to the color- balanced one results from the fact that, with the balance filter in place, we could increase the voltage on the lamp from 7v to 10v – thus increasing its temperature from ~ 2300 K to ~ 2800 K – without getting exposure times that were too short. The higher temperature increases the overall ratio of B-to-I flux and reduces the erosion of the short-wavelength side of the filter's principal passband. The rest of the improvement comes from the fact that, even with its own red leak, the color balance filter still reduces the throughput in the spectral region 6000 Å to 12.000 Å by a factor ranging from 4 to 20. Fig. 3 shows that the spectral response of the balanced B dome flat is a much better match to that of a typical star than is the unbalanced dome flat. Is it a better match to the star than the twilight-sky flat is? For wavelengths shortward of about 8800 Å the answer is "yes" – in spectral-energy distribution the color-balanced dome lamp matches starlight better than the twilight sky does. Longward of about 8800 Å the color-balanced dome lamp may still *look* like a better match, but remember: the plot is on a logarithmic vertical scale. In reality, 10^{-4} is more than nine times closer to 10^{-6} than it is to 10^{-3}. This means that if the CCD's sensitivity continues to fall off exponentially (or slower) to a point *much* longward of 8800 Å, twilight-sky flats might be more reliable even than color-balanced dome flats. (This conclusion is based solely on the spectral properties of the incident light, and neglects the possibility that the dome flats might not illuminate the detector uniformly – another reason why twilight flats might be better).

Similar comparisons for the "Mould" interference-filter V, R, and I passbands are shown in Figs. 4-6. We see that the advantage of using the color-balance filter quickly disappears as redder passbands are considered and, indeed, for R and I the color-balance filter does more harm than good. If we refer again to Fig. 2, we can see immediately why this happens: the color-balance filter suppresses the R and I flux even more than it does the red leak.

What are the consequences of using an incorrect flat-field frame? Since the flat-field correction merely scales one part of the image relative to another, and since the absolute throughput of the system will be determined by observations of standard stars made with the same device (and rectified by the same flat-field frame), using the wrong flat field should not introduce a net photometric error. What will happen is that the magnitude derived for a star will depend upon where that star fell on the detector. Thus, there will be additional scatter introduced into the photometry when stars from various places on the detector are measured. Of course, if the observer always places the standard stars at the center of the field, and then reduces program stars that come from all parts of the detector (such as in star-cluster observations), a systematic error can be introduced.

Here are my conclusions and recommendations: (1) *Know your filters and detector.* Don't be satisfied with statements like, "The red leak is negligible," or, "The detector dies beyond one micron."Demand to know how negligible is negligible and how dead is dead. You'll almost certainly find, as we have, that the people responsible for the instrument will be helpful and cooperative, but they are invariably busy and may only do extraordinary tests when they are specifically requested. (2) *Color-balance filters (for dome flats)* may *do more harm than good.* Since 1986 Kitt Peak has introduced a new set of "Mould" filters with red leaks apparently less than 0.01% out to 12.000 Å. With these, the color-balance filters will probably do more good than harm, but with the previous filters the opposite may have been the case. In which situation are you? You probably won't know until you've followed recommendation 1. (3) *Illuminate your dome flats with diffuse sunlight.* Nature has provided us with a 5700 K light bulb, with a spectral-energy distribution remarkably like that of a typical star. It has lots of flux for those troublesome U and narrow-band filters, too. Why not open the dome slit a few inches and let some of that free light illuminate your dome screen? (Turn the slit away from direct sunlight and check the screen for shadows before starting. Exposure levels will vary on days with scattered clouds, but this may not be a serious problem if enough frames are taken.) (4) *If you get a cloudy night, rejoice!* You can spend the time taking every sort of calibration frame you can think of: dome flats with balance filters, dome flats without balance filters, dome flats to low and high exposure levels – the works! If you gain a deeper understanding of your instrument, and can therefore reduce your systematic errors by 30%, that's as good as a second clear night!

2 Shutter-timing errors

When you start a CCD integration, the instrument-control computer sends a signal to the camera head instructing it to open its mechanical shutter. Some time later the shutter begins to open, and at some still later time it is completely open. After the prearranged exposure time has elapsed, the computer sends another signal to the camera telling it to close the shutter; after some time the shutter actually begins to close, and eventually it is completely closed. These various delays are typically of order many milliseconds, and their net effect is to produce a difference, δ, between the presumed and actual exposure times. The shutter timing error, δ, may be either positive or negative. Since it is produced by events occurring just at the beginning and the end of the exposure, and since the computer's own clock should be accurate to microseconds, at worst, δ should be nearly independent of total integration time. This shutter-timing error could become significant if sub-percent accuracy is desired for bright stars requiring exposure times of order one second. In principal, δ could be measured by the instrument scientist and then allowed for in the camera-control program, but since different shutters with different time-constants may be controlled by the same computer, this is often not done.

One way to check for shutter timing errors is to take dome flat-field exposures with different integration times, say, one and ten seconds. (Since we will use only the ratio of these frames, it does not matter whether the frame is absolutely uniformly illuminated, or whether there is a red leak.) In order to improve the signal-to-noise ratio of the short exposures and to avoid any non-linearity in the detector, I use the focussing sequence which allows me to open and close the shutter many times before reading out the chip. Thus, I make ten one-second exposures before reading out, and compare this frame with a single ten-second exposure. If many 10 x 1^s and 1 x 10^s exposures are taken alternately, any fluctuations in the dome lamp's brightness can be averaged out. Repeat determinations on different days check the reliability of the results.

Define R as the mean brightness of a 10 x 1^s exposure divided by the mean brightness of a 1 x 10^s exposure. Then

$$R = \frac{10 \cdot (1 + \delta)}{10 + \delta} \quad \Rightarrow \quad \delta = \frac{10 \cdot (R - 1)}{10 - R}$$

During 1988 June, I determined the shutter errors of the RCA5 detector on CTIO's 4m telescope. On two consecutive days I got values of $\delta = 12 \pm 1$ msec and 9 ± 1 msec. The agreement isn't perfect, but it's adequate to correct my shortest exposure times to better than 0.1%.

However, there is no reason to stop here. In fact, these mechanical shutters tend to open in one place first and to close in the same place last: with iris shutters, for instance, the center of the field tends to get longer exposures than the corners. By taking the ratios of the 10 x 1^s and 1 x 10^s images, rather than of their mean intensities, we can create a two-dimensional map $\delta(x, y)$:

$$\delta(x, y) = \frac{10 \cdot (R(x, y) - 1)}{10 - R(x, y)}$$

An example of such a map derived from the same 1988 June CTIO data is shown in Fig. 7, where the contour levels represent loci of constant δ, with the values indicated in milliseconds. One way to use this map would be to enter it with the known (x, y) coordinates of some star, and read out from it a shutter correction for that position. There is a still cleverer way to use it, however. Say we have some frame which has a nominal exposure time of two seconds. Take our two-dimensional map of δ and add a constant value of 2; now we have a map of $(2 + \delta)$. Divide our program image by this map and multiply it by 2. Now we have a flat, uniform exposure time of exactly two seconds everywhere in the frame. This correction can easily be applied when the raw data are flat-fielded, and can be ignored thenceforth.

3 Mean airmass for long integrations

In many cases, especially for data taken at Cassegrain foci, integration times can be very long – up to an hour or more. The program field is moving across the sky this whole time, so what should be used as the effective air mass to correct the observation for atmospheric extinction? The simplest way out would be simply to use the air mass at mid-observation; a slightly more sophisticated approach would be to average the initial and final air masses. But, as Jeff Pier has pointed out (1982, A.J. 87, 1515), neither of these is satisfactory. Consider the case of an exposure which passes through the meridian at mid-observation: the air mass at mid-observation is the minimum air mass for the entire observation, and the average of the initial and final air masses is the maximum air mass for the observation – clearly neither of these represents a sensible average. What we really want is the mean air mass that correctly predicts the total number of photons accumulated by the detector during the course of the exposure. In other words, recognizing that the instantaneous apparent magnitude of a star, v, is related to its extra-atmospheric magnitude, V, and to the instantaneous value of the air mass, X, by

$$v = V + K \cdot X,$$

we want to determine the mean air mass $\overline{X}$ such that

$$(t_1 - t_0) \cdot 10^{-0.4 K \overline{X}} = \int_{t_0}^{t_1} 10^{-0.4 K X(t)} dt$$

This is a very unpleasant equation, because it indicates that the effective mean air mass of the observation depends upon the extinction coefficient, K, which is an unknown. In fact, it is in order to determine K that we want to compute the air mass in the first place. Furthermore, the effective extinction coefficient depends somewhat on the color of the star being studied, $K \approx k_0 + k_1 (B - V)$, which means that in principle we may have to compute different mean air masses for stars of different colors recorded in the same frame!

However, not all is lost. For mathematical simplicity, let's convert this to powers of e rather than powers of 10. Also, we recognize that $\overline{X}$ is a constant, even if it is unknown, so we can bring it to the right side of the equation and into the integral:

$$1 = \frac{1}{t_1 - t_0} \int_{t_0}^{t_1} e^{-0.921\,K(X(t)-\overline{X})}dt.$$

If we approximate this integral using Simpson's 1/3-rule, we obtain

$$1 = \frac{1}{6} \left[e^{-0.921\,K\,(X_0-\overline{X})} + 4e^{-0.921\,K\,(X_{1/2}-\overline{X})} + e^{-0.921\,K\,(X_1-\overline{X})} \right],$$

where X_0, $X_{1/2}$, and X_1 are the instantaneous air masses at the beginning, the middle, and the end of the observation, respectively. Simpson's rule is very good; the error term is $\left[\frac{(t_1-t_0)^4}{2880} \cdot \frac{\partial^4 X}{\partial t^4} \right]$. Now, taking a first-order Taylor expansion of this we get

$$\begin{aligned}
1 &= \frac{1}{6} \left\{ 1 - 0.921\,K\,(X_0 - \overline{X}) + 4 \left[1 - 0.921\,K\,(X_{1/2} - \overline{X}) \right] \right. \\
&\quad \left. + 1 - 0.921\,K\,(X_1 - \overline{X}) + 0 \left(0.4\,K^2\,\overline{(X - \overline{X})^2} \right) \right\}
\end{aligned}$$

Note that $\frac{1+4+1}{6} = 1$, which cancels the 1 on the left side, and then we can divide through by $0.921\,K$, ultimately arriving at

$$\overline{X} = \frac{X_0 + 4X_{1/2} + X_1}{6} + 0 \left(0.08\,K\,\overline{(X - \overline{X})^2} \right)$$

The first part of this equation, up to the plus sign, is our adopted approximation for the mean air mass of the exposure. It is very conveniently *independent* of the extinction coefficient, K. The error term at the far right is pleasantly small: for reasonable values of $K \sim 0.25$ magnitude and $X_{max} - X_{min} < 0.25$, the error is of order 10^{-4}.

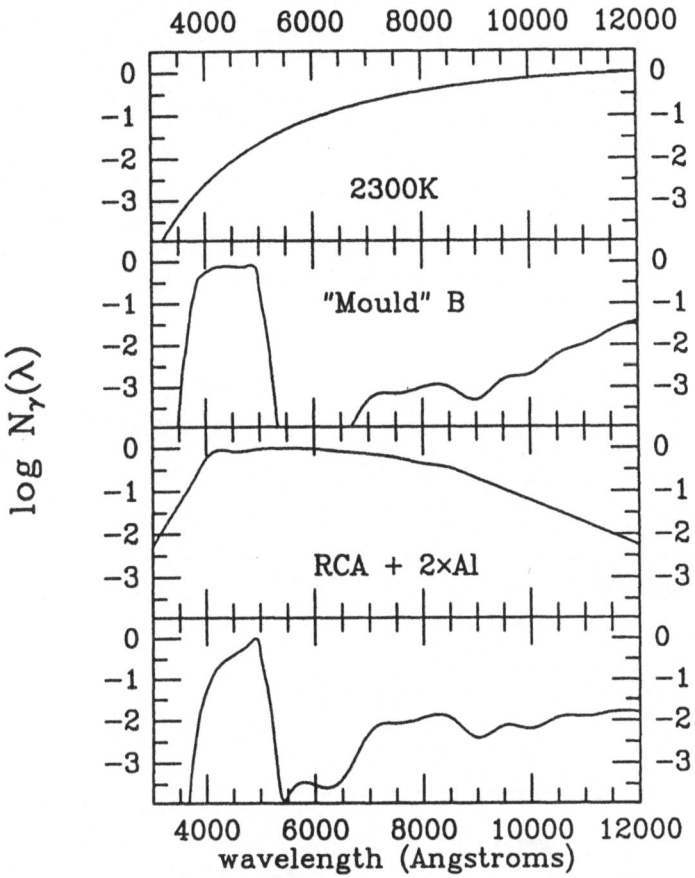

Figure 1

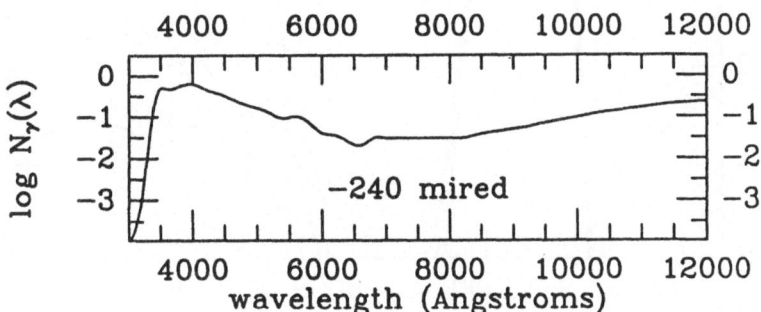

Figure 2

642

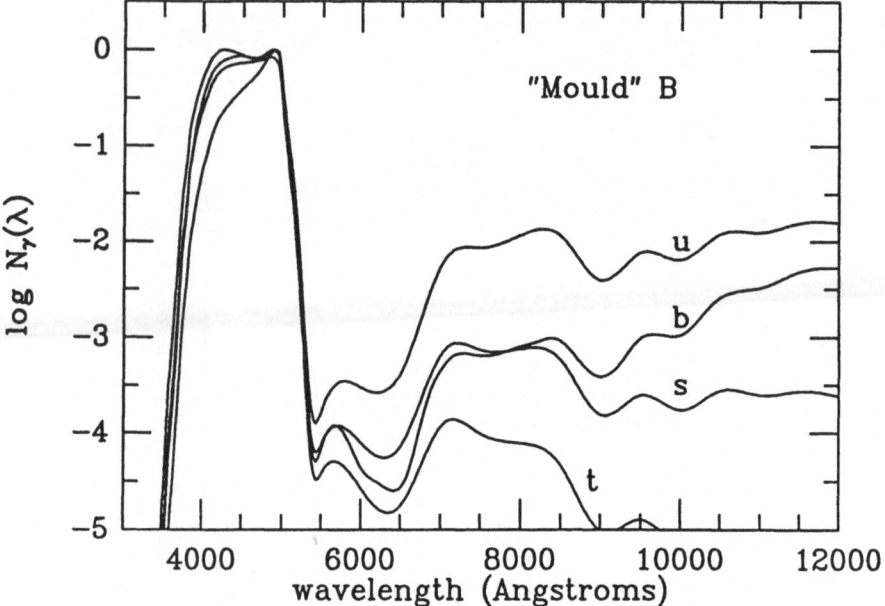

Figure 3

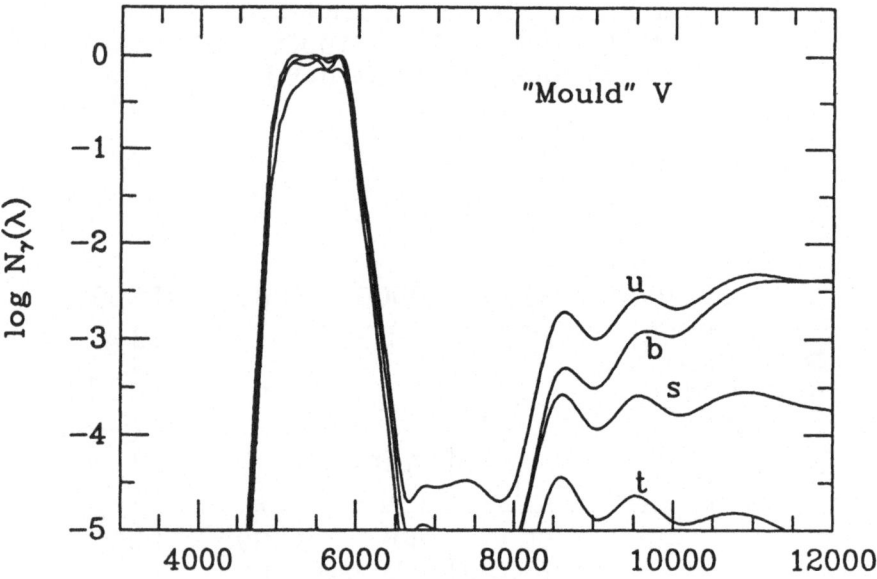

Figure 4

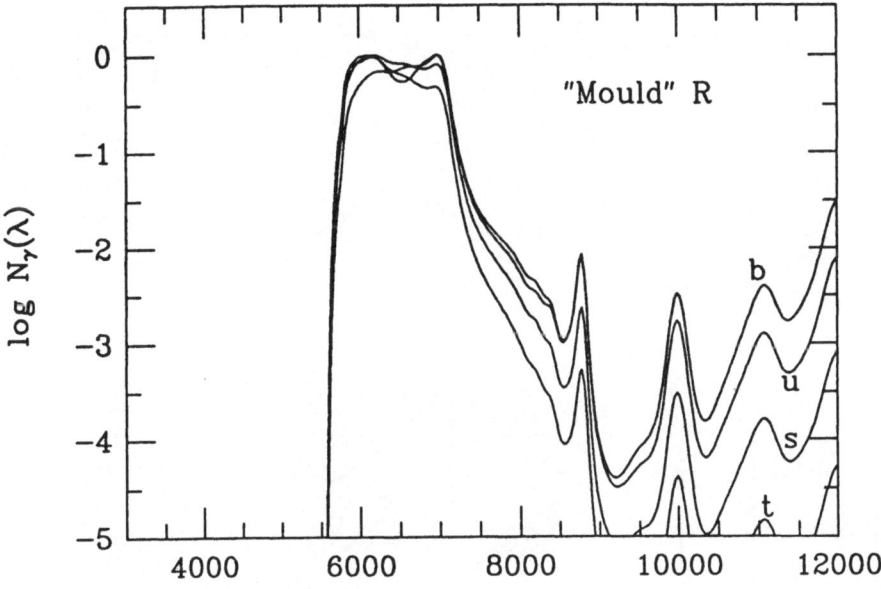

Figure 5

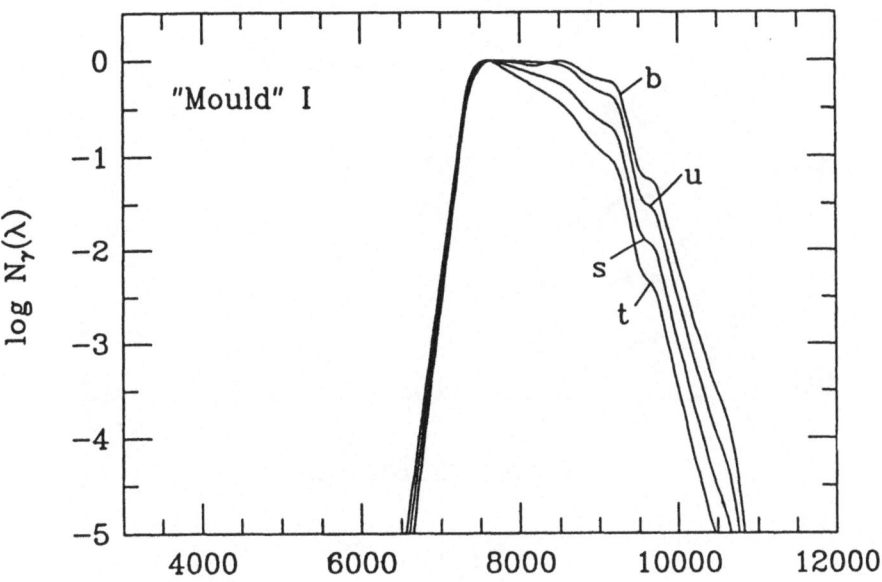

Figure 6

644

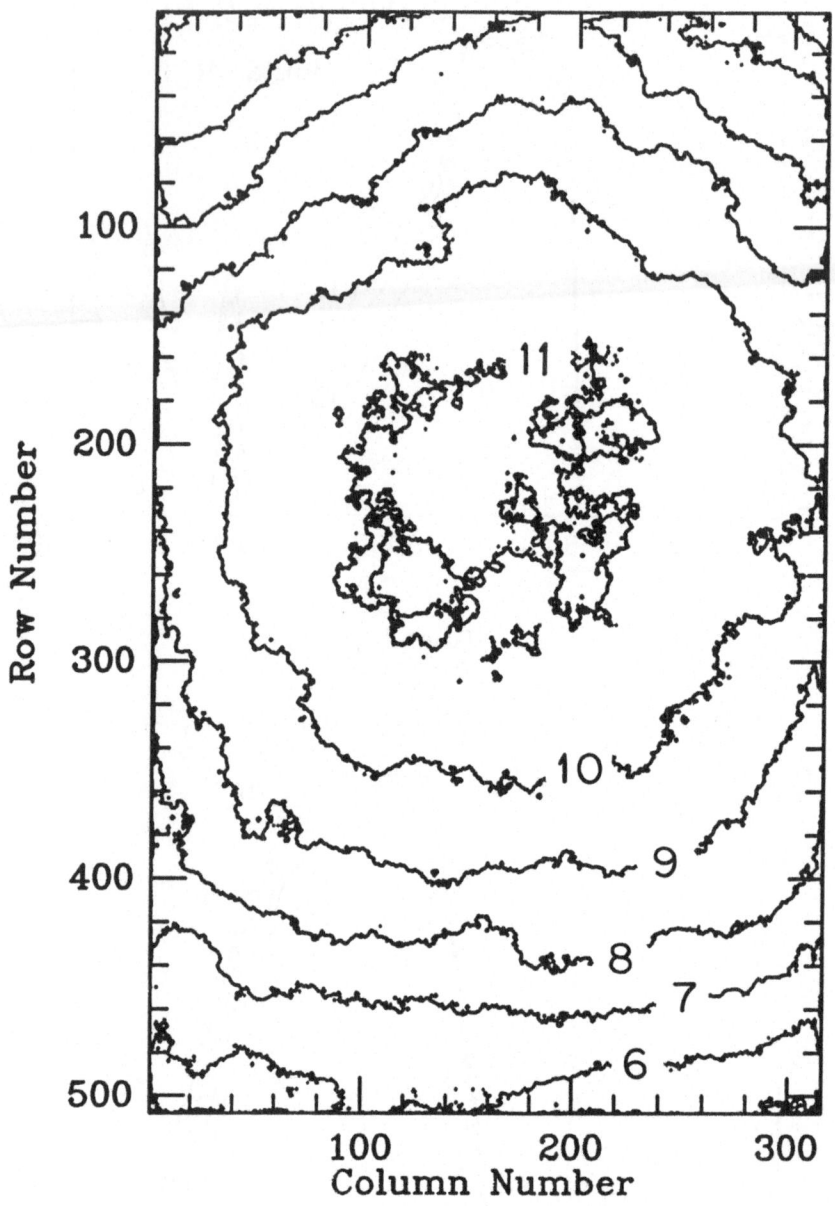

Figure 7

CCD DATA TAKING MODES AND FLATFIELDING PROBLEMS

S. Djorgovski
California Institute of Technology
Pasadena, CA 91125, USA.

M. Dickinson
Astronomy Department, University of California
Berkeley, CA 94720, USA.

ABSTRACT. We briefly review some problems in the flatfielding of CCD images, in the context of three data-taking modes: stare, short scan, and drift scan. The principal sources of flatfielding imperfections are: (1) mismatch in the spectra of the astronomical sources of interest and the flatfield illumination; (2) a variety of low-level additive errors; and (3) nonlinearities of the CCD response. Residual flatfielding errors are probably the limiting factor in high-precision astronomical photometry. Flatfielding accuracies of the order of $1 - 2\%$ per pixel are commonly achieved; with some effort, accuracies of $\sim 0.1\%$ can be reached; higher accuracies require a substantial effort, or improvements in the quality of CCD chips. In general, scanning data taking modes outperform the "standard" stare mode.

The CCD revolution in astronomy is comparable in its impact to the introduction of astronomical photography a century ago. There is hardly a field of astronomy where the new optical and near-optical imaging and spectroscopy with CCD's has not brought vast qualitative and quantitative advances. Most of the success is due to the high quantum efficiency (QE) and benign noise properties of these devices, as well as to the relative ease of obtaining linear, calibrated data in a digital form with them. However, this ease is somewhat deceptive; few observers bother to extract astronomical data from their raw CCD images with the accuracy of which the devices are capable. Worse yet, inadequate preprocessing procedures can degrade the quality of the data, or introduce systematic errors which are not reflected in nominal Poissonian error-bars. For many applications, high photometric accuracy is not required, or even pursued, but it is achievable with a modicum of effort.

For the general lore about astronomical CCD's, their charms and problems, the reader is referred to the reviews by Mackay (1986) and Djorgovski (1984), the proceedings edited by Baluteau and D'Odorico (1986), the thesis by Wright (1982), many *S.P.I.E.* proceedings on astronomical instrumentation (e.g., volumes 264, 290, 445, 501, and 627), the issues 8 – 10 of the vol. 26 of *Optical Engineering* (1987), and references therein. Many application papers often give useful advice on processing procedures. For example, the paper by Stetson and Harris (1988) is

645

D. McNally (ed.), Highlights of Astronomy, Vol. 8, 645–650.

a model of a crowded-field stellar photometry study. The paper by Tyson (1988; and references therein) addresses faint galaxy counts at the limits of present-day technology. A truly interested reader should also try to get on the preprint mailing list of James Janesick from the JPL.

Different sources of errors will dominate in different applications (e.g, faint galaxy counts vs. standard star exposures). For example, in short exposures, readout noise and charge transfer inefficiencies will define the noise floor, whereas flatfielding errors may dominate in the high-signal regime. One should thus optimize one's data-taking and processing techniques for the particular problem at hand. There is no universal best solution for all CCD observations, although some hygienic practices are always commendable. In this paper, we deal with the CCD flatfielding and its limitations in general.

We define flatfielding as the removal of sensitivity variations of a CCD detector on all spatial scales. Different flatfield images can be used to remove the sensitivity or illumination variations at different scales: pixel-to-pixel, global gradients and vigneting, or the scales in-between. The *intra*-pixel variations (if important) must be modeled. Such variations, generally due to the CCD architecture and illumination patterns, are seen in the GEC CCD's (Wright 1982), but mercifully not in the popular "HST/Galileo" TI 800^2 devices (Gunn, priv. comm.) In general, when observing, it is a worthwhile goal to expose one's object(s) not just on a single area of the CCD, but *multiply*, on various different pixels or portions of the CCD, so that residual variations partly average out or cancel. This may be achieved in a number of ways discussed below.

There are three principal modes of CCD data taking: (1) *Stare* or *Steady*, in which the telescope and the detector are in a stable pointing, tracking the object of interest. This is the standard mode, and the easiest to implement, but it gives the worst results. An expensive (in terms of observing and data reduction overhead), but highly beneficial variation on it is a mode of *Multiple Exposures with Shifting and Addition*; if the device is good, and enough independent exposures are taken, the flatfielding may be unnecessary (Tyson 1988). (2) *Drift Scan*, in which either the telescope or the CCD moves, and the charge is continuously read out, row by row, at a synchronized rate so that the charge image and the optical image always coincide (cf. Wright 1982, or Hall and Mackay 1984). Not only is the flatfielding reduced to a 1-dimensional problem (column-to-column response) by averaging the object signal over all rows of the device, but the method also removes the pesky interference fringes that occur in many systems. The bad side of this technique, from a photometric point of view, is that it is sensitive to seeing, sky brightness, and transparency variations; also, it is efficient only in long scan strips, due to the overhead "ramp-up" period during which not all pixels read out have been exposed to the sky for the same period of time. (3) *Short Scan*, invented by Roger Lynds (KPNO), is like (2), but the CCD is scanned over only part of its extent, typically 30 – 50 rows. The ramp-up is lost, but the ramp-down is kept, and is at least partly usable. This method has practically all the advantages of drift scanning with none of its drawbacks. If the telescope is capable of precision offset guiding, or a moving-stage CCD camera is available, this mode may be the optimal one in many applications, yet to date it is still not widely known or used.

The principal difficulties in CCD flatfielding come from any spectral mismatch between the data and the flatfield. This is the result of two facts: (1) QE response curves differ somewhat from one pixel to another, e.g., due to impurities or thickness variations; and (2) any finite-bandpass image is really a sum of monochromatic images, each with its own interference fringe pattern, weighted by the spectrum of the incident light and the bandpass response. This is easily demonstrated by comparing the flatfields obtained with two different narrow-band filters. Consequently, broader bands are harder to flatfield, and the wavelength dependence of the flatfield response also varies spatially on the CCD. This suggests that the future of CCD-based photometry may be in intermediate or narrow band systems in which the placement of the bandpasses is motivated by some astrophysical reason (e.g., Gunn-Thuan, Stromgren, etc.), rather than the sensitivity of obsolete photocathodes. Moreover, in pixels where the object signal is comparable to, or greater than the sky foreground (i.e., for almost anything brighter than $\sim 21^m$, and in particular for the *standard stars*), this color-dependent variation will vary from one object to another, and can the coefficients of instrumental color equations. This can be a real problem in stellar photometry where the PSF is undersampled, e.g., with HST, and some iterative scheme using different color flatfields may be necessary.

The most common type of flatfields are out-of-focus images of the telescope dome, illuminated by an incandescent lamp. These provide a high signal, and thus low pixel-to-pixel Poisson noise, but this may be highly deceptive; generally, the use of dome flats leads to the worst systematic flatfielding problems due to the frequency response variations discussed above. Some of the difficulties may be caused by red leaks in one's filters (Stetson 1988), but some are simply due to the mismatch between the flatfield illumination spectrum and that of the sky and astronomical objects. Using high-temperature light bulbs and/or blue filters for the dome illumination can help. Dome flatfields often have a residual low spatial frequency variation or a gradient due to an uneven illumination, which can be removed with the use of some kind of a sky flatfield. These can be dawn or twilight sky exposures, which have a high signal and the correct illumination, but colors which may be just as wrong as those of dome flats. Better than this are median-filtered "blank sky" exposures, or sky median stacks. In general, some combination of the dome and sky flats is used in most CCD imaging work. Similar techniques apply for both stare and short scan modes. Typical resulting accuracies are on the order of 0.2 to 2% above expected Poissonian errors, on all spatial scales, and often the flatfielding is worse in the bluer bands.

The sky median stacks are obtained using a large number (at least 5) of "blank" sky (i.e., free of any bright or extended sources) images, scaled to the same sky mean, with all discernible objects removed or flagged. If there are N such images, a data cube is formed, and at each row and column (i, j), a set of corresponding (non-flagged) pixels from all N images is extracted: $I_1(i,j)$, ... $I_N(i,j)$. The median, or the average of a few values around the median of that set is then an estimate of the unpolluted sky signal at the given row and column (i, j). The quality of the median stack frame, and its pixel-to-pixel Poissonian noise are determined by the number and exposure times of the sky fields used to construct it. Its accuracy is limited by the unremoved faint stars and galaxies, and PSF wings, typically at a $\sim 1\%$ level, over the scales of tens to hundreds of pixels. A good sky median stack is the "ideal" flatfield for the situations in which the sky signal dominates (i.e., any faint objects work). In practice, it is hard to accumulate enough signal in a given night to have a median sky stack usable as a flatfield on its own, but median stacks for the run are possible. One subtlety worth noting is that one should avoid, if it is at all possible, dividing an object frame by a median sky flat which *includes* that object frame as part of its constituent stack. This will result in some fraction of pixels in the object frame being in effect divided *by themselves* during the flatfielding process; a distinctive and undesirable spike results in the histogram of resulting pixel values. This can have unpleasant consequences for photometry, and the sky determination.

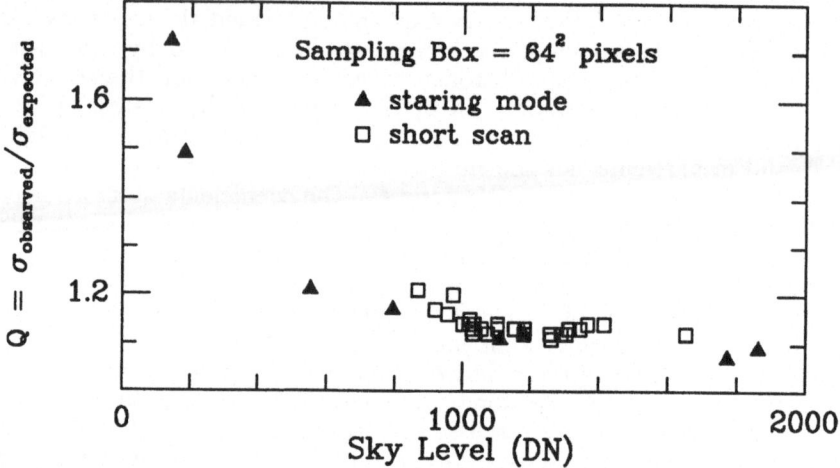

Figure 1. Dependence of the "excess noise" parameter Q on the mean signal level. The data are from a TI 800×800 CCD at the prime focus of the KPNO 4-m telescope, obtained in the R band, and flatfielded with the simple dome flats. The expected noise (Poisson + readout) is $\sim 2.2\%$ at 500 DN, and $\sim 1.1\%$ at 1000 DN ($1 \text{ DN} \simeq 4.15e^-$). The triangles are the data obtained in the stare mode, and the squares are the short scans.

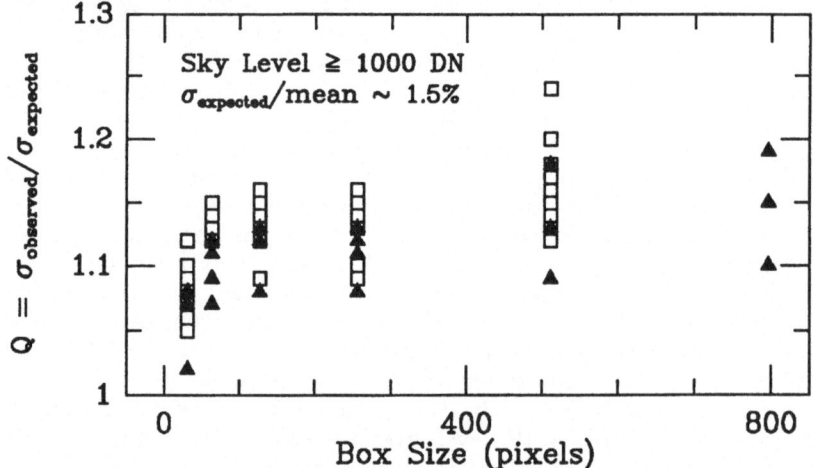

Figure 2. Dependence of the parameter Q on the size of the subimage used to evaluate the data histograms. The data and the symbols are as in Fig. 1, except that the frames are limited to those with mean sky level ≥ 1000 DN, and the expected noise is $\sim 1.5\%$ for all data points. Note that the residual noise is still $\sim 0.2\%$ above the Poissonian at most spatial scales.

A good tool for the investigations of flatfielding efficiency are the data histograms. In uncrowded fields, they can be well modeled as Gaussians with dispersion σ^2. Let us define the "excess noise" parameter, $Q = \sigma_{observed}/\sigma_{expected}$, where the $\sigma_{expected}$ accounts for the Poissonian and readout noise. Figure 1 shows dependence of the parameter Q on the mean signal level, for CCD subimages of size 64^2 pixels. The rise at the low light level is probably due to the charge transfer inefficiency and additive errors. Leveling-off at the higher signal levels suggests color mismatching or nonlinearities. Figure 2 shows the dependence of Q on the spatial scale: for a pure Poissonian noise, one would expect Q to go down as more pixels are sampled; instead, it is nearly constant. This suggests that the residual flatfielding noise is of the $1/f$ type, and a simple reasoning of the type $(S/N) \sim \sqrt{t_{exposure}}$ is likely to be incorrect.

In the case of drift scans, the column-to-column response can be derived from the data itself, e.g., as column medians, iteratively corrected for any sky brightness variations (row-to-row). This is not feasible if there are large, extended objects in the field, e.g., bright galaxies. An alternative is to generate a column vector from another sky exposure, or from a dome flat, but the color of the latter can be wrong. For faint object imaging in good photometric conditions, the data can be corrected to almost the Poissonian limit accuracy, better than 0.5%, and perhaps as good as 0.1% above the Poissonian noise.

Flatfielding of spectroscopic CCD data is in principle simpler than that for direct images, because all the pixels are illuminated by a monochromatic signal, and as long as there are no shifts between the pixels and the Ångstroms, even the interference fringes stay anchored. An additional problem is presented by the uneven width of the entrance apertures or the slit, but it can be easily corrected using a co-called slit function, derived from sky exposures; cf. Djorgovski and Spinrad (1983) for more details.

Most spectrum-related problems in CCD flatfielding are multiplicative errors, which can in principle be corrected by division by an appropriate correction image. The more insidious limits on flatfielding accuracy are caused by the *additive errors*, which are much harder to correct. Obviously, presence of additive signals in the raw charge image and/or the flatfield will perturb their division and give an incorrect astronomical object image. There are several possible sources of such additive errors: bias and dark current variations; charge transfer problems, unremoved interference fringes (from the night sky or emission-line objects), scattered light from bright sources, electronic drifts, and possibly polarization dependences of the CCD response. Some of these are fixed properties of a given CCD device, which can be mapped and then removed before flatfielding, e.g., deferred charge columns (Baum, Thomsen, and Kreidl 1981). Others, e.g., bias and dark variations, or electronics artifacts, are preventable through a good hardware design and maintenance, and can be easily monitored. But some problems, e.g., the charge transfer inefficiency, may be scene-dependent and difficult to remove.

The best way of coping with additive errors may be reference sky frame subtraction. Baum, Thomsen, and Morgan (1986) achieved flatfielding accuracies of the order of 0.2% with this technique. However, the fundamental limit comes is set by the faint stars and galaxies which are by necessity present in the reference sky frames. The procedure is familiar to IR observers, and is essential to the use of all present and near-future IR imaging devices.

Finally, CCD's are highly, but not perfectly linear-response detectors. The measured non-linearities of *most* types of astronomical CCD's (TI's are better than

RCA's) are less than 1%, perhaps about 0.01% for most part of their dynamic range (well below saturation level). Still, ideally the mean flatfield signal level should be close to the mean level of the signal of interest. More insidious is quantum efficiency hysteresis (Griffiths 1985). This is minimized with the new flashgate technology: platinum coating, oxygen soaking, UV flooding, and other magick; some elixirs and physics are described by Janesick *et al.* (1986, 1987).

Other sources of flatfielding errors may include moving dust specks, and their out-of-focus shadows. The problem can be rather bad in the cameras with many optical surfaces and a considerable flexure. If computer time is no objection, and there are flatfields taken at different times and telescope pointings, one can try to separate the time-dependent and time-independent factors in a flatfield. A much better alternative is to have a well-designed and clean CCD camera.

How well can one do? The commonly used methods (e.g., sky-corrected dome flats) generally achieve flatfielding accuracies on the order of 0.5 – 2% in for a broad-band image taken with either steady or short scan modes. Sky medians are hard to make better than about 1 – 2%, because of the residual faint objects or PSF wings. In *narrow* bands, with dome or dawn/twilight sky flatfields, and with reference sky subtraction, one can reach $\sim 0.1\%$. Self-flatfielded drift scans may approach the Poissonian limits, but that has been tested only at the 0.1 – 0.5% level so far; besides, drift scans are not practical for most standard star observations. Thus, it appears that the present limits are reaching a plateau of flatfielding accuracy at a $\sim 0.1\%$ level above the expected Poissonian noise, on all spatial scales. For averaging over many pixels, one may be able to do slightly better (0.0005^m?). However, in order to push the accuracy further, it may be necessary to obtain *better*, and not merely bigger, CCD's.

REFERENCES:

Baluteau, J.-P., and D'Odorico, S. (editors) 1986, Proceedings of the ESO-OHP Workshop on *The Optimization of the Use of CCD Detectors in Astronomy*, ESO Conference Proceedings No. 25.

Baum, W., Thomsen, B., and Kreidl, T.J. 1981, *Proc. S.P.I.E.* **290**, 24.

Baum, W., Thomsen, B., and Morgan, B. 1986, *Astrophys. J.* **301**, 83.

Djorgovski, S., and Spinrad, H. 1983, in *Proceedings of the AAS/OSA Joint Topical Meeting on Information Processing in Astronomy and Optics*, June 1983, St. Paul, Minn., p. ThB2-1. AAS/OSA publication.

Djorgovski, S. 1984, in *Proceedings of the Workshop on Improvements to Photometry*, eds. W. Borucki and A. Young, p. 152. NASA CP-2350.

Griffiths, R. 1985, *STScI Newsletter*, October 1985, p. 5.

Hall, P., and Mackay, C. 1986, *M.N.R.A.S.* **210**, 979.

Janesick, J., Elliot, T., Daud, T., and Campbell, D. 1986, *Proc. S.P.I.E.* **627**, 543.

Janesick, J., Campbell, D., Elliot, T., and Daud, T. 1987, *Opt. Eng.* **26**, 852.

Mackay, C. 1986, *Ann. Rev. Astron. Astrophys.* **22**, 255.

Stetson, P. 1988, DAO preprint; and this volume.

Stetson, P., and Harris, W. 1988, *Astron. J.* **96**, 909.

Tyson, J.A. 1988, *Astron. J.* **96**, 1.

Wright, J.F. 1982, Ph.D. Thesis, University of Cambridge.

HIGH PRECISION CROWDED FIELD PHOTOMETRY

Peter Linde
Lund Observatory
Box 43
S-221 00 Lund, Sweden

ABSTRACT. Methods have been developed for high precision photometry in crowded stellar fields. The procedure includes the following steps:

- Determination of a two-dimensional point spread function
- Definition of groups of stars with mutually overlapping images
- Determination of local background for each star group
- Simultaneous fitting of point spread functions, one for each star in the group

The Lund approach emphasizes interactivity. Many fundamental procedures are facilitated by the use of an optimised image display. Extensive modelling has been done to study the influence of various error sources.

1 Derivation of the point spread function

The Lund method uses a numerical point spread function (PSF), which is derived from the image to be processed. The following steps are taken:

- Selection of stars

- Editing of star images

- Centering of stars

- Normalisation and summation of star images

- PSF quality check

Selection of suitable PSF candidate stars, up to a maximum of twelve, is made by pointing in the stellar field displayed on the screen. The stars are extracted into a set of subimages. In cases where disturbances are seen, single pixel or group pixel editing is performed. Cursors of different geometric shapes are used to identify disturbing pixels. The pixel values are either replaced with a mean of the local neighbourhood or marked for avoidance in following processing steps. The group editing uses a PSF for fitting to neighbouring disturbing stellar images, followed by a subtraction. In this case, the necessary PSF often has to be approximately derived in a previous step.

For star images of typical CCD observations, the sampling is such that subpixel centering is necessary. This step necessitates a resampling of the subimages. The edit and centering

D. McNally (ed.), Highlights of Astronomy, Vol. 8, 651–656.
© *1989 by the IAU.*

functions can optionally be done automatically, using an iterative algorithm known as $\kappa - \sigma$ clipping (Newell, 1979). By cross-correlating the selected subimages, pixel values above a preset threshold can be clipped. Recentering and clipping continues until no further changes are introduced.

Remaining pixels in any subimage, differing more than two standard deviations from the mean, are discarded in the summation process. A weight is given to each participating star before summation. Optionally, a filtering algorithm can be applied to the data in order to suppress any remaining noise.

The quality of the resulting PSF is estimated from a careful study of residual images. These are made by subtracting the scaled PSF from the participating individual star subimages. Sometimes, the PSF generation procedure needs to be repeated a few times before the residuals are at a minimum.

2 Definition of initial star positions

Although it is attractive to use automatic procedures to locate stars and groups in a crowded field, such detection becomes extremely difficult in very crowded fields. Frequently, faint stars lie in the wings of much brighter stellar images, or conglomerates of faint objects lie near the background noise level. It is clear that undetected group members may seriously affect the measurements of other stars in a group. Therefore, the Lund method emphasizes an interactive approach for very crowded fields. The star position and group definitions are made by pointing in a computer image. The image is displayed on two monitors. One shows a normal grey-scale version and the other shows a pseudo-colour version using a histogram equalised, 16 level, colour sequence. This allows for efficient simultaneous interpretation of both large and small intensity variations in the image.

3 Determination of the background

The determination of the background for each measured star is in many cases highly essential for the final precision in the photometry. During the development of the Lund system, several methods have been tried:

- treating the background as a free parameter in the fitting procedure

- interactive determination of local backgrounds for each group

- interactive determination of a background image

- automatic determination from the histogram distribution of the neighbourhood

The first method gave inferior results in very crowded fields due to influence of non-group members. The second method proved to be too laborious. The last two methods are included in the current algorithms. A background image can be computed using a two-dimensional spline fit to interactively determined background points. The background image is subtracted from the target image. In the final reduction procedure the background then becomes a constant.

Alternatively, an automatic background determination mode is available in the final reduction procedure. Pixel values in a selectable range outside the outer perimeter of each star group are used to compute a background value. Both a clipped mean value and a mode value can be chosen.

The actual choice of background determination method depends on the degree of image crowding and the background properties in the image. In simple cases, the automatic method works well with either clipped mean or mode values. In more crowded cases, the mode values give better results. In very difficult circumstances, the separately computed background image is the preferred choice.

4 Fitting the point spread function

The Lund method uses a multi-fitting technique to calculate position and intensity of each member of a star group. The algorithm creates a model for each group by setting up one copy of the PSF per member star. All PSF:s are then fitted simultaneously, with three free parameters per star. The iteration terminates when residuals between the model and the real group are minimised.

The program runs in two phases. In the first phase, pixels that may disturb the final fitting procedure are identified and excluded by means of dynamic checking of residuals. Such pixels could belong to a wing of some close stellar image, or to a CCD defect. This exclusion procedure is iterated until only relevant pixel data remain.

In the second phase, the final fitting is done using a non-linear least-squares minimising technique. The method is a modified version of the Gauss-Newton algorithm (Gill and Murray, 1978). After each iteration, all variables are changed simultaneously. The method is unsensitive to initial values of the variables and the convergence is usually fast. The stability is also good, although problems may arise in special circumstances.

5 Simulations

In the Lund system, several facilities are available to produce simulated images. With a single command it is possible to produce crowded field test images with arbitrary stellar distributions, PSFs, noise levels, geometries, etc. The stellar magnitude distribution is controlled by astrophysical parameters, such as luminosity functions, colour-magnitude diagrams, etc. Figure 1 shows an example of a simulated V image with realistic noise properties. Figure 2 shows an "ideal" colour-magnitude diagram which would result from a perfect reduction, having photon noise as the only error source in the corresponding B and V images.

The images are subsequently used as input to the crowded field analysis program. A synthetic output image is generated from the results of the program. The difference between the input and the output images, the residual image, is used to examine the quality of the results. An example is seen in Figure 3. Finally, Figure 4 shows an output colour-magnitude diagram, derived using results from both a V and a B image reduction. This illustrates the reliability of the reduction procedure in astrophysical terms.

This approach has been used to study different error sources connected to crowded field photometry, including the following:

- different degree of crowding

- image noise

- various PSF sampling levels

- undetected stars

- different PSF shapes

- different fitting radii

- different background determination methods

Although the various effects are interdependent, efforts were made to separate the effects as much as possible. Thus, optimal conditions for the non-varied parameters were maintained. For example, preknowledge of stellar positions was utilised for the initial coordinates, the PSF used to generate the images was the same used by the analysis program, etc.

6   Summary of results

A brief summary of the results will be given here. A full description is available in Linde (1988).

In the first test, analysis was made of noise-free images with increasing degree of crowding. The resulting colour-magnitude diagrams showed almost no increase of spread at the crowding level illustrated in Figure 1.

Photon noise was then added to the same set of images. Figures 1-4 illustrate this experiment. Some spread, in addition to the expected, is seen in the corresponding colour-magnitude diagram. It should be noted that the data shown in Figure 4 is complete and unedited. From studying residuals in Figure 3, several outliers may be removed. The increase of spread for more crowded fields is larger in the noise case than in the noise-free case. This indicates that coupling exists between increased crowding and increased noise.

Undersampling was tested by using a PSF with a FWHM of 1.8 pixels. Although residuals became slightly more noticeable, the algorithm still worked reliably in the great majority of cases. The main problem for undersampled images is the definition of the PSF.

Undetected stars is the most serious problem in crowded field photometry. A comparison was made between "ideal" coordinate input and the interactive initial coordinate determination method described above. Several stars went unnoticed and showed up in the residual image. The magnitude determinations of neighbouring stars became strongly affected in many cases. However, since their positions were also disturbed, they could often be identified as unreliable in the position matching needed between different exposures. Clearly, many unfavourable cases could be solved by rerunning with improved input data.

A deliberately erroneous PSF was used to test the influence of incorrect PSF shape. The resulting differential image exhibited a considerably increased level of residuals, with a corresponding increase in spread in the colour-magnitude diagram. The main problem revealed by this test was the increased instability of the fitting procedure. In addition, a systematic error in the form of a shift towards brighter magnitudes was noticed.

The test image backgrounds were not complex enough to test the effects of different background estimators. Actually, all three background estimation methods, discussed in section 3, worked equally well.

7   Speed and positional accuracy

The computer used was a Hewlett-Packard 9000/530 with floating point accelerator. For groups with less than five members, the computing time was about 0.5 minutes per star. For a group with fifteen stars, the time rises to four minutes per star. Although the program allows a large number of stars per group, it is currently not practical to have more than twenty at a time. As a byproduct of the reductions, precise star positions are also determined. For images with typical noise, about 70 % of correlated position determinations lie within 0.05 pixels.

8 Comparison with other methods

Several algorithms for crowded field photometry have been described in the literature. These include the ROMAPHOT package (Buonanno et al., 1983), algorithms by Penny (1976) and Irwin (1986), which all utilise analytically defined PSFs. The DAOPHOT package (Stetson, 1987) uses an approach similar to the Lund method regarding the derivation and usage of numerical PSFs. However, in very dense fields, automatic algorithms tend to give misleading results. This is where interactive elements are of greatest importance. The Lund algorithms provide interactivity in the most critical phases of crowded field photometry. They include careful preparation of the input data as well as qualitative (visual) and quantitative quality checks of the output data.

9 References

Buonanno, R., Buscema, G., Corsi, C.E., Ferraro, I. and Iannicola, G., *Astron. Astrophys*, **126**, p. 278, 1983.

Gill, P.E., and Murray, W. *SIAM Journal on Numerical Analysis*, **15**, p. 977, 1978.

Irwin, M.J., In *Data Analysis in Astronomy*, Eds. di Gesù, V., Scarsi, L., Crane, P., Friedman, J.H., Levialdi, S., Plenum Press, New York-London, p. 439, 1985.

Linde, P., *Image analysis in Astronomy*, thesis, Lund Observatory 1988.

Newell, B., In *Image Processing in Astronomy*, Eds. G. Sedmak, M. Cappaccioli, R.J. Allen, Osservatorio Astronomico di Trieste, p. 100, 1979.

Penny, A.J., In *IAU Colloquium 40, Astronomical Applications of Image Detectors with Linear Response*, eds. M. Duchesne and G. Lelievre (Paris: Paris-Meudon Observatory), p. 49-1, 1976.

Stetson, P.B., *Publ. Astron. Soc. Pacific*, **99**, p. 191, 1 987.

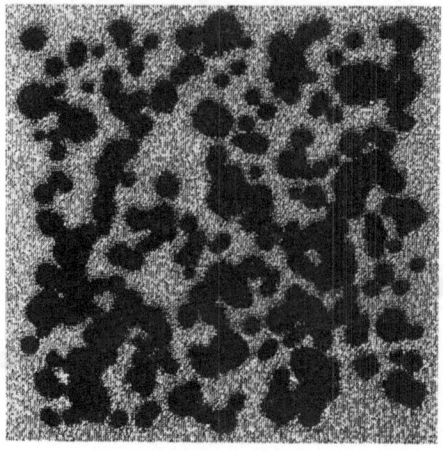

Figure 1. An artificial V field with 337 stars. Image size is 200*200 pixels. Noise according to Poisson statistics has been added. The FWHM size of the point spread function used for generating the stars is 3.6 pixels.

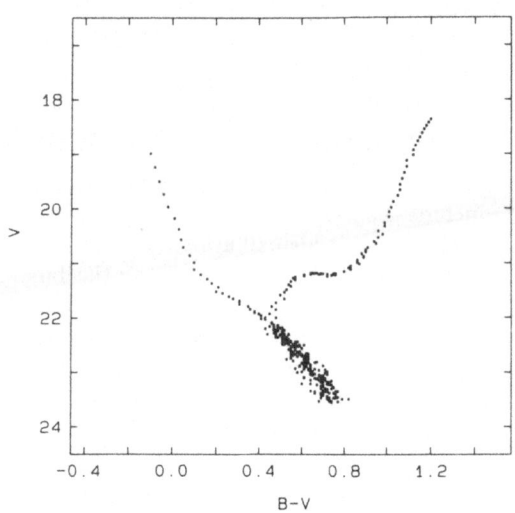

Figure 2. A colour-magnitude diagram as it would appear if the image shown in Figure 1 (and its B image counterpart) had been reduced with no error source except the photon noise.

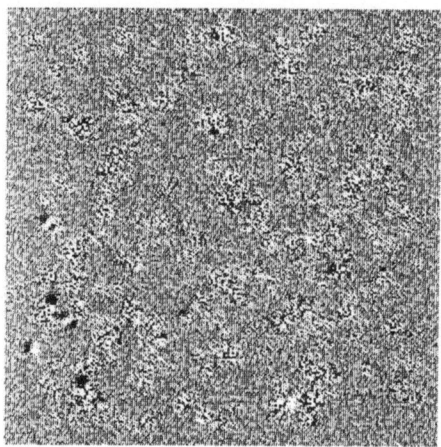

Figure 3. A differential image showing residuals between the field shown in Figure 1 and the measured data.

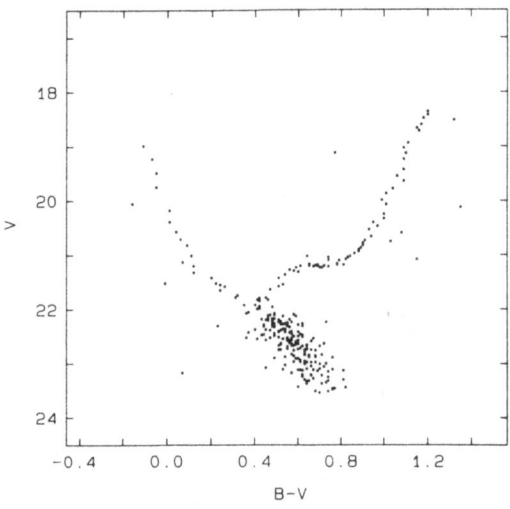

Figure 4. The resulting (raw) colour-magnitude diagram from analysis of the image shown in Figure 1 (and its B counterpart).

ANALYTICAL APPROXIMATIONS OF LONG-EXPOSURE POINT SPREAD FUNCTIONS AND THEIR USE

O.Bendinelli[1], G.Parmeggiani[2], F.Zavatti[1]

[1] Dipartimento di Astronomia, Universita' di Bologna
[2] Osservatorio Astronomico di Bologna

ABSTRACT. The observed light distribution in long exposure star images (PSF) may be fitted equally well by a variety of models. But dealing with undersampled star images, only the use of the multi-Gaussian model allows the correct model parameters estimation, taking into account integration on pixel surface, image off-centering and background behaviour. It is also shown that the convolution of a spherical source with the multi-Gaussian and Moffat's models gives in practice the same result.

1   Introduction

It is known that seeing theory does not give a simple description for the light distribution in long exposure star images, as done for short exposure images by Fried (1966). Nevertheless, its behaviour is well known empirically as it appears from the pioneering works by de Vaucouleurs (1948, 1958) and the more recent ones by King (1971) and Kormendy (1973). The PSF results nearly Gaussian at the centre, with extended wings following approximatively first an exponential and after an inverse power law fading in the background. The energy encircled in these zones clearly depends on the sky condition during the observations, typical values being of the order of 60, 30 and 10 per cent of the total. It must be stressed that the external zone, containing a small fraction of the total light spreaded over a very large region is generally negligible in observations reduction, excepting particular cases as detection of a faint star imbedded in the halo of a much brighter one or determination of the surface brightness of a distant galaxy with a very bright nucleus (see Schweizer, 1979 and Djorgovski, 1984). In any case, the extreme wings of the PSF are too poorly known, the only information available being the 3 values of the power -2.0, -1.5 and -1.7, due respectively to King (1971), Kormendy (1973) and Capaccioli and de Vaucouleurs (1979).

2   PSF empirical models

Among the variety of empirical smooth fitting functions proposed to represent long-exposures PSFs, the more widely employed are the Moffat (1969) distribution, a central truncated Gaussian overlapping an exponential wing (King, 1971), a generalization of the Lorentzian (Franz, 1973) , a mixture of Gaussians (Brown, 1974) and the sum of a Gaussian and an exponential convolved with the same Gaussian (Lauer, 1985). All these approximations give good fits of oversampled star images, i.e. those characterized by a scale length larger than

657

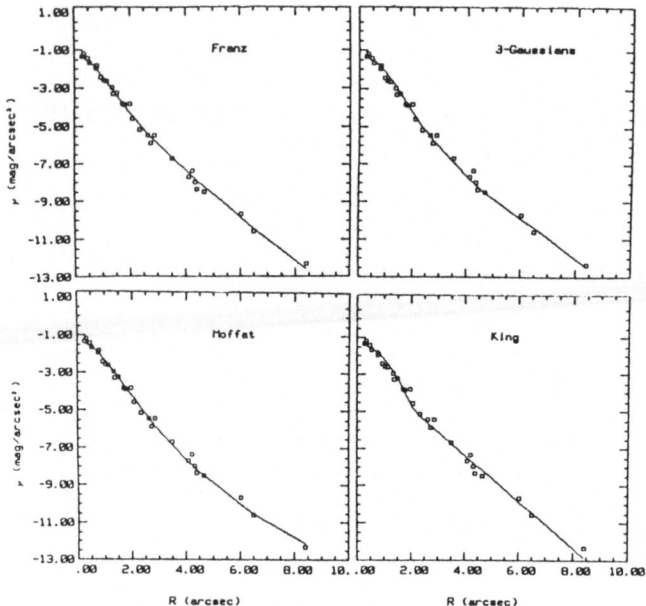

Figure 1: Fitting of King's (1971) star profile by different models (solid line)

the pixel size for CCD observations or the microdensitometer spot for photografic plates. This is shown in Fig.1, where the composite profile by King (1971) has been fitted by four of the above models.

But a serious problem arises in PSF determination from undersampled images (i.e.,those with seeing scale length smaller than pixel or spot size), since neither the integration on the pixel surface nor the off-centering of the image centre from the central pixel (i.e., that of local maximum intensity) can be neglected. The reason of this resides in the analytical properties of models which must be pointed out. First, none of them is linear in the parameters, the worse being from this point of view that by Lauer, in which the convolution of the Gaussian and the Exponential is not expressible in finite terms, and that by Franz where the identifiability conditions are severely violated. In spite of its simplicity, also the King's model is not useful owing to the discontinuity point. Secondly, excepting the multi-Gaussian model, variables are not separable when the pixel (or the spot) integration is taken into account. So, only Moffat's and Brown's models can be used in practice for more sophisticated applications.

3 PSF determination from CCD frames

CCD frames with pixel size greater than the dispersion parameter of the Gaussian seeing do not allow easy determination of the PSF for two main reasons. The first is that the apparent radial brightness profile of a star image is smoother than the true seeing profile, being integrated on the pixel surface, the second that the star image, even with constant background, is skewed, unless its centre coincides with the pixel centre. This means that the accurate knowledge of the seeing may be obtained only by simultaneous estimate of all the parameters (including background ones) of the image model, and taking into account the integration limits, which are off-centering dependent. In practice the centering algorithms of Van Altena and Auer (1975) and Chiu (1977) are improved by using the analytical properties of Gaussians and the Newton-Gauss regularized method (NGR) to secure the convergence of the iterative process leading to the parameters estimate. As shown in Bendinelli et al. (1987), the marginal distribution $f_x(x_n)$ of the light intensity (i.e., the column pixel sums) may be expressed by

$$f_x(x_n) = 0.5 \sum_i a_i \{ erf[(x_n + 1 - x_c)/\sigma_i \sqrt{2}]$$

$$-erf[(x_n - 1 - x_c)/\sigma_i \sqrt{2}]\} + Mb + M(x_n - x_c)\Delta_x b, \qquad (1)$$

which is a suitable form for the application of the NGR method since it is easily differentiable with respect to the 2N+3 parameters a_i, σ_i, x_c, b, $\Delta_x b$. A similar equation can evidently be derived for $f_y(y_m)$ in rewriting Eq.(1). If both the parameters estimates are in reasonable agreement, the PSF is circularly symmetric even though the apparent star image is elliptical due to background gradients. On the contrary, a more sophisticated seeing model could be taken, such as a sum of bivariate Gaussian distributions with dispersion ellipses rotated with respect to the x, y axes (see e.g. Chiu, 1977). If one tries to use Moffat's model, the analogous of Eq.(1) cannot be obtained because the model itself is not integrable in finite terms. As shown in Bendinelli et al. (1988), approximate values of Moffat's shape parameters can be derived by the sequence of integrated luminosities over squares surfaces with n pixels sizes $L(r_n)$, expressed by

$$L(r_n) = L_T[1 - (1 + r^2/\alpha^2)^{1-\beta}] + A_n f, \qquad (2)$$

which enables us to estimate the parameters by the Newton-Gauss regularized method since differentiable with respect to parameters α, β, f, L_T. As far as the off-centering is concerned, it can be derived, before the search of shape parameters, from the moments of the marginal pixel intensity distributions (see Chiu, 1977). The fits of a star image in the M 31 field (obtained by S.G. Djorgovski at the Kitt Peak 4-m telescope with a TI 800x800 CCD, pixel size of 0.298 arcsec) by the above mentioned methods are shown in Fig.2. It is evident that the brightness profiles computed with both models agree reasonably well with the observed data. The models diverge out of about 4.5 arcsec (i.e. roughly speaking 3 σ_3 and 7 α), but it should be stressed that the integrated luminosity of Moffat model from 7 α to the infinity, see Eq.(2), is less than one hundreth of the total, so that the true brightness distribution in extreme wings is irrelevant for any reasonable application (see, for instance, the next section).

4 Convolution of a spherical source

In a series of papers (see Bendinelli et al., 1986, and references therein) it has been shown that convolution-deconvolution of a spherical source with the PSF, approximated by a sum of weighted Gaussians, can be performed by the monodimensional integral equation

$$f(r) = \sum_{i=1}^{N} (a_i/\sigma_i^2) exp(-r^2/2\sigma_i^2 \mathrm{x} \int_0^{+\infty} exp(-\rho^2/2\sigma_i^2)I_0(r\rho/\sigma_i^2)\phi(\rho)d\rho, \qquad (3)$$

relating the true brightness distribution $\phi(r)$ and the observed one f(r). Let us assume Moffat's PSF approximation, then convolution is expressed by the double integral equation

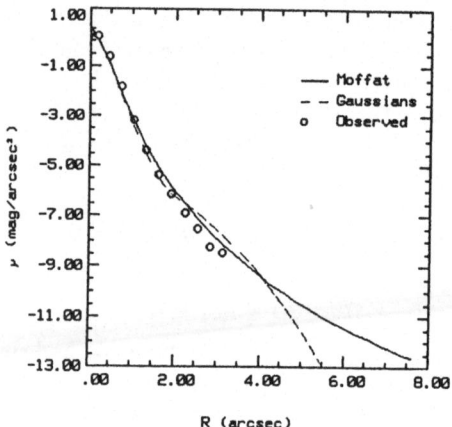

Figure 2: Fits of a star image in M31 field

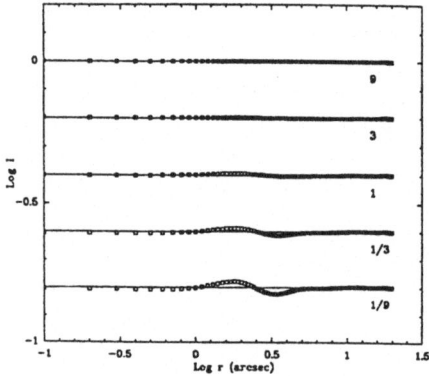

Figure 3: differences in convolved profiles using Moffat and multi-Gaussian approximations

$$f(r) = [2(\beta - 1)/\pi\alpha^2] \int_0^{+\infty} \rho F(\rho) \int_0^\pi [1 + (\rho^2 + r^2 - 2\rho r cos\theta)/\alpha^2]^{-\beta} d\rho d\theta, \qquad (4)$$

which requires, to be computed or inverted up to the radial distance where left-hand side term become negligible, about a factor four in time more than Eq.(3). To prove the substantial equivalence of Eqs.(3) and (4), a set of empirical King models has been choosen. They are characterized by the same concentration index $c = 2.25$ and r_c varying from 9 to 0.11 arcseconds, to simulate distance effect or smaller and smaller intrinsic size. Resulting differences in convolved profiles using Eqs.(3) or (4) are shown in Fig.3.

From the figure some main conclusions can be drawn: $i)$ dealing with large sources ($r > 1$") both PSF approximations give practically the same results, $ii)$ for small sources ($r < 1$") local differences of the order of 1 mag arcsec^{-2} between PSF approximations may cause about 0.1 mag arcsec^{-2} in convolved profiles, $iii)$ convolution effects become negligible for both PSF approximations at comparable distances from the center, so we must be reasonably confident that the outer profile of distant sources is not an artefact of seeing convolution.

5 Concluding remarks

In conclusion, it seems that parameters of both the multi-Gaussian and Moffat models can be calculated taking into account the finite pixel size, with practically the same accuracy. The use of one model depends on the particular research field. For instance in astrometry the multi-Gaussians is surely preferable, giving also the off-centering, while in extragalactic astronomy the other should be used in order to avoid artefacts of PSF wings on the appearance of distant sources, as discussed for instance by Schweizer (1981) and Djorgovski (1984).

6 References

Auer,L.H., Van Altena,W.F.:1978, *Astron. J.*, **83**, 531.

Bendinelli,O., Parmeggiani,G., Zavatti,F.: 1986, *Astrophys. J.*, **308**, 611.

Bendinelli,O., Parmeggiani,G., Piccioni,A., Zavatti,F.,:1987, *Astron. J.*, **48**, 1095.

Bendinelli,O., Parmeggiani,G., Zavatti,F.,:1988, *J. Astrophys. Astron.*, **9**, 17.

Brown,G.S.:1974, *University of Texas Pub.Astr.*, **No. 11.**

Capaccioli,M., de Vaucouleurs,G.: 1979, *Astrophys. J. Suppl. Ser.*, **52**, 465.

Chiu,L.T.G.:1977, *Astron. J.*, **82**, 842.

de Vaucouleurs,G.: 1948, *Ann. Astrophys.*, **11**, 247.

de Vaucouleurs,G.: 1958, *Astrophys. J.*, **128**, 465.

Djorgovski,S.:1984, *J. Astrophys. Astron.*, **4**, 271.

Franz,O.G.:1973, *J. R. Astr. Soc. Can.*, **67**, 81.

Fried,D.L.:1966,*J. Opt. Soc. Am.*,**56**,1372.

King,I.R.:1971, *Publ. Astron. Soc. Pac.*, **83**, 199.

Kormendy,J.:1973, *Astron. J.*, **78**, 255.

Lauer,T.R.:1985, *Astrophys. J. Suppl. Ser.*, **57**, 473.

Moffat,A.F.J.:1969, *Astron. Astrophys.*, **3**, 455.

Schweizer,F.:1979, *Astrophys. J.*, **233**, 23.

Schweizer,F.:1981, *Astron. J.*, **86**, 662.

Van Altena,V.F., Auer,L.H.:1975, in *Image Processing Techniques in Astronomy*, (Dordrecht:Reidel Pub. Co.), p.411.

PHOTOMETRIC DATA ARCHIVES

C. Jaschek
CDS, Observatoire Astronomique
Strasbourg / France

Astronomy being an observational science, it is clear that archiving must be an important part of our professional activities, because not preserving our observations means that we are building "on sand".

Archiving photometric observations, and specifically CCD-type photoelectric observations, means essentially that we keep record of:

1. field center with coordinates and equinox, dimensions of the field,

2. the dates of the observations,

3. the technical details about the system we observed in (filters, receivers),

4. the details of the reduction process (sky background, extinction),

5. the names of the observers, the telescope they used and its geographical location.

Probably you will smile at such a list of details, but I would bet that except in a very few cases, much of such details are unavailable in the large majority of presently published papers providing photoelectric observations. Let me just quote a few consequences of its omission.

One of the most serious errors is the omission of observing dates. This is unforgivable for a professional astronomer and prevents many possible uses of the data – what if the star is later recognized as variable, eclipsing binary or nova?

Lack of filter specifications prevents use of practically all photometry done between the 1920's and the 1950's – observers measured with great care something down to a 1% level, but we do not know what they measured. Thirty years of observations lost!

Later on we learned that even it is not enough to tell the system, since individual filters may deviate considerably from average transmission curves, causing all kinds of secondary effects.

Reductions are performed nowadays usually at the telescope – this is fine but it would be better to know what procedure was used in order to get a real idea about the precision to be expected, even if the author claims ± 0m002.

Clearly such a list could be extended, but I think the principle is clear – we must have behind us sufficient details so that our observations may be re-used. Astronomy is full of examples of observations which are used for purposes very different from the ones foreseen by its authors. For instance we would be grateful to Hipparchus and to Ptolemy if they

D. McNally (ed.), Highlights of Astronomy, Vol. 8, 663–664.
© 1989 by the IAU.

had left traces of how they set up the magnitude system, although they did not consider it something important – they were only interested if the number of stars was invariable or not. Their star positions were also used to derive proper motions of the stars, something whose very existence they denied.

What are thus the minimum requirements for a photoelectric archive? I have given at the start a list which I think contains the basic items. I shall simply add that of course each object must have an identifier *and* coordinates, to prevent that an error in one item invalidates the use of the observation. And then of course attention must be paid to engineering data of the telescope.

Sometimes the objection is made that observing dates are not very popular with magazines which prefer to "gain space" by omitting the column – but then leave half a blank page at the end of the paper. Similarly other "details" of my list are also left out "for editorial reasons". Although I doubt that editors are that harsh, observers have no excuse for not providing these data to the observatory archives, which *must* become a permanent feature of all modern observatories.

Observatory archives should be started right away, if they do not exist.

7. STAR CLUSTERS IN THE MAGELLANIC CLOUDS

Chairman: P. Demarque

Supporting Commissions:

28, 35, 37

A report of this Meeting will appear in Transactions of the International Union, Volume XXB, as part of the report of Commission 37, Star Clusters and Associations.

ADDITIONAL JOINT COMMISSION MEETING

SYSTEMATIC OBSERVATIONS OF THE SUN

A Joint Commission Meeting in honour of

HELEN DODSON PRINCE

Chairmen and Editors: J.C. Pecker and P. Wilson

Supporting Commissions:

10, 12

COMMISSIONS 10 and 12
SYSTEMATIC OBSERVATIONS OF THE SUN

Jean-Claude Pecker
Laboratoire d'Astrophysique Théorique
du Collège de France
98 bis Bd Arago
75014 Paris - France

Peter Wilson
University of Sydney
Department of Applied Maths.
Sydney NSW 2007
Australia

This meeting was dedicated to Helen DODSON-PRINCE who has contributed so much to the field. Unfortunately, she was unable to attend the General Assembly, but a telegram has been approved by acclamations and sent to her, expressing to our colleague and friend the deep admiration and the friendly feelings of the solar community, as gathered in Baltimore.

SUMMARY

One can say that, from the one-day discussion, there emerged two challenges, one scientific and the other one science-political.

The scientific challenge has several aspects. Firstly, the purely MHD treatment of solar magnetic fields cannot yet provide a satisfactory diagnosis of solar synoptic magnetic observations. Secondly, the convective structure of the Sun and its relation to magnetic phenomena is still uncertain. Models of both doughnut-shaped toroidal cells and banana cells have been proposed but it is far from clear whether one or other should be preferred at all times and latitudes or whether transitions may occur from high to low latitudes and at different phases of the solar cycle. Finally, the equatorward migration of solar phenomena oven an 18-22 year period, which has been referred to as the "extended cycle", has been contrasted with suggestions that the cyclic phenomena may operate, to some extent, independently at low and high latitudes.

The science-political challenge is to organize and coordinate a Solar Synoptic Network (SSN) of observatories located at appropriate longitudes and dedicated to the daily pursuit of synoptic studies of active phenomena, both to follow the physics of individual centers and to record consistently the long-term variations of the observable

D. McNally (ed.), Highlights of Astronomy, Vol. 8, 671–678.
© *1989 by the IAU.*

parameters of the cycle.

1. OBSERVATIONS (morning session)

The observational papers covered a wide range of phenomena.

1.1. P. McINTOSH compared the properties of Hα synoptic charts and synoptic magnetic maps generated by solar magnetographs. Data from 23 years were used to argue that the large-scale magnetic fields originate from giant-cell, convectively-driven sources of magnetic flux, and that they are continually renewed by weak flux eruption rather than by diffusion from sunspots. He found that long-lived patterns of polarity create boundaries which are preferred locations of the stronger active regions, and the centers of certain of these patterns are preferred locations for coronal holes which play an important role in the cycle. He suggested that the cycle can now be defined independently of sunspots by observing the changes in rotation rate, wavenumber, and pattern scale of large-scale magnetic fields, and that poleward-moving and equatorward-moving patterns co-exist among the patterns obtained from Hα synoptic charts.

1.2 H. SNODGRASS then discussed the large-scale velocity fields. He outlined the problem of differentiating between the several components of the large-scale velocity fields (i.e. rotational, longitude-dependent azimuthal, meridional, and radial flows). He noted the problem of the limb-effect, whereby the nett blue-shift of lines observed near disk center results in an apparent red-shift of lines near the limb. He then discussed the different indicators of rotation and differential rotation, noting that the supergranule rate is significantly greater than that obtained from other tracers and from spectroscopic data. Finally, he showed that contour plots of the greater-than-average shear zones correspond closely to the butterfly diagram in sunspot latitudes, but extend backwards in time to higher latitudes, so that the preceding component of the butterfly diagram is overlapped, forming a "herringbone diagram".

1.3. Z. MOURADIAN discussed systematic local deviations from the mean differential rotation. Using filaments as tracers, he showed that in many cases they are not drawn out by the differential rotation but mark limited regions of rigid rotation which rotate about a central or "pivot point" during several solar rotation. He pointed out that activity centers frequently appear close to pivot points, whereas filaments exhibiting the normal differential rotation show no similar correlation. He argued that the activity centers may emerge as a consequence of local dynamos and that the pivot point may be the surface signature of the associated velocity fields.

1.4. Z. MOURADIAN then presented a contribution on <u>magnetic tracers</u> of convective patterns, on behalf of E. Ribes. Using observations of the meridional motions of young sunspots and of predominantly east-west oriented filaments, she has the suggested existence of azimuthal convective rolls, having mean plasma velocities 15-20 m s^{-1}, lifetimes 2-3 years, typical sizes 200,000 km and exhibiting alternate magnetic polarities. The rolls first appear some two years after sunspot minimum at mid-latitudes, following the appearance of major activity phenomena, and subsequently appear at lower latitudes as the emerging active regions trace out the butterfly diagram. She suggested that the rolls reach the polar regions shortly after sunspot maximum and contribute to the polarity reversals.

1.5. K. HARVEY compared the statistical properties of <u>ephemeral active regions</u> (ERs) (areas in the range of 0.3-5.0 sq.deg.) and small active regions (ARs) (areas $\geq$ 2.5 sq.deg.). Although the number distributions of the two populations are different, they vary with the cycle in a similar manner, so that the differences are maintained. She found that, while ERs occur at all latitudes throughout the cycle, there is statistical evidence for a distinct high-latitude population in which the emerging ERs tend to adopt orientations consistent with the Hale-Nicholson law for the following cycle. Further, she found that small ARs, exhibiting a clear tendency to a similar orientation, appeared at mid-latitudes as early as 1982, suggesting that the new cycle 22 was in operation at that time.

1.6. This suggestion was supported by R. ALTROCK, who reviewed the <u>coronal green line emission</u> data of Trellis and others during cycle 19 and more recent results from cycle 20 and 21 obtained at the N.S.O. When plotted on synoptic charts, these data show a close correspondence with the butterfly diagram at low latitudes and a high-latitude component which, from the maximum to sunspot latitudes and appears to connect smoothly with the low-latitude component. However, in the N.S.O. data, Altrock found that, at the minimum between cycles 20 and 21, a bifurcation occurred and a high-latitude branch proceeded from mid-latitudes towards the poles, closely parallelling the "rush to the poles" of the polar crown filaments. He has also discussed the apparent occurrence of a double maximum in the total emission when averaged over several cycles, but concluded that this was not a real property of individual cycles.

1.7. P. SIMON presented a review by himself and J.-P. LEGRAND describing the behaviour across the cycle of <u>coronal holes, solar wind streams and geomagnetic disturbances</u>. These long series of solar and solar-related data can be used to study the cyclic behaviour of the toroidal and poloidal components of the solar magnetic field. Estimating the strength of the toroidal field from the sunspot number at maximum and the poloidal component from the maximum intensity of the solar dipole and the thickness of the helioneutral sheet, he defined an integrated magnetic field cycle. The poloidal component begins at mid-latitudes after sunspot maximum and builds towards the maximum of the solar dipole which occurs at the commencement of the following sunspot cycle and is followed by the development of the associated toroidal field. Thus the duration of the magnetic cycle is 17-19 years but its periodicity is ~ 11 years.

1.8. G. ALISSANDRAKIS then reviewed the radio-data, coming from interferometric E-W Nançay data, and from 2-dimensional images (Culgoora, Nançay, Clark Lake), as a source of information from a large range of altitudes. These data show well the coronal holes and emission regions. In particular, Earth's rotation aperture synthesis maps (Nançay, 169 MHz) show that solar emission comes from coronal holes, arch regions associated with the neutral magnetic line, and weak type I continuum, and the correlation with K-coronameter synoptic charts is good. The maps at metric and decametric wavelengths are the only means of studying the physical conditions of coronal features and their long-term variability from disk observations. They permit studies of the correlations between coronal structures and their chromospheric and photospheric counterparts and can be used to determine the position of the base of the heliosheet and the physical conditions therein.

1.9. Finally, H. NECKEL reported on variations of photospheric features, describing a long-term program, started in 1986 by himself and D. Labs at Kitt Peak, to record regularly during cycle 22, the solar limb-darkening at selected continuum wavelengths and high-resolution spectra (FTS) in the 330-400 nm band. The 1986-7 results and earlier observations (1981, CLV of 2 nm-wide spectral bands) revealed unexpected short-term variations of the intensities along the observed diameter, with amplitudes up to 1 % in the continuum, and up to 2 % in the 2 nm spectral bands. Time scales range from minutes to days. The spectra show related changes in line profiles and in line wavelengths.

1.10. The observational session concluded with the announcement of the following poster presentations by their authors: P. PETROPOULOS, in a poster with X. POULAKIS, described the distribution of solar flares for the period 1986-7. M.H. GOKHALE commented about sunspot activity as originated from interference of solar global magnetic oscillations. SIVARAMAN presented a review and a discussion of the long series of synoptic observations of the Sun at Kodaikanal. Finally, J. PAP has shown how the solar irradiance variations (as measured by SMM/ACRIM) are related to different active events, notably to emerging new activity.

2. MODELLING IMPLICATIONS (afternoon session)

2.1. P. WILSON discussed <u>the implications of high-latitude data</u>. He described two fundamentally different approaches to an understanding of cyclic phenomena; in one, a magnetic wave is generated by an unknown mechanism below the convection zone and, in the other, a dynamo operates within or at the base of the convection zone. He pointed out that, in the first approach, the surface phenomena are the incidental by-products of the mechanism, which remains "obscured" by the convection zone. However, in the second, the surface fields and their subsurface connections are, through their interaction with the convective and rotational motions, an essential part of the mechanism. Clearly, observations assume a greater significance in the latter approach.

Wilson then surveyed the high-latitude data which, on synoptic charts, exhibit negative latitude gradients and appear to run smoothly into the corresponding low-latitude patterns. Since the latter coincide with the butterfly diagram for the next cycle and the high-latitude ERs and small ARs tend to exhibit orientations appropriate to that cycle, these data suggest that the high-latitude components are part of an extended overlapping cycle, in which case relaxation models such as that of Babcock are excluded.

However, Wilson noted an alternative view in which the low-and high-latitude components operate independently of each other, having opposite magnetic signatures and this view was later argued by P. Gilman. Wilson stressed that the resolution of this question is of considerable importance, pointing out that the first view supports the concept of a propagating dynamo wave in the form of magnetic toroids (two per hemisphere), as first proposed by Parker. He also showed that a doughnut roll model of the fundamental mode of convection yields the appropriate direction of propagation (towards the equator). Nevertheless, it is well known that some phenomena (e.g., the polar crown) propagate polewards, which suggests that a more complex system of dynamo waves is operating. Wilson argued that the polar field reversals may provide a crucial test for the various models and that careful observations of the details of the reversals should have a high priority.

2.2. P. GILMAN then presented a review on "<u>theory and observations of the solar cycle, and global circulation</u>". He pointed out that global convection models for the solar convection zone do produce about the right differential rotation. The driving convection takes the form of rolls with north-south axes near the equator, which transport angular momentum towards the equator from high latitudes, and he showed that axisymmetric meridional circulations are very weak in such models.

However, these models also predict angular velocity decreasing with depth, and nearly constant on cylinders concentric with the rotation axis, and Gilman pointed out that recent estimates of interior rotation of the Sun, derived from oscillation measurements, indicate there may be very little radial gradient of angular velocity in the convection zone. This raises problems for the global convection models and suggests an interior rotation rate intermediate between the maximum and minimum surface rates. He referred to recent work by Cherilynn Ann Morrow which suggests a near balance of torques between the convection zone and interior, with low latitudes of the convection zone attempting to pull the interior ahead, while high latitudes are pulling it back. There is therefore a cycling of angular momentum between the convection zone and the interior, with a necessary flow of angular momentum from low latitudes to high within a transition zone just below the convection zone.

Gilman noted that the so-called α-ω dynamos, applied to the solar convection zone, can produce the correct solar butterfly diagram only if the angular velocity increases with depth within the convection zone, whereas global convection models predict instead a decrease with depth. Thus the oscillation measurements which indicate zero gradient, along with other considerations, led him to suggest that the seat of the solar dynamo is at the interface between the convection zone and the interior. He described how strong radial gradients of rotation in the interface region, of opposite sign in low and high latitudes, as inferred from the oscillation measurements, coupled with helicity pumped into this region from the convection zone above, could drive a solar dynamo with the appropriate butterfly diagram.

Gilman raised questions about the evidence for an extended solar cycle. In particular, he argued that the torsional oscillation measurements in low and high latitudes may not indicate a true migration from high latitudes to low. He also displayed recent data from Mt. Wilson (provided by Ulrich) suggesting the possibility of a longitude dependence of the torsional signal, and thus raising doubts about its axisymmetric nature. He introduced the question of magnetic contamination of the torsional oscillation signal and emphasized how small the torsional oscillation signal is as a Doppler shift -about 0.1 mÅ. Gilman also argued that evidence for an extended cycle in coronal data took insufficient account of poleward migration of coronal features, and put forward an alternative explanation of the equatorward migration in terms of changes in time of the "tilted dipole" in the large-scale solar magnetic fields. He also gave his opinion that the evidence for an extended cycle in the ephemeral regions data was a very minor and uncertain effect, suggesting instead that the occurrence of ephemeral regions seems to be nearly uniform with latitude and varies over all latitudes by a factor of 2 or 3 during the solar cycle. He suggested that dynamo theories should concentrate on explaining these large amplitude effects, rather than the much less certain signals suggesting an extended cycle.

2.3. There followed a vigorous discussion in which several speakers took issue with some of Gilman's arguments. Snodgrass asserted that the problem of the magnetic contamination of the Doppler signal had been resolved. Altrock rejected the criticism of the coronal data and pointed out that no mechanism had been put forward to support the tilted dipole model and that it was not supported by the data. Clearly the question is still to be conclusively resolved.

3. FUTURE WORK (afternoon session, second part)

W. LIVINGSTON first presented the case for the establishment of a Solar Synoptic Network of observatories (or a SSN) situated at appropriate longitudes, which would work towards the continuous study of the evolution of solar activity phenomena and large-scale features over periods which are not limited by the observing day at a single observatory.

An open discussion followed during which all participants agreed on such a need in particular Gurtovenko, Ai, Hiei and Sivaraman; and most of them openly regretted the closure of the operations of several solar observatories, including the radio interferometers in Culgoora and Clark Lake, at sensible longitudes. In particular, K. ZWAAN noted that, in order to get a SSN started, it is essential to restrict combined observing programs to the data that are really needed to follow the variable Sun and to back-up other programs. He emphasized that truly synoptic programs should be distinguished from campaign-type programs that also require a network of facilities, but that explore either transient phenomena or phenomena with time constant less than one year and that do not need full-disk data. As to the synoptic program, the features to be observed should be rated according to priorities and the required time resolution established in a modest fashion. Campaing-type programs may use some or all of the SSN instruments and may involve many of the scientists taking part in the synoptic program. In this way, SSN may be, according to Zwaan, a stimulus for a very broad sector of solar physics. But a first priority in setting up a SSN is a draft for a contained and truly synoptic program. It was agreed, in the discussion, that all astronomers interested in the SSN should contact Livingston. The participants of the meeting stressed the importance of the resolution voted by Commissions 10 and 12 concerning the pursuit and development of solar observations. The need to continue to observe solar eclipses has also been expressed, notably by E. HIEI.

In conclusion, and coming back to the confrontation between theory and observations of the solar cycle, L. PATERNO (who had chaired section 2) noted that the knowledge of the structure and evolution of the Sun is of overwhelming importance for understanding stellar phenomena: at the present, we have to fit more than two observational constraints of luminosity and radius, which are easily fitted, by varying two free parameters (as the helium abundance and mixing-length ratio) of the theory of convection, in the classic evolutionary models of the Sun. Paterno then noted that the recent large flows of good quality helioseismological data and the consequent very accurate determination of the p-mode spectrum, together with the features of the internal dynamics, impose new constaints to be satisfied. The comparison of the calculated eigenfrequencies with the observed ones indicates, indeed, that the present models of the Sun are basically wrong in the very inner and outer layers. The inversion of the helioseismological data has shown that the interior of the Sun rotates essentially as a rigid body, except perhaps in the very inner core, where the p-mode probing fails. The consequences of this discovery raise very important questions concerning the possible existence of a magnetic field in the Sun's core, the history of the solar angular momentum and the location of the source of the solar activity. Paterno therefore emphasized the need to turn towards three-dimensional models of the solar structure and, by consequence, of the stellar structure, in cases where stars exhibit magnetic activity and star spots. In order to achieve this goal, Paterno noted that highly accurate g-mode observations from space are needed. They would represent the key for penetrating the details of the structure and dynamics of the inner core, essential for understanding the physical mechanisms operating in the stars.

The meeting was then adjourned.

ADDITIONAL CONTRIBUTIONS

THE MICROWAVE BACKGROUND RADIATION: RECENT ADVANCES AND NEW PROBLEMS

G. DE ZOTTI and L. TOFFOLATTI
Astronomical Observatory and Department of Astronomy
Vicolo dell'Osservatorio, 5
I-35122 Padova
Italy

ABSTRACT. The substantially improved intensity measurements at wavelengths longward of the intensity peak of the microwave background are fully consistent with a Planck spectrum. The most precise data disagree with non-relativistic comptonization models for the large submillimeter excess observed by the Nagoya-Berkeley collaboration. The interpretation of such excess as dust emission at high redshifts also faces severe difficulties. Reported anisotropies on scales of several degrees and of tens of arcsec may be contributed, at least in part, by discrete sources. Just because the best experiments at *cm* wavelengths have already got close to the source confusion limit, they also provide interesting information on the large scale distribution of radio sources. Polarimetry may be decisive in clarifying the origin of observed fluctuations.

1. Spectrum

Spectral measurements reported in the last three years are summarized in Table I and displayed in Fig. 1. In the range 10 *cm* to a few *mm*, a key factor in improving the accuracy of the results has been a careful investigation of the atmospheric emission (Danese and Partridge 1988, and references therein). As a bonus, these investigations also led to a better understanding of the atmospheric emission itself. For example, Danese and Partridge were able to determine accurately the non-resonant O_2 contribution which was previously uncertain by a factor of 2.

There is still room for significant improvements, particularly longward of $\lambda \simeq$ 10 *cm*, where Bose-Einstein–like and bremsstrahlung distortions are expected to show up. A big problem here is the accurate modelling of the emission from our own galaxy (cf. Sironi et al. 1988), as well as the subtraction of the essentially isotropic component due to unresolved discrete sources. The conventional approach (extrapolation of the low–frequency measurements of the extragalactic radio background) ignores the effect of the bimodal distribution of spectral indices of radio-sources. On the other hand, the very deep source counts, now available at several radio frequencies, allow direct estimates of (or, strictly speaking, lower limits to) contributions of radio sources to the extragalactic background (see Table II).

At short wavelengths the major event was the rocket experiment by the Nagoya–Berkeley collaboration (Matsumoto et al. 1988), which detected a large excess in the Wien region, similar to that reported earlier by Gush (1981). Although the authors were very careful in controlling or ruling out many possible sources of spurious signals, the results certainly need confirmation. A possible test can be provided by measurements of the dipole anisotropy at submillimeter wavelengths:

681

D. McNally (ed.), Highlights of Astronomy, Vol. 8, 681–688.

TABLE I. Recent measurements of the MWB brightness temperature

References	λ (cm)	T_{MWB} (K)
Sironi et al 1988	50.0	2.98 ± 0.55
Levin et al 1988	21.3	2.11 ± 0.38
Sironi and Bonelli 1986	12.0	2.79 ± 0.15
De Amici et al 1988	8.1	2.59 ± 0.13
Mandolesi et al 1986b	6.3	2.70 ± 0.07
Kogut et al 1988	3.0	2.61 ± 0.06
Johnson and Wilkinson 1987 ..	1.2	2.783 ± 0.025
De Amici et al (1985)	0.909	2.81 ± 0.12
Bersanelli (Smoot et al 1987) ..	0.333	2.60 ± 0.10
Peterson et al 1985	0.351	2.80 ± 0.16
Meyer and Jura 1985	0.264	2.70 ± 0.04
Crane et al. 1988	0.264	$2.796^{+0.018}_{-0.040}$
Peterson et al 1985	0.198	$2.95^{+0.11}_{-0.12}$
Peterson et al 1985	0.148	2.92 ± 0.10
Meyer and Jura 1985	0.132	2.76 ± 0.20
Palazzi et al. 1988	0.132	2.67 ± 0.10
Matsumoto et al 1988	0.116	2.799 ± 0.018
Peterson et al 1985	0.114	$2.65^{+0.09}_{-0.10}$
Peterson et al 1985	0.100	$2.55^{+0.14}_{-0.18}$
Matsumoto et al 1988	0.0709	2.955 ± 0.017
Matsumoto et al 1988	0.0481	3.175 ± 0.027

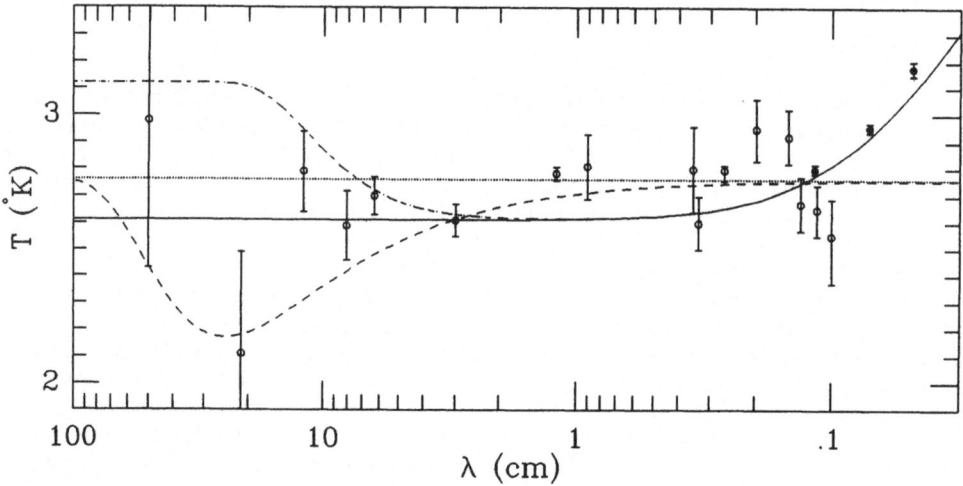

Figure 1. Comparison of the data in Table I with potential distortions of the MWB spectrum: non-relativistic comptonization with $y = 0.025$ (solid line); bremsstrahlung distortion for the same value of y, $z_{heating} = 10^4$, and $\Omega_{baryon} = 1$ (dot-dashed line); Bose-Einstein distortion with chemical potential $\mu_0 = 1 \times 10^{-2}$ and $\Omega_{baryon} = 0.1$.

in the presence of the reported rise of the brightness temperature, the Compton-Getting effect would lead to a decrease of the dipole amplitude (Danese and De Zotti 1981). This expectation is not borne out by the measurements of Halpern et

TABLE II. Contributions of discrete sources to the sky brightness temperature

$\nu\ (GHz)$	$\lambda\ (cm)$	$T_{sources}\ (K)$
0.178	168.5	21.8
0.408	73.5	2.585
0.6	50.0	.88
0.8	37.5	.40
1.4	21.4	.094
2.5	12.0	.020

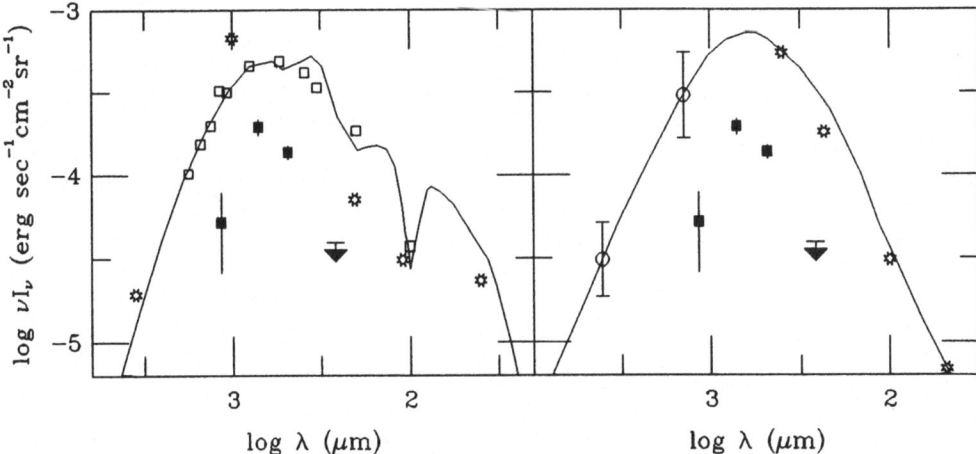

Figure 2. Comparison of the Matsumoto et al. (1988) excess (filled squares and arrow) with observed shapes of dust emission spectra. The panel on the left hand side shows the shape of the spectrum of the 3-kpc ring in NGC 1068 (open squares) and of regions of massive star formation in our Galaxy (stars); for references on the data and on the model (solid line), see Rowan-Robinson (1987). The data in other panel refer to the isolated dark cloud L 810 (Chini et al. 1986). Dust emission spectra have been arbitrarily redshifted and scaled in flux to ease the comparison with the data by Matsumoto et al. (1988).

al. (1988), although the disagreement is only at the 2σ level.

If the sub-mm excess is real, two possible explanations come immediately to mind, and indeed have already been discussed by Hayakawa et al. (1987), Smoot et al. (1988), and others.

A first possibility is comptonization by non-relativistic electrons. Since the excess energy is $\Delta\epsilon/\epsilon \simeq 10\%$, the comptonization parameter is $y \simeq \frac{1}{4}(\Delta\epsilon/\epsilon) \simeq 0.025$. The corresponding comptonized spectrum is shown in Fig. 1. To obtain a good fit to the Nagoya–Berkeley data we would actually need a slightly larger comptonization parameter ($y \simeq 0.028$, Hayakawa et al. 1987). But a larger y would further enhance the disagreement with the two most accurate results at longer wavelengths.

Kawasaki and Sato (1987) claim that a good fit may be obtained with comptonization of photons produced by radiative decay of massive weakly interacting particles having appropriate mass, lifetime, and number density. On the other hand, they apparently don't use the right electron temperature as a function of time. For example, after the end of the decay process, their formula yields $T_e \simeq T_R(1 + y)$, while the correct result is $T_e \simeq T_R(1 + 5.4y)$. In addition, particle decay is further severely constrained by consideration of the stability of late stages of stellar evolution.

The difficulty with the data at $\lambda \geq 0.264$ cm is somewhat relieved in the case of comptonization by relativistic electrons (Wright 1979), such as those invoked by some models for a bremsstrahlung origin of the X–ray background (Barcons and Fabian 1988). However, a spectrum of this kind does not provide a good fit to the Nagoya–Berkeley data (Hayakawa et al. 1987); in particular, it tends to conflict with the tight upper limit at 0.26 mm.

An alternative possibility (Hayakawa et al. 1987; Hogan and Bond 1988) is reradiation from dust heated by pregalactic stars or by massive starbursts during early phases of galaxy evolution. This picture, in addition to energetic problems (Lacey and Field 1988), has serious difficulties in accounting for the shape of the observed excess. After subtraction of a blackbody at $T = 2.762 \pm 0.013$ K (the weighted mean of data at $\lambda \geq 0.264$ mm listed in Table I) we end up with a remarkably narrow residual spectrum, well fitted by a single temperature dust emission. As illustrated by Fig. 2, the observed dust emission spectra in a variety of astrophysical settings are substantially broader: very special conditions are apparently required for the dust to be strictly isothermal. In addition, it appears to be difficult to avoid the further smearing out due to the redshift distribution of the emitting dust.

2. Isotropy

2.1. LARGE ANGULAR SCALES

The dipole anisotropy is now measured with great accuracy. The data from radiometers carried aloft by balloons, by the Berkeley group (Lubin et al. 1985) and by the Princeton group (Fixen et al. 1983), each covering $\simeq 85\%$ of the sky, have been combined into a map which is better connected and has a sky coverage increased to $\simeq 90\%$ (Lubin and Villela 1986). The average dipole amplitude is $\Delta T = 3.26 \pm 0.23$ mK, and the average solar velocity direction is $\alpha = 11^h.25 \pm 0^h.15$, $\delta = -5°.6 \pm 2°.0$. Both results are in excellent agreement with those of the *Relict* experiment aboard the *Prognoz 9* satellite (Strukov et al. 1987) which has scanned the whole celestial sphere: $\Delta T = 3.16 \pm 0.12$ mK, $\alpha = 11^h.28 \pm 0^h.16$, $\delta = -7°.5 \pm 2°.5$.

As is well known, the microwave dipole is a crucial benchmark for studies of large-scale streaming motions (see, e.g., Lynden-Bell et al. 1988).

No signal was found for higher harmonics. The quadrupole limits of both balloon experiments are $\Delta T/T \leq 7 \times 10^{-5}$. An harmonic analysis of *Relict* data yields $\Delta T/T \leq 3 \times 10^{-5}$. Even tighter constraints are obtained from the variance analysis for specific fluctuation spectra (Strukov et al. 1988); for example, for a Zeldovich spectrum $(\Delta T/T)_{quadrupole} \leq 1.6 \times 10^{-5}$ at the 95% confidence.

2.2. SMALL ANGULAR SCALES

Recent results on anisotropies at intermediate and small angular scales are listed in Table III.

Significant fluctuations with an observed standard deviation of 3.7×10^{-5} on scales of $8°$–$10°$ have been reported by the Jodrell–IAC group (Davies et al. 1987), at $\lambda \simeq 2.87$ cm. While recognizing that structure in the radio continuum emission from our galaxy may contribute appreciably, the authors argue that a substantial

TABLE III. Small scale anisotropies

Reference	λ (cm)	θ	$\frac{\Delta T}{T}$
Davies et al. (1987)	3	$8°$	3.7×10^{-5}
Klypin et al. (1987)	0.8	$6°$	$\leq 5.6 \times 10^{-6}$
Mandolesi et al. (1986a)	3.	$2° - 6°$	$\leq 5 - 7 \times 10^{-4}$
Uson and Wilkinson (1985)	1.5	$4'.5$	$\leq 2 \times 10^{-5}$
Readhead et al. (1988)	1.5	$7'$	$\leq 1.5 \times 10^{-5}$
Martin and Partridge (1988)	6	$18'' - 80''$	$(1.7 \pm 0.5) \times 10^{-4}$
Martin and Partridge (1988)	6	$36'' - 160''$	$(1.3 \pm 0.2) \times 10^{-4}$
Hogan and Partridge (1988)	2	$18''$	$(2 \pm 1) \times 10^{-4}$
Fomalont et al. (1988)	6	$12''$	$\leq 8.5 \times 10^{-4}$
Fomalont et al. (1988)	6	$18''$	$\leq 1.2 \times 10^{-4}$
Fomalont et al. (1988)	6	$30''$	$\leq 0.8 \times 10^{-4}$
Fomalont et al. (1988)	6	$60''$	$\leq 0.6 \times 10^{-4}$

part of the signal is probably intrinsic. Fluctuations of similar amplitude on an angular scale of 6° were previously observed by Melchiorri et al. (1981) at sub-mm wavelengths; again, the interpretation of the result depends on an uncertain correction for the galactic contribution. More recently, the Jodrell–IAC group (Watson et al. 1988) has obtained new results on an angular $\simeq 5°$. They were able to identify all features in the scan which have temperature excesses $\Delta T/T > 3 \times 10^{-5}$ with beam–diluted known sources, with the exception of a peak in the direction $\alpha \simeq 14^h\ 40^m$, $\delta \simeq 40°$; they note that this is the direction of the Bootes void as well as of an unusual string of galaxies. No signal of this amplitude has been reported by the *Relict* experiment, operating at $\lambda = 8\ mm$.

High sensitivity observations by several groups have led to remarkably tight upper limits on arcminute scales. At the same time, improved methods have been developed to compare the experimental results with theoretical predictions (Bond and Efstathiou 1987; Vittorio and Juskiewicz 1987). The analysis of Lasenby and Kaiser (1988) has shown that the upper limit originally reported by Uson and Wilkinson (1985), and quoted in Table III, should be revised upwards by a factor ≥ 2.

On still smaller angular scales, around or below $1'$, a signal has definitely been detected. Its interpretation, however, is still controversial. While Martin and Partridge (1988) argue that discrete sources cannot explain all the variance they see, Fomalont et al. (1988) come to the opposite conclusion. Franceschini et al. (1988) confirm that the counts of known classes of radio sources fall short by a factor $\simeq 3$ to account for the fluctuation amplitude measured by Martin and Partridge. On the other hand, the results of the VLA survey at $\lambda \simeq 20\ cm$ by Mitchell and Condon (1985) may already suggest that true counts are substantially higher than predicted by current models, and could saturate the observed sky variance. The discrete source contribution to fluctuations could be further enhanced by clustering.

Careful reviews of the quantitative predictions for anisotropies of the microwave background for a wide variety of theories of the formation of structure in the universe have recently been presented by Kaiser and Silk (1986), Bond (1988), and Hogan and Bond (1988).

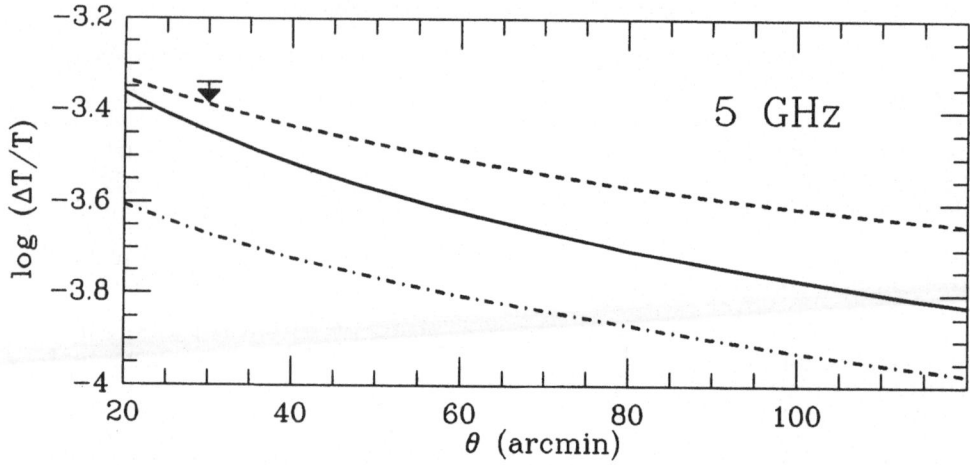

Figure 3. Contribution of discrete sources to temperature fluctuations, as a function of the angular scale. The solid line refers to a Poisson distribution of sources (Franceschini et al. 1988). The dashed and the dot-dashed lines show the additional contributions due to clustering described by a power-law $\xi(r)$ (see text) with $r_0 = 40\ Mpc$ and $r_0 = 20\ Mpc$ ($H_0 = 50$), respectively, and $r_{max} = 2r_0$. The upper limit is from Lasenby and Davies (1983).

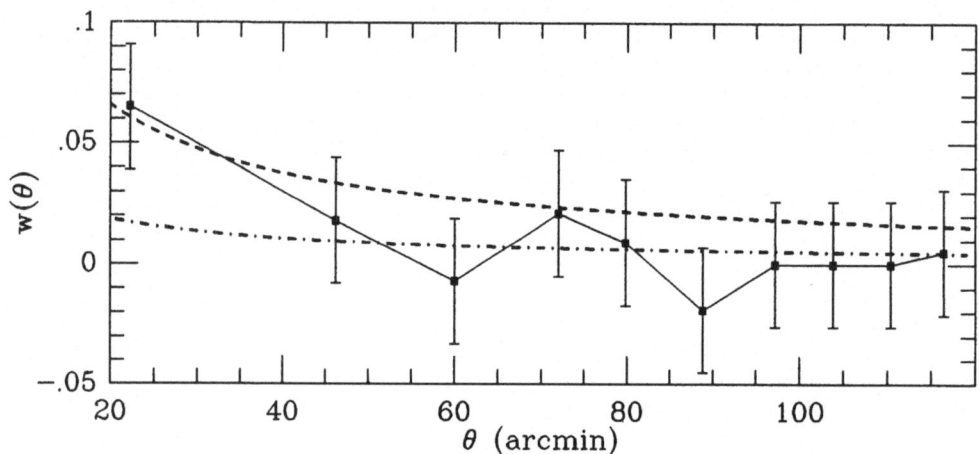

Figure 4. Comparison of the angular correlation functions corresponding to the same choices for $\xi(r)$ as in Fig. (3) with the data of Langston (1988).

2.3. SUNYAEV-ZELDOVICH EFFECT

The Sunyaev-Zeldovich dips on scales of a few arcmin, in the directions of three rich clusters of galaxies were the first small scale anisotropies for which detection at a high level of significance has been reported (Birkinshaw 1987). On the other hand, many years of experimental work have revealed several astrophysical and instrumental effects that may distort the results (cf. Partridge et al. 1987 and references therein); a full exploitation of the rich astrophysical information provided by the S-Z effect is probably still premature.

2.4. FLUCTUATIONS DUE TO DISCRETE SOURCES

As already mentioned, small scale temperature fluctuations due to the non–uniform space distribution of discrete sources may seriously hinder searches for primordial anisotropies, except for a relatively narrow frequency range (Franceschini et al. 1988).

On the other hand, fluctuations due to sources are themselves interesting quantities to measure. For example, they may be informative on the large–scale distribution of radio–sources, which, in turn, may reflect the large–scale distribution of matter in general. In Fig. 3 we compare the Poisson contributions to $\Delta T/T$ with the additional fluctuations expected if radio–sources have a power–law two–point correlation function, $[\xi(r) = (r/r_o)^{-1.8}]$, cut down at the radius r_{max}. For $r_0 \geq$ 30–40 Mpc the fluctuations are dominated by the effect of clustering, and the upper limit set by Lasenby and Davies (1983) already implies that r_0 cannot be much larger than 40 Mpc. This constraint is significantly stronger than that obtained from the study of the two–point angular correlation function of sources in the MG II survey (Langston 1988; see Fig. 4).

3. Polarization

Until recently, both observational and theoretical studies dealt with polarization on large angular scales, introduced by anisotropic expansion (Tolman 1985, and references therein).

As first pointed out by Kaiser (1983), however, a significant degree of polarization is associated with small scale anisotropies induced by adiabatic perturbations. Detailed studies (Milaneschi and Valdarnini 1986; Bond and Efstathiou 1987) have shown that, for a variety of models for primordial perturbations, the predicted $\Delta T_p/T$ range from $\simeq 2\%$ to $\simeq 10\%$ of intensity fluctuations. The ratio $(\Delta T_p/T)/(\Delta T/T)$ peaks at small angular scales. As pointed out by Milaneschi and Valdarnini (1986), this can be easily understood since long wavelength perturbations contribute to temperature anisotropies through potential fluctuations, but there are not enough scatterings for them to generate polarization (in dealing with very large perturbations, recombination can be treated as instantaneous). Similarly, $(\Delta T_p/T)/(\Delta T/T)$ decreases as the spectral index n of the primordial perturbation spectrum $(\delta^2 \propto k^n)$ decreases, since the relative power on large scales increases.

Polarimetry may thus be decisive in determining the origin of anisotropies, in the presence of confusion from faint sources.

Limits on linear and circular polarization (Stokes parameters Q, U, and V) for scales $18" \leq \theta \leq 160"$ were recently reported by Partridge et al. (1988). They range from $(\Delta T_p/T) \leq 4 \times 10^{-5}$ to $(\Delta T_p/T) \leq 2.2 \times 10^{-4}$, and are about three times lower than upper limits on intensity fluctuations set on the same angular scales.

Acknowledgements. We are indebted to L. Danese and A. Franceschini for many useful discussions. Work supported in part by MPI and CNR (through GNA and PSN).

References

Barcons, X., and Fabian, A.C. 1988, *M.N.R.A.S.*, **230**, 139.

Birkinshaw, M. 1987, in *Observational Cosmology*, A. Hewitt et al. (eds.), p. 83.
Bond, J.R. 1988, in *Large Scale Structures of the Universe*, Kluwer, p. 93.
Bond, J.R., and Efstathiou, G. 1987, *M.N.R.A.S.*, **226**, 655.
Chini, R., et al. 1986, in *Light on Dark Matter*, F.P. Israel (ed.), Reidel, p. 29.
Crane, P., et al. 1988, in *Proc. Third ESO/CERN Symposium*, in press.
Danese, L., and De Zotti, G. 1981, *Astr. Ap.*, **94**, L33.
Danese, L., and Partridge, R.B. 1988, *Ap. J.*, submitted.
Davies, R.D., et al. 1987, *Nature*, **326**, 462.
De Amici, G., et al. 1988, *Ap. J.*, **329**, 556.
De Amici, G., Smoot, G.F., Friedman, S.D., and Witebsky, C. 1985, *Ap. J.*, **298**, 710.
Fixen, D.J., Cheng, E.S., and Wilkinson, D.T. 1983, *Phys. Rev. Letters*, **50**, 620.
Fomalont, E.B., et al. 1988, *A.J.*, **96**, 1187.
Franceschini, A., Toffolatti, L., Danese, L., and De Zotti, G. 1988, *Ap. J.*, submitted.
Gush, H.P. 1981, *Phys. Rev. Letters*, **47**, 745.
Halpern, M., et al. 1988, *Ap. J.*, **332**, 596.
Hayakawa, S., et al. 1988, *Publ. Astron. Soc. Japan*, **39**, 941.
Hogan, C.J., and Bond, J.R. 1988, in *The Post-Recombination Universe*, N. Kaiser and
     A.N. Lasenby (eds.), Kluwer, p. 141.
Hogan, C., and Partridge, R.B. 1988, preprint.
Johnson, D.G., and Wilkinson, D.T. 1987, *Ap. J. Letters*, **313**, L1.
Kaiser, N. 1983, *M.N.R.A.S.*, **202**, 1169.
Kaiser, N., and Silk, J. 1986, *Nature*, **324**, 529.
Kawasaki, M., and Sato, K. 1987, *Publ. Astr. Soc. Japan*, **39**, 837.
Klypin, A.A., et al. 1987, *Pis'ma Astron. Zh.*, **13**, 259 [*Sov. Astron. Letters*, **13**, 194]
Kogut, A., et al. 1988, *Ap. J.*, **325**, 1.
Lacey, C.G., and Field, G.B. 1988, *Ap. J. Letters*, **330**, L1.
Langston, G. 1988, in *Large Scale Structure and Motions in the Universe*, in press.
Lasenby, A.N., and Davies, R.D. 1983, *M.N.R.A.S.*, **203**, 1137.
Lasenby, A.N., and Kaiser, N. 1988, preprint.
Levin, S.M., et al. 1988, *Ap. J.*, in press.
Lubin, P., Villela, T. 1986, in *Galaxy Distances and Deviations from Universal Expansion*,
     B.F. Madore and R.B. Tully (eds.), Reidel, P. 169.
Lubin, P., Villela, T., Epstein, G., and Smoot, G.F. 1985, *Ap. J. Letters*, **298**, L1.
Lynden-Bell, D., et al. 1988, *Ap. J.*, **326**, 19.
Mandolesi, N., et al. 1986a, *Nature*, **319**, 751.
Mandolesi, N., et al. 1986b, *Ap. J.*, **310**, 561.
Martin, H.M., and Partridge, R.B. 1988, *Ap. J.*, **324**, 794.
Matsumoto, T., et al. 1988, *Ap. J.*, **329**, 567.
Melchiorri, F., et al. 1981, *Ap. J. Letters*, **250**, L1.
Meyer, D.M., and Jura, M. 1985, *Ap. J.*, **297**, 119.
Milaneschi, E., and Valdarnini, R. 1986, *Astr. Ap.*, **162**, 5.
Mitchell, K.J., and Condon, J.J. 1985, *A. J.*, **90**, 1957.
Palazzi, et al. 1988, in *Proc. Third ESO/CERN Symposium*, in press.
Partridge, R.B., Nowakowski, J., and Martin, H.M. 1988, *Nature*, **331**, 146.
Partridge, R.B., Perley, R.A., Mandolesi, N., and Delpino, F. 1987, *Ap. J.*, **317**, 112.
Peterson, J.B., Richards, P.L., and Timusk, T. 1985, *Phys. Rev. Letters*, **55**, 332.
Readhead, A.C.S., et al. 1988, in *Large Scale Structures of the Universe*, Kluwer, p. 37.
Rowan-Robinson, M. 1987, in *Star Formation in Galaxies*, C.J. Lonsdale (ed.), p. 133.
Sironi, G., and Bonelli, G. 1986, *Ap. J.*, **311**, 418.
Sironi, G., et al. 1988, *Ap. J.*, submitted.
Smoot, G.F., et al. 1987, *Ap. J. Letters*, **317**, L45.
Smoot, G.F., et al. 1988, *Ap. J.*, **331**, 653.
Strukov, I.A., et al. 1987, *Pis'ma Astron. Zh.*, **13**, 163 [*Sov. Astron. Letters*, **13**, 65]
Tolman, B.W. 1985, *Ap. J.*, **290**, 1.
Uson, J.M., and Wilkinson, D.T. 1985, *Nature*, **312**, 427.
Vittorio, N., and Juskiewicz, R. 1987, *Ap. J. Letters*, **314**, L29.
Watson, R.A., et al. 1988, in *Large Scale Structure and Motions in the Universe*, in press.
Wright, E.L. 1979, *Ap. J.*, **232**, 348.

Submillimeter Spectrum of the Cosmic Background Radiation

Toshio Matsumoto
Department of Astrophysics, Nagoya University
Chikusa-ku, Nagoya, Japan 464-01

Recent observations of the spectrum of the 3K cosmic background radiation (CBR) indicate that the CBR spectrum is consistent with a blackbody spectrum of $T = 2.74 \pm 0.02$K (Smoot et al. 1987). These measurements, however, were carried out in the Rayleigh-Jeans part of the spectrum, while theories predict spectral distortion in the Wien part. Therefore, we tried to observe the submillimeter spectrum of the CBR with a liq.He cooled radiometer onboard a sounding rocket. The experiment was a collaboration between Nagoya University and University of California, Berkeley.

The winston light concentrater had a 4 cm aperture with a 7.6° beam. A flared baffle at the top of the horn resulted in very low side lobe response. The photometer was designed to observe sky brightness in 6 filter bands between $100\mu m$ and 1mm simultaneously (Lange et al. 1987, Sato et al. 1987). Since the whole optical system was cooled down to 1.0K, instrumental emission was negligible. Careful absolute calibration in the laboratory provided small systematic errors, especially for 3 long wavelength channels.

The sounding rocket, K-9M-80, having the radiometer onboard was launched on 1987 February 23, 0:00 JST (February 22, 15:00 UT), from the Kagoshima Space Center of the Institute of Space and Astronautical Science, Japan (Matsumoto et al.1988). The rocket reached an apogee of 317km at 287s after launch. The lid of the cryostat was opened at 150s and the observation commenced. The rocket axis which coincides with the optical axis of the radiometer precessed with a period of 10s around 30° ± 1.5° full cone angle on the sky passing through the zenith. The center of the precession cone pointed towards $l = 203° \pm 2°$, $b = 33° \pm 2°$. Just after the lid opened, the temperature of the photometer increased a little and environmental emission was observed in the short wavelength channels. After 200s, however, the transient phenomena disappeared and all signals became steady. Fig.1 shows the average sky brightness of the observed region of the sky.

D. McNally (ed.), Highlights of Astronomy, Vol. 8, 689–691.
© *1989 by the IAU.*

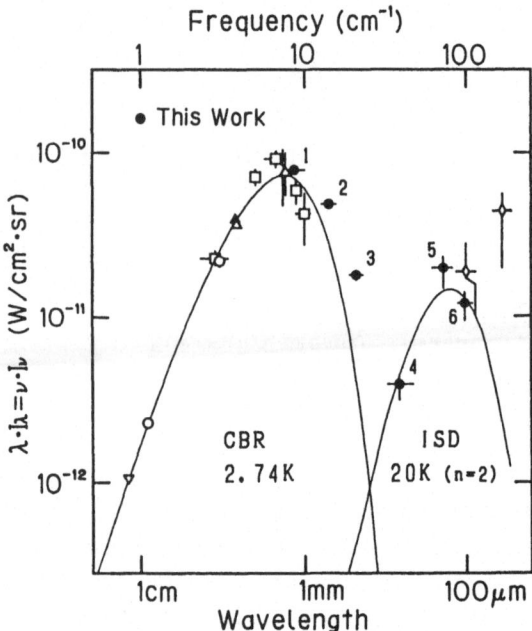

Figure 1: The observed spectrum of the astrophysical background. Filled circles indicate our data. Vertical and horizontal bars indicate 1σ errors and bandwidths, respectively.

A characteristic feature of Fig.1 is an excess submillimeter brightness above a blackbody spectrum of 2.74K. Signals of the three long wavelength channels were spatialy isotropic, indicating their cosmological origin. This is the first significant detection of a distortion of the CBR spectrum in the Wien region.

The spectrum of the three short wavelength channels shows a feature expected for thermal emission of the diffuse interstellar dust (ISD). The spatial distributions of these channels indicate a clear correlation with column density of neutral hydrogen, N(HI), as shown in Fig.2. Assuming that a correlated brightness with N(III) is of galactic origin, then ISD emission was well fit by the thermal emission of the dust, assuming $T_d = 19 \pm 3K$ and emissivity index of 2. Ratios of IR-brightness to N(HI) were also obtained. This ratio at $102\mu m$, $0.65 \pm 0.14 MJy.sr^{-1}/10^{20}H$, is consistent with that obtained by Terebey and Fich (1986) for IRAS but is a little smaller than that by Boulanger and Perault (1988). The galactic emission thus obtained, however, cannot explain all the observed brightness. In the $102\mu m$ band, residual emission could be attributed to zodiacal emission, but a significant isotropic emission $(5 \pm 2 \times 10^{-12}W.cm^{-2}.sr^{-1})$ remains in the $137\mu m$ band. The origin of this emission is not certain, but may be related to isotropic emission observed by IRAS at $100\mu m$ (Boulanger and Perault 1988).

The most important result of this experiment is an excess submillimeter CBR

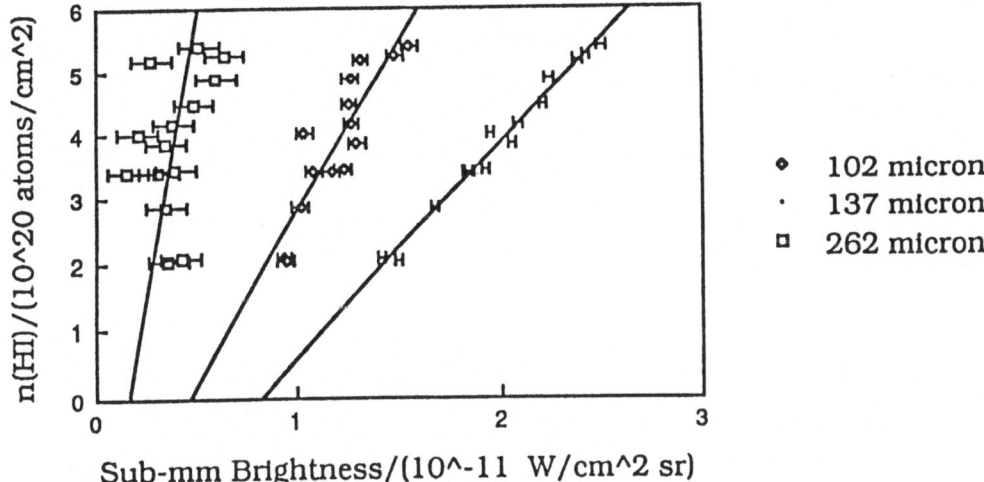

Figure 2: The correlation of N(HI) with submillimeter brightness. The horizontal error bars indicate 1σ statistical errors only.

brightness. The energy density of this excess amounts to 10% of a 2.74K CBR, which provides a serious restriction on baryonic energy generation at high redshift. Two physical processes which explain the observed spectrum are proposed (Hayakawa et al. 1987). One is Compton scattering in intergalactic hot plasma. Another is dust emission, which was generated and heated by Pop.III stars in the early universe, superposed on the CBR spectrum. Future advanced studies of the submillimeter CBR will provide a more definite scenario for understanding the origin and evolution of the early universe.

References
Boulanger,F and Perault,M. 1988, *Ap.J.*, **330**, 964.
Hayakawa,S. et al. 1987, *Pub.Astr.Soc.Japan*, **39**, 941.
Lange,A. et al. 1987, *Appl.Optics*, **26**,401.
Matsumoto,T. et al. 1988, *Ap.J*, **329**, 567.
Sato,S. et al. 1987, *Appl.Optics*, **26**, 410.
Smoot,G. et al. 1987, *Ap.J.Letters*, **317**, L45.
Terebey,S and Fich,M 1986, *Ap.J.Letters*, **309**, L73.

The status of Big Bang nucleosynthesis in July 1988

Hubert Reeves   Section d'Astrophysique C.E.N.S. Saclay Gif F - 91191
France

Institut d'Astrophysique de Paris 75014 Paris

Important progresses have been made in two fronts in the few past years.

On the theoretical ground, we have realized the importance of the quark-hadron phase transition as possibly influencing in a major way the yields of the primordial nucleosynthesis isotopes.

On the observational ground, the status of lithium as a bona-fide cosmological observable has been confirmed and its primordial abundance can be evaluated with increased confidence.

The quark-hadron phase transition.

The physics of the quark-hadron phase transition (or transitions, since there is a chiral transition and a confinement transition) is presently the object of intense studies (Iso et al 1986) (Satz 1985,1987) (Leutwyler, 1988) etc). Many of the parameters of the transitions are still poorly known despite the vigourous effort being made in QCD calculations on networks.

The relevant parameters, as far as BBN is concerned, are the following. First: the order of the transitions. There are strong indications that in the baryonic density range of the BBN, the transitions are *first order*, leading to nucleation and to bubbles of high density matter in a low density background (or the inverse).

Second : the critical temperature T_c of the transitions. Again in the density range of cosmological interest, the transitions appear to occur at the same T_c and to be simultaneous. QCD calculations give a range of $150 \text{ MeV} < T_c < 250 \text{ MeV}$. Recent chiral perturbation calculations have been published which quote a narrower range of $180 \text{ MeV} < T_c < 220 \text{ MeV}$ (Gasser and Leutwyler, 1987, 1988)

The density contrast R between the high and low baryonic density phases can be computed, assuming chemical potential equilibrium between the two phases (Sale and Matthews 1986) (Applegate and Hogan 1985) (Applegate, Hogan and Sherrer 1987), (Alcock et al 1987) (Fuller et al 1987) (Kapusta and Olive 1988). The result depends strongly on the value of the assumed critical temperature. At low T_c, the computed value of R is larger than ten, decreasing gradually at higher T_c. Thus, through the value of R, the lithium abundance is related to the value of T_c. In reality the gradual hadronization of the quark sea leads to a distribution of baryonic densities.

As the universe cools from T_c, at approximately 20 μsec, to one MeV, at one second, the contrast R between the phases is maintained. The neutron to proton ratio (n / p), governed by weak processses, is the same in all phases. It is given by the

693

D. McNally (ed.), Highlights of Astronomy, Vol. 8, 693–696.
© *1989 by the IAU.*

Boltzmann formula of mass-action. Below one MeV the weak processes are no more in thermal equilibrium . The neutrons diffuse from high density phases into low density phases, changing both their density and their n / p ratio.

It may be a long time before we get definite results on the effect of the quark-hadron phase transitions on the formation rate of the cosmological nuclides. Nevertheless recents calculations , especially those taking into account the effect of neutron back-diffusion during nucleosynthesis (Terasawa and Sato 1988, Kurki-Suonio et al (1988) already give us the general trends.

We may expect the final results to be quite comparable with the results of homogeneous density calculations . Our present ignorance of the exact values of many relevant parameters of the Q-H transition can be assimilated to corresponding uncertainties on the final results . These uncertainties are likely to decrease as more detailed studies of the transition become available.

Taking into account these various effects and the associated uncertainties in the parameters of the transitions, various calculations have been made of the primordial nucleosynthesis yields of the cosmological isotopes . The present state of the art can be summarized in the following sentences.

Given the all the uncertainties, including those related to the determination of the Hubble parameter, the range of Ω b goes from 0.1 to 0.01. This is *appreciably larger* than in the case of a homogeneous density universe. This does not appear to be large enough to allow the baryons to close the universe($\Omega_b < 1$).

The cosmic density of luminous matter (stars and X-ray cluster gas) is $\Omega_L = 0.01$ whitin a factor of two ,while the density of clustered matter needed to account for the stability of clusters of galaxy or large scale motions is $\Omega_G = 0.1$ to 0.2.

Thus, within the uncertainties , at one end of the scale the baryonic matter could be entirely luminous (no baryonic dark matter) while at the other end of the scale the clustered matter could be entirely baryonic (no non-baryonic dark matter).

The comparison between the present calculations and the cosmic abundances suggest that the contrast R between the phases is unlikely to be larger than ten. This result is in agreement with the best estimate of the critical temperature of the phase transition (180 MeV < T_c < 220 MeV) leading to R $\approx$ 7.

The promotion of ^{7}Li to the status of "cosmological observable".

Thanks to a number of new developppments, both observational and theoretical, the isotope lithium-7 has gradually been promoted to the status of a *bona fide* cosmological observable. First because its BBN contribution can now be evaluated more properly . Second because its potential message has been clarified , in relation with baryon inhomogeneities (as stressed, for example , by the anaysis of the physics of the quark-hadron phase transition. (Witten 1984)).

One of the most important event in observational cosmology in recent years has been the discovery of lithium in PopII stars by the Spite (1983 a and b) . Later, a number of other observations have confirmed their data and added a wealth of new measurements (Spite et al 1985) (Hobbs and Duncan 1987) (Rebolo, Molaro and Beckman 1987) . The full data and its significance are best presented as a function of the hydrogen to iron ratio. This parameter is a measure of the importance of stellar nucleosynthesis on a galactic scale The abundance of lithium has remained almost constant (within a factor of two) while iron grows from .0003 to 0.1 of the solar value. The message is clear : *the lithium in this range is not mostly produced by stellar processes* . There exist a primordial component which dominates the stellar contribution all through this range.

Observations have shown that this component is mostly made of ^7Li : ($^7Li / ^6Li$ > 10 , Maurice et al 1984). This, however, is not a very tell-taling result since in typical stellar outer layers ^6Li is thermonuclearly destroyed one hundred times faster than ^7Li.

More interesting results come from beryllium and boron, two elements which are produced in Galactic Cosmic Rays (Meneguzzi et al 1971) but not in BBN (Wagoner et al 1967). The rate of formation of lithium (both isotopes) is , to better than a factor of two , the same as the rate of formation of boron (both isotopes). It is approximately ten times larger than the rate of formation of beryllium (Reeves and Meyer, 1978, Walker et al 1985). Furthermore, lithium, at all relevant temperature, is destroyed faster than beryllium and boron by stellar processes . Thus the abundance of beryllium gives an estimate of the GCR contributed lithium in a star while the abundance of the boron gives an upper limit to the GCR contribution.

Beryllium has been recently detected in Pop II stars (Rebolo , Molaro, and Beckman 1988) with hydrogen ratio $^9Be / H = 2 \times 10^{-12}$. The corresponding GCR lithium is only one tenth of the observed Pop II abundance.

A search for boron (Molaro 1987) in a Pop II star (HD 140283) has yielded an upper limit of $B / Li < 0.04$. The corresponding upper limit to the ^7Li isotope is $(7 / H)_{GCR} < 10^{-11}$, at least an order of magnitude smaller than the PopII observations Thus we may conclude that *the primordial lithium component is not the result of hypothetical primordial cosmic rays* (Montmerle 1977). The only other process known to us to generate lithium-7 in interesting amount is BBN and we may thus conclude that the lithium in Pop II stars mostly is of cosmological origin.

But , in order to recover the primordial abundance relevant to BBN model calculations ,we must face the question of possible depletion of this element by stellar surface processes.

The mean abundance of lithium in Pop II stars is $^7Li / H = 1.6 \times 10^{-10}$ with a dispersion of a factor of two (Rebolo et al 1987). This small dispersion is the main argument in favor of the hypothesis that this lithium abundance has not suffered much depletion by processes associated with the stellar surfaces (Michaud 1986) . Following the same logic, it appears reasonable to estimate that *the fractional depletion should not be larger as the observed dispersion*. A thorough study of surface depletion processes by Delyannis et al (1988)has given a similar result . In consequence we estimate an initial value of $^7Li / H = 2.0^{+0.5} \times 10^{-10}$.

Vauclair (1987, 1988), studying the effect of rotational mixing on stellar lithium, has argued that the depletion may have been larger and that the primordial value may be the same as the Pop 1 value ($^7Li / H = 1.0 \times 10^{-9}$) . It remains to be seen if the theory will be able to reproduce the small abundance dispersion displayed in figure 1. More work is being done on this subject.

Bibliography

Alcock, C.R., Fuller, G.M., and Mathews, G.J., Astrophys. J. **320**, 439, 1987

Applegate, J.H., and Hogan, C. 1985 Phys . Rev. **D31** 3037.

Applegate. J.H., Hogan, C.,and Sherrer R.J. 1987a Phys Rev. , **D35** , 1151.

Cayrel. R., I.A.U Symposium no 126 Cambridge 1986.

Cayrel . R., Proceedings of the Alpbach Summer school 1988.

Delyannis, C., Demarque, P., Kawaler, S., Krauss, L., and Romanelli, P. preprint, (1988).

Fuller ,G.M., Mathews G.J., and Alcock, C.R. Phys. ReV . **D37** 1380 1988

696

Gasser , J and Leutwyler, H., Light quarks at low temperature. Phys Lett B 184, 83 , 1988.

Gasser , J and Leutwyler, H., Thermodynamics of chiral symmetry . Phys Lett B 188, 477 , 1987.

Hobbs, L.M., and Duncan, D.K., Ap.J. 317, 796, 1987. (Li in halo stars)

Iso, K., Kodama, H., and Sato, K., Phys. Lett. 337, 169B, 1986.

Kapusta,J.I., and Olive K.A. 1988 preprint

Kurki-Suonio, H., Matzner , M.M., Centrella, J.M., Rothman, T., and Wilson, J.R., , preprint submitted to Physical Review D 1988 preprint

Leutwyler, H., QCD: low temperature expansion and finite -size effects. Proceedings of the Seillac Conference to appear in April 88.

Maurice, E., Spite., F., and Spite , M., Astron. Astrophys132, 278 , 1984 (Li-7 / Li -6 ratio in old stars.)

Meneguzzi,M., Audouze, J., and Reeves , H., 1971, Astr. Ap., 15, 337.

Michaud, G., Ap. J. 302, 650, 1986

Molaro. P. Astr. Ap. 183, 394 1987 (boron in Pop II)

Montmerle 1977 thesis Université de Paris.

Rebolo, R., Beckman,.J., and Molaro, P., Astron, Astrophys. 172 L17 1987 (Li in G 64 -12)

Rebolo, R., Molaro, P., Abia, C., and Beckman, J.E., Astron. Astrophys., 193, 193-201, 1988. (Be-9 in Pop II)

Reeves, 1971, p . 256 , American Physical Society Meeting , Porto Rico Dec 1971

Reeves . H., Ann Rev Astron Astrophys. 12,437, 1974

Reeves, H. Varenna School "Confrontations between Observations and Theories in Cosmology" July 1987

Reeves, H., Delbourgo-Salvador, P , Audouze, J. , and Salatti, P. to appear in European Journal of Physics august 88

Reeves, H., and Meyer, J.P., 1978, Astrophys. J. 226, 613.

Reeves, H., and Richer, J., 1988 in preparation.

Sale, K.E. and Mathews , G.J. 1986 Ap.J. 309 L1

Satz ,H., Ann. Rev. Nucl. Sci. 35 1985

Satz,H., Proceedings of the Strasbourg Symposium on the quark-hadron phase transition. July 1987

Spite, M and Spite, F. 1983a Nature 297, 483.

Spite, F. and Spite. M. 1983b Astr. Ap ., 115, 337.

Spite, F., Spite, M., Peterson, R.C., and Chafee, F.H., Astron. Astrophys. 172 L9 1987, (Li in metal poor halo stars).

Spite F. and M., 1985 (^7Li / ^6Li > 10)

Terasawa, N., and Sato, K., preprint (1988)

Vauclair , S., Procedings of Meudon IAU Symposium July 1987

Vauclair , S., 1988 preprint.

Walker , T.P., Mathews, G.J. and Viola, V.E. 1985, Ap.J., 299,, 745.

Wagoner, R. V., Fowler, W. A., and Hoyle, F., Ap.J. 148, 3, 1967.

Witten E., 1984 Phys. Rev. D30 272

Yang , J., Turner, M.S., Steigman, G., Schramm, D.N., and Olive, K, 1984, Ap.J. , 281, 493

AUTHOR INDEX

Abt, H.	73	Dickinson, M.	645
Adler, D.S.	587	Djorgovski, S.	645
Alexeyeva, L.N.	229	Dluzhnevskaya, O.B.	84
Alissandrakis, G.	674	d'Odorico, S.	629
Altrock, R.	673	Dollfus, A.	295
Andersen, J.	145	Downes, D.	555, 589
Anosova, J.P.	143	Dudley, J.	92, 97
Aoki, S.	475	Duncombe, R.L.	482
Asseo, E.	429	Dupraz, C.	581
		Durisen, R.H.	133
Babadzhanov, P.B.	287		
Bahcall, N.A.	435	Eckart, A.	589
Baldwin, J.E.	549	Eddy, J.A.	503
Baum, W.A.	631	Einaudi, G.	529
Bell, R.A.	365	Ekers, R.D.	551
Bender, P.	470		
Bendinelli, O.	657	Feissel, M.	478
Benn, C.R.	81, 85	Ferrari, A.	417
Benz, A.O.	539	Field, G.B.	567
Bester, M.	563	Fransson, C.	223
Bicknell, G.V.	409		
Black, J.H.	331	Garcia-Burillo, S.	575
Blundell, R.	575	Geary, J.C.	623
Bodo, G.	417	Genzel, R.	589
Boss, A.P.	123	Ghosh, S.K.	345
Bouton, E.	90	Gilman, P.	675
Brownlee, D.E.	281	Gokhale, M.H.	674
Butler, L.	78	Grabhorn, R.	133
		Greenberg, J.M.	241
Campbell, B.	109	Guélin, M.	575
Capitaine, N.	474		
Casoli, F.	581	Hachisu, I.	175
Catchpole, R.M.	185	Halbwachs, J.-L.	81
Cayrel de Strobel, G.	609, 611	Hanami, H.	217
Cernicharo, J.	575	Harris, A.I.	589
Christou, J.C.	561	Harris, H.C.	631
Chupp, E.L.	199	Hartquist, T.W.	375
Combes, F.	581	Harvey, K.	673
Cornwell, T.J.	547	Heere, K.R.	345
		Henkel, C.	597, 601
Danchi, W.C.	563	Hidayat, B.	153
Davis, J.	545	Hiei, E.	513, 678
de Greve, J.P.	149	Hjalmarson, A.	579
Demarque P.	665	Höflich, P.	207
de Narbonne, A.-M.M.	91	Hollweg, J.V.	517
Dermott, S.F.	259	Hong, S.S.	267
Despois, D.	575	Hughes, J.A.	488
de Zotti, G.	681	Hunter, D.A.	631

Illingworth, G.D.	449	Obrubov, Yu.V.	287
Irvine, W.M.	339	Ochsenbein, F.	86
		Olsson-Steel, D.	313
Jaffe, D.T.	589	Omont, A.	357
Jaschek, C.O.	76, 83, 663		
Jones, J.H.	631	Paczynski, B.	167
		Pagani, L.	581
Karovska, M.	193	Pap, J.	674
Kato, M.	175	Papaliolios, C.	193
Kinoshita, H.	472	Parmeggiani, G.	657
Koechlin, L.	193	Parthasarathy, M.	399
Kozai, Y.	251	Paterno, L.	678
Kreidl, T.J.	631	Pecker, J.C.	671
Krolik, J.H.	161	Pelletier, G.	429
Kuperus, M.	535	Petropoulos, P.	674
Kwon, S.M.	267	Phillips, T.G.	591
		Poulakis, X.	674
Lasota, J.-P.	173	Prasad, S.S.	345
Latham, D.W.	103	Prieur, J.-L.	6291
Lasker, B.M.	78		
Lastovica, E.	90	Rees, M.	45
Lebovitz, N.R.	129	Reeves, H.	693
Legrand, J.-P.	673	Rey-Watson, J.	77
Léna, P.J.	571	Ridgway, S.T.	559
Liller, W.	183	Roberts, W.W., Jr	587
Linde, P.	651	Rosner, R.	521
Livingston, W.	677	Rubio, M.	599
Longair, M.S.	455	Rufener, F.	617
Lundqvist, P.	223	Rushton, M.	85
Marsden, B.G.	79	Saio, H.	175
Massaglia, S.	417	Sakurai, T.	513
Mathieu, R.D.	111	Sanders, D.B.	591, 603
Matsumoto, T.	689	Sargent, A.I.	591
McAlister, H.A.	569	Schmidt, M.	33
McCullough, P.R.	563	Schmutz, W.	215
McIntosh, P.	672	Schwan, H.	482
Merkle, F.	565	Scoville, N.Z.	591, 605
Millar, T.J.	369	Seidelmann, P.K.	465
Mirabel, I.F.	603	Shames, P.M.B.	78
Mitton, S.	71	Shao, M.	469
Miyama, S.M.	127	Sharp, C.M.	207
Moran, J.M.	553	Shobbrook, R.M.	82, 92
Morando, B.	484	Simon, M.	117
Moroz, V.I.	17	Simon, P.	673
Mouradian, Z.	672, 673	Sivaraman, K.R.	674
Mukai, T.	305	Snodgrass, H.	672
Murray, C.A.	481	Sofue, Y.	583
		Sol, H.	429
Nagasawa, M.	213	Souchay, J.	472
Neckel, H.	674	Spite, F.M.	86
Nicholson, P.D.	259	Spite, M.	609
Nisenson, P.	193	Standley, C.	193

Standish, E.M., Jr.	476	Wahr, J.	473
Stark, A.A.	593	Weaver, H.A.	387
Steppe, H.	575	Weiler, E.J.	441
Stetson, P.B.	635	West, R.M.	3
Stevens-Rayburn, S.	89	Whittle, M.	423
Strobel, D.F.	395	Wielebinski, R.	595
Swarup, G.	557	Wiklind, T.	601
		Wild, W.	589
Taam, R.E.	155	Wilkins, G.A.	74, 75, 88
Tarafdar, S.P.	345	Williams, D.A.	383
Taylor, J.H.	471	Wilson, P.	671, 675
Toffolatti, L.	681	Withbroe, G.	525
Tohline, J.E.	137	Woodgate, B.E.	445
Townes, C.H.	563		
Trauger, J.	443	Yallop, B.D.	480
Trimble, V.	177	Yang, S.	133
Trussoni, E.	417	Yoshida, T.	217
van der Hucht, K.A.	153	Zappala', V.	273
van Dishoeck, E.F.	323	Zavatti, F.	657
Verbunt, F.	139	Zuckerman, B.	119
Vermeulen, R.C.	403		
Verter, F.	585	Zwaan, K.	677